Introduction to Health and Safety in Construction

endorsed by

nebosh

This publication is endorsed by NEBOSH as offering high quality support for the delivery of NEBOSH qualifications. NEBOSH endorsement does not imply that this publication is essential to achieve a NEBOSH qualification, nor does it mean that this is the only suitable publication available to support NEBOSH qualifications. No endorsed material will be used verbatim in setting any NEBOSH examination and all responsibility for the content remains with the publisher. Copies of official specifications for all NEBOSH qualifications may be found on the NEBOSH website – www.nebosh.org.uk

Introduction to Health and Safety in Construction

The Handbook for Construction Professionals and Students on NEBOSH and Other Construction Courses

Third edition

Phil Hughes MBE MSc, FCIOSH, RSP, Chairman NEBOSH 1995–2001

Ed Ferrett PhD, BSc (Hons Eng), CEng, MIMechE, MIEE, MIOSH, Deputy Chairman NEBOSH

ELSEVIER

AMSTERDAM • BOSTON • HEIDELBERG • LONDON • NEW YORK • OXFORD • PARIS • SAN DIEGO • SAN FRANCISCO • SINGAPORE • SYDNEY • TOKYO
Butterworth-Heinemann is an imprint of Elsevier

Butterworth-Heinemann is an imprint of Elsevier
The Boulevard, Langford Lane, Kidlington, Oxford, OX5 1GB
30 Corporate Drive, Suite 400, Burlington, MA 01803, USA

First edition 2005
Reprinted 2005, 2009
Second edition 2007
Third edition 2008
Reprinted 2009

Notice
No responsibility is assumed by the publisher for any injury and/or damage to persons or property as a matter of products liability, negligence or otherwise, or from any use or operation of any methods, products, instructions or ideas contained in the material herein.
Because of rapid advances in the medical sciences, in particular, independent verification of diagnoses and drug dosages should be made

British Library Cataloguing in Publication Data
A catalogue record for this book is available from the British Library

Library of Congress Cataloguing in Publication Data
A catalogu record for this book is available from the Library of Congress

ISBN: 978-1-85617-521-0

For information on all Butterworth-Heinemann publications
visit our website at http://books.elsevier.com

Typeset by Charon Tec Ltd., A Macmillan Company. (www.macmillansolutions.com)

Printed and bound in China
09 10 11 12   12 11 10 9 8 7 6 5 4 3 2

Working together to grow
libraries in developing countries

www.elsevier.com | www.bookaid.org | www.sabre.org

ELSEVIER   BOOK AID International   Sabre Foundation

Contents

Contents

Contents

Contents

Contents

Preface

The *Introduction to Health and Safety in Construction* has quickly established itself, in its first 3 years, as the standard text for students taking the NEBOSH National Certificate in Construction Safety and Health and for those taking other courses in building or construction. It is also of great value to those working in the construction industry at all levels, particularly construction site managers and foremen. As it has become a significant work of reference for managers with health and safety responsibilities, it is a matter of prime importance that it should be kept up-to-date, as far as is possible, with new legislation and recent developments.

There has been concern over a number of years at the poor record of health and safety in the construction industry, particularly in 2006/07 when there was a 28% increase in construction fatalities. The legal health and safety requirements for all places of work are numerous and complex; it is the intention of the authors to offer an introduction to the subject for all those who have the maintenance of good health and safety standards as part of their employment duties or those who are considering the possibility of a career as a health and safety professional. Health and safety is well recognized as an important component of the activities of any organization, not only because of the importance of protecting people from harm but also because of the growth in the direct and indirect costs of accidents. These costs have increased higher than the rate of retail price inflation by a considerable amount in the last few years as the number of civil claims and awards have risen each year. It is very important that basic health and safety legal requirements are clearly understood by all organizations whether public or private, large or small. A good health and safety performance is normally only achieved when health and safety is effectively managed so that significant risks are identified and reduced by adopting appropriate high quality control measures.

This third edition of *Introduction to Health and Safety in Construction* is based on the QCA (Qualification and Curriculum Authority) accredited NEBOSH Certificate in Construction Safety and Health syllabus as revised in July 2007. It has been developed specifically for students who are studying for that NEBOSH National Certificate course in Construction Safety and Health but will also be useful for those studying a variety of building and construction courses, such as the Higher National Certificate and Diploma. It was felt appropriate to produce a text book which mirrored the Construction Certificate syllabus in its revised unitized form and in a single volume to the required breadth and depth. The syllabus, which follows the general pattern for health and safety management set by the Health and Safety Executive in their guidance HSG 65, is risk and management based so it does not start from the assumption that health and safety is best managed by looking first at the causes of failures. Fortunately, failures such as accidents and ill-health are relatively rare and random events in most work places. A full copy of the syllabus and guide can be obtained from NEBOSH.

The most significant change since the second edition has been the introduction of the unitized NEBOSH National Certificate in Construction Safety and Health. The course has been divided into three distinct units, each of which is assessed separately. The three units are NGC1 – Management of health and Safety; NCC1 – Managing and controlling hazards in construction activities; and NCC2 – Construction health and safety practical application.

This development offers the opportunity for additional and more flexible course formats and students may now study parallel courses (in, say, general health and safety and fire) without repeating the management unit. Students who decide to take individual units will, on passing, receive a Unit Certificate. However, it has necessitated the need for a new chapter (Chapter 9) on construction law and management to deal with those construction topics that were not in the original management syllabus.

There have also been some changes to the syllabus and the main ones are as follows:

➤ scope and application of CDM 2007
➤ safety in the use of lifting and moving equipment

➤ additional substances that pose construction health hazards

➤ further controls of substances that can cause cancer, asthma or damage to genes

➤ arrangements for site inductions

➤ construction activities on public highways and road marking equipment

➤ asbestos awareness training and surveys

➤ the use of crash decks when working at height

➤ the use of freezing equipment during excavation work

➤ the protection of migrant workers and

➤ the safe storage and disposal of incompatible wastes.

This third edition has been produced in order to update the health and safety legislation, with particular regard to changes in legislation relating to fire – the Regulatory Reform Fire Safety Order 2004. This removes the requirement for fire certificates and revokes the Fire Precautions Act 1971 and the Fire Precautions (Work Place) Requirements. This additional information enables the book to be useful to students undertaking the new NEBOSH Certificate in Fire Safety and Risk Management.

Other important changes in health and safety legislation, which are included in this edition, are the Control of Asbestos Regulations, the Hazardous Waste (England and Wales) Regulations and the new Construction (Design and Management) Regulations.

Some more recent examination questions have been added at the end of each chapter. Many questions cover the contents of more than one chapter. It is recommended that students have sight of the published examiners' reports, which are available from NEBOSH. These reports not only provide an excellent guide on the expected answers to the questions but also indicate areas of student misunderstanding.

We have been asked about the allocation of marks to each question but since the marks awarded to each part of a question can vary, only general guidance can be given. All one- or two-part questions are 8 mark questions with a minimum of 2 marks awarded to any part. Questions with three or more parts are 20 mark questions with a minimum of 4 marks for each part.

The book is also intended as a useful reference guide for managers and directors with health and safety responsibilities and for safety representatives. Chapter 21 summarizes all the most commonly used Acts and Regulations. It was written to provide an easily accessible reference source for students during and after the course and many others in industry and commerce such as managers, supervisors and safety representatives.

There is considerable commonality between the NEBOSH Construction Certificate, the NEBOSH Certificate in Fire Safety and Risk Management and the NEBOSH National General Certificate, and this commonality is reflected in the book. There is also some overlap of topics between syllabus elements and chapters and the authors have drawn attention to this when it occurs by indicating where the fullest account of the topic may be found.

Finally, since one of the objectives of the book is to provide a handbook for the use of any person who has health and safety as part of their responsibilities, we thought that it would be helpful to add a few useful topics which are outside the syllabus. These include managing occupational road safety (Chapter 11), fast track settlement of compensation claims following the Woolfe reforms (Chapter 8) and demolition using explosives (Chapter 20). We have also used the opportunity of a new edition to include a chapter on the international aspects of construction health and safety (Chpater 22). This will be useful for those who work overseas.

We hope that you find this new edition to be useful.

Phil Hughes MBE
Ed Ferrett

Acknowledgements

Throughout the book, definitions used by the relevant legislation, the Health and Safety Commission, the Health and Safety Executive and advice published in Approved Codes of Practice or various Health and Safety Commission/ Executive publications have been utilized. Most of the references produced at the end of each Act or Regulation summary in Chapter 21, are drawn from the HSE Books range of publications.

At the end of each chapter, there are some examination questions taken from recent NEBOSH National General Certificate and Construction Certificate papers. Some of the questions may include topics which are covered in more than one chapter. However, most of the answers to these questions are to be found within the chapter at whose end the questions are listed. NEBOSH publishes a very full examiners' report after each public examination which give further information on each question. Most accredited NEBOSH training centres will have copies of these reports and further copies may be purchased directly from NEBOSH. The authors would like to thank NEBOSH for giving them permission to use these questions.

The authors' grateful thanks go to Liz Hughes for her proofreading and patience and to Jill Ferrett for encouraging Ed to keep at the daunting task of text preparation. The authors are particularly grateful to Liz for the excellent study guide that she has written for all NEBOSH students. A study guide is included as Chapter 23 of this book which mature students who have not taken formal examinations for a number of years have found very useful.

The author would also like to acknowledge the contribution made by Hannah Ferrett for the help that she gave during the research for the book and with some of the word processing. The advice given on the specimen practical application by Neil Caddy, a Construction Health and Safety Consultant from Truro and a NEBOSH Construction examiner, is also gratefully acknowledged.

We would like to thank Teresa Budworth, the Chief Executive of NEBOSH, for her support during this third edition, the staff of NEBOSH and various HSE staff for their generous help and advice. Finally we would like to thank Stephen Vickers, the immediate past Chief Executive of NEBOSH, for his encouragement at the beginning of the project and Doris Funke and all the production team at Elsevier who have worked hard to translate our dream into reality.

About the authors

Phil Hughes is a well-known UK safety professional with over 30 years' worldwide experience as Head of Environment Health and Safety at two large multinationals, Courtaulds and Fisons. Phil started in health and safety with the Factory Inspectorate in the Derby District in 1969 and moved to Courtaulds in 1974. He joined IOSH in that year and became Chairman of the Midland Branch, National Treasurer and then President in 1990–1991. Phil was very active on the NEBOSH Board for over 10 years and served as Chairman from 1995 to 2001. He is also a Professional Member of the American Society of Safety Engineers and has lectured widely throughout the world. Phil received the RoSPA Distinguished Service Award in May 2001 and became a Director and Trustee of RoSPA in 2003. He received an MBE in the New Year's Honours List 2005 for Services to Health and Safety.

Dr Edward Ferrett is an experienced health and safety consultant who has practised for over 20 years. He spent 30 years in higher and further education, retiring as the Head of the Faculty of Technology of Cornwall College in 1993. Since then he has been an independent consultant to several public and private sector organizations, the Regional Health and Safety Adviser for the Government Office (West Midlands) and Vice Chair of NEBOSH. Edward was the Chair of West of Cornwall Primary Care NHS Trust for 6 years until 2006.

He has delivered many health and safety courses and is a manager of NEBOSH courses at the Cornwall Business School. He is currently an Inspector for the British Accreditation Council for Independent Further and Higher Education. Ed is a Chartered Engineer and a Chartered Member of IOSH.

List of principal abbreviations

Most abbreviations are defined within the text. Abbreviations are not always used if it is not appropriate within the particular context of the sentence. The most commonly used ones are as follows:

ACL	Approved carriage list
ACM	Asbestos containing material
ACOP	Approved Code of Practice
AIB	Asbestos Insulation Board
APAU	Accident Prevention Advisory Unit now Operations Unit
ARCA	Asbestos Removal Contractors Association
BA	Breathing apparatus
BAT	Best available techniques
BRE	Building Research Establishment
BSI	British Standards Institution
CAR	Control of Asbestos Regulations
CBI	Confederation of British Industry
CD	Consultative document
CDM	Construction (Design and Management) Regulations
CECA	Civil Engineering Contractors Association
CEN	Comite Europeen de Normalisation
CENELEC	Comite Europeen de Normalisation Electrotechnique
CHIP	Chemicals (Hazard Information and Packaging) Regulations
CIB	Chartered Institute of Building
CIRA	Construction Industry Research and Information Association
CLAW	Control of Lead at Work Regulations
CONIAC	Construction Industry Advisory Committee
CORGI	Council for Registered Gas Installers
COSHH	Control of Substances Hazardous to Health Regulations
dB(A)	Decibel (A – weighted)
dB(C)	Decibel (C – weighted)
DSE	Display screen equipment
DSEAR	Dangerous Substances and Explosive Atmospheres Regulations
EAV	Exposure action value
EC	European Community
ELV	Exposure limit value
EMAS	Employment Medical Advisory Service
EPA	Environmental Protection Act
EU	European Union
HAV	Hand–arm vibration
HSAC	Health and Safety Advice Centre
HSC	Health and Safety Commission
HSCER	Health and Safety (Consultation with Employers) Regulations

List of principal abbreviations

HSE	Health and Safety Executive
HSL	Health and Safety Laboratory
HSW etc Act	Health and Safety at Work etc Act
IAC	Industry Advisory Committee
ILO	International Labour Office
IOSH	Institution of Occupational Safety and Health
LEAL	Lower exposure action level
LOLER	Lifting Operations and Lifting Equipment Regulations
LPG	Liquefied petroleum gas
MCG	Major Contractors Group
MHOR	Manual Handling Operations Regulations
MHSW	Management of Health and Safety at Work Regulations
MOT	Ministry of Transport (still used for vehicle tests)
NAWR	Control of Noise At Work Regulations
NEBOSH	National Examination Board in Occupational Safety and Health
NVQ	National Vocational Qualification
OHSAS	Occupational health and safety assessment series
OSH	Occupational safety and health
PPE	Personal protective equipment
ppm	Parts per million
PUWER	Provision and Use of Work Equipment Regulations
RCD	Residual current device
REACH	Registration Evaluation and Authorization of Chemicals
RIDDOR	Reporting of Injuries, Diseases and Dangerous Occurrences Regulations
ROES	Representative(s) of employee safety
RoSPA	Royal Society for the Prevention of Accidents
RPE	Respiratory protective equipment
RRFSO	Regulatory Reform Fire Safety Order
RTA	Road traffic accident
SFARP	So far as reasonably practicable
SPL	Sound pressure level
STEL	Short term exposure limit
SWL	Safe working load
SWP	Safe working pressure
TLV	Threshold limit value
TUC	Trades Union Congress
TWA	Time-weighted average
UEAL	Upper exposure action level
UK	United Kingdom
VAWR	Vibration at Work Regulations
WAHR	Work at Height Regulations
WBV	Whole body vibration
WEL	Workplace exposure limit
WHO	World Health Organization
WRULD	Work-related upper limb disorder
WWMRegs	The Woodworking Machines Regulations

Illustrations credits

Figure 1.5   Source HSE. © Crown copyright material is reproduced with the permission of the Controller of HMSO and Queen's Printer for Scotland.

Figure 1.12   From HSG65 *Successful Health and Safety Management* (HSE Books 1997) ISBN 0717612767. © Crown copyright material is reproduced with the permission of the Controller of HMSO and Queen's Printer for Scotland.

Figure 2.7   From HSG149 *Backs for the Future: Safe Manual Handling in Construction* (HSE Books 2000) ISBN 0717611221. © Crown copyright material is reproduced with the permission of the Controller of HMSO and Queen's Printer for Scotland.

Figure 3.7   Courtesy of Stocksigns.

Figure 4.3   From HSG57 *Seating at Work* (HSE Books 1998) ISBN 0717612317. © Crown copyright material is reproduced with the permission of the Controller of HMSO and Queen's Printer for Scotland.

Figure 4.4   From HSG168 *Fire Safety in Construction Work* (HSE Books 1997) ISBN 0717613321. © Crown copyright material is reproduced with the permission of the Controller of HMSO and Queen's Printer for Scotland.

Figure 4.7   From HSG48 *Reducing Error and Influencing Behaviour* (HSE Books 1999) ISBN 0717624528. © Crown copyright material is reproduced with the permission of the Controller of HMSO and Queen's Printer for Scotland.

Figure 5.1   From HSG149 *Backs for the Future: Safe Manual Handling in Construction* (HSE Books 2000) ISBN 0717611221. © Crown copyright material is reproduced with the permission of the Controller of HMSO and Queen's Printer for Scotland.

Figure 6.1   Reproduced with permission from *The Argus*, Brighton.

Figure 6.3   Courtesy of Stocksigns.

Figure 6.4   Courtesy of Stocksigns.

Figure 6.5   Courtesy of Stocksigns.

Figure 6.6   Courtesy of Stocksigns.

Figure 6.7   Courtesy of Stocksigns.

Figure 6.8   Courtesy of Stocksigns.

Illustrations credits

Figure 6.9	Courtesy of Stocksigns.
Figure 6.10	Courtesy of Stocksigns.
Figure 6.11	Courtesy of Stocksigns.
Figure 6.12	Courtesy of Stocksigns.
Figure 6.13	From HSG150 (rev1) *Health and Safety in Construction* (HSE Books 2001) ISBN 0717621065. © Crown copyright material is reproduced with the permission of the Controller of HMSO and Queen's Printer for Scotland.
Figure 6.14	From HSG150 (rev1) *Health and Safety in Construction* (HSE Books 2001) ISBN 0717621065. © Crown copyright material is reproduced with the permission of the Controller of HMSO and Queen's Printer for Scotland.
Figure 6.16	From HSG110 *Seven Steps to Successful Substitution of Hazardous Substances* (HSE Books 1994) ISBN 0717606953. © Crown copyright material is reproduced with the permission of the Controller of HMSO and Queen's Printer for Scotland.
Figure 6.17	From PUWER 1998. *Provision and Use of Work Equipment Regulations 1998: Open Learning Guidance* (HSE Books 1999) ISBN 0717624595. © Crown copyright material is reproduced with the permission of the Controller of HMSO and Queen's Printer for Scotland.
Figure 6.18	Cover of INDG98 *Permit-to-Work Systems* (HSE 1998) ISBN 0717613313. © Crown copyright material is reproduced with the permission of the Controller of HMSO and Queen's Printer for Scotland.
Figure 7.1	From HSG65 *Successful Health and Safety Management* (HSE Books 1997) ISBN 0717612767. © Crown copyright material is reproduced with the permission of the Controller of HMSO and the Queen's Printer for Scotland.
Figure 7.2	From *Guide to Measuring Health and Safety Performance* (HSE Books 2001). © Crown copyright material is reproduced with the permission of the Controller of HMSO and Queen's Printer for Scotland.
Figure 8.4	From BI510 *Accident Book* (HSE Books 2003) ISBN 0717626032. © Crown copyright material is reproduced with the permission of the Controller of HMSO and Queen's Printer for Scotland.
Figure 8.5	From HSE31 (rev1) *RIDDOR Explained: Reporting of Injuries, Diseases and Dangerous Occurrences Regulations* (HSE Books 1999) ISBN 0717624412. © Crown copyright material is reproduced with the permission of the Controller of HMSO and Queen's Printer for Scotland.
Figure 9.4	Front cover from HSG150 (3rd edition) *Health and Safety in Construction* (HSE Books 2001) ISBN 0717621065. © Crown copyright material is reproduced with the permission of the Controller of HMSO and Queen's Printer for Scotland.
Appendix 9.1	From L23 *Manual Handling* (HSE Books 2004) ISBN 071762832X. © Crown copyright material is reproduced with the permission of the Controller of HMSO and Queen's Printer for Scotland.
Figure 10.2	From HSG151 *Protecting the Public – Your Next Move* (HSE Books 1997) ISBN 0717611485. © Crown copyright material is reproduced with the permission of the Controller of HMSO and Queen's Printer for Scotland.
Figure 10.3	From HSG150 (rev1) *Health and Safety in Construction* (HSE Books 2001) ISBN 0717621065. © Crown copyright material is reproduced with the permission of the Controller of HMSO and Queen's Printer for Scotland.

Figure 10.4	From HSG185 *Health and Safety in Excavations* (HSE Books 1999) ISBN 0717615634. © Crown copyright material is reproduced with the permission of the Controller of HMSO and Queen's Printer for Scotland.
Figure 10.5	From HSG150 *Health and Safety in Construction* Edition 3 HSE Books 2006, ISBN 0-717661822 2006. © Crown copyright material is reproduced with the permission of the Controller of HMSO and Queen's Printer for Scotland.
Figure 11.1	From HSG155 *Slips and Trips* (HSE Books 1996) ISBN 0717611450. © Crown copyright material is reproduced with the permission of the Controller of HMSO and Queen's Printer for Scotland.
Figure 11.4	Courtesy of Speedy Hire.
Figure 11.5a	From HSG144 *The Safe Use of Vehicles on Construction Sites* (HSE Books 2001) ISBN 071761610X. © Crown copyright material is reproduced with the permission of the Controller of HMSO and Queen's Printer for Scotland.
Figure 11.7	From HSG150 *Health & Safety in Construction* (HSE Books 2001) ISBN 0717621065. © Crown copyright material is reproduced with the permission of the Controller of HMSO and Queen's Printer for Scotland.
Figure 11.10	From HSG144 *The Safe Use of Vehicles on Construction Sites* (HSE Books 1998) ISBN 071761610X. © Crown copyright material is reproduced with the permission of the Controller of HMSO and Queen's Printer for Scotland.
Figure 11.12a	Courtesy of Stocksigns
Figure 11.12b	Courtesy of Stocksigns
Figure 11.13	Courtesy of Stocksigns
Figure 11.18	Courtesy of Stocksigns
Figure 12.4	From L23 *Manual Handling Operations – Guidance on Regulations* (HSE Books 2004) ISBN 071762823X. © Crown copyright material is reproduced with the permission of the Controller of HMSO and Queen's Printer for Scotland.
Figure 12.5	From HSG149 *Backs for the Future: Safe Manual Handling in Construction* (HSE Books 2000) ISBN 0717611221. © Crown copyright material is reproduced with the permission of the Controller of HMSO and Queen's Printer for Scotland.
Figure 12.6	From HSG115 *Manual Handling Solutions You Can Handle* (HSE Books 2001) ISBN 0717606937. © Crown copyright material is reproduced with the permission of the Controller of HMSO and Queen's Printer for Scotland.
Figure 12.8a	From HSG149 *Backs for the Future: Safe Manual Handling in Construction* (HSE Books 2000) ISBN 0717611221. © Crown copyright material is reproduced with the permission of the Controller of HMSO and Queen's Printer for Scotland.
Figure 12.8b	From HSG149 *Backs for the Future: Safe Manual Handling in Construction* (HSE Books 2000) ISBN 0717611221. © Crown copyright material is reproduced with the permission of the Controller of HMSO and Queen's Printer for Scotland.

Figure 13.31   From L114 *Safe Use of Woodworking Machinery* (HSE Books 1998). ISBN 0717616304 © Crown copyright material is reproduced with the permission of the Controller of HMSO and Queen's Printer for Scotland.

Figure 13.32   From WIS17 (rev1) *Safety in the Use of Hand-fed Planing Machines*. Woodworking Sheet No. 17 (revised) (HSE Books 2000) © Crown copyright material is reproduced with the permission of the Controller of HMSO and Queen's Printer for Scotland.

Figure 13.33   From WIS18 (rev1) *Safe Use of Vertical Spindle Moulding Machines*. Woodworking Sheet No. 18 (revised) (HSE Books 2001) © Crown copyright material is reproduced with the permission of the Controller of HMSO and Queen's Printer for Scotland.

Figure 13.34   Courtesy of Winget.

Figure 13.35   Courtesy of Speedy Hire.

Figure 13.36   Courtesy of BOMAG.

Figure 13.39   Courtesy of TS Thomas and Sons Ltd.

Figure 14.2   Courtesy of Stocksigns.

Figure 14.4a   Courtesy of Speedy Hire.

Figure 14.4b   Courtesy of Speedy Hire.

Figure 14.6   Courtesy of DeWalt.

Figure 14.7   From *Essentials of Health and Safety* (HSE Books 1999) ISBN 071760716X. © Crown copyright material is reproduced with the permission of the Controller of HMSO and Queen's Printer for Scotland.

Figure 14.8   From *Essentials of Health and Safety* (HSE Books 1999) ISBN 071760716X. © Crown copyright material is reproduced with the permission of the Controller of HMSO and Queen's Printer for Scotland.

Figure 14.10   From INDG236 *Maintaining Portable Electrical Equipment in Offices* (HSE Books 1996) ISBN 0717612724. © Crown copyright material is reproduced with the permission of the Controller of HMSO and Queen's Printer for Scotland.

Figure 14.11a   From HSG185 *Health and Safety in Excavations* (HSE Books 1999) ISBN 0717615634. © Crown copyright material is reproduced with the permission of the Controller of HMSO and Queen's Printer for Scotland.

Figure 14.11b   From HSG185 *Health and Safety in Excavations* (HSE Books 1999) ISBN 0717615634. © Crown copyright material is reproduced with the permission of the Controller of HMSO and Queen's Printer for Scotland.

Figure 15.1   Amended from HSG168 *Fire Safety in Construction Work* (HSE Books 1997) ISBN 0717613321. © Crown copyright material is reproduced with the permission of the Controller of HMSO and Queen's Printer for Scotland.

Figure 15.7   From HSG168 *Fire Safety in Construction Work* (HSE Books 1997) ISBN 0717613321. © Crown copyright material is reproduced with the permission of the Controller of HMSO and Queen's Printer for Scotland.

Figure 15.8a   Courtesy of Guardian Safety Products.

Illustrations credits

Figure 15.8b	From HSG168 *Fire Safety in Construction Work* (HSE Books 1997) ISBN 0717613321. © Crown copyright material is reproduced with the permission of the Controller of HMSO and Queen's Printer for Scotland.
Figure 15.9	From HSG168 *Fire Safety in Construction Work* (HSE Books 1997) ISBN 0717613321. © Crown copyright material is reproduced with the permission of the Controller of HMSO and Queen's Printer for Scotland.
Figure 15.10	From HSG168 *Fire Safety in Construction Work* (HSE Books 1997) ISBN 0717613321. © Crown copyright material is reproduced with the permission of the Controller of HMSO and Queen's Printer for Scotland.
Figure 15.11	From HSG168 *Fire Safety in Construction Work* (HSE Books 1997) ISBN 0717613321. © Crown copyright material is reproduced with the permission of the Controller of HMSO and Queen's Printer for Scotland.
Figure 16.2	Reprinted from *Anatomy and Physiology in Health and Illness*, ninth edition, Waugh and Grant, pages 240 and 248, 2002, with permission from Elsevier.
Figure 16.3	Reprinted from *Anatomy and Physiology in Health and Illness*, ninth edition, Waugh and Grant, page 9, 2002, with permission from Elsevier.
Figure 16.4	Reprinted from *Anatomy and Physiology in Health and Illness*, ninth edition, Waugh and Grant, page 8, 2002, with permission from Elsevier.
Figure 16.5	Reprinted from *Anatomy and Physiology in Health and Illness*, ninth edition, Waugh and Grant, page 340, 2002, with permission from Elsevier.
Figure 16.6	Reprinted from *Anatomy and Physiology in Health and Illness*, ninth edition, Waugh and Grant, page 363, 2002, with permission from Elsevier.
Figure 16.7	From HSG150. (3rd edition) *Health and Safety in Construction* (HSE Books 2006) ISBN 07176 6182 2 © Crown copyright material is reproduced with the permission of the controller of HMSO and Queen's printer for Scotland.
Figure 16.9	Courtesy of Draeger Safety UK Limited.
Figure 16.10	From HSG54 *Maintenance, Examination and Testing of Local Exhaust Ventilation* (HSE Books 1998) ISBN 0717614859. © Crown copyright material is reproduced with the permission of the Controller of HMSO and Queen's Printer for Scotland.
Figure 16.11	From HSG202 *General Ventilation in the Workplace* (HSE Books 2000) ISBN 0717617939. © Crown copyright material is reproduced with the permission of the Controller of HMSO and Queen's Printer for Scotland.
Figure 16.12a	From HSG53 *The Selection, Use and Maintenance of Respiratory Protective Equipment* (HSE Books 1998) ISBN 0717615375. © Crown copyright material is reproduced with the permission of the Controller of HMSO and Queen's Printer for Scotland.
Figure 16.12b	From HSG53 *The Selection, Use and Maintenance of Respiratory Protective Equipment* (HSE Books 1998) ISBN 0717615375. © Crown copyright material is reproduced with the permission of the Controller of HMSO and Queen's Printer for Scotland.
Figure 16.12c	From HSG53 *The Selection, Use and Maintenance of Respiratory Protective Equipment* (HSE Books 1998) ISBN 0717615375. © Crown copyright material is reproduced with the permission of the Controller of HMSO and Queen's Printer for Scotland.

Figure 16.13	Courtesy of Draper.
Figure 16.15	From HSG213 *Introduction to Asbestos Essentials* (HSE Books 2001) ISBN 071761901X. © Crown copyright material is reproduced with the permission of the Controller of HMSO and Queen's Printer for Scotland.
Figure 17.1	From HSG121 *A Pain in Your Workplace* (HSE Books 1994) ISBN 0717606686. © Crown copyright material is reproduced with the permission of the Controller of HMSO and Queen's Printer for Scotland.
Figure 17.3	From INDG175 (rev1) *Health Risks from Hand-Arm Vibration* (HSE Books 1998) ISBN 0717615537. © Crown copyright material is reproduced with the permission of the Controller of HMSO and Queen's Printer for Scotland.
Figure 17.4	From HSG170 *Vibration Solutions* (HSE Books 1997) ISBN 0717609545. © Crown copyright material is reproduced with the permission of the Controller of HMSO and Queen's Printer for Scotland.
Figure 17.6	Courtesy of Speedy Hire.
Figure 17.8a	Reprinted from *Anatomy and Physiology in Health and Illness*, Ninth edition, Waugh and Grant, page 195, 2002, with permission from Elsevier.
Figure 17.8b	Reprinted from *Anatomy and Physiology in Health and Illness*, Ninth edition, Waugh and Grant, page 195, 2002, with permission from Elsevier.
Figure 17.10	From L108 *Reducing Noise at Work* (HSE Books 1998) ISBN 0717615111. © Crown copyright material is reproduced with the permission of the Controller of HMSO and Queen's Printer for Scotland.
Figure 17.11	Source: *Heat Stress Card*, published by the Occupational Safety & Health Administration, USA.
Figure 17.13	From INDG69 (rev) *Violence at Work: A Guide for Employers* (HSE Books 2000) ISBN 0717612716. © Crown copyright material is reproduced with the permission of the Controller of HMSO and Queen's Printer for Scotland.
Figure 18.2	Courtesy of Speedy Hire.
Figure 18.3	From HSG33 *Health & Safety in Roof Work* (HSE Books 1998) ISBN 0717614255. © Crown copyright material is reproduced with the permission of the Controller of HMSO and Queen's Printer for Scotland.
Figure 18.5	From HSG33 *Health & Safety in Roof Work* (HSE Books 1998) ISBN 0717614255. © Crown copyright material is reproduced with the permission of the Controller of HMSO and Queen's Printer for Scotland.
Figure 18.6	From HSG33 *Health & Safety in Roof Work* (HSE Books 1998) ISBN 0717614255. © Crown copyright material is reproduced with the permission of the Controller of HMSO and Queen's Printer for Scotland.
Figure 18.7	From HSG150. (3rd edition) *Health and Safety in Construction* © Crown copyright material is reproduced with the permission of the controller of HMSO and Queen's printer for Scotland.
Figure 18.8a	From HSG150. (3rd edition) *Health and Safety in Construction* © Crown copyright material is reproduced with the permission of the controller of HMSO and Queen's printer for Scotland.
Figure 18.8b	From HSG150. (3rd edition) *Health and Safety in Construction* © Crown copyright material is reproduced with the permission of the controller of HMSO and Queen's printer for Scotland.

Figure 19.6	From HSG185 *Health and Safety in Excavations* (HSE Books 1999) ISBN 0717615634. © Crown copyright material is reproduced with the permission of the Controller of HMSO and Queen's Printer for Scotland.
Figure 19.7	From HSG185 *Health and Safety in Excavations* (HSE Books 1999) ISBN 0717615634. © Crown copyright material is reproduced with the permission of the Controller of HMSO and Queen's Printer for Scotland.
Figure 19.8	From HSG185 *Health and Safety in Excavations* (HSE Books 1999) ISBN 0717615634. © Crown copyright material is reproduced with the permission of the Controller of HMSO and Queen's Printer for Scotland.
Figure 19.11	Courtesy of Speedy Hire.
Figure 19.12	Courtesy of Speedy Hire.
Figure 20.1	Source: *Approved Code of Practice for Demolition*, 1994, ISBN: 0477035582, published by Occupational Safety and Health Service, Department of Labour, Wellington, New Zealand.
Figure 20.2	Courtesy of Stocksigns.
Figure 20.5	Courtesy of Stocksigns.
Figure 21.1	From INDG350 *The idiot's guide to CHIP: Chemicals (Hazard Information and Packaging for Supply) Regulations 2002* (HSE Books 2002) ISBN 0717623335. © Crown copyright material is reproduced with the permission of the Controller of HMSO and Queen's Printer for Scotland.
Figure 21.2	From L23 *Manual Handling Operations – Guidance on Regulations* (HSE Books 2004) ISBN 071762823X. © Crown copyright material is reproduced with the permission of the Controller of HMSO and Queen's Printer for Scotland.
Figure 22.5	From HSG65 *Successful Health and Safety Management* (HSE Books 1997) ISBN 07176 12767 © Crown copyright material is reproduced with the permission of the controller of HMSO and Queen's printer for Scotland.

Health and safety foundations

1

1.1 Introduction

Occupational health and safety is relevant to all branches of industry, business and commerce including traditional industries, information technology companies, the National Health Service, care homes, schools, universities, leisure facilities and offices. It is particularly important for the construction industry.

The purpose of this chapter is to introduce the foundations on which appropriate health and safety management systems may be built. Occupational health and safety affects all aspects of work. In a low hazard organization, health and safety may be supervised by a single competent manager. In a high hazard manufacturing plant, many different specialists, such as engineers (electrical, mechanical and civil), lawyers, medical doctors and nurses, trainers, work planners and supervisors, may be required to assist the professional health and safety practitioner in ensuring that there are satisfactory health and safety standards within the organization.

Construction is the largest industry in the UK and accounts for 8% of its gross domestic product. It employs 10% of the working population and has an annual turnover of £250 billion. The construction industry has a world reputation for the quality of its work but it remains one of the most dangerous in the UK. In 2004/05, the fatal injury rate (per 100,000 workers) was 3.4 while the industrial average was 0.8. In 2006/07, there was a 28% increase in fatalities in the industry, which accounted for 32% of all notifiable fatal injuries.

In response to the 'Revitalising Health and Safety' campaign launched by the Health and Safety Commission (HSC) and the Government in June 2000, the construction industry set itself a target to reduce the rate of fatal and major injury to its workers by 40% in 2004/05 and by 66% in 2009/10. The construction client who commissions

the work is a very important agent in the drive for improved health and safety standards. He should insist on evidence of a good health and safety record and performance of a contractor at the tendering stage, and ensure that health and safety standards are being met on site. He should also require that all the people working on the site are properly trained for their particular job.

There are many obstacles to the achievement of good standards. The pressure of production or performance targets, financial constraints and the complexity of the organization are typical examples of such obstacles. However, there are some powerful incentives for organizations to strive for high health and safety standards. These incentives are moral, legal and economic.

Corporate responsibility, a term used extensively in the 21st century world of work, covers a wide range of issues. It includes the effects that an organization's business has on the environment, human rights and Third World poverty. Health and safety in the workplace is an important corporate responsibility issue.

Corporate responsibility has various definitions. However, broadly speaking it covers the ways in which organizations manage their core business to add social, environmental and economic value in order to produce a positive, sustainable impact on both society and the business itself. Terms such as 'corporate social responsibility', 'socially responsible business', and 'corporate citizenship' all refer to this concept.

The UK's Health and Safety Executive's (HSE) mission is to ensure that the risks to health and safety of workers are properly controlled. In terms of corporate responsibility, they are working to encourage organizations to:

➤ improve health and safety management systems to reduce injuries and ill-health
➤ demonstrate the importance of health and safety issues at board level

1

Figure 1.1 At work.

➤ report publicly on health and safety issues within their organization, including their performance against targets.

The HSE believe that effective management of health and safety:

➤ is vital to employee well-being
➤ has a role to play in enhancing the reputation of businesses and helping them achieve high-performance teams
➤ is financially beneficial to business.

1.2 Some basic definitions

Before a detailed discussion of health and safety issues can take place, some basic occupational health and safety definitions are required:

Health – the protection of the bodies and minds of people from illness resulting from the materials, processes or procedures used in the workplace.

Safety – the protection of people from physical injury. The borderline between health and safety is ill-defined and the two words are normally used together to indicate concern for the physical and mental well-being of the individual at the place of work.

Welfare – the provision of facilities to maintain the health and well-being of individuals at the workplace. Welfare facilities include washing and sanitation arrangements, the provision of drinking water, heating, lighting, accommodation for clothing, seating (when required by the work activity), eating and rest rooms. First aid arrangements are also considered as welfare facilities.

Occupational or work-related ill-health – is concerned with those illnesses or physical and mental disorders that are either caused or triggered by workplace activities. Such conditions may be induced by the particular work activity of the individual or by activities of others in the workplace. The time interval between exposure and the onset of the illness may be short (e.g. asthma attacks) or long (e.g. deafness or cancer).

Environmental protection – arrangements to cover those activities in the workplace which affect the environment (in the form of flora, fauna, water, air and soil) and, possibly, the health and safety of employees and others. Such activities include waste and effluent disposal and atmospheric pollution.

Accident – defined by the HSE as 'any unplanned event that results in injury or ill-health of people, or damage or loss to property, plant, materials or the environment or a loss of a business opportunity'. Other authorities define an accident more narrowly by excluding events that do not involve injury or ill-health. This book will always use the HSE definition.

Near miss – is any incident that could have resulted in an accident. Knowledge of near misses is very important since research has shown that, approximately, for every ten 'near miss' events at a particular location in the workplace, a minor accident will occur.

Dangerous occurrence – is a 'near miss' which could have led to serious injury or loss of life. Dangerous occurrences are defined in the Reporting of Injuries, Diseases and Dangerous Occurrences Regulations 1995 (often known as RIDDOR) and are always reportable to the Enforcement Authorities. Examples include the collapse of a scaffold or a crane or the failure of any passenger carrying equipment.

Hazard and risk – a **hazard** is the *potential* of a substance, activity or process to cause harm. Hazards take many forms including, for example, chemicals, electricity and working from a ladder. A hazard can be ranked relative to other hazards or to a possible level of danger.

A **risk** is the *likelihood* of a substance, activity or process to cause harm. A risk can be reduced and the hazard controlled by good management.

It is very important to distinguish between a *hazard* and a *risk* – the two terms are often confused and activities such as construction work are called *high risk* when they are *high hazard*. Although the hazard will continue to be high, the risks will be reduced as controls are implemented. The level of risk remaining when controls have been adopted is known as the *residual risk*. There should only be *high residual risk* where there is poor health and safety management and inadequate control measures.

1.3 The legal framework for health and safety

1.3.1 *Sub-divisions of law*

There are two sub-divisions of the law that apply to health and safety issues: criminal law and civil law.

Criminal law

Criminal law consists of rules of behaviour laid down by the government or the state and, normally, enacted by Parliament through Acts of Parliament. These rules or Acts are imposed on the people for the protection of the people. Criminal law is enforced by several different Government Agencies who may prosecute individuals (and Organizations) for contravening criminal laws. It is important to note that, except for very rare cases, only these Agencies are able to decide whether to prosecute an individual or not.

An individual who breaks criminal law is deemed to have committed an offence or crime and, if he is prosecuted, the court will determine whether he is guilty or not. If he is found guilty, the court could sentence him to a fine or imprisonment. Due to this possible loss of liberty, the level of proof required by a criminal court is very high and is known as proof 'beyond reasonable doubt', which is as near certainty as possible. While the prime object of a criminal court is the allocation of punishment, the court can award compensation to the victim or injured party. One example of criminal law is the Road Traffic Acts, which are enforced by the police. However, the police are not the only criminal law enforcement agency. The Health and Safety at Work etc Act (HSW Act) is another example of criminal law and this is enforced either by the Health and Safety Executive (HSE) or Local Authority Environmental Health Officers (EHO). Other agencies which enforce criminal law include the Fire Authority, the Environment Agency, Trading Standards and Customs and Excise.

There is one important difference between procedures for criminal cases in general and criminal cases involving health and safety. The prosecution in a criminal case has to prove the guilt of the accused beyond reasonable doubt. While this obligation is not totally removed in health and safety cases, section 40 of the HSW Act 1974 transferred, where there is a duty to do something 'so far as is reasonably practicable' or 'so far as is practicable' or 'use the best practicable means', the onus of proof to the accused to show that there was no better way to discharge his duty under the Act. However, when this burden of proof is placed on the accused, they need only satisfy the court on the balance of probabilities that what they are trying to prove has been done.

Civil law

Civil law concerns disputes between individuals or individuals and companies. An individual sues another individual or company to address a civil wrong or tort (or delict in Scotland). The individual who brings the complaint to court is known as the claimant or plaintiff (pursuer in Scotland) and the individual or company who is being sued is known as the defendant (defender in Scotland).

The civil court is concerned with liability and the extent of that liability rather than guilt or non-guilt. Therefore, the level of proof required is based on the 'balance of probability', which is a lower level of certainty than that of 'beyond reasonable doubt' required by the criminal court. If a defendant is found to be liable, the court would normally order him to pay compensation and possibly costs to the plaintiff. However, the lower the balance of probability, the lower the level of compensation awarded.

In extreme cases, where the balance of probability is just over 50%, the plaintiff may 'win' his case but lose financially because costs may not be awarded and the level of compensation low. The level of compensation may also be reduced through the defence of **contributory negligence**, which is discussed later in Section 1.8 of this chapter.

For cases involving health and safety, civil disputes usually follow accidents or illnesses and concern negligence or a breach of statutory duty. The vast majority of cases are settled 'out of court'. While actions are often between individuals, where the defendant is an employee who was acting in the course of his employment during the alleged incident, the defence of the action is transferred to his employer – this is known as **vicarious liability**. The civil action then becomes one between the individual and an employer.

The Employers' Liability (Compulsory Insurance) Act makes it a legal requirement for all employers to have employers' liability insurance. This ensures that any employee, who successfully sues his employer following an accident, is assured of receiving compensation irrespective of the financial position of the employer.

1.4 The legal system in England and Wales

The description which follows applies to England and Wales (and, with a few minor differences, to Northern Ireland).

Only the court functions concerning health and safety are mentioned. Figure 1.2 shows the court hierarchy in schematic form.

1.4.1 *Criminal law*

Magistrates Courts

Most criminal cases begin and end in the Magistrates Courts. Health and safety cases are brought before the court by enforcement officers (HSE or Local Authority EHO) and they are tried by a bench of three lay magistrates (known as Justices of the Peace) or a single district judge. The lay magistrates are members of the public, usually with little previous experience of the law, whereas a district judge is legally qualified.

The Magistrates Court has limited powers with a maximum fine of £5,000 (for employees) to £20,000 (for employers) or for those who ignore prohibition notices. Magistrates are also able to imprison for up to 6 months for breaches of enforcement notices. The vast majority of health and safety criminal cases are dealt with in the Magistrates Court.

Crown Court

The Crown Court hears the more serious cases, which are passed to them from the Magistrates Court – normally because the sentences available to the magistrates are felt to be too lenient. Cases are heard by a judge and jury, although some cases are heard by a judge alone. The penalties available to the Crown Court are an unlimited

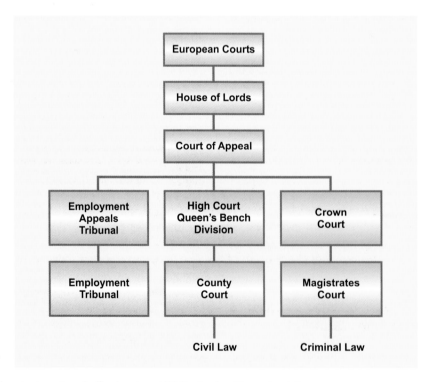

Figure 1.2 The main court system in England and Wales for health and safety.

fine and for breaches of enforcement notices up to 2 years imprisonment The Crown Court also hears appeals from the Magistrates Court.

Appeals from the Crown Court are made to the Court of Appeal (Criminal Division) who may then give leave to appeal to the most senior court in the country – the House of Lords. The most senior judge at the Court of Appeal is the Lord Chief Justice.

1.4.2 *Civil law*

County Court

The lowest court in civil law is the County Court, which only deals with minor cases (for compensation claims of up to £50,000 if the High Court agrees). Cases are normally heard by a judge sitting alone. For personal injury claims of less than £5,000, a small claims court is also available.

High Court

Many health and safety civil cases are heard in the High Court (Queen's Bench Division) before a judge only. It deals with compensation claims in excess of £50,000 and acts as an appeal court for the County Court.

Appeals from the High Court are made to the Court of Appeal (Civil Division). The House of Lords receives appeals from the Court of Appeal or, on matters of law interpretation, directly from the High Court. The most senior judge at the Court of Appeal is the Master of the Rolls.

The judges in the House of Lords are called Law Lords and are sometimes called upon to make judgements on points of law, which are then binding on lower courts. Such judgements form the basis of Common Law, which is covered later.

Other courts – Employment Tribunals

These were established in 1964 and primarily deal with employment and conditions of service issues, such as unfair dismissal. However, they also deal with appeals over health and safety enforcement notices, disputes between recognized safety representatives and their employers and cases of unfair dismissal involving health and safety issues. There are usually three members who sit on a Tribunal. These members are appointed and are often not legally qualified. Appeals from the Tribunal may be made to the Employment Appeal Tribunal or, in the case of enforcement notices, to the High Court. Appeals from Tribunals can only deal with the clarification of points of law.

1.5 The legal system in Scotland

Scotland has both criminal and civil courts but prosecutions are initiated by the procurator-fiscal rather than the HSE. The lowest criminal court is called the District

Court and deals with minor offences. The Sheriff Court has a similar role to that of the Magistrates Court (for criminal cases) and the County Court (for civil cases), although it can deal with more serious cases involving a sheriff and jury.

The High Court of Judiciary, in which a judge and jury sit, has a similar role to the Crown Court and appeals are made to the Court of Criminal Appeal. The Outer House of the Court of Session deals with civil cases in a similar way to the English High Court. The Inner house of the Court of Session is the Appeal Court for civil cases.

For both appeal courts, the House of Lords is the final court of appeal.

There are Industrial Tribunals in Scotland with the same role as Employment Tribunals in England and Wales.

1.6 European Courts

There are two European Courts – the European Court of Justice and the European Court of Human Rights.

The European Court of Justice, based in Luxembourg, is the highest court in the European Union (EU). It deals primarily with community law and its interpretation. It is normally concerned with breaches of community law by Member States and cases may be brought by other Member States or institutions. Its decisions are binding on all Member States. There is currently no right of appeal.

The European Court of Human Rights, based in Strasbourg, is not directly related to the EU – it covers most of the countries in Europe including the EU Member States. As its title suggests, it deals with human rights and fundamental freedoms. With the introduction of the Human Rights Act 1998 in October 2000, many of the human rights cases are heard in the UK.

1.7 Sources of law (England and Wales)

There are two sources of law: common law and statute law.

1.7.1 *Common law*

Common law dates from the 11th century when William I set up Royal Courts to apply a uniform (common) system of law across the whole of England. Prior to that time, there was a variation in law, or the interpretation of the same law, from one town or community to another. Common law is based on judgements made by courts (or strictly judges in courts). In general, courts are bound

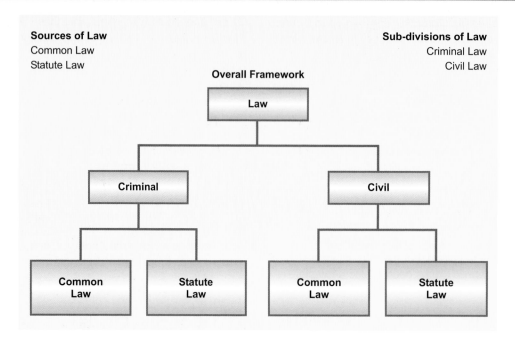

Sources of Law
Common Law
Statute Law

Sub-divisions of Law
Criminal Law
Civil Law

Figure 1.3 Sub-divisions and sources of law.

by earlier judgements on any particular point of law – this is known as 'precedent'. Lower courts must follow the judgements of higher courts. Hence judgements made by the Law Lords in the House of Lords form the basis of most of the common law currently in use.

In health and safety, the legal definition of negligence, duty of care and terms such as 'practicable' and 'as far as is reasonably practicable' are all based on legal judgements and form part of the common law. Common law also provides the foundation for most civil claims made on health and safety issues.

1.7.2 *Statute law*

Statute law is law which has been laid down by Parliament as Acts of Parliament. In health and safety, an Act of Parliament, the HSW Act 1974, lays down a general legal framework. Specific health and safety duties are, however, covered by Regulations or Statutory Instruments – these are also examples of statute law. If there is a conflict between statute and common law, statute law takes precedence. However, as with common law, judges interpret statute law usually when it is new or ambiguous. Although for health and safety, statute law is primarily the basis of criminal law, there is a tort of breach of statutory duty which can be used when a person is seeking compensation following an accident or illness. Breaches of the HSW Act 1974 cannot be used for civil action but breaches of most of the Regulations produced by the Act may give rise to civil actions.

1.7.3 *The relationship between the sub-divisions and sources of law*

The two sub-divisions of law may use either of the two sources of law (as shown in figure 1.3). For example, murder is a common law crime. In terms of health and safety, however, criminal law is only based on statute law, whereas civil law may be based on either common law or statute law.

In summary, criminal law seeks to protect everyone in society whereas civil law seeks to protect recompense the individual citizen.

1.8 Common law torts and duties

1.8.1 *Negligence*

The only tort (civil wrong) of real significance in health and safety is negligence. Negligence is the lack of reasonable care or conduct which results in the injury, damage (or financial loss) of or to another. Whether the act or omission was reasonable is usually decided as a result of a court action or out-of-court settlement.

There have been two important judgements that have defined the legal meaning of negligence. In 1856, negligence was judged to involve *actions or omissions* and the need for *reasonable and prudent* behaviour. In 1932, Lord Atkin said,

You must take reasonable care to avoid acts or omissions which you reasonably foresee would be likely to injure your neighbour. Who then, in law is my neighbour? The answer seems to be persons who are so closely and directly affected by my act that I ought reasonably to have them in contemplation as being so affected when I am directing my mind to the acts or omissions which are called in question.

It can be seen, therefore, that for negligence to be established, it must be reasonable and foreseeable that the injury could result from the act or omission. In practice, the Court may need to decide whether the injured party is the neighbour of the perpetrator. A collapsing scaffold could easily injure a member of the public, who could be considered a neighbour to the scaffold erector.

An employee who is suing his employer for negligence, needs to establish the following three criteria:

1. A duty was owed to him by his employer since the incident took place during the course of his employment.
2. There was a breach of that duty because the event was foreseeable and all reasonable precautions had not been taken.
3. The breach resulted in the specific injury, disease, damage and/or loss suffered.

These tests should also be used by anyone affected by the employer's undertaking (such as contractors and members of the public) who is suing the employer for negligence.

If the employer is unable to defend against the three criteria, two further partial defences are available. It could be argued that the employee was fully aware of the risks that were taken by not complying with safety instructions (known as *volenti non fit injuria* or 'the risk was willingly accepted'). This defence is unlikely to be totally successful because courts have ruled that employees have not accepted the risk voluntarily since economic necessity forces them to work.

The second possible defence is that of 'contributory negligence' where the employee is deemed to have contributed to the negligent act. This partial defence, if successful, can significantly reduce the level of compensation (up to 80% in some cases).

Any negligence claim must be made within a set time period (normally three years).

1.8.2 *Duties of care*

Several judgements have established that employers owe a duty of care to each of their employees. This duty cannot be assigned to others, even if a consultant is employed to advise on health and safety matters or if the employees are sub-contracted to work with another employer. These duties may be sub-divided into five groups. Employers must:

1. provide a safe place of work, including access and egress
2. provide safe plant and equipment
3. provide a safe system of work
4. provide safe and competent fellow employees
5. provide adequate levels of supervision, information, instruction and training.

Employer duties under common law are often mirrored in statute law. This, in effect, makes them both common law and statutory duties. These duties apply even if the employee is working at a third party premises or if he has been hired by his employer to work for another employer.

The requirements of a safe workplace, including the maintenance of floors and the provision of walkways and safe stairways, for example, are also contained in the Workplace (Health, Safety and Welfare) Regulations.

The requirement to provide competent fellow employees includes the provision of adequate supervision, instruction and training. As mentioned earlier, employers are responsible for the actions of their employees (**vicarious liability**) provided that the action in question took place during the normal course of their employment.

1.9 Levels of statutory duty

There are three principal levels of statutory duty, which form a hierarchy of duties. These levels are used extensively in health and safety statutory (criminal) law but have been defined by judges under common law. The three levels of duty are absolute, practicable and reasonably practicable.

1.9.1 *Absolute duty*

This is the highest level of duty and often occurs when the risk of injury is so high that injury is inevitable unless safety precautions are taken. It is a rare requirement regarding physical safeguards, although it was more common before 1992 when certain sections of the Factories Act 1961 were still in force. No assessment of risk is required but the duty is absolute and the employer has no choice but to undertake the duty. The verbs used in the Regulations are 'must' and 'shall'.

An example of this is Regulation 11(1) of the Provision and Use of Work Equipment Regulations concerning contact with a rotating stock bar which projects beyond a headstock of a lathe. Although this duty is absolute, it may still be defended using, for example, the argument that 'all reasonable precautions and all due diligence' were taken. This particular defence is limited to certain health and safety regulations such as The Electricity at Work Regulations and The Control of Substances Hazardous to Health.

Some of the health and safety management requirements contained in health and safety law place an absolute duty on the employer. The need for written safety policies and risk assessments when employee numbers rise above a basic threshold are examples of this.

1.9.2 *Practicable*

This level of duty is more often used than the absolute duty as far as the provision of safeguards is concerned and, in many ways, has the same effect. A duty that 'the employer ensure, so far as is practicable, that any control measure is maintained in an efficient state' means that if the duty is technically possible or feasible then it must be done irrespective of any difficulty, inconvenience or cost. Examples of this duty may be found in the Provision and Use of Work Equipment Regulations (Regulation 11(2) (a and b)).

1.9.3 *Reasonably practicable*

This is the most common level of duty in health and safety law and was defined by Judge Asquith in Edwards v. the National Coal Board (1949) as follows:

> *'Reasonably practicable' is a narrower term than 'physically possible', and seems to me to imply that computation must be made by the owner in which the quantum of risk is placed on one scale and the sacrifice involved in the measures necessary for averting the risk (whether in time, money or trouble) is placed in the other, and that, if it be shown that there is a gross disproportion between them – the risk being insignificant in relation to the sacrifice – the defendants discharge the onus on them.*

In other words, if the risk of injury is very small compared to the cost, time and effort required to reduce the risk, then no action is necessary. It is important to note that money, time and trouble must 'grossly outweigh' not balance the risk. This duty requires judgement on the part of the employer (or his adviser) and clearly needs a risk assessment to be undertaken with conclusions noted. Continual monitoring is also required to ensure that risks do not increase. There are numerous examples of this level of duty, including the Manual Handling

Operations Regulations and The Control of Substances Hazardous to Health Regulations. If organizations follow 'good practice', defined by the HSE as 'those standards which satisfy the law', then they are likely to satisfy the 'reasonably practicable' test. Approved codes of practice and guidance published by the HSE describe and define many of these standards.

The term 'suitable and sufficient' is used to define the scope and extent required for health and safety risk assessment and may be interpreted in a similar way to 'reasonably practicable'. More information is given on this definition in Chapter 5.

1.10 The influence of the EU on health and safety

As Britain is part of the EU, much of the health and safety law originates in Europe. Proposals from the European Commission may be agreed by member states. The member states are then responsible for making them part of their domestic law.

In Britain itself and in much of Europe, health and safety law is based on the principle of risk assessment described in chapter 5. The main role of the EU in health and safety is to harmonize workplace and legal standards and remove barriers to trade across member states. A directive from the EU is legally binding on each member state and must be incorporated into the national law of each member state. Directives set out specific minimum aims which must be covered within the national law. Some states incorporate Directives more speedily than others.

Directives are proposed by the European Commission, comprising 25 commissioners, who are citizens of each of the member states. The proposed Directives are sent to the European Parliament, which is directly elected from the member states. The European Parliament may accept, amend or reject the proposed Directives. The proposed Directives are then passed to the Council of Ministers,

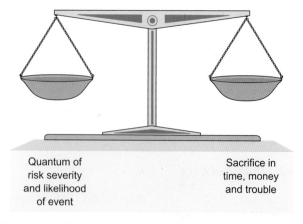

| Quantum of risk severity and likelihood of event | Sacrifice in time, money and trouble |

Figure 1.4 Diagrammatic view of 'reasonably practicable'.

who may accept the proposals on a qualified majority vote, unless the European Parliament has rejected them, in which case, they can only be accepted on a unanimous vote of the Council. The Council of Ministers consists of one senior Government minister from each of the member states.

The powers of the EU in health and safety law are derived from the Treaty of Rome 1957 and the Single European Act 1986. For health and safety, the Single European Act added two additional Articles, 100A and 118A. The 1997 Treaty of Amsterdam renumbered them as Articles 95 and 138 respectively. Article 95 is concerned with health and safety standards of equipment and plant and its Directives are implemented in the UK by the Department of Trade and Industry.

Article 138 is concerned with minimum standards of health and safety in employment and its Directives are implemented by the HSC/HSE.

The objective of the Single European Act 1986 is to produce a 'level playing field' for all member states so that goods and services can move freely around the EU without any one state having an unfair advantage over another. The harmonization of health and safety requirements across the EU is one example of the 'level playing field'.

The first introduction of an EU Directive into UK Health and Safety law occurred on 1 January 1993 when a Framework Directive on Health and Safety management and five daughter Directives were introduced using powers contained in the HSW Act 1974 (Figure 1.6).

Figure 1.5 Health and Safety Law poster – must be displayed or brochure given to employees. Specimen.

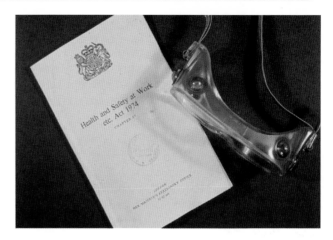

Figure 1.6 Health and Safety at Work Act.

These directives, known as the European Six Pack, covered the following areas:

1. Management of Health and Safety at Work
2. Workplace
3. Provision and Use of Work Equipment
4. Manual Handling
5. Personal Protective Equipment
6. Display Screen Equipment.

Since 1993, several other EU Directives have been introduced into UK law. Summaries of the more common ones are given in Chapter 21.

1.11 The Health and Safety at Work Act 1974

1.11.1 *Background to the Act*

The HSW Act resulted from the findings of the Robens Report, published in 1972. Earlier legislation had tended to relate to specific industries or workplaces. This resulted in over 5 m workers being unprotected by any health and safety legislation. Contractors and members of the public were generally ignored. The law was more concerned with the requirement for plant and equipment to be safe rather than the development of parallel arrangements for raising the health and safety awareness of employees.

A further serious problem was the difficulty that legislation had in keeping pace with developments in technology. For example, following a court ruling in 1955 which, in effect, banned the use of grinding wheels throughout industry, it took 15 years to produce the Abrasive Wheels Regulations 1970 to address the problem raised by the 1955 court judgement (John Summers and Sons Ltd v. Frost). In summary, health and safety legislation before 1974 tended to be reactive rather than proactive.

Lord Robens was asked, in 1970, to review the provision made for the health and safety of people at work. His report produced conclusions and recommendations upon which the HSW Act 1974 was based. The principal recommendations were as follows:

➤ There should be a single Act that covers all workers and that Act should contain general duties which should 'influence attitudes'.

➤ The Act should cover all those affected by the employer's undertaking such as contractors, visitors, students and members of the public.

➤ There should be an emphasis on health and safety management and the development of safe systems of work. This would involve the encouragement of employee participation in accident prevention. (This was developed many years later into the concept of the health and safety culture).

➤ Enforcement should be targeted at 'self-regulation' by the employer rather than reliance on prosecution in the courts.

These recommendations led directly to the introduction of the HSW Act in 1974.

1.11.2 *An overview of the Act*

Health and Safety Commission

The HSW Act established the HSC (Health and Safety Commission) and gave it the responsibility to draft new Regulations and to enforce them either through its executive arm, known as the HSE, or through the Local Authority EHO. The HSC has equal representation from employers, trade unions and special interest groups.

On 1st April 2008, the HSC and HSE merged into a single body called the Health and Safety Executive with the HSC functions being performed by a non-executive board acting within the HSE.

Regulations

The HSW Act is an **Enabling Act** which allows the Secretary of State to make further laws (known as regulations) without the need to pass another Act of Parliament. Regulations are law, approved by Parliament. These are usually made under the HSW Act, following proposals from the HSC. This applies to regulations based on EC Directives as well as 'home-grown' ones. It is a criminal offence to breach a regulation and any breaches may result in enforcement action as explained later in this chapter (Section 1.11.4).

The HSW Act, and general duties in the Management Regulations, aim to help employers to set goals, but leave them free to decide how to control hazards and risks which they identify. Guidance and Approved Codes of Practice give advice, but employers are free to take other routes to achieving their health and safety goals, so long as they do what is reasonably practicable. But some hazards are so great, or the proper control measures so expensive, that employers cannot be given discretion in deciding what to do about them. Regulations identify these hazards and risks and set out specific action that must be taken. Often these requirements are absolute – employers have no choice but to follow them and there is no qualifying phrase of 'reasonably practicable' included.

Some regulations apply across all organizations – the Manual Handling Regulations would be an example. These apply wherever things are moved by hand or bodily force. Equally, the Display Screen Equipment Regulations apply wherever visual display units are used at work. Other regulations apply to hazards unique to specific industries, such as mining or construction.

Wherever possible, the HSC will set out the regulations as goals, and describe what must be achieved, but not how it must be done.

Sometimes it is necessary to be *prescriptive*, and to spell out what needs to be done in detail, since some standards are absolute. For example, all mines should have two exits; contacts with live electrical conductors should be avoided. Sometimes European law requires prescription.

Some activities or substances are so dangerous that they have to be licensed, for example, explosives and asbestos removal. Large, complex installations or operations require 'safety cases', which are large-scale risk assessments subject to scrutiny by the Regulator. An example would be the recently privatized railway companies. They are required to produce safety cases for their operations.

Approved Code of Practice

An Approved Code of Practice (ACoP) is produced for most sets of regulations by the HSC/HSE and attempts to give more details on the requirements of the regulations. It also attempts to give the level of compliance needed to satisfy the regulations. ACoPs have a special legal status (sometimes referred to as quasi-legal). The relationship of an ACoP to a regulation is similar to the relationship of the Highway Code to the Road Traffic Acts. A person is never prosecuted for contravening the Highway Code but can be prosecuted for contravening the Road Traffic Acts. If a company is prosecuted for a breach of health and safety law and it is proved that it has not followed the relevant provisions of the ACoP, a court can find them at fault, unless the company can show that it has complied with the law in some other way.

Since most health and safety prosecutions take place in a Magistrates Court, it is likely that the lay magistrates will consult the relevant ACoP as well as the regulations when dealing with a particular case. Therefore, in practice, an employer must have a good reason for not adhering to an ACoP.

Codes of Practice generally are only directly legally binding if:

➤ the regulations or Act indicates that they are; for example The Safety Signs and Signals Regulations Schedule 2 specify British Standard Codes of practice for alternative hand signals, or
➤ they are referred to in an Enforcement Notice.

Guidance

Guidance, which has no formal legal standing, comes in two forms – legal and best practice. The Legal Guidance series of booklets is issued by the HSC and/or the HSE to cover the technical aspects of health and safety regulations. These booklets generally include the Regulations and the ACoP, where one has been produced.

Best practice guidance is normally published in the HSG series of publications by the HSE. Examples of best practice guidance books include *Health and Safety in Construction* HSG150 and *Lighting at Work* HSG38.

An example of the relationship between these three forms of requirement/advice can be shown using a common problem found throughout industry and commerce – minimum temperatures in the workplace.

Regulation 7 of the Workplace (Health, Safety and Welfare) Regulations states '(1) During working hours, the temperature in all workplaces inside buildings shall be reasonable'.

The ACoP states 'The temperature in workrooms should normally be at least 16 degrees Celsius unless much of the work involves severe physical effort in which case the temperature should be at least 13 degrees Celsius. . . .'

It would, therefore, be expected that employers would not allow their workforce to work at temperatures below those given in the ACoP unless it was **reasonable** for them to do so (if, for example, the workplace was a refrigerated storage unit).

Best practice guidance to cover this example is given in HSG194 (*Thermal Comfort in the Workplace*), in which possible solutions to the maintenance of employee welfare in low temperature environments is given.

Health and Safety Executive (HSE)

The prime function of the HSE is to monitor, review and enforce health and safety law and to produce codes of practice and guidance. However, the HSE also undertakes many other activities, such as the compilations of health and safety statistics, leading national health and safety campaigns, investigations into accidents or complaints, visiting and advising employers and the production of a very useful website.

1.11.3 *General duties and key sections of the Act*

A summary of the HSW Act is given in Chapter 21 (21.4) – only an outline will be given here.

Section 2: Duties of employers to employees

To ensure, so far as is reasonably practicable, the health, safety and welfare of all employees. In particular:

➤ safe plant and systems of work
➤ safe use, handling, transport and storage of substances and articles
➤ provision of information, instruction, training and supervision
➤ safe place of work, access and egress
➤ safe working environment with adequate welfare facilities
➤ a written safety policy together with organizational and other arrangements (if there are 5 or more employees)
➤ consultation with safety representatives and formation of safety committees where there are recognized trade unions.

Section 3: Duties of employers to others affected by their undertaking

A duty to safeguard those not in their employment but affected by the undertaking. This includes members of the public, contractors, patients, customers and students.

Section 4: Duties of landlords or owners

To ensure that means of access and egress are safe for those using the premises.

Section 6: Duties of suppliers

Persons who design, manufacture, import or supply any article or substance for use at work must ensure, so far as is reasonably practicable, that they are safe and without risk to health. Articles must be safe when they are set, cleaned, used and maintained. Substances must be without risk to health when they are used, handled, stored or transported. This requires that information must be supplied on the safe use of the articles and substances. There may be a need to guarantee the required level of safety by undertaking tests and examinations.

Section 7: Duties of employees

Two main duties:

1. to take reasonable care for the health and safety of themselves and others affected by their acts or omissions

Figure 1.7 Employees at work.

2. to co-operate with the employer and others to enable them to fulfil their legal obligations.

Section 8

No person is to misuse or interfere with safety provisions (sometimes known as the 'horseplay section').

Section 9

Employees cannot be charged for health and safety requirements such as personal protective equipment.

Section 37: Personal liability of directors

Where an offence is committed by a corporate body with the consent or connivance of, or is attributable to any neglect of, a director or other senior officer of the body, both the corporate body and the person are liable to prosecution.

1.11.4 *Enforcement of the Act*

Powers of inspectors

Inspectors under the Act work either for the HSE or the Local Authority. Local Authorities are responsible for retail and service outlets such as shops (retail and wholesale), restaurants, garages, offices, residential homes, entertainment and hotels. The HSE are responsible for all other work premises including the Local Authorities themselves. Both groups of inspectors have the same powers. The detailed powers of inspectors are given in Chapter 21. In summary an inspector has the right to:

➤ enter premises at any reasonable time, accompanied by a police officer, if necessary
➤ examine, investigate and require the premises to be left undisturbed

➤ take samples, photographs and, if necessary, dismantle and remove equipment or substances
➤ require the production of books or other relevant documents and information
➤ seize, destroy or render harmless any substance or article
➤ issue enforcement notices and initiate prosecutions.

An inspector may issue a formal caution when an offence has been committed but it is deemed that the public interest does not require a prosecution. Formal cautions are not normally considered if the offender has already had a formal caution.

Figure 1.8 The inspector inspects.

Enforcement notices

There are two types of enforcement notices.

Improvement Notice – This identifies a specific breach of the law and specifies a date by which the situation is to be remedied. An appeal must be made to the Employment Tribunal within 21 days. The notice is then suspended until the appeal is either heard or withdrawn. There are five main grounds for an appeal:

1. the inspector interpreted the law incorrectly.
2. the inspector exceeded his powers.
3. the breach of the law is admitted but the proposed solution is not practicable or reasonably practicable.
4. The time allowed to comply is too short.
5. The breach of the law is admitted but the breach is so insignificant that the notice should be cancelled.

Prohibition Notice – This is used to halt an activity which the inspector feels could lead to a serious personal injury. The notice will identify which legal requirement is being or is likely to be contravened. The notice takes effect as soon as it is issued. As with the improvement notice, an appeal may be made to the Employment Tribunal but, in this case, the notice remains in place during the appeal process.

There are two forms of prohibition notice:

1. an immediate prohibition notice – this stops the work activity immediately until the specified risk is reduced
2. a deferred prohibition notice – this stops the work activity within a specified time limit.

Penalties

(a) Magistrates Court (or Lower Court)
The Magistrates Court deals with summary offences.

For health and safety offences under sections 2–6 of the HSW Act, employers and others (self-employed, manufacturers and suppliers) may be fined up to £20,000.

For all other breaches of the HSW Act and breaches of health and safety regulations made under the Act, the maximum fine is £5,000. Thus the maximum fine that may be imposed on an employ under section 7 is £5,000.

For failure to comply with an enforcement notice or a court order, anybody may be fined up to £20,000 or imprisoned for up to 6 months.

(b) Crown Court (or Higher Court)
The Crown Court deals with indictable offences.

Fines are unlimited in the Crown Court for all health and safety offences and imprisonment is possible for up to 2 years for failure to comply with an enforcement notice or a court order.

(c) Both Courts
For the contravention of licensing requirements for nuclear installations, asbestos removal and the storage and manufacture of explosives, the lower court can fine up to £5,000 whereas the higher court can impose an unlimited fine and/or 2 years imprisonment.

The courts also have the power to disqualify directors who are convicted of health and safety offences for up to 5 years at the lower court and 15 years at the higher court.

Finally, it is important to stress that imprisonment for health and safety offences is rare. From 1996 until 2005, only 5 people were sent to prison for health and safety offences.

Summary of the actions available to an inspector
Following a visit by an inspector to premises, the following actions are available to an inspector:

➤ take no action
➤ give verbal advice
➤ give written advice
➤ formal caution
➤ serve an improvement notice
➤ serve a prohibition notice
➤ prosecute.

In any particular situation, more than one of these actions may be taken.

The decision to prosecute will be made either when there is sufficient evidence to provide a realistic possibility of conviction or when prosecution is in the public interest.

Work-related deaths are investigated by the police jointly with the HSE or Local Authority to ascertain whether a charge of manslaughter (culpable homicide in Scotland) is appropriate. If the police decide not to pursue such a charge, the HSE or Local Authority will continue the investigation under health and safety law.

In 2008, the Corporate Manslaughter and Corporate Homicide Act came into force. The Act clarifies the criminal liabilities of companies and other large organizations where serious failures in health and safety management systems led to a fatality. Prosecutions will be of the corporate body and not individuals. The liability of directors and board members under health and safety law is unaffected. The corporate body itself and individuals can still be prosecuted under the HSW Act for other health and safety offences.

Finally, the **Health and Safety (Information for Employees) Regulations** require that the Approved Poster entitled 'Health and safety Law – what you should know' is displayed or the Approved Leaflet is distributed. This information tells employees in general terms about the requirements of health and safety law. Employers must also inform employees of the local address of the enforcing authority (either the HSE or the local authority) and the Employment Medical Advisory Service. This should be marked on the poster or supplied with the leaflet.

From 1 July 2000 the latest version of the poster must be displayed or distributed. The revised text focuses on the modern framework of general duties of the employer and employee, supplemented by some basic information on health and safety management and risk assessment. It includes two additional boxes: one for details of trade union or other safety representatives and one for competent persons appointed to assist with health and safety and their responsibilities. The poster is shown in Figure 1.5.

1.12 The Management of Health and Safety at Work Regulations 1999

As mentioned earlier, on 1 January 1993, following an EC Directive, the Management of Health and Safety at

Work Regulations became law in the UK. These regulations were updated in 1999 and are described in detail in Chapter 21. In many ways the regulations were not introducing concepts or replacing the 1974 Act – they simply reinforced or amended the requirements of the HSW Act. Some of the duties of employers and employees were re-defined.

1.12.1 *Employers' duties*

Employers must:

➤ undertake suitable and sufficient written risk assessments when there are more than four employees
➤ put in place effective arrangements for the planning, organization, control, monitoring and review of health and safety measures in the workplace (including health surveillance). Such arrangements should be recorded if there are five or more than four employees
➤ employ (to be preferred) or contract competent persons to help them comply with health and safety duties
➤ develop suitable emergency procedures. Ensure that employees and others are aware of these procedures and can apply them
➤ provide health and safety information to employees and others, such as other employers, the self-employed and their employees who are sharing the same workplace and parents of child employees or those on work experience
➤ co-operate in health and safety matters with other employers who share the same workplace
➤ provide employees with adequate and relevant health and safety training
➤ provide temporary workers and their contract agency with appropriate health and safety information
➤ protect new and expectant mothers and young persons from particular risks
➤ under certain circumstances, as outlined in Regulation 6, provide health surveillance for employees.

The information that should be supplied by employers under the regulations is:

➤ risks identified by any risk assessments including those notified to an employer by other employers sharing the same workplace
➤ the preventative and protective measures that are in place
➤ the emergency arrangements and procedures and the names of those responsible for the implementation of the procedures.

Finally, it is important to note that the regulations outline the principles of prevention which employers and the self-employed need to apply so that health and safety risks are addressed and controlled. These principles are discussed in detail in Chapter 6.

1.12.2 *Employees' duties*

Employees must:

➤ use any equipment or substance in accordance with any training or instruction given by the employer
➤ report to the employer any serious or imminent danger
➤ report any shortcomings in the employer's protective health and safety arrangements.

1.13 Role and function of external agencies

The Health and Safety Commission, Health and Safety Executive and the Local Authorities (a term used to cover County, District and Unitary Councils) are all external agencies which have a direct role in the monitoring and enforcement of health and safety standards. There are, however, three other external agencies which have a regulatory influence on health and safety standards in the workplace (see Figure 1.9).

1.13.1 *Fire and rescue authority*

The Fire and Rescue Authority is situated within a single or group of local authorities and is normally associated with rescue, fire fighting and the offering of general advice. It has also been given powers to enforce fire precautions within places of work under fire safety law. The powers of the Authority are very similar to those of the HSE on health and safety matters. The Authority issues Alteration Notices to work places and conducts routine and random fire inspections (often to examine fire risk assessments). It should be consulted during the planning stage of proposed building alterations when such alterations may affect the fire safety of the building (e.g. means of escape).

The Fire and Rescue Authority can issue both Enforcement and Prohibition Notices. The Authority can prosecute for offences against fire safety law.

1.13.2 *The Environment Agency (Scottish Environmental Protection Agency)*

The Environment Agency was established in 1995 and was given the duty to protect and improve the environment.

Figure 1.9 The main external agencies that impact on the work place.

It is the regulatory body for environmental matters and has an influence on health and safety issues.

It is responsible for authorizing and regulating emissions from industry.

Other duties and functions of the agency include:

➤ ensuring effective controls of the most polluting industries
➤ monitoring radioactive releases from nuclear sites
➤ ensuring that discharges to controlled waters are at acceptable levels
➤ setting standards and issuing permits for the collection, transporting, processing and disposal of waste (including radioactive waste)
➤ enforcement of the Producer Responsibility Obligations (Packaging Waste) Regulations. These resulted from an EC Directive which seeks to minimize packaging and recycle at least 50% of it
➤ enforcement of the Waste Electrical and Electronic Equipment (WEEE) Directive and its associated directives.

The Agency may prosecute in the criminal courts for the infringement of environmental law – in one case a fine of £4 million was imposed.

1.13.3 *Insurance companies*

Insurance companies play an important role in the improvement of health and safety standards. Since 1969, it has been a legal requirement for employers to insure against liability for injury or disease to their employees arising out of their employment. This is called employers' liability insurance and was covered earlier in this chapter.

Certain public sector organizations are exempted from this requirement because any compensation is paid from public funds. Other forms of insurance include fire insurance and public liability insurance (to protect members of the public).

Premiums for all these types of insurance are related to levels of risk, which is related to standards of health and safety. In recent years, there has been a considerable increase in the number and size of compensation claims and this has placed further pressure on insurance companies.

Insurance companies are becoming effective health and safety regulators by weighing the premium offered to an organization according to its safety and/or fire precaution record.

1.14 Sources of information on health and safety

When anybody, whether a health and safety professional, a manager or an employee, is confronted with a health and safety problem, they will need to consult various items of published information to ascertain the scale of the problem and its possible remedies. The sources of this information may be internal to the organization and/or external to it.

Internal sources, which should be available within the organization include:

➤ accident and ill-health records and investigation reports

Figure 1.10 Good standards prevent harm and save money.

➤ absentee records
➤ inspection and audit reports undertaken by the organization and by external organizations such as the HSE
➤ maintenance, risk assessment (including COSHH) and training records
➤ documents which provide information to workers
➤ any equipment examination or test reports.

External sources, which are available outside the organization, are numerous and include:

➤ health and safety legislation
➤ HSC/HSE publications, such as Approved Codes of Practice, guidance documents, leaflets, journals, books and their website
➤ International (e.g. ILO), European and British standards
➤ health and safety magazines and journals
➤ information published by trade associations, employer organizations and trade unions
➤ specialist technical and legal publications
➤ information and data from manufacturers and suppliers
➤ the Internet and encyclopaedias.

Many of these sources of information will be referred to throughout this book.

1.15 Moral, legal and financial arguments for health and safety management

1.15.1 *Moral arguments*

The moral arguments are supported by the occupational accident and disease rates.

Accident rates
Accidents at work can lead to serious injury and even death. Although accident rates are discussed in greater detail in later chapters, some trends are shown in Tables 1.1–1.4 below. A major accident is a serious accident typically involving a fracture of a limb or a 24-hour stay in a hospital. An 'over 3-day accident' is an accident which leads to more than 3 days off work. Statistics are collected on all people who are injured at places of work not just employees.

It is important to note that since 1995 suicides and trespassers on the railways have been included in the HSE figures – this has led to a significant increase in the figures. Table 1.1 details accidents involving all people at a place of work over a 4-year period. In 2005/06, there were 384 fatal injuries to members of the public and about 66% of these were due to acts of suicide or trespass on the railways.

Table 1.1 Accidents involving all people at a place of work

Injury	2002/03	2003/04	2004/05	2005/06
Death	623	606	593	596
Major	41,985	45,516	46,018	45,230
Over 3 day	129,135	130,247	122,922	118,645

Table 1.2 shows, for the year 2004/05, the breakdown in accidents between employees, self-employed and members of the public. Table 1.3 shows the figures for employees only. It shows that whilst there has been a decline in fatalities, major accidents have increased.

Table 1.4 gives an indication of accidents in different employment sectors.

Table 1.2 Accidents for different groups of people for 2004/05

	Deaths	Major	Over 3 day
Total	593	46,018	122,922
Employees	172 (29%)	30,451	121,779
Self-employed	51 (9%)	1,251	1,143
Members of the public	370 (62%)	14,316	n/a

Table 1.3 Accidents involving employees in the workplace

Injury	2002/03	2003/04	2004/05	2005/06
Death	183	168	172	160
Major	28,113	30,666	30,451	28,605
Over 3 day	128,184	129,143	121,779	117,471

These figures indicate that there is a need for health and safety awareness even in occupations which many would consider very low hazard such as schools, the health services and hotels. In fact over 70% of all deaths occur in the service sector and manufacturing is considerably safer than construction and agriculture.

In 2005/06, slips and trips accounted for 38% of major accidents whereas 41% of over 3-day injuries were caused by handling, lifting and carrying.

Although there has been a decrease in fatalities over recent years, there is still a very strong moral case for improvement in health and safety performance.

Table 1.4 Accidents (per 100,000 workers) to all workers in various employment sectors (2005/06)

Sector	Deaths	Major
Agriculture	4.6	211.7
Construction	3.5	310.2
Transport	3.1	398.2
Manufacturing	1.4	180.3
Retail and wholesale	1.3	257.6
Hotel & catering	0.1	65.4
Health services	0.1	70.8

Disease rates

Work-related ill-health and occupational disease can lead to absence from work and, in some cases, to death. Such occurrences may also lead to costs to the state (the Industrial Injuries Scheme) and to individual employers (sick pay and, possibly, compensation payments).

In 2004/05, a major survey was undertaken into self-reported work-related illness. In that year there were 2.0 million people in the UK suffering from work-related illness, of whom 600,000 were new cases in that year. This led to 28 million working days lost compared to 6 million due to workplace injury. The top four illnesses were:

1. musculoskeletal disorders   1,010,000
2. stress related   509,000
3. lower respiratory disease   137,000
4. deafness, tinnitus and other ear problems   74,000

Due to musculoskeletal disorders, 11.6 million working days were lost causing each sufferer to have 21 days off work on average and 12.8 million working days were lost due to stress, depression and anxiety causing each sufferer to have 31 days off work on average. Recent research has shown that one in five people who are on sickness leave from work for 6 weeks, will stay off work and leave paid employment.

Ill-health statistics are estimated either from self-reported cases (Tables 1.5 and 1.6) or from data supplied by hospital specialists and occupational physicians (Table 1.7). The number of self-reported cases will always be much higher than those seeking advice from hospital-based specialists.

1.15.2 *Legal arguments*

The legal arguments, concerning the employer's duty of care in criminal and civil law, have been covered earlier.

Table 1.5 Self-reported work-related illness for 2005/06

Type of illness	Prevalence (thousands)
Musculoskeletal disorders of the back	437
Musculoskeletal disorders of the neck or upper limbs	374
Musculoskeletal disorders of the lower limbs	209
Musculoskeletal disorders in total	**1,020**
Stress, depression or anxiety	420
Breathing or lung problems	156
Hearing problems	68
Other (e.g. Skin or Heart problems)	294
Total	**1,958**

Table 1.6 Ill-health self-reported statistics (thousands) for all industries

	2001/02	2003/04	2004/05
Musculoskeletal Disorders	1,102	1,108	1,012
Stress, depression or anxiety	548	557	509
Breathing or lung problems	166	183	137
Hearing problems	88	81	74
Skin problems	38	31	29
Heart/circulatory problems	79	66	56
Headaches and/or eyestrain	53	37	31
Infectious disease	31	28	28
Other illnesses	171	142	130
Total	**2,276**	**2,233**	**2,006**

Some statistics on legal enforcement indicate the legal consequences resulting from breaches in health and safety law. There have been some very high compensation awards for health and safety cases in the civil courts and fines in excess of £100,000 in the criminal courts.

Table 1.7 Average annual incidence of new work-related illness reported by specialist doctors for 2003/05

Mental health (stress, anxiety)	7,200
Musculoskeletal disorders	6,400
Dermatitis and other skin disorders	3,500
Respiratory disease	3,200
Infections	1,300
Hearing loss	400
Total	**22,000**

Table 1.8 Number of enforcement notices issued over a 3-year period by the HSE

Year	Improvement notice	Prohibition notice
2003/04	6,798	4,537
2004/05	5,186	3,285
2005/06	3,787	2,596

Table 1.9 Number of prosecutions by the HSE over a 3-year period

Year	Offences prosecuted	Convictions
2003/04	1,720	1,317
2004/05	1,320	1,025
2005/06	1,012	741

Table 1.8 shows the number of enforcement notices served over a 3-year period by the HSE. Most notices are served in the manufacturing sector followed by construction and agriculture. Local Authorities serve 40% of all improvement notices and 20% of all prohibition notices.

Table 1.9 shows the number of prosecutions by the HSE over the same 3-year period and it can be seen that approximately 70% of those prosecuted are convicted. There was a reduction in the number of prosecutions over the 3-year period but provisional figures for 2006/07 show an increase of 20% in the number of prosecutions. HSE present 80% of the prosecutions and the remainder are presented by Local Authority EHO. Most of these prosecutions are for infringements of the

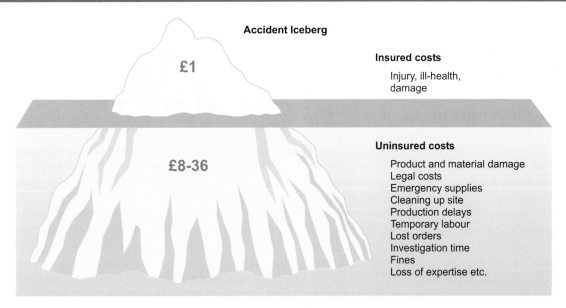

Accident Iceberg

£1

£8-36

Insured costs

Injury, ill-health,
damage

Uninsured costs

Product and material damage
Legal costs
Emergency supplies
Cleaning up site
Production delays
Temporary labour
Lost orders
Investigation time
Fines
Loss of expertise etc.

Figure 1.11 Insured and uninsured costs.

Construction Regulations and the Provision and Use of Work Equipment Regulations.

There are clear legal reasons for sound health and safety management systems.

1.15.3 *Financial arguments*

Costs of accidents

Any accident or incidence of ill-health will cause both direct and indirect costs and incur an insured and an uninsured cost. It is important that all of these costs are taken into account when the full cost of an accident is calculated. In a study undertaken by the HSE, it was shown that indirect costs or hidden costs could be 36 times greater than direct costs of an accident. In other words, the direct costs of an accident or disease represent the tip of the iceberg when compared to the overall costs (Figure 1.11).

Direct costs

These are costs which are directly related to the accident and may be insured or uninsured.

Insured direct costs normally include:

➤ claims on employers and public liability insurance
➤ damage to buildings, equipment or vehicles
➤ any attributable production and/or general business loss.

Uninsured direct costs include:

➤ fines resulting from prosecution by the enforcement authority
➤ sick pay
➤ some damage to product, equipment, vehicles or process not directly attributable to the accident (e.g. caused by replacement staff)

➤ increases in insurance premiums resulting from the accident
➤ any compensation not covered by the insurance policy due to an excess agreed between the employer and the insurance company
➤ legal representation following any compensation claim.

Indirect costs

These are costs which may not be directly attributable to the accident but may result from a series of accidents. Again these may be insured or uninsured.

Insured indirect costs include:

➤ a cumulative business loss
➤ product or process liability claims
➤ recruitment of replacement staff.

Uninsured indirect costs include:

➤ loss of goodwill and a poor corporate image
➤ accident investigation time and any subsequent remedial action required
➤ production delays
➤ extra overtime payments
➤ lost time for other employees, such as a First Aider, who attend to the needs of the injured person
➤ the recruitment and training of replacement staff
➤ additional administration time incurred
➤ first aid provision and training
➤ lower employee morale possibly leading to reduced productivity.

Some of these items, such as business loss, may be uninsurable or too prohibitively expensive to insure. Therefore, insurance policies can never cover all of the costs of an accident or disease either because some items are not covered by the policy or the insurance excess is greater than the particular item cost.

1.16 The framework for health and safety management

Most of the key elements required for effective health and safety management are very similar to those required for good quality, finance and general business management. Commercially successful organizations usually have good health and safety management systems in place. The principles of good and effective management provide a sound basis for the improvement of health and safety performance (see Figure 1.12).

HSE, in HSG 65, Successful Health and Safety Management, have identified six key elements involved in a successful health and safety management system. The following chapters will describe and discuss this framework in detail. The six elements are:

1. *Policy*: A clear health and safety policy contributes to business efficiency and continuous improvement throughout the operation. The demonstration of senior management involvement provides evidence to all stakeholders that responsibilities to people and the environment are taken seriously. The policy should state the intentions of the organization in terms of clear aims, objectives and targets.

2. *Organizing*: A well-defined health and safety organization offering a shared understanding of the organization's values and beliefs, at all levels of the organization, is an essential component of a positive health and safety culture. An effective organization will be noted for good staff involvement and participation; high quality communications; the promotion of competency; and the empowerment and commitment of all employees to make informed contributions.

3. *Planning and implementing*: A clear health and safety plan involves the setting and implementation of performance standards, targets and procedures through an effective heath and safety management system. The plan is based on risk assessment methods to decide on priorities and set objectives for the effective control or elimination of hazards and the reduction of risks. Measuring success requires the establishment of practical plans and performance targets against which achievements can be identified.

4. *Measuring performance*: This includes both active (sometimes called proactive) and reactive monitoring to see how effectively the health and safety management system is working. Active monitoring involves looking at the premises, plant and substances plus the people, procedures and systems. Reactive monitoring discovers through investigation of accidents

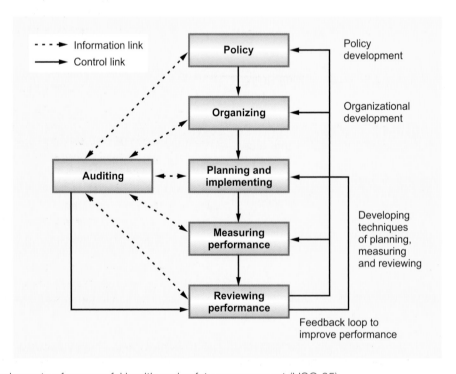

Figure 1.12   Key elements of successful health and safety management (HSG 65).

and incidents why controls have failed. It is also important to measure the organization against its own long-term goals and objectives.

5. *Reviewing performance*: The results of monitoring and independent audits should be systematically reviewed to evaluate the performance of the management system against the objectives and targets established by the health and safety policy. It is at the review stage that the objectives and targets set in the health and safety policy may be changed. Changes in the health and safety environment in the organization, such as an accident, should also trigger a performance review – this is discussed in more detail in Chapter 2. Performance reviews are not only required by the HSW Act but are part of any organization's commitment to continuous improvement. Comparisons should be made with internal performance indicators and the external performance indicators of similar organizations with exemplary practices and high standards.

6. *Auditing*: An independent and structured audit of all parts of the health and safety management system reinforces the review process. Such audits may be internal and external (the differences are discussed in Chapters 16 and 18). The audit assesses compliance with the health and safety management arrangements and procedures. If the audit is to be really effective, it must assess both the compliance with stated procedures and the performance in the workplace. It will identify weaknesses in the health and safety policy and procedures and identify unrealistic or inadequate standards and targets.

The conclusions from an audit of an organization's health and safety performance should be included in the annual report for discussion at Board meetings. This is considered best corporate practice.

1.17 Practice NEBOSH questions for Chapter 1

1. **Explain**, using an example in **EACH** case, the meaning of the following terms:
   (a) 'hazard'
   (b) 'risk'.

2. (a) **Outline** the main functions of:
      (i) criminal law
      (ii) civil law.
   (b) **Explain** the principal differences between common law and statute law.

3. **Outline** the **FOUR** key differences between civil law and criminal law.

4. An employer has common law duty of care for the health, safety and welfare of their employees. **Giving** an example in **EACH** case, **identify** what the employer must provide in order to fulfil this common law duty of care.

5. (a) **Define** the term 'negligence'.
   (b) **Outline** the **THREE** standard conditions that must be met for an employee to prove a case of negligence against an employer in seeking compensation for injuries following an accident at work. For **EACH** condition **EXPLAIN** its meaning.

6. (a) **Explain THREE** possible defences to a civil law claim of negligence.
   (b) **State** the circumstances, in which an employer may be held vicariously liable for the negligence of an employee.

7. **Explain** the meaning of the phrase 'so far as is reasonably practicable'.

8. **State** the general and specific duties of employers to their employees under section 2 of the Health and Safety at Work etc Act 1974.

9. **Outline** the general duties placed on employees by:
   (a) the Health and Safety at Work etc Act 1974
   (b) the Management of Health and Safety at Work Regulations 1999.

10. With respect to section 6 of the Health and Safety at Work etc Act 1974, **outline** the general duties of designers, manufacturers and suppliers of articles and substances for use at work to ensure that they are safe and without risk.

11. With reference to the Management of Health and Safety at Work Regulations 1999:
    (a) **outline** the information that an employer must provide to his employees
    (b) **identify FOUR** classes of persons, other than his own employees, to whom an employer must provide health and safety information
    (c) **identify** the specific circumstances when health and safety training should be given to employees.

12. **Explain** the meaning, status and roles of:
    (a) health and safety regulations
    (b) HSC Approved Codes of Practice
    (c) HSE guidance.

13. **Outline**, with an example of each, the differences between health and safety regulations and HSC approved codes of practice.

14. **List** the powers given to inspectors appointed under the Health and Safety at Work etc Act 1974.

15. (a) **Outline FOUR** powers available to an inspector when investigating an accident on a construction site.
    (b) **Identify** the **TWO** types of enforcement notice that may be served by an inspector and **state** the conditions that must be satisfied before **EACH** type of notice is served.

16. (a) **Explain**, using a relevant example, the circumstances under which a health and safety inspector may serve an improvement notice.
    (b) **Identify** the time period within which an appeal may be lodged against an improvement notice **AND state** the effect that the appeal will have on the notice.
    (c) **State** the penalties for contravening the requirements of an improvement notice when heard **BOTH** summarily **AND** on indictment.

17. (a) **Explain**, giving an example in each case, the circumstances under which a health and safety inspector may serve:
    (i) an improvement notice
    (ii) a prohibition notice.
    (b) **State** the effect on each type of enforcement notice of appealing against it.

18. **Outline** ways in which the Health and Safety Executive can influence the health and safety performance of an organization.

19. **Outline** the sources of published information that may be consulted when dealing with a health and safety problem at work.

20. (a) **Outline** the purpose of employer liability insurance.

(b) **Outline SIX** costs of a workplace accident that might be uninsured.

21. (a) **Explain** why it is necessary for sub-contractors to have employers' liability insurance.
    (b) **Outline SIX** costs of a workplace accident which would not be covered by such insurance.

22. Replacement or repair of damaged equipment is a cost that an organization may incur following an accident at work. **List EIGHT** other possible costs to the organization following a workplace accident.

23. **List EIGHT** possible costs to an organization when employees are absent due to work-related ill-health.

24. **Outline** reasons for maintaining good standards of health and safety within an organization.

25. (a) **Draw** a flowchart to show the relationships between the six elements of the health and safety management model in HSE's 'Successful Health and Safety Management' (HSG 65).
    (b) **Outline** the part that **EACH** element of the HSG 65 model plays within the health and safety management system.

26. (a) **Draw** a flowchart to **identify** the main components of the health and safety management system described in the HSE publication 'Successful Health and Safety Management' (HSG 65).
    (b) **Outline any TWO** components of the health and safety management system identified in (a).

27. **Outline** the main components of a health and safety management system.

28. **Outline** the economic benefits that an organization may obtain by implementing a successful health and safety management system.

Policy

2

2.1 Introduction

Every construction organization should have a clear policy for the management of health and safety so that everybody associated with the organization is aware of its health and safety aims and objectives. For a policy to be effective, it must be honoured in the spirit as well as the letter. A good health and safety policy will also enhance the performance of the organization in areas other than health and safety, help with the personal development of the workforce and reduce financial losses. It is important that each construction site throughout the organization is aware of the policy.

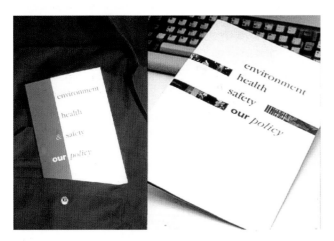

Figure 2.1 Well-presented policy documents.

2.2 Legal requirements

Section 2(3) of the Health and Safety at Work Act and the Employers' Health and Safety Policy Statements (exception) Regulations 1975 requires employers, with five or more employees, to prepare and reviue on a regular basis a written health and safety policy together with the necessary organization and arrangements to carry it out and to bring the policy and any revision of it to the notice of their employees.

This does not mean that organizations with four or less employees do not need to have a safety policy – it simply means that it does not have to be written down. The number of employees is the maximum number at any one time whether they are full time, part time or seasonal.

This obligation on employers was introduced for the first time by the HSW Act and is related to the reliance in the Act on self-regulation by employers to improve health and safety standards rather than on enforcement alone. A good health and safety policy involves the development, monitoring and review of the standards needed to address and reduce the risks to health and safety produced by the organization.

The law requires that the written health and safety policy should include the following three sections:

1. a health and safety policy statement of intent which includes the health and safety aims and objectives of the organization
2. the health and safety organization detailing the people with health and safety responsibilities and their duties
3. the health and safety arrangements in place in terms of systems and procedures.

The Management of Health and Safety at Work Regulations also requires the employer to 'make and give effect to such arrangements as are appropriate, having regard to the nature of his activities and the size of his undertaking, for the effective planning, organization, control, monitoring and review of the preventative and protective measures'. It further requires that these arrangements must be recorded when there are five or more employees.

When an inspector from the Health and Safety Executive (HSE) or Local Authority visits an establishment, it is very likely that they will wish to see the health and safety policy as an initial indication of the management attitude and commitment to health and safety. There have been instances of prosecutions being made due to the absence of a written health and safety policy. (Such cases are, however, usually brought before the courts because of additional concerns.)

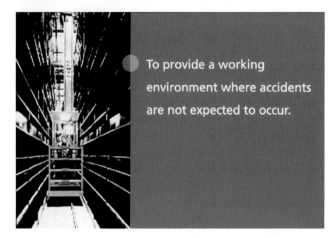

To provide a working environment where accidents are not expected to occur.

Figure 2.2 Part of a policy commitment.

2.3 Key elements of a health and safety policy

2.3.1 *Policy statement of intent*

The health and safety policy statement of intent is often referred to as the health and safety policy statement or simply (and incorrectly) as the health and safety policy. It should contain the aims (which are not measurable) and objectives (which are measurable) of the organization or company. Aims will probably remain unchanged during policy revisions whereas objectives will be reviewed and modified or changed each year. The statement should be written in clear and simple language so that it is easily understandable. It should also be fairly brief and broken down into a series of smaller statements or bullet points.

The statement should be signed and dated by the most senior person in the organization. This will demonstrate management commitment to health and safety and give authority to the policy. It will indicate where ultimate responsibility lies and the frequency with which the policy statement is reviewed.

The policy statement should be written by the organization and not by external consultants since it needs to address the specific health and safety issues and hazards within the organization. In large organizations, it may be necessary to have health and safety policies for each department and/or site with an overarching general policy incorporating the individual policies. Such an approach is often used by local authorities and multinational companies.

The following points should be included or considered when a health and safety policy statement is being drafted:

➤ the aims, which should cover health and safety, welfare and relevant environmental issues

➤ the position of the senior person in the organization or company who is responsible for health and safety (normally the chief executive)

➤ the names of the Health and Safety Adviser, any safety representatives or other health and safety competent persons

➤ a commitment to the basic requirements of the Health and Safety at Work Act (access, egress, safe plant and systems of work, use, handling, transport and handling of articles and substances, information, training and supervision)

➤ a commitment to the additional requirements of the Management of Health and Safety at Work Regulations (risk assessment, emergency procedures, health surveillance and employment of competent persons)

➤ duties towards the wider general public and others (contractors, customers, students, etc.)

➤ the principal hazards in the organization

➤ specific policies of the organization (smoking policy, violence to staff, etc.)

➤ a commitment to employee consultation possible using a safety committee or plant council

➤ duties of employees (particularly those defined in the Management of Health and Safety at Work Regulations)

➤ specific performance targets for the immediate and long-term future.

Health and safety **performance targets** are an important part of the statement of intent because:

➤ they indicate that there is management commitment to improve health and safety performance

➤ they motivate the workforce with tangible goals resulting, perhaps, in individual or collective rewards

➤ they offer evidence during the monitoring, review and audit phases of the management system.

The type of target chosen depends very much on the areas which need the greatest improvement in the organization. The following list, which is not exhaustive, shows common health and safety performance targets:

> a specific reduction in the number of accidents, incidents (not involving injury) and cases of work-related ill-health (perhaps to zero)
> a reduction in the level of sickness absence
> a specific increase in the number of employees trained in health and safety
> an increase in the reporting of minor accidents and 'near miss' incidents
> a reduction in the number of civil claims
> no enforcement notices from the HSE or Local Authority
> a specific improvement in health and safety audit scores
> the achievement of a nationally recognized health and safety management standard, such as OHSAS 18001 (see chapter 22).

The policy statement of intent should be posted on prominent notice boards at every site and throughout the workplace and brought to the attention of all employees at induction and refresher training sessions. It can also be communicated to the workforce during team briefing sessions, at 'toolbox' talks which are conducted at the workplace or directly by email, intranet, newsletters or booklets. It should be a permanent item on the agenda for health and safety committee meetings where it and related targets should be reviewed at each meeting.

2.3.2 Organization of health and safety

This section of the policy defines the names, positions and duties of those within the organization or company who have a specific responsibility for health and safety. Therefore, it identifies those health and safety responsi-

bilities and the reporting lines through the management structure. This section will include the following groups together with their associated responsibilities:

> directors and senior managers (responsible for setting policy, objectives and targets)
> supervisors (responsible for checking day-to-day compliance with the policy)
> safety advisers (responsible for giving advice during accident investigations and on compliance issues)
> other specialists, such as an occupational nurse, chemical analyst and an electrician (responsible for giving specialist advice on particular health and safety issues)
> safety representatives (responsible for representing employees during consultation meetings on health and safety issues with the employer)
> employees (responsible for taking reasonable care of the health and safety of themselves and others who may be affected by their acts or omissions)
> fire marshals (responsible for the safe evacuation of the building in an emergency)
> first-aiders (responsible for administering first aid to injured persons).

For smaller organizations, some of the specialists mentioned above may well be employed on a consultancy basis.

For the health and safety organization to work successfully, it must be supported from the top (preferably at Board level) and some financial resource made available.

It is also important that certain key functions are included in the organization structure. These include:

> accident investigation and reporting
> health and safety training and information
> health and safety monitoring and audit
> health surveillance
> monitoring of plant and equipment and its maintenance and risk assessment

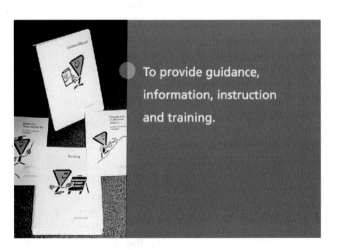

To provide guidance, information, instruction and training.

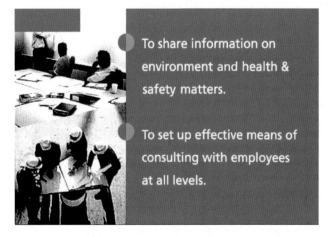

To share information on environment and health & safety matters.

To set up effective means of consulting with employees at all levels.

Figure 2.3 Good information, training and working with employees is essential.

➤ liaison with external agencies
➤ management and/or employee safety committees – the management committee will monitor day-to-day problems and any concerns of the employee health and safety committee.

The role of the health and safety adviser is to provide specialist information to managers in the organization and to monitor the effectiveness of health and safety procedures. The adviser is not 'responsible' for health and safety or its implementation; that is the role of the line managers.

Finally the job descriptions, which define the duties of each person in the health and safety organizational structure, must not contain responsibility overlaps or blur chains of command. Each individual must be clear about his/her responsibilities and the limits of those responsibilities.

2.3.3 *Arrangements for health and safety*

The arrangements section of the health and safety policy gives details of the specific systems and procedures used to assist in the implementation of the policy statement. This will include health and safety rules and procedures and the provision of facilities such as a first aid room and wash rooms. It is common for risk assessments (including COSHH, manual handling and personal protective equipment (PPE) assessments) to be included in the arrangements section particularly for those hazards referred to in the policy statement. It is important that arrangements for fire and other emergencies and for information, instruction, training and supervision are also covered. Local codes of practice (e.g. for fork lift drivers) should be included.

The following list covers the more common items normally included in the arrangements section of the health and safety policy:

➤ employee health and safety code of practice
➤ accident and illness reporting and investigation procedure
➤ emergency procedures, first aid
➤ fire drill procedure
➤ procedures for undertaking risk assessments
➤ control of exposure to specific hazards (noise, vibration, radiation, manual handling, hazardous substances, etc.)
➤ machinery safety (including safe systems of work, and lifting and pressure equipment)
➤ electrical equipment (maintenance and testing)
➤ maintenance procedures
➤ permits to work procedures
➤ use of PPE
➤ monitoring procedures including health and safety inspections and audits

➤ procedures for the control and safety of contractors and visitors
➤ provision of welfare facilities
➤ training procedures and arrangements
➤ catering and food hygiene procedures
➤ arrangements for consultation with employees
➤ terms of reference and constitution of the safety committee
➤ procedures and arrangements for waste disposal.

The three sections of the health and safety policy are usually kept together in a health and safety manual and copies distributed around the organization.

2.4 Review of health and safety policy

It is important that the health and safety policy is monitored and reviewed on a regular basis. For this to be successful, a series of benchmarks needs to be established. Such benchmarks, or examples of good practice, are defined by comparison with the health and safety performance of other parts of the organization or the national performance of the occupational group of the organization. The HSE publish an annual report, statistics and a bulletin all of which may be used for this purpose. Typical benchmarks include accident rates per employee and accident or disease causation. The targets set by the HSE Construction Industry Advisory Committee (CONIAC), described in Chapter 9, are the most appropriate benchmarks to use. Chapter 9 also describes the new Strategic Forum Health and Safety Task Group recently set up by the Government – this should also develop some targets and benchmarks.

There are several reasons to review the health and safety policy. The more important reasons are:

➤ significant organizational changes may have taken place
➤ there have been changes in personnel
➤ there have been changes in legislation
➤ the monitoring of risk assessments or accident/incident investigations indicate that the health and safety policy is no longer totally effective
➤ enforcement action has been taken by the HSE or Local Authority
➤ a sufficient period of time has elapsed since the previous review.

A positive promotion of health and safety performance will achieve far more than simply preventing accidents and ill-health. It will:

➤ support the overall development of personnel
➤ improve communication and consultation throughout the organization

Safety, quality and production will receive equal priority

Policy Statement View

Figure 2.4 Sound policy but not put into practice – blocked fire exit.

➤ minimize financial losses due to accidents and ill-health and other incidents

➤ directly involve senior managers in all levels of the organization

➤ improve supervision, particularly for young persons and those on occupational training courses

➤ improve production processes

➤ improve the public image of the organization or company.

It is apparent, however, that some health and safety policies appear to be less than successful. There are many reasons for this. The most common are:

➤ the statements in the policy and the health and safety priorities are not understood by or properly communicated to the workforce

➤ minimal resources are made available for the implementation of the policy

➤ too much emphasis on rules for employees and too little on management policy

➤ a lack of parity with other activities of the organization (such as finance and quality control) due to mistaken concerns about the costs of health and safety and the effect of those costs on overall performance. This attitude produces a poor health and safety culture

➤ lack of senior management involvement in health and safety, particularly at board level

➤ employee concerns that their health and safety issues are not being addressed or that they are not receiving adequate health and safety information. This can lead to low morale amongst the workforce and, possibly, high absenteeism

➤ high labour turnover

➤ inadequate or no PPE (Personal Protective Equipment)

➤ unsafe and poorly maintained machinery and equipment

➤ a lack of health and safety monitoring procedures.

In summary, a successful health and safety policy is likely to lead to a successful organization or company. A checklist for assessing any safety policy has been produced by the HSE and has been reproduced in Appendix 2.1.

2.5 Practice NEBOSH questions for Chapter 2

1. (a) **Outline** the circumstances under which an employer must, by law, prepare a written health and safety policy.

    (b) **Explain** the purposes of **EACH** of the following sections of a health and safety policy document:

    (i) 'statement of intent'

    (ii) 'organization'

    (iii) 'arrangements'.

2. (a) **State** the legal requirements whereby employers must prepare a written statement of their health and safety policy.

   (b) **Outline** the various methods for communicating the contents of a health and safety policy to a workforce.

3. **Explain** why a health and safety policy should be signed by the most senior person in an organization, such as a Managing Director or Chief Executive.

4. (a) **Identify** the typical content of the 'statement of intent' section of an organization's health and safety policy document.

   (b) **Outline** the factors that may indicate that health and safety standards within an organization do not reflect the objectives within the 'statement of intent'.

5. **Identify SIX** categories of persons who may be shown in the 'organization' section of a health

and safety policy document **AND state** their likely general or specific health and safety responsibilities.

6. **Outline** the issues that are typically included in the arrangements section of a health and safety document.

7. **Outline** the key areas that should be addressed in the 'arrangements' section of a health and safety policy document.

8. (a) **Explain** why it is important for an organization to set targets in terms of its health and safety performance.

   (b) **Outline SIX** types of target that an organization might typically set in relation to health and safety.

9. **Outline** the circumstances that would require a health and safety policy to be reviewed.

Appendix 2.1 – Health and safety policy checklist

The following checklist is intended as an aid to the writing and review of a safety policy. It is derived from the booklet *Writing a safety policy statement* published by the HSE in booklet HSC 6.

General policy and organization

➤ Does the statement express a commitment to health and safety and are your obligations towards your employees made clear?

➤ Does it say which senior manager is responsible for seeing that it is implemented and for keeping it under review, and how this will be done?

➤ Is it signed and dated by you or a partner or senior director?

➤ Have the views of managers and supervisors, safety representatives and of the safety committee been taken into account?

➤ Were the duties set out in the statement discussed with the people concerned in advance, and accepted by them, and do they understand how their performance is to be assessed and what resources they have at their disposal?

➤ Does the statement make clear that co-operation on the part of all employees is vital to the success of your health and safety policy?

➤ Does it say how employees are to be involved in health and safety matters, for example, by being consulted, by taking part in inspections, and by sitting on a safety committee?

➤ Does it show clearly how the duties for health and safety are allocated and are the responsibilities at different levels described?

➤ Does it say who is responsible for the following matters (including deputies where appropriate)?

   ➤ reporting investigations and recording accidents

   ➤ fire precautions, fire drill, evacuation procedures

   ➤ first aid

   ➤ safety inspections

   ➤ the training programme

   ➤ ensuring that legal requirements are met, for example regular testing of lifts and notifying accidents to the health and safety inspector.

Arrangements that need to be considered

➤ keeping the workplace, including staircases, floors, ways in and out, washrooms, etc. in a safe and clean condition by cleaning, maintenance and repair

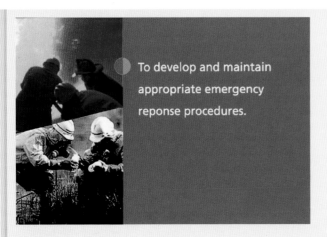

Figure 2.5   Emergency procedures.

To develop and maintain appropriate emergency reponse procedures.

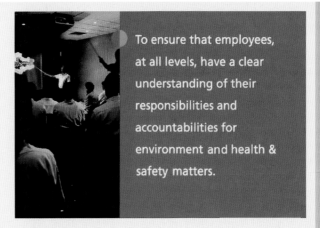

Figure 2.6   Responsibilities.

To ensure that employees, at all levels, have a clear understanding of their responsibilities and accountabilities for environment and health & safety matters.

> the requirements of the Work at Height Regulations
> Any suitable and sufficient risk assessments.

Plant and substances

> maintenance of equipment such as tools, ladders, etc. Are they in a safe condition?
> maintenance and proper use of safety equipment such as helmets, boots, goggles, respirators, etc.
> maintenance and proper use of plant, machinery and guards
> regular testing and maintenance of lifts, hoists, cranes, pressure systems, boilers and other dangerous machinery, emergency repair work, and safe methods of doing it
> maintenance of electrical installations and equipment
> safe storage, handling and, where applicable, packaging, labelling and transport of dangerous substances
> controls of work involving harmful substances such as lead and asbestos
> the introduction of new plant, equipment or substances into the workplace by examination, testing and consultation with the workforce.

Other hazards

> noise problems – wearing of hearing protection, and control of noise at source
> vibration problems – hand–arm and whole-body control techniques and personal protection
> preventing unnecessary or unauthorized entry into hazardous areas
> lifting of heavy or awkward loads

> protecting the safety of employees against assault when handling or transporting the employer's money or valuables
> special hazards to employees when working on unfamiliar sites, including discussion with site manager where necessary
> control of works transport, for example fork lift trucks by restricting use to experienced and authorized operators or operators under instruction (which should deal fully with safety aspects).

Figure 2.7   Vacuum operated coping stone placer.

Emergencies

> ensuring that fire exits are marked, unlocked and free from obstruction
> maintenance and testing of fire-fighting equipment, fire drills and evacuation procedures

(Continued)

Appendix 2.1 – *Continued*

➤ First aid, including name and location of person responsible for first aid and deputy, and location of first aid box.

Communication

➤ giving your employees information about the general duties under the Health and Safety at Work Act and specific legal requirements relating to their work

➤ giving employees necessary information about substances, plant, machinery, and equipment with which they come into contact

➤ discussing with contractors, before they come on site, how they plan to do their job, whether they need any equipment from your organization to help them, whether they can operate either in a segregated area or only when part of the plant is shut down and, if not, what hazards they may create for your employees and vice versa.

Training

➤ training employees, supervisors and managers to enable them to work safely and to carry out their health and safety responsibilities efficiently.

Supervising

➤ supervising employees so far as necessary for their safety – especially young workers, new employees and employees carrying out unfamiliar tasks.

Keeping check

➤ regular inspections and checks of the workplace, machinery appliances and working methods.

Organizing for health and safety

3

3.1 Introduction

This chapter is about managers in construction organizations, setting out clear responsibilities and lines of communications for everyone in the enterprise. The chapter also covers the legal responsibilities that exist between duty holders under the Construction (Design and Management, CDM) Regulations; between people who control premises and those who use them; and between contractors and those who hire them; and the duties of suppliers, manufacturers and designers of articles and substances for use at work.

Chapter 2 is concerned with policy, which is an essential first step. The policy will only remain as words on paper, however good the intentions, until there is an effective organization set up to implement and monitor its requirements.

The policy sets the direction for health and safety within the enterprise and forms the written intentions of the principals or directors of the business. The organization needs to be clearly communicated and people need to know what they are responsible for in their day-to-day operations. A vague statement that 'everyone is responsible for health and safety' is misleading and fudges the real issues. Everyone is responsible but management has more responsibilities. There is no equality of responsibility under law between those who provide direction and create policy and those who are employed to follow. Principals, or employers in terms of the Health and

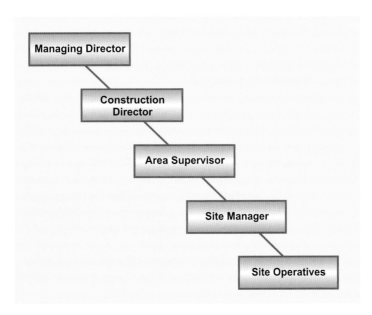

Figure 3.1 Everyone from senior managers down has health and safety responsibilities.

Safety at Work etc Act (HSW Act), have substantially more responsibility than employees.

Some policies are written so that most of the wording concerns strict requirements laid on employees and only a few vague words cover managers' responsibilities. Generally, such policies do not meet the HSW Act or the Management of Health and Safety at Work Regulations, which require an effective policy with a robust organization and arrangements to be set up.

3.2 Control

Like all management functions, establishing control and maintaining it day in day out is crucial to effective health and safety management. Managers, particularly at senior levels, must take proactive responsibility for controlling issues that could lead to ill-health, injury or loss. A nominated senior manager at the top of the organization needs to oversee policy implementation and monitoring. The nominated person will need to report regularly to the most senior management team and will be a director or principal of the organization.

Health and safety responsibilities will need to be assigned to line managers and expertise must be available, either inside or outside the enterprise, to help them achieve the requirements of the HSW Act and the regulations made under the Act. The purpose of the health and safety organization is to harness the collective enthusiasm, skills and effort of the entire workforce with managers taking key responsibility and providing clear direction. The prevention of accidents and ill-health through management systems of control becomes the focus rather than looking for individuals to blame after the incident occurs.

The control arrangements should be part of the written health and safety policy. Performance standards will need to be agreed and objectives set which link the outputs required to specific tasks and activities for which individuals are responsible. For example, the objective could be to carry out a workplace inspection once a week to an agreed checklist and rectify faults within three working days. The periodic, say annual, audit would check to see if this was being achieved and if not the reasons for non-compliance with the objective.

People should be held accountable for achieving the agreed objectives through existing or normal procedures such as:

➤ job descriptions, which include health and safety responsibilities plus
➤ performance appraisal systems, which look at individual contributions plus

➤ arrangements for dealing with poor performance plus
➤ where justified, the use of disciplinary procedures.

Such arrangements are only effective if health and safety issues achieve the same degree of importance as other key management concerns and a good performance is considered to be an essential part of career and personal development.

3.3 Employers' responsibilities

Employers have duties under both criminal and civil law. The civil law duties are covered in Chapter 1. The general duties of employers under the HSW Act relate to:

➤ the health, safety and welfare at work of employees and other workers, whether part-time, casual, temporary, homeworkers, or on work experience or Government Training Schemes or on site as contractors – i.e. anyone working under their control or direction
➤ the health and safety of anyone who visits or uses the workplace or site
➤ the health and safety of anyone who is allowed to use the organization's equipment
➤ the health and safety of those affected by the work activity, for example neighbours, and the general public.

Other duties on employers are covered in Chapter 1 and the summary of the HSW Act is given in Chapter 21.

3.4 Employees' responsibilities

Employees have specific responsibilities under the HSW Act, which are:

➤ to take reasonable care for the health and safety of themselves and of other persons who may be affected by their acts or omissions at work. This involves the same wide group that the employer has to cover, not just the people on the next desk or bench
➤ to cooperate with employers in assisting them to fulfil their statutory duties
➤ not to interfere with deliberately or misuse anything provided, in accordance with health and safety legislation, to further health and safety at work.

3.5 Organizational health and safety responsibilities

In addition to the legal responsibilities on management, there are many specific responsibilities imposed by each organization's health and safety policy. The appendices to this chapter show a typical summary of the health and safety responsibilities and accountability of each level of the line organization to provide an understanding of how health and safety responsibilities and accountability are integrated within the total organization. The responsibilities cover directors, senior managers, site managers, supervisors and employees. Many organizations will not fit this exact structure but most will have those who direct, those who manage or supervise and those who have no line responsibility but have responsibilities to themselves and fellow workers.

Because of the special role and importance of directors, these are covered here in detail.

3.5.1 *Directors' responsibilities*

The Chairman of the Health and Safety Commission said at the launch of the guidance on Directors' responsibilities:

> *Health and safety is a boardroom issue. Good health and safety reflects strong leadership from the top and that is what we want to see. The company whose chairperson or chief executive is the champion of health and safety sends the kind of message which ensures good performance on the ground.*
>
> *Those who are at the top have a key role to play, which is why Boards are being asked to nominate one of their members to be a 'health and safety' director. But appointing a health and safety director or department does not absolve the Board from its collective responsibility to lead and oversee health and safety management.*

Directors' Responsibility for Health and Safety INDG 343 published in 2002 sets out the following action points for directors:

➤ The Board needs to accept formally and publicly its collective role in providing health and safety leadership in its organization.

➤ Each member of the Board needs to accept their individual role in providing health and safety leadership for their organization.

➤ The Board needs to ensure that all Board decisions reflect its health and safety intentions, as articulated in the health and safety policy statement. It is important for Boards to remember that, although health and safety functions can (and should) be delegated, legal responsibility for health and safety rests with the employer.

➤ The Board needs to recognize its role in engaging the active participation of workers in improving health and safety.

➤ The Board needs to ensure that it is kept informed of, and alert to, relevant health and safety risk management issues. The Health and Safety Commission recommends that boards appoint one of their number to be the 'Health and Safety Director'.

Directors need to ensure that the Board's health and safety responsibilities are properly discharged. The Board will need to:

➤ carry out an annual review of health and safety performance

➤ keep the health and safety policy statement up-to-date with current Board priorities and review the policy at least every year

➤ ensure that there are effective management systems for monitoring and reporting on the organization's health and safety performance

➤ ensure that any significant health and safety failures and their investigation are communicated to Board members

➤ ensure that when decisions are made the health and safety implications are fully considered

➤ ensure that regular audits are carried out to check that effective health and safety risk management systems are in place.

By appointing a 'Health and Safety Director' there will be a board member who can ensure that these health and safety risk management issues are properly addressed, both by the Board and more widely throughout the organization.

The Chairman and/or Chief Executive have a critical role to play in ensuring risks are properly managed and that the Health and Safety Director has the necessary competence, resources and support of other Board members to carry out their functions. Indeed, some Boards may prefer to see all the health and safety functions assigned to their Chairman and/or Chief Executive. As long as there is clarity about the health and safety responsibilities and functions, and the Board properly addresses the issues, this is acceptable.

The health and safety responsibilities of all Board members should be clearly articulated in the organization's statement of health and safety policy and arrangements. It is important that the role of the Health and Safety Director should not detract either from the responsibilities of other directors for specific areas of health and safety risk management or from the health and safety responsibilities of the Board as a whole.

3.6 Role and functions of health and safety and other advisers

3.6.1 *Competent person*

One or more competent persons must be appointed to help managers comply with their general duties under health and safety law. The essential point is that managers should have access to expertise to help them fulfil the legal requirements. However, they will always remain as advisers and do not assume responsibility in law for health and safety matters. This responsibility always remains with line managers and cannot be delegated to an adviser whether inside or outside the organization. The appointee could be:

➤ an employer if they are sure they know enough about what to do. This may be appropriate in a small low hazard business
➤ one or more employees, as long as they have sufficient time and other resources to do the task properly
➤ someone from outside the organization who has sufficient expertise to help.

If an employer decides to seek outside help they should check to see if an existing employee is competent to assist. Many health and safety issues can be tackled by people with an understanding of current best practice and an ability to judge and solve problems. Some help is needed long term, other help for a one-off short period. There is a wide range of experts available for different types of health and safety problems. For example:

➤ engineers for specialist ventilation or chemical processes
➤ occupational hygienists for assessment and practical advice on exposure to chemical (dust, gases, fumes, etc.), biological (viruses, fungi, etc.) and physical (noise, vibration, etc.) agents
➤ occupational health professionals for medical examinations and diagnosis of work-related disease, pre-employment and sickness advice, health education, etc.
➤ ergonomists for advice on suitability of equipment, comfort, physical work environment, work organization, etc.
➤ physiotherapists for treatment and prevention of musculoskeletal disorders, etc.

➤ radiation protection advisers for advice on compliance with the Ionising Radiation Regulations 1999
➤ health and safety practitioners for general advice on implementation of legislation, health and safety management, risk assessment, control measures and monitoring performance, etc.

Figure 3.2 Safety practitioner at the front line.

3.6.2 *Health and safety adviser*

Status and **competence** are essential to the role of health and safety and other advisers. They must be able to advise management and employees or their representatives with authority and independence. They need to be able to advise on:

➤ creating and developing health and safety policies. These will be for existing activities plus new acquisitions or processes
➤ the promotion of a positive health and safety culture. This includes helping managers to ensure that an effective health and safety policy is implemented
➤ health and safety planning. This will include goal-setting, deciding priorities and establishing adequate systems and performance standards. Short- and long-term objectives need to be realistic
➤ advising on the competence of planning supervisors, sub-contractors
➤ day-to-day implementation and monitoring of policy and plans. This will include accident and incident investigation, reporting and analysis
➤ performance reviews and audit of the whole health and safety management system.

3.6.3 *To do this properly, health and safety advisers need to*

➤ have proper training and be suitably qualified – for example NEBOSH Diploma, relevant degree and, where appropriate, a Chartered Safety and Health

Practitioner, NEBOSH Certificate in construction safety and health for small, medium hazard construction businesses such as painting and decorating or replacement window contractors

➤ keep up-to-date information systems on such topics as civil and criminal law, health and safety management and technical advances

➤ know how to interpret the law as it applies to their own organization

➤ actively participate in the establishment of organizational arrangements, systems and risk control standards relating to hardware and human performance. Health and safety advisers will need to work with management on matters such as legal and technical standards

➤ undertake the development and maintenance of procedures for reporting, investigating, recording and analysing accidents and incidents

➤ develop and maintain procedures to ensure that senior managers get a true picture of how well health and safety is being managed (where a benchmarking role may be especially valuable). This will include monitoring, review and auditing

➤ be able to present their advice independently and effectively.

3.6.4 *Relationships within the organization*

For health and safety advisers to operate effectively within the organization they must support the provision of authoritative and independent advice. They should report directly to directors on matters of policy and have the authority to stop work if it contravenes agreed standards and puts people at risk of injury.

The health and safety adviser is responsible for professional standards and systems. They may also have line management responsibility for other health and safety professionals, in a large group of companies or on a large and/or high-hazard site.

3.6.5 *Relationships outside the organization*

Health and safety advisers also have a function outside their own organization. They provide the point of liaison with a number of other agencies including the following:

➤ environmental health officers and licensing officials
➤ architects and consultants
➤ HSE and the Fire and Rescue Authorities
➤ the police
➤ HM Coroner or the Procurator Fiscal
➤ local authorities
➤ insurance companies
➤ contractors
➤ clients and sub-contractors

➤ the public
➤ equipment suppliers
➤ the media
➤ general practitioners
➤ IOSH and occupational health specialists and services.

3.7 Persons in control of premises

Section 4 of the HSW Act requires that 'Persons in control of **non-domestic** premises' take such steps as are reasonable in their position to ensure that there are no risks to the health and safety of people who are not employees but use the premises. This duty extends to:

➤ people entering the premises to work
➤ people entering the premises to use machinery or equipment
➤ access and exit from the premises
➤ common corridors, stairs, hoists, work platforms and storage areas.

Those in control of premises are required to take a range of steps depending on the likely use of the premises and the extent of their control and knowledge of the actual use of the premises.

Figure 3.3 NEBOSH are in control here.

3.8 Self-employed

The duties of the self-employed under the HSW Act are fairly limited. The Revitalising Health and Safety Strategy document expressed concern about whether the HSW Act sufficiently covers the area, in view of the huge growth in the use of contractors and sub-contractors throughout the UK. Under the HSW Act the self-employed are:

➤ responsible for their own health and safety
➤ responsible to ensure that others who may be affected are not exposed to risks to their health and safety.

These responsibilities are extended by the Management Health and Safety at Work Regulations, which require self-employed people to:

➤ carry out risk assessment
➤ cooperate with other people who work in the premises and, where necessary, in the appointment of a health and safety coordinator
➤ provide comprehensible information to other people's employees working in their undertaking.

3.9 The supply chain

3.9.1 *Introduction*

Market leaders in every industry are increasing their grip on the chain of supply. They do so by monitoring rather than managing, and also by working more closely with suppliers. The result of this may be that suppliers or contractors are absorbed into the culture of the dominant firm, while avoiding the costs and liabilities of actual management. Powerful procurement departments emerge, to define and impose the necessary quality standards and guard the lists of preferred suppliers.

The trend in many manufacturing businesses is to involve suppliers in a greater part of the manufacturing process so that much of the final production is the assembly of prefabricated sub-assemblies. This is particularly true of the automotive and aircraft industries. This is good practice as it:

➤ involves the supplier in the design process
➤ reduces the number of items being managed within the business
➤ reduces the number of suppliers
➤ improves quality management by placing the onus on suppliers to deliver fully checked and tested components and systems.

In retail, suppliers are even given access to daily sales and forecasts of demand which would normally be considered as highly confidential information. In the process, the freedom of local operating managers to pick and choose suppliers is reduced. Even though the responsibility to do so is often retained, it is strongly qualified by centrally imposed rules and lists, and assistance or oversight.

Suppliers have to be:

➤ trusted
➤ treated with fairness in a partnership

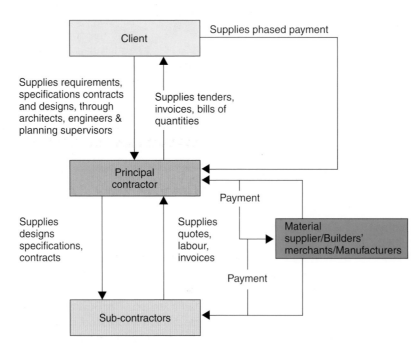

Figure 3.4 Typical supply chain.

➤ given full information to meet the demands being placed on them.

In these conditions, suppliers and contractors looking for business with major firms need greater flexibility and wider competence than previously. This often implies increased size and perhaps mergers, though in principle, bids could be and perhaps are made by loose partnerships of smaller firms organized to secure such business. Figure 3.4 shows a typical supply chain.

3.9.2 *Advantages of good supply chain management*

Reduction of waste

This is an important objective of any business and involves not only waste of materials but also time-related waste. Examples of waste are:

➤ unwanted materials due to over ordering, damage or incorrect specifications
➤ extraneous activities like double handling, for example between manufacturer, builder's merchant and the site
➤ re-working, re-fitting due to poor quality, design, storage or manufacture
➤ time wasted such as waiting for supplies due to excessive time from ordering to delivery or early delivery long before they are needed.

Faster reaction

A well-managed supply chain should be able to respond rapidly to changing requirements. Winter conditions require very different materials than during warmer or drier seasons. A contractor may have to modify plans rapidly and suppliers may need to ramp up or change production at short notice.

Reduction in accidents

A closer relationship between client, designers, principal contractors, and suppliers of services and products can result not only in a safer finished product but, in construction, a safer method of erection. If more products are pre-assembled in ideal factory conditions and then fixed in place on site it is often safer than utilizing a full assembly approach in poor weather conditions on temporary work platforms. Examples include made-up roof trusses and prefabricated doors and windows already fitted to their frames.

3.9.3 *Legislation*

The HSW Act Section 6 places a duty on everyone in the supply chain, from the designer to the final installer of articles of plant or equipment for use at work or any article of fairground equipment, to:

➤ ensure that the article will be safe and without risks to health at all times when it is being set, used, cleaned or maintained
➤ carry out any necessary testing and examination to ensure that it will be safe, and
➤ provide adequate information about its safe setting, use, cleaning, maintenance, dismantling and disposal.

There is an obligation on designers or manufacturers to do any research necessary to prove safety in use. Erectors or installers have special responsibilities to make sure that the plant or equipment is safe to use when handed over.

Similar duties are placed on manufacturers and suppliers of substances for use at work to ensure that the substance is safe when properly used, handled, processed, stored or transported, to provide adequate information and do any necessary research, testing or examining.

Where articles or substances are imported, the suppliers' obligations outlined above attach to the importer, whether a separate importing business or the user himself.

Often items are obtained through hire purchase, leasing or other financing arrangements with the ownership of the items being vested with the financing organization. Where the financing organization's only function is to provide the money to pay for the goods, the supplier's obligations do not attach to them.

3.9.4 *Information for customers*

The quality movement has drawn attention to the need to ensure that there are processes in place which ensure quality, rather than just inspecting and removing defects when it is too late. In much the same way, organizations need to manage health and safety rather than acting when it is too late.

Customers need information and specifications from the manufacturer or supplier – especially where there is a potential risk involved for them. When deciding what the supplier needs to pass on, careful thought is required about the health and safety factors associated with any product or service.

This means focusing on four key questions and then framing the information supplied so that it deals with each one. The questions are:

➤ Are there any inherent dangers in the product or service being passed on – what could go wrong?
➤ What can the manufacturer or supplier do while working on the product or service to reduce the chance of anything going wrong later?

➤ What can be done at the point of handover to limit the chances of anything going wrong?

➤ What steps should customers take to reduce the chances of something going wrong? What precisely would they need to know?

Depending on what is being provided to customers, the customer information may need to comply with the following legislation:

➤ Supply of Machinery (Safety) Regulations 1992 and amendment 1994

➤ Provision and Use of Work Equipment Regulations

➤ Control of Substances Hazardous to Health Regulations

➤ Chemicals (Hazard Information and Packaging for Supply) Regulations 2002 & 2005 amendments.

This list is not exhaustive.

3.9.5 *Buying problems*

Examples of problems that may arise when purchasing include:

➤ second-hand equipment which does not conform to current safety standards

➤ starting to use new substances which do not have safety data sheets

➤ machinery which, while well guarded for operators, may pose risks for a maintenance engineer

➤ office chairs which do not provide adequate back support.

A risk assessment should be done on any new product, taking into account the likely life expectancy (delivery, installation, use, cleaning, maintenance, disposal, etc.). The supplier should be able to provide the information needed to do this. This will help the purchaser make an informed decision on the total costs because the risks will have been identified as will the precautions needed to control those risks. A risk assessment will still be needed for a CE-marked product. The CE marking signifies the manufacturer's declaration that the product conforms to relevant European Directives. Declarations from reputable manufacturers will normally be reliable. However, purchasers should be alert to fake or inadequate declarations and technical standards which may affect the health and safety of the product despite the CE marking. The risk assessment is still necessary to consider how and where the product will be used, what effect it might have on existing operations and what training will be required.

Employers have some key duties when buying plant and equipment:

➤ They must ensure that work equipment is safe, suitable for its purpose and complies with the relevant legislation. This applies equally to equipment which is adapted to be used in ways for which it was not originally designed.

➤ When selecting work equipment, they must consider existing working conditions and health and safety issues.

➤ They must provide adequate health and safety information, instructions, training and supervision for operators. Manufacturers and suppliers are required by law to provide information that will enable safe use of the equipment, substances, etc. and without risk to health.

Some of the issues that will need to be considered when buying in product or plant include:

➤ ergonomics – risk of work-related upper limb disorders (WRULD)

➤ manual handling needs

➤ access/egress

➤ storage, for example of chemicals

➤ risk to contractors when decommissioning old plant or installing new plant

➤ hazardous materials – provision of extraction equipment or personal protective equipment

➤ waste disposal

➤ safe systems of work

➤ training

➤ machinery guarding

➤ emissions from equipment/plant, such as noise, heat or vibration.

3.9.6 *Supply chains in construction*

Recent studies by the Building Research Establishment and The Logistics Business Ltd have shown that there is a long way to go before the construction sector matches the best of manufacturing and retail businesses in their management of supply chains.

This is partly due to the 'one-off' approach applied to building design and the British and others' love of traditional buildings using small components glued or fixed together on site both for the structure (e.g. brickwork) and services (e.g. electrics and plumbing).

Understanding customer needs

The studies found a number of problems in construction which included the following:

➤ There is a failure to see the chain of customer–supplier relationships.

➤ The customer needs and service levels are not defined and monitored and, hence, improved.

Supplier partnerships

➤ There are generally too many suppliers.

➤ The relationships are adversarial rather than managed through service level agreements.

➤ Supplier selection is driven by price, rather than by value.

➤ There is a tendency to secrecy rather than sharing of good quality information.

Control of materials and information

➤ No effort is made to reduce material waste, reduce double handling or improve efficiency generally through an understanding of where goods are and what has to be done to them.

➤ Sources of wasted time are generally not well understood.

➤ Waste is not monitored and no targets are set for improvement.

➤ Care must be taken with second-hand equipment which may not conform to current standards.

3.9.7 *Influencing change*

The attitude of the end client or principal contractor is extremely important in determining the way in which the

Figure 3.5 Inadequate chair – take care when buying second-hand.

supply chain acts. For example, supermarkets demand absolute compliance with legislation and codes of practice in matters of food hygiene, because this is paramount to their business. They have rigorous systems of monitoring and inspection to ensure that agreed procedures are being followed. Unfortunately similar regimes do not exist for the health and safety of people at work. The HSE are concerned to encourage larger organizations to influence the standards adopted by smaller sub-contractors.

Companies want to ensure that they engage safe contractors. This involves a close attention to detail from the pre-contract stage to the end of the project. Large firms often have their own approved list of contractors and suppliers. To get on this list will usually involve a demonstration that the supplier has all the necessary policies and procedures in place and has a good health and safety track record. Failure to meet the client's requirements because of accidents or enforcement action may mean a loss of business and even exclusion from the approved list. Some clients or principal contractors may want to audit a supplier to check compliance.

Firms applying to tender may not always understand that large firms will usually make inquiries of other firms with experience of their health and safety performance. Many large firms regard health and safety performance as a good indicator of the general competence of a smaller firm, precisely because it is an aspect that is often neglected. An adverse report will, at the very least, mean that a bidder must do even better on other aspects to succeed and, in these circumstances, companies generally make special arrangements to ensure that their safety standards are met by the contractors.

3.10 Contractors

3.10.1 *Introduction*

The use of contractors has always been significant in the construction industry but is increasing as many companies turn to outside resources to supplement their own staff and expertise. A contractor is anyone who is brought in to work who is not an employee. Contractors are used for electrical, plumbing, cladding, brick laying, carpentry, roofing, repairs, installation, asbestos removal, demolition, cleaning, security, health and safety and many other tasks. Sometimes there are several contractors on site at any one time. Companies need to think about how their work may affect each other and how they interact with the client if on an occupied site.

Where the CDM Regulations apply there are specific responsibilities between duty holders which go beyond the normal responsibilities between a client, contractors

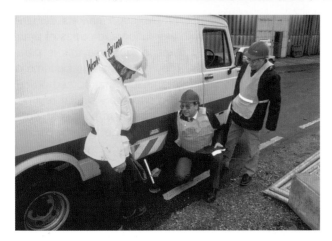

Figure 3.6 Contractors at work.

and sub-contractors. This is covered later. First, the general client/contractor relationship is considered.

3.10.2 *Legal considerations*

The HSW Act applies to all work activities. It requires employers to ensure, so far as is reasonably practicable, the health and safety of:

➤ their employees
➤ other people at work on their site, including contractors
➤ members of the public who may be affected by their work.

All parties to a contract have specific responsibilities under health and safety law, and these cannot be passed on to someone else:

➤ Employers are responsible for protecting people from harm caused by work activities. This includes the responsibility not to harm contractors and sub-contractors on site.
➤ Employees and contractors have to take care not to endanger themselves, their colleagues or others affected by their work.
➤ Contractors also have to comply with the HSW Act and other health and safety legislation. Clearly, when contractors are engaged, the activities of different employers do interact. So cooperation and communication are needed to make sure all parties can meet their obligations.
➤ Employees have to cooperate with their employer on health and safety matters, and not do anything that puts them or others at risk.
➤ Employees must be trained and clearly instructed in their duties.

➤ Self-employed people must not put themselves in danger, or others who may be affected by what they do.
➤ Suppliers of chemicals, machinery and equipment have to make sure their products or imports are safe, and provide information on this.

The Management of Health and Safety at Work Regulations apply to everyone at work and encourage employers to take a more systematic approach to dealing with health and safety by:

➤ assessing the risks which affect employees and anyone who might be affected by the site occupier's work, including contractors
➤ setting up emergency procedures
➤ providing training
➤ cooperating with others on health and safety matters, for example, contractors who share the site with an occupier
➤ providing temporary workers, such as contractors, with health and safety information.

The principles of cooperation, coordination and communication between organizations underpin the Management of Health and Safety at Work Regulations and the CDM Regulations covered in chapter 9. See later Section 3.11 on joint occupation of premises. For more information on the Management of Health and Safety at Work Regulations, read the summary in Chapter 21.

3.10.3 *Contractor selection*

The selection of the right contractor for a particular job is probably the most important element in ensuring that the risks to the health and safety of everybody involved on the activity and people in the vicinity are reduced as far as possible. Ideally, selection should be made from a list of approved contractors who have demonstrated that they are able to meet the client's requirements.

The selection of a contractor has to be a balanced judgement with a number of factors taken into account. Fortunately, a contractor who works well and meets the client's requirements in terms of the quality and timeliness of the work is likely also to have a better than average health and safety performance. Cost, of course, will have to be part of the judgement but may not provide any indication of which contractor is likely to give the best performance in health and safety terms. In deciding which contractor should be chosen for a task, the following should be considered:

➤ registration with either the HSE or Environmental Health Department of a Local Authority
➤ do they have an adequate health and safety policy?

➤ can they demonstrate that the person responsible for the work is competent?

➤ can they demonstrate that competent safety advice will be available?

➤ do they monitor the level of accidents at their work site?

➤ do they have a system to assess the hazards of a job and implement appropriate control measures?

➤ will they produce a method statement, which sets out how they will deal with all significant risks?

➤ do they have guidance on health and safety arrangements and procedures to be followed?

➤ details of competence certification, particularly when working with gas or electricity may be involved

➤ details of insurance arrangements in force at the time of the contract

➤ do they have effective monitoring arrangements?

➤ do they use trained and skilled staff who are qualified where appropriate? (Judgement will be required, as many construction workers have had little or no training except training on the job.) Can the company demonstrate that the employees or other workers used for the job have had the appropriate training and are properly experienced and, where appropriate, qualified?

➤ details of emergency procedures, including fire precautions, for contractor employees

➤ details of any previous accidents or incidents reported under RIDDOR

➤ details of accident reporting procedure

➤ details of previous work undertaken by the contractor

➤ can they produce good references indicating satisfactory performance?

SITE SAFETY

Under the Health & Safety at Work Act 1974 all persons entering this site must comply with all regulations under this Act.
All visitors must report to the site office and obtain permission to proceed on to the site or any work area.
Safety signs and procedures must be observed and personal protection and safety equipment must be used at all times.

 Construction work in progress.
Parents are advised to warn children of the dangers of entering this site.

 Safety helmets must be worn

 Unauthorised entry to this site is strictly forbidden.

Figure 3.7 Site safety rules.

3.10.4 *Control of contractors*

On being selected, contractors should be expected to:

➤ familiarize themselves with those parts of the health and safety plan which affect them and their employees and/or sub-contractors

➤ cooperate with the principal contractor in his health and safety duties to contractors

➤ comply with their legal health and safety duties.

On arrival at the site, sub-contractors should ensure that:

➤ they report to the site office and the site manager

➤ they abide by any site rules, particularly in respect of personal protective equipment

➤ the performance of their work does not place others at risk

➤ they are familiar with the first aid and accident reporting arrangements of the principal contractor

➤ they are familiar with all emergency procedures on the site

➤ any materials brought onto the site are safely handled, stored and disposed of in compliance, where appropriate, with the current Control of Substances Hazardous to Health Regulations

➤ they adopt adequate fire precaution and prevention measures when using equipment which could cause fires

➤ they minimize noise and vibration produced by their equipment and activities

➤ any ladders, scaffolds and other means of access are erected in conformity with good working practice and any statutory requirements

➤ any welding or burning equipment brought onto the site is in safe operating condition and used safely with a suitable fire extinguisher to hand

➤ any lifting equipment brought onto the site complies with the current Lifting Operation and Lifting Equipment Regulations

➤ all electrical equipment complies with the current Electricity at Work Regulations

➤ connection to the electricity supply is from a point specified by the principal contractor and is by proper cables and connectors. For outside construction work, only 110V equipment should be used

➤ any restricted access to areas on the site is observed

➤ welfare facilities provided on site are treated with respect

➤ any vehicles brought onto the site observe any speed, condition or parking restriction.

The control of sub-contractors can be exercised by monitoring them against the criteria listed above and by regular site inspections. On completion of the contract, the work should be checked to ensure that the agreed standard has been reached and that any waste material has been removed from the site.

When working on an occupied site, contractors, their employees, and sub-contractors and their employees, should not be allowed to commence work on any site without authorization signed by the Client Company contact. The authorization should clearly define the range of work that the contractor can carry out and set down any special requirements, for example protective clothing, fire exits to be left clear, isolation arrangements.

Permits will be required for operations, such as hot work. All contractors should keep a copy of their authorization at the place of work. A second copy of the authorization should be kept at the site and be available for inspection.

The Client Company contact signing the authorization will be responsible for all aspects of the work of the contractor. The contact will need to check as a minimum the following:

➤ that the correct contractor for the work has been selected
➤ that the contractor has made appropriate arrangements for supervision of staff
➤ that the contractor has received and signed for a copy of the contractor's safety rules
➤ that the contractor is clear what is required, the limits of the work and any special precautions that need to be taken
➤ that the contractor's personnel are properly qualified for the work to be undertaken.

The Client Company contact should check whether sub-contractors will be used. They will also require authorization, if deemed acceptable. It will be the responsibility of the Client Company contact or the principal contractor, if under CDM 2007 Regulations, to ensure that sub-contractors are properly supervised. See Chapters 9 and 21 for further information on CDM 2007.

The Client Company contact will be responsible for ensuring that there is adequate and clear communication between different contractors and Company personnel where this is appropriate.

3.10.5 *Safety rules for contractors*

In the conditions of contract there should be a stipulation that the contractor and all of their employees adhere to the contractor's safety rules. Contractor's safety rules should contain as a minimum the following points:

➤ **health & safety:** that the contractor operates to at least the minimum legal standard and conforms to accepted industry good practice
➤ **supervision:** provides a good standard of supervision of their own employees
➤ **sub-contractors:** that they may not use sub-contractors without prior written agreement from the company
➤ **authorization:** that each employee must carry an authorization card issued by the Company at all times while on site.

Figure 3.8 Rules at site entrance.

3.10.6 *Example of rules for contractors*

Contractors engaged by a client's organization to carry out work on its premises will:

➤ familiarize themselves with that part of the organization's safety policy as affects them and will ensure that appropriate parts of the policy are communicated to their employees, and any sub-contractors and employees of sub-contractors who will do work on the premises

➤ co-operate with the organization in its fulfilment of its health and safety duties to contractors and take the necessary steps to ensure the like co-operation of their employees

➤ comply with their legal and moral health, safety and food hygiene duties

➤ ensure the carrying out of their work on the organization's premises in such a manner as not to put either themselves or any other persons on or about the premises at risk

➤ where they wish to avail themselves of the organization's first aid arrangements/facilities while on the premises, ensure that written agreement to this effect is obtained prior to first commencement of work on the premises

➤ where applicable, and requested by the organization, supply a copy of its statement of policy, organization and arrangements for health and safety written for the purposes of compliance with The Management of Health and Safety at Work Regulations 1999 and section 2(3) of the Health and Safety at Work Act 1974

➤ abide by all relevant provisions of the organization's safety policy, including compliance with health and safety rules

➤ ensure that on arrival at the premises, they and any other persons who are to do work under the contract, report to reception or their designated organization contact.

Without prejudice to the requirements stated above, contractors, sub-contractors and employees of contractors and sub-contractors will, to the extent that such matters are within their control, ensure:

➤ the safe handling, storage and disposal of materials brought onto the premises

➤ that the organization is informed of any hazardous substances brought onto the premises and that the relevant parts of the Control of Substances Hazardous to Health Regulations in relation thereto are complied with

➤ that fire prevention and fire precaution measures are taken in the use of equipment which could cause fires

➤ that steps are taken to minimize nois produced by their equipment and activiti

➤ that scaffolds, ladders and other such access for work at height, are erected and accordance with statutory requirements and g working practice

➤ that any welding or burning equipment brought onto the premises is in safe operating condition and used in accordance with all safety requirements

➤ that any lifting equipment brought onto the premises is adequate for the task and has been properly tested/certified

➤ that any plant and equipment brought onto the premises is in safe condition and used/operated by competent persons

➤ that, for vehicles brought onto the premises, any speed, condition or parking restrictions are observed

➤ that compliance is made with the relevant requirements of the Electricity at Work Regulations

➤ that connection(s) to the organization's electricity supply is from a point specified by its management and is by proper connectors and cables

➤ that they are familiar with emergency procedures existing on the premises

➤ that welfare facilities provided by the organization are treated with care and respect

➤ that access to restricted parts of the premises is observed and the requirements of food safety legislation are complied with

➤ that any major or lost-time accident or dangerous occurrence on the organization's premises is reported as soon as possible to their site contact

➤ that where any doubt exists regarding health and safety requirements, advice is sought from the site contact.

The foregoing requirements *do not* exempt contractors from their statutory duties in relation to health and safety, but are intended to assist them in attaining a high standard of compliance with those duties.

3.11 Joint occupation of premises

The Management of Health and Safety at Work Regulations specifically state that where two or more employers share a workplace – whether on a temporary or a permanent basis – each employer shall:

➤ co-operate with other employers

➤ take reasonable steps to co-ordinate between other employers to comply with legal requirements

➤ take reasonable steps to inform other employers where there are risks to health and safety.

ople involved should
ements adopted are
ployer controls the
uld help to assess
y necessary control
olling employer the
joint arrangements
h as appointing a

Figure 3.9 Working together.

3.12 Consultation with the workforce

It is important to gain the co-operation of all employees
if a successful health and safety culture is to become
established. This co-operation is best achieved by
consultation. The easiest and potentially the most
effective method of consultation is the health and
safety committee. It will realize its full potential if its
recommendations are seen to be implemented and
both management and employee concerns are freely
discussed. It will not be so successful if it is seen as a
talking shop.

The committee should have stated objectives, which
mirror the objectives in the organization's health and
safety policy statement, and its own terms of reference.
Terms of reference could include the following:

➤ the study of accident and notifiable disease statistics to enable reports to be made of recommended
remedial actions
➤ the examination of health and safety audit and statutory inspection reports
➤ the consideration of reports from the external
enforcement agency
➤ the review of new legislation, approved codes
of practice and guidance, and its effect on the
organization
➤ the monitoring and review of all health and safety
training and instruction activities in the organization
➤ the monitoring and review of health and safety
publicity and communication throughout the
organization.

There are no fixed rules on the composition of the health
and safety committee except that it should be representative of the whole organization. It should have representation from the workforce and the management including
at least one senior manager.

There are two pieces of legislation that cover health
and safety consultation with employees and both are
summarized in Chapter 21.

3.12.1 The Safety Representatives and Safety Committees Regulations

These regulations only apply to those organizations that
have recognized trade unions for collective bargaining purposes. The recognized trade union may appoint
safety representatives from among the employees and
notify the employer in writing.

The safety representative has several functions (not
duties) including:

➤ representing employees in consultation with the
employer
➤ investigating potential hazards and dangerous
occurrences
➤ investigating the causes of accidents
➤ investigating employee complaints relating to health,
safety and welfare
➤ making representations to the employer on health,
safety and welfare matters
➤ carrying out inspections
➤ representing employees at the workplace in consultation with enforcing inspectors
➤ receiving information
➤ attending safety committee meetings.

Representatives must be allowed time off with pay to
fulfil these functions and to undergo health and safety
training. They must also be allowed to inspect the workplace at least once a quarter or sooner if there is has
been a substantial change in the conditions of work.

Finally, if at least two representatives have requested in writing that a safety committee be set up, the employer has 3 months to comply.

3.12.2 *The Health and Safety (Consultation with Employees) Regulations*

These regulations were introduced so that employees working in organizations without recognized trade unions would still need to be consulted on health and safety matters. All employees must now be consulted either on an individual basis (e.g. very small companies) or through safety representatives elected by the workforce, known as ROES (Representative(s) of Employee Safety). The guidance to these regulations emphasizes the difference between informing and consulting. Consultation involves listening to the opinion of employees on a particular issue and taking it into account before a decision is made. Informing employees means providing information on health and safety issues such as risks, control systems and safe systems of work.

The functions of these representatives are similar to those under the Safety Representatives and Safety Committees Regulations as are the rights to health and safety training (i.e. time off and costs covered by the employer).

The employer must consult on the following:

➤ the introduction of any measure or change which may substantially affect employees' health and safety
➤ the arrangements for the appointment of competent persons to assist in following health and safety law
➤ any information resulting from risk assessments or their resultant control measures which could affect the health, safety and welfare of employees
➤ the planning and organization of any health and safety training required by legislation
➤ the health and safety consequences to employees of the introduction of new technologies into the workplace.

However, the employer is not expected to disclose information if:

➤ it violates a legal prohibition
➤ it could endanger national security
➤ it relates specifically to an individual without their consent
➤ it could harm substantially the business of the employer or infringe commercial security
➤ it was obtained in connection with legal proceedings.

Finally, an employer should ensure that the representative receives reasonable training in health and safety at the employer's expense, is allowed time during working hours to perform his duties and provides other facilities and assistance that the representative should reasonably require.

3.13 Practice NEBOSH questions for Chapter 3

1. **Explain** the health and safety responsibilities of a Managing Director of a medium-sized construction company.

2. **Outline** the main health and safety responsibilities of the following persons within a medium-sized construction company:
   (a) Managing Director
   (b) Construction Director/Manager
   (c) Site Supervisor/Manager
   (d) Site Operatives.

3. **Outline** the factors that will determine the level of supervision that an employee should receive when newly employed on a construction site.

4. (a) **Explain** the meaning of the term 'competent person'.
   (b) **Outline** the organizational factors that may cause a person to work unsafely even though they are competent.

5. **Outline** a procedure designed to ensure the health and safety of visitors to work premises.

6. **Outline** the main factors to be considered when assessing the health and safety competence of a sub-contractor.

7. **Outline** the checks that could be made in assessing the health and safety competence of a contractor.

8. A contractor has been engaged by a manufacturing company to undertake extensive maintenance work on the interior walls of a factory workshop.
   (a) **State** the legal duties that the manufacturing company owes the contractor's employees under the Health and Safety at Work etc Act 1974.
   (b) **Outline** the information relevant to health and safety that should be provided before work commences by:
   (i) the manufacturing company to the contractor; **AND**
   (ii) the contractor to the manufacturing company.

(c) **Describe** additional procedural measures that the manufacturing company should take to help ensure the health and safety of their own and the contractor's employees.

9. (a) **State** the circumstances under which an employer must establish a health and safety committee.

   (b) **Give SIX** reasons why a health and safety committee may prove to be ineffective in practice.

10. (a) A new construction scheme has been set up with an integral project-based health and safety committee.

    (i) **Outline** the benefits to the scheme of having a health and safety committee.

    (ii) **Outline** the reasons why such a committee may prove to be ineffective.

   (b) **Identify** a range of methods that a principal contractor can use to provide health and safety information direct to individual workers on a project.

11. **Outline** the factors that may determine the effectiveness of a safety committee.

12. Site safety committees are becoming more common on larger construction projects. **Outline** the topics that may typically be included on the agenda of a site safety committee.

13. In relation to the Safety Representatives and Safety Committees Regulations 1977, **outline**:

   (a) the rights and functions of a trade union appointed safety representative

   (b) the facilities that an employer may need to provide to safety representatives.

14. With respect to the Safety Representatives and Safety Committees Regulations 1977, **state** when a safety representative is legally entitled to inspect the workplace.

15. By reference to the Safety Representatives and Safety Committees Regulations 1977:

   (a) **explain** the circumstances under which an employer must form a health and safety committee;

(b) **identify TWO** reasons why a safety representative may have cause to complain to an Employment Tribunal and **state** the period of time within which such a complaint must be made.

16. In relation to the Health and Safety (Consultation with Employees) Regulations 1996, **identify**:

   (a) the health and safety matters on which employers have a duty to consult their employees

   (b) four types of information that an employer is not obliged to disclose to an employee representative.

17. With reference to the Health and Safety (Consultation with Employees) Regulations 1996:

   (a) **explain** the difference between consulting and informing,

   (b) **outline** the health and safety matters on which employers must consult their employees,

   (c) **outline** the entitlements of representatives of employee safety who have been elected under the Regulations.

18. In relation to the Health and Safety (Consultation with Employees) Regulations 1996, **outline** the types of facility that an employer may need to provide to representatives of employee safety.

19. For high standards of health and safety to be achieved, it is important that site management provides information to, and consults with, all workers on a project.

   (a) **Explain** the meaning of, and the differences between, "informing" and "consulting" on health and safety matters.

   (b) **Outline** various means by which health and safety information can be provided to site operatives.

   (c) **Describe** the opportunities that might be available to site management for consulting with workers on a construction site.

Appendix 3.1 – Typical organizational responsibilities

Senior managers

- are responsible and held accountable for the relevant organization's health and safety performance
- develop strong, positive health and safety attitudes among those employees reporting directly to them
- provide guidance to their management team
- establish minimum acceptable health and safety standards within the organization
- provide necessary staffing and funding for the health and safety effort within the organization
- authorize necessary major expenditures
- evaluate, approve, and authorize health and safety-related projects developed by members of the organization's health and safety advisers
- review and approve policies, procedures, and programmes developed by group staff
- maintain a working knowledge of areas of health and safety that are regulated by governmental agencies
- keep health and safety in the forefront by including it as a topic in business discussions
- review health and safety performance statistics and provide feedback to the management team
- review and act upon major recommendations submitted by outside loss-prevention consultants
- make health and safety a part of site tours by making specific remarks about observations of acts or conditions that fall short of or exceed health and safety standards
- personally investigate fatalities and major property losses
- personally chair the organization's main health and safety committee (if applicable).

Site managers

- are responsible and held accountable for their site's health and safety performance
- establish, implement, and maintain a formal, written site health and safety programme, encompassing applicable areas of loss prevention, that is consistent with the organization's health and safety policy
- establish controls to ensure uniform adherence to health and safety programme elements. These controls should include corrective action and follow-up

- develop, by action and example, a positive health and safety attitude and a clear understanding of specific responsibilities in each member of management
- approve and adopt local health and safety policies, rules, and procedures
- chair the site health and safety committee
- personally investigate fatalities, serious lost workday case injuries, and major property losses
- review monthly health and safety activity reports and performance statistics
- review lost workday cases injury/illness investigation reports
- review health and safety reports submitted by outside agencies
- attend site health and safety audits regularly to appraise the programme's effectiveness
- review as appropriate the health and safety programme's effectiveness and make adjustments where necessary
- evaluate the functional performance of the health and safety staff and provide guidance or training where necessary
- personally review, sign and approve corrective action planned for lost-time accidents
- monitor staff's progress toward achieving their health and safety goals and objectives.

Supervisors

- are responsible and held accountable for their group's health and safety performance
- conduct health and safety meetings for employees at least monthly
- enforce method statements and/or safe job procedures
- report to site manager and act upon any weaknesses in method statements and/or safe job procedures, as revealed by health and safety risk assessments and observations
- report jobs not covered by safe job procedures
- review unsafe acts and conditions and direct daily health and safety activities to correct the causes
- instruct employees in health and safety rules and regulations; make records of instruction; and enforce all health and safety rules and regulations
- make daily inspections of assigned work areas and take immediate steps to correct unsafe or unsatisfactory conditions; report to the manager

those conditions that cannot be immediately corrected; instruct employees on housekeeping standards

➤ instruct employees that tools/equipment are to be inspected before each use; make spot checks of tools'/equipment's condition

➤ instruct each new employee personally on job health and safety requirements in assigned work areas

➤ enforce the organization's/site's medical recommendations with respect to an employee's physical limitations. Report on employee's apparent physical limitations to their manager; and request physical examination of the employee

➤ enforce personal protective equipment requirements; make spot checks to determine that hard hats and other PPE are being used; and periodically appraise condition of equipment

➤ see that, in a case of serious injury, the injured employee receives prompt medical attention; isolate the area or shut down equipment, as necessary; and immediately report to the site manager the facts regarding the employee's accident or illness and any action taken. In serious incident cases, the supervisor determines the cause, takes immediate steps to correct an unsafe condition, and isolates area and/or shuts down the equipment, as necessary. They immediately report facts and action taken to the manager

➤ make thorough investigation of all accidents, serious incidents, and cases of ill-health involving employees in assigned work areas; immediately after the accident, prepare a complete accident investigation report and include recommendations for preventing recurrence

➤ Check changes in operating practices, procedures, and conditions at the start of each shift/day and before relieving the 'on-duty' supervisor (if applicable), noting health and safety-related incidents that have occurred since their last working period

➤ make, at the start of each shift/day, an immediate check to determine absentees. If site injury is claimed, an immediate investigation is instituted and the site manager is notified

➤ make daily spot checks and take necessary corrective action regarding housekeeping, unsafe acts or practices, unsafe conditions, job procedures and adherence to health and safety rules

➤ attend all scheduled and assigned health and safety training meetings

➤ instruct personally or provide on-the-job instruction on safe and efficient performance of assigned jobs

➤ act on all employee health and safety complaints and suggestions

➤ maintain, in their assigned area, health and safety signs and notice boards in a clean and legible condition.

Employees

➤ are responsible for their own safety and health

➤ ensure that their actions will not jeopardize the safety or health of other employees

➤ are alert to observe and correct, or report, unsafe practices and conditions

➤ maintain a healthy and safe place to work and co-operate with managers in the implementation of health and safety matters

➤ make suggestions to improve any aspect of health and safety

➤ maintain an active interest in health and safety

➤ learn and follow the operating procedures and health and safety rules and procedures for safe performance of the job

➤ follow the established procedures if accidents occur.

Appendix 3.2 – Checklist for supply chain health and safety management

This checklist is taken from the HSE leaflet INDG 268 Working Together: Guidance on Health and Safety for Contractors and Suppliers 2002. It is a reminder of the topics that might need to be discussed with people who individual contractors may be working with.

It is not intended to be exhaustive and not all questions will apply at any one time, but it should help people to get started.

Responsibilities

➤ What are the hazards of the job?
➤ Who is to assess particular risks?
➤ Who will co-ordinate action?
➤ Who will monitor progress?

The job

➤ Where is it to be done?
➤ Who with?
➤ Who is in charge?
➤ How is the job to be done?
➤ What other work will be going on at the same time?
➤ How long will it take?
➤ What time of day or night?
➤ Do you need any permit to do the work?

The hazards and risk assessments

Site and location
Consider the means of getting into and out of the site and the particular place of work – are they safe? and

➤ Will any risks arise from environmental conditions?
➤ Will you be remote from facilities and assistance?
➤ What about physical/structural conditions?
➤ What arrangements are there for security?

Substances

➤ What supplier's information is available?
➤ Is there likely to be any microbiological risk?
➤ What are the storage arrangements?
➤ What are the physical conditions at the point of use? Check ventilation, temperature, electrical installations, etc.
➤ Will you encounter substances that are not supplied, but produced in the work, for example fumes from hot work during dismantling plant? Check how much, how often, for how long, method of work, etc.
➤ What are the control measures? For example, consider preventing exposure, providing engineering controls, using personal protection (in that order of choice).

➤ Is any monitoring required?
➤ Is health surveillance necessary, for example for work with sensitizers? (Refer to health and safety data sheet.)

Plant and equipment

➤ What are the supplier/hirer/manufacturer's instructions?
➤ Are any certificates of examination and test needed?
➤ What arrangements have been made for inspection and maintenance?
➤ What arrangements are there for shared use?
➤ Are the electrics safe to use? Check the condition of power sockets, plugs, leads and equipment. (Don't use damaged items until they have been repaired.)
➤ What assessments have been made of noise levels?

People

➤ Is information, instruction and training given, as appropriate?
➤ What are the supervision arrangements?
➤ Are members of the public/inexperienced people involved?
➤ Have any disabilities/medical conditions been considered?

Emergencies

➤ What arrangements are there for warning systems in case of fire and other emergencies?
➤ What arrangements have been made for fire/ emergency drills?
➤ What provision has been made for first aid and fire fighting equipment?
➤ Do you know where your nearest fire exits are?
➤ What are the accident reporting arrangements?
➤ Are the necessary arrangements made for availability of rescue equipment and rescuers?

Welfare

Who will provide:

➤ shelter
➤ food and drinks
➤ washing facilities
➤ toilets (male and female)
➤ clothes changing/drying facilities?

There may be other pressing requirements which make it essential to re-think health and safety as the work progresses.

Promoting a positive health and safety culture

4

Introduction

In 1972, the Robens report recognized that the introduction of health and safety management systems was essential if the ideal of self-regulation of health and safety by industry was to be realized. It further recognized that a more active involvement of the workforce in such systems was essential if self-regulation was to work. Self-regulation and the implicit need for health and safety management systems and employee involvement was incorporated into the Health and Safety at Work Act 1974.

Since the introduction of the Act, health and safety standards have improved considerably but there have been some catastrophic failures. One of the worst was the fire on the off-shore oil platform, Piper Alpha, in 1988 when 167 people died. At the subsequent inquiry, the concept of a safety culture was defined by the Director General of the Health and Safety Executive (HSE) at the time, J R Rimington. This definition has remained as one

> "Safety is, without doubt, the most crucial investment we can make. And the question is not what it costs us, but what it saves."
>
> Robert E McKee, Chairman and Managing Director, Conoco (UK) Ltd

Figure 4.1 Safety investment.

of the main checklists for a successful health and safety management system.

4.2 Definition of a health and safety culture

The health and safety culture of an organization may be described as the development stage of the organization in health and safety management at a particular time. HSG65 gives the following definition of a health and safety culture:

> *The safety culture of an organization is the product of individual and group values, attitudes, perceptions, competencies and patterns of behaviour that determine the commitment to, and the style and proficiency of, an organization's health and safety management.*
>
> *Organizations with a positive safety culture are characterized by communications founded on mutual trust, by shared perceptions of the importance of safety and by confidence in the efficacy of preventive measures.*

There is concern among some health and safety professionals that many health and safety cultures are developed and driven by senior managers with very little input from the workforce. Others argue that this arrangement is sensible because the legal duties are placed on the employer. A positive health and safety culture needs the involvement of the whole workforce just as a successful quality system does. There must be a joint commitment in terms of attitudes and values. The workforce must believe that the safety measures put in place will

51

be effective and followed even when financial and performance targets may be affected.

4.3 Safety culture and safety performance

4.3.1 *The relationship between health and safety culture and health and safety performance*

The following elements are the important components of a positive health and safety culture:

➤ leadership and commitment to health and safety throughout and at all levels of the organization
➤ acceptance that high standards of health and safety are achievable as part of a long-term strategy formulated by the organization
➤ a detailed assessment of health and safety risks in the organization and the development of appropriate control and monitoring systems
➤ a health and safety policy statement outlining short- and long-term health and safety objectives. Such a policy should also include codes of practice and required health and safety standards
➤ relevant employee training programmes and communication and consultation procedures
➤ systems for monitoring equipment, processes and procedures and the prompt rectification of any defects
➤ the prompt investigation of all incidents and accidents and reports made detailing any necessary remedial actions.

If the organization adheres to these elements, then a basis for a good performance in health and safety will have been established. However, to achieve this level of performance, sufficient financial and human resources must be made available for the health and safety function at all levels of the organization.

All managers, supervisors and members of the governing body (e.g. directors) should receive training in health and safety and be made familiar during training sessions with the health and safety targets of the organization. The depth of training undertaken will depend on the level of competence required of the particular manager. Managers should be accountable for health and safety within their departments and be rewarded for significant improvements in health and safety performance. They should also be expected to discipline employees within their departments who infringe health and safety policies or procedures.

4.3.2 *Important indicators of a health and safety culture*

There are several outputs or indicators of the state of the health and safety culture of an organization. The most important are the numbers of accidents, near misses and occupational ill-health cases occurring within the organization or on the construction site.

While the number of accidents may give a general indication of the health and safety culture, a more detailed examination of accidents and accident statistics is normally required. A calculation of the rate of accidents enables health and safety performance to be compared between years and organizations.

The simplest measure of accident rate is called the incident rate and is defined as:

$$\frac{\text{Total number of accidents}}{\text{Number of persons employed}} \times 1,000$$

or the total number of accidents per 1,000 employees.

A similar measure (per 100,000) is used by the Health and Safety Commission (HSC) in its annual report on national accident statistics and enables comparisons to be made within an organization between time periods when employee numbers may change. It also allows comparisons to be made with the national occupational or industrial group relevant to the organization.

There are four main problems with this measure which must be borne in mind when it is used:

➤ There may be a considerable variation over a time period in the ratio of part-time to full-time employees.
➤ The measure does not differentiate between major and minor accidents and takes no account of other incidents, such as those involving damage but no injury (although it is possible to calculate an incidence rate for a particular type or cause of accident).
➤ There may be significant variations in work activity during the periods being compared.
➤ Under-reporting of accidents will affect the accuracy of the data.

Subject to the above limitations, an organization with a high accident incidence rate is likely to have a negative or poor health and safety culture.

There are other indications of a poor health and safety culture or climate. These include:

➤ a high sickness, ill-health and absentee rate among the workforce
➤ the perception of a blame culture
➤ high staff turnover leading to a loss of momentum in making health and safety improvements

➤ no resources (in terms of budget, people or facilities) made available for the effective management of health and safety

➤ a lack of compliance with relevant health and safety law and the safety rules and procedures of the organization

➤ poor selection procedures and management of contractors

➤ poor levels of communication, cooperation and control

➤ a weak health and safety management structure

➤ either a lack or poor levels of health and safety competence

➤ high insurance premiums.

In summary, a poor health and safety performance within an organization is an indication of a negative health and safety culture.

4.3.3 Factors affecting a health and safety culture

The most important factor affecting the culture is the commitment to health and safety from the top of an organization. This commitment may be shown in many different ways. It not only needs to have a formal aspect in terms of an organizational structure, job descriptions and a health and safety policy, but it also needs to be apparent during crises or other stressful times. The health and safety procedures may be circumvented or simply forgotten when production or other performance targets are threatened.

Structural reorganization or changes in market conditions will produce feelings of uncertainty among the workforce which, in turn, will affect the health and safety culture.

Poor levels of supervision, health and safety information and training are very significant factors in reducing health and safety awareness and, therefore, the culture. In the construction industry, the poor selection of contractors has a major negative effect on the culture.

Finally, the degree of consultation and involvement with the workforce in health and safety matters is crucial for a positive health and safety culture. Most of these factors may be summed up as human factors.

4.4 Human factors and their influence on safety performance

4.4.1 Human factors

Over the years, there have been several studies undertaken to examine the link between various accident

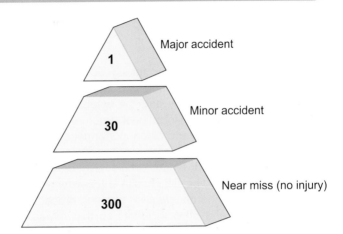

Figure 4.2 Heinrich's accidents/incidents ratio.

types, graded in terms of their severity, and near misses. One of the most interesting was conducted in the USA by H W Heinrich in 1950. He looked at over 300 accidents/incidents and produced the ratio illustrated in Figure 4.2.

This study indicated that for every 10 near misses, there will be an accident. While the accuracy of this study may be debated and other studies have produced different ratios, it is clear that if near misses are continually ignored, an accident will result. Further, the HSE Accident Prevention Unit has suggested that 90% of all accidents are due to human error and 70% of all accidents could have been avoided by earlier (proactive) action by management. It is clear from many research projects that the major factors in most accidents are human factors.

The HSE has defined human factors as, 'environmental, organizational and job factors, and human and individual characteristics which influence behaviour at work in a way which can affect health and safety'.

In simple terms, the health and safety of people at work are influenced by:

➤ the organization
➤ the job
➤ personal factors.

These are known as human factors since they each have a human involvement. The personal factors which differentiate one person from another are only one part of those factors – and not always the most important.

Each of these elements will be considered in turn.

4.4.2 The organization

The organization is the company or corporate body and has the major influence on health and safety. It must have its own positive health and safety culture and produce an environment in which it:

➤ manages health and safety throughout the organization, including the setting and publication of a health and safety policy and the establishment of a health and safety organizational structure

➤ measures the health and safety performance of the organization at all levels and in all departments. The performance of individuals should also be measured. There should be clear health and safety targets and standards and an effective reporting procedure for accidents and other incidents so that remedial actions may be taken

➤ motivates managers within the organization to improve health and safety performance in the workplace in a proactive rather than reactive manner.

The HSE has recommended that an organization needs to provide the following elements within its management system:

➤ a clear and evident commitment from the most senior manager downwards, which provides a climate for safety in which management's objectives and the need for appropriate standards are communicated and in which constructive exchange of information at all levels is positively encouraged

➤ an analytical and imaginative approach identifying possible routes to human factor failure. This may well require access to specialist advice

➤ procedures and standards for all aspects of critical work and mechanisms for reviewing them

➤ effective monitoring systems to check the implementation of the procedures and standards

➤ incident investigation and the effective use of information drawn from such investigations

➤ adequate and effective supervision with the power to remedy deficiencies when found.

It is important to recognize that there are often reasons for these elements not being present resulting in weak management of health and safety. The most common reason is that individuals within the management organization do not understand their roles – or their roles have never been properly explained to them. The higher that a person is within the structure the less likely it is that he has received any health and safety training. Such training at board level is rare.

Objectives and priorities may vary across and between different levels in the structure leading to disputes which affect attitudes to health and safety. For example, a warehouse manager may be pressured to block walkways so that a large order can be stored prior to dispatch.

Motivations can also vary across the organization, which may cause health and safety to be compromised. The production controller will require that components of a product are produced as near simultaneously as possible so that final assembly of them is performed as quickly as possible. However, the health and safety adviser will not want to see safe systems of work compromised.

In an attempt to address some of these problems, the HSC produced guidance in 2001 on the safety duties of company directors. Each director and the board, acting collectively, will be expected to provide health and safety leadership in the organization. The board will need to ensure that all its decisions reflect its health and safety intentions and that it engages the workforce actively in the improvement of health and safety. The board will also be expected to keep itself informed of changes in health and safety risks. (See Chapter 3 for more details on directors' responsibilities.)

The following simple checklist may be used to check any organizational health and safety management structure. Does the structure have:

➤ an effective health and safety management system?
➤ a positive health and safety culture?
➤ arrangements for the setting and monitoring of standards?
➤ adequate supervision?
➤ effective incident reporting and analysis?
➤ learning from experience?
➤ clearly visible health and safety leadership?
➤ suitable team structures?
➤ efficient communication systems and practices?
➤ adequate staffing levels?
➤ suitable work patterns?

HSG 48 Reducing Error and Influencing Behaviour, gives the following causes for failures in organizational and management structures:

➤ poor work planning, leading to high work pressure
➤ lack of safety systems and barriers
➤ inadequate responses to previous incidents
➤ management based on one-way communications
➤ deficient coordination and responsibilities
➤ poor management of health and safety
➤ poor health and safety culture.

Organizational factors play a significant role in the health and safety of the workplace. However, this role is often forgotten when health and safety is being reviewed after an accident or when a new process or piece of equipment is introduced.

4.4.3 The job

Jobs may be highly dangerous or present only negligible risk of injury. Health and safety is an important element during the design stage of the job and any equipment, machinery or procedures associated with the job. Method study helps to design the job in the most cost effective way and ergonomics helps to design the job with health and safety in mind. Ergonomics is the science of matching equipment and machines to man rather than the other way round. An ergonomically designed machine will ensure that control levers, dials, metres and switches are sited in a convenient and comfortable position for the machine operator. Similarly, an ergonomically designed workstation will be designed for the comfort and health of the operator. Chairs, for example, will be designed to support the back properly throughout the working day.

Physically matching the job and any associated equipment to the person will ensure that the possibility of human error is minimized. It is also important to ensure that there is mental matching of the person's information and decision-making requirements. A person must be capable, either through past experience or through specific training, of performing the job with the minimum potential for human error.

The major considerations in the design of the job, which would be undertaken by a specialist, have been listed by the HSE as follows:

➤ the identification and detailed analysis of the critical tasks expected of individuals and the appraisal of any likely errors associated with those tasks
➤ evaluation of the required operator decision-making and the optimum (best) balance between the human and automatic contributions to safety actions (with the emphasis on automatic whenever possible)
➤ application of ergonomic principles to the design of man–machine interfaces, including displays of plant and process information, control devices and panel layout
➤ design and presentation of procedures and operating instructions in the simplest terms possible
➤ organization and control of the working environment, including the workspace, access for maintenance, lighting, noise and heating conditions
➤ provision of the correct tools and equipment
➤ scheduling of work patterns, including shift organization, control of fatigue and stress and arrangements for emergency operations
➤ efficient communications, both immediate and over a period of time.

For some jobs, particularly those with a high risk of injury, a job safety analysis should be undertaken to check that all necessary safeguards are in place. All jobs should carry a job description and a safe system of work for the particular job. The operator should have sight of the job description and be trained in the safe system of

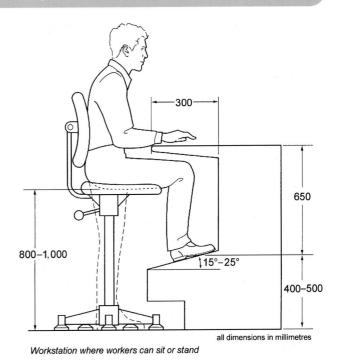

Workstation where workers can sit or stand

Figure 4.3 Well-designed workstation for sitting or standing.

work before commencing the job. More information on both these latter items is given in Chapter 6.

The following simple checklist may be used to check that the principal health and safety considerations of the job have been taken into account:

➤ Have the critical parts of the job been identified and analysed?
➤ Have the employee's decision-making needs been evaluated?
➤ Has the best balance between human and automatic systems been evaluated?
➤ Have ergonomic principles been applied to the design of equipment displays, including displays of plant and process information, control information and panel layouts?
➤ Has the design and presentation of procedures and instructions been considered?
➤ Has the guidance available for the design and control of the working environment, including the workspace, access for maintenance, lighting, noise and heating conditions, been considered?
➤ Have the correct tools and equipment been provided?
➤ Have the work patterns and shift organization been scheduled to minimize their impact on health and safety?
➤ Has consideration been given to the achievement of efficient communications and shift handover?

Figure 4.4 Most construction rubbish can burn. Make sure that it is swept up and removed from the site as soon as possible.

HSG48 gives the following causes for failures in job health and safety:

- illogical design of equipment and instruments
- constant disturbances and interruptions
- missing or unclear instructions
- poorly maintained equipment
- high workload
- noisy and unpleasant working conditions.

It is important that health and safety monitoring of the job is a continuous process. Some problems do not become apparent until the job is started. Other problems do not surface until there is a change of operator or a change in some aspect of the job.

It is very important to gain feedback from the operator on any difficulties experienced because there could be a health and safety issue requiring further investigation.

4.4.4 Personal factors

Personal factors, which affect health and safety, may be defined as any condition or characteristic of an individual which could cause or influence him to act in an unsafe manner. They may be physical, mental or psychological in nature. Personal factors, therefore, include issues such as attitude, motivation, training and human error and their interaction with the physical, mental and perceptual capability of the individual.

These factors have a significant effect on health and safety. Some of them, normally involving the personality of the individual, are unchangeable but others, involving skills, attitude, perception and motivation can be changed, modified or improved by suitable training or other measures. In summary, the person needs to be matched to the job. Studies have shown that the most common personal factors which contribute to accidents are low skill and competence levels, tiredness, boredom, low morale and individual medical problems.

It is difficult to separate all the physical, mental or psychological factors because they are interlinked. However, the three most common factors are psychological factors – attitude, motivation and perception.

Attitude is the tendency to behave in a particular way in a certain situation. Attitudes are influenced by the prevailing health and safety culture within the organization, the commitment of the management, the experience of the individual and the influence of the peer group. Peer group pressure is a particularly important factor among young people and health and safety training must be designed with this in mind by using examples or case studies that are relevant to them. Behaviour may be changed by training, the formulation and enforcement of safety rules and meaningful consultation – **attitude** change often follows.

Motivation is the driving force behind the way a person acts or the way in which people are stimulated to act. Involvement in the decision-making process in a meaningful way will improve motivation as will the use of incentive schemes. However, there are other important influences on motivation such as recognition and promotion opportunities, job security and job satisfaction. Self-interest, in all its forms, is a significant motivator and personal factor.

Perception is the way in which people interpret the environment or the way in which a person believes or understands a situation. In health and safety, the perception of hazards is an important concern. Many accidents occur because people do not perceive that there is a risk. There are many common examples of this, including the use of personal protective equipment PPE (such as hard hats) and guards on drilling machines and the washing of hands before meals. It is important to understand that when perception leads to an increased health and safety risk, it is not always caused by a conscious decision of the individual concerned. The stroboscopic effect caused by the rotation of a drill at certain speeds under fluorescent lighting will make the drill appear stationary. It is a well-known phenomenon, especially among illusionists, that people will often see what they expect to see rather than reality. Routine or repetitive tasks will reduce attention levels leading to the possibility of accidents.

Other personal factors which can affect health and safety include physical stature, age, experience, health, hearing, intelligence, language, skills, level of competence and qualifications.

Finally, memory is an important personal factor since it is influenced by training and experience. The efficiency of memory varies considerably between people and during the lifetime of an individual. The overall health of a person can affect memory as can personal crises. Due to these possible problems with memory, important safety instructions should be available in written as well as verbal form.

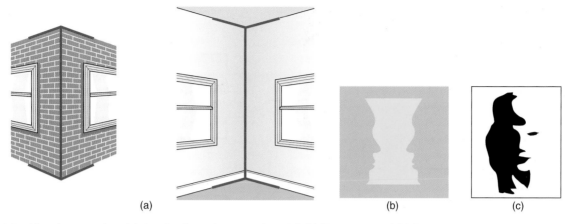

Figure 4.5   Visual perception. (a) Are the lines the same length? (b) Face or vase? (c) Face or saxophone player.

The following checklist given in HSG48 may be used to check that the relevant personal factors have been covered:

➤ Has the job specification been drawn up and included age, physique, skill, qualifications, experience, aptitude, knowledge, intelligence and personality?
➤ Have the skills and aptitudes been matched to the job requirements?
➤ Have the personnel selection policies and procedures been set up to select appropriate individuals?
➤ Has an effective training system been implemented?
➤ Have the needs of special groups of employees been considered?
➤ Have the monitoring procedures been developed for the personal safety performance of safety critical staff?
➤ Have fitness for work and health surveillance been provided where it is needed?

Figure 4.6   Motivation and activity.

➤ has counselling and support for ill-health and stress been provided?

Personal factors are the attributes that employees bring to their jobs and may be strengths or weaknesses. Negative personal factors cannot always be neutralized by improved job design. It is, therefore, important to ensure that personnel selection procedures should match people to the job. This will reduce the possibility of accidents or other incidents.

4.5   Human errors and violations

Human failures in health and safety are classified as either errors or violations. An error is an unintentional deviation from an accepted standard, while a violation is a deliberate deviation from the standard (see Figure 4.7).

4.5.1   *Human errors*

Human errors fall into three groups – slips, lapses and mistakes, which can be further sub-divided into rule-based and knowledge-based mistakes.

Slips and lapses
These are very similar in that they are caused by a momentary memory loss often due to lack of attention or loss of concentration. They are not related to levels of training, experience or motivation and they can usually be reduced by redesigning the job or equipment or minimizing distractions.

Slips are failures to carry out the correct actions of a task. Examples include the use of the incorrect switch, reading the wrong dial or selecting the incorrect component for an assembly. A slip also describes an action taken too early or too late within a given working procedure.

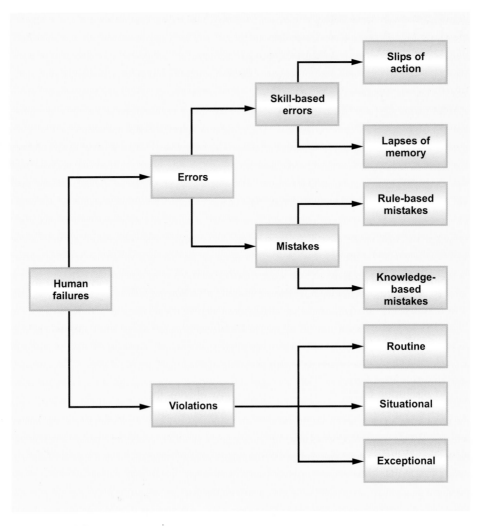

Figure 4.7 Types of human failure.

Lapses are failures to carry out particular actions which may form part of a working procedure. A forklift truck driver leaving the keys in the ignition lock of his truck is an example of a lapse as is the failure to replace the petrol cap on a car after filling it with petrol. Lapses may be reduced by redesigning equipment so that, for example, an audible horn indicates the omission of a task. They may also be reduced significantly by the use of detailed checklists.

Mistakes

Mistakes occur when an incorrect action takes place but the person involved believes the action to be correct. A mistake involves an incorrect judgement. There are two types of mistakes – rule-based and knowledge-based.

Rule-based mistakes occur when a rule or procedure is remembered or applied incorrectly. These mistakes usually happen when, due to an error, the rule that is normally used no longer applies. For example, a particular job requires the counting of items into groups of ten followed by the adding together of the groups so that the total number of items may be calculated. If one of the groups is miscounted, the final total will be incorrect even though the rule has been followed.

Knowledge-based mistakes occur when well-tried methods or calculation rules are used inappropriately. For example, the depth of the foundations required for a particular building was calculated using a formula. The formula, which assumed a clay soil, was used to calculate the foundation depth in a sandy soil. The resultant building was unsafe.

The HSE has suggested the following points to consider when the potential sources of human errors are to be identified:

➤ What human errors can occur with each task? There are formal methods available to help with this task.
➤ What influences are there on performance? Typical influences include time pressure, design of controls, displays and procedures, training and experience, fatigue and levels of supervision.

➤ What are the consequences of the identified errors? What are the significant errors?

➤ Are there opportunities for detecting each error and recovering it?

➤ Are there any relationships between the identified errors? Could the same error be made on more than one item of equipment due, for example, to the incorrect calibration of an instrument?

Errors and mistakes can be reduced by the use of instruction, training and relevant information. However, communication can also be a problem, particularly at shift handover times. Environmental and organizational factors, involving workplace stress will also affect error levels.

The following steps are suggested to reduce the likelihood of human error:

➤ Examine and reduce the workplace stressors (noise, poor lighting and so on) which increase the frequency of errors.

➤ Examine and reduce any social or organizational stressors (insufficient staffing levels, peer pressure and so on).

➤ Design plant and equipment to reduce error possibilities – poorly designed displays, ambiguous instructions.

➤ Ensure that there are effective training arrangements

➤ simplify any complicated or complex procedures.

➤ Ensure that there is adequate supervision, particularly for inexperienced or young trainees.

➤ Check that job procedures, instructions and manuals are kept up to date and are clear.

➤ Include the possibility of human error when undertaking the risk assessment.

➤ Isolate the human error element of any accident or incident and introduce measures to reduce the risk of a repeat.

➤ Monitor the effectiveness of any measures taken to reduce errors.

4.5.2 *Violations*

There are three categories of violation – routine, situational and exceptional.

Routine violation occurs when the breaking of a safety rule or procedure is the normal way of working. It becomes routine not to use the recommended procedures for tasks. An example of this is the regular speeding of forklift trucks in a warehouse so that orders can be fulfilled on time.

There are many reasons given for routine violation – the job would not be completed on time, lack of supervision and enforcement or, simply, the lack of knowledge of the procedures. Routine violations can be reduced by regular monitoring, ensuring that the rules are actually necessary or redesigning the job.

The following features are very common in many workplaces and often lead to routine violations:

➤ poor working posture due to poor ergonomic design of the workstation or equipment

➤ equipment difficult to use and/or slow in response

➤ equipment difficult to maintain or pressure on time available for maintenance

➤ procedures unduly complicated and difficult to understand

➤ unreliable instrumentation and/or warning systems

➤ high levels of noise and other poor aspects to the environment (fumes, dusts, humidity)

➤ associated PPE either inappropriate, difficult and uncomfortable to wear or ineffective due to lack of maintenance.

Situational violations occur when particular job pressures at particular times make rule compliance difficult. They may happen when the correct equipment is not available or weather conditions are adverse. A common example is the use of a ladder rather than a scaffold for working at height to replace window frames in a building. Situational violations may be reduced by improving job design, the working environment and supervision.

Exceptional violations rarely happen and usually occur when a safety rule is broken to perform a new task. Good examples are the violations which can occur during the operations of emergency procedures, such as fires or explosions. These violations should be addressed in risk assessments and during training sessions for emergencies (e.g. fire training).

Everybody is capable of making errors. It is one of the objectives of a positive health and safety culture to reduce them and their consequences as much as possible.

4.6 The development of a positive health and safety culture

No single section or department of an organization can develop a positive health and safety culture on its own. There needs to be commitment by the management, the promotion of health and safety standards, effective communication within the organization, cooperation from and with the workforce and an effective and developing training programme. Each of these topics will be examined in turn to show their effect on improving the health and safety culture in the organization.

4.6.1 *Commitment of management*

As mentioned earlier, there needs to be a commitment from the very top of the organization and this commitment will, in turn, produce higher levels of motivation and commitment throughout the organization. Probably the best indication of this concern for health and safety is shown by the status given to health and safety and the amount of resources (money, time and people) allocated to health and safety.

The management of health and safety should form an essential part of a manager's responsibility and they should be held to account for their performance on health and safety issues. Specialist expertise should be made available when required (e.g. for noise assessment), either from within the workforce or by the employment of external contractors or consultants.

Health and safety should be discussed on a regular basis at management meetings at all levels of the organization. If the organization employs sufficient people to make direct consultation with all employees' difficulties, there should be a health and safety committee at which there is employee representation. In addition, there should be recognized routes for anybody within the organization to receive health and safety information or have their health and safety concerns addressed.

The health and safety culture is enhanced considerably when senior managers appear regularly at all levels of an organization whether it be the shop floor, the hospital ward or the general office and are willing to discuss health and safety issues with staff. A visible management is very important for a positive health and safety culture.

Finally, the positive results of a management commitment to health and safety will be the active involvement of all employees in health and safety, the continuing improvement in health and safety standards and the subsequent reduction in accident and occupational ill-health rates. This will lead, ultimately, to a reduction in the number and size of compensation claims.

HSG48 makes some interesting suggestions to managers on the improvements that may be made to health and safety which will be seen by the workforce as a clear indication of their commitment. The suggestions are as follows:

➤ Review the status of the health and safety committees and health and safety practitioners. Ensure that any recommendations are acted upon or implemented.

➤ Ensure that senior managers receive regular reports on health and safety performance and act on them.

➤ Ensure that any appropriate health and safety actions are taken quickly and are seen to have been taken.

➤ Develop any action plans in consultation with employees, based on a shared perception of hazards and risks; the plans should be workable and continually reviewed.

4.6.2 *The promotion of health and safety standards*

For a positive health and safety culture to be developed, everyone within the organization needs to understand the standards of health and safety expected by the organization and the role of the individual in achieving and maintaining those standards. Such standards are required to control and minimize health and safety risks.

Standards should clearly identify the actions required of people to promote health and safety. They should also specify the competencies needed by employees and they should form the basis for measuring health and safety performance.

Health and safety standards cover all aspects of the organization. Typical examples include:

➤ the design and selection of premises
➤ the design and selection of plant and substances (including those used on site by contractors)
➤ the recruitment of employees and contractors
➤ the control of work activities, including issues such as risk assessment
➤ competence, maintenance and supervision
➤ emergency planning and training
➤ the transportation of the product and its subsequent maintenance and servicing.

Having established relevant health and safety standards, it is important that they are actively promoted within the organization by all levels of management. The most effective method of promotion is by leadership and example. There are many ways to do this such as:

➤ the managers involved in workplace inspections and accident investigations
➤ the use of PPE (e.g. goggles and hard hats) by all managers and their visitors in designated areas
➤ ensuring that employees attend specialist refresher training courses when required (e.g. first aid and forklift truck driving)
➤ full cooperation with fire drills and other emergency training exercises
➤ comprehensive accident reporting and prompt follow up on recommended remedial actions.

The benefit of good standards of health and safety will be shown directly in less lost production, fewer accidents and compensation claims and lower insurance premiums. It may also be shown in higher product quality and better resource allocation.

An important and central necessity for the promotion of high health and safety standards is health and safety competence. What is meant by 'competence'?

4.6.3 *Competence*

The word 'competence' is often used in health and safety literature. One definition, made during a civil case in 1962, stated that a competent person is:

> *a person with practical and theoretical knowledge as well as sufficient experience of the particular machinery, plant or procedure involved to enable them to identify defects or weaknesses during plant and machinery examinations, and to assess their importance in relation to the strength and function of that plant and machinery.*

This definition concentrates on a manufacturing rather than service industry requirement of a competent person.

Regulation 7 of the Management of Health and Safety at Work Regulations 1999 requires that 'every employer shall employ one or more competent persons to assist him in undertaking the measures he needs to take to comply with the requirements and prohibitions imposed upon him by or under the relevant statutory provisions'. In other words, competent persons are required to assist the employer in meeting his obligations under health and safety law. This may mean a health and safety adviser in addition to, say, an electrical engineer, an occupational nurse and a noise assessment specialist. The number and range of competent persons will depend on the nature of the business of the organization.

It is recommended that competent employees are used for advice on health and safety matters rather than external specialists (consultants). It is recognized, however, that if employees, competent in health and safety, are not available in the organization, then an external service may be enlisted to help. The key is that management and employees need access to health and safety expertise.

The regulations do not define 'competence' but do offer advice to employers when appointing health and safety advisers who should have:

> ➤ a knowledge and understanding of the work involved, the principles of risk assessment and prevention and current health and safety applications
> ➤ the capacity to apply this to the task required by the employer in the form of problem and solution identification, monitoring and evaluating the effectiveness of solutions and the promotion and communication of health and safety and welfare advances and practices.

Such competence does not necessarily depend on the possession of particular skills or qualifications. It may only be required to understand relevant current best practice, be aware of one's own limitations in terms of experience and knowledge and be willing to supplement existing experience and knowledge. However, in more complex or technical situations membership of a relevant professional body and/or the possession of an appropriate qualification in health and safety may be necessary.

It is important that any competent person employed to help with health and safety has evidence of relevant knowledge, skills and experience for the tasks involved. The appointment of a competent person as an adviser to an employer does not absolve the employer from his responsibilities under the Health and Safety at Work Act and other relevant statutory provision.

Finally, it is worth noting that the requirement to employ competent workers is not restricted to those having a health and safety function but covers the whole workforce.

Competent workers must have sufficient training, experience, knowledge and other qualities to enable them properly to undertake the duties assigned to them.

4.7 Effective communication

Many problems in health and safety arise due to poor communication. It is not just a problem between management and workforce – it is often a problem the other way or indeed at the same level within an organization. It arises from ambiguities or, even, accidental distortion of a message.

The Health and Safety (Information for Employees) Regulations require that the latest version of the approved health and safety poster be displayed prominently in the workplace. Additional information on these regulations is given in Chapter 1.

There are three basic methods of communication in health and safety – verbal, written and graphic.

Verbal communication is the most common. It is communication by speech or word of mouth. Verbal communication should only be used for relatively simple pieces of information or instruction. It is most commonly used in the workplace, during training sessions or at meetings.

There are several potential problems associated with verbal communication. The speaker needs to prepare the communication carefully so that there is no confusion about the message. It is very important that the recipient is encouraged to indicate their understanding of the communication. There have been many cases of accidents occurring because a verbal instruction has not been understood. There are several barriers to this understanding from the point of view of the recipient, including language and dialect, the use of technical language and abbreviations, background noise and distractions, hearing problems, ambiguities in the message, mental weaknesses and learning disabilities, lack of interest and attention.

Having described some of the limitations of verbal communication, it does have some merits. It is less formal, enables an exchange of information to take place quickly and the message to be conveyed as near to the workplace as possible. Training or instructions that are delivered in this way are called toolbox talks and can be very effective.

Written communication takes many forms from the simple memo to the detailed report.

A memo should contain one simple message and be written in straightforward and clear language. The title should accurately describe the contents of the memo. In recent years, emails have largely replaced memos since it has become a much quicker method to ensure that the message gets to all concerned (although a recent report has suggested that many people are becoming overwhelmed by the number of emails which they receive!). The advantage of memos and emails is that there is a record of the message after it has been delivered. The disadvantage is that they can be ambiguous or difficult to understand or, indeed, lost within the system.

Reports are more substantial documents and cover a topic in greater detail. The report should contain a detailed account of the topic and any conclusions or recommendations. The main problem with reports is that they are often not read properly due to the time constraints on managers. It is important that all reports have a summary attached so that the reader can decide whether it needs to be read in detail.

The most common way in which written communication is used in the workplace is the noticeboard. For a noticeboard to be effective, it needs to be well positioned within the workplace and there needs to be a regular review of the notices to ensure that they are up to date and relevant.

There are many other examples of written communications in health and safety, such as employee handbooks, company codes of practice, minutes of safety committee meetings and health and safety procedures.

Graphic communication is communication by the use of drawings, photographs or DVDs. It is used either to impart health and safety information (e.g. fire exits) or health and safety propaganda. The most common forms of health and safety propaganda are the poster and the video. Both can be used very effectively as training aids since they can retain interest and impart a simple message. Their main limitation is that they can become out of date fairly quickly or, in the case of posters, become largely ignored.

There are many sources of health and safety information which may need be consulted before an accurate communication can be made. These include regulations, judgements, approved codes of practice, guidance, British and European standards, periodicals, case studies and HSE publications.

4.8 Health and safety training

Health and safety training is a very important part of the health and safety culture and it is also a legal requirement, under the Management of Health and Safety at Work Regulations 1999 and other regulations, for an employer to provide such training. Training is required on recruitment, at induction or on being exposed to new or increased risks due to:

➤ being transferred to another job or given a change in responsibilities
➤ the introduction of new work equipment or a change of use in existing work equipment
➤ the introduction of new technology
➤ the introduction of a new system of work or the revision of an existing system of work
➤ an increase in the employment of more vulnerable employees (young or disabled persons)
➤ particular training required by the organization's insurance company (e.g. specific fire and emergency training).

Additional training may well be needed following a single or series of accidents or near misses, the introduction of new legislation, the issuing of an enforcement notice or as a result of a risk assessment or safety audit.

It is important during the development of a training course that the target audience is taken into account. If the audience are young persons, the chosen approach must be capable of retaining their interest and any illustrative examples used must be within their experience. The trainer must also be aware of external influences, such as peer pressures, and use them to advantage. For example, if everybody wears PPE then it will be seen as the thing to do. Levels of literacy and numeracy are other important factors.

The way in which the training session is presented by the use of DVDs, power point, case studies, lectures or small discussion groups needs to be related to the material to be covered and the backgrounds of the trainees. Supplementary information in the form of copies of slides and additional background reading is often useful. The environment used for the training sessions is also important in terms of room layout and size, lighting and heating.

Attempts should be made to measure the effectiveness of the training by course evaluation forms issued at the time of the session, by a subsequent refresher session and by checking for improvements in health and safety performance (such as a reduction in specific accidents).

There are several different types of training; these include induction, job specific, supervisory and management, and specialist. Informal sessions that take place at the place of work are known as 'toolbox talks'. Such sessions should only be used to cover a limited number of issues. They can become a useful route for employee consultation.

4.8.1 *Induction training*

Induction training should always be provided to new employees, trainees and contractors. While such training covers items such as pay, conditions and quality, it must also include health and safety. It is useful if the employee signs a record to the effect that training has been received. This record may be required as evidence should there be a subsequent legal claim against the organization.

Most induction training programmes would include the following topics:

➤ the health and safety policy of the organization including a summary of the organization and arrangements including employee consultation
➤ a brief summary of the health and safety management system including the name of the employee's direct supervisor and safety representative, and source of health and safety information
➤ the employee responsibility for health and safety including any general health and safety rules (e.g. smoking prohibitions)
➤ the accident reporting procedure of the organization, the location of the accident book and the location of the nearest first aider
➤ the fire and other emergency procedures including the location of the assembly point
➤ a summary of any relevant risk assessments and safe systems of work
➤ the location of welfare, canteen facilities and rest rooms.

Additional items, which are specific to the organization, may need to be included such as:

➤ internal transport routes and pedestrian walkways (e.g. forklift truck operations)
➤ the correct use of PPE and maintenance procedures
➤ manual handling techniques and procedures
➤ details of any hazardous substances in use and any procedures relating to them (e.g. health surveillance).

There should be some form of follow-up with each new employee after 3 months to check that the important messages have been retained. This is sometimes called a refresher course, although it is often better done on a one-to-one basis.

It is very important to stress that the content of the induction course should be subject to constant review and updated following an accident investigation, new legislation, changes in the findings of a risk assessment or the introduction of new plant or processes.

4.8.2 *Job-specific training*

Job-specific training ensures that employees undertake their jobs in a safe manner. Such training, therefore, is a form of skill training and is often best done 'on the job' – sometimes known as 'toolbox training'. Details of the safe system of work or, in more hazardous jobs, a permit to work system, should be covered. In addition to normal safety procedures, emergency procedures and the correct use of PPE need to be included. The results of risk assessments are very useful in the development of this type of training. It is important that any common causes of human errors (e.g. discovered as a result of an accident investigation), any standard safety checks or maintenance requirements are addressed.

It is common for this type of training to follow an operational procedure in the form of a checklist which the employee can sign on completion of the training. The new employee will still need close supervision for some time after the training has been completed.

4.8.3 *Supervisory and management training*

Supervisory and management health and safety training follows similar topics to those contained in an induction training course but will be covered in more depth. There will also be a more detailed treatment of health and safety law. There has been considerable research over the years into the failures of managers that have resulted in accidents and other dangerous incidents. These failures have included:

➤ lack of health and safety awareness, enforcement and promotion (in some cases there has been an encouragement to circumvent health and safety rules)
➤ lack of consistent supervision of and communication with employees
➤ lack of understanding of the extent of the responsibility of the supervisor.

It is important that all levels of management, including the board, receive health and safety training. This will not only keep everybody informed of health and safety legal requirements, accident prevention techniques and changes in the law, but also encourage everybody to monitor health and safety standards during visits or tours of the organization.

4.8.4 *Specialist training*

Specialist health and safety training is normally needed for activities that are not related to a specific job but more to an activity. Examples include first aid, fire prevention, forklift truck driving, overhead crane operation, scaffold inspection and statutory health and safety inspections. These training courses are often provided by specialist organizations and successful participants are awarded certificates. Details of two of these courses will be given here by way of illustration.

Fire prevention training courses include the causes of fire and fire spread, fire and smoke alarm systems, emergency lighting, the selection and use of fire extinguishers and sprinkler systems, evacuation procedures, high risk operations and good housekeeping principles.

A forklift truck driver's course would include the general use of the controls, loading and unloading procedures, driving up or down an incline, speed limits, pedestrian awareness (particularly in areas where pedestrians and vehicles are not segregated), security of the vehicle when not in use, daily safety checks and defect reporting, refuelling and/or battery charging and emergency procedures.

Details of other types of specialist training appear elsewhere in the text.

Training is a vital part of any health and safety programme and needs to be constantly reviewed and updated. Many health and safety regulations require specific training (e.g. manual handling, PPE and display screens). Additional training courses may be needed when there is a major reorganization, a series of similar accidents or incidents or a change in equipment or a process. Finally, the methods used to deliver training must be continually monitored to ensure they are effective.

4.9 Internal influences

There are many influences on health and safety standards; some are positive and others negative. No business, particularly a small business, is totally divorced from its suppliers, customers and neighbours. This section considers the internal influences on a business, including management commitment, production demands, communication, competence and employee relations.

4.9.1 *Management commitment*

Managers, particularly senior managers, can give powerful messages to the workforce by what they do for health and safety. Managers can achieve the level of health and safety performance that they demonstrate they want to achieve. Employees soon get the negative message if directors disregard safety rules and ignore written policies to get urgent orders to the customer or to avoid personal inconvenience. It is what they do that counts not what they say. In Chapter 3 managers' organizational roles were listed showing the ideal level of involvement for senior managers. Depending on the size and geography of the organization, senior managers should be personally involved in:

> ➤ health and safety inspections or audits
> ➤ meetings of the central health and safety or joint consultation committees
> ➤ involvement in the investigation of accidents, ill-health and incidents. The more serious the incident the more senior the manager who takes an active part in the investigation.

It may well be, following a particularly serious incident, that disciplinary action or the use of penalties by the project client will be required. It is important that published disciplinary procedures have been formulated, discussed with the workforce and implemented if necessary. Penalties also have a role both at the client/principal contractor and principal contractor/sub-contractor interfaces.

4.9.2 *Production/service demands*

Managers need to balance the demands placed by customers and action required to protect the health and safety of their employees. How this is achieved has a strong influence on the standards adopted by the organization. The delivery driver operating to near impossible delivery schedules and the manager agreeing to the strident demands of a large customer regardless of the risks to employees involved, are typical dilemmas facing workers and managers alike. The only way to deal with these issues is to be well organized and agree with the client/employees ahead of the crisis how they should be prioritized.

Rules and procedures should be intelligible, sensible and reasonable. They should be designed to be followed under normal production or service delivery conditions. If following the rules involves very long delays or impossible production schedules, they should be revised rather than ignored by workers and managers alike until it is too late and an accident occurs. Sometimes the safety rules are simply used in a court of law, in an attempt to defend the company concerned, following the accident.

Managers who plan the impossible schedule or ignore a safety rule to achieve production or service demands, are held responsible for the outcomes. It is acceptable to balance the cost of the action against the level of risk being addressed but it is never acceptable to ignore safety rules or standards simply to get the work done. Courts are not impressed by managers who put profit considerations ahead of safety requirements.

4.9.3 *Communication*

Communications are covered in depth earlier in the chapter but clearly will have significant influence on health and safety issues. This will include:

➤ poorly communicated procedures that will not be understood or followed
➤ poor verbal communications which will be misunderstood and will demonstrate a lack of interest by senior managers
➤ missing or incorrect signs may cause accidents rather than prevent them
➤ managers who are nervous about face-to-face discussions with the workforce on health and safety issues will have a negative effect.

Managers and supervisors should plan to have regular discussions to learn about the problems faced by employees and discuss possible solutions. Some meetings, like the safety committee, are specifically planned for safety matters, but this should be reinforced by discussing health and safety issues at all routine management meetings. Regular one-to-one talks should also take place in the workplace, preferably to a planned theme or safety topic, to get specific messages across and get feedback from employees.

4.9.4 *Competence*

Competent people, who know what they are doing and have the necessary skills to do the task correctly and safely, will make the organization effective. Competence can be bought in through recruitment or consultancy but is often much more effective if developed from among employees. It demonstrates commitment to health and safety and a sense of security for the work force. The loyalty that it develops in the workforce can be a significant benefit to safety standards. Earlier in this chapter there is more detail on how to achieve competence.

4.9.5 *Employee representation*

Enthusiastic, competent employee safety representatives given the resources and freedom to fulfil their function effectively can make a major contribution to good health and safety standards. They can provide the essential bridge between managers and employees. People are more willing to accept the restrictions that some precautions bring if they are consulted and feel involved, either directly in small workforces or through their safety representatives. See earlier in Chapter 3 for the details.

4.10 External influences

The role of external organizations is set out in Chapter 1. Here the influence they can have is briefly discussed, including societal expectations, legislation and enforcement, insurance companies, trade unions, economics and commercial stakeholders.

4.10.1 *Societal expectations*

Societal expectations are not static and tend to rise over time, particularly in a wealthy nation like the UK. For example, the standards of safety accepted in a motor car 50 years ago would be considered to be totally inadequate at the beginning of the 21st century. We expect safe, quiet, comfortable cars that do not break down and which retain their appearance for many thousands of miles. Industry should strive to deliver these same high standards for the health and safety of employees or service providers. The question is whether societal expectations are as great an influence on workplace safety standards as they are on product safety standards. Society can influence standards through:

➤ people only working for good employers. This is effective in times of low unemployment
➤ national and local news media highlighting good and bad employment practices
➤ schools teaching good standards of health and safety
➤ the purchase of fashionable and desirable safety equipment, like trendy crash helmets or mountain bikes
➤ only buying products from responsible companies. The difficulty of defining what is responsible has been partly overcome through ethical investment criteria but this is possibly not widely enough understood to be a major influence
➤ watching TV and other programmes which improve safety knowledge and encourage safe behaviour from an early age. It will be interesting to see if 'Bob the Builder' and all the books, DVDs and toys produced to enhance sales will have an effect on future building and DIY health and safety standards.

4.10.2 *Legislation and enforcement*

Good comprehensible legislation should have a positive effect on health and safety standards. Taken together legislation and enforcement can affect standards by:

➤ providing a level to which every employer has to conform
➤ insisting on minimum standards which also enhances people's ability to operate and perform well
➤ providing a tough, visible threat of getting shut down or a heavy fine
➤ stifling development by being too prescriptive. For example, woodworking machines did not develop quickly in the 20th century, partly because the regulations were so detailed in their requirements that new designs were not feasible
➤ providing well presented and easily read guidance for specific industries at reasonable cost or free.

On the other hand a weak enforcement regime can have a powerful negative effect on standards.

4.10.3 *Insurance companies*

Insurance companies can influence health and safety standards mainly through financial incentives. Employers' liability insurance is a legal requirement in the UK and therefore all employers have to obtain this type of insurance cover. Insurance companies can influence standards through:

➤ discounting premiums to those in the safest sectors or best individual companies
➤ insisting on risk reduction improvements to remain insured. This is not very effective where competition for business is fierce
➤ encouraging risk reduction improvements by bundling services into the insurance premium
➤ providing guidance on standards at reasonable cost or free.

4.10.4 *Trade unions*

Trade unions can influence standards by:

➤ providing training and education for members
➤ providing guidance and advice cheaply or free to members
➤ influencing governments to regulate, enhance enforcement activities and provide guidance
➤ influencing employers to provide high standards for their members. This is sometimes confused with financial improvements, with health and safety getting a lower priority
➤ encouraging members to work for safer employers

➤ helping members to get proper compensation for injury or ill-health if it is caused through their work.

4.10.5 *Economics*

Economics can play a major role in influencing health and safety standards. The following ways are the most common:

➤ lack of orders and/or money can cause employers to try to ignore health and safety requirements
➤ if employers were really aware of the actual and potential cost of accidents and fires they would be more concerned about prevention. The HSE believe that the ratio between insured and uninsured costs of accidents is between 1:8 and 1:36 (see *The Costs of Accidents at Work*, HSE Books, HSG96)
➤ perversely when the economy is booming activity increases and, particularly in the building industry, accidents can sharply increase. The pressures to perform and deliver for customers can be safety averse
➤ businesses that are only managed on short-term performance indicators seldom see the advantage of the long-term gains that are possible with a happy, safe and fit work force.

The cost of accidents and ill-health, in both human and financial terms, needs to be visible throughout the organization so that all levels of employee are encouraged to take preventative measures.

4.10.6 *Commercial stakeholders*

A lot can be done by commercial stakeholders to influence standards. This includes:

➤ insisting on proper arrangements for health and safety management at supplier companies before they tender for work or contracts
➤ checking on suppliers to see if the workplace standards are satisfactory
➤ encouraging ethical investments
➤ considering ethical standards as well as financial when banks provide funding
➤ providing high quality information for customers
➤ insisting on high standards to obtain detailed planning permission (where this is possible)
➤ providing low cost guidance and advice.

4.11 Practice NEBOSH questions for Chapter 4

1. **Outline** the factors that might contribute towards a positive health and safety culture within a construction company.

2. **Describe FIVE** components of a positive health and safety culture.

3. **Outline FOUR** indicators of a poor health and safety culture within a construction company.

4. **Outline** the ways in which the health and safety culture of an organization might be improved.

5. **Outline** the factors that might cause the safety culture within an organization to decline.

6. (a) **Define** the term 'accident incidence rate'.
   (b) **Outline** how information on accidents could be used to promote health and safety in the workplace.

7. **Outline** the personal factors that might place an individual at a greater risk of harm while at work.

8. (a) **Explain**, using an example, the meaning of the term 'attitude'.
   (b) **Outline THREE** influences on the attitude towards health and safety of employees within an organization.
   (c) **Outline** the ways in which employers might motivate their employees to comply with health and safety procedures.

9. (a) **Explain** the meaning of the term 'motivation'.
   (b) Other than lack of motivation, **outline SIX** reasons why employees may fail to comply with safety procedures at work.
   (c) **Outline** ways in which employers may motivate their employees to comply with health and safety procedures.

10. A health and safety audit of a long-term construction project has identified a general lack of compliance with procedures.
    (a) **Describe** the possible reasons for procedures not being followed.
    (b) **Outline** the practical measures that could be taken to motivate site workers to comply with health and safety procedures.

11. (a) **Explain** the meaning of the term 'perception'.
    (b) **Outline** the factors relating to the individual that may influence a person's perception of an occupational risk.
    (c) **Outline** ways in which employees' perceptions of hazards in the workplace might be improved.

12. (a) Giving a practical example, **explain** the meaning of the term 'human error'.
    (b) **Outline** *individual* (or *personal*) factors that may contribute to human errors occurring at work.

13. **Describe**, using practical examples, **FOUR** types of human error that can lead to accidents in the workplace.

14. (a) **Outline** ways of reducing the likelihood of human error in a construction workplace.
    (b) **Give FOUR** reasons why the seriousness of a hazard may be underestimated by a construction worker exposed to it.
    (c) **Outline** ways in which managers can motivate employees to work safely.

15. **Outline** the practical means by which a manager could involve employees in the improvement of health and safety in the workplace.

16. Following a significant increase in accidents, a health and safety campaign is to be launched within an organization to encourage safer working by employees.
    (a) **Outline** how the organization might ensure that the nature of the campaign is effectively communicated to, and understood by, employees.
    (b) **Explain** why it is important to use a variety of methods to communicate health and safety information in the workplace.
    (c) Other than poor communications, **describe** the organizational factors that could limit the effectiveness of the campaign.

17. **Give** reasons why a verbal instruction may not be clearly understood by an employee.

18. **Explain** why employees may *fail* to comply with safety procedures at work.

19. **Outline FOUR** advantages and **FOUR** disadvantages of using 'propaganda' posters to communicate health and safety information to the workforce.

20. (a) **Identify FOUR** types of health and safety information that might usefully be displayed on a noticeboard within a workplace.
    (b) **Explain** how the effectiveness of noticeboards as a means of communicating health and safety information to the workforce can be maximized.

21. **Identify** a range of methods that an employer can use to provide health and safety information directly to individual employees.

22. (a) **Identify** the **TWO** means by which employers may provide information to employees in order

to comply with the Health and Safety Information for Employees Regulations 1989.

(b) **Outline** the categories of information provided to employees by the means identified in (a).

23. **Explain** how induction training programmes for new employees can help to reduce the number of accidents in the workplace.

24. A contractor has been engaged to undertake building maintenance work in a busy warehouse.

   Outline the issues that should be covered in an induction programme for the contractor's employees.

25. **Describe:**
   (a) the purpose of induction training and the issues that should typically be covered by such training.
   (b) the nature of 'toolbox talks', including their purpose, mode of presentation and typical subject areas.

26. A site manager has been tasked to provide site induction training to the employees of a contractor about to start work on site. **List SIXTEEN** items that should be addressed in the session.

27. **Describe** how the principal contractor on a large construction project might ensure that all persons on site have received adequate training and information to enable them to carry out their work safely.

28. With reference to the Management of Health and Safety at Work Regulations 1999, **identify:**
   (a) the particular matters on which employees should receive health and safety information
   (b) the specific circumstances when health and safety training should be given to employees.

29. **Outline** the circumstances which may suggest that a review of health and safety training is needed within an organization.

30. An independent audit of an organization has concluded that employees have received insufficient health and safety training.
   (a) **Describe** the factors that should be considered when developing an extensive programme of health and safety training within an organization.
   (b) **Outline** the various measures that might be used to assess the effectiveness of such training,
   (c) **Give FOUR** reasons why it is important for an employer to keep a record of the training provided to each employee.

31. **Outline** reasons why an employee might require additional health and safety training at a later stage of employment within an organization.

32. **Describe TWO** internal and **TWO** external influences on the health and safety culture of an organization.

Risk assessment

5

5.1 Introduction

Risk assessment is an essential part of the planning stage of any health and safety management system. Health and Safety Executive (HSE), in the publication HSG65 *Successful Health and Safety Management*, states that the aim of the planning process is to minimize risks.

> *Risk assessment methods are used to decide on priorities and to set objectives for eliminating hazards and reducing risks. Wherever possible, risks are eliminated through selection and design of facilities, equipment and processes. If risks cannot be eliminated, they are minimized by the use of physical controls or, as a last resort, through systems of work and personal protective equipment.*

5.2 Legal aspects of risk assessment

The general duties of employers to their employees in section 2 of the Health and Safety at Work Act 1974 imply the need for risk assessment. This duty was also extended by section 3 of the Act to anybody else affected by activities of the employer – contractors, visitors, customers or members of the public. However, the Management of Health and Safety at Work Regulations are much more specific concerning the need for risk assessment. The following requirements are laid down in those Regulations:

> *the risk assessment shall be 'suitable and sufficient' and cover both employees and non-employees affected by the employer's*

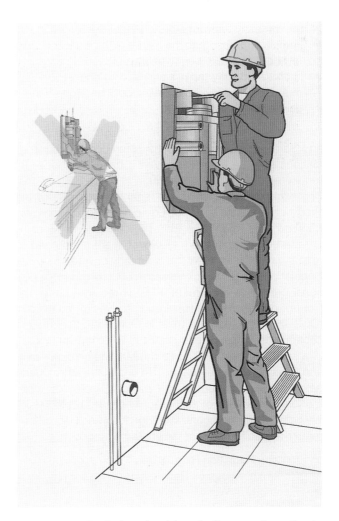

Figure 5.1 Reducing the risk – finding an alternative to fitting a wall-mounted boiler.

undertaking [e.g. contractors, members of the public, students, patients, customers];

every self-employed person shall make a 'suitable and sufficient' assessment of the risks to which they or those affected by the undertaking may be exposed;

any risk assessment shall be reviewed if there is reason to suspect that it is no longer valid or if a significant change has taken place;

where there are 5 or more employees, the significant findings of the assessment shall be recorded and any specially at risk group of employees identified. [This does not mean that employers with 4 or less employees need not undertake risk assessments.]

The term 'suitable and sufficient' is important since it defines the limits to the risk assessment process. A suitable and sufficient risk assessment should:

➤ identify the significant risks and ignore the trivial ones
➤ identify and prioritize the measures required to comply with any relevant statutory provisions
➤ remain appropriate to the nature of the work and valid over a reasonable period of time.
➤ it should identify the risk arising or in connection with the work. The level of detail should be proportionate to the risk.

When assessing risks under the Management of Health and Safety at Work Regulations, reference to other regulations may be necessary even if there is no specific requirement for a risk assessment in those Regulations. For example, reference to the legal requirements of the Provision and Use of Work Equipment Regulations will be necessary when risks from the operation of machinery are being considered. However, there is no need to repeat a risk assessment if it is already covered by other Regulations (e.g. a risk assessment considering personal protective equipment (PPE) is required under the Control of Substances Hazardous to Health regulations (COSHH) Regulations so there is no need to undertake a separate risk assessment under the PPE Regulations).

Apart from the duty under the Management of Health and Safety at Work Regulations to undertake a health and safety risk assessment of any person (employees, contractors or members of the public) who may be affected by the activities of the organization, the following Regulations require a specific risk assessment to be made:

➤ Ionizing Radiation Regulations
➤ Control of Asbestos Regulations
➤ Electricity at Work Regulations
➤ Noise at Work Regulations
➤ Manual Handling Operations Regulations
➤ Health and Safety (Display Screen Equipment) Regulations

➤ Personal Protective Equipment at Work Regulations
➤ Confined Space Regulations
➤ The Control of Vibration at Work Regulations and also the Regulatory Reform (Fire Safety) Order
➤ Control of Lead at Work Regulations
➤ Control of Substances Hazardous to Health Regulations.

A detailed comparison of the risk assessments required for most of these and more specialist regulations is given in the HSE Guide to Risk Assessment Requirements, INDG218.

5.3 Forms of risk assessment

There are two basic forms of risk assessment.

A quantitative risk assessment attempts to measure the risk by relating the probability of the risk occurring to the possible severity of the outcome and then giving the risk a numerical value. This method of risk assessment is used in situations where a malfunction could be very serious (e.g. aircraft design and maintenance or the petrochemical industry).

The more common form of risk assessment is the qualitative assessment, which is based purely on personal judgement and is normally defined as high, medium or low. Qualitative risk assessments are usually satisfactory since the definition (high, medium or low) is normally used to determine the time frame in which further action is to be taken.

The term 'generic' risk assessment is sometimes used and describes a risk assessment which covers similar activities or work equipment in different departments, sites or companies. Such assessments are often produced by specialist bodies, such as trade associations. If used, they must be appropriate to the particular job and they will need to be extended to cover additional hazards or risks.

5.4 Some definitions

Some basic definitions were introduced in Chapter 1 and those relevant to risk assessment are reproduced here.

Electricity is an example of a high hazard since it has the potential to kill a person. The risk associated with electricity – the likelihood of being killed on coming into contact with an electrical device – is, hopefully, low.

5.4.2 *Occupational or work-related ill-health*

This is concerned with those illnesses or physical and mental disorders that are either caused or triggered by

5.4.1 *Hazard and risk*

A hazard is the *potential* of a substance, activity or process to cause harm. Hazards take many forms including, for example, chemicals, electricity and the use of a ladder. A hazard can be ranked relative to other hazards or to a possible level of danger.

A risk is the *likelihood* of a substance, activity or process to cause harm. Risk (or, strictly, the level of risk) is also linked to the severity of its consequences. A risk can be reduced and the hazard controlled by good management.

It is very important to distinguish between a hazard and a risk – the two terms are often confused and activities often called high risk are, in fact, high hazard. There should only be high residual risk where there is poor health and safety management and inadequate control measures.

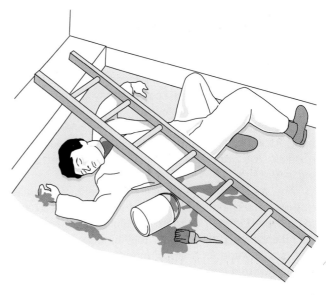

Figure 5.2   Accident at work.

Examples include the collapse of a scaffold or a crane or the failure of any passenger-carrying equipment.

In 1969, F E Bird collected a large quantity of accident data and produced a well-known triangle (Figure 5.3).

Figure 5.3   F E Bird's well-known accident triangle.

It can be seen that damage and near miss accidents occur much more frequently than injury accidents and are, therefore, a good indicator of risk. The study also shows that most accidents are predictable and avoidable.

workplace activities. Such conditions may be induced by the particular work activity of the individual or by activities of others in the workplace. The time interval between exposure and the onset of the illness may be short (e.g. asthma attacks) or long (e.g. deafness or cancer).

5.4.3 *Accident*

This is defined by the HSE as 'any unplanned event that results in injury or ill-health of people, or damage or loss to property, plant, materials or the environment or a loss of a business opportunity'. Other authorities define an accident more narrowly by excluding events that do not involve injury or ill-health. This book will always use the HSE definition.

5.4.4 *Near miss*

This is any incident that could have resulted in an accident. Knowledge of near misses is very important since research has shown that, approximately, for every 10 'near miss' events at a particular location in the workplace, a minor accident will occur.

5.4.5 *Dangerous occurrence*

This is a 'near miss' which could have led to serious injury or loss of life. Dangerous occurrences are defined in the Reporting of Injuries, Diseases and Dangerous Occurrences Regulations 1995 (often known as RIDDOR) and are always reportable to the Enforcement Authorities.

5.5 The objectives of risk assessment

The main objective of risk assessment is to determine the measures required by the organization to comply with relevant health and safety legislation and, thereby, reduce the level of occupational injuries and ill-health. The purpose is to help the employer or self-employed person to determine the measures required to comply with their legal statutory duty under the Health and Safety at Work Act 1974 or its associated Regulations. The risk assessment will need to cover all those who may be at risk, such as customers, contractors and members of the public. In the case of shared work-places, an overall risk assessment may be needed in partnership with other employers.

In Chapter 1, the moral, legal and financial arguments for health and safety management were discussed in detail. The important distinction between the direct and indirect costs of accidents is reiterated here.

Any accident or incidence of ill-health will cause both direct and indirect costs and incur an insured and an uninsured cost. It is important that all of these costs are taken into account when the full cost of an accident is calculated. In a study undertaken by the HSE, it was shown that indirect costs or hidden costs could be 36 times greater than direct costs of an accident. In other words, the direct costs of an accident or disease represent the tip of the iceberg when compared with the overall costs (*The Cost of Accidents at Work* HSG96).

Direct costs are costs that are directly related to the accident. They may be insured (claims on employers' and public liability insurance, damage to buildings, equipment or vehicles) or uninsured (fines, sick pay, damage to product, equipment or process).

Indirect costs may be insured (business loss, product or process liability) or uninsured (loss of goodwill, extra overtime payments, accident investigation time, production delays).

There are many reasons for the seriousness of a hazard not to be obvious to the person exposed to it. It may be that the hazard is not visible (radiation, certain gases and biological agents) or have no short term effect (work-related upper limb disorders). The common reasons include lack of attention, lack of experience, the wearing of PPE, sensory impairment and inadequate information, instruction and training.

5.6 Accident categories

There are several categories of accident and each has a particular significance in the construction industry. All of these accident categories are discussed in considerable detail in later chapters. The principal categories are as follows:

➤ contact with moving machinery or material being machined
➤ struck by moving, flying or falling object
➤ hit by a moving vehicle
➤ struck against something fixed or stationary
➤ injured while handling, lifting or carrying
➤ slips, trips and falls on the same level

➤ falls from a height
➤ trapped by something collapsing
➤ drowned or asphyxiated
➤ exposed to, or in contact with, a harmful substance
➤ exposed to fire
➤ exposed to an explosion
➤ contact with electricity or an electrical discharge
➤ injured by an animal
➤ physically assaulted by a person
➤ other kind of accident.

5.7 Health risks

Risk assessment is not only concerned with injuries in the workplace but also needs to consider the possibility of occupational ill-health. Health risks fall into the following four categories:

➤ chemical (e.g. paint solvents, exhaust fumes)
➤ biological (e.g. bacteria, pathogens)
➤ physical (e.g. noise, vibrations)
➤ psychological (e.g. occupational stress).

There are two possible health effects of occupational ill-health.

They may be **acute**, which means that they occur soon after the exposure and are often of short duration, although in some cases emergency admission to hospital may be required.

They may be **chronic,** which means that the health effects develop with time. It may take several years for the associated disease to develop and the effects may be slight (mild asthma) or severe (cancer).

Health risks are discussed in more detail in Chapters 16 and 17.

5.8 The management of risk assessment

Risk assessment is part of the planning and implementation stage of the health and safety management system recommended by the HSE in its publication HSG65. All aspects of the organization, including health and safety management, need to be covered by the risk assessment process. This will involve the assessment of risk in areas such as maintenance procedures, training programmes and supervisory arrangements. A general risk assessment of the organization should reveal the significant hazards present and the general control measures that are in place. Such a risk assessment should be completed first and then followed by more specific risk assessments that examine individual work activities.

HSE has produced a free leaflet entitled *Five Steps to Risk Assessment*. It gives practical advice on assessing risks and recording the findings and is aimed at small- and medium-sized companies in the service and manufacturing sectors. The five steps are:

➤ Look for the hazards.
➤ Decide who might be harmed, and how.
➤ Evaluate the risks and decide whether existing precautions are adequate or more should be done.
➤ Record the significant findings.
➤ Review the assessment and revise it if necessary.

Each of these steps will be examined, in turn, in the next section.

Finally, it is important that the risk assessment team is selected on the basis of its competence to assess risks in the particular areas under examination in the organization. The team leader or manager should have health and safety experience and relevant training in risk assessment. It is sensible to involve the appropriate line manager, who has responsibility for the area or activity being assessed, as a team member. Other members of the team will be selected on the basis of their experience, their technical and/or design knowledge and any relevant standards or regulations relating to the activity or process. At least one team member must have communication and report writing skills. It is likely that team members will require some basic training in risk assessment.

5.9 The risk assessment process

The HSE approach to risk assessment (5 steps) will be used to discuss the process of risk assessment. It is, however, easier to divide the process into six elements:

➤ hazard identification
➤ persons at risk
➤ evaluation of risk level
➤ risk controls (existing and additional)
➤ record of risk assessment findings
➤ monitoring and review.

Each element will be discussed in turn.

5.9.1 Hazard identification

Hazard identification is the crucial first step of risk assessment. Only significant hazards, which could result in serious harm to people, should be identified. Trivial hazards should be ignored.

A tour of the area under consideration by the risk assessment team is an essential part of hazard identification as is consultation with the relevant section of the workforce.

A review of accident, incident and ill-health records will also help with the identification. Other sources of information include safety inspection, survey and audit reports, job or task analysis reports, manufacturers' handbooks or data sheets and approved codes of practice and other forms of guidance.

Hazards will vary from workplace to workplace but the checklist in Appendix 5.1 shows the common hazards that are significant in many workplaces. Many questions in the NEBOSH examinations involve several common hazards found in most workplaces.

It is important that unsafe conditions are not confused with hazards, during hazard identification. Unsafe conditions should be rectified as soon as possible after observation. Examples of unsafe conditions include missing machine guards, faulty warning systems and oil spillage on the workplace floor.

5.9.2 Persons at risk

Employees and contractors who work full time at the site or workplace are the most obvious groups at risk and it will be necessary to check that they are competent to perform their particular tasks. However, there may be other groups who spend time in or around the workplace. These include young workers, trainees, new and expectant mothers, cleaners, contractor and maintenance workers and members of the public. Members of the public will include visitors, patients, students or customers as well as passers by. On a construction site, the persons at risk could be site operatives, surveyors, transport drivers, building inspectors, other visitors and the general public.

The risk assessment must include any additional controls required due to the vulnerability of any of these groups, perhaps caused by inexperience or disability. It must also give an indication of the numbers of people from the different groups who come into contact with the hazard and the frequency of these contacts.

5.9.3 Evaluation of risk level

During most risk assessment it will be noted that some of the risks posed by the hazard have already been addressed or controlled. The purpose of the risk assessment, therefore, is to reduce the remaining risk. This is called the **residual risk**.

The goal of risk assessment is to reduce all residual risks to as low a level as reasonably practicable. In a relatively complex workplace, this will take time so that a system of ranking risk is required – the higher the risk level the sooner it must be addressed and controlled.

For most situations, a **qualitative** risk assessment will be perfectly adequate. (This is certainly the case for NEBOSH Certificate candidates and is suitable for use during the practical assessment.) During the risk assessment, a judgement is made as to whether the risk level is high, medium or low in terms of the risk of somebody being injured. This designation defines a timetable for remedial actions to be taken thereby reducing the risk. High-risk activities should normally be addressed in days, medium risks in weeks and low risks in months or in some cases no action will be required. It will usually be necessary for risk assessors to receive some training in risk level designation.

A **quantitative** risk assessment attempts to quantify the risk level in terms of the likelihood of an incident and its subsequent severity. Clearly the higher the likelihood and severity, the higher the risk will be. The likelihood depends on such factors as the control measures in place, the frequency of exposure to the hazard and the category of person exposed to the hazard. The severity will depend on the magnitude of the hazard (voltage, toxicity, etc.). HSE suggest in HSG65 a simple 3×3 matrix to determine risk levels.

Likelihood of occurrence	Likelihood level
Harm is certain or near certain to occur	High 3
Harm will often occur	Medium 2
Harm will seldom occur	Low 1

Severity of harm	Severity level
Death or major injury (as defined by RIDDOR)	Major 3
3-day injury or illness (as defined by RIDDOR)	Serious 2
All other injuries or illnesses	Slight 1

Risk = Severity × Likelihood

Likelihood	Severity		
	Slight 1	Serious 2	Major 3
Low 1	Low 1	Low 2	Medium 3
Medium 2	Low 2	Medium 4	High 6
High 3	Medium 3	High 6	High 9

Thus:

    6–9 High risk
    3–4 Medium risk
    1–2 Low risk

It is possible to apply such methods to organizational risk or to the risk that the management system for health and safety will not deliver in the way in which it was expected or required. Such risks will add to the activity or occupational risk level. In simple terms, poor supervision of an activity will increase its overall level of risk.

A risk management matrix has been developed which combines these two risk levels as shown below.

		Occupational risk level		
		Low	Medium	High
Organizational risk level	Low	L	L	M
	Medium	L	M	H
	High	M	H	Unsatisfactory

L Low risk   **M** Medium risk   **H** High risk

Whichever type of risk evaluation method is used, the level of risk simply enables a timetable of risk reduction to be formulated to an acceptable and tolerable level. The legal duty requires that all risks should be reduced to as low as is reasonably practicable.

5.10 Risk control measures

The next stage in the risk assessment is the control of the risk. In established workplaces, some control of risk will already be in place. The effectiveness of these controls needs to be assessed so that an estimate of the residual risk may be made. Many hazards have had specific acts, regulations or other recognized standards developed to reduce associated risks. Examples of such hazards are fire, electricity, lead and asbestos. The relevant legislation and any accompanying approved codes of practice or guidance should be consulted first and any recommendations implemented. Advice on control measures may also be available from trade associations, trade unions or employers' organizations.

Where there are existing preventative measures in place, it is important to check that they are working properly and that everybody affected has a clear understanding of the measures. It may be necessary to strengthen existing procedures, for example, by the introduction of a permit to work system. More details on the principles of control are contained in Chapter 6.

5.11 Hierarchy of risk control

When assessing the adequacy of existing controls or introducing new controls, a hierarchy of risk controls should be considered. The Management of Health and Safety at Work Regulations Schedule 1 specifies the general principles of prevention which are set out in the European Council Directive. These principles are:

1. avoiding risks
2. evaluating the risks which cannot be avoided
3. combating the risks at source
4. adapting the work to the individual, especially as regards the design of the workplace, the choice of work equipment and the choice of working and production methods, with a view, in particular, to alleviating monotonous work and work at a predetermined work-rate and to reducing their effects on health
5. adapting to technical progress
6. replacing the dangerous by the non-dangerous or the less dangerous
7. developing a coherent overall prevention policy which covers technology, organization of work, working conditions, social relationships and the influence of factors relating to the working environment
8. giving collective protective measures priority over individual protective measures and
9. giving appropriate instruction to employees.

These principles are not exactly a hierarchy but must be considered alongside the usual hierarchy of risk control, which is as follows:

➤ elimination
➤ substitution
➤ engineering controls (e.g. isolation, insulation and ventilation)
➤ reduced or limited time exposure
➤ good housekeeping
➤ safe systems of work
➤ training and information
➤ PPE
➤ welfare
➤ monitoring and supervision
➤ reviews.

See chapter 6 for more information.

5.12 Prioritization of risk control

The prioritization of the implementation of risk control measures will depend on the risk rating (high, medium and low) but the time scale in which the measures are introduced will not always follow the ratings. It may be convenient to deal with a low-level risk at the same time as a high-level risk or before a medium-level risk. It may also be that work on a high-risk control system

is delayed due to a late delivery of an essential component – this should not halt the overall risk reduction work. It is important to maintain a continuous programme of risk improvement rather than slavishly following a predetermined priority list. Many risk assessors seem unable to distinguish between priorities and timescales. The highest priority may require a complex solution and may not be able to be undertaken for several months whereas lower priorities, such as sweeping the floor, may be undertaken relatively quickly.

5.13 Record of risk assessment findings

It is very useful to keep a written record of the risk assessment even if there are less than five employees in the organization. For an assessment to be 'suitable and sufficient' only the significant hazards and conclusions need be recorded. The record should also include details of the groups of people affected by the hazards and the existing control measures and their effectiveness. The conclusions should identify any new controls required and a review date. The HSE booklet *Five Steps to Risk Assessment* provides a very useful guide and examples of the detail required for most risk assessments.

There are many other possible layouts which can be used for the risk assessment record. One example is given in Appendix 5.2.

The written record provides excellent evidence to a health and safety inspector of compliance with the law. It is also useful evidence if the organization should become involved in a civil action.

The record should be accessible to employees and a copy kept with the safety manual containing the safety policy and arrangements.

5.14 Monitoring and review

As mentioned earlier, the risk controls should be reviewed periodically. This is equally true for the risk assessment as a whole. Review and revision may be necessary when conditions change as a result of the introduction of new machinery, processes or hazards. There may be new information on hazardous substances or new legislation. There could also be changes in the workforce, for example, the introduction of trainees. The risk assessment only needs to be revised if significant changes have taken place since the last assessment was done. An accident or incident or a series of minor ones provides a good reason for a review of the risk assessment. This is known as the post-accident risk assessment.

5.15 Special cases

There are several groups of persons who require an additional risk assessment due to their being more 'at risk' than other groups. Three such groups will be considered – young persons, expectant and nursing mothers and disabled workers.

5.15.1 *Young persons*

There were 21 fatalities of young people at work in 2002/2003 and more recent figures have not shown much of an improvement. A risk assessment involving young people needs to consider the particular vulnerability of young persons in the workplace. Young workers clearly have a lack of experience and awareness of risks in the workplace, a tendency to be subject to peer pressure and a willingness to work hard. Many young workers will be trainees or on unpaid work experience. An amendment to the Health and Safety at Work Act enabled trainees on Government sponsored training schemes to be treated as employees as far as health and safety is concerned. The Management of Health and Safety at Work Regulations 1999 defines a young person as anybody under the age of 18 years and stipulates that a special risk assessment must be completed which takes into account their immaturity and inexperience and the assessment must be completed before the young person starts work. If the young person is of school age (16 years or less), the parents or guardian of the child should be notified of the outcome of the risk assessment and details of any safeguards which will be used to protect the health and safety of the child. The following key elements should be covered by the risk assessment:

➤ details of the work activity, including any equipment or hazardous substances
➤ details of any prohibited equipment or processes
➤ details of health and safety training to be provided
➤ details of supervision arrangements.

Induction training is important for young workers and such training should include site rules, restricted areas, prohibited machines and processes, fire precautions and details of any further training related to their particular job. At induction, they should be introduced to their mentor and given close supervision, particularly during the first few weeks of their employment.

Generally, young workers under the age of 18 years should not be given tasks which:

➤ exceed their mental or physical capabilities
➤ expose them to toxic or carcinogenic substances
➤ expose them to radiation or extremes of temperature, noise or vibration
➤ involve the use of hazardous equipment or machinery.

A guide is available to employers who organize site visits for young people. The guide (www.safevisits.org.uk) gives practical advice and a checklist that can be used during the organization of the visit.

More detailed guidance is available from HSE Books – HSG165 *Young people at work* and HSG199 *Managing health and safety on work experience.*

5.15.2 *Expectant and nursing mothers*

The Management of Health and Safety at Work Regulations 1999 incorporates the Pregnant Workers Directive from the EU. If any type of work could present a particular risk to expectant or nursing mothers, the risk assessment must include an assessment of such risks. Should these risks be unavoidable, then the woman's working conditions or hours must be altered to avoid the risks. The alternatives are for her to be offered other work or be suspended from work on full pay. The woman must notify the employer in writing that she is pregnant, or has given birth within the previous 6 months and/or is breastfeeding.

Pregnant workers should not be exposed to chemicals such as pesticides and lead or to biological hazards, such as hepatitis. Female agricultural workers, veterinaries or farmer's wives who are pregnant, should not assist with lambing so that any possible contact with ovine chlamydia is avoided.

Typical factors which might affect such women are:

➤ manual handling
➤ chemical or biological agents
➤ ionizing radiation
➤ passive smoking
➤ lack of rest room facilities
➤ temperature variations
➤ ergonomic issues related to prolonged standing, sitting or the need for awkward body movement
➤ issues associated with the use and wearing of PPE
➤ working excessive hours
➤ stress and violence to staff.

Detailed guidance is available in *New and expectant mothers at work*, HSG122, HSE Books.

5.15.3 *Workers with a disability*

Organizations have been encouraged for many years to employ workers with disabilities and to ensure that their premises provide suitable access for such people. From a health and safety point of view, it is important that workers with a disability are covered by special risk assessments so that appropriate controls are in place to protect them. For example, employees with a hearing problem will need to be warned when the fire alarm sounds or a forklift truck approaches. Special vibrating signals or flashing lights may be used. Similarly workers in wheelchairs will require a clear, wheelchair-friendly route to a fire exit and onwards to the assembly point. Safe systems of work and welfare facilities need to be suitable for any workers with disabilities.

The final stage of the goods and services provisions in Part III of the Disability Discrimination Act came into force on 1 October 2004. The new duties apply to service providers where physical features make access to their services impossible or unreasonable for people with a disability. The Act requires equal opportunities for employment and access to workplaces to be extended to all people with disabilities.

5.15.4 *Lone workers*

People who work alone, like those in small workshops, remote areas of a large site, social workers, sales personnel or mobile maintenance staff, should not be at more risk than other employees. It is important to consider whether the risks of the job can be properly controlled by one person. Other considerations in the risk assessment include the following:

➤ Does the particular workplace present a special risk to someone working alone?
➤ Is there safe egress/exit from the workplace?
➤ Can all the equipment and substances be safely handled by one person?
➤ Is violence from others a risk?
➤ Would women and young persons be specially at risk?
➤ Is the worker medically fit and suitable for working alone?
➤ Are special training and supervision required?

For details of further precautions for lone workers, see Section 6.6 of Chapter 6.

5.15.5 *Construction site visitors and trespassers*

Both authorized and unauthorized visitors to a construction site are vulnerable to an accident particularly if the trespassers are children.

All authorized visitors should be given a hard hat, briefed on health and safety issues and supervised throughout the period of their visit. Walkways on the site should avoid hazards from workers at height, vehicles or excavation work.

During non-working hours, the site and the equipment on it should be secured against trespassers. This will involve the use of security gates and boarding ladders, and emptying fuel from vehicles.

Figure 5.4 A lone worker – special arrangements required.

Site security issues are dealt with in detail in Chapter 10.

5.16 Practice NEBOSH questions for Chapter 5

1. (a) **Explain** the meaning of the term 'risk' as used in occupational safety and health.
   (b) **Outline** the logical steps to take in managing risks at work.

2. (a) **Identify** the particular requirements of regulation 3 of the Management of Health and Safety at Work Regulations 1999 in relation to an employer's duty to carry out risk assessments.
   (b) **Outline** the factors that should be considered when selecting individuals to assist in carrying out risk assessments in the workplace.

3. A new method for concrete shuttering is to be used on a construction site, for which a general risk assessment is to be undertaken using HSE's 'five steps' approach.
   (a) **Outline** the steps that should be used in carrying out the risk assessment, identifying the key issues that will need to be considered at each stage.
   (b) **Describe** the criteria that must be met for the assessment to be deemed 'suitable and sufficient'.

(c) **Identify** the circumstances that may give rise to a need for the assessment to be reviewed at a later stage.

4. (a) **Explain**, using an example, the meaning of the term 'risk'.
   (b) **Outline** the factors that should be considered when selecting individuals to assist in carrying out risk assessments on a construction site.

5. With respect to undertaking general risk assessments on activities within a workplace:
   (a) **Outline** the **FIVE** key stages of the risk assessment process, identifying the issues that would need to be considered at **EACH** stage.
   (b) **Identify FOUR** items of information from within and **FOUR** items of information from outside the organization that may be useful when assessing the activities.
   (c) **State** the legal requirements for recording workplace risk assessments.

6. (a) **Identify** the '**FIVE**' steps involved in the assessment of risk from workplace activities (as described in HSE's 'Five steps to risk assessment' (INDG163)).
   (b) **Explain** the criteria that should be applied to help develop an action plan to prioritize the control of health and safety risks in the workplace.

7. **Outline** the hazards that might be encountered in a busy hotel kitchen.

8. **Identify SIX** hazards that might be considered when assessing the risk to health and safety of a multi-storey car park attendant.

9. **Outline** the hazards that might be encountered by a gardener employed by a local authority parks department.

10. **Outline** the content of a training course for staff who are required to assist in carrying out risk assessments.

11. **Outline FOUR** sources of information that might be consulted when assessing the risk from a substance that has not previously been subject to risk assessment.

12. An employer has agreed to accept a young person on a work experience placement for 1 week. **Outline** the factors that the employer should consider prior to the placement.

13. (a) **Identify FOUR** 'personal' factors that may place young persons at a greater risk from workplace hazards.
    (b) **Outline FOUR** measures that could be taken to minimize the risks to young persons in the workplace.

14. **Outline** the factors that may increase risks to pregnant employees.

15. (a) **Outline THREE** work activities that may present a particular risk to pregnant women.
    (b) **Outline** the actions that an employer may take when a risk to a new or expectant mother cannot be avoided

16. An employee who works on a production line has notified her employer that she is pregnant. **Outline** the factors that the employer should consider when undertaking a specific risk assessment in relation to this employee.

17. **Outline** the issues to be considered to ensure the health and safety of disabled workers in the workplace.

18. **Identify** the factors to be considered to ensure the health and safety of persons who are required to work on their own away from the workplace.

19. **Outline** the issues that should be considered to ensure the health and safety of cleaners employed in a school out of normal working hours.

Appendix 5.1 – Hazard checklist

The following checklist may be helpful:

1. **Equipment/mechanical**
   entanglement
   friction/abrasion
   cutting
   shearing
   stabbing/puncturing
   impact
   crushing
   drawing-in
   air or high pressure fluid injection
   ejection of parts
   pressure/vacuum
   display screen equipment
   hand tools.
2. **Transport**
   site vehicles
   mechanical handling
   people/vehicle interface.
3. **Access**
   slips, trips and falls
   falling or moving objects
   obstruction or projection
   working at height
   confined spaces
   excavations.
4. **Handling/lifting**
   manual handling
   mechanical handling.
5. **Electricity**
   fixed installation
   portable and tools equipment.
6. **Chemicals**
   dust/fume/gas
   toxic
   irritant
   sensitizing
   corrosive
   carcinogenic
   nuisance.

7. **Fire and Explosion**
   flammable materials/gases/ liquids
   explosion
   means of escape/alarms/detection.
8. **Particles and dust**
   inhalation/ingestion
   abrasion of skin or eye.
9. **Radiation**
   ionizing
   non-ionizing.
10. **Biological**
    bacterial
    viral
    fungal.
11. **Environmental**
    noise
    vibration
    light
    humidity
    ventilation
    temperature
    overcrowding.
12. **The individual**
    individual not suited to work
    long hours
    high work rate
    violence to staff
    unsafe behaviour of individual
    stress
    pregnant/nursing women
    young people.
13. **Other factors to consider**
    poor maintenance
    lack of supervision
    lack of training
    lack of information
    inadequate instruction
    unsafe systems.

Appendix 5.2 – Example of a risk assessment record

General Health and Safety Risk Assessment	No
Firm/Company	Department
Contact Name	Nature of Business
Telephone Number	

Principal Hazards
Risks to employees and members of the public could arise due to the following hazards:
1. hazardous substances
2. electricity
3. fire
4. dangerous occurrences or other emergency incidents
5. ?

Persons at Risk
Employees, contractors, and members of the public.

Main Legal Requirements
1. Health and Safety at Work Act 1974 – Sections 2 and 3
2. Management of Health and Safety at Work Regulations 1999
3. Noise at Work Regulations 1989
4. Common Law Duty of Care
5. ?

Significant Risks
➤ acute and chronic health problems caused by the use or release of hazardous substances
➤ injuries to employees and members of the public due to equipment failure such as electric shock
➤ injuries to employees and members of the public from slips, trips and falls
➤ injuries to employees and members of the public caused by fire
➤ ?

Consequences
fractures, bruising, smoke inhalation and burns, acute and chronic health problems and death

Existing Control Measures
Possible examples include:
1. All sub-contractors are vetted prior to appointment.
2. All hazardous and harmful materials are identified and the risks to people assessed. COSHH assessments are provided and the appropriate are controls implemented. Health surveillance is provided as necessary.
3. Fire risk assessment has been produced. Fire procedures are in place and all employees are trained to deal with fire emergencies. A carbon dioxide fire extinguisher is available at every worksite.
4. A minimum of flammable substances are used on the premises, no more than a half day's supply at a time. Kept in fire resistant store.
5. No smoking is allowed on the premises.
6. Manual handling is kept to a minimum. Where there is a risk of injury manual handling assessments are carried out.
7. Method statements are used for complex and/or hazardous jobs and are followed at all times.
8. All accidents on or around the site are reported and investigated by management. Any changes found necessary are quickly implemented. All accidents reportable under RIDDOR 1995, are reported to the HSE on form F2508.
9. At least 1 qualified First Aider is available during working hours.
10. ?

Residual Risk i.e. after controls are in place.
Severity.......................   **Likelihood**......................   **Residual Risk**..................

Information
Details of various relevant HSE and trade publications.

Comments from Line Manager   **Comments from Risk Assessor**

Signed......................   **Date**...........   **Signed**......................   **Date**...........

Review Date

Principles of control

6

6.1 Introduction

The control of risks is essential to secure and maintain a healthy and safe construction site or workplace which complies with the relevant legal requirements. Hazard identification and risk assessment are covered in Chapter 5 and these together with appropriate risk control measures form the core of the HSG65 'implementing and planning' section of the management model. Chapter 1 covers this in more detail.

Figure 6.1 When controls break down.

In industry today, including the construction industry, safety is controlled through a combination of engineered measures such as the provision of safety protection, that is guarding and warning systems, and operational measures in training, safe work practices, operating procedures and method statements, along with management supervision.

These measures (collectively) are commonly known in health and safety terms as **control measures**. Some of these more common measures will be explained in more detail later.

This chapter concerns the principles that should be adopted when deciding on suitable measures to eliminate or control both acute and chronic risks to the health and safety of people at work on construction operations. The principles of control can be applied to both health risks and safety risks, although health risks have some distinctive features that require a special approach.

Chapters 10 to 20 deal with specific workplace hazards and controls subject by subject.

The principles of prevention now enshrined in the Management of Health and Safety at Work (MHSW) Regulations need to be used jointly with the hierarchy of control methods which give the preferred order of approach to risk control.

When risks have been analysed and assessed, decisions can be made about workplace precautions.

All final decisions about risk control methods must take into account the relevant legal requirements, which establish minimum levels of risk prevention or control. Some of the duties imposed by the HSW Act and the relevant statutory provisions are **absolute** and must be complied with. Many requirements are, however, qualified by the words **so far as is reasonably practicable**, or **so far as is practicable**. These require an assessment of cost, along with information about relative costs, effectiveness and reliability of different control measures. Further guidance on the meaning of these three expressions is provided in Chapter 1.

6.2 Principles of prevention

The MHSW Regulations Schedule 1 specifies the general principles of prevention which are set out in article 6(2) of the European Council Directive 89/391/EEC. For the first time the principles have been enshrined directly in Regulations which state, at regulation 4, that *Where an employer implements any preventative measures he shall do so on the basis of the principles specified in Schedule 1.* These principles are:

1. Avoiding risks
   This means, for example, trying to stop doing the task or using different processes or doing the work in a different safer way.
2. Evaluating the risks which cannot be avoided
   This requires a risk assessment to be carried out.
3. Combating the risks at source
   This means that risks, such as a dusty work atmosphere, are controlled by removing the cause of the dust rather than providing special protection; or that holes in floors are covered or repaired rather than putting up a sign.
4. Adapting the work to the individual, especially as regards the design of the workplace, the choice of work equipment and the choice of working and production methods, with a view, in particular, to alleviating monotonous work and work at a predetermined work-rate and to reducing their effect on health.
   This will involve consulting those who will be affected when site layouts, construction methods, safety method statements and safety procedures are designed. The control individuals have over their work should be increased, and time spent working at predetermined speeds and in monotonous work should be reduced where it is reasonable to do so.
5. Adapting to technical progress
   It is important to take advantage of technological and technical progress, which often gives clients, designers, contractors and other employers the chance to improve both safety and working methods. With the Internet and other international information sources available a very wide knowledge, going beyond what is happening in the UK or Europe, will be expected by the enforcing authorities and the courts.
6. Replacing the dangerous by the non-dangerous or the less dangerous
   This involves substituting, for example, equipment or substances with non-hazardous or less hazardous substances.
7. Developing a coherent overall prevention policy which covers technology, organization of work, working conditions, social relationships and the influence of factors relating to the working environment.
   Health and safety policies should be prepared and applied by reference to these principles.
8. Giving collective protective measures priority over individual protective measures
   This means giving priority to control measures which make the construction workplace safe for everyone working there, so giving the greatest benefit; for example, removing hazardous dust by exhaust ventilation rather than providing a filtering respirator to an individual worker. This is sometimes known as a 'Safe Place' approach to controlling risks.
9. Giving appropriate instruction to employees
   This involves making sure that employees are fully aware of company policy, risk assessments, method statements, safety procedures, good practice, official guidance, any test results and legal requirements. This is sometimes known as a 'Safe Person' approach to controlling risks where the focus is on individuals. A properly set up health and safety management system should cover and balance both a Safe Place and Safe Person approach.

6.3 General control measures

6.3.1 *Hierarchy of control*

When assessing the adequacy of existing controls or introducing new controls, a hierarchy of control should be considered. The principles of prevention in the MHSW Regulations is not a hierarchy but a list of prevention principles which must all be considered when controlling risks. However, there is a preferred hierarchy of control following these principles and those found in HSG65. The 2007 NEBOSH Construction syllabus uses basically the same hierarchy with a few minor differences. These are shown in Table 6.1 alongside the HSG65 hierarchy of risk control principles.

The hierarchy reflects that risk elimination and risk control by the use of physical engineering controls and safeguards can be more reliably maintained than those which rely solely on people. These concepts are now written into the Control of Substances Hazardous to Health (COSHH) and Management of Health and Safety at Work (MHSW) regulations. Where a range of control measures are available, it will be necessary to weigh up the relative costs of each against the degree of control each provides, both in the short and long term. Some control measures, such as eliminating a risk by choosing a safer alternative substance or machine, provide a high degree of control and are reliable. Physical safeguards such as guarding a machine or enclosing a hazardous process need to be maintained. In making decisions about risk control, it will therefore be necessary to consider the degree of control and the reliability of the control measures along with the costs of both providing and maintaining the measure.

Table 6.1 Hierarchy of control

2007 NEBOSH construction certificate	HSG65 summary of risk control principles
Avoidance of risks	**Eliminate risk** by substituting the dangerous by the inherently less dangerous for example:
Elimination of hazards or substitution for something less hazardous	➤ Use less hazardous substances ➤ Substitute a type of machine which is better guarded to make the same product ➤ Avoid the use of certain processes.
Reducing or limiting the duration of exposure to the hazard	**Combat risks** at source by engineering controls and giving collective protective measures priority, for example:
Isolation/segregation	➤ Separate the operator from the risk of exposure to a known hazardous substance by enclosing the process ➤ Protect the dangerous parts of a machine by guarding ➤ Design process machinery and work activities to minimize the release, of or to suppress or contain, airborne hazards
Engineering controls	➤ Design machinery which is remotely operated and to which materials are fed automatically, thus separating the operator from danger areas.
Safe systems of work	**Minimize risk** by:
Training and information	➤ designing suitable systems of working ➤ using personal protective clothing and equipment; this should only be used as a last resort.
Personal protective equipment	
Welfare	
Monitoring and supervision	

6.3.2 Avoidance of risks by elimination or substitution

This is the best and most effective way of reducing risks by avoiding a hazard and its associated risks. For example avoid working at height by installing a permanent working platform with stair access. Avoid entry into a confined space by for example using a sump pump in a pit which is removed by a lanyard for maintenance. Eliminate the fire risks from tar boilers by using bitumen which can be applied cold.

Substitution describes the use of a less hazardous form of a substance or process. There are many examples of substitution such as the use of water based rather than oil based paints, the use of asbestos substitutes and the use of compressed air as a power source rather than electricity to reduce both electrical and fire risks. Use mechanical excavators instead of hand digging.

In some cases it is possible to change the method of working so that risks are reduced. For example use rods to clear drains instead of strong chemicals; use the roof scaffold to fix the gutters instead of ladders. Sometimes the pattern of work can be changed so that people can do things in a more natural way; for example when placing bricks and mortar for bricklaying consider whether people are right or left handed; encourage people in site offices to take breaks from computer screens by getting up to photocopy, fetch files or print documents.

Care must be taken not to forget about additional hazards which may be involved or to introduce additional risks as a result of a substitution.

6.3.3 Reduced time exposure

This involves reducing the time during the working day that the employee is exposed to the hazard, either by giving the employee other work or rest periods. It is normally only suitable for the control of health hazards associated with, for example, noise, vibration, excessive heat or cold, display screens and hazardous substances. However, it is important to note that for many hazards there are short-term exposure limits as well as normal workplace exposure limits (WELs) over an 8-hour period (see Chapter 16). Short-term limits must not be exceeded during the reduced time exposure intervals.

It cannot be argued that a short time of exposure to a dangerous part of a machine is acceptable. However it is possible to consider short bouts of intensive work with rest periods when employees are engaged in heavy labour such as manual digging when machines are not permitted into the confines of the space or buried services.

6.3.4 Isolation/segregation

Controlling risks by isolating them or segregating people and the hazard is an effective control measure used in many instances. For example separating vehicles and pedestrians on building sites, providing separate walk-ways for the public on road repairs, providing warm rooms on sites or noise refuges in noisy processes.

The principle of isolation is usually followed with the storage of highly flammable liquids or gases which are put into open air or well ventilated compounds away from other hazards such as sources of ignition and peo-ple who may be at risk from a fire or explosion.

6.3.5 *Engineering controls*

This describes the control of risks by means of engineer-ing design rather than a reliance on preventative actions by the employee. There are several ways of achieving such controls:

1. Control the risks at the source (e.g. the use of more efficient dust filters or the purchase of less noisy equipment).
2. Control the risk of exposure by:
   - ➤ *isolating* the equipment by the use of an enclosure, a barrier or a guard
   - ➤ *insulating* any electrical or temperature hazard
   - ➤ *ventilate* away any hazardous fumes or gases either naturally or by the use of extractor fans and hoods.

6.3.6 *Safe systems of work*

Operating procedures or safe systems of work are prob-ably the most common form of control measure used in industry today and may be the most economical and, in some cases, the only practical way of managing a par-ticular risk. They should allow for methodical execution of tasks. The development of safe operating procedures should address the hazards that have been identified in the risk assessment. The system of work describes the safe method of performing the job or activity. A safe sys-tem of work is a requirement of the Health and Safety at Work Act and is dealt with in detail later.

If the risks involved in the task are high or medium, the details of the system should be in writing and should be communicated to the employee formally in a training session. Systems for low risk activities may be conveyed verbally. There should be records that the employee (or contractor) has been trained or instructed in the safe system of work and that they understand it and will abide by it.

6.3.7 *Training*

Training helps people acquire the skills, knowledge and attitudes to make them competent in the health and safety aspects of their work. There are generally two types of safety training:

- ➤ Specific safety training (or on-the-job training) which aims at tasks where training is needed due to the specific nature of such tasks. This is usually a job for supervisors who, by virtue of their authority and close daily contact, are in a position to convert safety generalities to the everyday safe practice procedures that apply to individual tasks, machines, tools and processes.
- ➤ Planned training, such as general safety training, induction training, management training, skill train-ing or refresher courses, that are planned by the organization, and relate to managing risk through policy, legislative or organizational requirements, that are common to all employees.

Before any employee can work safely they must be shown safe procedures for completing their tasks. The purpose of safety training should be to improve the safety awareness of employees, and show them how to perform their jobs employing acceptable safe behaviour.

See Chapter 4 for more details on health and safety training.

6.3.8 Information

Organizations need to ensure that they have effective arrangements for:

- ➤ identifying and receiving relevant health and safety information from outside the organization including:
  - ➤ ensuring that pertinent health and safety infor-mation is communicated to all people in the organization who need it
  - ➤ ensuring that relevant information is commu-nicated to people outside the organization who require it
  - ➤ encouraging feed-back and suggestions from employees on health and safety matters.

Anyone who is affected by what is happening in the work-place will need to be given safety information. This does not only apply to staff. It can also apply to visitors, mem-bers of the public and contractors.

Information to be provided for people in a workplace include:

- ➤ who is at risk and why
- ➤ how to carry out specific tasks safely
- ➤ correct operation of equipment
- ➤ emergency action
- ➤ accident and hazard reporting procedures; and
- ➤ the safety responsibilities of individual people.

Information can be provided in a variety of ways. These include safety signs, posters, newsletters, memos, emails, personal briefings, meetings, tool box talks, formal training, written safe systems of work and written health and safety arrangements. For more details see Chapters 1, 4 and 9.

6.3.9 Safety signs

A summary of the Safety Signs and Signals Regulations is given in Chapter 21. All general health and safety signs used in the workplace must include a pictorial symbol categorized by shape, colour and graphic image.

All workplaces need to display safety signs of some kind but deciding what is required can be confusing. Here are the basic requirements for the majority of small premises or sites like small construction sites, canteens, shops, small workshop units and offices. This does not cover any signs which food hygiene law may require.

Most requirements are covered by the Health and Safety (Safety Signs and Signals) Regulations 1996. These require signs wherever a risk has not been controlled by other means. For example, if a wet area of floor is cordoned off, a warning sign will not be needed, because the barrier will keep people out of the danger area. Signs are not needed where the sign would not reduce the risk or the risk is insignificant.

The following signs are typical of some of the ones most likely to be needed in these premises. Others may be necessary, depending on the hazards and risks present.

(i) Overhead obstacles/construction site

Figure 6.3 Overhead obstacles.

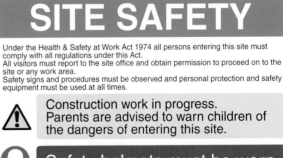

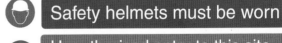

Figure 6.4 Construction site signs.

⊘	**Prohibition** A red circular band with diagonal cross bar on a white background, the symbol within the circle to be black denoting a safety sign that indicates that a certain behaviour is prohibited.
△	**Warning** A yellow triangle with black border and symbol within the yellow area denoting a safety sign that gives warning of a hazard.
●	**Mandatory** A blue circle with white symbol denoting a sign that indicates that a specific course of action must be taken.
▪	**Safe Condition** A green oblong or square with symbol or text in white denoting a safety sign providing information about safe conditions.
▪	**Fire Equipment** A red oblong or square with symbol in white denoting a safety sign that indicates the location of fire fighting equipment.

Figure 6.2 Colour categories and shapes of signs.

(ii) Wet floors Signs need to be used wherever a slippery area is not cordoned off. Lightweight stands holding double-sided signs are readily available.

Figure 6.5 Wet floors.

(iii) Chemical storage Where hazardous chemicals are stored, apart from keeping the store locked, a suitable warning notice should be posted if it is considered this would help to reduce the risk of injury for example:

Figure 6.6 Chemicals.

(iv) Fire safety signs The Regulations apply in relation to general fire precautions. The guidance under the Fire Safety Order requires signs to comply with BS5499-4 & 5 and the Safety Signs and Signals Regulations.

Since 24 December 1998 the older, text-only 'fire exit' signs should have been supplemented or replaced with pictogram signs. Fire safety signs complying with BS5499-4 & 5 already contain a pictogram and do not require changing.

Figure 6.7 Fire safety signs.

(v) Fire action signs These and other fire safety signs, such as fire extinguisher location signs shown below, will be needed.

Figure 6.8 Fire action signs.

(vi) First Aid
Signs showing the location of first aid facilities will be needed. Advice on the action to take in the case of electric shock is no longer a legal requirement but is recommended.

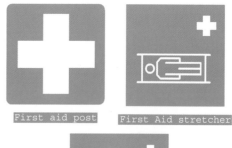

Figure 6.9 First aid signs.

(vii) LPG cylinder stores LPG cylinder stores should have the following sign.

Figure 6.10 LPG sign.

(viii) *No smoking* Areas substantially enclosed should have the following sign under smokefree legislation (see Chapter 17 for more details).

Figure 6.11 Smokefree no-smoking sign.

(ix) *Fragile roofs* Signs should be erected at roof access points and at the top of outside walls where ladders may be placed.

Figure 6.12 Fragile roof sign.

(x) *Obstactes and dangerous locations* For example: low head-height, tripping hazard. Use tape of alternating yellow and black diagonal stripes.

(xi) *Other signs and posters*
➤ Health and safety law – What you should know (there is a legal requirement to display this poster or distribute equivalent leaflet)

➤ Certificate of Employer's Liability Insurance (there is a legal requirement to display this)
➤ Scalds and burns are common in kitchens. A poster showing recommended action is advisable, for example 'First Aid for Burns'

(xii) *Sign checklist*
Existing signs should be checked to ensure that:

➤ they are the correct signs
➤ they carry the correct warning symbol where appropriate
➤ they are relevant to the hazard
➤ they are easily understood
➤ they are suitably located and not obscured
➤ they are clean, durable and weatherproof where necessary
➤ illuminated signs have regular lamp checks
➤ they are used when required (e.g. 'Caution wet floor' signs) and
➤ they are obeyed or effective.

6.3.10 *Personal protective equipment*

Personal protective equipment (PPE) should only be used as a last resort. There are many reasons for this. The most important limitations are that PPE:

➤ only protects the person wearing the equipment not others nearby
➤ relies on people wearing the equipment at all times
➤ must be used properly
➤ must be replaced when it no longer offers the correct level of protection. This last point is particularly relevant when respiratory protection is used.

The benefits of PPE are that:

➤ it gives immediate protection to allow a job to continue while engineering controls are put in place
➤ in an emergency it can be the only practicable way of effecting rescue or shutting down plant in hazardous atmospheres
➤ it can be used to carry out work in confined spaces where alternatives are impracticable. But it should never be used to allow people to work in dangerous atmospheres, which are, for example, enriched with oxygen or explosive.

See Chapter 16 for more details on PPE.

6.3.11 *Welfare*

Welfare facilities include general workplace ventilation, lighting and heating and the provision of drinking water,

Figure 6.13 Good dust control for a chasing operation. A dust mask is still required for complete protection.

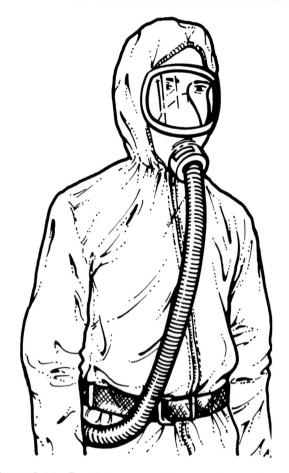

Figure 6.14 Respiratory protection and disposable overalls are needed when working in higher levels of asbestos dust.

sanitation and washing facilities. There is also a requirement to provide eating and rest rooms. Risk control may be enhanced by the provision of eye washing and shower facilities for use after certain accidents. Within this area of welfare, first aid is an important service, which should be available.

Good housekeeping is a very cheap and effective means of controlling risks. It involves keeping the workplace clean and tidy at all times and maintaining good storage systems for hazardous substances and other potentially dangerous items. The risks most likely to be influenced by good housekeeping are fire and slips, trips and falls.

See Chapter 17 for more information on the work environment.

6.3.12 *Monitoring and supervision*

All risk control measures, whether they rely on engineered or human behavioural controls, must be monitored for their effectiveness and supervised to ensure that they have been applied correctly. Competent people who have a sound knowledge of the equipment or process should undertake monitoring. Checklists are useful to ensure that no significant factor is forgotten. Any statutory inspection or insurance company reports should be checked to see whether any areas of concern were highlighted and if any recommendations were implemented. Details of any accidents, illnesses or other incidents will give an indication on the effectiveness of the risk control measures. Any emergency arrangements should be tested during the monitoring phase including first-aid provision.

It is crucial that the operator should be monitored to ascertain that all relevant procedures have been understood and followed. The operator may also be able to suggest improvements to the equipment or system of work. The supervisor is an important source of information during the monitoring process.

Where the organization is involved with shift work, it is essential that the risk controls are monitored on all shifts to ensure the uniformity of application.

The effectiveness and relevance of any training or instruction given should be monitored.

Periodically the risk control measures should be reviewed. Monitoring and other reports are crucial for the review to be useful. Reviews often take place at safety committee and/or at management meetings. A serious accident or incident should lead to an immediate review of the risk control measures in place.

6.4 Controlling health risks

6.4.1 Types of health risk

The principles of control for health risks are the same as those for safety. However, the nature of health risks can make the link between work activities and employee ill-health less obvious than in the case of injury from an accident.

The COSHH Amendment Regulations 2004 sets out the principles of control (see Box 6.1).

Figure 6.16 shows a route map for achieving adequate control.

Unlike safety risks, which can lead to immediate injury, the results of daily exposure to health risks may not manifest itself for months, years and, in some cases, decades. Irreversible health damage may occur before any symptoms are apparent. It is, therefore, essential to develop a preventive strategy to identify and control risks before anyone is exposed to them.

Risks to health from work activities include:

➤ skin contact with irritant substances, leading to dermatitis, etc.
➤ inhalation of respiratory sensitizers, triggering immune responses such as asthma
➤ badly designed workstations requiring awkward body postures or repetitive movements, resulting in upper limb disorders, repetitive strain injury and other musculoskeletal conditions
➤ noise levels which are too high, causing deafness and conditions such as tinnitus
➤ too much vibration, for example, from hand-held tools leading to hand–arm vibration syndrome and circulatory problems
➤ exposure to ionizing and non-ionizing radiation including ultraviolet in the sun's rays, causing burns, sickness and skin cancer
➤ infections ranging from minor sickness to life-threatening conditions, caused by inhaling or being contaminated with microbiological organisms
➤ stress causing mental and physical disorders.

Figure 6.15 Health risk – checking on the contents.

Some illnesses or conditions, such as asthma and back pain, have both occupational and non-occupational causes and it may be difficult to establish a definite causal link with a person's work activity or their exposure to particular agents or substances. But, if there is evidence that shows the illness or condition is prevalent among the type of workers to which the person belongs or among workers exposed to similar agents or substances, it is likely that their work and exposure has contributed in some way.

6.4.2 Control hierarchy for exposure to substances hazardous to health

The following shows how the general principles and hierarchy can be applied to substances hazardous to health, which come under the COSHH Regulations:

➤ Change the process or task so that there is no need for the hazardous substance.
➤ Replace the substance with a safer material.
➤ Use the substance in a safer form, for example, in liquid or pellets to prevent dust from powders. The HSE has produced a useful booklet on *7 steps to successful substitution of hazardous substances* HSG110, HSE Books, which should be consulted.
➤ Totally enclose the process.
➤ Partially enclose the process and use local exhaust ventilation to extract the harmful substance.
➤ Provide high quality general ventilation, if individual exposures do not breach the exposure limits.
➤ Use safe systems of work and procedures to minimize exposures, spillage or leaks.
➤ Reduce the number of people exposed or the duration of their exposure.

Box 6.1   Principles of good practice for the control of exposure to substances hazardous to health

(a) Design and operate processes and activities to minimize emission, release and spread of substances hazardous to health.

(b) Take into account all relevant routes of exposure – inhalation, skin absorption and ingestion – when developing control measures.

(c) Control exposure by measures that are proportionate to the health risk.

(d) Choose the most effective and reliable control options which minimize the escape and spread of substances hazardous to health.

(e) Where adequate control of exposure cannot be achieved by other means, provide, in combination with other control measures, suitable PPE.

(f) Check and review regularly all elements of control measures for their continuing effectiveness.

(g) Inform and train all employees on the hazards and risks from the substances with which they work and the use of control measures developed to minimize the risks.

(h) Ensure that the introduction of control measures does not increase the overall risk to health and safety.

If none of the above control measures prove is to be adequate on its own, PPE should be provided. This is a last resort and is only permitted by the COSHH Regulations in conjunction with other means of control if they prove inadequate. Special control requirements are needed for carcinogens, which are set out in the Carcinogen Approved Code of Practice.

6.4.3   *Assessing exposure and health surveillance*

Some aspects of health exposure will need input from specialist or professional advisers, such as occupational health hygienists, nurses and doctors. However,

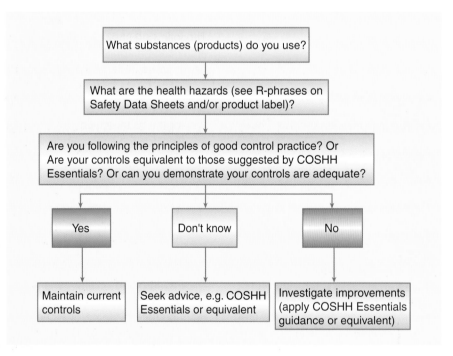

Figure 6.16   Route map for adequate control for Small and Medium Enterprises (SMEs)/non-experts.

considerable progress can be made by taking straightforward measures such as:

- consulting the workforce on the design of workplaces
- talking to manufacturers and suppliers of substances and work equipment about minimizing exposure
- enclosing machinery to cut down dust, fume and noise
- researching the use of less hazardous substances
- ensuring that employees are given appropriate information and are trained in the safe handling of all the substances and materials to which they may be exposed.

To assess health risks and to make sure that control measures are working properly, it may be necessary, for example, to measure the concentration of substances in air to make sure that exposures remain within the assigned WELs. Sometimes health surveillance of workers who may be exposed will be needed. This will enable data to be collected to check control measures and for early detection of any adverse changes to health. Health surveillance procedures available include biological monitoring for bodily uptake of substances, examination for symptoms and medical surveillance – which may entail clinical examinations and physiological or psychological measurements by occupationally qualified registered medical practitioners. The procedure chosen should be suitable for the case concerned. Sometimes a method of surveillance is specified for a particular substance, for example, in the COSHH ACoP. Whenever surveillance is undertaken, a health record has to be kept for the person concerned.

Health surveillance should be supervised by a registered medical practitioner or, where appropriate, it should be done by a suitably qualified person (e.g. an occupational nurse). In the case of inspections for easily detectable symptoms like chrome ulceration or early signs of dermatitis, health surveillance should be done by a suitably trained responsible person. If workers could be exposed to substances listed in Schedule 6 of the COSHH Regulations, medical surveillance, under the supervision of an HSE employment medical adviser or a doctor appointed by HSE, is required.

6.5 Safe systems of work

6.5.1 *What is a safe system of work?*

A safe system of work has been defined as:

The integration of personnel, articles and substances in a laid out and considered method of working which takes proper

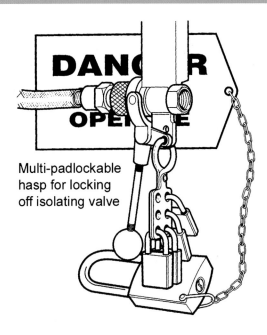

Multi-padlockable hasp for locking off isolating valve

Figure 6.17 Multi-padlocked hasp for locking off an isolation valve.

account of the risks to employees and others who may be affected, such as visitors and contractors, and provides a formal framework to ensure that all of the steps necessary for safe working have been anticipated and implemented.

In simple terms, a safe system of work is a defined method for doing a job in a safe way. It takes account of all foreseeable hazards to health and safety and seeks to eliminate or minimize these. Safe systems of work are normally formal and documented, for example, in written operating procedures but, in some cases, they may be verbal.

The particular importance of safe systems of work stems from the recognition that most accidents are caused by a combination of factors (plant, substances, lack of training, supervision, etc.). Hence prevention must be based on an integral approach and not one which only deals with each factor in isolation. The adoption of a safe system of work provides this integral approach because an effective safe system:

- is based on looking at the job as a whole
- starts from an analysis of all foreseeable hazards (e.g. physical, chemical, health)
- brings together all the necessary precautions, including design, physical precautions, training, monitoring, procedures and PPE.

The operations covered may be simple or complex, routine or unusual.

Whether the system is verbal or written, and whether the operation it covers is simple or complex, routine or unusual, the essential features are forethought and planning – to ensure that all foreseeable hazards are identified and controlled. In particular, this will involve scrutiny of:

➤ the sequence of operations to be carried out

➤ the equipment, plant, machinery and tools involved

➤ chemicals and other substances to which people might be exposed in the course of the work

➤ the people doing the work – their skill and experience

➤ foreseeable hazards (health, safety, environment), whether to the people doing the work or to others who might be affected by it

➤ practical precautions which, when adopted, will eliminate or minimize these hazards

➤ the training needs of those who will manage and operate under the procedure

➤ monitoring systems to ensure that the defined precautions are implemented effectively.

It follows from this that the use of safe systems of work is in no way a replacement for other precautions, such as good equipment design, safe construction and the use of physical safeguards. However, there are many situations where these will not give adequate protection in themselves, and then a carefully thought-out and properly implemented safe system of work is especially important. This is usually called a health and safety method statement (or method statement for short).

6.5.2 *Legal requirements*

The HSW Act section 2 requires employers to provide safe plant and systems of work. In addition, many Regulations made under the Act, such as the Provision and Use of Work Equipment Regulations, require information and instruction to be provided to employees and others. In effect, this is also a more specific requirement to provide safe systems of work. Many of these safe systems, information and instructions will need to be in writing.

6.5.3 *Health and safety method statements*

Method statements are not required by law, but they are an effective and practical management tool for higher risk activities. The Construction (Design and Management) Regulations require both a Construction phase Health and Safety Plan and a Health and Safety File.

Method statements will make compliance with these Regulations easier.

A method statement draws together the information compiled about various hazards and the ways in which they are to be controlled for any particular task.

A method statement is a formal written document detailing how an activity or task is to be undertaken. It pays particular attention to the health, safety and welfare implications in carrying out such an activity or task.

The method statement should take into account the conclusions of risk assessments made under the Management, Control of Substances Hazardous to Health and the Manual Handling Operations Regulations. It can also help other contractors working on a site to understand the effects work will have on them. It can help the principal contractor to draw up the *construction phase safety plan.*

As a general rule, if potentially hazardous activities are to be undertaken then method statements should be prepared.

If the work is to be carried out by sub-contractors then they should prepare and issue the method statement.

Typical work which will require method statements includes:

➤ erection and dismantling of design scaffolding, temporary support systems, form work and false work

➤ demolition work

➤ excavation work below 1.2 m

➤ refurbishment work, which may affect the structural stability of such a structure

➤ roof work

➤ erection of structures

➤ work on high voltage electrical equipment

➤ entry into confined spaces

➤ hot work

➤ work involving highly flammable liquids such as petrol.

The extent and detail of a method statement will depend upon the size and/or complexity of the work, activity or task to be undertaken. A method statement should contain the following:

➤ management arrangements, including identified persons with authority

➤ detailed sequence of work operations in a chronological order

➤ drawings and/or technical information

➤ detailed information on plant, equipment, substances, etc.

➤ inspection and monitoring controls

➤ risk assessments

➤ emergency procedures and systems

> arrangements for delivery, stacking, storing and movement of logistics on site
> details of site features, layout and access, which may affect the method of working
> procedures for changing or departing from the method statement.

It must be remembered that the method statement is a dynamic document and must be adhered to and kept up-to-date.

Permits to work are used as a control to implement a method statement or safe system of work for a particular high-risk task. See Section 6.7 for guidance on permits to work.

6.5.4 *Development of safe systems/method statements*

Role of competent person
The competent person appointed under the MHSW Regulations and/or the CDM Coordinator appointed under the CDM2007 Regulations should assist managers to draw up guidelines for safe systems of work including method statements, with suitable forms and should advise management on the adequacy of the safe systems produced.

Role of managers
Primarily management is responsible to provide safe systems of work. Managers and employees know the detailed way in which the task should be carried out. On many construction sites the responsibility will be with the Principal Contractor to ensure that their employees and their sub-contractor have risk assessments, safe systems of work and method statements where necessary.

Management are responsible to ensure that employees are adequately trained in a specific safe system of work and are competent to carry out the work safely. Managers need to provide sufficient supervision to ensure that the system of work is followed and the work is carried out safely. The level of supervision will depend on the experience of the particular employees concerned and the complexity and risks of the task.

Principal Contractors will need to monitor sub-contractors to check that they are providing suitable safe systems of work, have trained their employees and are carrying out the tasks in accordance with the safe systems.

Role of employees/consultation
Many people operating a piece of plant or a construction process are in the best position to help with the preparation of safe systems of work. Consultation with those employees who will be exposed to the risks, either directly or through their representatives, is also a legal requirement. The importance of discussing the proposed system with those who will have to work under it, and those who will have to supervise its operation, cannot be emphasized enough.

Employees have a responsibility to follow the safe system of work.

Analysis
The safe system of work or method statement should be based on a thorough analysis of the job or operation to be covered by the system. The way this analysis is done will depend on the nature of the job/operation.

If the operation being considered is a new one involving high loss potential, the use of formal hazard analysis techniques such as HAZOP (Hazard and Operability study), FTA (Fault Tree Analysis) or Failure Modes and Effects Analysis should be considered.

However, where the potential for loss is lower, a more simple approach, such as JSA (Job Safety Analysis), will be sufficient. This will involve three key stages:

1. identification of the key steps in the job/operation – what activities will the work involve?
2. analysis and assessment of the risks associated with each stage – what could go wrong?
3. definition of the precautions or controls to be taken – what steps need to be taken to ensure the operation proceeds without danger, either to the people doing the work, or to anyone else?

The results of this analysis are then used to draw up the safe operating procedure or method statement. (See Appendix 6.1 for a suitable form.)

Introducing controls
There are a variety of controls that can be adopted in safe systems of work. They can be split into the following three basic categories:

1. *Technical*: these are engineering or process type controls which engineer out or contain the hazard so that the risks are acceptable. For example exhaust ventilation, a machine guard, dust respirator.
2. *Procedural*: these are ways of doing things to ensure that the work is done according to the procedure, legislation or cultural requirements of the organization. For example, a supervisor must be involved, the induction course must be taken before the work commences, a particular type of form or a person's signature must be obtained before proceeding and the names of the workforce must be recorded.

3. *Behavioural*: these are controls which require a certain standard of behaviour from individuals or groups of individuals. For example no smoking is permitted during the task, hard hats must be worn, all lifts are to be in tandem between two workers.

6.5.5 *Preparation of safe systems/method statements*

A checklist for use in the preparation of safe systems of work or method statements is set out as follows:

➤ What is the work to be done?
➤ What are the potential hazards?
➤ Is the work covered by any existing instructions or procedures? If so, to what extent (if any) do these need to be modified?
➤ Who is to do the work?
➤ What are their skills and abilities – is any special training needed?
➤ Under whose control and supervision will the work be done?
➤ Will any special tools, protective clothing or equipment (e.g. breathing apparatus) be needed? Are they ready and available for use?
➤ Are the people who are to do the work adequately trained to use the above?
➤ What isolations and locking-off will be needed for the work to be done safely?
➤ Is a permit to work required for any aspect of the work?
➤ Will the work interfere with other activities? Will other activities create a hazard to the people doing the work?
➤ Have other departments been informed about the work to be done, where appropriate?
➤ How will the people doing the work communicate with each other?
➤ Have possible emergencies and the action to be taken been considered?
➤ Should the emergency services be notified?
➤ What are the arrangements for handover of the plant/equipment to the client at the end of the work?
➤ Do the planned precautions take account of all foreseeable hazards?
➤ Who needs to be informed about or receive copies of the safe system of work/method statement?
➤ What arrangements will there be to see that the agreed system is followed and that it works in practice?
➤ What mechanism is there to ensure that the safe system of work/method statement stays relevant and up-to-date?

Appendix 6.2 gives an example of a safety method statement form. The form should be produced electronically in the form of a table so that each section can be enlarged as required.

6.5.6 *Documentation*

Safe systems of work/method statements should be properly documented.

Whatever method is used, all written systems of work/method statements should be signed by the relevant managers to indicate approval or authorization.

As far as possible, systems should be written in a non-technical style and should specifically be designed to be as intelligible and user-friendly as possible.

6.5.7 *Communication and training*

People doing work or supervising work must be made fully aware of the laid-down safe systems that apply. The preparation of safe systems/method statements will often identify a training need that must be met before the system can be implemented effectively.

In addition, people should receive training in how the system is to operate. This applies not only to those directly involved in doing the work but also to supervisors/managers who are to oversee it.

In particular, the training might include:

➤ why a safe system is needed
➤ what is involved in the work
➤ the hazards which have been identified
➤ the precautions which have been decided and, in particular:
  ➤ the isolations and locking-off required, and how this is to be done
  ➤ details of the permit-to-work system, if applicable
  ➤ any monitoring (e.g. air testing) which is to be done during the work, or before it starts
  ➤ how to use any necessary PPE
  ➤ if on a client's site the local site's relevant rules and procedures
  ➤ emergency procedures.

6.5.8 *Monitoring safe systems/method statements*

Safe systems of work should be monitored to ensure that they are effective in practice and relevant to the actual task being undertaken. This will involve:

➤ reviewing and revising the systems themselves, to ensure they stay up-to-date
➤ inspection to identify how fully they are being implemented.

In practice, these two things go together, since it is likely that a system that is out of date will not be fully implemented by the people who are intended to operate it.

6.6 Lone workers

People who work by themselves without close or direct supervision are found in many work situations. In some cases they are the sole occupant of small workshops or work positions; they may work in remote sections of a large construction site; they may work out of normal hours, like cleaners or security personnel; they will often be working away from their main base as sub-contract installers, maintenance people, and drivers.

There is no general legal reason why people should not work alone but there may be special risks which require two or more people to be present, for example, during entry into a confined space in order to effect a rescue. It is important to ensure that a lone worker is not put at any higher risk than other workers. This is achieved by carrying out a specific risk assessment and introducing special protection arrangements for their safety. People particularly at risk, like young people or women, should also be considered. People's overall health and suitability to work alone should be taken into account.

Procedures may include:

➤ periodic visits from the supervisor to observe what is happening
➤ regular voice contact between the lone worker and the supervisor
➤ automatic warning devices to alert others if a specific signal is not received from the lone worker
➤ other devices to raise the alarm, which are activated by the absence of some specific action
➤ checks that the lone worker has returned safely home or to their base
➤ special arrangements for first aid to deal with minor injuries; this may include mobile first-aid kits
➤ arrangements for emergencies should be established and employees trained.

6.7 Permits to work

6.7.1 *Introduction*

Safe systems of work/method statements are crucial in work such as the maintenance of chemical and other processing plants where the potential risks are high and the careful coordination of activities and precautions is essential to safe working. In this situation and others of similar risk potential, the safe system of work is likely to be implemented by formal controls through a permit-to-work procedure.

The permit-to-work procedure is a specialized system for ensuring that potentially very dangerous work (e.g. entry into process plant and other confined spaces) is done safely and the safe system or work/method statement is properly and formally implemented.

Although it has been developed and refined by the chemical industry, the principles of permit-to-work procedures are equally applicable to the management of complex risks in other industries such as construction. In many cases contractors will be required to follow the permit-to-work procedure operated on a client's site; in others they may be required to set up their own system.

Its fundamental principle is that certain defined operations are prohibited without the specific permission of a responsible manager, this permission being only granted once stringent checks have been made to ensure that all necessary precautions have been taken and that it is safe for work to go ahead.

The people doing the work take on responsibility for following and maintaining the safeguards set out in the permit, which will define the work to be done (no other work being permitted) and the timescale in which it must be carried out.

Figure 6.18 Permit to work.

To be effective, the permit system requires the training needs of those involved to be identified and met, and monitoring procedures to ensure that the system is operating as intended.

6.7.2 *Permit-to-work procedures*

The permit-to-work procedure allows certain categories of high risk-potential work to be done with the specific permission of an authorized manager. This permission (in the form of the permit to work) will only be given if the laid-down precautions are in force and have been checked.

The permit document will typically specify:

➤ what work is to be done
➤ the plant/equipment involved, and how it is identified
➤ who is authorized to do the work
➤ the steps which have already been taken to make the plant safe
➤ potential hazards which remain, or which may arise as the work proceeds
➤ the precautions to be taken against these hazards
➤ for how long the permit is valid
➤ that the equipment is released to those who are to carry out the work.

In accepting the permit, the person in charge of doing the authorized work normally undertakes to take/maintain whatever precautions are outlined in the permit. The permit will also include spaces for:

➤ signature certifying that the work is complete
➤ signature confirming re-acceptance of the plant/ equipment.

6.7.3 *Principles*

Permit systems must adhere to the following eight principles:

1. Wherever possible, and especially with routine jobs, hazards should be eliminated so that the work can be done safely without requiring a permit to work.
2. Although the site manager may delegate the responsibility for the operation of the permit system, the overall responsibility for ensuring safe operation rests with him/her.
3. The permit must be recognized as the master instruction which, until it is cancelled, overrides all other instructions.
4. The permit applies to everyone on site, including sub-contractors.

5. Information given in a permit must be detailed and accurate. It must state:
   (a) which plant/equipment has been made safe and the steps by which this has been achieved
   (b) what work may be done
   (c) the time at which the permit comes into effect.
6. The permit remains in force until the work has been completed and the permit is cancelled by the person who issued it, or by the person nominated by management to take over the responsibility (e.g. at the end of a shift or during absence).
7. No work other than that specified is authorized. If it is found that the planned work has to be changed, the existing permit should be cancelled and a new one issued.
8. Responsibility for the plant must be clearly defined at all stages.

6.7.4 *Work requiring a permit*

The nature of permit-to-work procedures will vary in its scope depending on the job and the risks involved. However, a permit-to-work system is unlikely to be needed where, for example:

(a) the assessed risks are low and can be controlled easily
(b) the system of work is very simple and
(c) other work being done nearby cannot affect the work concerned in, say, a confined space entry, or a welding operation.

However, where there are high risks and the system of work is complex and other operations may interfere, a formal permit to work should be used.

The main types of permit and the work to be covered by each are identified below. Appendix 6.3 illustrates the essential elements of a permit form with supporting notes on its operation.

General permit

The general permit should be used for work such as:

➤ alterations to or overhaul of plant or machinery where mechanical, toxic or electrical hazards may arise
➤ work on or near overhead crane tracks
➤ work on pipelines with hazardous contents
➤ work with asbestos-based materials
➤ work involving ionizing radiation
➤ work at height where there are exceptionally high risks
➤ excavations to avoid underground services.

Confined space permit

Confined spaces include chambers, tanks (sealed and open-top), vessels, furnaces, ducts, sewers, manholes, pits, flues, excavations, boilers, reactors and ovens.

Many fatal accidents have occurred where inadequate precautions were taken before and during work involving entry into confined spaces. The two main hazards are the potential presence of toxic or other dangerous substances and the absence of adequate oxygen. In addition, there may be mechanical hazards (entanglement on agitators) and raised temperatures. The work to be carried out may itself be especially hazardous when done in a confined space; for example, cleaning using solvents, cutting/welding work. Should the person working in a confined space get into difficulties for whatever reason, getting help in and getting the individual out may prove difficult and dangerous.

Stringent preparation, isolation, air testing and other precautions are therefore essential and experience shows that the use of a confined space entry permit is essential to confirm that all the appropriate precautions have been taken.

The Confined Spaces Regulations are summarized in Chapter 21. They detail the specific controls that are necessary when people enter a confined space. For more details on confined spaces, see Chapter 19.

Work on high voltage apparatus (including testing) and work on high voltage apparatus (over about 600 V) is potentially high risk. Hazards include:

➤ possibly fatal electric shock/burns to the people doing the work
➤ electrical fires/explosions
➤ consequential danger from disruption of power supply to safety-critical plant and equipment.

In view of the risk, this work must only be done by suitably trained and competent people acting under the terms of a high voltage permit.

Hot work

Hot work is potentially hazardous as a:

➤ source of ignition in any plant in which highly flammable materials are handled
➤ cause of fires in all locations, regardless of whether highly flammable materials are present.

Hot work includes cutting, welding, brazing, soldering and any process involving the application of a naked flame. Drilling and grinding should also be included where a flammable atmosphere is potentially present. In high-risk areas hot work may also involve any equipment or procedure that produces a spark of sufficient energy to ignite highly flammable substances.

Hot work should therefore be done under the terms of a hot work permit, the only exception being where hot work is done in a designated area suitable for the purpose.

Figure 6.20 A hot work permit is essential except in designated areas.

6.7.5 *Responsibilities*

The effective operation of the permit system requires the involvement of many people. The following specific responsibilities can be identified.

(*Note*: all appointments, definitions of work requiring a permit, etc. must be in writing. All the categories of people identified below should receive training in the operation of the permit system as it affects them.)

Figure 6.19 Entering a confined space.

Site manager

➤ has overall responsibility for the operation and management of the permit system

➤ appoints a senior manager (normally the engineering manager on a client's site) to act as senior authorized person.

Senior authorized person

➤ is responsible to the site manager for the operation of the permit system

➤ defines the work on the site which requires a permit

➤ ensures that people responsible for this work are aware that it must only be done under the terms of a valid permit

➤ appoints all necessary authorized persons

➤ appoints a deputy to act in his/her absence.

Authorized persons

➤ issue permits to competent persons and retain copies

➤ personally inspect the site to ensure that the conditions and proposed precautions are adequate and that it is safe for the work to proceed

➤ accompany the competent person to the site to ensure that the plant/equipment is correctly identified and that the competent person understands the permit

➤ cancel the permit on satisfactory completion of the work.

Competent persons

➤ receive permits from authorized persons

➤ read the permit and make sure they fully understand the work to be done and the precautions to be taken

➤ signify their acceptance of the permit by signing both copies

➤ comply with the permit and make sure those under their supervision similarly understand and implement the required precautions

➤ on completion of the work, return the permit to the authorized person who issued it and sign the clearance or hand back section.

Operatives

➤ read the permit and comply with its requirements, under the supervision of the competent person.

Specialists

A number of permits require the advice/skills of specialists in order to operate effectively. Such specialists may include chemists, electrical engineers, health and safety advisers and fire officers. Their role may involve:

➤ isolations within their discipline – for example electrical work

➤ using suitable techniques and equipment to monitor the working environment for toxic or flammable materials, or for lack of oxygen

➤ giving advice to managers on safe methods of working.

Specialists must not assume responsibility for the permit system. This lies with the site manager and the senior authorized person.

Contractors

The permit system should be applied to sub-contractors in the same way as to direct employees of the main contractor.

The contractor must be given adequate information and training on the permit system, the restrictions it imposes and the precautions it requires.

6.8 Emergency procedures

6.8.1 Introduction

Most of this chapter is about the principles of control to prevent accidents and ill-health. Emergency procedures, however, are about control procedures and equipment to limit the damage to people and property caused by an incident. Local fire and rescue authorities will often be involved and are normally prepared to give advice to employers.

Under Regulation 8 of the MHSW Regulations, procedures must be established and set in motion when necessary to deal with serious and imminent danger to persons at work. Necessary links must be maintained with local authorities, particularly with regard to first aid, emergency medical care and rescue work.

Although fire is the most common emergency likely to be faced, there are many other possibilities which should be considered including:

➤ gas explosion
➤ electrical burn or electrocution
➤ escape of toxic gases or fumes
➤ discovery of dangerous dusts like asbestos in the atmosphere
➤ bomb warning
➤ large vehicle crashing into the premises

- collapse of a permanent structure during construction or demolition
- collapse of a temporary structure such as a scaffold
- aircraft crash if near a flight path
- spread of highly infectious disease
- severe weather with high winds and flooding.

For fire emergencies see Chapter 15.

6.8.2 *Supervisory duties*

A member of the site staff should be nominated to supervise and coordinate all emergency arrangements. This person should be in a senior position or at least have direct access to a senior manager. Senior members of the staff should be appointed as departmental fire wardens, with deputies for every occasion of absence, however brief. They should ensure that the following precautions are taken, i.e. that:

- everyone on site can be alerted to an emergency
- everyone on site knows what signal will be given for an emergency and knows what to do
- someone who has been trained in what to do is on site and ready to coordinate activities
- emergency routes are kept clear, signed and adequately lit
- there are arrangements for calling the fire and rescue services and to give them special information about high hazard work, for example in tunnels or confined spaces
- there is adequate access to the site for the emergency services and this is always kept clear
- suitable arrangements for treating and recovering injured people are set up
- someone is posted to the site entrance to receive and direct the emergency services.

6.8.3 *Assembly and roll call*

Assembly points should be established for use in the event of evacuation. It should be in a position, preferably under cover, which is unlikely to be affected at the time of the emergency. In some cases it may be necessary to make mutual arrangements with the client or occupiers of nearby premises.

In the case of small sites, a complete list of the names of all staff should be maintained so that a roll call can be made if evacuation becomes necessary.

In those premises where the number of staff would make a single roll call difficult, each area fire warden should maintain a list of the names of employees and sub-contractors in their area. Roll call lists must be updated regularly.

6.9 First aid at work

6.9.1 *Introduction*

People at work can suffer injuries or fall ill. It doesn't matter whether the injury or the illness is caused by the work they do. What is important is that they receive immediate attention and that an ambulance is called in serious cases. First aid at work (FAW) covers the arrangements employers must make to ensure this happens. It can save lives and prevent minor injuries becoming major ones.

The Health and Safety (First-Aid) Regulations 1981 requires employers to provide adequate and appropriate equipment, facilities and personnel to enable first aid to be given to employees if they are injured or become ill at work.

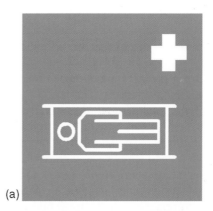

(a)

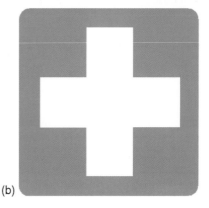

(b)

Figure 6.21 (a) First aid and stretcher sign; (b) first-aid sign.

What is adequate and appropriate will depend on the circumstances in a particular site.

The minimum first-aid provision on any work site is:

➤ a suitably stocked first-aid box
➤ an appointed person to take charge of first-aid arrangements.

It is also important to remember that accidents can happen at any time. First-aid provision needs to be available at all times people are at work.

Some small sites will only need to make the minimum first-aid provision. However, construction work is normally considered to be high risk and greater provision is usually necessary. The following checklist covers the points that should be considered.

6.9.2 Aspects to consider

The risk assessments carried out under the MHSW and COSHH regulations should show whether there are any specific risks in the workplace. The following should be considered:

➤ Are there hazardous substances, dangerous tools and equipment, dangerous manual handling tasks, electrical shock risks, dangers from neighbours or animals?
➤ Are there different levels of risk in parts of the site?
➤ What is the accident and ill-health record, and type and location of incidents?
➤ What is the total number of persons likely to be on site?
➤ Are there young people, or pregnant or nursing mothers on site, employees with disabilities or special health problems?
➤ Are the facilities widely dispersed with several buildings or compact in a multi-storey building?
➤ What is the pattern of working hours; does it involve night work?
➤ Is the site remote from emergency medical services?
➤ Do employees travel a lot or work alone?
➤ Do any employees work at sites occupied by other employers?
➤ Are members of the public regularly on site?

6.9.3 Impact on first-aid provision if risks are significant

First aiders may need to be appointed if risks are significant.

This will involve a number of factors which must be considered, including:

➤ training for first aiders
➤ additional first-aid equipment and the contents of the first-aid box
➤ siting of first-aid equipment to meet the various demands on the site. For example, provision of equipment in each area or on several floors of a large site. There needs to be first-aid provision at all times during working hours
➤ informing local medical services of the site and its risks
➤ any special arrangements that may be needed with the local emergency services.

If sub-contractors are employed the principal/main contractor needs to consider:

➤ sharing arrangements with sub-contractors
➤ information telling people who the appointed person or first aider is
➤ ensuring that everyone on site knows through information and suitable signage what arrangements have been made. A notice in the site hut is a good way of doing this with a sign outside.

Although there are no legal responsibilities for non-employees, the HSE strongly recommends that they are included in any first-aid provision.

6.9.4 Contents of the first-aid box

There is no standard list of items to put in a first-aid box. It depends on what the employer assesses the needs to be. Where there is no special risk in the workplace, a minimum stock of first-aid items is required; these are listed in Table 6.2. There should be sufficient stock to cover all persons on site.

Tablets or medicines should not be kept in the first-aid box. Table 6.2 is a suggested contents list only; equivalent but different items will be considered acceptable.

6.9.5 Appointed persons

As a minimum each site must have an appointed person, who is someone that is appointed by management to:

➤ take charge when someone is injured or falls ill. This includes calling an ambulance if required
➤ look after the first-aid equipment, for example, keeping the first-aid box replenished
➤ keeping records of treatment given.

Table 6.2 Contents of first-aid box – low risk

Stock for up to 50 persons:	
A leaflet giving general guidance on first aid, for example HSE leaflet *Basic advice on first aid at work*.	
➤ Medical adhesive plasters	40
➤ Sterile eye pads	4
➤ Individually wrapped triangular bandages	6
➤ Safety pins	6
➤ Individually wrapped: medium sterile unmedicated wound dressings	8
➤ Individually wrapped: large sterile unmedicated wound dressings	4
➤ Individually wrapped wipes	10
➤ Paramedic shears	1
➤ Pairs of latex gloves	2
➤ Sterile eyewash if no clean running water	2

Appointed persons should never attempt to give first aid for which they are not competent. Short emergency first-aid training courses are available. Remember that an appointed person should be available at all times when people are at work on site – this may mean appointing more than one. The training should be repeated every 3 years to keep up-to-date.

6.9.6 *Qualifications and numbers of people*

A first aider is someone who has undergone an HSE approved training course in administering FAW and holds a current FAW certificate. Lists of local training organizations are available from the local environmental officer or HSE offices. The training should be repeated every 3 years to maintain a valid certificate and keep the first aider up-to-date.

It is not possible to give hard and fast rules on when or how many first aiders or appointed persons might be needed. This will depend on the circumstances of each particular organization or worksite. Table 6.3 offers suggestions on how many first aiders or appointed persons might be needed in relation to categories of risk and number of employees. The details in the table are suggestions only; they are not definitive, nor are they a legal requirement.

To ensure cover at all times when people are at work and where there are special circumstances, such as remoteness from emergency medical services, shift work, or sites with several separate buildings, there may need to be more first-aid personnel than set out in Table 6.3. Provision must be sufficient to cover for absences.

Employees must be informed of the first-aid arrangements. Putting up notices telling workers who and where the first aiders or appointed persons are and where the first-aid box is will usually be enough. Special

Table 6.3 Number of first-aid personnel

Category of risk	Numbers employed at any location	Suggested number of first-aid personnel
Lower risk		
For example shops and offices, libraries	Fewer than 50 50–100 More than 100	At least one appointed person At least one first aider One additional first aider for every 100 employed
Medium risk		
For example light engineering and assembly work, food processing, warehousing	Fewer than 20 20–100 More than 100	At least one appointed person At least one first aider for every 50 employed (or part thereof) One additional first aider for every 100 employed
Higher risk		
For example most construction, slaughterhouses, chemical manufacture, extensive work with dangerous machinery or sharp instruments	Fewer than 5 5–50 More than 50	At least one appointed person At least one first aider One additional first aider for every 50 employed

Source: HSE.

arrangements will be needed for employees with reading or language difficulties.

6.9.7 *Changes to first-aid requirements*

In June 2006 the HSE indicated that their review of the First Aid Regulations had decided to retain approval of the training courses within the HSE but make changes to the structure of the course. The intention is that in future, employers will be able to send suitable employees on either a 6 hour (minimum) emergency first aid at work (EFAW) or an 18 hour (minimum) FAW course, based on the findings of their first-aid needs assessment (see Figure 6.22). After 3 years, first aiders will need to complete another course (either a 6 hour EFAW or 12 hour FAW requalification course, as appropriate) to obtain a new certificate. Within any 3 year certification period, first aiders should complete two annual refresher

courses, covering basic life support/skills updates, that will each last for at least 3 hours.

The likely date for the introduction of these courses is 2009. The HSE is aware that FAW training providers require an adequate lead-in period. There are a number of additional issues to be clarified before a date can be finalized. Firstly, whether the 3 day FAW course needs to be trialled in a pilot study prior to its introduction. Secondly, consideration will be given to any arrangements that need to be put in place for the approval and monitoring of those providers wishing to offer the EFAW course and whether these arrangements should be extended to those offering annual refresher training.

HSE will revise the guidance for employers in its publication: *First aid at work: The Health and Safety (First-Aid) Regulations 1981 Approved Code of Practice and Guidance (L74)*. This will help employers understand

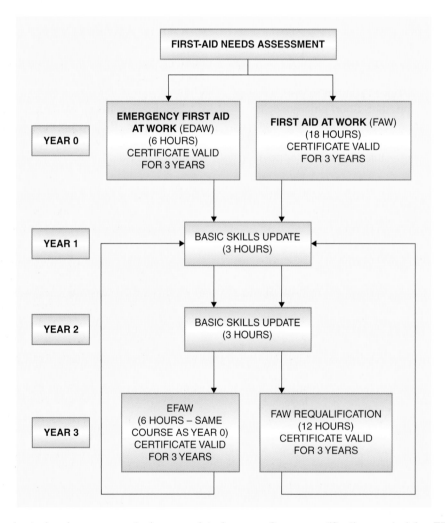

Figure 6.22 Flow chart showing courses to be completed over a 3 year certification period for EFAW and FAW. The dotted line indicates the route to be taken in subsequent years after completion of the relevant course at year 3.

the new training regime and will provide more detailed information on how to conduct a first-aid needs assessment. HSE recognizes that this is crucial in ensuring employers send prospective first aiders on the course that most closely meets the needs of their workplace. The guidance will also strongly emphasize that HSE will consider it as good practice for first aiders to undertake annual refresher training. Once the guidance has been drafted, it will be subject to a consultation process with stakeholders, as appropriate.

Summary of the HSE's review is that:

➤ HSE recognizes the First aid at Work Council (FWC) as the lead industry body with representation from all sectors of the FAW training provider industry

➤ HSE will not transfer first-aid approval and monitoring section (FAAMS) to the FWC and will retain it in-house

➤ HSE has made some modifications to the structure of FAW training courses since its Position Statement was published

➤ HSE needs to resolve a number of issues before a date for the introduction of the new courses is announced and an adequate lead-in period will be given for FAW training providers.

Any further significant developments will be published on the first-aid web pages of HSE's website.

6.10 Practice NEBOSH questions for Chapter 6

1. With respect to the management of risk within the workplace:
   (a) **explain** the meaning of the term 'hierarchy of control'
   (b) **outline**, with examples, the standard hierarchy that should be applied with respect to controlling health and safety risks in the workplace.

2. Approximately one quarter of all accidents on construction sites are associated with plant, machinery and vehicles. **Outline** the elements of a strategy designed to prevent such accidents.

3. **Outline** the possible effects on health and safety of poor standards of housekeeping in the workplace.

4. **Name** and **describe FOUR** classes of safety sign prescribed by the Health and Safety (Safety Signs and Signals) Regulations 1996.

5. **State** the shape and colour, and give a relevant example, of each of the following types of safety sign:
   (a) prohibition
   (b) warning
   (c) mandatory
   (d) emergency escape or first aid.

6. **Explain** why Personal Protective Equipment PPE should be considered as a last resort in the control of occupational health hazards.

7. (a) **Explain** the purposes of a method statement.
   (b) **Identify** the main factors relevant to health and safety that should be included in a method statement.

8. (a) **Explain** the meaning of the term 'safe system of work'.
   (b) **Describe** the enforcement action that could be taken by an enforcing authority when a safe system of work has not been implemented.

9. **Outline** the factors that should be considered when developing a safe system of work.

10. **Identify EIGHT** sources of information that might usefully be consulted when developing a safe system of work for a construction activity.

11. Due to an increase in knife related accidents amongst hotel kitchen staff that use sharp tools in the preparation of food for the restaurant, a safe system of work is to be developed to minimize the risk of injury to this group of employees.
    (a) **Identify** the legal requirements under which the employer must provide a safe system of work.
    (b) **Describe** the issues to be addressed when developing the safe system of work for the hotel kitchen staff that use knives as part of their work.
    (c) **Outline** the ways in which the employer could motivate the hotel kitchen staff to follow the safe system of work.

12. (a) **Define** the term 'permit-to-work system'.
    (b) **Outline THREE** types of work situation that may require a permit-to-work system, giving reasons in EACH case for the requirement.
    (c) **Outline** the specific details that should be included in a permit to work for entry into a confined space.

13. (a) **Identify TWO** specific work activities for which a permit-to-work system might be needed.

(b) **Outline** the key elements of a permit-to-work system.

14. **Outline** the issues to be addressed in a training session on the operation of a permit-to-work system.

15. **Outline** the factors that should be considered when preparing a procedure to deal with a workplace emergency.

16. **Outline** the reasons why employees may fail to comply with safety procedures at work.

17. With reference to the Confined Space Regulations 1997:
    (a) **Define** the meaning of 'confined space', giving **TWO** workplace examples.
    (b) **Outline** specific hazards associated with working in confined spaces.
    (c) **Identify FOUR** 'specified risks' that may arise from work in a confined space.

18. **Outline** the precautions that should be taken in order to ensure the safety of employees undertaking maintenance work in an underground storage vessel.

19. **Outline** the factors that should be considered when preparing a procedure to deal with a workplace emergency.

20. (a) **Identify FOUR** types of emergency procedure that an organization might need to have in place.
    (b) **Explain** why visitors to a workplace should be informed of the emergency procedures.

21. (a) **Identify THREE** types of emergency in the workplace for which employees may need to be evacuated.
    (b) **Explain** why it is important to develop workplace procedures to enable the safe evacuation of employees during an emergency.

22. (a) **Identify** the **TWO** main functions of first-aid treatment.
    (b) **Outline** the factors to consider when making an assessment of first-aid provision in a workplace.

23. **Outline** the factors to consider when making an assessment of first-aid provision in a workplace.

24. An out-of-town business park, consisting of several self-contained units, is currently under construction. **List** the factors to be considered in deciding whether first-aid arrangements for the site are adequate.

25. (a) **Identify** the **TWO** main functions of first-aid treatment.
    (b) **Outline** the factors to consider when making an assessment of first-aid provision in a workplace.

26. **Identify** the factors that should be considered when assessing the adequacy of first-aid arrangements on a site.

27. Construction is shortly to commence on an out-of-town supermarket development. **Outline** the factors to be considered when deciding upon adequate first-aid arrangements for the site.

Appendix 6.1 – Job safety analysis form

JOB SAFETY ANALYSIS

Job	Date
Department	Carried out by

Description of job

Legal requirements and guidance

Task steps	Hazards	Consequence	Severity	Risk C×S	Controls

Safe system of work

Job Instruction

Training requirements

Review date

Appendix 6.2 – Example of a safety method statement form

HEALTH AND SAFETY METHOD STATEMENT

Company: ....

Site: ....   Job No: ....

In respect of:	
Statement prepared by:	
Date:	
Initial Risk Assessments included:	
Client's Name:	
Client's Address:	
Site Address:	
Date for Commencement of Work:	
Date for Completion of Work:	
Personnel Carrying Out Work:	
Equipment Being Used (Make? Model? Reg. No?)	
How Site will be Accessed:	
Site Details: (Parking, Fuel Depot, Equipment Store, Welfare Facilities, Emergency Procedures)	
Hazards on Site: (e.g. Power Lines, Buildings, Demolition, Hazardous Processes, Underground Services)	
Standards Required	
Planning Procedures	
Supervision	
How Task will Proceed: Safe System of Work	

Signed: ....   Date: ....

Appendix 6.3 – Essential elements of a permit-to-work form

2. Should have any reference to other relevant permits or isolation certificates.	1 Permit Title	2 Permit Number
	3 Job Title	
	4 Plant identification	
5. Description of work to be done and its limitations.	5 Description of Work	
6. Hazard identification including residual hazards and hazards introduced by the work.	6 Hazard Identification	
7. Precautions necessary – person(s) who carry out precautions, e.g. isolations, should sign that precautions have been taken.	7 Precautions necessary.	7 Signatures . . . .
8. Protective equipment needed for the task.	8 Protective Equipment	
9. Authorization signature confirming that isolations have been made and precautions taken, except where these can only be taken during the work. Date and time duration of permit.	9 Authorization	
10. Acceptance signature confirming understanding of work to be done, hazards involved and precautions required. Also confirming permit information has been explained to all workers involved.	10 Acceptance	
11. Extension/Shift handover signature confirming checks have been made that plant remains safe to be worked on, and new acceptance/ workers made fully aware of hazards/ precautions. New time expiry given.	11 Extension/Shift handover	
12. Hand back signed by acceptor certifying work completed. Signed by issuer certifying work completed and plant ready for testing and re-commissioning.	12 Hand back	
13. Cancellation certifying work tested and plant satisfactorily re-commissioned.	13 Cancellation	

Source: HSE

Monitoring review and audit

7

7.1 Introduction

This chapter concerns the monitoring of health and safety performance, including both positive measures like inspections and negative measures like injury statistics. It is about reviewing progress to see if something better can be done and auditing to ensure that what has been planned is being implemented.

Measurement is a key step in any management process and forms the basis of continuous improvement. If measurement is not carried out correctly, the effectiveness of the health and safety management system is undermined and there is no reliable information to show managers how well the health and safety risks are controlled.

Managers should ask key questions to ensure that arrangements for health and safety risk control are in place, comply with the law as a minimum, and operate effectively. There are two basic types of monitoring:

Proactive or active monitoring, by taking the initiative before things go wrong, involves routine inspections and checks to make sure that the standards and policies are being implemented and that controls are working.

Reactive monitoring, after things go wrong, involves looking at historical events to learn from mistakes and see what can be put right to prevent a recurrence.

The Health and Safety Executive's (HSE's) experience is that organizations find health and safety performance measurement a difficult subject. They struggle to develop health and safety performance measures which are not based solely on injury and ill-health statistics.

7.2 The traditional approach to measuring health and safety performance

Senior managers often measure company performance by using, for example, percentage profit, return on investment or market share. A common feature of the measures would be that they are generally positive in nature – which demonstrates achievement – rather than negative, which demonstrates failure.

Yet, if senior managers are asked how they measure their companies' health and safety performance, it is likely that the only measure would be accident or injury statistics. While the general business performance of an organization is subject to a range of positive measures, for health and safety it too often comes down to one negative measure, injury and ill-health statistics – measures of failures.

Health and safety differs from many areas measured by managers because improvement in performance means fewer outcomes from the measure (injuries or ill-health) rather than more. A low-injury or ill-health rate trend over years is still no guarantee that risks are being controlled and that incidents will not happen in the future. This is particularly true in organizations where major hazards are present but there is a low probability of accidents.

There is no single reliable measure of health and safety performance. What is required is a 'basket' of measures, providing information on a range of health and safety issues.

There are some significant problems with the use of injury and ill-health statistics in isolation:

➤ There may be under-reporting – focusing on injury and ill-health rates as a measure, especially if a reward system is involved, can lead to non-reporting to keep up performance.
➤ It is often a matter of chance whether a particular incident causes an injury, and they may not show whether or not a hazard is under control; luck or a reduction in the number of people exposed may produce a low injury/accident rate rather than good health and safety management.
➤ An injury is the particular consequence of an incident and often does not reflect the potential severity; for example, an unguarded machine could result in a cut finger or an amputation.
➤ People can be absent from work for reasons which are not related to the severity of the incident.
➤ There is evidence to show that there is little relationship between 'occupational' injury statistics (e.g. slips, trip and falls) and the reasons for the lack of control of major accident hazards (e.g. loss of containment of flammable or toxic material).
➤ A small number of accidents may lead to complacency.
➤ Injury statistics demonstrate outcomes not causes.

Because of the potential shortcomings related to the use of accident/injury and ill-health data as a single measure of performance, more proactive or 'up stream' measures are required. These require a systematic approach to deriving positive measures and how they link to the overall risk control process, rather than a quick fix based on things that can easily be counted, such as the number of training courses or number of inspections, which has limited value. The resultant data provide no information on how the figure was arrived at, whether it is 'acceptable' (i.e. good/bad) or the quality and effectiveness of the activity. A more disciplined approach to health and safety performance measurement is required. This needs to develop as the health and safety management system develops.

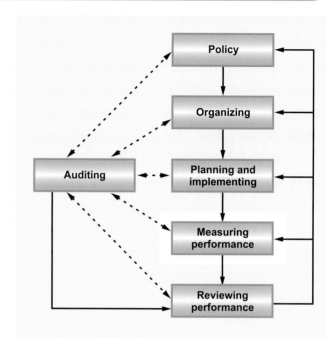

Figure 7.1 The health and safety management system.

as much part of a health and safety management system as financial, production or service delivery management. The HSG65 framework for managing health and safety, discussed at the end of Chapter 1 and illustrated in Figure 7.1, shows where measuring performance fits within the overall health and safety management system.

The main purpose of measuring health and safety performance is to provide information on the progress and current status of the strategies, processes and activities employed to control health and safety risks. Effective measurement not only provides information on what the levels are but also why they are at this level, so that corrective action can be taken.

7.3.2 *Answering questions*

Health and safety monitoring or performance measurement should seek to answer such questions as:

➤ Where is the position relative to the overall health and safety aims and objectives?
➤ Where is the position relative to the control of hazards and risks?
➤ How does the organization compare with others?
➤ What is the reason for the current position?
➤ Is the organization getting better or worse over time?

7.3   Why measure performance?

7.3.1 *Introduction*

> *You can't manage what you can't measure –*
> *Drucker*

Measurement is an accepted part of the 'plan-do-check-act' management process. Measuring performance is

Is the management of health and safety doing the right things?

Is the management of health and safety doing things right consistently?

Is the management of health and safety proportionate to the hazards and risks?

Is the management of health and safety efficient?

Is an effective health and safety management system in place across all parts of the organization?

Is the culture supportive of health and safety, particularly in the face of competing demands?

These questions should be asked at all management levels throughout the organization. The aim of monitoring should be to provide a complete picture of an organization's health and safety performance.

7.3.3 Decision-making

The measurement information helps in deciding:

where the organization is in relation to where it wants to be

what progress is necessary and reasonable in the circumstances

how that progress might be achieved against particular restraints (e.g. resources or time)

priorities – what should be done first and what is most important

effective use of resources.

7.3.4 Addressing different information needs

Information from the performance measurement is needed by a variety of people. These will include directors, senior managers, line managers, supervisors, health and safety professionals, and employees or safety representatives. They each need information appropriate to their position and responsibilities within the health and safety management system.

For example, what the chief executive officer of a large organization needs to know from the performance measurement system will differ in detail and nature from the information needs of the manager of a particular site.

A coordinated approach is required so that individual measuring activities fit within the general performance measurement framework.

Although the primary focus for performance measurement is to meet the internal needs of an organization, there is an increasing need to demonstrate to external stakeholders (regulators, insurance companies, shareholders, suppliers, sub-contractors, members of the public, etc.) that arrangements to control health and safety risks are in place, operating correctly and effectively.

7.4 What to measure

7.4.1 Introduction

In order to achieve an outcome of no injuries or work-related ill-health, and to satisfy stakeholders, health and safety risks need to be controlled. Effective risk control is founded on an effective health and safety management system. This is illustrated in Figure 7.2.

7.4.2 Effective risk control

The health and safety management system comprises three levels of control (see Figure 7.2):

Level 1 – the key elements of the health and safety management system: the management arrangements (including plans and objectives) necessary to organize, plan, control and monitor the design and implementation of RCSs.

Level 2 – risk control systems (RCSs): the basis for ensuring that adequate workplace precautions are provided and maintained

Level 3 – effective workplace precautions provided and maintained to prevent harm to people who are exposed to the risks

The health and safety culture must be positive to support each level.

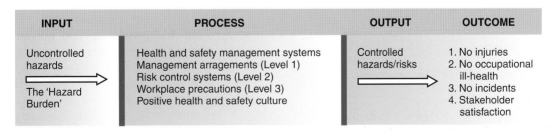

INPUT	PROCESS	OUTPUT	OUTCOME
Uncontrolled hazards The 'Hazard Burden'	Health and safety management systems Management arragements (Level 1) Risk control systems (Level 2) Workplace precautions (Level 3) Positive health and safety culture	Controlled hazards/risks	1. No injuries 2. No occupational ill-health 3. No incidents 4. Stakeholder satisfaction

Figure 7.2 Health and safety management system.

Performance measurement should cover all the elements of Figure 7.2 and be based on a balanced approach which combines:

➤ **input: monitoring** the scale, nature and distribution of hazards created by the organization's activities – measures of the hazard burden
➤ **process: active monitoring** of the adequacy, development, implementation and deployment of the health and safety management system and the activities to promote a positive health and safety culture – measures of success
➤ **outcomes: reactive monitoring** of adverse outcomes resulting in injuries, ill-health, loss and accidents with the potential to cause injuries, ill-health or loss – measures of failure.

7.5 Measuring failure – reactive monitoring

So far, this chapter has concentrated on measuring activities designed to prevent the occurrence of injuries and work-related ill-health (active monitoring). Failures in risk control also need to be measured (reactive monitoring), to provide opportunities to check performance, learn from failures and improve the health and safety management system.

Reactive monitoring arrangements include systems to identify and report:

➤ injuries and work-related ill-health (details of the incident rate calculation are given in Chapter 4)
➤ other losses such as damage to property
➤ incidents, including those with the potential to cause injury, ill-health or loss (near misses)
➤ hazards and faults
➤ weaknesses or omissions in performance standards and systems, including complaints from employees and enforcement action by the authorities.

Guidance on investigating these incidents is given in Chapter 8.

7.6 Proactive or active monitoring – how to measure performance

7.6.1 Introduction

The measurement process can gather information through:

➤ direct observation of conditions and of people's behaviour (sometimes referred to as unsafe acts and unsafe conditions monitoring)

➤ talking to people to elicit facts and their experiences as well as gauging their views and opinions
➤ examining written reports, documents and records.

These information sources can be used independently or in combination. Direct observation includes inspection activities and the monitoring of the work environment (e.g. temperature, dust levels, solvent levels, noise levels) and people's behaviour. Each Risk Control System (RCS) should have a built-in monitoring element that will define the frequency of monitoring; these can be combined to form a common inspection system.

7.6.2 Inspection

General

This may be achieved by developing a checklist or inspection form that covers the key issues to be monitored in a particular department or area of the organization within a particular period. It might be useful to structure this checklist using the 'four Ps' (note that the examples are not a definitive list):

➤ **premises**, including:
   ➤ access/escape
   ➤ housekeeping
   ➤ services like gas and electricity
   ➤ working environment
   ➤ fire precautions.
➤ **plant and substances**, including:
   ➤ machinery guarding
   ➤ tools and equipment
   ➤ local exhaust ventilation
   ➤ use/storage/separation of materials/chemicals.
➤ **procedures**, including:
   ➤ safe systems of work
   ➤ method statements
   ➤ permits to work
   ➤ use of personal protective equipment
   ➤ procedures followed.
➤ **people**, including:
   ➤ health surveillance
   ➤ people's behaviour
   ➤ training and supervision
   ➤ appropriate authorized person.

It is essential that people carrying out an inspection do not in any way put themselves or anyone else at risk. Particular care must be taken with regard to safe access. In carrying out these safety inspections, the safety of people's actions should be considered, in addition to the safety of the conditions they are working in – a ladder might be in perfect condition but it has to be used properly too.

Figure 7.3 Poor conditions – inspection needed.

Key points in becoming a good observer

To improve health and safety performance, managers and supervisors must eliminate unsafe acts by observing them, taking immediate corrective action, and following up to prevent recurrence. To become a good observer, they must improve their observation skills and must learn how to observe effectively. Effective observation includes the following key points:

- Be selective.
- Know what to look for.
- Practice.
- Keep an open mind.
- Guard against habit and familiarity.
- Do not be satisfied with general impressions.
- Record observations systematically.

Observation techniques

In addition, to become a good observer, a person must:

- stop for 10–30 seconds before entering a new area to ascertain where employees are working
- be alert for unsafe practices that are corrected as soon as you enter an area
- observe activity – do not avoid the action
- remember ABBI – look Above, Below, Behind, Inside
- develop a questioning attitude to determine what injuries might occur if the unexpected happened and how the job might be accomplished more safely. Ask 'why?' and 'what could happen if . . .?'
- use all senses: sight, hearing, smell, touch
- maintain a balanced approach. Observe all phases of the job
- be inquisitive
- observe for ideas – not just to determine problems
- recognize good performance.

Daily/weekly/monthly safety inspections

These will be aimed at checking conditions in a specified area against a suitable checklist drawn up by local site management. It will cover specific items, such as the guards at particular machines, whether access or agreed routes are clear, whether scaffold inspections are in date, whether ladders are secured, and whether fire extinguishers are in place. The checks should be carried out by site staff who should sign off the checklist. It should not last more than half an hour, perhaps less. This is not a specific hazard spotting operation, but there should be a space on the checklist for the inspectors to note down any particular problems encountered.

Reports from inspections

Some of the items arising from safety inspections will have been dealt with immediately; other items will require action by specified people. Where there is some doubt about the problem, and what exactly is required, advice should be sought from the site safety adviser or external expert. A brief report of the inspection and any resulting action list should be submitted to the safety committee. While the committee may not have the time available to consider all reports in detail, it will want to be satisfied that appropriate action is taken to resolve all matters; it will be necessary for the committee to follow up the reports until all matters are resolved.

Inspection standards

In order to get maximum value from inspection checklists, they should be designed so that they require objective rather than subjective judgements of conditions. For example, asking the people undertaking a general inspection of the workplace to rate housekeeping as good or bad, begs questions as to what does good and bad mean, and what criteria should be used to judge this. If good housekeeping means there is no rubbish left on the floor; all waste bins are regularly emptied and not overflowing; floors are swept each day and cleaned once a week; and decorations should be in good condition with no peeling paint, then this should be stated. Adequate expected standards should be provided in separate notes so that those who inspect know the standards that are required.

The checklist or inspection form should facilitate:

➤ the planning and initiation of remedial action, by requiring those doing the inspection to rank deficiencies in priority order (those actions which are most important rather than those which can be easily done quickly)

➤ identifying those responsible for taking remedial actions, with sensible timescales to track progress on implementation

➤ periodic monitoring to identify common themes which might reveal underlying problems in the system

➤ management information on the frequency or nature of the monitoring arrangements.

Appendix 7.1 gives examples of poor workplaces which can be used for practice exercises. Appendix 7.2 shows a basic list of construction inspection issues, which can be adapted to any particular organization's needs. A policy checklist is given in Chapter 2.

7.6.3 *Safety sampling*

Safety sampling is a useful technique that helps organizations to concentrate on one particular area or subject at a time. A specific area is chosen which can be inspected in about 30 minutes. A checklist is drawn up to facilitate the inspection, looking at specific issues. These may be different types of hazard; they may be unsafe acts or conditions noted; they may be proactive, good behavioural practices noted.

The inspection team or person then carries out the sampling at the same time each day or week in the specified period. The results are recorded and analysed to see if the changes are good or bad over time. Of course, defects noted must be brought to the notice of the appropriate person for action on each occasion.

7.7 Who should monitor performance?

Performance should be measured at each management level from directors downwards. It is not sufficient to monitor by exception so that unless problems are raised it is assumed to be satisfactory. Senior managers must satisfy themselves that the correct arrangements are in place and working properly. Responsibilities for both active and reactive monitoring must be laid down and managers need to be personally involved in making sure that plans and objectives are met and compliance with standards is achieved. Although systems may be set up with the guidance of safety professionals, managers should be personally involved and given sufficient training to be competent to make informed judgements about monitoring performance.

Other people, like safety representatives, will also have the right to inspect the workplace. Each employee should be encouraged to inspect their own workplace frequently to check for obvious problems and rectify them if possible or report hazards to their supervisors.

Specific statutory (or thorough) examinations of, for example, lifting equipment or pressure vessels, have to be carried out at intervals laid down in written schemes by competent persons – usually specially trained and experienced inspection/insurance company personnel.

Excavations, scaffolds, working platforms and other equipment for working at height require specific statutory inspections, normally carried out by competent people on the site. For details see Chapters 18–20.

7.8 Frequency of monitoring and inspections

This will depend on the level of risk and any statutory inspection requirement. Directors may be expected to examine the premises formally at an annual audit, while site supervisors may be expected to carry out inspection each week. Senior managers should regularly monitor the health and safety plan to ensure that objectives are being met and to make any changes to the plan as necessary.

Data from reactive monitoring should be considered by senior managers at least once a month. In most organizations serious events would be closely monitored as they happen.

7.9 Report writing

There are three main aims to the writing of reports and they are all about communication. A report should aim to:

➢ get a message through to the reader
➢ make the message and the arguments clear and easy to understand
➢ make the arguments and conclusions persuasive.

Communication starts with trying to get into the mind of the reader, imagining what would most effectively catch the attention, what would be most likely to convince, what will make this report stand out among others.

A vital part of this is presentation; so while a handwritten report is better than nothing if time is short, a well-organized, typed report is very much clearer. To the reader of the report, who may well be very busy with a great deal of written information to wade through, a clear, well-presented report will produce a positive attitude from the outset, with instant benefit to the writer.

Five factors which help to make reports effective are:

1. structure
2. presentation of arguments
3. style
4. presentation of data
5. how the report itself is presented.

7.9.1 Structure

The structure of a report is the key to its professionalism. Good structuring will:

➢ help the reader to understand the information and follow the arguments contained in the report
➢ increase the writer's credibility
➢ ensure that the material contained in the report is organized to the best advantage.

The following list shows a frequently used method of producing a report, but always bear in mind that different organizations use different formats. It is important to check with the organization requesting the report in case their in-house format differs:

➢ title page
➢ summary
➢ contents list
➢ introduction
➢ main body of the report
➢ conclusions
➢ recommendations
➢ appendices
➢ references.

1. Title page
This will contain:

➢ a title and often a subtitle
➢ the name of the person or organization to whom the report is addressed
➢ the name of the writer(s) and their organization
➢ the date on which the report was submitted.

As report writing is about communication, it is a good idea to choose a title that is eye-catching and memorable as well as being informative.

2. Summary
Limit the summary to between 150 and 500 words. Do not include any evidence or data. This should be kept for the main report. Include the main conclusions and principal recommendations and place them near the front of the report.

3. Contents list
Put the contents list near the beginning of the report. Short reports do not need a list but if there are several headings it does help the reader to grasp the overall content of the report in a short time.

4. Introduction
The introduction should contain the following:

➢ information about who commissioned the report and when
➢ the reason for the report
➢ objectives of the report
➢ terms of reference
➢ preparation of the report (type of data, research undertaken, subjects interviewed, etc.)
➢ methodology used in any analysis

➤ problems and the methods used to tackle them
➤ details about consultation with clients, employees, etc.

There may be other items that are specific to the report.

5. The main body of the report

This part of the report should describe, in detail, what was discovered (the facts), the significance of these discoveries (analysis) and their importance (evaluation). Graphs, tables and charts are often used at this stage in the report. These should have the function of summarizing information rather than giving large amounts of detail. The more detailed graphics should be made into appendices.

To make it more digestible, this part of the report should be divided into sections, using numbered headings and sub-headings. Very long and complex reports will need to be broken down into chapters.

6. Conclusions

The concluding part of the report should be a reasonably detailed 'summing up'. It should give the conclusions arrived at by the writer and explain why the writer has reached these conclusions.

7. Recommendations

The use of this section depends on the requirements of the person commissioning the report. If recommendations are required, provide as few as possible, to retain a clear focus. Report writers are often asked only to provide the facts.

8. Appendices

This part of the report should contain sections that may be useful to a reader who requires more detail. Examples would be the detailed charts, graphs and tables mentioned in 'the main body of the report', any questionnaires used in constructing the evidence, lists, forms, case studies and so on. The appendices are the background material of the report.

9. References

If any books, papers or journal articles have been used as source material, it should be acknowledged in a reference section. There are a number of accepted referencing methods used by academics.

Because the reader is likely to be a person with some degree of expertise in the subject, a report must be reliable, credible, relevant and thorough. It is therefore important to avoid emotional language, opinions presented as facts and arguments that have no supporting evidence. To make a report more persuasive, the writer needs to:

➤ present the information clearly
➤ provide reliable evidence
➤ present arguments logically
➤ avoid falsifying, tampering with or concealing facts.

Expertise in an area of knowledge means that distortions, errors and omissions will quickly be spotted by the reader and the presence of any of these will cast doubt on the credibility of the whole report.

Reports are usually used as part of a decision-making process. If this is the case, clear, unembellished facts are needed. Exceptions to this would be where the report is a proposal document or where a recommendation is specifically requested. Unless this is the case, it is better not to make recommendations.

A report should play a key role in organizing information for the use of decision makers. It should review a complex and/or extensive body of information and make a summary of all the important issues.

It is relatively straightforward to produce a report, as long as the writer keeps to a clear format. Using the format described here, it should be possible to tell the reader as clearly as possible:

➤ what happened
➤ what it cost
➤ what the result was, etc.

There may be a request for a special report and this is likely to be longer and more difficult to produce. Often it will relate to a 'critical incident' and the decision makers will be looking for information to help them:

➤ decide whether this is a problem or an opportunity
➤ decide whether to take action
➤ decide what action, if any, to take.

Finally, report writing should be kept simple. Nothing is gained, in fact much is lost in the use of long, complicated sentences, jargon and official-sounding language. When the report is finished, it is helpful to run through it with the express intention of simplifying the language and making sure that it says what was intended in a clear and straightforward way.

KEEP IT SHORT AND SIMPLE

7.10 Review and audit

7.10.1 *Audits – purpose*

The final steps in the health and safety management control cycle are auditing and performance review. Organizations need to be able to reinforce, maintain and develop the ability to reduce risks. The 'feedback loop' produced by this final stage in the process enables them to do this and to ensure continuing effectiveness of the health and safety management system.

Audit is a business discipline which is frequently used, for example, in finance, environmental matters and quality. It can equally well be applied to health and safety.

The term is often used to mean inspection or other monitoring activity. Here, the following definition is used, which complies with HSG65:

> *The structured process of collecting independent information on the efficiency, effectiveness and reliability of the total health and safety management system and drawing up plans for corrective action.*

Over time, it is inevitable that control systems will decay and they may even become obsolete as things change. Auditing is a way of supporting monitoring by providing managers with information. It will show how effectively plans and the components of health and safety management systems are being implemented. In addition, it will provide a check on the adequacy and effectiveness of the management arrangements and RCSs.

Auditing is critical to a health and safety management system, but it is not a substitute for other essential parts of the system. Companies need systems in place to manage cash flow and pay the bills – this cannot be managed through an annual audit. In the same way, health and safety needs to be managed on a day-to-day basis and for this organizations need to have systems in place. A periodic audit will not achieve this.

The aims of auditing should be to establish that the three major components of a safety management system are in place and operating effectively. It should show that:

➤ appropriate management arrangements are in place
➤ adequate RCSs exist, are implemented, and consistent with the hazard profile of the organization
➤ appropriate health and safety precautions are in place.

Where the organization is spread over a number of sites, the management arrangements linking the centre with the business units and sites should be covered by the audit.

There are a number of ways in which this can be achieved and some parts of the system do not need auditing as often as others. For example, an audit to verify the implementation of RCSs would be made more frequently than a more overall audit of the capability of the organization or of the management arrangements for health and safety. Critical RCSs, which control the principle hazards of the business, would need to be audited more frequently. Where there are complex workplace precautions, it may be necessary to undertake technical audits. An example would be engineering construction projects for chemical process plants.

A well-structured auditing programme will give a comprehensive picture of the effectiveness of the health and safety management system in controlling risks. Such a programme will indicate when and how each component part will be audited. Managers, safety representatives and employees, working as a team, will effectively widen involvement and cooperation needed to put together the programme and implement it.

The process of auditing involves:

➤ gathering information from all levels of an organization about the health and safety management system
➤ making informed judgements about its adequacy and performance.

7.10.2 *Gathering information*

Decisions will need to be made about the level and detail of the audit before starting to gather information about the health and safety management of an organization. Auditing involves sampling, so initially it is necessary to decide how much sampling is needed for the assessment to be reliable. The type of audit and its complexity will relate to its objectives and scope, to the size and complexity of the organization and to the length of time that the existing health and safety management system has been in operation.

Information sources of interviewing people, looking at documents and checking physical conditions are usually approached in the following order:

Preparatory work

➤ Meet with relevant managers and employee representatives to discuss and agree the objectives and scope of the audit.
➤ Prepare and agree the audit procedure with managers.
➤ Gather and consider documentation.

On-site

➤ interviewing
➤ review and assessment of additional documents
➤ observation of physical conditions and work activities.

Conclusion

➤ Assemble the evidence.
➤ Evaluate the evidence.
➤ Write an audit report.

7.10.3 *Making judgements*

It is essential to start with a relevant standard or bench-mark against which the adequacy of a health and safety management system can be judged. If standards are not clear, assessment cannot be reliable. Audit judgements should be informed by legal standards, HSE guidance and applicable construction industry standards. HSG65 sets out benchmarks for management arrangements and for the design of RCSs. This book follows the same concepts.

Auditing should not be seen as a fault-finding activity. It should make a valuable contribution to the health and safety management system and to learning. It should recognize achievement as well as highlighting areas where more needs to be done.

Scoring systems can be used in auditing along with judgements and recommendations. This can be seen as a useful way to compare sites or monitor progress over time. However, there is no evidence that quantified results produce a more effective response than the use of qualitative evidence. Indeed, the introduction of a scoring system can, the HSE believes, have a negative effect as it may encourage managers to place more emphasis on high-scoring questions which may not be as relevant to the development of an effective health and safety management system.

To achieve the best results, auditors should be competent people who are independent of the area and of the activities being audited. External consultants can be used or staff from other areas of the organization. An organization can use its own auditing system or one of the proprietary systems on the market or, since it is unlikely that any ready-made system will provide a perfect fit, a combination of both. With any scheme, cost and benefits have to be taken into account. Common problems include:

➤ systems which are too general in their approach. These may need considerable work to make them fit the needs and risks of the organization
➤ systems which are too cumbersome for the size and culture of the organization
➤ scoring systems may conceal problems in underlying detail
➤ organizations may design their management system to gain maximum points rather than using one which suits the needs and hazard profile of the business.

HSE encourages organizations to assess their health and safety management systems using in-house or proprietary schemes but without endorsing any particular one.

7.10.4 *Performance review*

When performance is reviewed, judgements are made about its adequacy and decisions are taken about how and when to rectify problems. The feedback loop is needed by organizations so that they can see whether the health and safety management system is working as intended. The information for review of performance comes from audits of RCSs and workplace precautions, and from the measurement of activities. There may be other influences, both internal and external, such as reorganization, new legislation or changes in current good practice. These may result in the necessity to redesign or change parts of the health and safety management system or to alter its direction or objectives.

Performance standards need to be established which will identify the systems requiring change, responsibilities, and completion dates. It is essential to feed back the information about success and failure so that employees are motivated to maintain and improve performance.

In a review, the following areas will need to be examined:

➤ the operation and maintenance of the existing system
➤ how the safety management system is designed, developed and installed to accommodate changing circumstances.

Reviewing is a continuous process. It should be undertaken at different levels within the organization. Responses will be needed as follows:

➤ by first-line supervisors or other managers to remedy failures to implement workplace precautions which they observe in the course of routine activities
➤ to remedy sub-standard performance identified by active and reactive monitoring
➤ to the assessment of plans at individual, departmental, site, group or organizational level
➤ to the results of audits.

The frequency of review at each level should be decided upon by the organization and reviewing activities should be devised which will suit the measuring and auditing activities. The review will identify specific remedial actions which establish who is responsible for implementation and set deadlines for completion.

7.11 Practice NEBOSH questions for Chapter 7

1. **Outline** the ways in which an organization can monitor its health and safety performance.

2. **Identify EIGHT** measures that could be used by an organization in order to monitor its health and safety performance.

3. **Outline TWO** reactive measures and **TWO** proactive (or active) measures that can be used in monitoring an organization's health and safety performance.

4. **Outline FOUR** proactive (or active) monitoring methods that can be used in assessing the health and safety performance of an organization.

5. **Outline FOUR** proactive (or active) monitoring methods that can be used in assessing the health and safety performance on a construction project.

6. An employer intends to implement a programme of regular workplace inspections following a workplace accident.
   (a) **Outline** the factors that should be considered when planning such inspections.
   (b) **Outline THREE** additional proactive (active) methods that could be used in the monitoring of health and safety performance.
   (c) **Identify** the possible costs to the organization as a result of the accident.

7. **Identify** a range of measures that can be used to monitor a construction company's health and safety performance, clearly distinguishing between proactive and reactive measures.

8. **Identify** the main topic areas that should be included in a planned health and safety inspection of a workplace.

9. **Explain** how the following may be used to improve safety performance within an organization:
   (a) accident data
   (b) safety inspections.

10. **Outline** the ways in which an organization can monitor its health and safety performance.

11. **Outline** the factors that would determine the frequency with which health and safety inspections should be undertaken in the workplace.

12. (a) **Define** the term 'safety survey'.
    (b) **Outline** the issues which should be considered when a safety survey of a workplace is to be undertaken.

13. **Outline** the *strengths* **AND** *weaknesses* of using a checklist to complete a health and safety inspection of a workplace.

14. **Outline** the topics that should be included in a health and safety audit.

15. **Outline** the reasons why an organization should review and monitor its health and safety performance.

16. A health and safety audit of an organization has identified a general lack of compliance with procedures.
    (a) **Describe** the possible reasons for procedures not being followed.
    (b) **Outline** the practical measures that could be taken to motivate employees to comply with health and safety procedures.

17. **List** the documents that are likely to be examined during a health and safety audit.

18. **Identify** the *advantages* **AND** *disadvantages* of carrying out a health and safety audit of an organization's activities by:
    (a) an internal auditor
    (b) an external auditor.

Appendix 7.1 – Workplace inspection exercises

The following four examples are given of workplaces with numerous inadequately controlled hazards. They can be used to practise workplace inspections and risk assessments.

Figure 7.4 Construction site.

Figure 7.5 Road repair.

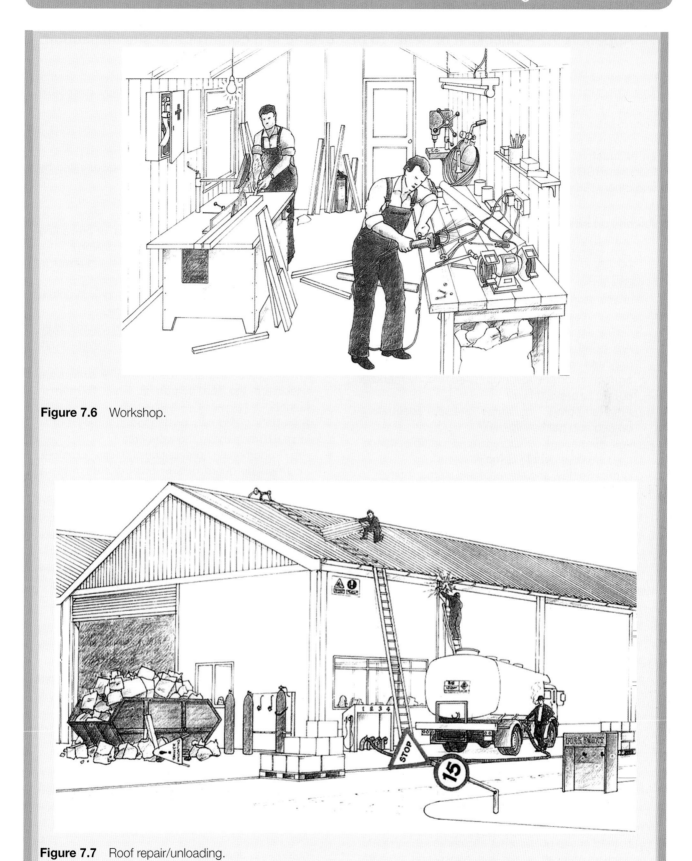

Figure 7.6 Workshop.

Figure 7.7 Roof repair/unloading.

Appendix 7.2 – Checklist of items to be covered in a construction site inspection

This checklist identifies some of the hazards most commonly found on construction sites. The questions it asks are intended to help you decide whether your site is a safe and healthy place to work. **It is not an exhaustive list**. More detailed information can be found in HSG150 *Health and safety in construction* and other HSE publications.

Access on site

➤ Can everyone reach their place of work safely; for example, are roads, gangways, passageways, passenger hoists, staircases, ladders and scaffolds in good condition?

➤ Are access routes free from obstructions and clearly signposted?

➤ Are holes and openings securely guard railed, provided with an equivalent standard of edge protection or provided with fixed, clearly marked covers to prevent falls?

➤ Are temporary structures stable, adequately braced and not overloaded?

➤ Will permanent structures remain stable during any refurbishment or demolition work?

➤ Is the site tidy, and are materials stored safely?

➤ Are there proper arrangements for collecting and disposing of waste materials?

➤ Is the work area adequately lit?

➤ Is lighting adequate especially when work is carried on after dark outside or inside buildings?

Welfare

➤ Have suitable and sufficient numbers of toilets been provided and are they kept clean and properly lit?

➤ Are there clean wash basins, hot and cold (or warm) water, soap and towels or means of drying?

➤ Are washbasins large enough to wash up to the elbow?

➤ Is suitable clothing provided for those who have to work in wet, dirty or otherwise adverse conditions?

➤ Are there facilities for changing, drying and storing clothes?

➤ Are drinking water and cups provided?

➤ Is there a site hut or other accommodation where workers can sit, make hot drinks and prepare food?

➤ Are welfare facilities easily and safely accessible to all who need to use them?

Scaffolds

➤ Are scaffolds erected, altered and dismantled by competent persons?

➤ Is there safe access to the scaffold platform?

➤ Are all uprights provided with base plates (and, where necessary, timber sole plates) or prevented in some other way from slipping or sinking?

➤ Are all the uprights, ledgers, braces and struts in position?

➤ Is the scaffold secured to the building or structure in enough places to prevent collapse?

➤ Are there double guard rails and toe boards or an equivalent standard of protection at every edge to prevent falling?

➤ Where guard rails and toe boards or similar are used: are the toe boards at least 150mm in height?

➤ Are upper guard rails positioned at a height of at least 910mm above the work area?

➤ Are additional precautions, for example, intermediate guard rails or brick guards in place to ensure that there is no unprotected gap of more than 470mm between the toe board and upper guard rail?

➤ Are the working platforms fully boarded and are the boards arranged to avoid tipping or tripping?

➤ Are there effective barriers or warning notices in place to stop people using an incomplete scaffold (e.g. where working platforms are not fully boarded)?

➤ Has the scaffold been designed and constructed to cope with the materials stored on it and are these distributed evenly?

➤ Does a competent person inspect the scaffold regularly, for example, at least once a week if the working platform is 2m or above in height or at suitable intervals if less than 2m, and always after it has been substantially altered or damaged and following extreme weather?

➤ Are the results of inspections recorded and kept?

➤ Have proprietary tower scaffolds been inspected and are they erected and used in accordance with suppliers' instructions?

➤ Are the wheels of tower scaffolds locked and outriggers deployed when in use and are the platforms empty when they are moved?

Powered access equipment

➤ Has the equipment been erected/installed by a competent person?

➤ Is fixed equipment (e.g. mast climbers) rigidly connected to the structure against which it is operating?

➤ Does the working platform have adequate secure guard rails and toe boards or other barriers to prevent people and materials failing off?

➤ Have precautions been taken to prevent people being struck by:
  ➤ the moving platform,
  ➤ projections from the building or
  ➤ falling materials,
    ➤ for example, by a barrier or fence around the base?

➤ Are the operators trained and competent?

➤ Is the safe working load clearly marked?

➤ Is the equipment inspected by a competent person?

➤ Is the power supply isolated and the equipment secured at the end of the working day?

Ladders

➤ Does the risk assessment conclude that ladders are the right way to do the job? Don't work from a ladder if there is a safer way using more suitable equipment.

➤ Are all ladders in good condition?

➤ Are they correctly inclined at a ratio of 1 out to 4 up?

➤ Are they secured to prevent them slipping sideways or outwards?

➤ Do ladders rise a sufficient height above their landing place (about 5 rungs or 1m? If not, are there other handholds available?

➤ Are the ladders positioned so that users don't have to overstretch or climb over obstacles to work?

➤ Does the ladder rest against a solid surface and not on fragile or insecure materials?

Roof work

➤ Are there enough barriers and is there other edge protection to stop people or materials failing from roofs?

➤ Do the roof battens provide safe hand and foot holds? If not, are crawling ladders or boards provided and used?

➤ During industrial roofing, are precautions taken (e.g. nets) to stop people failing from the leading edge of the roof or from fragile or partially fixed sheets which could give way?

➤ Where nets are used have they been rigged safely by a competent person?

➤ Have fragile surfaces been identified (e.g. fibre, cement sheets or roof lights)?

➤ Are suitable barriers, guard rails or covers, etc. provided where people pass or work near fragile material such as asbestos cement sheets and roof lights?

➤ Are crawling boards provided where work on fragile materials cannot be avoided?

➤ Are people excluded from the area below the roof work? If this is not possible, have additional precautions been taken to stop debris failing onto them?

Traffic, vehicles and plant

➤ Are vehicles and pedestrians kept separate wherever possible? If not are:
  ➤ they separated as much as possible and then barriers used?
  ➤ people told about the problem and what to do about it?
  ➤ warning signs displayed?

➤ Have one-way systems or turning points been provided to minimize the need for reversing?

➤ Where vehicles have to reverse, are they controlled by properly trained signallers?

➤ Are vehicles maintained; do the steering, handbrake and footbrake work properly?

➤ Have drivers received proper training and are they competent for the vehicles and plant they are operating?

➤ Are loads properly secured?

➤ Are passengers prevented from riding in dangerous positions and only on vehicles designed to carry them?

➤ Are vehicles and plant prevented from operating on dangerous slopes?

➤ Can zero tail swing excavators be used or is there adequate clearance around slewing vehicles?

Hoists

➤ Has the hoist and its equipment been installed by a competent person?

(Continued)

Appendix 7.2 – *Continued*

➤ Is the hoist protected by a substantial enclosure to prevent someone from being struck by any moving part of the hoist or falling down the hoistway?

➤ Are gates provided at all landings, including ground level?

➤ Are the gates kept shut except when the platform is at the landing?

➤ Are the controls arranged so that the hoist can be operated from one position only?

➤ Is the hoist operator trained and competent?

➤ Is the hoist's safe working load or rated capacity clearly marked?

➤ If the hoist is for materials only, is there a warning notice on the platform or cage to stop people riding on it?

➤ Does the hoist have a current report of thorough examination and a record of inspection by a competent person?

➤ Are the results of inspection recorded?

Cranes and lifting appliances

➤ Is the crane suitable for the job?

➤ Has the lift been properly planned by an 'Appointed person'?

➤ Is the crane on a firm level base? Are the riggers properly set?

➤ Are the safe working loads and corresponding radii known and considered before any lifting begins?

➤ Who is the appointed 'crane supervisor' responsible for controlling the lifting operation on site?

➤ Is the load secure?

➤ Are all operators trained and competent?

➤ Have arrangements been made to make sure that the driver can see the load or has a signaller been provided to help?

➤ Has the signaller/slinger been trained to give signals and to attach loads correctly?

➤ Do the operator and slinger find out the weight and centre of gravity of the load before trying to lift it?

➤ Does the crane have a current report of thorough examination and a record of inspection by a competent person?

➤ Are people stopped from walking beneath a raised load?

Excavations

➤ Is an adequate supply of timber, trench sheets, props or other supporting material made available before excavation work begins, or has the excavation been sloped or battered back to a safe angle?

➤ If the sides of the excavation are sloped back or battered, is the angle of batter sufficient to prevent collapse?

➤ Is this material strong enough to support the sides?

➤ Is a safe method used for putting in the support, i.e. one that does not rely on people working within an unsupported trench?

➤ Is there safe access to the excavation (e.g. by a sufficiently long, secured ladder)?

➤ Are there guard rails or other equivalent protection to stop people falling in?

➤ Are properly secured stop blocks or other safeguards provided to prevent vehicles falling in?

➤ Does the excavation affect the stability of neighbouring structures or services?

➤ Are materials, spoil or plant stored away from the edge of the excavation in order to reduce the likelihood of a collapse of the side?

➤ Is the excavation regularly inspected by a competent person, for example, at the start of every shift; and after any accidental collapse or event likely to have affected its stability?

Manual handling

➤ Has the risk of manual handling injuries been assessed?

➤ Can lighter or smaller sizes of material be used, for example, cement in 25 kg bags and so avoid the repetitive laying of building blocks weighing more than 20 kg?

➤ Are hoists, telehandlers, wheelbarrows and other plant or equipment used so that manual lifting and handling of heavy objects is kept to a minimum?

➤ Have people been instructed and trained how to lift safely?

Tools and machinery

➤ Are the right tools and machinery being used for the job?

➤ Are all dangerous parts guarded (e.g. exposed gears, chain drives, projecting engine shafts)?

➤ Are guards secured and in good repair?

➤ Are tools and machinery maintained in good repair and are all safety devices operating correctly?

➤ Are all operators trained and competent?

Fire and emergencies

General

➤ Have emergency procedures been developed, for example, evacuating the site in case of fire or rescue from a confined space?

➤ Are people on site aware of the procedures?

➤ Is there a means of raising the alarm and does it work?

➤ Is there a way of contacting emergency services from site?

➤ Are there adequate escape routes and are these kept clear?

➤ Is there adequate first aid provision?

Fire

➤ Is the quantity of flammable materials, liquids and gases on site kept to a minimum?

➤ Are there proper storage areas for flammable materials, flammable liquids and gases?

➤ Are containers and cylinders returned to ventilated stores at the end of the shift?

➤ Are suitable containers used for flammable liquids?

➤ If liquids are transferred from their original containers are the new containers suitable for flammable materials?

➤ Is smoking banned in areas where gases or flammable liquids are stored and used? Are other ignition sources also prohibited?

➤ Are gas cylinders, associated hoses and equipment properly maintained and in good condition?

➤ When gas cylinders are not in use, are the valves fully closed?

➤ Are adequate bins or skips provided for storing flammable and combustible waste?

➤ Is flammable and combustible waste removed regularly?

➤ Are the right number and type of fire extinguishers available and accessible?

Hazardous substances

➤ Have all harmful materials (e.g. asbestos, lead, solvents, paints cement and dust, been identified)?

➤ Have the risks to everyone who might be exposed to these substances been assessed?

➤ Has a check been made to see whether a licensed contractor is needed to deal with asbestos on site (most work with asbestos needs a license, although some limited work can be done with asbestos without a license)?

➤ Have precautions to prevent or control exposure to hazardous substances been identified and put into place by:

  ➤ doing the work in a different way, to remove the risk entirely;

  ➤ using less hazardous materials;

  ➤ using tools fitted with dust suppression (water lubrication) or extraction;

  ➤ providing suitable personal protective equipment?

➤ Have workers received information and training so they know what the risks are from the hazardous substances used and produced on site, and what they need to do to avoid the risks?

➤ Are there procedures to prevent contact with wet cement?

➤ Has health surveillance been arranged for people using certain hazardous substances like lead?

Noise

➤ Have workers received information and training so they know what the risks are from noise on site, and what they need to do to avoid the risks?

➤ Has workers' exposure to noise been assessed?

➤ Can the noise be reduced by using different working methods or selecting quieter plant, for example, by fitting breakers and other plant or machinery with silencers?

➤ Are barriers erected to reduce the spread of noise?

➤ Is work sequenced to minimize the number of people exposed to noise?

➤ Are others not involved in the work kept away from the source of noise?

➤ Is suitable hearing protection provided and worn in noisy areas?

➤ Have hearing protection zones been marked?

➤ Has health surveillance been arranged for people exposed to high levels of noise?

Hand–arm vibration

➤ Have workers received information and training so they know what the risks are from hand–arm

(Continued)

Appendix 7.2 – *Continued*

vibration (HAV) on site, and what they need to do to avoid the risks?

- Have risks to workers from prolonged use of vibrating tools such as breakers, angle grinders, hammer drills and chainsaws been identified and assessed?
- Has exposure to HAV been reduced as much as possible by selecting suitable work methods and equipment?
- Have reduced vibration tools been used whenever possible?
- Are vibrating tools properly maintained?
- Have arrangements for health surveillance for people exposed to high levels of HAV (especially when exposed for long periods) been put in place?

Electricity and other services

- Have all necessary services been provided on site before work begins and have existing services present on site been identified (e.g. electricity cables, gas and water mains) and effective steps taken to prevent danger from them?
- Is the supply voltage for tools and equipment the lowest necessary for the job (could battery operated tools and reduced voltage systems, e.g. 110 V, or even lower in wet conditions, be used)?
- Where mains voltage has to be used, are trip devices (e.g. residual current devices (RCDs)) provided for all equipment?
- Are RCDs protected from damage, dust and dampness and checked daily by users?
- Are cables and leads protected from damage by sheathing, protective enclosures or positioning away from causes of damage?
- Are all connections to the system properly made and are suitable plugs used?
- Is there an appropriate system of user checks, formal visual examinations by site managers and combined inspection and test by competent persons for all tools and equipment?
- Are scaffolders, roofers, etc., or cranes or other plant, working near or under overhead lines? Has the electricity supply been turned off, or have other precautions, such as 'goal posts' or taped markers been provided to prevent them contacting the lines?

- Have underground electricity cables and other services been located (with a locator and plans) and marked, and precautions for safe digging and other work been taken?

Confined spaces

- Is work done in confined spaces (including tanks, pits sewers and manholes) where there may be an inadequate supply of oxygen or the presence of dangerous or flammable gases, dusts or fumes?
- If so have all necessary precautions been taken including operating a formal permit to work system?

Protective clothing

- Has adequate personal protective equipment (e.g. hard hats, safety boots, gloves, goggles, and dust masks) been provided?
- Is the equipment in good condition and worn by all who need it?

Protecting the public

- Are the public fenced off or otherwise protected from the work?
- Are roadworks barriered off and lit, and is a safe alternative route provided?
- Are the public protected from falling material?
- Has a safe route been provided through roadworks for people with prams, wheelchair users, and visually impaired people?
- When work has stopped for the day:
  - are the gates secured?
  - is the perimeter fencing secure and undamaged?
  - are all ladders removed or their rungs boarded so that they cannot be used?
  - are excavations and openings securely covered or fenced off?
  - is all plant immobilized to prevent unauthorized use?
  - are bricks and materials safely stacked?
  - are flammable or dangerous substances locked away in secure storage places?

Incident investigation, recording and reporting

8

Introduction

This chapter is concerned with the recording of incidents and accidents at work; their investigation; the legal reporting requirements; and simple analysis of incidents to help managers benefit from the investigation and recording process.

Incidents and accidents rarely result from a single cause and many turn out to be complex. Most incidents involve multiple, interrelated causal factors. They can occur whenever significant deficiencies, oversights, errors, omissions or unexpected changes occur. Any one of these can be the precursor for an accident or incident. There is a value on collecting data on all incidents and potential losses as it helps to prevent more serious events. (See Chapter 5 for accident ratios and definitions.)

Incidents and accidents, whether they cause damage to property, or more seriously, injury and/or ill-health to people, should be properly and thoroughly investigated to allow an organization to take the appropriate action to prevent a recurrence. Good investigation is a key element to making improvements in health and safety performance.

Incident investigation is considered to be part of a **reactive monitoring system** because it is triggered after an event. The range of events includes:

➤ injuries and ill-health, including sickness absence
➤ damage to property, personal effects, work in progress, etc.
➤ incidents which have the potential to cause injury, ill-health or damage

➤ hazards
➤ deficiencies in performance standards.

Figure 8.1 A dangerous occurrence – fire.

Each type of event gives the opportunity to:

➤ check performance
➤ identify underlying deficiencies in management systems and procedures
➤ learn from mistakes and add to the corporate memory
➤ reinforce key health and safety messages
➤ identify trends and priorities for prevention
➤ provide valuable information if there is a claim for compensation
➤ help to meet legal requirements for reporting certain incidents to the authorities.

8.2 Reasons for incident/accident investigation

8.2.1 *Logic and understanding*

Incident/accident investigation is based on the logic that:

➤ all incidents/accidents have causes . . . eliminate the cause and eliminate future incidents
➤ the direct and indirect causes of an incident/accident can be discovered through investigation
➤ corrective action indicated by the causation can be taken to eliminate future incidents/accidents.

Investigation is not intended to be a mechanism for apportioning blame. There are often strong emotions associated with injury or significant losses. It is all too easy to look for someone to blame without considering the reasons why a person behaved in a particular way. Often short cuts to working procedures that may have contributed to the accident give no personal advantage to the person injured. The short cut may have been taken out of loyalty to the organization or ignorance of a safer method.

Valuable information and understanding can be gained from carrying out accident/incident investigations. These include:

➤ an understanding of how and why problems arose which caused the accident/incident
➤ an understanding of the ways people are exposed to substances or situations which can cause them harm
➤ a snapshot of what really happens, e.g. why people take short cuts or ignore safety rules
➤ identifying deficiencies in the control of risks in the organization.

8.2.2 *Legal reasons*

The legal reasons for conducting an investigation are:

➤ to ensure that the organization is operating in compliance with legal requirements
➤ that it forms an essential part of the MHSW Regulation 5 requirements to plan, organize, control, monitor and review health and safety arrangements
➤ to comply with the Woolf Report on civil action which changed the way cases are run. Full disclosure of the circumstances of an accident has to be made to the injured parties considering legal action.

The fact that a thorough investigation was carried out and remedial action taken would demonstrate to a court that a company has a positive attitude to health and safety. The investigation will also provide essential information for insurers in the event of an employer's liability or other claim.

8.2.3 *Benefits*

There are many benefits from investigating accidents/incidents. These include:

➤ the prevention of similar events occurring again. Where the outcomes are serious injuries the enforcing authorities are likely to take a tough stance if previous warnings have been ignored
➤ the prevention of business losses due to disruption immediately after the event, loss of production, loss of business through a lowering of reputation or inability to deliver and the costs of criminal and legal actions
➤ improvement in employee morale and general attitudes to health and safety particularly if they have been involved in the investigations
➤ improving management skills to improve health and safety performance throughout the organization.

The case for investigating near misses and undesired circumstances may not be so obvious but it is just as useful and much easier as there are no injured people to deal with. There are no demoralized people at work or distressed families and seldom a legal action to answer. Witnesses will be more willing to speak the truth and help with the investigation.

8.2.4 *What managers need to do*

Managers need to:

➤ communicate the type of accident and incident that needs to be reported
➤ provide a system for reporting and recording
➤ check that the proper reports are being made
➤ make appropriate records of accidents and incidents
➤ investigate all incidents and accidents reported
➤ analyse the events routinely to check for trends in performance and the prevalence of types of incident or injury
➤ monitor the system to make sure that it is working satisfactorily.

8.3 Which incidents/accidents should be investigated?

8.3.1 Injury/accident

Should every accident be investigated or only those that lead to serious injury? In fact the main determinant is the potential of the accident to cause harm rather than the actual harm resulting. For example, a slip can result in an embarrassing flailing of arms or, just as easily, a broken leg. The frequency of occurrence of the accident type is also important – a stream of minor cuts from paper needs looking into.

As it is not possible to determine the potential for harm simply from the resulting injury, the only really sensible solution is to investigate all accidents. The amount of time and effort spent on the investigation should, however, vary depending on the level of risk (severity of potential harm, frequency of occurrence). The most effort should be focused on significant events involving serious injury, ill-health or losses and events which have the potential for multiple or serious harm to people or substantial losses. These factors should become clear during the accident investigation and be used to guide how much time should be taken.

Table 8.1 has been developed by the HSE to help to determine the level of investigation which is appropriate. The worst case scenario for the consequences should be considered in using the table. A particular incident like a scaffold collapse may not have caused an injury but had the potential to cause major or fatal injuries.

- ➤ In a **minimal level** investigation, the relevant supervisor will look into the circumstances of the accident/incident and try to learn any lessons which will prevent future incidents.
- ➤ A **low level** investigation will involve a short investigation by the relevant supervisor or line manager into the circumstances and immediate underlying and root causes of the accident/incident, to try to prevent a recurrence and to learn any general lessons.
- ➤ A **medium level** investigation will involve a more detailed investigation by the relevant supervisor or line manager, the health and safety adviser and employee representatives and will look for the immediate, underlying and root causes.
- ➤ A **high level** investigation will involve a team-based investigation, involving supervisors or line managers, health and safety advisers and employee representatives. It will be carried out under the supervision of senior management or directors and will look for the immediate, underlying and root causes.

Table 8.1 Level of investigation required

Likelihood of recurrence	Potential worst injury consequences of accident/incident			
	Minor	Serious	Major	Fatal
Certain				
Likely				
Possible				
Unlikely				
Rare				
Risk	Minimal	Low	Medium	High
Investigation level	Minimal level	Low level	Medium level	High level

8.3.2 Incidents (including near miss)

Figure 8.2 Unsafe scaffold – accident waiting to happen.

8.4 Investigations and causes of incidents

8.4.1 Who should investigate?

Investigations should be led by line managers or other people with sufficient status and knowledge to make recommendations that will be respected by the organization. The person to lead many investigations will be the site manager or supervisor of the person/area involved because they:

➤ know about the situation
➤ know most about the employees
➤ have a personal interest in preventing further incidents/accidents affecting 'their' people, equipment, area, materials
➤ can take immediate action to prevent a similar incident
➤ can communicate most effectively with the other employees concerned
➤ can demonstrate practical concern for employees and control over the immediate work situation.

8.4.2 *When should the investigation be conducted?*

The investigation should be carried out as soon as possible after the incident to allow the maximum amount of information to be obtained. There may be difficulties which should be considered in setting up the investigation quickly – if, for example, the victim is removed from the site of the accident, or if there is a lack of a particular expert. An immediate investigation is advantageous because:

➤ factors are fresh in the minds of witnesses
➤ witnesses have had less time to talk (there is an almost automatic tendency for people to adjust their story of the events to bring it into line with a consensus view)
➤ physical conditions have had less time to change
➤ more people are likely to be available, for example, delivery drivers, contractors and visitors who will quickly disperse following an incident making contact very difficult
➤ there will probably be the opportunity to take immediate action to prevent a recurrence and to demonstrate management commitment to improvement
➤ immediate information from the person suffering the accident often proves to be most useful.

Consideration should be given to asking the person to return to the site for the accident investigation if they are physically able, rather than wait for them to return to work. A second option, although not as valuable, would be to visit the injured person at home or even in hospital (with their permission) to discuss the accident.

8.4.3 *Investigation method*

There are four basic elements to a sound investigation:

1. Collect facts about what has occurred.
2. Assemble and analyse the information obtained.

3. Compare the information with acceptable industry and company standards and legal requirements to draw conclusions.
4. Implement the findings and monitor progress.

Information should be gathered from all available sources, for example, witnesses, supervisors, physical conditions, hazard data sheets, written systems of work, training records. The amount of time spent should not, however, be disproportionate to the risk. The aim of the investigation should be to explore the situation for possible underlying factors, in addition to the immediately obvious causes of the accident. For example, in an accident involving a faulty scaffold it would not be sufficient to conclude that an accident occurred because the scaffold was incorrectly assembled. It is necessary to look into the possible underlying system failure that may have occurred which allowed the scaffold to be sub-standard.

Investigations have three facets, which are particularly valuable and can be used to check against each other:

1. direct observation of the scene, premises, workplace, relationship of components, materials and substances being used, possible reconstruction of events, and injuries or condition of the person concerned
2. documents including written instructions, training records, procedures, safe operating systems, risk assessments, policies, records of inspections or tests and examinations carried out
3. interviews (including written statements) with persons injured, witnesses, people who have carried out similar functions or examinations and tests on the equipment involved and people with specialist knowledge.

Immediate causes

A detailed investigation should look at the following factors as they can provide useful information about **immediate causes** that have been manifested in the incident/accident.

➤ Personal factors:
  ➤ behaviour of the people involved
  ➤ suitability of people doing the work
  ➤ training and competence.
➤ Task factors:
  ➤ workplace conditions and precautions or controls
  ➤ actual method of work adopted at the time
  ➤ ergonomic factors
  ➤ normal working practice either written or customary.

Root or underlying causes

A thorough investigation should also look at the following factors as they can provide useful information about

underlying causes that have been manifested in the incident/accident:

➤ Management and organizational factors:
  ➤ previous similar incidents
  ➤ supervision
  ➤ the control and coordination of work
  ➤ quality of the health and safety policy and procedures
  ➤ quality of consultation and cooperation of employees
  ➤ the adequacy and quality of communications and information, including to sub-contractors
  ➤ deficiencies in risk assessments, plans and control systems
  ➤ deficiencies in monitoring and measurement of work activities
  ➤ quality and frequency of reviews and audits.

8.4.4 *Investigation interview techniques*

It must be made clear at the outset and during the course of the interview that the aim is not to apportion blame but to discover the facts and use them to prevent similar accidents in the future.

A witness should be given the opportunity to explain what happened in their own way without too much interruption and suggestion. Questions should then be asked to elicit more information. These should be of the open type, which do not suggest the answer. Questions starting with the following words (Figure 8.3) are useful.

'Why' should not be used at this stage. The facts should be gathered first, with notes being taken at the end of the explanation. The investigator should then read them or give a summary back to the witness, indicating clearly that they are prepared to alter the notes, if the witness is not content with them.

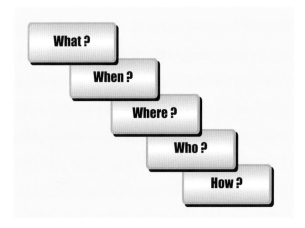

Figure 8.3 Questions to be asked in an investigation.

If possible, indication should be given to the witness about immediate actions that will be taken to prevent a similar occurrence and that there could be further improvements depending on the outcome of the investigation.

Accidents can often be very upsetting for witnesses, which should be borne in mind. This does not mean they will not be prepared to talk about what has happened. They may in fact wish to help, but questions should be sensitive; upsetting the witness further should be avoided.

8.4.5 *Comparison with relevant standards*

There are usually suitable and relevant standards which may come from the HSE, construction industry or the organization itself. These should be carefully considered to see if:

➤ suitable standards are available to cover legal standards and the controls required by the risk assessments
➤ suitable method statements were provided as necessary and followed the standards
➤ the standards were sufficient and available to the organization
➤ the standards were implemented in practice
➤ the standards were implemented; why was there a failure?
➤ there are any changes to the standards needed.

8.4.6 *Recommendations*

The investigation should have highlighted both immediate causes and root or underlying causes. Recommendations, both for immediate action and for longer-term improvements, should come out of this, but it may be necessary to ensure that the report goes further up the management chain if the improvements recommended require authorization, which cannot be given by the investigating team.

8.4.7 *Follow-up*

It is essential that a follow-up is made to check on the implementation of the recommendations. It is also necessary to review the effect of the recommendations to check whether they have achieved the desired result and whether they have had unforeseen 'knock-on' effects, creating additional risks and problems.

8.4.8 *Use of information*

The accident investigation should be used to generate recommendations but should also be used to generate safety awareness. The investigation report or a summary should therefore be circulated locally to relevant people and, when appropriate, summaries circulated throughout the organization. The accident does not need to have resulted in a 3-day lost time injury for this system to be used.

8.4.9 *Training*

A number of people will potentially be involved in accident investigation. For most of these people it will only be necessary on rare occasions. Training guidance and help will therefore be required. Training can be provided in accident investigation in courses run on site and also in numerous off-site venues. Computer-based training courses are also available. These are intended to provide refresher training on an individual basis or complete training at office sites, for example, where it may not be feasible to provide practical training.

8.4.10 *Investigation form*

Headings which could be used to compile an accident/incident investigation form are given below:

➤ date and location of accident
➤ name of anyone who was injured together with the type of injury and brief details avoiding details which would contravene the Data Protection Act
➤ circumstances of accident
➤ immediate cause of accident
➤ root or underlying cause of accident
➤ immediate action taken
➤ recommendation for further improvement
➤ report circulation list
➤ date of investigation
➤ signature of investigation team leader
➤ names of investigating team.

Follow-up
➤ were the recommendations implemented?
➤ were the recommendations effective?

An example of an investigation form using 16 different causes of accidents for analysis purposes is shown in Appendix 8.1.

8.5 Legal recording and reporting requirements

8.5.1 *Accident book*

Under the Social Security (Claims and Payments) Regulations 1979, regulation 25, employers must keep a record of accidents at premises where more than ten people are employed. Anyone injured at work is required to inform the employer and record information on the accident in an accident book, including a statement on how the accident happened.

The employer is required to investigate the cause and enter this in the accident book if they discover anything that differs from the entry made by the employee. The purpose of this record is to ensure that information is available if a claim is made for compensation.

The HSE produced a modified Accident Book BI 510 in May 2003 with notes on these Regulations and the Reporting of Injuries, Diseases and Dangerous Occurrences Regulations (RIDDOR) (see Figure 8.4), and which now complies with the Data Protection Act 1998. The Stationery Office also produces a similar accident book.

8.5.2 *RIDDOR*

RIDDOR requires employers and those in control of a site, to report certain more serious accidents and incidents to the HSE or other enforcing authority and to keep a record. There are no exemptions for small organizations. Whoever is in control of a site must report accidents to the self-employed and members of the public. A full summary of the Regulations is given in Chapter 21.

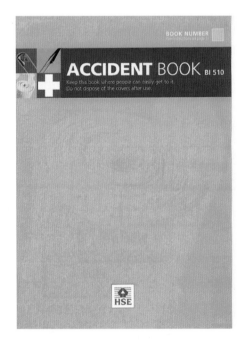

Figure 8.4 The Accident Book. BI 510 ISBN 978-0-7176-2603-2.

If a principal contractor has been appointed, contractors should promptly provide them with details of accidents, diseases or dangerous occurrences which are reportable under RIDDOR.

The reporting and recording requirements are as follows.

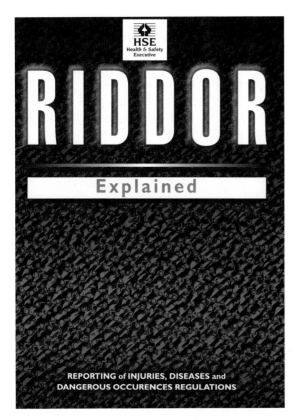

Figure 8.5 RIDDOR leaflet from HSE Books.

Death or major injury

If an accident occurs at work and:

➤ an employee, or self-employed person working on the premises is killed or suffers a major injury (including the effects of physical violence) (see Chapter 21 under RIDDOR for definition of major injury)
➤ a member of the public is killed or taken to hospital.

The responsible person must notify the enforcing authority without delay by the quickest practicable means, for example, telephone. They will need to give brief details about the organization, the injured person(s) and the circumstances of the accident and, within 10 days, the responsible person must also send a completed accident report form, F2508.

Over 3-day lost time injury

If there is an accident connected with work (including physical violence) and an employee, or self-employed person working on the premises, suffers an injury and is away from work or not doing their normal duties for more than 3 days (including weekends, rest days or holidays) but not counting the day of the accident, the responsible person must send a completed accident report form, F2508, to the enforcing authority within 10 days.

Disease

If a doctor notifies the responsible person that an employee suffers from a reportable work-related disease a completed disease report form, F2508A, must be sent to the enforcing authority. A summary is included in Chapter 21 and a full list is included with a pad of report forms. The HSE InfoLine or the Incident Contact Centre (ICC) can be contacted to check whether a particular disease is reportable.

Dangerous occurrence

If an incident happens which does not result in a reportable injury, but obviously could have done, it could be a Dangerous Occurrence as defined by a list in the Regulations (see Chapter 21 for a summary of Dangerous Occurrences). All Dangerous Occurrences must be reported immediately by, for example, telephone, to the enforcing authorities. The HSE InfoLine or the ICC can be contacted to check whether a dangerous occurrence is reportable.

A completed accident report form, F2508, must be sent to the enforcing authorities within 10 days but see information concerning the Incident Report Centre which follows.

8.5.3 *Whom to report to*

Until 2001 all reports had to be made to the local HSE office or local authority. This can still be done, but there is a centralized national system, called the ICC, which is a joint venture between the HSE and local authorities. All reports sent locally are now passed on to the ICC.

This means that employers no longer need be concerned about which authority to report to, as all incidents can be reported to the ICC directly. Reports can be sent by telephone, fax, the Internet, or by post. If reporting by the Internet or telephone a copy of the report is sent to the responsible person for correction, if necessary, and for their records.

➤ Postal reports should be sent to:
Incident Contact Centre
Caerphilly Business Park
Caerphilly
CF83 3GG
➤ For Internet reports go to:
www.riddor.gov.uk or link through
www.hse.gov.uk

➤ By telephone (charged at local rates) 0845 300 9923
➤ By fax (charged at local rates) 0845 300 9924
➤ By email: riddor@natbrit.com

8.6 Typical examples of incidents within the construction industry

8.6.1 *Accidents*

➤ falls while working at height on scaffolds, structure, roofs or ladders
➤ falling through holes in uncompleted structures or when lifting covers over holes
➤ being hit by falling objects from scaffolds or partly completed buildings
➤ nails through feet while walking over timber left lying around site

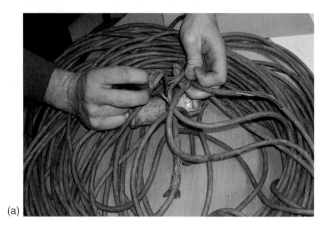

(a)

(b)

Figure 8.6 (a) Cable in a dangerous condition; (b) foot injury from upturned nails is all too common.

➤ dropping heavy items like bricks on unprotected feet
➤ hit by moving pieces of plant or inadequately guarded elevators
➤ crushed by overturning vehicle
➤ contact with live overhead power lines
➤ overcome in a confined space like a sewer.

8.6.2 *Diseases*

➤ severe dermatitis from wet cement
➤ occupational asthma from adhesives and solvents
➤ asbestosis- or asbestos-related cancer from stripping asbestos-containing materials
➤ severe back injury from lifting or manual handling of loads.

8.6.3 *Dangerous occurrences*

➤ collapse of a scaffold from overloading, incorrect design or high winds
➤ fires from welding, cutting or tar boilers
➤ overturning of cranes.

8.7 Internal systems for collecting and analysing incident data

8.7.1 *Introduction*

Managers need effective internal systems to know whether the organization is getting better or worse; to know what is happening and why; and to assess whether objectives are being achieved. Chapter 7 deals with monitoring generally, but here, the basic requirements of a collection and analysis system for incidents are discussed.

The incident report form (discussed earlier) is the basic starting point for any internal system. Each organization needs to lay down what the system involves and who is responsible to do each part of the procedure. This will involve:

➤ what type of incidents should be reported
➤ who completes the incident report form – normally the manager responsible for the investigation
➤ how copies should be circulated in the organization
➤ who is responsible to provide management measurement data
➤ how the incident data should be analysed and at what intervals
➤ the arrangements to ensure that action is taken on the data provided.

The data should seek to answer the following questions:

➤ Are failure incidents occurring, including injuries, ill-health and other loss incidents?
➤ Where are they occurring?
➤ What is the nature of the failures?
➤ How serious are they?
➤ What were the potential consequences?
➤ What are the reasons for the failures?
➤ How much has it cost?
➤ What improvements in controls and the management system are required?
➤ How do these issues vary with time?
➤ Is the organization getting better or worse?

8.7.2 *Type of incident*

Most organizations will want to collect data on:

➤ all injury accidents
➤ cases of ill-health
➤ sickness absence
➤ damage to property, personal effects and work in progress
➤ incidents with the potential to cause serious injury, ill-health or damage.

Not all of these are required by law, but this should not deter the organization that wishes to control hazards effectively.

8.7.3 *Analysis*

All the information, whether in accident books or report forms, will need to be analysed so that useful management data can be prepared. Many organizations look at analysis each month and annually. However, where there are very few incidents, quarterly may be sufficient. The health and safety information should be used alongside other business measures and should receive equal status.

There are several ways in which data can be analysed and presented. The most common ways are:

➤ by causation using the classification used on the RIDDOR form F2508. This has been used on the example accident report form (see Appendix 8.1)
➤ by nature of the injury, such as cuts, abrasions, asphyxiation, amputations
➤ by the part of the body affected, such as hands, arms, feet, lower leg, upper leg, head, eyes, back and so on. Sub-divisions of these categories could be useful if there were sufficient incidents
➤ by age and experience at the job

➤ by time of day
➤ by occupation or location of the job
➤ by type of equipment used.

There are a number of up-to-date computer recording programs which can be used to manipulate the data if significant numbers are involved. The trends can be shown against monthly, quarterly and annual past performance of, preferably, the same organization. If indices are calculated, such as Incident Rate, comparisons can be made nationally with HSE figures and with other similar organizations or businesses in the construction industry. This is really only of major value to larger organizations with significant numbers of events. (See Chapter 4 for more information on definitions and calculation of Incident Rates.)

The HSE produces annual bulletins of national performance plus a detailed statistical report, which can be used for comparisons. There are difficulties in comparisons across Europe and say, with the USA, where the definitions of accidents or time lost vary.

Reports should be prepared with simple tables and graphs showing trends and comparisons. Line graphs, bar charts and pie charts are all used quite extensively with good effect. All analysis reports should be made available to employees as well as managers. This can often be done through the health and safety committee and safety representatives, where they exist, or directly to all employees in small organizations. Other routine meetings, team briefings and notice boards can all be used to communicate the message.

It is particularly important to make sure that any actions recommended or highlighted by the reports are taken quickly and employees kept informed.

8.8 Compensation and insurance issues

Accidents arising out of the organization's activities resulting in injuries to people and incidents resulting in damage to property can lead to compensation claims. The second objective of an investigation should be to collect and record relevant information for the purposes of dealing with any claim. It must be remembered that, in the longer term, prevention is the best way to reduce claims and must be the first objective in the investigation. An overzealous approach to gathering information concentrating on the compensation aspect can, in fact, prompt a claim from the injured party where there was no particular intention to take this route before the investigation. Nevertheless, relevant information should be collected. Sticking to the collection of facts is usually the best approach.

As mentioned earlier the legal system in England and Wales changed dramatically with the introduction of the Woolf reforms in 1999. These reforms apply to injury claims. This change was feared by many because of the uncertainty and the fact that the pre-action protocols were very demanding.

The essence of the pre-legal action protocols is as follows:

1. 'Letter of claim' to be acknowledged within 21 days
2. 90 days from date of acknowledgement to either accept liability or deny. If liability is denied then full reasons must be given
3. Agreement to be reached on using a single expert.

The overriding message is that to comply with the protocols quick action is necessary. It is also vitally important that records are accurately kept and accessible.

Lord Woolf made it clear in his instructions to the judiciary that there should be very little leeway given to claimants and defendants who did not comply.

What has been the effect of these reforms on day-to-day activity?

At the present time there has been a dramatic drop in the number of cases moving to litigation. Whilst this was the primary objective of the reforms, the actual drop is far more significant than anyone anticipated. In large part this may simply be due to general unfamiliarity with the new rules and we can expect an increase in litigation as time goes by.

However, some of the positive effects have been:

➤ the elimination of speculative actions due to the requirement to fully outline the claimant's case in the letter of claim
➤ earlier and more comprehensive details of the claim allow a more focused investigation and response
➤ 'Part 36 offers' (payments into court) – these seem to be having greater effect in deterring claimants from pursuing litigation
➤ overall faster settlement.

The negative effects have principally arisen from failure to comply with the timescales particularly relating to the gathering of evidence and records and having no time therefore to construct a proper defence.

Appendix 8.2 provides a checklist of headings, which may assist in the collection of information. It is not expected that all accidents and incidents will be investigated in depth and a dossier with full information prepared. Judgement has to be applied as to which incidents might give rise to a claim and when a full record of information is required. All accident report forms should include the names of all witnesses as a

minimum. Where the injury is likely to give rise to lost time, a photograph(s) of the situation should be taken.

8.9 Practice NEBOSH questions for Chapter 8

1. (a) **Give FOUR** reasons why an organization should have a system for the internal reporting of accidents.
    (b) **Outline** factors that may discourage employees from reporting accidents at work.

2. An employee has been seriously injured after being struck by a reversing vehicle in a loading bay.
    (a) **Give FOUR** reasons why the accident should be investigated by the person's employer.
    (b) **Outline** the information that should be included in the investigation report.
    (c) **Outline FOUR** possible immediate causes and **FOUR** possible underlying (root) causes of the accident.

3. **State** the issues that should be included in a typical workplace accident reporting procedure.

4. An organization has decided to conduct an internal investigation of an accident in which an employee was injured following the collapse of storage racking.
    (a) **Outline FOUR** benefits to the organization of investigating the accident.
    (b) Giving reasons in **EACH** case, **identify FOUR** people who may be considered useful members of the investigation team.
    (c) Having defined the team, **outline** the factors that should be considered when planning the investigation.

5. (a) **Give TWO** typical examples of a 'near miss' on a construction site.
    (b) **Explain** why it is desirable to investigate and maintain records of 'near misses'.

6. **Explain** the differences between the *immediate* and the *root (underlying)* causes of an accident.

7. **Outline** the immediate and longer-term actions that should be taken following a major injury accident at work.

8. **Outline**:
    (a) the immediate; and
    (b) the longer-term actions
    that should be taken after an employee has been taken to hospital with a fractured leg following an accident on a construction site.

9. An employee slipped on a patch of oil on a warehouse floor and was admitted to hospital where he remained for several days. The oil was found close to a stack of pallets that had been left abandoned on the designated pedestrian walkway.
   (a) **Outline** the legal requirements for reporting the accident to the enforcing authority.
   (b) **Identify** the possible *immediate* **AND** *root* causes of the accident.
   (c) **Outline** ways in which management could demonstrate their commitment to improve health and safety standards in the workplace following the accident.

10. **Outline** the key points that should be covered in a training session for employees on the reporting of accidents/incidents.

11. With reference to the RIDDOR 1995:
   (a) **State** the legal requirements for reporting a fatality resulting from an accident at work to an enforcing authority.
   (b) **Outline THREE** further categories of work-related injury (other than fatal injuries) that are reportable.
   (c) **State** the requirements for reporting an 'over 3-day' injury.

12. With reference to the RIDDOR 1995:
   (a) **list FOUR** types of major injury
   (b) **outline** the procedures for reporting a major injury to an enforcing authority.

13. With reference to the RIDDOR 1995:
   (a) **Explain**, using **TWO** examples, the meaning of the term 'dangerous occurrence'.

   (b) **State** the requirements for reporting a dangerous occurrence.

14. (a) **List FOUR** types of dangerous occurrence (as specified in the RIDDOR 1995) likely to be encountered in the construction industry.
   (b) **Identify** the procedures for reporting a specified dangerous occurrence to the relevant enforcing authority.

15. (a) **Identify TWO** ill-health conditions that are reportable under the RIDDOR 1995.
   (b) **Outline** reasons why employers should keep records of occupational ill-health among employees.

16. **Outline** the procedures to be followed after an employee has been taken to hospital with a fractured leg following an accident on a construction site.

17. An employee has suffered a foot injury on a construction site. This has caused him to be absent from work for more than 3 days.
   Outline the procedures that should be followed in these circumstances.

18. An employee is claiming compensation for injuries received during an accident involving a forklift truck.
   Identify the documented information that the employer might draw together when preparing a possible defence against the claim.

Appendix 8.1 – Injury report form

To comply with the Data Protection Act 1998, personal details entered in an accident record must be kept confidential. The record sheets should be stored securely.

INCIDENT/ACCIDENT REPORT

INJURED PERSON: . . . .   Date of Accident: // 20 Time . . . . am/pm
POSITION: . . . .   Place of Incident: . . . .
DEPARTMENT: . . . .   Details of Injury: . . . .
Investigation carried out by: . . . .
Position: . . . .   Estimated Absence: . . . .

Brief details of Accident (A detailed report together with diagrams, photographs and any witness statements should be attached where necessary. Please complete all details requested overleaf.)

Immediate Causes	**Underlying Causes**

Conclusions (How can we prevent this kind of incident/accident occurring again?)

Action to be taken: **Completion Date: /   /20**

IMPORTANT

Please ensure that an accident investigation and report is completed and forwarded to Human Resources within 48 hours of the accident occurring.

Remember that accidents involving major injuries or dangerous occurrences have to be notified immediately by telephone to the HSE/Local authority.

Signature of manager making report: . . . .   Copies: Personnel Manager
Health & Safety Manager
Date: /   /20   Payroll Controller

INJURED PERSON: Surname . . . .   Forenames . . . .

Male/Female

Home address . . . .   Age . . . .

Employee ☐   Agency Temp ☐   Contractor ☐   Visitor ☐   Youth Trainee ☐   (Tick one box)

Kind of Accident Indicate what kind of accident led to the injury or condition (tick one box)

Contact with moving machinery or material being machined 1	Injured while handling lifting or carrying 5	Drowning or asphyxiation 9	Contact with electricity or an electrical discharge 13
Struck by moving, flying, or falling object 2	Slip, trip or fall on same level 6	Exposure to or contact with harmful substance 10	Injured by an animal 14
Struck by moving vehicle 3	Fall from height indicate approx distance of fall. . . . . . .m 7	Exposure to fire 11	Violence, physically assaulted by a person 15
Struck against something fixed or stationary 4	Trapped by something collapsing 8	Exposure to an explosion 12	Other kind of accident 16

Detail any machinery, chemicals, tools, etc. involved

Accident first reported to: Name . . . .

Position & . . . .   Dept . . . .

First Aid/medical attention by: First Aider Name . . . .   Dept . . . .

                                        Doctor Name . . . .

                                        Medical centre . . . .   Hospital . . . .

WITNESSES

Name	Position & Dept	Statement obtained (yes/no) Attach all statements taken
. . . .	. . . .	yes/no
. . . .	. . . .	yes/no
. . . .	. . . .	yes/no
. . . .	. . . .	yes/no

For Office use only

If relevant: Date reported to HSE/LA   a) by telephone   .../.../20

                                                    b) on form F2508   .../.../20

                    Date reported to organization's Insurers   .../.../20

Wᴇʀᴇ ᴛʜᴇ Rᴇᴄᴏᴍᴍᴇɴᴅᴀᴛɪᴏɴꜱ ᴇꜰꜰᴇᴄᴛɪᴠᴇ? **YES/NO**
Iꜰ ɴᴏ ꜱᴀʏ ᴡʜᴀᴛ ᴀᴄᴛɪᴏɴ ꜱʜᴏᴜʟᴅ ʙᴇ ᴛᴀᴋᴇɴ.

Appendix 8.2 – Information for insurance/compensation claims

Information for insurance/compensation purposes following accident or incident

Factual information needs to be collected where there is the likelihood of some form of claim either against the organization or by the organization (e.g. damage to equipment). This aspect should be considered as a second objective in accident investigation, the first being to learn from the accident to reduce the possibility of accidents occurring in the future.

Workplace claims

- accident book entry
- first aider report
- surgery record
- foreman/supervisor accident report
- safety representatives accident report
- RIDDOR report to HSE
- other communications between defendants and HSE
- minutes of Health and Safety Committee meeting(s) where accident/matter considered
- report to DSS
- documents listed above relative to any previous accident/matter identified by the claimant and relied upon as proof of negligence
- earnings information where defendant is employer.

Documents produced to comply with requirements of the Management of Health and Safety at Work Regulations 1999.

- pre-accident Risk Assessment
- post-accident Re-Assessment
- Accident Investigation Report prepared in implementing the requirements; Health Surveillance Records in appropriate cases; Information provided to employees
- documents relating to the employee's health and safety training.

WORKPLACE CLAIMS – Examples of disclosure where specific regulations apply

Section A – Workplace (Health, Safety and Welfare) Regulations 1992

- repair and maintenance records
- housekeeping records
- hazard warning signs or notices (Traffic Routes).

Section B – Provision and Use of Work Equipment Regulations 1998

- manufacturers' specifications and instructions in respect of relevant work equipment establishing its suitability to comply with Regulation 5
- maintenance log/maintenance records required; Documents providing information and instructions to employees; Documents provided to the employee in respect of training for use
- any notice, sign or document relied upon as a defence to alleged breaches dealing with controls and control systems.

Section C – Personal Protective Equipment at Work Regulations 1992

- documents relating to the assessment of the PPE
- documents relating to the maintenance and replacement of PPE
- record of maintenance procedures for PPE
- records of tests and examinations of PPE
- documents providing information, instruction and training in relation to the PPE
- instructions for use of Personal Protective Equipment to include the manufacturers' instructions.

Section D – Manual Handling Operations Regulations 1992

- manual Handling Risk Assessment carried out
- re-assessment carried out post-accident
- documents showing the information provided to the employee to give general indications related to the load and precise indications on the weight of the load and the heaviest side of the load if the centre of gravity was not positioned centrally
- documents relating to training in respect of manual handling operations and training records.

Section E – Health and Safety (Display Screen Equipment) Regulations 1992

- analysis of work stations to assess and reduce risks
- re-assessment of analysis of work stations to assess and reduce risks following development of symptoms by the claimant
- documents detailing the provision of training including training records
- documents providing information to employees.

Section F – Control of Substances Hazardous to Health Regulations

➤ risk assessments and any reviews
➤ copy labels from containers used for storage handling and disposal of carcinogens
➤ warning signs identifying designation of areas and installations which may be contaminated by carcinogens
➤ documents relating to the assessment of the Personal Protective Equipment
➤ documents relating to the maintenance and replacement of Personal Protective Equipment
➤ record of maintenance procedures for Personal Protective Equipment
➤ records of tests and examinations of Personal Protective Equipment

➤ documents providing information, instruction and training in relation to the Personal Protective Equipment
➤ instructions for use of Personal Protective Equipment to include the manufacturers' instructions
➤ air monitoring records for substances assigned a workplace exposure limit
➤ maintenance examination and test of control measures records
➤ monitoring records
➤ health surveillance records
➤ documents detailing information, instruction and training including training records for employees
➤ labels and Health and Safety data sheets supplied to the employers.

Construction law and management

9

9.1 Introduction

Management of health and safety in the construction industry is more important now than it has ever been. The number of construction projects continues to grow in terms of both new build and refurbishment. The labour force is also changing as more and more migrant workers join the industry. At the same time, the fatalities and injuries to the workforce, even when measured as an incidence rate, are increasing. So concerned are the Government and the Health and Safety Executive (HSE) with this trend that a new Framework for Action has been agreed between the Government, the HSE, the employers and the trade unions.

Figure 9.1   Building site entrance.

This chapter covers the scope and nature of construction activities and the moral, legal and financial consequences of failing to manage health and safety in the construction industry. Details of the amended Construction (Design and Management) (CDM) Regulations are given. Since these regulations now incorporate the Construction (Health, Safety and Welfare) Regulations (CHSW), they form the legal foundations for the management of health and safety in construction activities.

It is important to stress that this chapter must be read in conjunction with Chapter 1 which dealt with the health and safety foundations of law and management systems.

9.2 The scope and definition of construction

The scope of the construction industry is very wide. The most common activity is general building work, which is domestic, commercial or industrial in nature. This work may be new building work, such as a building extension or, more commonly, the refurbishment, renovation, alteration, maintenance or repair of existing buildings. The buildings may be occupied or unoccupied. Such projects may begin with a partial or total demolition of a structure which is a particularly hazardous operation. Larger civil engineering projects involving road and bridge building, water supply and sewage schemes and river and canal work all come within the scope of construction.

Most construction projects cover a range of activities such as site clearance, the demolition or dismantling of building structures or plant and equipment, the felling of trees and the safe disposal of waste materials. The work could involve hazardous operations, such as roof work or excavation, or contact with hazardous materials, such as asbestos or lead. The site activities will include the loading, unloading and storage of materials and site movements of vehicles and pedestrians. Finally, the construction processes themselves are often hazardous. These processes include fabrication, decoration, cleaning,

Figure 9.2 Ground clearance and foundations.

installation and the removal and maintenance of services (electricity, water, gas and telecommunications). Construction also includes the use of woodworking workshops together with woodworking machines and their associated hazards, painting and decorating and the use of heavy machinery. It will often require work to take place in confined spaces, such as excavations and underground chambers. At the end of most projects, the site is landscaped, which will introduce a new set of hazards.

9.3 Particular issues relating to the construction industry

Nearly all construction sites are temporary in nature and, during the construction process, are constantly changing. The workforce, itself, is transitory in nature and, at any given time, there are often many young people receiving training on site in the various construction trades. These trainees need supervision and structured training programmes. The level of numeracy and literacy among some of the workforce may make the communication of health and safety information difficult.

The progress of work on the site can be impeded by poor weather conditions leading to the possibility of contract penalties. Clients are usually keen to see the work finished as soon as possible and put additional pressure on the contractor. All these issues always lead to the temptation to compromise on health and safety issues, such as the provision of adequate welfare facilities or the safe re-routing of site traffic.

In recent years, some additional management issues have arisen with the increasing employment of migrant workers.

9.3.1 *Migrant workers in the construction industry*

It has been estimated that, in 2007, migrant labour accounted for 6% of the total construction workforce; around Greater London it represents 26% of the workforce. In the United Kingdom, five foreign workers were killed on construction sites in 2005/06 and a further five were killed in 2006/07. The average proportion of migrant worker deaths over the last 3 years was 8% of the total deaths in the industry.

It is not easy to define where the responsibility for health and safety lies – with the gangmaster or the construction employer. It is recommended that there should be liaison between the two so that all relevant issues, such as personal protective equipment (PPE), are covered. The HSE have held various events to familiarize migrants with the important aspects of the health

and safety regime in the United Kingdom and provided several translations of health and safety publications including the practical checklist in 'Health and Safety in Construction HSG150'. Construction Skills, the Sector Skills Council for construction, has developed web-based material to provide construction employers with information so that they can ensure that any migrants they employ are properly qualified, competent and safe. For many migrants there are cultural as well as language differences that need to be taken into account. Sometimes the cultural differences are more important than language differences.

Before migrants begin work, the following checks should be made by the employer:

➤ Are any special vocational skills required for the job? If so, ensure that any worker supplied has those qualifications or skills.
➤ Have they a reasonable command of written and spoken English?
➤ What information, instruction and training will need to be provided as a result of the risk assessment?

A special risk assessment is important in assessing the risks posed to migrant workers when on site. Particular issues include:

➤ the ability to speak and write English to a reasonable standard so that basic communication on site is possible
➤ the required basic competencies in terms of literacy, numeracy, skills and past work experience
➤ any physical and health attributes needed for the job
➤ the compatibility or equivalence of foreign vocational qualifications with those required in the UK
➤ the provision of necessary information, instruction, training and supervision
➤ the provision of safe equipment, substances and PPE
➤ the provision of any special welfare and emergency arrangements.

The risk assessment will need to be reviewed on a very regular basis to reflect changes in migrant nationalities and changes to work processes and procedures.

It is important to give migrant workers relevant induction and general health and safety training and to ensure that such training is fully understood. The key points are to ensure that:

➤ essential induction information is given
➤ any necessary job-related and/or vocational training is given
➤ consideration is given to the needs of workers with a poor command of English; the use of accredited

translation services will probably be needed together with visual non-verbal material, such as pictures, signs, videos and DVDs

➤ any safety critical documents, such as safe systems of work, permits-to-work, method statements and emergency procedures, are translated by a professional interpreter

➤ the workers know the identity of their supervisor and with whom they can raise any concerns about health and safety issues

➤ the workers have clearly understood all the information, instruction and training that they have received.

Finally, while it is not the responsibility of construction employers to provide transport to and from the site, it is reasonable for them to check that some form of safe transport is available to the migrant workers.

9.4 Moral, legal and financial arguments for health and safety management

9.4.1 Moral arguments

The moral arguments are reflected by the occupational accident and disease rates.

Accident rates

Accidents at work can lead to serious injury and even death. A major accident is a serious accident typically involving a fracture of a limb or a 24-hour stay in a hospital. An 'over 3-day accident' is an accident that leads to more than 3 days off work. Statistics are collected on all people who are injured at construction workplaces not just employees.

In 2006/07, the construction industry had the highest total of fatal injuries to workers (77) of all the main industrial sectors. This represented 31% of all fatal injuries of workers and a 28% increase on the number of construction industry fatalities in the previous year (2005/06). There was a concern that the increase in the number of migrant workers in the industry may be one of the causes of this increase. The sharpest increase in the number of fatalities (39) occurred in the repair, refurbishment and new-build housing sector of the industry.

Table 9.1 shows the number of accidents involving all site workers in the construction industry from 2002 to 2006.

In 2005/06, there were 59 fatal injuries to construction workers. Falls from height caused 24 deaths (41%) and being struck by a moving vehicle caused 8 deaths (14%). There were also more that 4,000 major accidents.

Table 9.1 Accidents involving all workers on site in the construction industry

Injury	2002/03	2003/04	2004/05	2005/06
Death	71	71	71	59
Major	4,721	4,728	4,486	4,419
Over 3 day	9,578	8,995	8,250	8,291

Table 9.2 shows, for the year 2005/06, the breakdown in accidents between employees, self-employed and members of the public.

Table 9.2 Accidents for different groups of people for 2005/06

	Deaths	Major	Over 3-day
Total	64	4,607	8,291
Employees	42 (66%)	3,677	7,492
Self-employed	17 (27%)	742	799
Members of the public	5 (7%)	188	n/a

Table 9.3 shows the figures for employees in the construction industry only.

Table 9.3 Accidents involving employees in the construction industry

Injury	2002/03	2003/04	2004/05	2005/06
Death	56	52	56	42
Major	4,031	3,978	3,760	3,677
Over 3 day	8,949	8,256	7,509	7,492

The main causes of fatal accidents in the construction industry in 2005/06 were:

➤ falling through roof lights and fragile roofs
➤ falling from ladders, scaffolds and other workplaces
➤ being struck by excavators, lift trucks or dumpers
➤ being struck by falling loads and equipment
➤ being crushed by collapsing structures.

In the same year, falls from height accounted for 25% of the major injuries in the construction industry. The occupations where the proportion of major injuries due to falls from height was higher included:

➤ painters and decorators (62%)
➤ bricklayers and masons (38%)
➤ scaffolders, stagers and steeplejacks (36%)
➤ electrical fitters (36%).

Injuries due to handling, lifting and carrying and being hit by moving/falling objects, each accounted for 16% of the major injuries in the construction industry.

Table 9.3 shows that fatalities, having fallen considerably to one of the lowest for the industry in 2003/04, increased again in 2004/05. Table 9.4 gives an indication of accidents in different employment sectors and includes accidents to members of the public in 2005/06.

These figures indicate that there is a need for health and safety awareness even in occupations which many would consider very low hazard, such as schools and health and social work. In fact, 70% of all deaths occur in the service sector and manufacturing appears safer than construction.

Although there has been a decrease in fatalities over recent years since 2000/01, there has been a large increase in 2006/07.

Table 9.4 Accidents to all people in various employment sectors (2005/06)

Sector	Deaths	Majors
Education	4 (1%)	5,746 (11%)
Health and social work	58 (9%)	4,150 (8%)
Service industries	437 (67%)	34,060 (62%)
Manufacturing	46 (7%)	5,554 (10%)
Construction	64 (10%)	4,607 (8%)
Agriculture	43 (6%)	597 (1%)
Total	652	54,714

Disease rates

Work-related ill-health and occupational diseases can lead to long absences from work and even death. In the construction industry, recent figures indicate that average prevalence rates for all illnesses are similar to those found in other industries with some notable exceptions.

Table 9.5 compares the annual cases seen by specialist doctors in the construction and all industries. The incidence rates show that asbestos-related diseases, musculoskeletal disorders and the effects of exposure to

Table 9.5 Annual cases and incidence rates for work related ill-health seen by the Health and Occupational Reporting Network Disease Specialists from 2002 to 2004

	Construction		All industries	
	Average annual cases	Rate per 100,000	Average annual cases	Rate per 100,000
Diffuse pleural thickening	399	19.3	976	1.5
Mesothelioma	278	13.4	819	1.0
All musculoskeletal disorders	214	10.0	2,020	7.0
Asbestosis	24	1.2	122	0.1
Dermatitis	143	7.0	1,687	6.0
Vibration white finger	86	4.1	171	0.3
Stress	28	1.0	2,223	8.0
Asthma	19	1.0	370	1.0
Occupational deafness	17	1.0	99	0.0
Infections	1	0.0	1,207	4.0

noise and vibration are far more prevalent in the construction industry. Diffuse pleural thickening of the lung lining is thought to be principally due to exposure to asbestos but can be caused by other diseases, such as tuberculosis, or certain drugs. One positive fact is that the construction industry has one of the lowest rates for stress.

In 2005/06 there were 3.2 million days lost in the construction industry due to work-related ill-health and workplace injuries. This was equivalent to 1.5 days per worker in construction compared to a rate for all industries of 1.3 days.

Not all of the injury and ill-health statistics for 2006/07, except fatalities, were available as this edition of the book went to press.

9.4.2 *Legal arguments*

In 2002, the Construction Division of the HSE undertook four construction blitzes across the country. They concentrated on falls from height and gave advance warning of their visits. Work was stopped on nearly 50% of sites visited. A number of prosecutions followed involving large as well as small companies. Thus there were some clear legal reasons for implementing sound health and safety management systems.

During the summer of 2007, the HSE conducted almost 1,600 inspections of construction sites and stopped work on 244 of them because they felt that there was a high risk of a serious injury. These inspections followed the publication of the fatality statistics for 2006/07 mentioned earlier in this chapter. The inspection emphasis was on working at height and keeping the site tidy so that accidents due to slips, trips and falls could be reduced. The HSE has made it clear that it is determined to continue its campaign to protect workers in the industry from serious injury.

The Head of Construction at the HSE said after the inspections:

'Nearly 1 in 3 construction refurbishment sites inspected put the lives of workers at risk. We stopped work on site immediately during 244 inspections because we felt there was a real possibility that life would be lost or ruined through serious injury. It is completely unacceptable that so many lives have been put at risk. Our inspectors were appalled at the apparent willingness to ignore basic safety precautions.

Figure 9.3   Bare wires on a concrete mixer wedged in with chipboard. Earth wire cut off.

'The simple fact is that despite knowing what they should be doing, too many people are prepared to allow bad practices to continue, even though last year 39 people died on refurbishment, repair and maintenance sites.

'We are determined to tackle this issue head on and will continue to take enforcement action against those rogues who flout safety precautions. Let me be clear to all those who put lives at risk – we will continue to carry out further inspections and will take all action necessary to protect workers, including closing sites and prosecution.

'Work at height remains the biggest concern. Over half of the enforcement action taken during this inspection initiative was against dangerous work at height, which last year led to the death of 23 workers.

Table 9.6 compares the number of enforcement notices served by the HSE on the construction industry with those given to all industries over a 3-year period. Although construction received only 11% of all improvement notices, it received 57% of all prohibition notices. In addition to these figures, Local Authorities, through their Environmental Health Officers, served 40% of all improvement notices and 20% of all prohibition notices.

Table 9.7 shows the number of prosecutions by the HSE over the same 3-year period. The HSE present 80% of all prosecutions and the remainder are presented by Local Authority Environmental Health Officers. For construction, most of these prosecutions are for infringements of the various Construction Regulations and the Provision and Use of Work Equipment Regulations. The fines quoted exclude all large fines so that it is easier to compare year with year. The average fine for each offence between 2003 and 2005 was £9,000.

9.4.3 *Financial arguments*

Costs of accidents

Any accident or incidence of ill-health causes both direct and indirect costs and incurs an insured and an uninsured cost. This is particularly the case in the construction industry where the cost of non-injury accidents is considerably higher than the costs of ill-health and injury accidents. A non-injury accident is an accident that causes damage or loss to property, plant, materials, the environment or a business opportunity.

Table 9.8 shows the range of costs for ill-health, injury and non-injury accidents in a study undertaken by the HSE in 1999. While the figures will have increased since then, the relative differences between them should have remained the same.

It can be seen that non-injury accident costs are 80% of the total costs and between 3 and 11 times the injury accident costs.

Table 9.6 Number of enforcement notices issued by the HSE over a 3-year period

Year	All industries		Construction	
	Improvement notice	Prohibition notice	Improvement notice	Prohibition notice
2002/03	8,140	5,184	778	2,804
2003/04	6,798	4,537	798	2,689
2004/05	5,167	3,278	548	1,933

Table 9.7 Number of prosecutions by the HSE over a 3-year period

Year	All industries			Construction		
	Offences prosecuted	Convictions	Average fine (£)	Offences prosecuted	Convictions	Average fine (£)
2003/04	1,720	1,317	6,524	617	418	6,174
2004/05	1,320	1,025	6,990	574	405	7,530
2005/06	1,012	741	6,219	448	303	4,553

Table 9.8 Range of accident costs in the construction industry

Ill-health (£ million)		Injury (£ million)		Non-injury (£ million)		Total (£ million)	
Min	Max	Min	Max	Min	Max	Min	Max
100	180	140	140	420	1,570	660	1,890

Direct costs and indirect costs

Direct costs are directly related to the accident. They may be insured (claims on employers' and public liability insurance, damage to buildings, equipment or vehicles) or uninsured (fines, sick pay, damage to product, equipment or process). **Indirect costs** may also be insured (business loss, product or process liability) or uninsured (loss of goodwill, extra overtime payments, accident investigation time, production delays).

It is important that all of these costs are taken into account when the full cost of an accident is calculated.

Section 1.15.3 covered this topic in detail and most of the costs mentioned are also relevant to construction. Probably one of the most significant costs for the construction industry is contract penalty costs.

9.5 The role of the regulator in the improvement of health and safety in the construction industry

The health and safety problem in the construction industry is its poor record when compared with the other parts of British industry. This performance deteriorated in 2000 and certain actions were taken by the HSE. A new construction division was launched in April 2002 and a new intervention strategy was developed. Clients and developers as well as construction sites have been targeted.

The Construction Industry Advisory Committee (CONIAC) to the Health and Safety Commission (HSC) has set a series of targets for the industry and these targets considerably exceed those set for the rest of industry. The targets, known as 'revitalizing targets', are:

➤ to reduce the incidence rate of fatalities and major injuries by 40% by 2004/05 and by 66% by 2009/10
➤ to reduce the incidence rate of cases of work-related ill-health by 20% by 2004/05 and by 50% by 2009/10 and

➤ to reduce the number of working days lost per 100,000 workers from work-related injury and ill-health by 20% by 2004/05 and by 50% by 2009/10.

They are ambitious targets since by aiming to reduce the incidence rate of major injuries and deaths by 66%, there will need to be 3,000 fewer reported in 2009/10 than there were in 1999/2000.

By 2004/05 the number of fatalities was 71, which were the same as for 2003/04, but the workforce had increased so that the fatal injury rate had fallen by 3%. The 40% reduction in fatalities target had therefore been met. However the reduction in the major injury and 3-day injury rates were both below the target of 40%.

As mentioned earlier in this chapter, the downward trend in fatalities was reversed in 2006/07, leading to 1,600 site inspections and the intervention of the Government. The Secretary of State for Work and Pensions convened a Construction Forum which was attended by the Government, the HSE, employers, trade unions, trade associations, contractors and suppliers. A Framework for Action was agreed to reduce the number of future fatalities and major accidents. Key areas singled out for urgent action were:

➤ sharing best practice – particularly in house building, repair and refurbishment projects
➤ ensuring that all construction workers receive induction training before starting work on a site
➤ raising competence levels of all site workers in the house building sector, enabling them to carry a Construction Skills Certification Scheme (CSCS) card, or be able to demonstrate their competence
➤ encouraging worker involvement, by ensuring that all projects include trade union and worker safety representatives
➤ tackling the problem of the informal economy and the use of casual labour.

A Strategic Forum Health and Safety Task Group is to oversee the implementation of the framework.

In contrast to the situation in the United Kingdom, the Republic of Ireland almost halved its construction fatalities in 2006 and increased its site inspections by 13%. The Irish Health and Safety Authority also introduced a safe system of work plans for house building, ground works and demolition. These plans, in the form of pictograms, must be completed and signed off by the site manager and shared with the rest of the site team.

The HSE has published on their website a list of topics that they will be checking during site visits. These include:

➤ competence and control of contractors and employees
➤ risk assessments, particularly for work at height
➤ communication of risk controls to the workforce
➤ selection, use, inspection and maintenance of equipment and
➤ welfare facilities and site tidiness.

Finally, the HSE have launched a 'Falls from Vehicles' campaign in the construction industry. Annually, approximately 200 workers in the construction industry are seriously injured and even lose their lives as a result of falling from a vehicle; 70% of all falls from vehicle injuries occur to non-drivers. Such accidents often occur during unloading of delivery vehicles.

9.6 Scope and application of the CDM Regulations

9.6.1 *Introduction*

The management of construction work, including the selection and control of contractors, is governed by the CDM Regulations 2007 known as CDM 2007. CDM Regulations apply to the whole of a construction project, from the initial feasibility study to the hand over of the completed structure to the customer.

CDM 2007 came into force on 6th April 2007. They replace both the Construction (Design and Management) Regulations 1994 (CDM94) and the Construction (Health, Safety and Welfare) Regulations 1996 (CHSW). The new Approved Code of Practice replaces the ACOP under CDM94.

The key aim of the CDM 2007 is to integrate health and safety into the management of the project and to encourage everyone involved to work together.

Figure 9.4 Design and management of construction work.

CDM 2007 is split into several main parts:

➤ Part 2 covers all construction projects and sets out general management duties.

➤ Part 3 sets out additional duties where the project is notifiable.

➤ Part 4 sets out the duties relating to health and safety on construction sites and Schedule 2, welfare arrangements. (Largely the former Construction (Health, Safety and Welfare) Regulations 1996).

A detailed summary of the CDM 2007 is given in Section 21.10. Table 9.9 shows a summary of the responsibilities of various duty holders.

9.6.2 Key differences between CDM 94 and CDM 2007

Regulations in CDM 2007 have been re-ordered to group duties together by duty holder and to show whether individual provisions apply to all projects, only notifiable projects, or only non-notifiable projects. (The definition of 'Notifiable' is unchanged, but the application provision relating to fewer than five workers on site has been removed.)

The construction health and safety requirements formerly in the Construction (Health, Safety and Welfare) Regulations have now been included within the CDM 2007 as Part 4 and Schedule 2.

Work for domestic clients no longer needs to be notified.

Under CDM 2007, a group of clients involved in a project can now elect one or more of its members to be the only client(s).

The 'CDM co-ordinator' is introduced to support and advise the client and to co-ordinate design and planning. Therefore, the role of the planning supervisor ceases to exist.

The appointment of a CDM co-ordinator, principal contractor and preparation of a written health and safety plan are only required for notifiable projects (but demolition work requires a written system of work).

Duty holders cannot accept an appointment/ engagement unless they are competent to carry it out. Additionally, they cannot arrange for, or instruct, anyone to carry out or manage design or construction work, unless that person is competent (or being supervised by someone who is competent).

Assessment and demonstration of competence is simplified. The new ACOP (L144 Managing Health and Safety in Construction) has specific sections on individual and corporate competence.

Everyone involved in a project has general co-operation and co-ordination duties (relating to others on the same or adjoining sites). This is a specific requirement to implement any preventive and protective measures,

on the basis of the principles of prevention specified in the MH&SW Regulations.

Clients now have the duty (which already exists in the HSW Act and MH&SW Regulations) to take reasonable steps to ensure that managerial arrangements made by duty holders (including time and other resources) enable construction work to be carried out without risk to health or safety. Clients have the duty to ensure that the arrangements are maintained and reviewed throughout the project.

Clients must tell designers and contractors how much time they have, before the start of work on site, for planning and preparing construction work.

9.6.3 Explanation of terms used in the CDM 2007 and ACOP

(a) **A Client** is an organization or individual for whom a construction project is carried out. Clients only have duties when the project is associated with a business or other undertaking (whether for profit or not). This can include, for example, local authorities, school governors, insurance companies and project originators or Private Finance Initiative (PFI) projects. Domestic clients are a special case and do not have duties under CDM 2007.

(b) **Construction work** See Chapter 21 for more information on the definition of construction.

(c) **Designer** Designers are those who have a trade or a business which involves them in:

➤ preparing designs for construction work, including variations; this includes preparing drawings, design details, specifications, bills of quantities and the specification (or prohibition) of articles and substances, as well as all the related analysis, calculations, and preparatory work or

➤ arranging for their employees or other people under their control to prepare designs relating to a structure or part of a structure.

it does not matter whether the design is recorded (e.g. on paper or a computer) or not (e.g. it is only communicated orally).

Designers may include:

➤ architects, civil and structural engineers, building surveyors, landscape architects, other consultants, manufacturers and design practices (of whatever discipline) contributing to, or having overall responsibility for, any part of the design (e.g. drainage engineers)

➤ anyone who specifies or alters a design, or who specifies the use of a particular method of work or material, such as a design manager, quantity surveyor who insists on specific material, or a client who stipulates a particular layout for a new building

Table 9.9 Summary of the responsibilities of various duty holders under CDM 2007

	All construction projects (Part 2 of the Regulations)	Additional duties for notifiable projects (Part 3 of Regulations)

Clients (excluding domestic clients)	➤ Check competence and resources of all appointees.
	➤ Ensure there are suitable management arrangements for the project including welfare facilities.
	➤ Allow sufficient time and resources for all stages.
	➤ Provide pre-construction information to designers and contractors.
		➤ Appoint CDM co-ordinator.*
		➤ Appoint principal contractor.*
		➤ Provide pre-construction information to CDM co-ordinator.
		➤ Make sure that the construction phase does not start unless there are suitable:
		  ➤ welfare facilities and
		  ➤ a construction phase plan in place.
		➤ Provide information relating to the health and safety file to the CDM co-ordinator.
		➤ Retain and provide access to the health and safety file.
		(*There must be a CDM co-ordinator and principal contractor until the end of the construction phase)

CDM Co-ordinators		➤ Advise and assist the client with his/her duties. ➤ Notify HSE. ➤ Co-ordinate health and safety aspects of design work and co-operate with others involved in the project. ➤ Facilitate good communication between client, designers and contractors. ➤ Liaise with principal contractor re ongoing design. ➤ Identify, collect and pass on pre-construction information. ➤ Prepare/update health and safety file.
Designers	➤ Eliminate hazards and reduce risks due to design. ➤ Provide information about remaining risks.	➤ Check client is aware of duties and CDM co-ordinator has been appointed. ➤ Check HSE has been notified. ➤ Provide any information needed for the health and safety file.
Principal contractors		➤ Plan, manage and monitor construction phase in liaison with contractors. ➤ Prepare, develop and implement a written plan and site rules. (Initial plan completed before the construction phase begins.) ➤ Give contractors relevant parts of the plan. ➤ Make sure suitable welfare facilities are provided from the start and maintained throughout the construction phase. ➤ Check competence of all their appointees. ➤ Ensure all workers have site inductions and any further information and training needed for the work. ➤ Consult with the workers. ➤ Liaise with CDM co-ordinator re ongoing design. ➤ Secure the site.
Contractors	➤ Plan, manage and monitor own work and that of workers. ➤ Check competence of all their appointees and workers. ➤ Train own employees. ➤ Provide information to their workers. ➤ Comply with requirements in Part 4 of the Regulations. ➤ Ensure there are adequate welfare facilities for their workers.	➤ Check client is aware of duties and a CDM co-ordinator has been appointed and HSE notified before starting work. ➤ Co-operate with principal contractor in planning and managing work, including reasonable directions and site rules. ➤ Provide details to the principal contractor of any contractor whom he engages in connection with carrying out the work. ➤ Provide any information needed for the health and safety file. ➤ Inform principal contractor of problems with the plan. ➤ Inform principal contractor of reportable accidents, diseases and dangerous occurrences.
Everyone	➤ Check own competence. ➤ Co-operate with others and co-ordinate work so as to ensure the health and safety of construction workers and others who may be affected by the work. ➤ Report obvious risks. ➤ Comply with requirements in Schedule 2 (ACOP says schedule 3 but believed a mistake) and Part 4 of the Regulations for any work under their control.	

➤ building service designers, engineering practices or others designing plant which forms part of the permanent structure (including lifts, heating, ventilation and electrical systems), for example a specialist provider of permanent fire extinguishing installations

➤ those purchasing materials where the choice has been left open, for example those purchasing building blocks and so deciding the weights that a bricklayer must handle

➤ contractors carrying out design work as part of their contribution to a project, such as an engineering contractor providing design, procurement and construction management services

➤ temporary works engineers, including those designing auxiliary structures, such as formwork, falsework, façade retention schemes, scaffolding and sheet piling

➤ interior designers, including shopfitters, who also develop the design

➤ heritage organizations, who specify how work is to be done in detail, for example providing detailed requirements to stabilize existing structures

➤ those determining how buildings and structures are altered (e.g. during refurbishment) where this has the potential for partial or complete collapse.

(d) **CDM co-ordinator as with e, f, g below** means the person appointed as the CDM co-ordinator under regulation 14(1) for notifiable projects only. The CDM co-ordinator provides clients with a key project adviser in respect of construction health and safety risk management matters. Their main purpose is to help clients to carry out their duties; to co-ordinate health and safety aspects of the design work and to prepare the health and safety file.

(e) **Principal contractor** is the contractor appointed by the client for notifiable projects. The principal contractor can be an organization or an individual, and is the main or managing contractor. A principal contractor's key duty is to co-ordinate and manage the construction phase to ensure the health and safety of everybody carrying out construction work, or who are affected by the work.

(f) **Contractor** is any person (including a client, principal contractor or other person referred to in the regulations) who, in the course or furtherance of a business, carries out or manages construction work.

(g) **Pre-construction health and safety information** must be provide by the client to designers and contractors and for notifiable projects the CDM co-ordinator, specific health and safety information needed to identify the hazards and risks associated with the design and construction work. The information needs to be identified, assembled and sent out in good time, so that those who need it when preparing a bid or preparing for work can decide what resources will be needed to enable design, planning and construction work to be organized and carried out properly.

The topics which should be addressed in the pre-construction information are given in Appendix 9.2

(h) **Construction Phase health and safety plan** Principal contractors must set out the way in which the construction phase will be managed and the key health and safety issues for the particular project must be set out in writing. The health and safety plan should set out the organization and arrangements that have been put in place to manage risks and co-ordinate the work on site. It should not be a repository for detailed generic risk assessments, records of how decisions were reached or detailed method statements, but it may, for example, set out when such documents will need to be prepared. It should be well focused, clear and easy for contractors and others to understand – emphasizing key points and avoiding irrelevant material. It is crucial that all relevant parties are involved and co-operate in the development and implementation of the plan as work progresses.

The plan must be tailored to the particular project. Generic plans will not satisfy the requirements of regulation 23. Photographs and sketches can greatly simplify and shorten explanations. It should also be organized so that relevant sections can easily be made available to designers and contractors.

The Construction phase health and safety plan for the initial phase of the construction work must be prepared before any work begins. Later details may need to be added as work progress and full designs become available.

See Appendix 9.3 for details of what should be included in the construction phase health and safety plan.

(i) **Health and safety file** this is a record of information for the client to retain which focuses on health and safety. It alerts those who are responsible for the structure and equipment in it to the significant health and safety risks that will need to be dealt with during subsequent use, construction, maintenance, cleaning, repair, alterations, refurbishment and demolition work. See later for further information on the file.

(j) **Method statement** is a written document laying out the work procedures and sequences of operations to ensure health and safety. It results from the risk assessment carried out for the task or operation and the control measures identified. If the risk is low, a verbal statement may suffice. More information is contained in Chapter 6

(k) **Notifiable work** Construction work (except for a domestic client) is notifiable to the HSE if it lasts longer than 30 working days or will involve more than 500 person-days of work. (For example, 50 people working for 10 days). Holidays and weekends do not count if no construction work takes place on those days. The CDM co-ordinator is responsible to make the notification as soon as possible after their appointment for a particular project. The notice must be displayed where it can be read by people working on the site. A F10(rev) may be used and is available from local HSE offices or can be completed online. This form does not have to be used as long as all the particulars shown in Schedule 1 (see Figure 21.2 in Chapter 21) are provided.

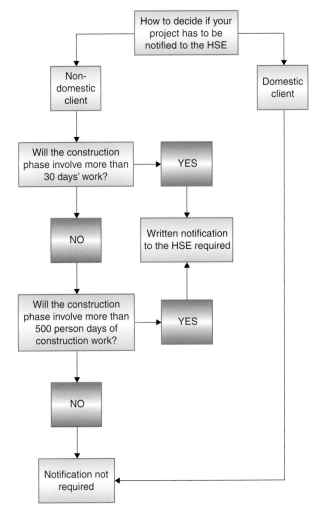

Figure 9.5 Notification requirements.

For notifiable projects, designers are prohibited from doing anything more than initial design work before the CDM co-ordinator has been appointed. In preparing or modifying a design they are required, so far as is reasonably practicable, to avoid risks to the health or safety of any person using a structure designed as a workplace. They must eliminate hazards which may give rise to risks and reduce risks from any remaining hazards.

9.6.4 *Part 1 non-notifiable project responsibilities of duty holders*

General

A summary of duties is shown in Table 9.9. Also see the flow diagram in Appendix 9.1. The ACOP for CDM 2007

covers the competence of each duty holder in some depth. This is an essential part of the effective management of construction projects. The ACOP should be consulted for further details.

Co-ordination and co-operation are the two main requirements for all members of the project team. However, there is no requirement for the appointment of a CDM co-ordinator or principal contractor, or a construction-phase health and safety plan for non-notifiable projects. For low risk projects a low key approach will suffice. However, for higher-risk projects, for example those involving demolition a more rigorous approach is required. The architect or lead designer should normally lead the co-ordination of the design work while the main contractor or builder should co-ordinate the construction work.

A brief summary plan may be all that is required in many cases, but where the risks are higher, something close to the construction phase plan will be needed. For example where work involves:

➤ structural alterations
➤ deep excavations, and those in unstable or contaminated ground
➤ unusual working methods or safeguards
➤ ionizing radiation or other significant health hazards
➤ close proximity to high voltage power lines
➤ a risk of falling into water which is, or may become, fast flowing
➤ diving
➤ explosives
➤ heavy or complex lifting operations.

For all **demolition work,** those in control of the work are required to produce a written plan showing how danger will be addressed.

Clients

The role of the client has been given a higher profile in CDM 2007 to ensure that the construction project has sufficient leadership. Clients are clearly accountable for the effect that their attitudes to health and safety has on those working on, or affected by, the project.

Clients are not expected to manage or plan projects themselves. They are not expected to develop substantial expertise in construction health and safety, unless this is central to their business. Clients must ensure that various things are done, but are not normally expected to do them alone. If help is needed this should be available from the competent person appointed under regulation 7 of the Management of Health and Safety at Work Regulation. Alternatively, clients could seek advice from someone who has acted as a CDM co-ordinator on other projects. (The appointment of a CDM co-ordinator is **not** a requirement for non-notifiable projects.)

Figure 9.6 (a) Notifiable over 30 days but domestic client.

Clients must make sure that:

➤ designers, contractors and others whom they propose to engage are competent or work under the supervision of competent people, and are adequately resourced and appointed early enough to fulfil their duties
➤ they allow sufficient time for each stage of the project
➤ they co-operate with others concerned as necessary to allow people to fulfil their duties
➤ they co-ordinate their own work with others to ensure the safety of those carrying out the construction work
➤ there are reasonable management arrangements in place throughout the project to ensure the work can be carried out (SFARP) safely and without risks to health (this does not mean managing the risk themselves as few clients have the expertise)
➤ contractors have made arrangements for suitable welfare facilities to be provided from the start and throughout the project

➤ any fixed workplaces which are to be constructed comply with the Workplace (Health Safety and Welfare) Regulations
➤ relevant information (pre-construction information) likely to be needed by designers, contractors or others to plan and manage the work is passed to them.

The client will need to ensure that arrangements are in place to ensure that:

➤ there is clarity as to the roles, functions and responsibilities of members of the project team
➤ those with duties under the regulations have sufficient time and resources to comply with their duties
➤ there is good communication, co-ordination and co-operation between members of the project team (e.g. between designers and contractors)
➤ designers are able to confirm that their designs (and any design changes) have taken account of the

Figure 9.6 (b) Non-notifiable domestic client but only 3 days' work.

requirements of Regulation 11 (designers' duties), and that the different design elements will work together in a way which does not create risks to the health and safety of those constructing, using or maintaining the structure

➤ that the contractor is provided with the pre-construction information

➤ contractors are able to confirm that health and safety standards on site will be controlled and monitored, and welfare facilities will be provided by the contractor from the start of the construction phase through to handover and completion.

Most of these arrangements will be made by others in the project team, such as designers and contractors. Before they start work, a good way of checking is to ask the relevant members of the team to explain their arrangements, or to ask for examples of how they will manage these issues during the life of the project. When discussing roles and responsibilities, on simple projects all that may be needed is a simple list of who does what.

Having made these initial checks before work begins, clients should monitor the project to ensure that the arrangements which have been made are maintained. For non-notifiable projects, only simple checks will be needed, for example:

➤ checking that there is adequate protection for the client's workers and/or members of the public

➤ checking to make sure that adequate welfare facilities have been provided by the contractor

➤ checking that there is good co-operation and communication between designers and contractors

➤ asking for confirmation from the contractor that the arrangements that they agreed to make have been implemented.

Most clients on non-notifiable projects should be able to carry out these checks for themselves.

Figure 9.7 Protection of the public in main shopping area.

Designers

CDM 2007 recognizes the key role the designers have in construction projects for health and safety. The aim of the Regulations is to ensure that the designers do not produce designs which cannot be constructed, used or maintained safely and with proper consideration for health issues. The amount of effort put in to eliminating hazards and reducing risk should be proportional to the degree of risk.

Designers should:

➤ make sure that they are competent and adequately resourced to address the health and safety issues likely to be involved in the design

➤ check that clients are aware of their duties

➤ when designing the project, avoid foreseeable risks to those involved in the construction and future use of the structure and, in so doing, they should eliminate hazards So Far As is Reasonably Possible (SFARP), and reduce risks associated with the hazards that remain

➤ provide adequate information about any significant risks associated with the design

➤ co-ordinate their work with that of others in order to improve the way in which risks are managed and controlled.

Designers need to consider risks to those carrying out the construction work including demolition; those cleaning and maintaining the structure (particularly windows or translucent panels); those who use the structure for a workplace; and those who may be affected by the work (e.g. the public or other customers).

Designers may not commence work in relation to the project unless their client is aware of their duties under the Regulations. This, in turn, will help to ensure that the client's requirements are clearly understood by encouraging discussion and co-operation.

These duties apply whenever designs are prepared which may be used in construction work in Great Britain. This includes concept design, competitions, bids for grants, modifications of existing designs and relevant work carried out as part of feasibility studies. It does not matter whether or not planning permission or funds have been secured; the project is notifiable or high risk; or the client is a domestic client.

Contractors

Contractors and their employees, those actually doing the construction work, are most at risk of injury and ill-health. They have a key role to play, in co-operation with other duty holders planning and managing the work, to ensure that risks are identified and are properly controlled.

Contractors may include utilities, specialist contractors, contractors nominated by the client and self-employed persons. Contractors are often sub-contractors to the principal contractor. Contractors may also have duties as designers if they are involved in designing elements of their work, such as precast concrete planks or curtain walling.

Anyone who directly employs or engages construction workers, or manages construction work, is a contractor for the purposes of these regulations. This includes companies that use their own workforce to carry out construction work on their own premises.

For all projects contractors must:

➤ check that the clients are aware of their duties
➤ satisfy themselves that they and anyone they employ or engage are competent and adequately resourced
➤ plan, manage and monitor their own work to make sure that workers under their control are safe from the start of their work
➤ ensure that any contractor whom he appoints or engages to work on the project is informed of the minimum amount of time which will be allowed for them to plan and prepare before starting work
➤ provide workers under their control (whether employed or self-employed) with any necessary information, including relevant aspects of other contractors' work, and site induction which they need to work safely, to report problems or to respond appropriately in an emergency
➤ ensure that any design work they do complies with the requirements on designers in regulation 11

Figure 9.8 Contractors at work.

➤ comply with any relevant requirement in part 4 and schedule 2 of the Regulations regarding health and safety on construction sites

➤ co-operate with others and co-ordinate their own work with others working on the project

➤ obtain specialist advice where necessary when planning high-risk work – for example, alterations that could result in a structural collapse or work on contaminated ground.

Contractors should always plan, manage, supervise and monitor their own work and that of their workers to ensure that it is carried out safely and that health risks are also addressed. The effort invested in this should reflect the risk involved and the experience and track record of the workers involved. Where contractors identify unsafe practices, they must take appropriate action to ensure health and safety.

9.6.5 *Additional things duty holders must do for notifiable projects*

Clients

For notifiable projects the client has the following additional duties:

➤ appoint a CDM co-ordinator to advise and assist with their duties and to co-ordinate the arrangements for health and safety during the planning phase

➤ appoint a principal contractor to plan and manage the construction work – preferably early enough for them to work with the designer on issues relating to buildability, usability and maintainability

➤ ensure that the construction phase does not start until the principal contractor has prepared a suitable health and safety plan and made arrangements for suitable welfare facilities to be present from the start of the work

➤ make sure the health and safety file is prepared, reviewed, or updated ready for handover at the end of the construction work. This must be kept available for any future construction work or to pass on to a new owner.

Early appointments are essential as the client is likely to have to rely on the advice from the CDM co-ordinator on the competence of appointees and the adequacy of the arrangements. If a client does not make these appointments they are liable for the work of the CDM co-ordinator and the principal contractor as well as not making the appointment.

CDM co-ordinators (notifiable projects only)

A new role of CDM co-ordinator has been created by CDM 2007. It is a wider role than that of the planning supervisor, which it replaces. The CDM co-ordinator is effectively the client's health and safety adviser for the project.

The role involves advising and assisting the client in undertaking the measures needed to comply with CDM 2007, in particular, the client's duties both at the start of the construction phase and during it.

CDM co-ordinators must:

➤ give suitable and sufficient advice and assistance to clients in order to help them to comply with their duties, in particular:
  ➤ the duty to appoint competent designers and contractors and
  ➤ the duty to ensure that adequate arrangements are in place for managing the project.
➤ notify the HSE about the project
➤ co-ordinate design work, planning and other preparation for construction where relevant to health and safety
➤ identify and collect the pre-construction information and advise the client if surveys need to be commissioned to fill significant gaps
➤ promptly provide, in a convenient form, to those involved with the design of the structure, and to every contractor (including the principal contractor) who may be or has been appointed by the client, such parts of the pre-construction information which are relevant to each
➤ manage the flow of health and safety information between clients, designers and contractors
➤ advise the client on the suitability of the initial construction phase plan and the arrangements made to ensure that welfare facilities are on site from the start
➤ produce or update a relevant, user friendly, health and safety file suitable for future use at the end of the construction phase.

The CDM co-ordinator should assist with the development of these arrangements, and should advise the clients on whether or not the arrangements are adequate. They should assist the client with decisions about for how much time a contractor will need to prepare before construction work begins. When advising and assisting the client, the following issues should be considered:

1. Is the client aware of their duties and do they understand what is expected of them?
2. Has the client prepared relevant information about the site in order to meet their duties under regulation 10?
3. Have the necessary appointments been made, and has the project been notified?
4. Is there an established project team who meet regularly to discuss and co-ordinate activities in relation to the project?

5. Are project team members clear about their roles and responsibilities?
6. Are there arrangements in place for co-ordinating design work and reviewing the design to ensure that the requirements in regulation 11 are being addressed?
7. Are there arrangements in place for dealing with late changes to the design, and for co-operating with contractors, so that problems are shared?
8. Has the principal contractor been given enough time to plan and prepare for the work, and mobilize for the start of the construction phase?
9. Has the principal contractor made arrangements for providing welfare facilities on site from the outset, and have they prepared a construction phase plan that addresses the main risks during the early stages of construction?
10. Are there suitable arrangements for developing the plan to cover risks that arise as the work progresses?
11. Has the format for the health and safety file been agreed, and are arrangements in place for collecting the information which it will contain?
12. Has the principal contractor put in place suitable arrangements for consulting with workers on site; for carrying out site induction; and for ensuring that workers are adequately trained and supervised?

Not all of these questions will need answers at the start of the project, and the arrangements will need to evolve as the project develops. The key thing is to plan ahead so that arrangements are in place before the risks that need managing materialize on site.

Designers

In addition to the duties outlined previously, when the project is notifiable, designers should:

➤ check that the client has appointed a CDM co-ordinator
➤ ensure that they do not start design work other than initial design work unless a CDM co-ordinator has been appointed
➤ co-operate with the CDM co-ordinator, principal contractor and with any other designers or contractors as necessary for each of them to comply with their duties; this includes providing any information needed for the pre-construction information or health and safety file.

For a notifiable project, designers need to ensure that a CDM co-ordinator has been appointed. If appointment has been done, then designers can assume that the client is aware of their duties.

Principal contractors (notifiable projects only)

The principal contractor is the key duty holder, who is required to ensure effective management of health and safety throughout the construction phase of the project. Their main duty is to properly plan, manage and co-ordinate work during the construction phase in order to ensure that hazards are identified and risks are properly controlled.

The principal contractor has a duty to liaise with all duty holders and the workforce. In particular:

➤ consulting with the workforce – directly or via their (sub) contractors
➤ co-operating with designers and CDM co-ordinators, particularly if any changes occur to the design
➤ ensuring clients are aware of their duties.

Whilst the principal contractor is under a duty to co-operate and have systems that allow and facilitate co-operation, the duty and responsibility for managing health and safety in the construction phase, lies clearly with the principal contractor.

Principal contractors must be competent to carry out the work they are engaged to do in a safe manner, and ensure they give proper consideration to the potential effects of their activities on everyone who may be affected. Principal contractors are required to demonstrate to the client that they have sufficient resources, including properly trained and experienced staff, to carry out the project.

It is essential that principal contractors are fully aware of the duties of other CDM duty holders, so that they know the level of information they may reasonably expect. Principal contractors must recognize that time is a resource and that they must be allowed to have reasonable time to plan activities with proper regard to health and safety.

Should design changes occur, the principal contractors must allow the CDM co-ordinator to carry out their duties, but must at all times retain responsibility for managing their activities and those of their contractors and sub-contractors. The principal contractor must be in control of the site for clear commercial responsibility as well as for health and safety reasons.

Principal Contractors must:

1. satisfy themselves that clients are aware of their duties, that a CDM co-ordinator has been appointed and the HSE notified before they start work
2. make sure that they are competent to address the health and safety issues likely to be involved in the management of the construction phase; attaining a NEBOSH National Certificate in Construction Safety and Health is recommended

3. ensure that the construction phase is properly planned, managed and monitored, with adequately resourced, competent site management appropriate to the risk and activity

4. ensure that every contractor who will work on the project is informed of the minimum amount of time which they will be allowed for planning and preparation before they begin work on site

5. ensure that all contractors are provided with the information about the project that they need to enable them to carry out their work safely and without risk to health. Requests from contractors for information should be met promptly

6. ensure safe working and co-ordination and co-operation between contractors

7. ensure that a suitable construction phase plan ('the plan') is:
   - ➤ prepared before construction work begins
   - ➤ developed in discussion with, and communicated to, contractors affected by it
   - ➤ implemented
   - ➤ kept up to date as the project progresses

8. satisfy themselves that the designers and contractors that they engage are competent and adequately resourced

9. ensure suitable welfare facilities are provided from the start of the construction phase

10. take reasonable steps to prevent unauthorized access to the site

11. prepare and enforce any necessary site rules

12. provide (copies of or access to) relevant parts of the plan and other information to contractors, including the self-employed, in time for them to plan their work

13. liaise with the CDM co-ordinator on design carried out during the construction phase, including design by specialist contractors, and its implications for the plan

14. provide the CDM co-ordinator promptly with any information relevant to the health and safety file

15. ensure that all the workers have been provided with suitable health and safety induction, information and training

16. ensure that the workforce is consulted about health and safety matters

17. display the project notification.

Figure 9.9 Barriets to prevent unauthorised access to construction work.

Contractors and the self-employed

In the case of notifiable projects, contractors must also:

1. check that a CDM co-ordinator has been appointed and HSE notified before they start work (having a copy of the notification of the project to HSE (form 10), is normally sufficient)
2. co-operate with the principal contractor, CDM co-ordinator and others working on the project or adjacent sites
3. tell the principal contractor about risks to others created by their work
4. provide details to the principal contractor of any contractor whom they engage in connection with carrying out the work
5. comply with any reasonable directions from the principal contractor, and with any relevant rules in the construction phase plan
6. inform the principal contractor of any problems with the plan or risks identified during their work that have significant implications for the management of the project
7. tell the principal contractor about accidents and dangerous occurrences
8. provide information for the health and safety file.

Contractors must co-operate with the principal contractor, and assist them in the development of the construction phase plan and its implementation. Where contractors identify shortcomings in the plan, the contractor should inform the principal contractor.

9.6.6 *Site induction, information and training*

Contractors must not start work on a construction site until they have been provided with basic information. This should include information from the client about any particular risks associated with the project (including information about existing structures where these are to be demolished or structurally altered), and from designers about any significant risks associated with the design.

Contractors must ensure, so far as is reasonably practicable, that every worker has:

➤ a suitable induction and
➤ any further information and training needed for the particular work.

Further advice on training and competence is given in the ACP.

Inductions are a way of providing workers with specific information about the particular risks associated with the site and the arrangements which have been put in place for their control. On non-notifiable sites, induction will need to be provided by the contractor, or by arrangement with the main contractor on site.

Induction is not intended to provide general health and safety training, but it should include site-specific issues as covered in Chapter 10.

9.6.7 *Workers*

Workers, along with all others involved in the life of a project, have duties to co-operate and to co-ordinate with others. The term 'worker' includes managers and supervisors.

Workers need to be involved as soon as possible and should:

➤ give feedback to their employer via the agreed consultation method
➤ provide input on risk assessments and developing a method statement
➤ work to the agreed method statement or approach their employer to discuss implementing any change or improvement
➤ use welfare facilities with respect
➤ keep tools and PPE in good condition
➤ be vigilant for hazards and risks and keep management
➤ be aware of arrangements and actions to take if a dangerous situation arises.

9.6.8 *Health and safety file (notifiable projects only)*

The CDM co-ordinator has the responsibility to prepare a suitable health and safety file or update it if one already exists.

Under the CDM94 a separate health and safety file was required for each project. CDM 2007 requires one file for each site, structure or, occasionally, group of structures (e.g. bridges along a road). The file can then be developed over time as information is added from different projects.

The health and safety file should contain the information needed to allow future construction work, including cleaning, maintenance, alterations, refurbishment and demolition to be carried out safely. Information in the file should alert those carrying out such work to risks, and should help them to decide how to work safely. The file should be useful to:

➤ clients, who have a duty to provide information about their premises to those who carry out work there
➤ designers during the development of further designs or alterations

➤ CDM co-ordinators preparing for construction work
➤ principal contractors and contractors preparing to carry out or manage such work.

The file should form a key part of the information that the client, or the client's successor, is required to provide for future construction projects under Regulation 10. The file should therefore be kept up to date after any relevant work or surveys.

The scope, structure and format for the file should be agreed between the client and CDM co-ordinator at the start of a project. There can be a separate file for each structure, one for an entire project or site, or one for a group of related structures. The file may be combined with the Building Regulations Log Book, or a maintenance manual provided that this does not result in the health and safety information being lost or buried. What matters is that people can find the information they need easily and that any differences between similar structures are clearly shown.

The potential practical value of the information contained in the file is also likely to increase as more clients make use of the Internet to share this information with designers and contractors. (For example, maintenance contractors could check what access equipment and parts they are likely to need to repair a fault before leaving for the site, saving them and their clients inconvenience, time and money.)

Clients, designers, principal contractors, other contractors and CDM co-ordinators all have legal duties in respect of the health and safety file:

➤ CDM co-ordinators must prepare, review, amend or add to the file as the project progresses, and give it to the client at the end of project.
➤ Clients, designers, principal contractors and other contractors must supply the information necessary for compiling or updating the file.
➤ Clients must keep the file to assist with future construction work.
➤ Everyone providing information should make sure that it is accurate, and provided promptly.

A file must be produced or updated (if one already exists) as part of all notifiable projects. For some projects (e.g. redecoration using non-toxic materials) there may be nothing of substance to record. Only information likely to be significant for health and safety in future work need be included. The NHBC Purchaser Manual provides suitable information for developers to give to householders. A file on the whole structure is not required if a project only involves a small amount of construction work on part of the structure.

The client should make sure that the CDM co-ordinator compiles the file. In some cases (e.g. design and build

contracts), it is more practical for the principal contractor to obtain the information needed for the file from the specialist contractors. In these circumstances the principal contractor can assemble the information and give it to the CDM co-ordinator as the work is completed.

It can be difficult to obtain information for the file after designers or contractors have completed their work. What is needed should be agreed in advance to ensure that the information is prepared and handed over in the required form and at the right time.

The contents of the health and safety file are given in the ACOP as follows:

➤ a brief description of the work carried out
➤ any residual hazards which remain and how they have been dealt with (e.g. surveys or other information concerning asbestos; contaminated land; water-bearing strata; buried services)
➤ key structural principles (e.g. bracing, sources of substantial stored energy – including pre- or post-tensioned members) and safe working loads for floors and roofs, particularly where these may preclude placing scaffolding or heavy machinery there
➤ hazardous material used (e.g. lead paint, pesticides, special coatings which should not be burnt off)
➤ information regarding the removal or dismantling of installed plant and equipment (e.g. any special arrangements for lifting, order of or other special instructions for dismantling)
➤ health and safety information about equipment provided for cleaning or maintaining the structure
➤ the nature, location and markings of significant services, including underground cables; gas supply equipment; firefighting services
➤ information and as-built drawings of the structure, its plant and equipment (e.g. the means of safe access to and from service voids or fire doors and compartmentalization).

9.6.8 *Health, safety and welfare requirements*

Part 4 of the Regulations relates to health and safety on construction sites. Schedule 2 covers welfare arrangements. These parts of CDM 2007 effectively cover the requirements previously contained in the now revoked Construction (Health, Safety and Welfare) Regulations 1996.

Chapter 21 covers these requirements in more detail in the summary of the Regulations. Most of the issues are covered in Chapters 10–20.

They deal with:

1. safe place of work
2. good order and site security
3. stability of structures

4. demolition and dismantling
5. explosives
6. excavations
7. cofferdams and caissons
8. reports of inspections
9. energy distribution installations
10. prevention of drowning
11. traffic routes
12. vehicles
13. prevention of risk from fire
14. emergency procedures
15. emergency routes and exits
16. fire detection and firefighting
17. fresh air
18. temperature and weather protection
19. lighting
20. sanitary conveniences
21. washing facilities
22. drinking water
23. changing rooms and lockers
24. facilities for rest.

9.7 Sources of information on health and safety in construction

9.7.1 Genereal sources of information

All the sources of information outlined in Chapter 1 (1.14) are relevant to the construction industry in addition to the following:

➤ construction legislation (specific Regulations are now the CDM 2007), and the Construction (Head Protection) Regulations 1909 (see Chapter 21).

➤ specific construction publications and reports from the HSC/E. See the HSE's micros site for Construction URL: http://www.hse.gov.uk/construction/index.htm

   Also see their infonet and news bulletin sign-up page URL: http://www.hse.gov.uk/construction/infonet.htm to receive regular free information on construction health and safety

➤ reports from trade organizations and construction forums, such as CITB (Construction Industry Training Board), CIRIA (Construction Industry Research and Information Association, see Section 9.7.1), CECA (Civil Engineering Contractors Association, see Section 9.7.2) and MCG (Major Contractors Group)

➤ CIOB (Chartered Institute of Building) the Health and Safety pages of the CIOB's website provide all the latest construction health and safety news, as well as several valuable resources for those working in the industry. http://www.ciob.org.uk/home

➤ CSCS (Construction Skills certification scheme) CSCS was established in 1995 with the objective of helping the construction industry to get quality up, accidents down and cowboy builders out. Since then a great deal of progress has been made. In May 2007 they presented the millionth CSCS card. That's a million people who have formally proved their competency and their awareness of health and safety, a positive move for them, their colleagues, employers and clients as well as the productivity and wellbeing of the industry. See their website: http://www.cscs.uk.com/RunScript.asp?page=1&p=ASP\Pg1.asp

➤ International Labour Organization http://www.ilo.org/global/lang–en/index.htm

➤ CEN, the European Committee for Standardization, was founded in 1961 by the national standards bodies in the European Economic Community and EFTA countries. Now CEN is contributing to the objectives of the European Union and European Economic Area with voluntary technical standards which promote free trade, the safety of workers and consumers, interoperability of networks, environmental protection, exploitation of research and development programmes, and public procurement. CEN is a non-profit-making technical organization set up under Belgian law. http://www.cen.eu/cenorm/businessdomains/businessdomains/isss/index.asp

➤ BSI, British Standards Institute, publish more than 2,000 new and revised standards every year, for all industry sectors. Standards are available from British Standards Online as either a hard copy or electronic format or from their Customer Services department: Tel: +44 (0)20 8996 9001, or email: orders@bsiglobal.com

➤ the Internet, a truly global resource for those in the construction industry. Only a very few links have been given here but each of the websites has other links to sources of information. A quick search will also provide numerous sources of specific information.

9.7.2 Construction health and safety forums

There are several construction health and safety forums to help people discuss and share best practice in the industry. This will assist in the review of standards to determine whether an organization is performing well.

   Three of the principal organizations are:

1. CIRIA – Construction Industry Research and Information Association
2. CECA – Civil Engineering Contractors Association
3. MCG – Major Contractors Group.

9.7.3 *CIRIA*

The Construction Industry Research and Information Association was founded in 1960. In recent times the organizations' name was shortened to CIRIA. Today, CIRIA is now known solely by its abbreviated name. CIRIA's mission is to improve the performance of those in the construction and related industries.

CIRIA works with the construction industry, government and academia to provide performance improvement products and services in the construction and related industries and currently engages with around 700 subscribing organizations. Activities include collaborative projects, networking, publishing, workshops, seminars and conferences. Each year CIRIA runs about 40 projects, holds over 90 events and publishes 25 best practice guides.

CIRIA does not answer technical enquiries. CIRIA provides consensus-driven best practice guidance through publications, seminars, training and multimedia solutions. Many of CIRIA's members and funders may be able to provide assistance.

Website: http://www.ciria.org.uk/faq.htm. Email: enquiries@ciria.org and assistance will be given where possible.

London office:

Classic House, 174–180 Old Street, London EC1V 9BP, UK. Tel: +44 (0) 20 7549 3300; Fax: +44 (0) 20 7253 0523

9.7.4 *CECA*

The Civil Engineering Contractors Association (CECA) was established in November 1996 at the request of contractors to represent the interests of civil engineering contractors registered in the UK as well as to provide a full range of services to members.

CECA promotes the positive contribution that the civil engineering industry makes to the nation. The Industry is an integral part of the economy and provides the transport and other infrastructure necessary for civilized life.

CECA's current membership is in excess of 200 civil engineering companies, which range in size from large and well-known national names to the medium- and smaller-sized company which may operate at a more regional level.

CECA's main objectives are:

➤ to ensure that CECA is and is recognized to be fully representative of British civil engineering contractors
➤ to promote and lobby effectively on behalf of civil engineering contractors
➤ to provide an effective range of services to CECA regional companies and CECA members.

Website: http://www.ceca.co.uk
Contact address: 55 Tufton Street, London SW1P 3QL. Tel: +44 (0) 20 7227 4620; Fax: +44 (0) 20 7227 4621

9.7.5 *MCG*

The Major Contractors Group (MCG) represents the UK's large construction companies. MCG's members carry out over £20 billion worth of construction work each year.

MCG represents the interests of major contractors to government and other decision-makers.

The MCG's Health and Safety Charter for member companies is as follows:

➤ Sharing common goals, the member companies of the Major Contractors Group (MCG) are committed to operating construction sites that provide a working environment, which is both safe and free from health hazards for all stakeholders within the construction industry and for members of the public.
➤ MCG member companies working with their supply chains will commit to:
  ➤ leading behavioural change on all our sites to eliminate accidents and incidence of ill-health
  ➤ a fully qualified workforce
  ➤ an effective site-specific induction process before anyone is allowed to work on site
  ➤ all workers being consulted on health and safety matters in a way that engages them in improving health and safety
  ➤ exchanging best practice and lessons learned in order to establish the root cause of incidents
  ➤ raising awareness and insisting on the highest standards of PPE
  ➤ publishing an annual report of progress made against the commitments of this Charter
  ➤ reducing the incidence rate of work-related ill-health in the construction industry by health surveillance, education, rehabilitation and reducing exposure. Companies will monitor progress by measuring the total number of days lost due to sickness absence.

Website: http://www.mcg.org.uk
Contact address: 55 Tufton Street, London, SW1P 3QL. Tel: +44 (0) 20 7227 4503; email: mcg@mcg.org.uk

9.8 Practice NEBOSH questions for Chapter 9

1. **Outline** the benefits to a construction company of maintaining high standards of health and safety.

2. **Outline** some of the pressures on managers of construction projects that could lead to accidents on construction sites.

3. A principal contractor needs to employ sub-contractors, some of whom may not have English as their first language.

   Outline how the principal contractor could ensure that all site operatives fully understand site rules with regard to health and safety.

4. The principal contractor on a major new-build project will require a number of sub-contractors to work for him. **Outline** what health and safety information should be requested from prospective contractors in order to aid the principal contractor to decide which sub-contractors he should appoint.

5. **Outline** the role of the HSE in improving the health and safety performance of the construction industry.

6. Falls from height are the major cause of accidents in the construction industry, particularly falls from ladders and scaffolds. **Outline SIX** possible reasons for such accidents.

7. **Outline** the principal duties of a client under the CDM Regulations.

8. (a) **List FOUR** duties of designers under the CDM Regulations.
   (b) **Explain**, with practical examples where appropriate, how failures on the part of designers to comply with their duties can cause health and safety problems during the construction phase of a project.

9. **Outline FOUR** duties of each of the following persons under the CDM Regulations:
   (a) the CDM co-ordinator
   (b) the principal contractor.

10. **Describe**, with relevant examples, the information that a principal contractor should provide to other contractors in order to fulfil his duties under the CDM Regulations.

11. A brick-built unit for the storage of packaging materials is to be constructed within a chemical processing plant.
    With respect to the proposed work, **outline** the areas to be addressed in the health and safety plan:
    (a) as received by the principal contractor from the CDM co-ordinator
    (b) as developed by the principal contractor.

12. (a) **Outline** the responsibilities of the parties involved in a construction project who have particular duties in relation to the preparation of a health and safety file.
    (b) **Identify** the types of information that may be appropriate to include in a health and safety file.

13. With reference to the CDM Regulations:
    (a) **identify** the circumstances under which a construction project must be notified to an enforcing authority
    (b) **outline** the duties of the client under the Regulations.

14. (a) **Outline** the main duties of a CDM co-ordinator under the CDM Regulations.
    (b) **Identify FOUR** items of information in the health and safety file for an existing building that might be needed by a contractor carrying out refurbishment work.

15. **Explain** the requirements to inform the HSE about the commencement of work on a construction project.

16. In relation to the CDM Regulations, **outline**:
    (a) the circumstances under which the main regulations do not apply to a construction project
    (b) the meaning of 'notifiable project'
    (c) the particulars that must be provided to the enforcing authority when a project is required to be notified.

17. **Outline** the sources of published information that may be consulted when dealing with a health and safety problem on a construction site.

Appendix 9.1 – Summary of application and notification under CDM 2007

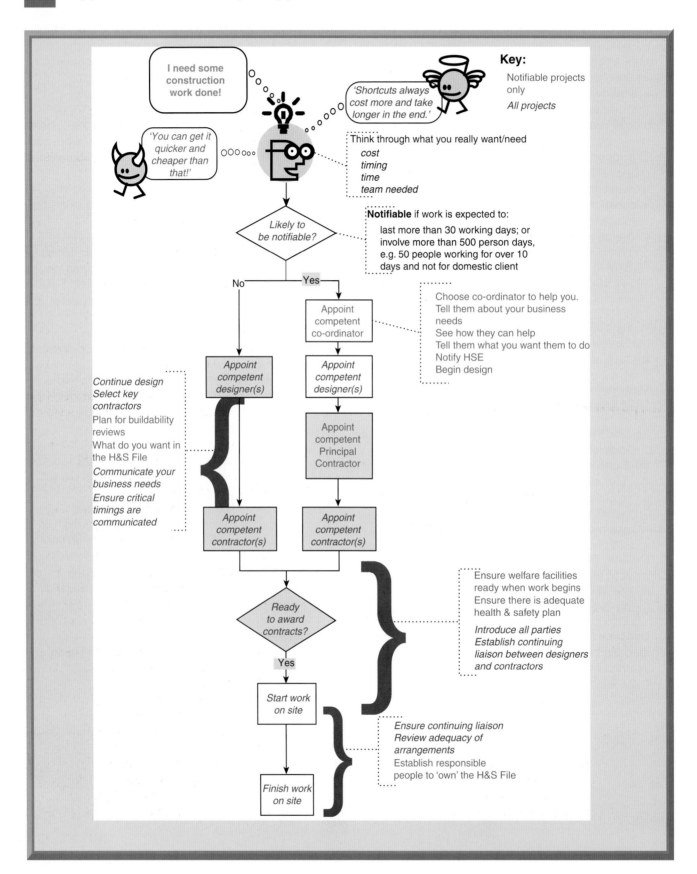

Appendix 9.2 – Pre-construction information

This is taken from Appendix 2 of the ACOP Managing health and safety in construction, L144

When drawing up the pre-construction information, each of the following topics should be considered. Information should be included where the topics are relevant to the work proposed. The pre-construction information provides information for those bidding for or planning work, and for the development of the construction phase plan. The level of detail in the information should be proportionate to the risks involved in the project.

Pre-construction information

1. Description of project
(a) project description and programme details including
  ➤ key dates (including planned start and finish of the construction phase); and
  ➤ the minimum time to be allowed between appointment of the principal contractor and instruction to commence work on site.
(b) details of client, designers, CDM co-ordinator and other consultants
(c) whether or not the structure will be used as a workplace (if so, the finished design will need to take account of the relevant requirements of the Workplace (Health, Safety and Welfare) Regulations 1992)
(d) extent and location of existing records and plans.

2. Client's considerations and management requirements
(a) arrangements for
  ➤ planning for and managing the construction work, including any health and safety goals for the project
  ➤ communications and liaison between client and others
  ➤ security of the site
  ➤ welfare provisions.
(b) requirements relating to the health and safety of the client's employees or customers or those involved in the project such as:
  ➤ site hoarding requirements
  ➤ site transport arrangements or vehicle movement restrictions

➤ client permit-to-work systems
➤ fire precautions
➤ emergency procedures and means of escape
➤ 'no-go' areas or other authorizations requirements for those involved in the project
➤ any areas the client has designated as confined spaces
➤ smoking and parking restrictions.

3. Environmental restrictions and existing on-site risks
(a) safety hazards, including:
  ➤ boundaries and access, including temporary access (e.g. narrow streets, lack of parking, turning or storage)
  ➤ any restrictions on deliveries or waste collection or storage
  ➤ adjacent land uses (e.g. schools, railway lines or busy roads)
  ➤ existing storage of hazardous materials
  ➤ location of existing services, particularly those that are concealed – water, electricity, gas, etc.
  ➤ ground conditions, underground structures or water courses where these might affect the safe use of plant (e.g. cranes, or the safety of ground work)
  ➤ information about existing structures – stability, structural form, fragile or hazardous materials, anchorage points for fall arrest systems (particularly where demolition is involved)
  ➤ previous structural modifications, including weakening or strengthening of the structure (particularly where demolition is involved)
  ➤ fire damage, ground shrinkage, movement or poor maintenance which may have adversely affected the structure
  ➤ any difficulties relating to plant and equipment in the premises, such as overhead gantries whose height restricts access
  ➤ health and safety information contained in earlier design, construction or 'as-built' drawings, such as details of pre-stressed or post-tensioned structures

(Continued)

Appendix 9.2 – *Continued*

(b) health hazards, including:
  ➤ asbestos, including results of surveys (particularly where demolition is involved)
  ➤ existing storage of hazardous material
  ➤ contaminated land, including results of surveys
  ➤ existing structures containing hazardous materials
  ➤ health risks arising from client's activities.

4. **Significant design and construction hazards**
  (a) significant design assumptions and suggested work methods, sequences or other control measures

(b) arrangements for co-ordination of ongoing design work and handling design changes
(c) information on significant risks identified during design
(d) materials requiring particular precautions.

5. **The health and safety file**
   description of its format and any conditions relating to its content.

Appendix 9.3 — Construction phase plan

This is taken from Appendix 3 of the ACOP Managing Health and Safety in Construction L144.

When drawing up the construction phase plan, employers should consider each of the following topics. Information should be included in the plan where the topic is relevant to the work proposed. The plan sets out how health and safety is to be managed during the construction phase. The level of detail should be proportionate to the risks involved in the project.

Construction phase plan

1. **Description of project**
   (a) project description and programme details including key dates
   (b) details of client, CDM co-ordinator, designers, principal contractor and other consultants
   (c) extent and location of existing records and plans that are relevant to health and safety on site, including information about existing structures when appropriate.

2. **Management of the work**
   (a) management structure and responsibilities
   (b) health and safety goals for the project and arrangements for monitoring and review of health and safety performance
   (c) arrangements for:
     ➤ regular liaison between parties on site
     ➤ consultation with the workforce

   ➤ exchange of design information between the client, designers, CDM co-ordinator and contractors on site
   ➤ handling design changes during the project
   ➤ the selection and control of contractors
   ➤ the exchange of health and safety information between contractors
   ➤ site security
   ➤ site induction
   ➤ on-site training
   ➤ welfare facilities and first aid
   ➤ the reporting and investigation of accidents and incidents including near misses
   ➤ the production and approval of risk assessments and written systems of work
   ➤ site rules
   ➤ fire and emergency arrangements.

3. **Arrangements for controlling significant site risks**
   (a) safety risks, including:
     ➤ delivery and removal of materials(including waste) and work equipment taking account of any risks to the public (e.g. during access to and egress from the site)
     ➤ dealing with services – water, electricity and gas, including overhead power lines and temporary electrical installations
     ➤ accommodating adjacent land use

- stability of structures whilst carrying out construction work, including temporary structures and existing unstable structures
- preventing falls
- work with or near fragile materials
- control of lifting operations
- the maintenance of plant and equipment
- work on excavations and work where there are poor ground conditions
- work on well, underground earthworks and tunnels
- work on or near water where there is a risk of drowning
- work involving diving
- work in a caisson or compressed air working
- work involving explosives
- traffic routes and segregation of vehicles and pedestrians

- storage of materials (particularly hazardous materials) and work equipment
- any other significant safety risks.

(b) health risks, including:
- the removal of asbestos
- dealing with contaminated land
- manual handling
- use of hazardous substances, particularly where there is a need for health monitoring
- reducing noise and vibration
- working with ionizing radiation; any other significant health risks.

4. **The health and safety file**
(a) layout and format
(b) arrangements for the collation and gathering of information
(c) storage of information.

Construction site issues – hazards and control

10

10.1 Introduction

The construction industry covers a wide range of activities ranging from large-scale civil engineering projects to very small house extensions. The construction industry has approximately 200,000 firms of whom only 12,000 employ more than seven people – many of these firms are much smaller. The use of sub-contractors is very common at all levels of the industry.

Over many years, the construction industry has had a poor health and safety record. In 1966, there were 292 fatalities in the industry and by 1995 this figure had reduced to 62 but by 2000/01 the figure had increased to 106. Over the last 25 years, 2,800 people have died on

Figure 10.1 Typical construction site.

construction sites. These figures include deaths of members of the public including children playing on construction sites. Although most of these fatalities (over 70%) were caused by falls from height, there are other significant causes. The Health and Safety Executive (HSE) has listed the main causes of fatalities as:

➤ falling through fragile roofs and rooflights
➤ falling from ladders, scaffolds and other workplaces
➤ being struck by excavators, forklift trucks or dumper trucks
➤ overturning vehicles
➤ being crushed by collapsing structures.

At a conference organized by the Health and Safety Commission in February 2001 to address the problem, it was noted that at least two construction workers were being killed each week. Targets were set to reduce the number of fatalities and major injuries by 40% over a 4-year period. All these statistics, however, ignore the large number of 'near miss' incidents, such as material falling from a height but not coming into contact with an employee, a machine overturning without causing injury and a trench collapse at a time when no one was working in it. A proper investigation of near miss incidents and the identification of the causes allow preventive action to be taken before an accident occurs. It also signals that all failures must be taken seriously whether or not they lead to injury and may yield a greater understanding of deficiencies within the health and safety management system. Studies have shown that approximately ten 'near miss' similar incidents will be followed by an accident if no remedial action is taken.

Due to the fragmented nature of the industry and its accident and ill-health record, the recent construction industry legal framework has concentrated on hazards associated with the industry, welfare issues and the need for management and control at all stages of a construction project.

Finally, it is important to note that the construction site may be inside a building (refurbishment work) as well as an open air site.

10.2 General hazards and controls

The Construction (Design and Management) (CDM) Regulations deal with the main hazards likely to be found on a construction site. In addition to these specific hazards, there will be the more general hazards (manual handling, electricity, noise, etc.), which are discussed in more detail in other chapters. The hazards and controls identified in the CDM 2007 are as follows.

Figure 10.2 Secure site access gate.

10.2.1 *Safe place of work*

Safe access to and egress from the site and the individual places of work on the site are fundamental to a good health and safety environment. This clearly requires that all ladders, scaffolds, gangways, stairways and passenger hoists are safe for use. It further requires that all excavations are fenced, the site is tidy and proper arrangements are in place for the storage of materials and the disposal of waste. The site needs to be adequately lit and secured against intruders, particularly children, when it is unoccupied. Such security will include:

➤ secure and locked gates with appropriate notices posted
➤ a secure and undamaged perimeter fence with appropriate notices posted
➤ all ladders either stored securely or boarded across their rungs
➤ all excavations covered
➤ all mobile plant immobilized, fuel removed, where practicable, and services isolated
➤ secure storage of all inflammable and hazardous substances
➤ a programme of visits to local schools to explain the dangers present on a construction site. This has been shown to reduce the number of child trespassers
➤ security patrols and closed circuit television – may need to be considered if unauthorized entry persists.

10.2.2 *Protection against falls*

Falls from a height are the most common cause of serious injury or death in the construction industry. These falls are normally from a height (see Chapter 8) but may also be on the same level. It is important that trip

174

hazards, such as building and waste materials, are not left on-site on walkways or roadways.

10.2.3 Fragile roofs

Roof work, particularly work on pitched roofs, is hazardous and requires a specific risk assessment and method statement prior to the commencement of work. Further details are given in Chapter 18.

10.2.4 Protection against falling objects

Construction workers and members of the public need to be protected from the hazards associated with falling objects. Both groups should be protected by the use of covered walkways or suitable netting to catch falling debris. Waste material should be brought to ground level by the use of chutes or hoists. Waste should not be thrown from scaffolding and only minimal quantities of building material should be stored on working platforms. The Construction (Head Protection) Regulations virtually mandate employers to supply head protection (hard hats) to employees whenever there is a risk of head injury from falling objects. (Sikhs wearing turbans are exempted from this requirement.) The employer is also responsible for ensuring that hard hats are properly maintained and replaced when they are damaged in any way. Self-employed workers must supply and maintain their own head protection. Employees must wear head protection when so directed, take good care of it and report any loss or defect. Visitors to construction sites should always be supplied with head protection and mandatory head protection signs should be displayed around the site.

Injuries due to falling materials are not limited to materials falling from height. Many fatalities have been caused by the collapse of walls, particularly retaining walls, and by the collapse of excavations.

10.2.5 Demolition

Demolition is one of the most hazardous construction operations and is responsible for more deaths and major injuries than any other activity. The management of demolition work is controlled by the CDM Regulations and requires a planning supervisor and a health and safety plan (as covered in Chapter 3). A more detailed discussion of demolition is given in Chapter 20.

10.2.6 Excavations

This topic is covered in more detail in Chapter 19. Excavations must be constructed so that they are safe environments for construction work to take place. They

must also be fenced and suitable notices posted so that neither people nor vehicles fall into them.

10.2.7 Prevention of drowning

Where construction work takes place over water, steps should be taken to prevent people falling into the water and rescue equipment should be available at all times.

Rescue and safety equipment must always be easily available and in good condition

Figure 10.3 Prevention of drowning.

10.2.8 Vehicles and traffic routes

All vehicles used on site should be regularly maintained and records kept. Only trained drivers should be allowed to drive vehicles and the training should be relevant to the particular vehicle (forklift truck, dumper truck, etc.). Vehicles should be fitted with reverse warning systems. HSE investigations have shown that in over 30% of dumper truck accidents, the drivers had little experience and no training. Common forms of accident include driving into excavations, overturning whilst driving up steep inclines and runaway vehicles which have been left unattended with the engine running. Many vehicles such as mobile cranes require regular inspection and test certificates.

Traffic routes and loading and storage areas need to be well designed with enforced speed limits (a maximum of 10 mph is the limit on most sites), good visibility and the separation of vehicles and pedestrians being considered. The use of one-way systems and separate site access gates for vehicles and pedestrians may be required. Finally, the safety of members of the public must be considered particularly where vehicles cross public footpaths.

10.2.9 *Fire and other emergencies*

Emergency procedures relevant to the site should be in place to prevent or reduce injury arising from fire, explosion, flooding or structural collapse. These procedures should include the location of fire points and assembly points, extinguisher provision, site evacuation, contact with the emergency services, accident reporting and investigation and rescue from excavations and confined spaces. There also needs to be training in these procedures at the induction of new workers and ongoing training for all workers.

10.2.10 *Electricity*

Electrical hazards are covered in detail in Chapter 14, and all the control measures mentioned apply on a construction site. However, due to the possibility of wet conditions, it is recommended that only 110 V equipment is used on site. If mains electricity is used (perhaps during the final fitting out of the building), residual current devices should be used with all electrical equipment. Where workers or tall vehicles are working near or under overhead power lines, either the power should be turned off or 'goal posts' or taped markers should be used to prevent contact with the lines. Similarly, underground supply lines should be located and marked before digging takes place.

10.2.11 *Noise*

Noisy machinery should be fitted with silencers. When machinery is used in a workshop (such as woodworking machines), a noise survey should be undertaken and, if the noise levels exceed the second action level, the use of ear defenders becomes mandatory. If it is necessary to shout when talking to a nearby colleague, then the noise level is excessive and ear defenders should be worn.

10.2.12 *Health hazards*

Health hazards are present on a construction site. These hazards include vibration, dust (including asbestos), cement, solvents and paints, and cleaners. A Control of Substances Hazardous to Health Regulations (COSHH) assessment is essential before work starts with regular updates as new substances are introduced. Copies of the assessment and the related safety data sheets should be kept in the site office for reference after accidents or fires. They will also be required to check that the correct personal protective equipment (PPE) is available.

A manual handling assessment should also be made to ensure that the lifting and handling of heavy objects is kept to a minimum. Additional health hazards may also be present when working near water whether the water is stagnant or not.

10.3 Initial site assessment

Before work commences, an initial site assessment should be undertaken. This assessment should not only consider hazards associated with the site itself but also the hazards which may well be present during the construction phase.

The previous and current use of the site will form the basis of the assessment. A greenfield site is likely to present fewer problems than a brownfield site, particularly when the site has been previously contaminated by earlier activities. Chemical contamination is common when the site formerly housed industrial plant. It is not uncommon in such situations for oils or other pollutants to be stored in underground storage chambers; and although normally the bulk of such substances will have been removed, residues will remain and present both environmental and cost problems to the construction developer who needs to understand the various remedial measures that may be used to prevent pollution of nearby streams. Remedial measures may involve the capping of the chamber or its excavation and the complete disposal of the pollutants (possibly including soil washing and thermal treatment). In many parts of the country, there are old mine workings, shafts and wells and in many cases there is no record of these on available public documents. A detailed site survey may well discover that such old workings exist and indicate that sufficient funds must be allocated to ensure that they are safely capped.

Knowledge of the history of the site is, therefore, important during the initial site assessment. Details will also be needed of any other activities, including non-construction ones, which are taking place at or near the site.

If the project involves **demolition and refurbishment**, then information on the occupancy of the premises or nearby premises needs to be ascertained. A structural survey should be made of the building to investigate the types of structural defect that may be apparent during a visual inspection and explain the causes of those defects. Typical defects include corrosion of steelwork, cracking due to fatigue in structural steel and welds, cracked brickwork and concrete, concrete degradation (flaking), dropped lintels and distorted doors or window frames, wood rot, evidence of timber infestation, dampness above the damp-proof course, cracks or dips in floors and ceilings, dips in the roof line, deformed structural members and bulging walls. Possible causes could include the quality of the initial design, the age of the building, lack of maintenance, poor quality materials, poor construction standards, solar radiation, damage from extreme weather conditions, vibration from machinery or traffic, chemical attack, overloading, accidental impact, ground settlement or subsidence, leaking drains, unauthorized modifications, fire damage, vandalism, the action of tree roots and pest infestation.

The provision of temporary supports where required, suitable access equipment and barriers or covers to prevent falls through floor openings and the need to carry out an assessment of the manual handling operations are additional factors to be considered. Asbestos, vermin infestation and exposure to lead from paintwork and pipes are possible health hazards and evidence of any of these must also be included in any assessment.

Also an emergency evacuation plan and the use of hot work permits may need to be considered at the initial site assessment stage.

A survey of the topography of the site, including the ground conditions, ground stability and type of ground or soil (e.g. gravel, sand or clay), is essential prior to the commencement of any construction work.

Access to the site, particularly for vehicles, is an important consideration especially if access is from a busy road or close to a school or hospital.

Finally **the nature of the surroundings** of the site should be included in the assessment and should include the following considerations:

- the proximity of roads, footpaths, railways, rivers and other waterways
- the details of any fencing or hedging and general vegetation
- the position and description of any trees which need to be felled and those that must be untouched (particularly if they are subject to a tree preservation order)
- the proximity of residential, commercial or industrial properties and concerns
- the proximity of schools, colleges, hospitals and playing fields
- the presence of overhead power lines and details of the responsible authority
- the presence of buried services including sewers and disused wells
- the availability of services, such as electricity, telephone lines and water, and drainage systems
- any evidence of fly tipping or other unauthorized activity
- measures required to prevent unauthorized access by vehicles.

10.4 Site controls

Many of the site controls have been covered earlier in this chapter (Section 10.2 – General hazards and controls). Site controls can be conveniently subdivided under four headings – site planning, site preparation, site security and the arrangements with the client and/or occupier of the premises.

10.4.1 *Site planning*

An essential element of site planning is the layout of the site. Due to a lack of forethought or planning, many sites rapidly degenerate into a muddled or confused state. This leads to a corresponding reduction in health and safety standards and an increase in accidents.

During the site planning stage, arrangements will be needed to ensure that there is safe access to and around the site for both pedestrians and vehicles. Care must be taken during the planning of traffic routes so that pedestrians and vehicles are kept apart as much as possible. It will also be necessary to provide suitable loading and unloading areas, which may need to be moved as the construction work progresses, and recognized parking areas.

Method statements (see Chapter 6) should be developed at this stage. They will detail how a specific job or process is to be undertaken together with the associated hazards and the necessary control measures that should be in place to ensure health and safety. If contractors are to be used, a pre-tender questionnaire should be submitted by them so that their competence may be determined. The method statement should be checked and approved by the employer or a responsible person, such as the site manager or supervisor, before the work starts. When the work has started, the employer is obliged, under section 3 of the Health and Safety at Work Act, to ensure adherence to the method statement and that a safe system of work is being used while the work is being undertaken.

Adequate lighting must be provided in all parts of the site especially along vehicle routes, although natural light should be made available whenever possible. When lighting failure could lead to serious risks of accidents, such as working on a tower scaffold, secondary lighting should be provided, perhaps in the form of torches. Suitable signs will be required on the site – these may be health and safety, warning or directional signs. Areas on the site will also need to be designated for site offices, materials and on-site machinery (fixed and mobile) storage and welfare facilities for the workforce. Arrangements for the day-to-day management and control of the site, monitoring, review and inspection of procedures and equipment and workforce supervision must be made during the planning stage.

If demolition or refurbishment is to be undertaken, then hazards associated with structural collapse, fire, working at height, lifting and carrying and electricity must be considered together with health hazards from lead, asbestos, vermin and insects. As mentioned in Chapter 20, a structural survey before work commences should reduce the risk of unintentional structural collapse. The issue of hot work permits, provision of firefighting equipment and suitable storage facilities for flammable substances should control the fire risk. Health risks can be controlled, for example, by the use of licensed contractors for the removal of asbestos, specialist contractors to eradicate vermin and the provision of suitable PPE.

Emergency arrangements need to be in place before work begins. These arrangements should deal with fire precautions, accident reporting and investigation and the provision of an accident book together with RIDDOR reporting forms.

Whatever the type of construction project, the disposal of any waste produced must be planned. There are four requirements for the transfer of waste from the site to final treatment or disposal:

1. The waste must be clearly identified. This will involve the determination of the chemical and physical properties of the material, the identification of any special problems associated with the waste, and ensuring that it is properly labelled with a written description.
2. The waste must be securely stored with procedures to deal with spillages and prevent access, particularly by scavengers or children.
3. The waste must be transferred only to an authorized person. Classes of authorized person include registered waste carriers, holders of waste management licences or exempt parties such as charities or waste collection authorities. Waste transfer notes must be transferred from one duty holder to the next along the waste chain. Checks are advised to ensure that prospective waste contractors are competent to deal with waste.
4. The adequate records of waste consignments must be kept for at least 2 years in the case of controlled waste, and 3 years for any hazardous waste. (These categories of waste are defined in Chapter 16 which includes additional possible legal requirements for waste disposal on construction sites.)

Waste skip selection should be made during the site planning process. The selected skip must be suitable for the particular job. The following points should be considered:

➤ sufficient strength to cope with its load
➤ stability while being filled

➤ a reasonable uniform load distribution within the skip at all times
➤ the immediate removal of any damaged skip from service and the skip inspected after repair before it is used again
➤ sufficient space around the skip to work safely at all times
➤ the skip should be resting on firm level ground
➤ the skip should never be overloaded or overfilled, and
➤ there must be sufficient headroom for the safe removal of the skip when it is filled.

Finally the HSE should be notified before work begins (using F10 (rev) or the details required by F10) when the project construction phase involves more than 30 days of work or involves more than 500 person-days of construction work. The CDM co-ordinator is responsible for this notification to the HSE. See Chapter 21 for details of the information required.

10.4.2 *Site preparation*

It is quite possible that specialist activities, such as lifting, piling or steelwork, will be carried out on site during the construction phase and controls will need to be in place to address associated risks. Only workers who are competent in the specialist activity should be employed or contracted. One of the ways of assessing competence is to check on relevant method statements supplied by the workers. When the activity involves working at height, then the safeguards outlined in Chapter 18 must be made available. Care must also be taken to ensure that other major hazards are addressed, such as noise and vibration hazards associated with piling work and crane hazards during lifting operations. Site rules for site workers should be formulated during the planning stage and a specimen set of such rules is given in Appendix 10.1.

The preparation of the site may well involve the removal or de-branching of trees using chainsaws and a wheeled excavator to remove the roots. These are particularly hazardous operations when there are overhead or buried power lines at the site.

Chainsaws must only be used by trained and competent operators using a safe system of work which fully describes the hazards involved and the measures required to reduce the risks. Such measures include the necessary PPE, for example goggles, hearing protection, and ballistic trousers or overalls; the provision of safe and secure access to higher branches such as ladders or cherry pickers; and the provision of fall arrest equipment such as harnesses. There must be a high level of supervision at all times. There is much more information on chainsaws and the necessary precautions when used, in Chapter 13 (13.9.4).

Figure 10.4 Precautions for overhead lines: 'goalpost' crossing points beneath lines to avoid contact by plant.

When site preparation requires the removal of contaminated topsoil, it is important to ensure that appropriate PPE, including respirators, is issued to the workforce and the suitable welfare facilities, including a decontamination unit, are available. A decontamination unit will have a dirty chamber where contaminated clothing can be left, a shower facility and a clean chamber where clean clothes will be kept. Arrangements will also be needed to ensure that eating, rest and first-aid areas are free from contamination and that emergency decontamination facilities are available.

Finally, high-voltage overhead power lines should always be made dead to prevent contact with the energized cable. If this is not possible, other control measures should be used such as height restrictions for site vehicles, signs, goalpost barriers and the use of banksmen. It is also necessary to locate and identify any underground services. It is essential that the electrical supply company is informed of the construction work and that sound emergency procedures are in place.

10.4.3 *Site security*

Site security is required to ensure that construction materials, equipment and accessories are protected and this is usually achieved by the erection of a strong perimeter fence and a lockable gate. However, it is also important to protect members of the public including trespass by children. This is particularly relevant for a refurbishment project when strong hoardings or secure fencing must be provided. There is also a need to ensure that the site itself is safe should children manage to trespass onto the site. This involves the removal of access ladders, providing trench supports, covering holes where possible, immobilizing plant, safe stacking of materials, locking away dangerous chemicals, isolating electrical supplies and locking off fuel storage tanks. Many of these problems may be avoided by visiting local schools to inform children of construction site dangers. Accidents to children on construction sites are a national concern and a widespread problem everywhere.

The provision of warning signs at the perimeter fence and suitable viewing points for the public to observe the construction work safely has been shown to reduce the incidents of site trespass.

Whilst storage is an important factor for site security, it is also needed for health and safety reasons. Cement, which is a hazardous substance, must be securely stored in a dry store. Gas cylinders, fuel and other flammable substances must be stored in a separate flameproof store (more details of these requirements are given in Chapter 15).

10.4.4 *Environmental considerations*

The public must also be protected from environmental hazards resulting from the work, such as excessive dust, noise and mud on the public highway. Excessive dust may be controlled by damping down haul roads regularly, the sheeting of lorry loads and imposing speed limits on roads close to the site. During road repair operations, dust should be prevented from entering surface drains and precautions taken to prevent or restrict the effects of fuel spillage.

Noise levels may be reduced by the regular maintenance of equipment and the erection of noise barriers.

Other controls include the use of quieter equipment and restricting the use of noisy machinery to late morning or early afternoon.

Regular road sweeping will reduce mud on surrounding roads. The problem will also be alleviated by washing vehicle wheels before they leave the site.

For any of these measures to be successful, the site workforce must be informed of the procedures at induction and by posters and signs around the site.

10.4.5 Arrangements with the client and/or occupier of the premises

Such arrangements will include site rules (see Appendix 10.1 for an example) for the protection of members of the public, visitors and other employees. They should also cover issues such as shared facilities (e.g. first-aid arrangements and the use of catering facilities) and the need for full co-operation from the occupier and his staff in the observation of health and safety rules.

When the construction work involves a public highway, measures should be introduced to ensure the safe passage of vehicles and pedestrians. This will involve the erection of warning signs, the use of cones with lead-in and exit tapers and an indication of a safety zone between the cones and the work area. Pedestrian safety will require the use of barriers and lighting and traffic may need to be controlled using traffic lights or stop/go boards. Certain neighbours, such as schools, hospitals or factories, may also need to be informed of the timing of some aspects of the work.

10.4.6 Arrangements for site induction

Properly organized and prepared site inductions for all site workers are very important. As mentioned in Chapter 9, site induction training is seen by the HSE as an essential element in the drive to reduce accidents on construction sites. Induction training in general was covered in Chapter 4 (4.8.1) and the issues raised there should be included in an induction event on a construction site. However, the following additional topics should also be covered:

➤ senior management commitment to health and safety
➤ site health and safety policy
➤ the outline of the project
➤ the managerial and consultation arrangements on site including the identity of their supervisor
➤ site health and safety plans
➤ site welfare facilities
➤ any site-specific health and safety risks, for example in relation to access, transport, site contamination, hazardous substances and manual handling

➤ those site method statements which are relevant to the workers at the induction
➤ the findings of any relevant risk assessments
➤ details of any hazardous operations, such as work at height, site vehicular movement and confined space working
➤ control measures on the site, including:
  ➤ any site rules
  ➤ any permit-to-work systems
  ➤ traffic routes
  ➤ security arrangements
  ➤ hearing protection zones
  ➤ arrangements for PPE, including what is needed, where to find it and how to use it
  ➤ manual handling procedures and precautions
  ➤ arrangements for housekeeping and materials storage
  ➤ facilities available, including welfare facilities
  ➤ emergency procedures, including fire precautions, the action to take in the event of a fire, escape routes, assembly points, responsible people and the safe use of any firefighting equipment.
➤ arrangements for first aid
➤ arrangements for reporting accidents and other incidents
➤ details of any planned training, such as 'toolbox' talks
➤ arrangements for consulting and involving workers in health and safety, including the identity and role of any:
  ➤ appointed trade union representatives
  ➤ representatives of employee safety
  ➤ safety committees.
➤ information about the individual's responsibilities for health and safety.

The quality of the information presented at the site meeting will determine the effectiveness of the induction training. Some use should be made of videos or DVDs in the presentation and the opportunity for interactive discussion given. A copy of the site rules, health and safety policy and a simple organizational chart should be issued to each worker at the induction meeting.

10.5 Provision of welfare facilities

The HSE have been concerned for some time at the poor standard of welfare facilities on many construction sites. Welfare arrangements include the provision of sanitary conveniences and washing facilities, drinking water, accommodation for clothing, facilities for changing clothing and facilities for rest and eating meals. First-aid provision is also a welfare issue and is covered in

Chapter 6. However, the scale of first-aid provision will be related to the size and complexity of the project, the number of workers on site at any given time and the proximity of emergency hospital facilities.

Sanitary and washing facilities (including showers if necessary) with an adequate supply of drinking water should be provided for everybody working on the site. Accommodation will be required for the changing and storage of clothes and rest facilities for break times. There should be adequate first-aid provision (an accident book) and protective clothing against adverse weather conditions. Welfare issues are covered by the CDM 2007 and its AC.P.

Detailed guidance is also available in HSG 150 Health and Safety in Construction. Welfare requirements are also covered in the work place (Health Safety and Welfare) Regulations and its AC.P.

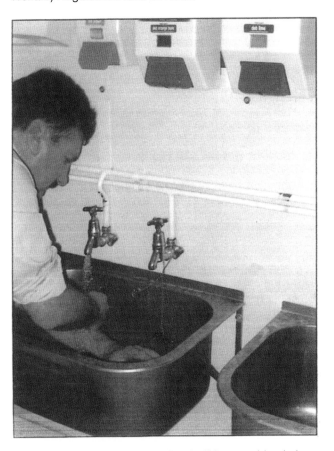

Figure 10.5 Welfare – washing facilities: washbasin large enough for people to wash their forearms.

Sanitary conveniences and washing facilities must be provided together and in proportion to the size of the workforce. The Approved Code of Practice of the Workplace (Health, Safety and Welfare) Regulations provides two tables offering guidance on the requisite number of water closets, wash stations and urinals for varying sizes of workforce (approximately one of

each for every 25 employees). Special provision should be made for disabled workers and there should normally be separate facilities for men and women. There should be adequate protection from the weather and only as a last resort should public conveniences be used. A good supply of warm water, soap and towels must be provided as close to the sanitary facilities as possible. Hand dryers are permitted but there are concerns about their effectiveness in drying hands completely and thus removing all bacteria. In the case of temporary or remote work sites, sufficient chemical closets and sufficient washing water in containers must be provided.

All such facilities should be well ventilated and lit and cleaned regularly.

Drinking water must be readily accessible to all the workforce. The supply of drinking water must be adequate and wholesome. Normally mains water is used and should be marked as 'drinking water' if water not fit for drinking is also available. On some sites, particularly during the early stages of construction, it may be necessary to provide bottled water or water in tanks. In this event, the water should be changed regularly and protected from contamination. Cups or mugs should be available near the tap unless a drinking fountain is provided.

Accommodation for clothing and facilities for changing clothing must be provided which is clean, warm, dry, well ventilated and secure. Where workers are required to wear special or protective clothing, arrangements should be such that the workers' own clothing is not contaminated by any hazardous substances. On smaller sites, the site office may be suitable for the storage of clothing, provided that it can be kept secure and used as a rest facility.

Facilities for rest and eating meals must be provided so that workers may sit down during break times in areas where they do not need to wear PPE. It should be possible for workers to make hot drinks and prepare food. Separate rooms should be provided for smokers and non-smokers. Facilities should also be provided for pregnant women and nursing mothers to rest. Arrangements must be in place to ensure that food is not contaminated by hazardous substances. Many fires have been caused by placing clothes on an electric heater. Damp clothes should not be positioned in contact with the heater. There should be adequate ventilation around the heater. If possible, the heater should be fitted with a high temperature cut-out device.

The location of the welfare facilities on the site is important. They should be located adjacent to each other and as close to the main working area as possible. It is likely that as the work progresses, the welfare facilities will need to be moved. It may be appropriate to use facilities in neighbouring premises. However, all such arrangements must be agreed in advance with the relevant

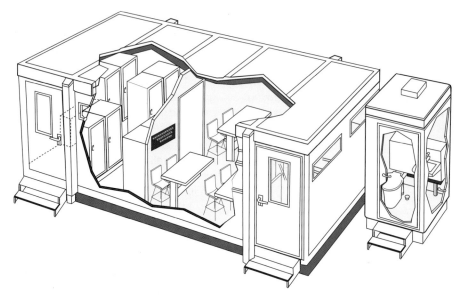

Figure 10.6 A wide range of portable welfare facilities like these are available. It may be possible when refurbishing buildings to use the facilities already on site.

neighbour. Reliance on public conveniences should be avoided due to their unreliability and lack of hot water for washing (normally).

Welfare arrangements for transient sites will be different to those for fixed sites. Transient sites include sites where the construction work is of short duration (up to 1 week in many cases), such as road works and emergency work. In these cases, welfare facilities (toilets, hand-washing basins, rest and eating arrangements) may be provided by a suitably equipped vehicle or by siting fixed installations at intervals along the length of the job.

HSE have produced two useful information leaflets on the provision of welfare facilities at fixed construction sites (CIS No. 18 (rev 1)) and transient construction sites (CIS No. 46).

10.6 Practice NEBOSH questions for Chapter 10

1. (a) **Describe** how a person new to a construction site could be made aware of the requirements to wear head protection.
   (b) **Outline** the requirements of the Construction (Head Protection) Regulations 1989.

2. **Outline** a system for the management and disposal of waste from a construction site.

3. Major refurbishment work, including the installation of new supports and joists for the first floor, is to be carried out on a two-storey Victorian building.

Identify the main hazards that may be encountered in carrying out this work and **outline** the precautions that should be taken to control them.

4. Extensive maintenance work is to be carried out on a piecemeal basis in a working warehouse. **List SIXTEEN** items that should be covered in a site induction briefing for the maintenance contractor's employees.

5. Site preparation work involves the removal of topsoil from land known to be contaminated with heavy metals. **Outline** the precautions required in relation to the transportation from site and disposal of the contaminated soil.

6. The preparation of a site requires the complete removal of some trees and the de-branching of others. An overhead power line crosses the site and buried cables may be present.
   Describe a safe system of work for:
   (a) cutting the timber using chainsaws
   (b) removing tree roots by means of a wheeled excavator.

7. Site preparation work involves the removal of topsoil from land known to be contaminated with heavy metals. **Outline** the specific requirements for this work in relation to:
   (a) Personal protective equipment and practices
   (b) welfare facilities.

8. Residents living close to a large construction site have complained about excessive noise levels

from the site, dust on their property and mud on residential streets.

(a) **Outline** the actions that could be taken to mitigate such nuisances.

(b) **Describe** the means by which site management could make contractors and operatives aware of any new procedures adopted.

9. **Outline EIGHT** precautions that should be considered to prevent accidents to children who might be tempted to gain access to a construction site.

10. **Identify** ways of minimizing the risk of accidents to children who might be tempted to gain access to a construction site.

11. **Outline** the issues that should be considered to ensure the health and safety of security personnel used on a construction site out of normal working hours.

12. **Explain**, with reference to practical examples, the legal requirements relating to the provision of welfare and first-aid facilities on construction sites.

13. **Outline** the requirements of the Construction (Design and Management) Regulations relating to sanitary and washing facilities on construction sites.

14. **Outline** the requirements of the Workplace (Health, Safety and Welfare) Regulations relating to the provision of welfare facilities.

15. A project involves the construction of an in situ concrete structure, including the installation of drains, on a brownfield site. Up to 60 people are likely to be involved in the work at any one time.

Identify the welfare facilities that should be provided for such work.

16. A project involves shallow excavations in land previously used as a landfill site.

Outline the welfare facilities that should be provided for such work.

Appendix 10.1 – A typical set of site rules

1. All contractors are required to comply with the guidance contained in the publication HSG150 *Health and Safety in Construction* as the basis upon which their construction activities are organized, planned and implemented. A copy will be provided on site for reference by contractors and employees.

2. Contractors must ensure that their employees are made aware of these site rules and fire, first-aid and emergency procedures.

3. Any person who is in doubt about safe working practices and procedures must contact their supervisor.

4. Any person becoming aware of any unsafe act, condition or faulty piece of equipment which they cannot put right themselves, must where possible warn others, isolate the hazard and immediately inform the person in charge. Where necessary, work should be stopped and, if appropriate the area evacuated.

5. All persons must report to the site manager when initially arriving on site.

6. Sub-contractors must advise the site manager of any employees under the age of 18 and provide a copy of their assessment of the risks specific to young persons.

7. Young persons on 'work experience' will not be permitted on site without the prior agreement of the site manager.

8. No person who is known to be under the influence of alcohol or illegal substances will be permitted to work.

9. The site manager has the authority to stop work if he is not satisfied with the provisions or arrangements made for health and safety whether in respect of unsafe practices, faulty plant and equipment or the conduct of employees.

10. All plant and equipment brought onto site must be in good condition and properly maintained.

11. Operators of plant and equipment must produce 'certificates of competence or training achievement' when requested or a letter from their employer confirming competence and authorization to operate the plant/equipment.

12. Plant and machinery must be immobilized and secured when the site is unattended.

13. The site manager must be advised before any 'hot work' is undertaken and an appropriate hot work permit form completed.

14. Entrance gates/fencing must be secured and locked when the site is unattended.

15. Contractors must supply details of risk assessment and method statements as required by the site manager.

16. Only 110 V electrical tools and equipment are to be used on site.

17. Hard hats must be worn in the designated areas.

18. Safety footwear must be worn at all times.

19. Debris and waste must be cleared away regularly, and dust (and noise) kept to a minimum. There must be no burning on site.

20. All accidents and incidents must be reported immediately to the site manager and entered in the accident book.

21. The statutory reporting of injuries is the responsibility of the individual employer of contracted workers. The site manager is to be provided with a copy of Form F2508.

22. Sub-contractors and employees must co-operate with the site manager in the investigation of any incident or accident.

Movement of people and vehicles – hazards and control

11

11.1 Introduction

People are most often involved in accidents as they walk around the construction site or when they come into contact with vehicles in or around the site. It is therefore important to understand the various common accident causes and the control strategies that can be employed to reduce them. Slips, trips and falls account for the majority of accidents to pedestrians and the more serious accidents between pedestrians and vehicles can often be traced back to excessive speed or other unsafe

vehicle practices, such as lack of driver training. Many of the risks associated with these hazards can be significantly reduced by an effective management system.

11.2 Hazards to pedestrians

The most common hazards to pedestrians at work are slips, trips and falls on the same level, falls from height, collisions with moving vehicles, being struck by moving, falling or flying objects and striking against fixed or stationary objects. Each of these will be considered in turn, including the conditions and environment in which the particular hazard may arise.

11.2.1 Slips, trips and falls on the same level

These are the most common of the hazards facing pedestrians and account for 30 per cent of all the major accidents and 20 per cent of over 3-day injuries reported to the HSE each year. It has been estimated that the annual cost of these accidents to the nation is £750 million with a direct cost to employers of £300 million. Slips and trips are the biggest single cause of reported injuries in the construction industry. There are over 1,000 major injuries on construction sites each year. Older workers, especially women, are the most severely injured group from falls resulting in fractures of the hips and/or femur. Civil compensation claims are becoming more common and costly to employers and such claims are now being made by members of the public who have tripped on uneven paving slabs on pavements or in shopping centres.

The Health and Safety Commission has been so concerned at the large number of such accidents that it has identified slips, trips and falls on the same level as

Figure 11.1 Tripping hazards.

Figure 11.2 Cleaning must be done carefully to prevent slipping.

a key risk area. The costs of slips, trips and falls on the same level are high to the injured employee (lost income and pain), the employer (direct and indirect costs including lost production) and to society as a whole in terms of health and social security costs.

Slip hazards are caused by:

➤ wet or dusty floors
➤ the spillage of wet or dry substances – oil, water, cement dust and fuel from site vehicles
➤ loose mats on slippery floors
➤ wet and/or icy weather conditions
➤ unsuitable footwear or floor coverings or sloping floors.

Trip hazards are caused by:

➤ obstacles, such as bricks, blocks or timber, left around the site
➤ loose floorboards or carpets
➤ obstructions, low walls, low fixtures on the floor
➤ cables or trailing leads across walkways or uneven surfaces. Leads to portable electrical hand tools and other electrical appliances (vacuum cleaners and overhead projectors). Raised telephone and electrical sockets are also a serious trip hazard (this can be a significant problem when the display screen workstations are re-orientated in an office)
➤ rugs and mats – particularly when worn or placed on a polished surface
➤ poor housekeeping – obstacles left on walkways and working platforms, construction waste and debris not removed regularly
➤ poor lighting levels – particularly near steps or other changes in level

➤ sloping or uneven floors – particularly where there is poor lighting or no handrails
➤ unsuitable footwear – shoes with a slippery sole or lack of ankle support.

The vast majority of major accidents involving slips, trips and falls on the same level result in dislocated or fractured bones.

11.2.2 *Falls from a height*

These are the most common cause of serious injury or death in the construction industry and the topic is covered in Chapter 18. These accidents are usually concerned with falls of greater than 2 m and often result in fractured bones, serious head injuries, loss of consciousness and death. Twenty-five per cent of all deaths at work and 19 per cent of all major accidents are due to falls from a height. Falls down staircases and stairways, through fragile roofs, off landings and stepladders and from vehicles, all come into this category. Injury, sometimes serious, can also result from falls below 2 m, for example, using swivel chairs for access to high shelves.

11.2.3 *Collisions with moving vehicles*

These can occur within the workplace premises or on the access roads around the building site. It is a particular problem where there is no separation between pedestrians and vehicles or where vehicles are speeding. Poor lighting, blind corners, the lack of warning signs and barriers at road crossing points also increase the risk of this type of accident. Eighteen per cent of fatalities at work are caused by collisions between pedestrians and moving vehicles with the greatest number occurring in the service sector (primarily in retail and warehouse activities). Collisions are also a significant cause of accidents in the construction industry.

11.2.4 *Being struck by moving, falling or flying objects*

This causes 18 per cent of fatalities at work and is the second highest cause of fatality in the construction industry. It also causes 15 per cent of all major and 14 per cent of over 3-day accidents. Moving objects include articles being moved, moving parts of machinery or conveyor belt systems. Flying objects are often generated by the disintegration of a moving part or a failure of a system under pressure. Falling objects are a major problem in construction (due to careless working at height) and in warehouse work (due to careless stacking of pallets on racking). The head is particularly vulnerable to these hazards. Items falling off high shelves and moving loads are also significant hazards in many sectors of industry.

Figure 11.3 Falling from a height – tower scaffold.

11.2.5 *Striking against fixed or stationary objects*

This accounts for over 1000 major accidents each year. Injuries are caused to a person either by colliding with a fixed part of the building structure, work in progress, a machine, a stationary vehicle or by falling against such objects. The head appears to be the most vulnerable part of the body to this particular hazard and this is invariably caused by the misjudgement of the height of an obstacle. Concussion in a mild form is the most common outcome and a medical check-up is normally recommended. It is a very common injury during maintenance operations when there is, perhaps, less familiarity with particular space restrictions around a machine.

Effective solutions to all these hazards need not be expensive, time consuming or complicated. Employee awareness and common sense combined with a good housekeeping regime will solve many of the problems.

11.3 Control strategies for pedestrian hazards

11.3.1 *Slips, trips and falls on the same level*

These may be prevented or, at least, reduced by several control strategies. These and all the other pedestrian hazards discussed should be included in the workplace risk assessments required under the Management of Health and Safety at Work Regulations by identifying slip or trip hazards, such as poor or uneven floor/pavement surfaces, badly lit stairways and puddles from leaking roofs. There is also a legal requirement in the CDM Regulations for all sites to be kept in good condition and in a reasonable state of cleanliness with no projecting

nails. Traffic routes must be so organized that people can move around the worksite safely.

The key elements of a health and safety management system are as relevant to these as to any other hazards:

➤ **planning** – remove or minimize the risks by using appropriate control measures and defined working practices (e.g. covering all trailing leads)
➤ **organization** – involve all workers and supervisors in the planning process by defining responsibility for keeping given areas of the site tidy and free from trip hazards
➤ **control** – record all cleaning and maintenance work. Ensure that anti-slip covers and cappings are placed on stairs, ladders, catwalks, kitchen floors and smooth walkways. Use warning signs when floor surfaces have recently been washed
➤ **monitoring and review** – carry out regular safety audits of cleaning and housekeeping procedures and include trip hazards in safety surveys. Check on accident records to see whether there has been an improvement or if an accident black spot may be identified.

Slip and trip accidents are a major problem for large retail stores both for customers and employees. The provision of non-slip flooring, a good standard of lighting and minimizing the need to block aisles during the re-stocking of merchandise are typical measures that many stores use to reduce such accidents. Other measures include the wearing of suitable footwear by employees, adequate handrails on stairways, the highlighting of any floor level changes and procedures to ensure a quick and effective response to any reports of floor damage or spillages. Good housekeeping procedures are essential. The design of the store layout and any associated warehouse can also ensure a reduction in all types of accident. Many of these measures are valid for a range of workplaces.

11.3.2 *Falls from a height*

Working at height is covered in detail in Chapter 18; however, there may be occasions using a fixed staircase (not covered by the Work at Height Regulations) when people can fall and be seriously injured.

Staircases are a source of accidents included within this category of falling from a height and the following design and safety features will help to reduce the risk of such accidents:

➤ adequate width of the stairway, depth of the tread and provision of landings and banisters or handrails and intermediate rails. The treads and risers should

Figure 11.4 Stairs leading from site accommodation.

always be of uniform size throughout the staircase and designed to meet Building Regulations requirements for angle of incline (i.e. steepness of staircase)

➤ provision of non-slip surfaces and reflective edging
➤ adequate lighting
➤ adequate maintenance
➤ special or alternative provision for disabled people (for example, personnel elevator at the side of the staircase).

Great care should be used when people are loading or unloading vehicles. As far as possible people should avoid and be deterred from climbing onto vehicles or their loads. For example, the sheeting of lorries should be carried out in designated places using properly designed access equipment.

11.3.3 Collisions with moving vehicles

These are best prevented by completely separating pedestrians and vehicles, and providing well marked, protected and laid out pedestrian walkways. People, particularly members of the public, should cross roads by designated and clearly marked pedestrian crossings. Suitable guardrails and barriers should be erected at entrances and exits from buildings and at 'blind' corners at the end of racking in warehouses. Particular care must be taken in areas where lorries are being loaded or unloaded. It is important that separate doorways are provided for pedestrians and vehicles and all such doorways should be provided with a vision panel and an indication of the safe clearance height, if used by vehicles. Finally, the enforcement of a sensible speed limit, coupled

where practicable, with speed governing devices, is another effective control measure.

11.3.4 Being struck by moving, falling or flying objects

This may be prevented by guarding or fencing the moving part (as discussed in Chapter 13) or by adopting the measures outlined for construction work (Chapter 18). Both construction workers and members of the public need to be protected from the hazards associated with falling objects. Both groups should be protected by the use of covered walkways or suitable netting to catch falling debris where this is a significant hazard. Waste material should be brought to ground level by the use of chutes or hoists. Waste should not be thrown from a height and only minimal quantities of building materials should be stored on working platforms. Appropriate personal protective equipment, such as hard hats or safety glasses, should be worn at all times when construction operations are taking place.

It is often possible to remove high-level storage in offices and provide driver protection on lift truck cabs in warehouses.

Storage racking is particularly vulnerable and should be strong and stable enough for the loads it has to carry. Damage from vehicles in a warehouse can easily weaken the structure and cause collapse. Uprights need protection, particularly at corners.

The following action can be taken to keep racking serviceable:

➤ Inspect it regularly and encourage workers to report any problems.
➤ Post notices with maximum permissible loads and never exceed the loading.

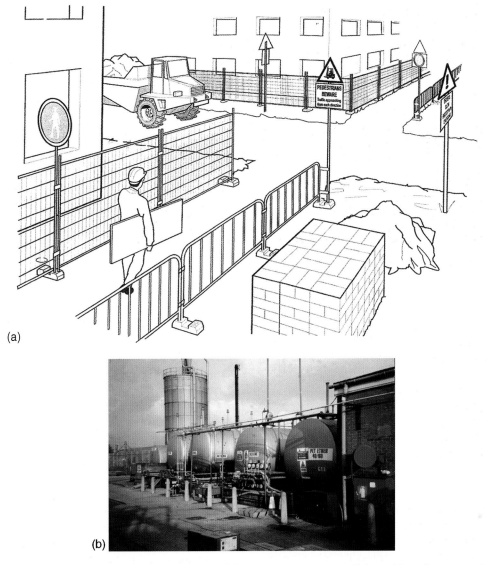

(a)

(b)

Figure 11.5 (a) Typical pedestrian/vehicle crossing area; (b) barriers to prevent collision with tank surrounds/bunds.

➤ Use good pallets and safe stacking methods.

➤ Band, box or wrap articles to prevent items falling.

➤ Set limits on the height of stacks and regularly inspect to make sure that limits are being followed.

➤ Provide instruction and training for staff and special procedures for difficult objects.

11.3.5 *Striking against fixed or stationary objects*

This can only be effectively controlled by:

➤ having good standards of lighting and housekeeping by ensuring that all waste debris and construction materials, particularly timber, are safely stored and/or removed from the site

➤ defining walkways and making sure they are used

➤ the use of awareness measures, such as training and information in the form of signs or distinctive colouring

➤ the use of appropriate personal protective equipment, in some cases, as discussed previously

➤ ensuring that all protruding nails or screws are either removed or flattened against the timber.

11.3.6 *General preventative measures for pedestrian hazards*

Minimizing pedestrian hazards and promoting good work practices requires a mixture of sensible planning, good housekeeping and common sense. Few of the required measures are costly or difficult to introduce

189

Figure 11.6 (a) A designated waste collection area; (b) unsafe stacks of heavy boxes.

and, although they are mainly applicable to slips, trips and falls on the same level and collisions with moving vehicles, they can be adapted to all types of pedestrian hazard. Typical measures include:

➤ Develop a safe workplace as early as possible and ensure that suitable floor surfaces and lighting are selected and vehicle and pedestrian routes are carefully planned. Lighting should not dazzle approaching vehicles nor should pedestrians be obscured by stored products. Lighting is very important where there are changes of level or stairways. Any physical hazards, such as low beams, vehicular movements or pedestrian crossings, should be clearly marked. Staircases need particular attention to ensure that they are slip resistant and the edges of the stairs marked to indicate a trip hazard.

➤ Consider pedestrian safety when re-orientating the workplace layout (e.g. the need to reposition lighting and emergency lighting).

➤ Adopt and mark designated walkways.

➤ Apply good housekeeping principles by keeping all areas, particularly walkways, as tidy as possible and ensure that any spillages are quickly removed.

➤ Ensure that all workers are suitably trained in the correct use of any safety devices (such as machine guarding or personal protective equipment) or cleaning equipment provided by the employer.

➤ Only use cleaning materials and substances that are effective and compatible with the surfaces being cleaned, so that additional slip hazards are not created.

➤ Ensure that a suitable system of maintenance, cleaning, fault reporting and repair are in place and working

effectively. Areas that are being cleaned must be fenced and warning signs erected. Care must also be taken with trailing electrical leads used with the cleaning equipment. Records of cleaning, repairs and maintenance should be kept.

➤ Ensure that all workers are wearing appropriate footwear with the correct type of anti-slip soles suitable for the type of flooring.

➤ Consider whether there are significant pedestrian hazards present in the area when any workplace risk assessments are being undertaken.

11.4 Hazards to the general public and the associated controls in construction activities, including public highways

Four chapters (10, 18, 19 and 20) deal in detail with the protection required for members of the public in many different construction scenarios. These hazards may be conveniently divided between those which are present inside the site and those which may occur outside the site. Inside the site, the public may be visitors (authorized or unauthorized) or joint occupiers of the site if it is a refurbishment project. The hazards from the construction work presented to the public outside the site could include materials falling from working platforms, and the operation of cranes and other lifting equipment.

The control measures begin with the posting of various warning signs and the provision of protected thoroughfares. Separate entrances for people and vehicles

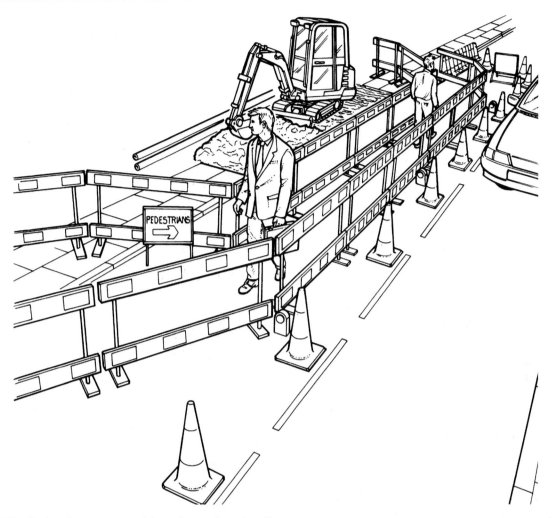

Figure 11.7 Pedestrians separated from the work and traffic.

are another good safeguard as is the provision of 2 m high perimeter fencing, with no clearance at ground level. This fence will provide good security even for short duration work. Security guards and/or cameras may be required on sites which are deemed particularly vulnerable to trespassers. Such trespassers are deterred from sites which are well lit and are surrounded by a high security fence. Site security is addressed in Chapter 10.

The protection of the public is a very important issue when the following groups of the public are concerned.

People with disabilities – particularly when pedestrian routes are affected by cable installation work or scaffolds on pavements. During the site assessment, pedestrian movements should be monitored to ascertain whether there is regular usage of the pavement by disabled people, the elderly or children in prams.

Children – this topic is covered in detail in Chapter 10. Local schools should be made aware of the proposed construction work and pupils warned of the hazards present on the site particularly when it is

not being worked. Work on school premises requires extra care and 2 m plywood panels may need to be fitted around the base of scaffolds to prevent access by children. If possible, such work should be done during weekends or school holiday periods. Deliveries should not occur at school start, finish or break times. Tunnels and fans (see Figure 18.13) may be required near playgrounds and entrances to prevent pupils, staff and parents being struck by falling objects.

Occupied premises – again this issue is covered in detail in Chapter 10. Where work is to be undertaken in or near occupied premises, such as hospitals, schools, office blocks, factories, shops or houses, it may be necessary to completely or partially evacuate the premises either during a hazardous part of the construction work or an emergency. In such cases, a detailed evacuation plan will be needed. For elderly residents or hospital patients, evacuation may present greater risks than the hazardous construction work itself.

Refurbishment of residential properties – the use of hazardous substances during damp course work or timber treatment may require the evacuation of local residents for a short period of time. A plan needs to be in place prior to the start of the work to deal with this eventuality. Good communication between the contractor, clients and residents is essential if risks are to be successfully controlled. Advice should be sought from the emergency services before work begins.

Refurbishment of commercial properties – this is covered in Chapter 10. The principal contractor will need to cooperate closely with the occupiers to manage the risks created by sharing the workplace as well as managing the risks created by his work. When work is to be done in areas which remain occupied, the risk assessment must clearly define the perimeter around the construction work area and detail the methods used to maintain the security of the site and the safety of members of the public passing around, under or over the site.

Healthcare premises – these include hospitals, clinics, surgeries and psychiatric hospitals. It is important that the principal contractor is informed of the nature of the work undertaken in these premises and whether his normal mode of operation will need to be modified. Unnecessary obstructions should be removed and adequate signage used to indicate hazards which are present. The accidental release of fungal spores or other organisms could adversely affect patients who are recovering from major surgery. It was decided to use a mobile elevating work platform at a site in a psychiatric hospital due to fears that patients might climb a traditional scaffold out of hours. Construction site vehicles and pedestrians should be separated from hospital traffic as much as possible during the contract.

House building – additional hazards are presented when the building programme takes place beside an established occupied housing estate. Site security, safe vehicular access through the estate to the site and trespass by children are all problems which must be addressed. An additional problem is created when a show house and sales office are located on the site. Prospective buyers must be protected from the hazards on the site.

Street works present particular hazards to pedestrians and many of these hazards are discussed in Chapters 10, 18, 19 and 20. Work on a pavement or road is hazardous for both members of the public and the construction workers. Pavements should be kept clear of tripping hazards, such as trailing cables. The site must be well lit at night. Road traffic past the site may also need to be controlled to protect the workforce. Members of the public and traffic vehicles must also be protected from the elbows of loaders, excavators and cranes which may swing into their path. More detailed advice is available from the Code of Practice 'Safety at street works and road works' related to the New Roads

and Street Works Act. The following points for the protection of pedestrians should be considered when work in streets or similar areas is being planned:

➤ temporary traffic controls
➤ signs for pedestrians and traffic
➤ cones or other barriers to designate the safety zone
➤ barriers to protect and guide the public around the site
➤ temporary walking surfaces for the public
➤ temporary lighting
➤ secure and safe storage of working materials
➤ arrangements for the movement of vehicles to and from the site
➤ provision of high visibility clothing for workers on the site
➤ arrangements for other hazards associated with the work at the site, such as noise, dust and buried services.

11.5 Mobile work equipment

11.5.1 Hazards of mobile work equipment

Many different kinds of vehicle are used in construction, including dumper trucks, heavy goods vehicles, all terrain vehicles and, perhaps the most common, the forklift truck.

Approximately 15 persons are killed annually following vehicle accidents on construction sites. There are also over 1000 major accidents (involving serious fractures, head injuries and amputations) caused by:

➤ collisions between pedestrians and vehicles
➤ people falling from vehicles
➤ people being struck by objects falling from vehicles
➤ people being struck by an overturning vehicle
➤ communication problems between vehicle drivers and employees or members of the public.

Figure 11.8 Telescopic materials handler.

A key cause of these accidents is the lack of competent and documented driver training. HSE investigations, for example, have shown that in over 30 per cent of dumper truck accidents on construction sites, the drivers had little experience and no training. Common forms of these accidents include driving into excavations, overturning while driving up steep inclines and runaway vehicles which have been left unattended with the engine running.

Risks of injuries to employees and members of the public involving vehicles could arise due to the following occurrences:

➤ poor maintenance with defective brakes, tyres and steering
➤ poor visibility because of dirty mirrors and windows or loads which obstruct the driver's view
➤ operating on rough ground or steep gradients which causes the mobile equipment to turn on its side 90° plus or roll over 180° or more
➤ carrying of passengers without the proper accommodation for them
➤ people being flung out as the vehicle overturns and being crushed by it
➤ being crushed under the wheels as the vehicle moves
➤ being struck by a vehicle or an attachment
➤ lack of driver training or experience
➤ underlying causes of poor management procedures and controls, safe working practices, information, instruction, training and supervision
➤ collision with other vehicles
➤ overloading of vehicles
➤ general vehicle movements and parking
➤ dangerous occurrences or other emergency incidents (including fire)
➤ access and egress from the buildings and the site.

The machines most at risk of roll over according to the HSE are:

➤ compact dumpers frequently used on construction sites
➤ agricultural tractors; and
➤ variable reach rough terrain trucks (telehandlers).

Vehicle operations need to be carefully planned so that the possibility of accidents is minimized.

There are several other more general hazardous situations involving pedestrians and vehicles. These include the following:

➤ reversing of vehicles, especially inside buildings
➤ poor road surfaces and/or poorly drained road surfaces
➤ roadways too narrow with insufficient safe parking areas

➤ roadways poorly marked out and inappropriate or unfamiliar signs used
➤ too few pedestrian crossing points
➤ the non-separation of pedestrians and vehicles
➤ lack of barriers along roadways
➤ lack of directional and other signs
➤ poor environmental factors, such as lighting, dust and noise
➤ ill-defined speed limits and/or speed limits which are not enforced
➤ poor or no regular maintenance checks
➤ vehicles used by untrained and/or unauthorized personnel
➤ poor training or lack of refresher training.

11.5.2 *Mobile work equipment legislation – PUWER 1998 Part III*

General

The main purpose of the mobile work equipment PUWER 1998 Part III, Regulations 25 to 30 is to require additional precautions relating to work equipment while it is travelling from one location to another or where it does work while moving. All appropriate sections of PUWER 1998 will also apply to mobile equipment as it does to all work equipment, for example dangerous moving parts of the engine would be covered by Part II Regulations 10, 11 and 12 of PUWER 1998. If the equipment is designed primarily for travel on public roads the Road Vehicles (Construction and Use) Regulations 1986 will normally be sufficient to comply with PUWER 1998.

Mobile equipment would normally move on wheels, tracks, rollers, skids, etc. Mobile equipment may be self-propelled, towed or remote controlled and may incorporate attachments. Pedestrian controlled work equipment, such as lawn mowers, is not covered by Part III.

Employees carried on mobile work equipment – Regulation 25

No employee may be carried on mobile work equipment unless:

➤ it is suitable for carrying persons; and
➤ it incorporates features to reduce risks as low as is reasonably practicable, including risks from wheels and tracks.

Where there is a significant risk of falling materials fit falling-object protective structures (FOPS).

Rolling over of mobile work equipment

Where there is a risk of overturning it must be minimized by:

➤ stabilizing the equipment;
➤ fitting a structure so that it only falls on its side;

Figure 11.9 Various types of construction plant with driver protection.

➤ fitting a structure which gives sufficient clearance for anyone being carried if it turns over further – ROPS (Roll-Over Protective Structure)
➤ a device giving comparable protection
➤ fitting a suitable restraining system for people if there is a risk of being crushed by rolling over.

Self-propelled work equipment

Where self-propelled work equipment may involve risks while in motion it shall have:

➤ facilities to prevent unauthorized starting
➤ (with multiple rail-mounted equipment) facilities to minimize the consequences of collision
➤ a device for braking and stopping
➤ (where safety constraints so require) emergency facilities for braking and stopping, in the event of failure of the main facility, which have readily accessible or automatic controls
➤ (where the driver's vision is inadequate) devices fitted to improve vision
➤ (if used at night or in dark places) appropriate lighting fitted; otherwise it shall be made sufficiently safe for use
➤ (if there is anything carried or towed that constitutes a fire hazard liable to endanger employees (particularly if escape is difficult such as from a tower crane)) appropriate fire-fighting equipment carried, unless it is sufficiently close by.

Rollover and falling object protection

Roll over protective structure (ROPS) is now becoming much more affordable and available for most types of mobile equipment where there is a high risk of turning over. Its use is spreading across most developed countries and even some less well-developed areas. A ROP is a cab or frame that provides a safe zone for the vehicle operator in the event of a rollover.

The ROPS frame must pass a series of static and dynamic crush tests. These tests examine the ability of the ROPS to withstand various loads to see if the protective zone around the operator remains intact in an overturn. A home made bar attached to a tractor axle or simple shelter from the sun or rain cannot protect the operator if the equipment overturns.

The ROPS must meet International Standards such as ISO 3471:1994. All mobile equipment safeguards should comply with the essential health and safety requirements of the Supply of Machinery (Safety) Regulations 1992 but need not carry a CE marking.

ROPS must also be correctly installed strictly following the manufacturers instructions and using the correct strength bolts and fixings. It should never be modified by drilling, cutting welding or other means as this may seriously weaken the structure.

Falling object protective structure (FOPS) is required where there is a significant risk of objects falling on the equipment operator or other authorized person using the mobile equipment. Canopies that protect against falling objects (FOPS applies) must be properly designed and certified for that purpose. Front loaders, working in woods or construction sites near scaffolding or buildings under construction and high bay storage areas are all scenarios where there is a risk of falling objects. Purchasers of equipment should check that any canopies fitted are FOPS. ROPS should never be modified by the user to fit a canopy without consultation with the manufacturers.

ROPS provides some safety during overturning but operatives must be confined to the protective zone of the ROPS. So where ROPS is fitted a suitable restraining system must be provided for all seats. The use of seat restraints could avoid accidents where drivers are thrown from machines, through windows or doors or around inside the cab. In agriculture and forestry 50 per cent of overturning accidents occur on slopes of less than 10° and 25 per cent on slopes of 5° or less. This means that seat restraints should be used most of the time that the vehicle is being operated.

11.6 Safe driving

Drivers have an important role to play in the safe use of mobile equipment. They should include the following in their safe working practice checklist. They must:

➤ make sure they understand fully the operating procedures and controls on the equipment being used

- only operate equipment for which they are trained and authorized
- never drive if abilities are impaired by for example alcohol, poor vision or hearing, ill-health or drugs whether prescribed or not
- use the seat restraints where provided
- know the site rules and signals
- know the safe operating limits relating to the terrain and loads being carried
- keep vehicles in a suitably clean and tidy condition with particular attention to mirrors and windows or loose items which could interfere with the controls
- drive at suitable speeds and following site rules and routes at all times
- only allow passengers when there are safe seats provided on the equipment
- park vehicles on suitable flat ground with the engine switched off and the parking brakes applied; use wheel chocks if necessary
- make use of visibility aids or a signaller when vision is restricted
- get off the vehicle during loading operations unless adequate protection is provided
- ensure that the load is safe to move

- do not get off vehicle until it is stationary, engine stopped and parking brake applied
- where practicable remove the operating key when getting off the vehicle
- take the correct precautions such as not smoking and switching off the engine when refuelling
- report any defects immediately.

11.7 Control strategies for mobile plant operations

Any control strategy involving vehicle operations will involve a risk assessment to ascertain where, on traffic routes, accidents are most likely to happen. It is important that the risk assessment examines both internal and external traffic routes, particularly when goods are loaded and unloaded from lorries. It should also assess whether designated traffic routes are suitable for the purpose and sufficient for the volume of traffic.

The following needs to be addressed:

- Traffic routes, loading and storage areas need to be well designed with enforced speed limits, good

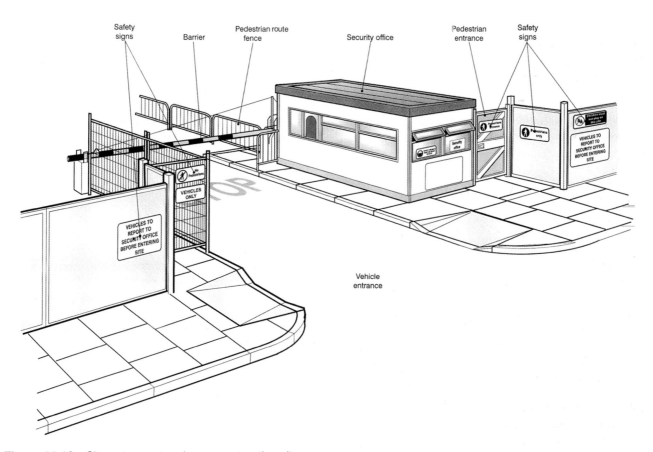

Figure 11.10 Site entrance to a large construction site.

visibility and the separation of vehicles and pedestrians whenever reasonably practicable.

➤ Environmental considerations, such as visibility, road surface conditions, road gradients and changes in road level, must also be taken into account.

➤ The use of one-way systems and separate site access gates for vehicles and pedestrians may be required.

➤ The safety of members of the public must be considered, particularly where vehicles cross public footpaths.

➤ All external roadways must be appropriately marked, particularly where there could be doubt on right of way, and suitable direction and speed limit signs erected along the roadways. While there may well be a difference between internal and external speed limits, it is important that all speed limits are observed.

➤ Induction training for all new employees must include the location and designation of pedestrian walkways and crossings, and the location of areas in the factory where pedestrians and forklift trucks use the same roadways.

➤ The identification of recognized and prohibited parking areas around the site should be given during these training sessions.

11.8 The management of vehicle movements

The movement of vehicles must be properly managed, as must vehicle maintenance and driver training. The development of an agreed code of practice for drivers, to which all drivers should sign up, and the enforcement of site rules covering all vehicular movements, are essential for effective vehicle management.

Figure 11.11 Dumper truck with roll over protection.

All vehicles should be subject to appropriate regular preventative maintenance programmes with appropriate records kept and all vehicle maintenance procedures properly documented. Consideration must be given to driver protection by fitting driver restraint (seat belts), falling object protective structure (FOPS) and roll-over or tip-over protective structure known as ROPS. (See the summary of the Provision and Use of Work Equipment Regulations in Chapter 21.) Many vehicles, such as mobile cranes, require regular inspection by a competent person and test certificates.

Certain vehicle movements, such as reversing, are more hazardous than others and particular safe systems should be set up. The reversing of lorries, for example, must be kept to a minimum (and then restricted to particular areas). Vehicles should be fitted with reversing warning systems as well as being able to give warning of approach. Refuges, where pedestrians can stand to avoid reversing vehicles, are a useful safety measure. Banksmen, who direct reversing vehicles, should also be alert to the possibility of pedestrians crossing in the path of the vehicle. Where there are many vehicle movements, consideration should be given to the provision of high visibility clothing.

Fire is often a hazard which is associated with many vehicular activities, such as battery charging and the storage of warehouse pallets. All batteries should be recharged in a separate, well-ventilated area.

As mentioned earlier, driver training, given by competent people, is essential. Only trained drivers should be allowed to drive vehicles and the training should be relevant to the particular vehicle (e.g. forklift truck, dumper truck, lorry). All drivers must receive specific training and instruction before they are permitted to drive vehicles. They must also be given refresher training at regular intervals. This involves a management system for assuring driver competence, which must include detailed records of all drivers with appropriate training dates and certification in the form of a driving licence or authorization. Competence and its definition are discussed in Chapter 4.

The HSE publications *Workplace Transport Safety. An Employers Guide HSG136*, and *The Safe Use of Vehicles on Construction Sites HSG144*, provide useful checklists of relevant safety requirements that should be in place when vehicles are used in a workplace.

11.9 Hazards and controls of vehicles on construction sites

11.9.1 *General*

Many of the hazards and controls required for vehicle movements on construction sites have been covered earlier in this chapter. Common vehicles found on construction sites include site dumper trucks, forklift trucks with telescopic handlers, all terrain and rough terrain vehicles, excavators and tipper lorries. On larger construction sites various types of earth moving equipment may also be used.

The most common hazards are those which cause loss of control of the vehicle, overturning and collision with structures on the site, pedestrians or other site vehicles. Such hazards often arise due to problems of site layout, such as the non-segregation of pedestrians and vehicles, and uneven road surfaces. Attempts to drive vehicles up inclines which are too steep have caused many fatalities on sites and the need for rollover protection and driver restraint systems (covered in detail in Chapter 13) cannot be overemphasized. The methods of construction can also present hazards, for example excavations, scaffolds and falsework, all of which should have additional protection fitted when sited near a roadway. Additional hazards from falling materials, noise, dust and poor maintenance also need to be addressed.

The key controls, therefore, are the provision of safe routes, the separation of pedestrians and vehicles, the restriction of the carriage of people on site and the prevention of falls into excavations by the erection of strong barriers. Other precautions involve the introduction of one-way systems, speed limits, adequate lighting at road junctions, separate entrances and exits to the site with adequate turning room, and the use of clear signs. All vehicles should be well maintained, fitted with flashing beacons or reversing alarms, not be overloaded and sheeted when loaded. Unintended vehicle movements must be prevented by the proper application of braking systems when the vehicle is stationary.

Clear routes should be set out across the site. These routes should avoid sharp bends, blind corners (unless suitably placed mirrors are located at the corner), narrow gaps, low head room and adverse cambers or steep gradients. Road surfaces should be inspected regularly and repaired using hardcore if necessary.

All vehicle drivers should be properly trained on each of the vehicles which they are expected to drive. Banksmen, wearing high visibility clothing, should be used to direct movement of lorries and excavators. Site rules should cover the use of vehicles on site and site management systems should include details of each driver and renewal dates for refresher training to ensure continued driver competency. A vehicle code of practice should be issued to all drivers. This code should include site safety rules, essential vehicle safety checks and reference to the hazards of driving near overhead power lines (see Chapter 10, Fig. 10.4).

A level parking area should be provided for vehicles and although they should be drained of fuel overnight this must not be done while the engine is still hot or running. Finally, mud should be cleaned from the wheels of a vehicle before it enters a public highway.

11.9.2 *Road works signs*

The control of vehicle movements during road works is governed by the Department for Transport Code of Practice 'Safety at street works and road works'. This gives detailed information on the temporary signs required and the order of these signs along the approach road. It also covers the way in which traffic cones should be set out with lead-in and exit tapers and it gives an indication of a safety zone between the cones and the work area. Means of controlling traffic, either by the use of traffic lights or stop/go boards, is also important, as is the provision of barriers for pedestrian safety.

The code of practice covers all situations from a footpath closure (see Figure 11.16) to high speed motorway lane closures and mobile work carried out from a vehicle like lane marking. The following gives an example of signage for advanced signs at road works.

It does not matter whether the works are small or large, on the ground or overhead; all street works require warning and information. In emergencies as much warning must be given as the circumstances permit, and full signing must be provided as quickly as possible.

High visibility clothing must be worn at all times. Jackets with sleeves must be worn on dual carriageway roads with a speed limit of 50 mph or above, unless operatives stay within the working space at all times.

Signs should be placed where they will be clearly seen, and cause minimum inconvenience to drivers, cyclists, pedestrians and other road users alike, and where there is a minimum risk of their being hit or knocked over by traffic. Where there is a grass verge the signs should normally be placed there; the placing of signs in the footway is permitted but in no circumstances must the footway width be reduced below 1 m. If there are already vehicles parked in the carriageway, place the advance signs so that they are not obscured.

a) Road Works Ahead

The **Road Works Ahead** sign is the first sign to be seen by the driver, so place it well before the works. Its size, the minimum distance from the start of the lead-in taper, and clear visibility distance will vary according to the type of road and its speed limit – see table on inside of back cover. The range of distances is given to allow the sign to be placed in the most convenient position bearing in mind available space and visibility for drivers. Do not simply choose the minimum distance – assess each site carefully.

Figure 11.12   (a) Road Works Ahead; (b) Road Narrows.

b) Road Narrows

A **Road Narrows Ahead** sign warns the driver which side of the carriageway is obstructed. Place it midway between the Road Works Ahead sign and the beginning of the lead-in taper.

On roads with speed limits of 50 mph or more, all advance signs should have plates giving the distance to the works in yards or miles not in metres.

c) Keep Right and Keep Left signs

Place **Keep Right**, or as appropriate, **Keep Left** signs at the beginning and end of the lead-in taper of cones. These signs must be the same size as the Road Works Ahead sign. **Make sure that the signs point in the correct direction**. Do not turn the sign frame on its side to make it point in the correct direction. These signs must NOT be used for directing pedestrians.

Figure 11.13   Keep Right/Left.

d) Cones and lamps

Place a line of traffic cones to guide traffic past the works and add road danger lamps in poor daytime visibility and bad weather. Where the traffic is faster the length of taper must be longer. Look at the table inside the back cover for details of positioning of cones and lamps.

Road danger lamps must be used at night on roads with a speed limit of 40 mph or above. On roads with a Lower Speed Limit, judgement may be used as to whether road danger lamps are needed, depending on the standard of street lighting.

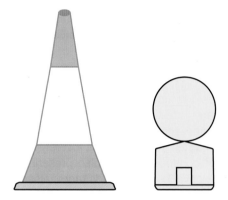

Figure 11.14   Cone and road danger lamp.

Road danger lamps must not be higher than 1.5 m above the road or 1.2 m where the speed limit is more than 40 mph.

The type of lamp to be used is as follows:

Type of road danger lamp	Conditions of use
Flashing lamp 55 to 150 flashes per minute	Only when ALL of the following conditions apply – the speed limit is 40 mph or less – the road danger lamp is within 50 m of a street lamp, and – the street lamp is illuminated
Steady lamp	On any road with or without street lighting

e) Barriers

Barriers may comprise separate portable post and plank systems, gate frames linked together, or semi-permanent constructions built to enclose the site.

There are several different requirements for the barrier planks associated with post and plank systems. The following explains the requirements and how they may be met using barrier planks which are red and white and manufactured in fully retro-reflective materials. (*Note*: Retro-reflective means that at night the material reflects light back to the light source Fig. 11.15.)

Figure 11.15   Red and white barrier rail.

Figure 11.16   Road works signs for footpath closure. Could be improved with walkway for pedestrians beside track. But only minor road, and pedestrians can cross to a good pavement opposite.

Barrier planks are required to carry out three functions:

1. As a **traffic barrier**. When a traffic lane is closed for works to take place, the regulations require this to be done with a retro-reflective red and white barrier plank placed across the lane. This is illustrated as a Traffic Barrier Lane Closed sign.
2. As a **pedestrian barrier**. Pedestrians must be separated from the works by barriers which are conspicuous and mounted as part of a portable fencing system. Pedestrian barrier planks may be of several different contrasting colours; yellow, white or orange colours are best detected by partially sighted people, but red and white is one of the acceptable combinations.
3. As a **tapping rail** for blind and partially sighted people. Tapping rails are placed as the bottom rail in a pedestrian fencing system. A red and white barrier plank may be used.

All barriers facing vehicular traffic should be of the fully retro-reflective red and white form. Red and white barrier planks do not have to be used for pedestrian barriers or tapping rails but, if they are, they must be retro-reflective. Other planks used for these purposes do not need to be retro-reflective.

f) Information board

An information board must be displayed at every site, except for mobile works and minor works which do not include excavation, (see Mobile works and minor works carried out from a vehicle). This board should be placed so that it does not obstruct footways or carriageways but can be read mainly by pedestrians, and possibly by drivers who have stopped.

Such boards are too detailed to be read easily by passing traffic. A board must give the name of the organization for which the works are being carried out, and a telephone number which can be contacted in emergencies. It may also contain other information that will be helpful in explaining to the public why the work is being done, who is doing it and how long it will take. Such additional information is to be encouraged where practical and could include some or all of the following: a brief description of the works, the name of the contractor and a message apologizing for inconvenience or delays. A completion date should normally be included if the works are expected to continue for more than a month.

Figure 11.17   Typical Information sign.

g) End sign

The End sign indicates not only the end of works but also the end of any temporary restrictions, including temporary speed limits, associated with the works.

If the permanent speed limit changes within the length of road covered by a temporary speed restriction, signs indicating the new speed limit must be provided on each side of the carriageway at the end of the works, in addition to the End sign.

An End sign must be placed beyond works that are 50 m or more in length, measured between the end of the lead-in taper and the beginning of the exit taper and beyond two or more adjacent sites.

But an End sign is not necessary on a road where ALL of the following conditions are met:

➤ there are no temporary speed limits or other traffic restrictions
➤ the speed limit is 30 mph or under
➤ there is a total two-way traffic flow of fewer than 20 vehicles counted over 3 minutes (400 veh/h)
➤ fewer than 20 heavy goods vehicles pass the works site per hour.

Figure 11.18   Road works ends sign.

h) Site layout

The layout of the work site (which must be marked off with cones) should include a working area, working space and a safety zone. The safety zone must not be used as a working area or for storing plant or materials.

i) Mobile works and minor works carried out from a vehicle

These include continuous **mobile operations**, as well as those which involve movement with periodic stops and short duration static works. They also include **minor works** which do not include excavations, involving the use of a single vehicle or a small number of vehicles.

Works in this category may omit the use of cones and a traffic barrier Lane Closed sign, provided that safe working methods are used. Single vehicle works must not be carried out on dual carriageways to which the national speed limit applies, unless they can be done at prevailing traffic speeds.

Using a single mobile vehicle or minor works with one or more vehicles – carry out the work when there is good visibility and **during periods of low risk**. (Consult the supervisor if work is to take place in the centre of the carriageway with traffic passing on both sides.)

Department for Transport – Safety at street works and road works basic requirements:

➤ The vehicle must be conspicuously coloured.
➤ The vehicle must have one or more roof mounted beacons operating.
➤ A Keep Right/Left sign must be displayed for drivers approaching on the same side of the carriageway, showing which side to pass. Vehicle mounted Keep Right/Left signs must be covered when the vehicle is travelling to and from the site. Do NOT simply turn the sign to point up or down.

Additional static signs – will be required when ANY of the following conditions apply:

➤ the works vehicle cannot be seen clearly because of hills, bends in the road, etc.
➤ stationary traffic may tail back
➤ there is not enough space for two-way traffic to pass the works vehicle
➤ the vehicle is slow moving or is required to make periodic stops.

See the *Department for Transport Code of Practice* for details. The work must be done by skilled and competent people.

11.9.3 *Check list*

Source: Department for Transport – Safety at street works and road works Code of Practice.

Before starting – General

Is high visibility clothing being worn by everyone on the site?
Are all signs, barriers, cones and lighting correctly placed?
Are signs obscured by bends, hills or dips in the road ?
Are advanced signs needed?
Will the site be safe at night or in wind, fog, snow or rain?
Are parked vehicles, trees or street furniture obscuring signs?
Is there enough road width remaining for two-way traffic?
Is traffic control with shuttle lane working required?
Are there any site specific risks requiring special guarding?
Has allowance been made for delivery and removal of materials?
Is the contact number displayed on the information board?

Before starting – Pedestrians

Are pedestrians given protected routes which are wide enough?
Are pedestrian routes clearly indicated?
If the footway is closed, is there an alternative route? If so, is it clearly marked?
Are there any special hazards for disabled pedestrians? If so, how can they be made safe?
If a temporary footway in the road is to be used, are ramps to the kerb provided where necessary?

Before starting – Traffic

Is type of traffic control right for work, traffic and speed?
Have any misleading permanent signs and road markings been covered?
Is there safe access to adjacent premises?
Have you a copy of portable traffic signals site approval?
Have you considered the needs of cyclists and horse riders?

When work is in progress

Does signing and guarding meet changing conditions?
Are signs, cones and lamps kept clean?
Can traffic control arrangements be improved to reduce traffic delays as conditions change?
Are the carriageway and footway being kept clear of mud and surplus equipment?
Are materials that are left on verges or lay-bys being properly guarded and lit?

When work is suspended

Will checks be made on signing, lighting and guarding?
Has the arrangement been changed to reflect conditions?

When work is finished

Have all signs, cones, barriers, and lamps been removed?
Have any covered permanent signs been restored?
Have the authorities been told the work is complete?

11.10 Managing occupational road safety

11.10.1 *Introduction*

This section has been added outside the NEBOSH certificate because it is an important area of concern recently given more prominence by the UK's HSE. It has been estimated that up to a third of all road traffic accidents involve somebody who is at work at the time. This may account for over 20 fatalities and 250 serious injuries every week. Some employers believe, incorrectly, that if they comply with certain road traffic law requirements, so that company vehicles have a valid MOT certificate, and drivers hold a valid licence, this is enough to ensure the safety of their employees, and others, when they are on the road. However, health and safety law applies to on-the-road work activities as it does to all work activities, and the risks should be managed effectively within a health and safety management system.

These requirements are in addition to the duties employers have under road traffic law, e.g. the Road Traffic Act and Road Vehicle (Construction and Use) Regulations, which are administered by the police and other agencies such as the Vehicle and Operator Services Agency.

Health and safety law does not apply to commuting, unless the employee is travelling from their home to a location which is not their usual place of work.

11.10.2 *Benefits of managing work-related road safety*

The true costs of accidents to organizations are nearly always higher than just the costs of repairs and insurance claims. The benefits of managing work-related road safety can be considerable, no matter what the size of the organizations. There will be benefits in the area of:

➤ *control*: costs, such as wear and tear and fuel, insurance premiums and claims can be better controlled
➤ *driver training and vehicle purchase*: better informed decisions can be made
➤ *lost time*: fewer days will be lost due to injury, ill-health and work rescheduling
➤ *vehicles*: fewer will need to be off the road for repair
➤ *orders*: fewer orders will be missed

➤ *key employees*: there is likely to be a reduction in driving bans.

11.10.3 *Managing occupational road risks*

Where work-related road safety is integrated into the arrangements for managing health and safety at work, it can be managed effectively. The main areas to be addressed are policy, responsibility, organization, systems and monitoring. Employees should be encouraged to report all work-related road incidents and be assured that punitive action will not be taken against them.

The risk assessment should:

➤ consider the use, for example, of air or rail transport as a partial alternative to driving
➤ attempt to avoid situations where employees feel under pressure
➤ make sure that maintenance work is organized to reduce the risk of vehicle failure
➤ insist that drivers and passengers are adequately protected in the event of an incident. Crash helmets and protective clothing for those who ride motorcycles and other two-wheeled vehicles should be of the appropriate colour and standard.
➤ ensure that company policy covers the important aspects of the Highway Code.

Figure 11.19 Occupational road risk.

11.10.4 *Evaluating the risks*

The following considerations can be used to check on work-related road safety management.

Figure 11.20   Concrete delivery.

The driver
Competency
➤ Is the driver competent, experienced and capable of doing the work safely?
➤ Is his or her licence valid for the type of vehicle to be driven?
➤ Is the vehicle suitable for the task or is it restricted by the driver's licence?
➤ Does recruitment procedure include appropriate pre-appointment checks?
➤ Is the driving licence checked for validity on recruitment and periodically thereafter?
➤ When the driver is at work, is he or she aware of company policy on work-related road safety?
➤ Are written instructions and guidance available?
➤ Has the company specified and monitored the standards of skill and expertise required for the circumstances for the job?

Training
Are drivers properly trained?
➤ Do drivers need additional training to carry out their duties safely?
➤ Does the company provide induction training for drivers?
➤ Are those drivers whose work exposes them to the highest risk given priority in training?

➤ Do drivers need to know how to carry out routine safety checks such as those on lights, tyres and wheel fixings?
➤ Do drivers know how to adjust safety equipment correctly, for example seat belts and head restraints?
➤ Is the headrest 3.8 cm (1.5 inches) behind the driver's head?
➤ Is the front of the seat higher than the back and are the legs 45° to the floor?
➤ Is the steering wheel adjustable and set low to avoid shoulder stress?
➤ Are drivers able to use anti-lock brakes (ABS) properly?
➤ Do drivers have the expertise to ensure safe load distribution?
➤ If the vehicle breaks down, do drivers know what to do to ensure their own safety?
➤ Is there a handbook for drivers?
➤ Are drivers aware of the dangers of fatigue?
➤ Do drivers know the height of their vehicle, both laden and empty?

Fitness and health
➤ The driver's level of health and fitness should be sufficient for safe driving.
➤ Drivers of HGVs must have the appropriate medical certificate.
➤ Drivers who are most at risk, should also undergo regular medicals. Staff should not drive, or undertake other duties, while taking a course of medicine that might impair their judgement.
➤ All drivers should have regular (every 2 years) eyesight tests. Recent research has shown many drivers have poor eyesight, which affects the standard of their driving.

The vehicle
Suitability
All vehicles should be fit for the purpose for which they are used. When purchasing new or replacement vehicles, the management should look for vehicles that are most suitable for driving and public health and safety. The fleet should be suitable for the job in hand. Where privately owned vehicles are used for work they should be insured for business use and have an appropriate MOT certificate (over 3 years old in the UK).

Condition and safety equipment
Are vehicles maintained in a safe and fit condition?
There will need to be:

➤ maintenance arrangements to acceptable standards
➤ basic safety checks for drivers
➤ a method of ensuring that the vehicle does not exceed its maximum load weight

Figure 11.21 Must have a licence valid for vehicle.

➤ reliable methods to secure goods and equipment in transit

➤ checks to make sure that safety equipment is in good working order

➤ checks on seatbelts and head restraints. Are they fitted correctly and functioning properly?

➤ drivers needing to know what action to take if they consider their vehicle is unsafe.

Ergonomic considerations

The health of the drivers, and possibly also their safety, may be put at risk through an inappropriate seating position or driving posture. Ergonomic considerations should therefore be considered before purchasing or leasing new vehicles. Information may need to be provided to drivers about good posture and, where appropriate, on how to set their seat correctly.

The load

For any lorry driving, most of the topics covered in this section are relevant. However, the load being carried is an additional issue. If the load is hazardous, emergency procedures (and possibly equipment) must be in place and the driver trained in those procedures. The load should be stacked safely in the lorry so that it cannot move during the journey. There must also be satisfactory arrangements for handling the load at either end of the journey.

The journey

Routes

Route planning is crucial. Safe routes should be chosen which are appropriate for the type of vehicle undertaking the journey wherever practicable. Motorways are the safest roads. Minor roads are suitable for cars, but they are less safe and could present difficulties for larger vehicles. Overhead restrictions, for example bridges, tunnels and other hazards such as level crossings, may present dangers for long and/or high vehicles so route planning should take particular account of these.

Scheduling

There are danger periods during the day and night when people are most likely, on average, to feel sleepy. These are between 2 am and 6 am and between 2 pm and 4 pm. Schedules need to take sufficient account of these periods. Where tachographs are carried, they should be checked regularly to make sure that drivers are not putting themselves and others at risk by driving for long periods without a break. Periods of peak traffic flow should be avoided if possible and new drivers should be given extra support while training.

Time

Has enough time been allowed to complete the driving job safely? A realistic schedule would take into account the type and condition of the road and allow the driver rest breaks. A non-vocational driver should not be expected to drive and work for longer than a professional driver. The recommendation of the Highway Code is for a 15 minute break every 2 hours.

➤ Are drivers put under pressure by the policy of the company? Are they encouraged to take unnecessary risks, for example exceeding safe speeds because of agreed arrival times?

➤ Is it possible for the driver to make an overnight stay? This may be preferable to having to complete a long road journey at the end of the working day.

➤ Are staff aware that working irregular hours can add to the dangers of driving? They need to be advised of the dangers of driving home from work when they are excessively tired. In such circumstances they may wish to consider an alternative, such as a taxi.

Distance

Managers need to satisfy themselves that drivers will not be put at risk from fatigue caused by driving excessive distances without appropriate breaks. Combining driving with other methods of transport may make it possible for long road journeys to be eliminated or reduced. Employees should not be asked to work an exceptionally long day.

Weather conditions

When planning journeys, sufficient consideration will need to be given to adverse weather conditions, such as snow, ice, heavy rain and high winds. Routes should be rescheduled and journey times adapted to take adverse weather conditions into consideration. Where poor weather conditions are likely to be encountered, vehicles should be properly equipped to operate, with, for example, anti-lock brakes.

Where there are ways of reducing risk, for example when driving a high-sided vehicle in strong winds with a light load, drivers should have the expertise to deal with the situation. In addition, they should not feel pressurized to complete journeys where weather conditions are exceptionally difficult and this should be made clear by management.

11.10.5   *Typical health and safety rules for drivers of cars on company business*

The following example shows typical rules that have been prepared for the use of car drivers.

Approximately 25 per cent of all road accidents are work-related accidents involving people who are using the vehicle on company business. Drivers are expected to understand and comply with the relevant requirements of the current edition of the Highway Code. The following rules have been produced to reduce accidents at work. Any breach of these rules will be a disciplinary offence:

➤ All drivers must have a current and valid driving licence.

➤ All vehicles must carry comprehensive insurance for use at work.

➤ Plan the journey in advance to avoid, where possible, dangerous roads or traffic delays.

➤ Use headlights in poor weather conditions and fog lights in foggy conditions.

➤ Use hazard warning lights if an accident or severe traffic congestion is approached (particularly on motorways).

➤ All speed limits must be observed but speeds should always be safe for the conditions encountered.

➤ Drivers must not drive continuously for more than 2 hours without a break of at least 15 minutes.

➤ Mobile phones, including hands-free equipment, must not be used whilst driving. They must be turned off during the journey and only used during the rest periods or when the vehicle is safely parked and the handbrake on.

➤ No alcohol must be consumed during the day of the journey until the journey is completed. Only minimal amounts of alcohol should be consumed on the day before a journey is to be made.

➤ No recreational drugs should be taken on the day of a journey. Some prescribed and over the counter drugs and medicines can also affect driver awareness and speed of reaction. Always check with a doctor or pharmacist to ensure that it is safe to drive.

11.10.6   *Further information*

The Highway Code The Stationery Office 2001 ISBN 0-11-552290-5. Can also be viewed on www.highwaycode.gov.uk

Managing occupational road risk Royal Society for the Prevention of Accidents; available from Edgbaston Park, 353 Bristol Road, Birmingham B5 7ST Tel: 0121 248 2000.

Driving at work; managing work-related road safety HSE INDG382 ISBN 0-7176-2824-8.

11.11   Practice NEBOSH questions for Chapter 11

1. (a) **Identify** the types of hazard that may cause slips or trips at work.

    (b) **Outline** how slip and trip hazards in the workplace might be controlled.

2. A cleaner is required to polish floors using a rotary floor polisher.
   (a) **Identify** the hazards that might be associated with this operation.
   (b) **Outline** suitable control measures that might be used to minimize the risk.

3. **List EIGHT** design features and/or safe practices intended to reduce the risk of accidents on staircases used as internal pedestrian routes within work premises.

4. **Outline** measures to be taken to prevent accidents when pedestrians are required to work in vehicle manoeuvring areas.

5. **Outline** the precautionary measures to be taken to avoid accidents involving reversing vehicles within a workplace.

6. **Outline** the preventive and protective measures needed to minimize the risk of injury to operators of mobile equipment due to the equipment overturning during use.

7. (a) **Outline** the requirements of the Construction (Design and Management) Regulations 2007 that relate to the use and movement of vehicles on a construction site.
   (b) **Describe** the elements of a safe system of work to be followed to ensure the safe use and movement of vehicles on a site where excavation work is taking place and where vehicles are entering the site either to deliver materials or to remove surplus material. Use sketches to illustrate your answer, in particular to show how vehicles and pedestrians on the site may be segregated.

8. **Outline** a hierarchy of measures to minimize the risks from reversing vehicles on a construction site.

9. A significant number of accidents associated with work on construction sites are associated with plant and vehicle movement.
   Outline the control measures that should be adopted to prevent such accidents occurring.

10. **Describe** the features of a well-designed and maintained traffic management system on a busy construction site.

11. Significant repairs and resurfacing to a road section is to be undertaken. Adjacent to the road is a primary school, a large warehouse and residential properties.
    (a) **Identify** the environmental factors that should be taken into account and **outline** how any detrimental environmental effects might be mitigated.
    (b) **Describe** the means by which the safe passage of vehicular and pedestrian traffic might be achieved.

Manual and mechanical handling hazards and control

12

12.1 Introduction

Figure 12.1 Handling on a construction site.

Until a few years ago, accidents caused by the manual handling of loads were the largest single cause of over 3-day accidents reported to the Health and Safety Executive (HSE). The Manual Handling Operations Regulations recognized this fact and helped to reduce the number of these accidents. However, accidents due to poor manual handling technique still accounts for over 25% of all reported accidents and in some occupational sectors, such as the health service, the figure rises above 50%. An understanding of the factors causing some of these accidents is essential if they are to be further reduced. Mechanical handling methods should always be used whenever possible, but they are not without their hazards, many of which have been outlined in Chapter 11. Much mechanical handling involves the use of lifting equipment, such as cranes and lifts which present specific hazards to both the users and bystanders. The risks from these hazards are reduced by thorough examinations and inspections as required by the Lifting Operations and Lifting Equipment Regulations (LOLER).

12.2 Manual handling hazards and injuries

The term 'manual handling' is defined as the movement of a load by human effort alone. This effort may be applied directly or indirectly using a rope or a lever. Manual handling may involve the transportation of the load or the direct support of the load including pushing, pulling, carrying, moving by using bodily force and, of course, straightforward lifting. Back injuries due to the lifting of heavy loads are very common and several million working days are lost each year as a result of such injuries.

Typical hazards of manual handling include:

> lifting a load which is too heavy or too cumbersome resulting in back injury
> poor posture during lifting or poor lifting technique resulting in back injury
> dropping a load, resulting in foot injury
> lifting sharp-edged or hot loads resulting in hand injuries.

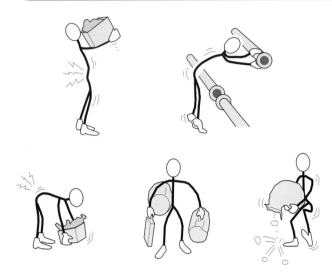

Figure 12.2 Manual handling – there are many potential hazards.

12.2.1 *Injuries caused by manual handling*

Manual handling operations can cause a wide range of acute and chronic injuries to workers. Acute injuries normally lead to sickness leave from work and a period of rest during which time the damage heals. Chronic injuries build up over a long period of time and are usually irreversible producing illnesses such as arthritic and spinal disorders. There is considerable evidence to suggest that modern life styles, such as a lack of exercise and regular physical effort, have contributed to the long-term serious effects of these injuries.

The most common injuries associated with poor manual handling techniques are all musculoskeletal in nature and are:

➤ muscular sprains and strains – caused when a muscular tissue (or ligament or tendon) is stretched beyond its normal capability leading to a weakening, bruising and painful inflammation of the area affected; such injuries normally occur in the back or in the arms and wrists

➤ back injuries – include injuries to the discs, situated between the spinal vertebrae (i.e. bones), and can lead to a very painful prolapsed disc lesion (commonly known as a slipped disc); this type of injury can lead to other conditions known as lumbago and sciatica (where pain travels down the leg)

➤ trapped nerve – usually occurring in the back as a result of another injury but aggravated by manual handling

➤ hernia – this is a rupture of the body cavity wall in the lower abdomen causing a protrusion of part of

the intestine; this condition eventually requires surgery to repair the damage

➤ cuts, bruising and abrasions – caused by handling loads with unprotected sharp corners or edges

➤ fractures – normally of the feet due to the dropping of a load. Fractures of the hand also occur but are less common

➤ Work-related upper limb disorders (WRULDs) – cover a wide range of musculoskeletal disorders, which are discussed in detail in Chapter 17

➤ rheumatism – this is a chronic disorder involving severe pain in the joints; it has many causes, one of which is believed to be the muscular strains induced by poor manual-handling lifting technique.

The sites on the body of injuries caused by manual handling accidents are shown in Figure 12.3.

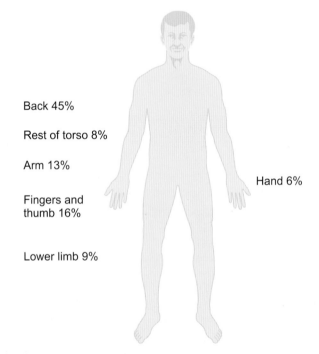

Back 45%

Rest of torso 8%

Arm 13%

Fingers and thumb 16%

Lower limb 9%

Hand 6%

Figure 12.3 Main injury sites caused by manual handling accidents.

In general, pulling a load is much easier for the body than pushing one. If a load can only be pushed, then pushing backwards using the back is less stressful on body muscles. Lifting a load from a surface at waist level is easier than lifting from floor level and most injuries during lifting are caused by lifting and twisting at the same time. If a load has to be carried, it is easier to carry it at waist level and close to the body trunk. A firm grip is essential when moving any type of load.

12.3 Manual handling risk assessments

12.3.1 Hierarchy of measures for manual handling operations

With the introduction of the Manual Handling Operations Regulations, the emphasis during the assessment of lifting operations changed from a simple reliance on safe lifting techniques to an analysis, using risk assessment, of the need for manual handling. The Regulations established a clear hierarchy of measures to be taken when an employer is confronted with a manual handling operation:

> Avoid manual handling operations so far as is reasonably practicable by either redesigning the task to avoid moving the load or by automating or mechanizing the operations
> If manual handling cannot be avoided, then a suitable and sufficient risk assessment should be made
> Reduce the risk of injury from those operations so far as is reasonably practicable, either by the use of mechanical handling or making improvements to the task, the load and the working environment.

The guidance given to the Manual Handling Operations Regulations (available in the HSE Legal series – L23) is a very useful document. It gives helpful advice on manual handling assessments and manual handling training. The advice is applicable to all occupational sectors. Appendix 12.1 gives an example of a manual handling assessment form.

In the construction industry, the typical manual handling tasks that occur on most sites are lifting building blocks, cement bags, doors, windows and lintels. Another common example occurs in road and street works – the lifting and manoeuvring of kerb stones and paving slabs. These examples can lead not only to a strained back and pulled muscles but also to trapped fingers and toes. This indicates the need for strong gloves and safety footwear. The handling of kerb stones is discussed in more detail in Section 12.3.4.

12.3.2 Manual handling assessments

The Regulations specify four main factors which must be taken into account during the assessment. These are the task, the load, the working environment and the capability of the individual who is expected to do the lifting.

The **task** should be analysed in detail so that all aspects of manual handling are covered including the use of mechanical assistance. The number of people involved and the cost of the task should also be considered.

Some or all of the following questions are relevant to most manual handling tasks:

> Is the load held or manipulated at a distance from the trunk? The further from the trunk, the more difficult it is to control the load and the stress imposed on the back is greater.
> Is a satisfactory body posture being adopted? Feet should be firmly on the ground and slightly apart and there should be no stooping or twisting of the trunk. It should not be necessary to reach upwards since this will place additional stresses on the arms, back and shoulders. The effect of these risk factors is significantly increased if several are present while the task is being performed.
> Are there excessive distances over which to carry or lift the load? Over distances greater than 10 m, the physical demands of carrying the load will dominate the operation. The frequency of lifting, and the vertical and horizontal distances the load needs to be carried (particularly if it has to be lifted from the ground and/or placed on a high shelf) are very important considerations.
> Is there excessive pulling and pushing of the load? The state of floor surfaces and the footwear of the individual should be noted so that slips and trips may be avoided.
> Is there a risk of a sudden movement of the load? The load may be restricted or jammed in some way.
> Is frequent or prolonged physical effort required? Frequent and prolonged tasks can lead to fatigue and a greater risk of injury.
> Are there sufficient rest or recovery periods? Breaks and/or the changing of tasks enable the body to recover more easily from strenuous activity.
> Is there an imposed rate of work on the task? This is a particular problem with some automated production lines and can be addressed by spells on other operations away from the line.
> Are the loads being handled while the individual is seated? In these cases, the legs are not used during the lifting processes and stress is placed on the arms and back.
> Does the handling involve two or more people? The handling capability of an individual reduces when he becomes a member of a team (e.g. for a three person team, the capability is half the sum of the individual capabilities). Visibility, obstructions and the roughness of the ground must all be considered when team handling takes place.

The **load** must be carefully considered during the assessment and the following questions asked:

> Is the load too heavy? The maximum load that an individual can lift will depend on the capability of

the individual and the position of the load relative to the body. There is therefore no safe load, but Figure 12.4 is reproduced from the HSE guidance, which does give some advice on loading levels. It does not recommend that loads in excess of 25 kg should be lifted or carried by a man (and this is only permissible when the load is at the level of and adjacent to the thighs). For women, the guideline figures should be reduced by about one-third.

➤ Is the load too bulky or unwieldy? In general, if any dimension of the load exceeds 0.75 m (2 ft), its handling is likely to pose a risk of injury. Visibility around the load is important. It may hit obstructions or become unstable in windy conditions. The position of the centre of gravity is very important for stable lifting – it should be as close to the body as possible.

➤ Is the load difficult to grasp? Grip difficulties will be caused by slippery surfaces, rounded corners or a lack of foot room.

➤ Are the contents of the load likely to shift? This is a particular problem when the load is a container full of smaller items, such as a sack full of nuts and bolts. The movement of people (e.g. in a nursing home) or animals (e.g. in a veterinary surgery) fall into this category as loads.

➤ Is the load sharp, hot or cold? Personal protective equipment may be required.

The **working environment** in which the manual handling operation is to take place, must be considered during the assessment. The following areas will need to be assessed:

➤ any space constraints which might inhibit good posture; such constraints include lack of headroom, narrow walkways and items of furniture

➤ slippery, uneven or unstable floors
➤ variations in levels of floors or work surfaces, possibly requiring the use of ladders
➤ extremes of temperature and humidity; these effects are discussed in detail in Chapter 17
➤ ventilation problems or gusts of wind
➤ poor lighting conditions.

Finally, the capability of the **individual** to lift or carry the load must be assessed. The following questions will need to be asked:

➤ Does the task require unusual characteristics of the individual (e.g. strength or height)? It is important to remember that the strength and general manual handling ability depends on age, gender, state of health and fitness.

➤ Are employees who might reasonably be considered to be pregnant or to have a health problem, put at risk by the task? Particular care should be taken to protect pregnant women or those who have recently given birth from handling loads. Allowance should also be given to any employee who has a health problem, which could be exacerbated by manual handling.

The assessment must be reviewed if there is reason to suspect that it is no longer valid or there has been a significant change to the manual handling operations to which it relates.

12.3.3 Reducing the risk of injury

This involves the introduction of control measures resulting from the manual handling risk assessment. The guidance

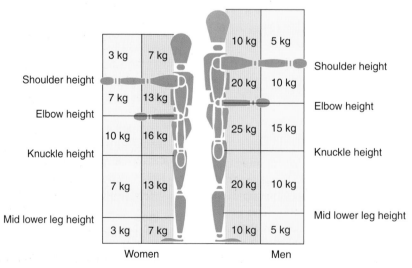

Figure 12.4 HSE guidance for manual lifting – recommended maximum weights.

Figure 12.5 Moving paving blocks.

to the Regulations (L23) and the HSE publication 'Manual Handling – solutions you can handle' (HSG115) contain many ideas to reduce the risk of injury from manual handling operations. An ergonomic approach, discussed in detail in Chapter 17, is generally required to design and develop the manual handling operation as a whole. The control measures can be grouped under five headings. However the first consideration, when it is reasonably practicable, is mechanical assistance.

Mechanical assistance involves the use of mechanical aids to assist the manual handling operation such as wheel barrows, hand-powered hydraulic hoists, specially adapted trolleys, hoist for lifting patients, roller conveyors and automated systems using robots.

Figure 12.6 A pallet truck.

The **task** can be improved by changing the layout of the workstation by, for example, storing frequently used loads at waist level. The removal of obstacles and the use

of a better lifting technique that relies on the leg rather than back muscles should be encouraged. When pushing, the hands should be positioned correctly. The work routine should also be examined to see whether job rotation is being used as effectively as it could be. Special attention should be paid to seated manual handlers to ensure that loads are not lifted from the floor whilst they are seated. Employees should be encouraged to seek help if a difficult load is to be moved so that a team of people can move the load. Adequate and suitable personal protective equipment should be provided where there is a risk of loss of grip or injury. Care must be taken to ensure that the clothing does not become a hazard in itself (e.g. the snagging of fasteners and pockets).

The **load** should be examined to see whether it could be made lighter, smaller or easier to grasp or manage. This could be achieved by splitting the load, the positioning of handholds or a sling, or ensuring that the centre of gravity is brought closer to the handler's body. Attempts should be made to make the load more stable and any surface hazards, such as slippery deposits or sharp edges, should be removed. It is very important to ensure that any improvements do not, inadvertently, lead to the creation of additional hazards.

The **working environment** can be improved in many ways. Space constraints should be removed or reduced. Floors should be regularly cleaned and repaired when damaged. Adequate lighting is essential and working at more than one level should be minimized so that hazardous ladder work is avoided. Attention should be given to the need for suitable temperatures and ventilation in the working area.

The **capability of the individual** is the fifth area where control measures can be applied to reduce the risk of injury. The state of health of the employee and his/her medical record will provide the first indication as to whether the individual is capable of undertaking the task. A period of sick leave or a change of job can make an individual vulnerable to manual handling injury. The Regulations require that the employee be given information and training. The information includes the provision, where it is reasonably practicable to do so, of precise information on the weight of each load and the heaviest side of any load whose centre of gravity is not centrally positioned. The training requirements are given in the next section.

In a more detailed risk assessment, other factors will need to be considered such as the effect of personal protective equipment and psychosocial factors in the work organization. The following points my need to be assessed:

➤ Does protective clothing hinder movement or posture?
➤ Is the correct personal protective equipment being worn?
➤ Is proper consideration given to the planning and scheduling of rest periods?
➤ Is there good communication between managers and employees during risk assessment or workstation design?
➤ Is there a mechanism in place to deal with sudden changes in the volume of workload?
➤ Have employees been given sufficient training and information?
➤ Has the worker any learning disabilities and, if so, has this been taken into account in the assessment?

An example of a manual handling assessment in construction would be the moving of a rolled steel joist (RSJ) beam. The assessment would need to consider the weight of the beam, the distance that it needed to be carried and the vertical height of the required lift. Other factors would include the condition of the floor and the space available to manoeuvre the beam into its final position.

Recent amendments to the Manual Handling Operations Regulations have emphasized that a worker may be at risk if he/she:

➤ is physically unsuited to carry out the tasks in question
➤ is wearing unsuitable clothing, footwear or other personal effects or
➤ does not have adequate or appropriate knowledge or training.

HSE has developed a Manual Handling Assessment Chart (MAC) to help with the assessment of common risks associated with lifting, carrying and handling. The MAC is available on the HSE website.

12.3.4 *Handling kerb stones*

The handling of kerb stones is one of the main causes of back problems and other musculoskeletal disorders. Standard kerb stones weigh approximately 67 kg and are made of pre-cast concrete. Thus the main hazards associated with the manual handling of kerb stones are the weight of the stones, unsuitable body posture and the frequency and repetitive nature of the work. The risk, therefore, of injury to workers who lay kerb stones by hand is high and employers must address all three hazards. HSE recommends a hierarchy of control measures in their information sheet, CIS No. 57, as follows:

1. Eliminate manual lifting at the design stage perhaps by dispensing with the need for a kerb.
2. Use total mechanization by ensuring that kerb stones are always lifted and laid mechanically (e.g. vacuum devices or a vehicle mounted crane and mechanical grab).
3. Use partial mechanization by ensuring that the maximum amount of kerb handling is done mechanically (e.g. off loading using a hoist).
4. Use manual handling when it is not possible to use any of the above methods; the risk of injury can be reduced if the workers are well trained, lighter kerb stones are used or devices which allow two people to share the lift are provided.

Many of the manual handling problems can be reduced at the design and manufacturing stages. Contractors should provide their workers with training in good handling techniques and the safe use of mechanical lifting equipment.

12.3.5 *Manual handling training*

Training alone will not reduce manual handling injuries – there still needs to be safe systems of work in place and the full implementation of the control measures highlighted in the manual handling assessment. The following topics should be addressed in a manual handling training session:

➤ types of injuries associated with manual handling activities
➤ the findings of the manual handling assessment
➤ the recognition of potentially hazardous manual handling operations
➤ the correct use of mechanical handling aids
➤ the correct use of personal protective equipment
➤ features of the working environment which aid safety in manual handling operations
➤ good housekeeping issues
➤ factors which affect the capability of the individual
➤ good lifting or manual handling technique as shown in Figure 12.7.

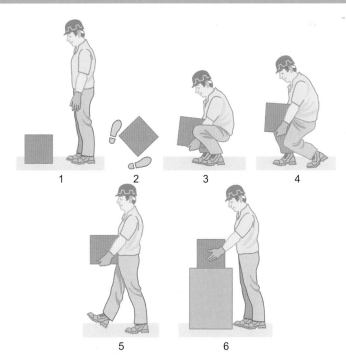

1. Check suitable clothing and assess load. Heaviest side to body.

2. Place feet apart – bend knees.

3. Firm grip – close to body slight bending of back, hips and knees at start.

4. Lift smoothly to knee level and then waist level. No further bending of back.

5. With clear visibility move forward without twisting. Keep load close to the waist. Turn by moving feet. Keep head up; do not look at load.

6. Set load down at waist level or to knee level and then floor.

Figure 12.7 The main elements of a good lifting technique.

Finally, it needs to be stressed that if injuries involving manual handling operations are to be avoided, planning, control and effective supervision are essential.

12.4 Safety in the use of lifting and moving equipment

An amendment to the LOLER included the positioning and installation of lifting equipment and the organization of lifting operations.

12.4.1 *Positioning and installation of lifting equipment*

Lifting equipment must be positioned and installed so as to reduce the risks, so far as is reasonably practicable, from:

➤ equipment or a load striking a person
➤ a load drifting, falling freely or being released unintentionally.

Lifting equipment should be positioned and installed to minimize the need to lift loads over people and to prevent crushing in extreme positions. It should be designed to stop safely in the event of a power failure and not release its load. Lifting equipment, which follows a fixed path, should be enclosed with suitable and substantial interlocked gates and any necessary protection in the event of power failure.

12.4.2 *The organization of lifting operations*

Every lifting operation, that is lifting or lowering of a load, shall be:

➤ properly planned by a competent person
➤ appropriately supervised
➤ carried out in a safe manner.

The person planning the operation should have adequate practical and theoretical knowledge and experience of planning lifting operations. The plan needs to address the risks identified by the risk assessment and identify the resources, the procedures and the responsibilities required so that any lifting operation is carried out safely. For routine simple lifts, a plan will normally be the responsibility of the people using the lifting equipment. For complex lifting operations, for example, where two cranes are used to lift one load, a written plan may need to be produced each time.

The planning should include the need to avoid suspending loads over occupied areas, visibility, the attaching/detaching and securing of loads, the environment, the location, the possibility of overturning, the proximity to other objects, any lifting of people and the pre-use checks required for the equipment.

213

12.4.3 *Summary of the requirements for lifting operations*

There are four general requirements for all lifting operations:

- strong, stable and suitable lifting equipment
- the equipment should be positioned and installed correctly
- the equipment should be visibly marked with the safe working load
- lifting operations must be planned, supervised and performed in a safe manner by competent people.

12.5 Types of mechanical handling and lifting equipment

There are four elements to mechanical handling, each of which can present hazards. These are handling equipment, the load, the workplace and the employees involved.

The **mechanical handling equipment** must be capable of lifting and/or moving the load. It must be fault-free, well maintained and inspected on a regular basis. The hazards related to such equipment include collisions between people and the equipment and personal injury from being trapped in moving parts of the equipment (such as belt and screw conveyors).

The **load** should be prepared for transportation in such a way as to minimize the possibility of accidents. The hazards will be related to the nature of the load (e.g. substances which are flammable or hazardous to health) or the security and stability of the load (e.g. collapse of bales or incorrectly stacked pallets).

The **workplace** should be designed so that, whenever possible, workers and the load are kept apart. If, for example, an overhead crane is to be used, then people should be segregated away or barred from the path of the load.

The **employees** and others and any other people who are to use the equipment must be properly trained and competent in its safe use.

12.5.1 *Conveyors and elevators*

Conveyors transport loads along a given level which may not be completely horizontal, whereas elevators move loads from one level or floor to another.

There are three common forms of **conveyor** – belt, roller and screw conveyors. The most common hazards and preventative measures are:

- the in-running nip, where a hand is trapped between the rotating rollers and the belt; protection from this hazard can be provided by nip guards and trip devices
- entanglement with the power drive requiring the fitting of fixed guards and the restriction of loose clothing which could become caught in the drive
- loads falling from the conveyor; this can be avoided by edge guards and barriers
- impact against overhead systems; protection against this hazard may be given by the use of bump caps, warning signs and restricted access
- contact hazards prevented by the removal of sharp edges, conveyor edge protection and restricted access
- manual handling hazards
- noise and vibration hazards.

Screw conveyors, often used to move very viscous substances, must be provided with either fixed guards or covers to prevent accidental access. People should be prohibited from riding on belt conveyors, and emergency trip wires or stop buttons must be fitted and operational at all times.

Elevators are used to transport goods between floors, such as the transportation of building bricks to upper storeys during the construction of a building or the transportation of grain sacks into the loft of a barn. Guards should be fitted at either end of the elevator and around the power drive. The most common hazard is injury due to loads falling from elevators. There are also potential manual handling problems at both the feed and discharge ends of the elevator.

12.5.2 *Fork-lift trucks*

The most common form of mobile handling equipment is the fork-lift truck. It comes from the group of vehicles known as lift trucks, and can be used in factories, on construction sites and on farms. The term fork-lift truck is normally applied to the counterbalanced lift truck, where the load on the forks is counterbalanced by the weight of the vehicle over the rear wheels. The reach truck is designed to operate in narrower aisles in warehouses and enables the load to be retracted within the wheelbase. The very narrow aisle (VNA) truck does not turn within the aisle to deposit or retrieve a load. It is often guided by guides or rails on the floor. Other forms of lift truck include the pallet truck and the pallet stacker truck, both of which may be pedestrian or rider controlled.

On construction sites, fork-lift trucks of various sizes (including rough terrain, see Figure 12.10) are used to lift loads on to scaffolding working platforms. The loading bay for such working platforms must be constructed to a higher specification than an ordinary scaffold with standards fitted closer together than is normal with extra bracing provided for greater rigidity.

There are many hazards associated with the use of fork-lift trucks. These include:

➤ overturning – manoeuvring at too high a speed (particularly cornering), wheels hitting an obstruction such as a kerb, sudden braking, poor tyre condition leading to skidding, driving forwards down a ramp, movement of the load, insecure, excessive or uneven loading, incorrect tilt or driving along a ramp
➤ overloading – exceeding the rated capacity of the machine

➤ collisions – particularly with warehouse racking which can lead to a collapse of the whole racking system
➤ silent operation of the electrically powered fork-lift truck – can make pedestrians unaware of its presence
➤ uneven road surface – due perhaps to potholes following a heavy storm – can cause the vehicle to overturn and/or cause musculoskeletal problems for the driver
➤ overhead obstructions – a problem for inexperienced drivers (overhead power lines pose a particular problem)

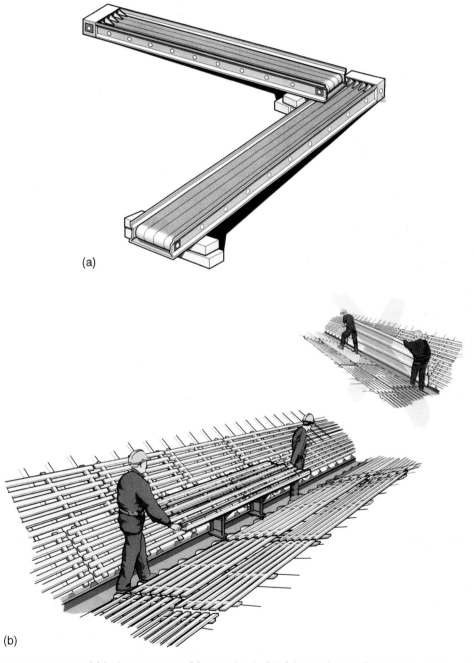

(a)

(b)

Figure 12.8   Conveyor systems: (a) belt conveyors; (b) a method of safely moving roofing sheets along a roof valley;

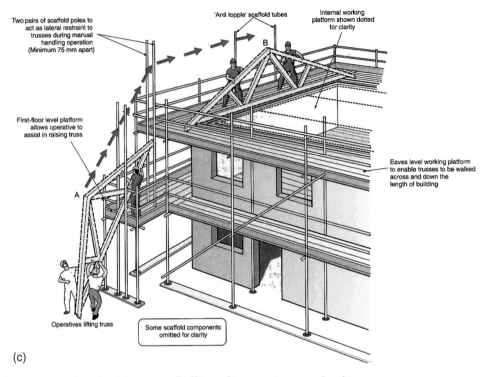

Figure 12.8 (c) a suggested method for manually lifting of trusses to eaves level.

Figure 12.9 A brick elevator.

➤ loss of load – shrink wrapping or sheeting will reduce this hazard

➤ inadequate maintenance leading to mechanical failure

Figure 12.10 Rough terrain counter-balanced lift truck.

➤ use as a work platform

➤ speeding – strict enforcement of speed limits is essential

➤ poor vision around the load

➤ pedestrians – particularly when pedestrians and vehicles use the same roadways; warning signs, indicating the presence of fork-lift trucks, should be posted at regular intervals

➤ dangerous stacking or de-stacking technique – this can destabilize a complete racking column

➤ carrying passengers – this should be a disciplinary offence

➤ battery charging – presents an explosion and fire risk

➤ fire – often caused by poor maintenance resulting in fuel leakages or engine/motor burn-out, or through using a fork-lift truck in areas where flammable liquids or gases are used and stored

➤ lack of driver training.

If fork-lift trucks are to be used outside, visibility and lighting, weather conditions and the movement of other vehicles become additional hazards.

There are also the following physical hazards:

➤ noise – caused by poor silencing of the power unit

➤ exhaust fumes – should only be a problem when the maintenance regime is poor

➤ vibrations – often caused by a rough road surface or wide expansion joints; badly inflated tyres will exacerbate this problem

➤ manual handling – resulting from manoeuvring the load by hand or lifting batteries or gas cylinders

➤ ergonomic – musculoskeletal injuries caused by soft tyres and/or undulating road surface or holes or cracks in the road surface (e.g. expansion joints).

Regular and documented maintenance by competent mechanics is essential. However, the driver should undertake the following checks at the beginning of each shift:

➤ condition of tyres and correct tyre pressures

➤ effectiveness of all brakes

➤ audible reversing horn and light working properly

➤ lights, if fitted, working correctly

➤ mirrors, if fitted, in good working order and properly set

➤ secure and properly adjusted seat

➤ correct fluid levels, when appropriate

➤ fully charged batteries, when appropriate

➤ correct working of all lifting and tilting systems.

A more detailed inspection should be undertaken by a competent person within the organization on a weekly basis to include the mast and the steering gear. Driver training is essential and should be given by a competent trainer. The training session must include the site rules covering items such as the fork-lift truck driver code of practice for the organization, speed limits, stacking procedures and reversing rules. Refresher training should be provided at regular intervals and a detailed record kept of all training received. Table 12.1 illustrates some key requirements of fork-lift truck drivers and the points listed should be included in most codes of practice.

Finally, care must be taken with the selection of drivers, including relevant health checks and previous experience.

Table 12.1 Safe driving of lift trucks

Drivers must:
➤ drive at a suitable speed to suit road conditions and visibility ➤ use the horn when necessary (at blind corners and doorways) ➤ always be aware of pedestrians and other vehicles ➤ take special care when reversing (do not rely on mirrors) ➤ take special care when handling loads which restrict visibility ➤ travel with the forks (or other equipment fitted to the mast) lowered ➤ use the prescribed lanes ➤ obey the speed limits ➤ take special care on wet and uneven surfaces ➤ use the handbrake, tilt and other controls correctly ➤ take special care on ramps ➤ always leave the truck in a state which is safe and discourages unauthorized use (brake on, motor off, forks down, key out).
Drivers must not:
➤ operate in conditions in which it is not possible to drive and handle loads safely (e.g. partially blocked aisles) ➤ travel with the forks raised ➤ use the forks to raise or lower persons unless a purpose-built working cage is used ➤ carry passengers ➤ park in an unsafe place (e.g. obstructing emergency exits) ➤ turn round on ramps ➤ drive into areas where the truck would cause a hazard (flammable substance store) ➤ allow unauthorized use.

Drivers should be at least 18 years of age and their fitness to drive should be re-assessed at 3-yearly intervals.

12.5.3 Other forms of lifting equipment

In construction work, there are many different types of vehicles and plant used to lift and move loads. These include excavators, telehandlers, rough terrain vehicles, dumper trucks and hoists, all of which are covered in more detail elsewhere in this book (Chapters 11, 13, 18 and 19).

A weekly inspection of a 360° wheeled excavator would include the condition of the tyres, wheels, steering controls, lights, brakes and hydraulic systems. Periodically, the seat rigidity and the proper functioning of seat restraints, instrumentation and controls, bucket condition and other attachments need to be inspected. Appendix 12.2 gives an example of a typical risk assessment used for an excavator.

Another of these pieces of plant will be discussed in a little more detail here (see Figure 12.11). The small dumper truck is widely used on all sizes of construction site.

Compact dumper trucks are involved in about 30% of construction transport accidents. The three main causes of such accidents are:

➤ overturning on slopes and at the edges of excavations
➤ poorly maintained braking systems
➤ driver error due to lack of training and/or inexperience.

Some of the hazards associated with these vehicles are collisions with pedestrians, other vehicles or structures, such as scaffolding. They can be struck by falling materials and tools or be overloaded. The person driving the truck can be thrown from the vehicle, come into contact with moving parts on the truck, suffer the effects of whole body vibrations due to driving over potholes in the roadway and suffer from the effects of noise and dust.

The precautions that can be taken to address these hazards include the use of authorized, trained, competent and supervised drivers only. As with so many other construction operations, risks should be assessed, safe systems of work followed and drivers forbidden from

Figure 12.11 Lifting roof trusses.

Figure 12.12 Typical jib tower crane.

taking short cuts. The following site controls should also be in place:

- designated traffic routes and signs
- speed limits
- stop blocks used when the vehicle is stationary
- proper inspection and maintenance procedures
- procedures for starting, loading and unloading the vehicle
- provision of roll-over protective structures (ROPS) and seat restraints
- provision of falling-object protective structures (FOPS) when there is a risk of being hit by falling materials
- visual and audible warning of approach
- where necessary, hearing protection.

For other forms of mobile construction equipment, the risk to people from the overturning of the equipment must always be safeguarded. This can usually be achieved by the avoidance of working on steep slopes, the provision of stabilizers and ensuring that the load carried does not affect the stability of the equipment/vehicle.

The other types of lifting equipment to be considered are cranes (mobile overhead and jib), lifts and hoists and lifting tackle. A sample risk assessment for the use of lifting equipment is given in Appendix 12.3.

Cranes may be either a jib crane, a tower crane or an overhead gantry travelling crane. The safety requirements are similar for each type. All cranes need to be properly designed, constructed, installed and maintained. They must also be operated in accordance with a safe system of work. They should only be driven by authorized persons who are fit and trained. Each crane is issued with a certificate by its manufacturer giving details of the SWL. The SWL must **never** be exceeded and should be marked on the crane structure. If the SWL is variable, as with a jib crane (the SWL decreases as the operating radius increases), an SWL indicator should be fitted. Care should be taken to avoid sudden shock loading since this will impart very high stresses to the crane structure. It is also very important that the load is properly shackled and all eyebolts tightened. Safe slinging should be included in any training programme. All controls should be clearly marked and be of the 'hold to run' type.

Large cranes, which incorporate a driving cab, often work in conjunction with a banksman, who will direct the lifting operation from the ground. It is important that banksmen are trained so that they understand recognized crane signals.

During the planning of a lifting operation close to a public highway, there should be liaison with the police and local authority over any road closures and the lift should be timed to take place when there will be a minimum number of people about. Arrangements may be needed to give advanced warning to motorists and other road users and appropriate signs and traffic controls used.

The basic principles for the safe operation of cranes are as follows. **For all cranes**, the driver must:

- undertake a brief inspection of the crane and associated lifting tackle each time before it is used
- check that all lifting accessory statutory inspections are in place
- check that tyre pressures, where appropriate, are correct
- ensure that loads are not left suspended when the crane is not in use
- before a lift is made, ensure that nobody can be struck by the crane or the load
- ensure that loads are never carried over people
- ensure good visibility and communications with the signaller
- lift loads vertically – cranes must not be used to drag a load
- travel with the load as close to the ground as possible
- switch off power to the crane when it is left unattended.

For mobile jib cranes, the following points should be considered:

- Each lift must be properly planned, with the maximum load and radius of operation known.

➤ Overhead obstructions or hazards must be identified; it may be necessary to protect the crane from overhead power lines by using goal posts and bunting to mark the safe headroom.
➤ The ground on which the crane is to stand should be assessed for its load-bearing capacity.
➤ If fitted, outriggers should be used.

The principal reasons for crane failure, including loss of load, are:

➤ overloading or incorrect use of lifting gear
➤ automatic safe load indicator not working
➤ poor slinging of load
➤ insecure or unbalanced load
➤ loss of load
➤ overturning
➤ collision with another structure or overhead power lines
➤ collision or struck by a vehicle
➤ strong winds
➤ foundation failure
➤ structural failure of the crane
➤ operator error
➤ lack of maintenance and/or regular inspections
➤ no signaller used when driver's view is obscured
➤ incorrect signals given.

During lifting operations using cranes, it must be ensured that:

➤ the driver has good visibility
➤ there are no pedestrians below the load by demarcating the area of lift and using barriers, if necessary
➤ there is adequate space for the installation, manoeuvring and operation of the crane
➤ all pedestrians are re-routed during the lifting operation
➤ everyone within the lifting area is wearing a safety helmet
➤ an audible warning is given prior to the lifting operation.

If lifting takes place in windy conditions, tag lines may need to be attached to the load to control its movement.

The Health and Safety (Safety Signs and Signals) Regulations and BS 7121 give examples of signals to be used when guiding cranes. The signalman should wear one or more appropriate distinctive items (a jacket, helmet, sleeves or armbands or carry bats). The distinctive items should be all of the same colour and for the exclusive use of the signalman. Figure 12.13 illustrates the basic signals.

The basic signals are as follows:

(a) General signals
  ➤ start operations
  ➤ stop interruption or end of movement
  ➤ end of the operation.
(b) Vertical movements
  ➤ raise
  ➤ lower
  ➤ vertical distance.
(c) Horizontal movements
  ➤ move forwards
  ➤ move backwards
  ➤ right

Meaning	Description	Illustration
General signals		
START Attention Start of Command	both arms are extended horizontally with the palms facing forwards	
STOP Interruption End of movement	the right arm points upwards with the palm facing forwards	
END of the operation	both hands are clasped at chest height	
Vertical movements		
RAISE	the right arm points upwards with the palm facing forwards and slowly makes a circle	
LOWER	the right arm points downwards with the palm facing inwards and slowly makes a circle	
VERTICAL DISTANCE	the hands indicate the relevant distance	

Figure 12.13 Minimum requirements for hand signals.

Meaning	Description	Illustration

Horizontal movements

MOVE FORWARDS	both arms are bent with the palms facing upwards, and the forearms make slow movements towards the body	
MOVE BACKWARDS	both arms are bent with palms facing downwards, and the forearms make slow movements away from the body	
RIGHT to the signalman's	the right arm is extended more or less horizontally with the palm facing downwards and slowly makes small movements to the right	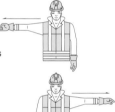
LEFT to the signalman's	the left arm is extended more or less horizontally with the palm facing downwards and slowly makes small movements to the left	
HORIZONTAL DISTANCE	the hands indicate the relevant distance	

Danger

DANGER Emergency stop	both arms point upwards with the palms facing forwards	
QUICK	all movements faster	
SLOW	all movements slower	

Figure 12.13 Continued.

> left
> horizontal distance.
(d) Danger
> danger emergency stop
> quick – all movements faster
> slow – all movements slower.

Finally, if a crane is hired to lift a load, the following items should be undertaken before the lift takes place:

> the suitability of the crane, in particular its load carrying capacity
> the validity of its documentation relating to inspection (test certificate) and maintenance

> an assessment of the area where the lifting is to take place including the ground conditions and the proximity of overhead power lines
> the competence of the driver, slinger and banksman
> the co-ordination of the lifting operation
> consultation with others affected by the work.

Following a spate of recent accidents involving cranes, an industry forum has been set up to influence health and safety throughout the construction industry. It recommends that site-specific inductions are given to everyone involved with tower crane erection, operation and dismantling with the emphasis on suitable risk assessment.

A **lift or hoist** incorporates a platform or cage and is restricted in its movement by guides. Hoists are generally used in industrial settings (e.g. construction sites and garages), whereas lifts are normally used inside buildings. Lifts and hoists may be designed to carry passengers and/or goods alone. They should be of sound mechanical construction and have interlocking doors or gates, which must be completely closed before the lift or hoist moves. Passenger carrying lifts must be fitted with an automatic braking system to prevent overrunning, at least two suspension ropes, each capable alone of supporting the maximum working load, and a safety device which could support the lift in the event of suspension rope failure. Maintenance procedures must be rigorous, recorded and only undertaken by competent persons. It is very important that a safe system of work is employed during maintenance operations to protect others, such as members of

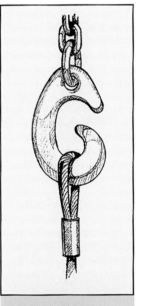
A hook designed to prevent displacement of the load.

A hook with a safety catch.

Figure 12.14 Specially designed safety hooks.

the public, from falling down the lift shaft and other hazards. More information on hoists is given in Chapter 18.

Other **items of lifting tackle**, usually used with cranes, include chain slings and hooks, wire and fibre rope slings, eyebolts and shackles. Special care should be taken when slings are used to ensure that the load is properly secured and balanced. Lifting hooks should be checked for signs of wear and any distortion of the hook. Shackles and eyebolts must be correctly tightened. Slings should always be checked for any damage before they are used and only competent people should use them. Wire rope slings are often found on construction sites and widely misused and become defective. Common problems are kinking of the wire and broken wire strands; the distortion (into an oval shape) of the hook caused by stretching due to overloading; damage to thimbles and ferrules; and 'birdcaging' of the wire strands.

Training and instruction in the use of lifting tackle is essential and should include regular inspections of the tackle, in addition to the mandatory thorough examinations. Finally, care should be taken when these items are being stored between use.

12.6 Requirements for the statutory examination of lifting equipment

The LOLER 1998 specify examinations required for lifting equipment by using two terms – an inspection and a thorough examination. Both terms are defined by the HSE in guidance accompanying various regulations.

An **inspection** is used to identify whether the equipment can be operated, adjusted and maintained safely so that any defect, damage or wear can be detected before it results in unacceptable risks. It is normally performed by a competent person appointed by the employer (often an employee).

A **thorough examination** is a detailed examination, which may involve a visual check, a disassembly and testing of components and/or an equipment test under operating conditions. Such an examination must normally be carried out by a competent person who is independent of the employer. The examination is usually carried out according to a written scheme and a written report is submitted to the employer.

A detailed summary of the thorough examination and inspection requirements of the LOLER is given in Chapter 21.

A thorough examination of lifting equipment should be undertaken at the following times:

➤ before the equipment is used for the first time
➤ after it has been assembled at a new location
➤ at least every 6 months for equipment for lifting persons or a lifting accessory

➤ at least every 12 months for all other lifting equipment including the lifting of loads over people
➤ in accordance with a particular examination scheme drawn up by an independent competent person
➤ each time that exceptional circumstances, which are likely to jeopardize the safety of the lifting equipment, have occurred (such as severe weather).

The person making the thorough examination of lifting equipment must:

➤ notify the employer forthwith of any defect which, in their opinion, is or could become dangerous
➤ as soon as practicable (within 28 days) write an authenticated report to the employer and any person who leased or hired the equipment.

The Regulations specify the information that should be included in the report.

The initial report should be kept for as long as the lifting equipment is used (except for a lifting accessory which need only be kept for 2 years). For all other examinations, a copy of the report should be kept until the next thorough examination is made or for 2 years (whichever is the longer). If the report shows that a defect exists, which could lead to an existing or imminent risk of serious personal injury, a copy of the report must be sent, by the person making the thorough examination, to the appropriate enforcing authority.

The equipment should be inspected at suitable intervals between thorough examinations. The frequency and the extent of the inspection are determined by the level of risk presented by the lifting equipment. A report or record should be made of the inspection which should be kept until the next inspection. Unless stated otherwise, lifts and hoists should be inspected every week.

12.7 Practice NEBOSH questions for Chapter 12

1. (a) **Outline** the main requirements of the Manual Handling Operations Regulations 1992.
   (b) **List** the possible indications of a manual handling problem in a workplace.

2. **Outline** the factors to be considered when undertaking a manual handling assessment of a task that involves lifting buckets of water out of a sink.

3. (a) **List TWO** types of injury that may be caused by the incorrect manual handling of loads.
   (b) **Outline** a good manual handling technique that could be adopted by a person required to lift a load from the ground.

4. An RSJ is to be used in the support of the second floor of a town house that is being refurbished. The layout of the house and the arrangement of the supports are such that mechanical means cannot be used either to transport the RSJ from the ground floor or to lift it into its final position.

     Outline the factors that would need to be considered when undertaking a manual handling assessment of the task.

5. Concrete blocks are to be moved by hand on a construction site.

     Outline the key issues to be addressed prior to this activity being carried out.

6. Concrete building blocks are to be moved on a construction site.

     Outline the key issues to be addressed if the blocks are to be moved by mechanical and manual means. **Give** practical examples within your answer.

7. (a) **Identify** the main risks to health and safety associated with kerb-laying.
    (b) **Outline** the steps to be taken to minimize the risks identified in (a).

8. **Outline** the health and safety considerations when a fork-lift truck is to be used to unload palletized goods from a vehicle parked in a factory car park.

9. **List EIGHT** items to be included on a checklist for the routine inspection of a fork-lift truck at the beginning of a shift.

10. **List** the ways in which a fork-lift truck may become unstable while in operation.

11. **Identify** the weekly maintenance inspection requirements for 360° wheeled excavator.

12. **Identify** the particular items that should be included in a periodic inspection of a 360° tracked excavator.

13. **Outline** the hazards associated with small dumper trucks.

14. The Provision and Use of Work Equipment Regulations (PUWER) 1998 require that work equipment used in hostile environments is inspected at suitable intervals.

     Identify the items on a small dumper truck that should be the subject of such an inspection.

15. **Outline** a procedure for the safe lifting of a load by a crane, having ensured that the crane has been correctly selected and positioned for the job.

16. An engineering workshop uses an overhead gantry crane to transport materials.
    (a) **Identify TWO** reasons why loads may fall from this crane.
    (b) **Outline** precautions to prevent accidents to employees working at ground level when overhead cranes are in use.

17. BS 7121 'Safe Use of Cranes' describes the signals that can be given by a signalman to the driver of a crane.
    (a) **List** eight these signals.
    (b) **Describe**, or **sketch**, **FOUR** different types of signal identified in (a).

18. **List** the main items to be checked by the person appointed to have overall control of a lifting operation to be carried out with the use of a crane hired for the purpose.

19. **Outline** the factors that might cause a mobile crane to overturn during use.

20. **Describe**, or **sketch**, **EIGHT** types of hand signal that can be given by a signaller to the driver of a crane during a lifting operation (as specified by BS 7121 'Safe Use of Cranes').

21. **Describe FOUR** defects that might be identified in a wire rope sling during routine inspection prior to use.

22. **Outline** the requirements relating to examinations and inspections of lifting equipment used to lift persons.

Appendix 12.1 – Manual handling of loads assessment checklist

Section A – Preliminary assessment	
Task description: Factors beyond the limits of the guidelines?	Is an assessment needed? (i.e. is there a potential risk of injury, and are the factors beyond the limits of the guidelines?) Yes/No
If 'YES' continue. If 'NO' no further assessment required.	
Tasks covered by this assessment: Location: People involved: Date of assessment:	Diagram and other information
Section B – See detailed analysis form	
Section C – Overall assessment of the risk of injury? Low/Med/High	
Section D – Remedial action needed:	
Remedial action, in priority order: A B C D F G H	
Date action should be completed:	Date for review:
Assessor's name: Signature:	

Manual Handling Risk Assessment—detailed analysis				
Task description		Employee		
		I.D. Number		
Risk Factors				
A. Task characteristics	Yes/No	Risk Level		Current Controls
		H	M	L	See guidance
1. Loads held away from trunk?				
2. Twisting?				

3. Stooping?			
4. Reaching upwards?			
5. Extensive vertical movements?			
6. Long carrying distances?			
7. Strenuous pushing or pulling?			
8. Unpredictable movements of loads?			
9. Repetitive handling operations?			
10. Insufficient periods of rest/recovery?			
11. High work rate imposed?			
B. Load characteristics			
1. Heavy?			
2. Bulky?			
3. Difficult to grasp?			
4. Unstable/unpredictable?			
5. Harmful (sharp/hot)?			
C. Work environment characteristics			
1. Postural constraints?			
2. Floor suitability?			
3. Even surface?			
4. Thermal/humidity suitability?			
5. Lighting suitability?			
D. Individual characteristics			
1. Unusual capability required?			
2. Hazard to those with health problems?			
3. Hazard to pregnant workers?			
4. Special information/training required?			
E. Other factors to consider			
1. Movement or posture hindered by protective clothing?			
2. Absence of correct/suitable PPE?			
3. Lack of planning and scheduling of basics/rest basics?			

(Continued)

225

Appendix 12.1 – *Continued*

4. Poor communication between managers and employees?					
5. Sudden changes in workload or seasonal changes in volume without mechanisms to deal with change?					
6. Lack of training and/or information?					
7. Any learning disabilities?					
Any further action needed?	**Yes/No**				
Details:					

Appendix 12.2 – A typical risk assessment for an excavator to be used for lifting

INITIAL RISK ASSESSMENT	Excavators used for lifting		
SIGNIFICANT HAZARDS	**Low**	**Medium**	**High**
1. Unplanned release or dropping of load		✓	
2. Boom striking overhead obstruction		✓	
3. Persons struck by machine boom or other part			✓
4. Objects falling from boom/bucket		✓	
5.			

ACTION ALREADY TAKEN TO REDUCE THE RISKS:

Compliance with:

Lifting Operations and Lifting Equipment Regulations (LOLER)
Provision and Use of Work Equipment Regulations (PUWER)
Health and Safety Executive Guidance Note – Excavators used as cranes
British Standards – Guide to selection and use of lifting slings for multipurposes
BS – Code of Practice for safe use of wire rope slings for general purposes CDM Regulations

Planning:

Excavators to be used as cranes must be designed for the purpose, and be fitted with sling attachments and check valves. A current test certificate and thorough examination must be available for the excavator. If any exemption certificates apply they should also be available. Excavator safe working load will be greater than the foreseeable weight of loads to be lifted.

Physical:

No persons are allowed to stand or work within operating radius without the operator's permission. Loads must not be slued over personnel, vehicle cabins or huts. A banksman is to be used where driver's vision is impaired or if operating in congested areas. SWL will be clearly marked on excavator, and a table of SWLs will be clearly visible to the driver; it will not be exceeded. The machine will be on a firm, level base, and the lifts will be carried out with the boom parallel to the machine tracks or wheels.

Managerial/Supervisory:

Certification of drivers must be checked; drivers must be over 18. Lifting operations will be supervised to ensure the stability of the machine and the load. Lifting operations will be restricted to those directly associated with excavations. All operatives working within the boom's radius will wear head protection.

Training:

Driver training to CITB standard is required. Operator training for earth-moving machinery. Excavator driving by non-certificate holding operatives is not permitted; this also applies to sub-contractors and the self-employed.

Date of Assessment	Assessment made by ....
Risk Re-Assessment Date	Site Manager's Comments:

Appendix 12.3 – A typical risk assessment for the use of lifting equipment

INITIAL RISK ASSESSMENT	Use of lifting equipment		
SIGNIFICANT HAZARDS	Low	Medium	High
1. Unintentional release of load			✓
2. Unplanned movement of load	✓		
3. Damage to equipment	✓		
4. Crush injuries to personnel		✓	
5.			

ACTION ALREADY TAKEN TO REDUCE THE RISKS:

Compliance with:

Lifting Operations and Lifting Equipment Regulations (LOLER)
Safety Signs and Signals Regulations (SSSR)
Provision and Use of Work Equipment Regulations (PUWER)
British Standard – Specification for flat woven webbing slings
BS – Guide to selection and use of lifting slings for multipurposes

Planning:

Copies of statutory thorough examinations of lifting equipment will be kept on site. Before selection of lifting equipment, the above standards will be considered as well as the weight, size, shape and centre of gravity of the load. Lifting equipment is subject to the planned maintenance programme.

Physical:

All items of lifting equipment will be identified individually and stored so as to prevent physical damage or deterioration. Safe working loads of lifting equipment will be established before use. Packing will be used to protect slings from sharp edges on the load. All items of lifting equipment will be visually examined for signs of damage before use. Ensuring the eyes of strops are directly below the appliance hook, and that tail ropes are fitted to larger loads will check swinging of the load. Banksmen will be used where the lifting equipment operator's vision is obstructed. Approved hand signals will be used.

Managerial/Supervisory:

Only lifting equipment that is in date for statutory examination will be used. Manufacturer's instructions will be checked to ensure that methods of sling attachment and slinging arrangements generally are correct.

Training:

Personnel involved in the slinging of loads and use of lifting equipment will be required to be trained to CITB or equivalent standard. Supervisors will be trained in the supervision of lifting operations.

Date of Assessment	Assessment made by ....
Risk Re-Assessment Date	Site Manager's Comments: ....

Work equipment hazards and control

13

13.1 Introduction

This chapter covers the scope and main requirements for work equipment as covered by Parts II and III of the Provision and Use of Work Equipment Regulations (PUWER). The requirements for the supply of new machinery are also included. Summaries of PUWER and The Supply of Machinery (Safety) Regulations are given in Chapter 21. The safe use of hand tools, hand-held power tools and the proper safe-guarding of a small range of machinery used in construction work are included.

Figure 13.1 Typical machinery safety notice.

Any equipment used by an employee at work is generally covered by the term 'Work Equipment'. The scope is extremely wide and includes hand tools, power tools, ladders, photocopiers, site plant, lifting equipment, forklift trucks and motor vehicles (which are not privately owned). Virtually anything used to do a job of work, including employees' own equipment, is covered. The uses covered include starting or stopping the equipment, repairing, modifying, maintaining, servicing, cleaning and transporting.

Employers and the self-employed must ensure that work equipment is suitable; maintained; inspected if necessary; provided with adequate information and instruction; and only used by people who have received sufficient training.

13.2 Suitability of work equipment and CE marking

13.2.1 *Standards and requirements*

When work equipment is provided it has to conform to standards which cover its supply as a new or second-hand piece of equipment and its use in the workplace. This involves:

- ➤ its initial integrity
- ➤ the place where it will be used
- ➤ the purpose for which it will be used.

There are two groups of law that deal with the provision of work equipment:

- ➤ one deals with what manufacturers and suppliers have to do. This can be called the 'supply' law. One of the most common of these is the Supply of Machinery (Safety) Regulations, which requires manufacturers and suppliers to ensure that machinery is safe when supplied and has CE marking. Its **primary** purpose is to prevent barriers to trade across the EU, and not to protect people at work.
- ➤ the other deals with what the users of machinery and other work equipment have to do. This can be called the 'user' law, PUWER, and applies to most pieces of work equipment. Its **primary** purpose is to protect people at work.

$$\mathsf{C}\,\mathsf{E}$$

Figure 13.2 CE mark.

Under 'user' law employers have to provide safe equipment of the correct type; ensure that it is correctly used; and maintain it in a safe condition. When buying new equipment the 'user' has to check that the equipment complies with all the 'supply' law that is relevant. **The user must check that the machine is safe before it is used**.

Most new equipment, including machinery in particular, must have 'CE' marking when purchased. 'CE' marking is a claim by the manufacturer that the equipment is safe and that they have met relevant supply law. If this is done properly manufacturers will have to do the following:

➤ find out about the health and safety hazards (trapping, noise, crushing, electrical shock, dust, vibration, etc.) that are likely to be present when the machine is used
➤ assess the likely risks
➤ design out the hazards that result in risks or, if that is not possible
➤ provide safeguards (e.g. guarding dangerous parts of the machine, providing noise enclosures for noisy parts) or, if that is not possible
➤ use warning signs on the machine to warn of hazards that cannot be designed out or safeguarded against (e.g. 'noisy machine' signs).

Manufacturers also have to:

➤ keep information, explaining what they have done and why, in a technical file
➤ fix CE marking to the machine where necessary, to show that they have complied with all the relevant supply laws
➤ issue a 'Declaration of Conformity' for the machine (see Figure 13.3). This is a statement that the machine complies with the relevant essential health and safety requirements or with the example that underwent type-examination. A declaration of conformity must:
  ➤ state the name and address of the manufacturer or importer into the EU
  ➤ contain a description of the machine, its make, its type and serial number
  ➤ indicate all relevant European Directives with which the machinery complies
  ➤ state details of any notified body that has been involved
  ➤ specify which Standards have been used in the manufacture (if any)
  ➤ be signed by a person with authority to do so.
➤ provide the buyer with instructions to explain how to install, use and maintain the machinery safely.

This can help to decide which equipment may be suitable, particularly if buying a standard piece of equipment 'off the shelf'.

If buying a more complex or custom-built machine the buyer should discuss their requirements with potential suppliers. For a custom-built piece of equipment, there is the opportunity to work with the supplier to design out

Before buying new equipment the buyer will need to think about:

➤ where and how it will be used
➤ what will it be used for
➤ who will use it (skilled employees, trainees)
➤ what risks to health and safety might result
➤ how well health and safety risks are controlled by different manufacturers.

the causes of injury and ill-health. Time spent now on agreeing the necessary safeguards, to control health and safety risks, could save time and money later.

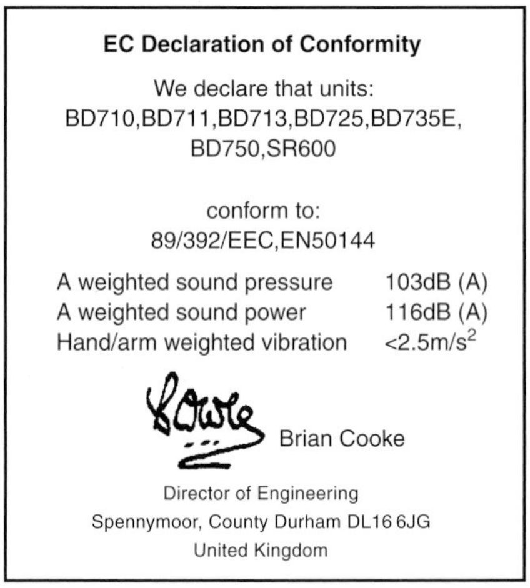

EC Declaration of Conformity

We declare that units:
BD710,BD711,BD713,BD725,BD735E,
BD750,SR600

conform to:
89/392/EEC,EN50144

A weighted sound pressure	103dB (A)
A weighted sound power	116dB (A)
Hand/arm weighted vibration	<2.5m/s^2

Brian Cooke

Director of Engineering
Spennymoor, County Durham DL16 6JG
United Kingdom

Figure 13.3 Typical certificate of conformity.

Note: Sometimes equipment is supplied via another organization, for example, an importer, rather than direct from the manufacturer, so this other organization is referred to as the supplier. It is important to realize that the supplier may not be the manufacturer.

When the equipment has been supplied the buyer should look for CE marking, check for a copy of the Declaration of Conformity and that there is a set of instructions in English on how the machine should be used and most important of all, **check to see that it is safe when in use**.

13.2.2 *Limitations of CE marking*

CE marking is not a guarantee that the machine is safe. It is a claim by the manufacturer that the machinery complies with the law. CE marking has many advantages if done properly, for example:

➤ it allows a common standard across Europe
➤ it provides a means of selling to all European Community Member States without barriers to trade

➤ it ensures that instructions and safety information is supplied in a fairly standard way in most languages in the EC

➤ it has encouraged the use of diagrams and pictorials which are common to all languages

➤ it allows for independent type-examination for some machinery like woodworking machinery which has not been made to an EC harmonized standard, (identified by an EN marking before the standard number (e.g. BS EN ....).

Clearly there are disadvantages as well, for example:

➤ instruction manuals have become very long; sometimes two volumes are provided, because of the number of languages required

➤ translations can be very poor and disguise the proper meaning of the instruction

➤ manufacturers can fraudulently put on the CE marking

➤ manufacturers may make mistakes in claiming conformity with safety laws.

13.3 Use and maintenance of equipment with specific risks

Figure 13.4 Using a bench-mounted abrasive wheel.

Some pieces of work equipment involve specific risks to health and safety where it is not possible to control adequately the hazards by physical measures alone, for example, the use of a bench-mounted circular saw or an abrasive wheel. In all cases the hierarchy of controls should be adopted to reduce the risks by:

➤ eliminating the risks or, if this is not possible

➤ taking physical measures such as guards to control the risks, but if the risks cannot be adequately controlled

➤ taking appropriate measures, such as a safe system of work.

PUWER regulation 7 restricts the use of such equipment to the persons designated to use it. These people need to have received sufficient information, instruction and training so that they can carry out the work, using the equipment, safely.

Repairs, modifications, maintenance or servicing is also restricted to designated persons. A designated person may be the operator if they have the necessary skills and have received specific instruction and training. Another person specifically trained to carry out a particular maintenance task, for example, dressing an abrasive wheel, may not be the operator but designated to do this type of servicing task on a range of machines.

13.4 Information, instruction and training

People using and maintaining work equipment, where there are residual risks that cannot be sufficiently reduced by physical means, require enough information, instruction and training to operate safely.

The information and instructions are likely to come from the manufacturer in the form of operating and maintenance manuals. It is up to the employer to ensure that what is provided is easily understood, set out logically with illustrations and standard symbols where appropriate. The information should normally be in good plain English but other languages may be necessary in some cases.

The extent of the information and instructions will depend on the complexity of the equipment and the specific risks associated with its use. They should cover:

➤ all safety and health aspects
➤ any limitations on the use of the equipment
➤ any foreseeable problems that could occur
➤ safe methods to deal with the problems

any relevant experience with the equipment that would reduce the risks or help others to work more safely, should be recorded and circulated to everyone concerned.

Everyone who uses and maintains work equipment needs to be adequately trained. The amount of training required will depend on:

➤ the complexity and level of risk involved in using or maintaining the equipment

➤ the experience and skills of the person doing the work, whether it is normal use or maintenance.

Training needs will be greatest when a person is first recruited but will also need to be considered:

➤ when working tasks are changed, particularly if the level of risk changes
➤ if new technology or new equipment is introduced
➤ where a system of work changes
➤ when legal requirements change
➤ periodically to update and refresh people's knowledge and skills.

Supervisors and managers also require adequate training to carry out their function, particularly if they only supervise a particular task occasionally. The training and supervision of young persons is particularly important because of their relative immaturity, unfamiliarity with a working environment and awareness of existing or potential risks. Some Approved Codes of Practice, for example on the *Safe Use of Woodworking Machinery* restrict the use of high-risk machinery, so that only young persons with sufficient maturity and competence, who have finished their training, may use the equipment unsupervised.

13.5 Maintenance and inspection

13.5.1 *Maintenance*

Figure 13.5 Typical maintenance notice.

Work equipment needs to be properly maintained so that it continues to operate safely and in the way it was designed to perform. The amount of maintenance will be stipulated in the manufacturers' instructions and will depend on the amount of use, the working environment and the type of equipment. High speed, high-hazard machines, which are heavily used in an adverse environment like salt water, may require very frequent maintenance, whereas a simple hand tool, like a shovel, may require very little.

Maintenance management schemes can be based around a number of techniques designed to focus on those parts which deteriorate and need to be maintained to prevent health and safety risks. These techniques include:

➤ **Preventative planned maintenance** – which involves replacing parts and consumables or making necessary adjustments at preset intervals, normally set by the manufacturer, so that there are no hazards created by component deterioration or failure. Vehicles are normally maintained on this basis.
➤ **Condition-based maintenance** – this involves monitoring the condition of critical parts and carrying out maintenance whenever necessary to avoid hazards which could otherwise occur.
➤ **Breakdown-based maintenance** – here maintenance is only carried out when faults or failures have occurred. This is only acceptable if the failure does not present an immediate hazard and can be corrected before the risk is increased. If, for example, a bearing overheating can be detected by a monitoring device, it is acceptable to wait for the overheating to occur as long as the equipment can be stopped and repairs carried out before the fault becomes dangerous to persons employed.

In the context of health and safety, maintenance is not concerned with operational efficiency but only with avoiding risks to people. It is essential to ensure that maintenance work can be carried out safely. This will involve:

➤ competent well-trained maintenance people
➤ the equipment being made safe for the maintenance work to be carried out. In many cases the normal safeguards for operating the equipment may not be sufficient as maintenance sometimes involves going inside guards to observe and subsequently adjust, lubricate or repair the equipment. Careful design allowing adjustments, lubrication and observation from outside the guards, for example, can often eliminate the hazard. Making equipment safe will usually involve disconnecting the power supply and then preventing anything moving, falling or starting during the work. It may also involve waiting for equipment to cool or warm up to room temperature
➤ a safe system of work being used to carry out the necessary procedures to make and keep the equipment safe and perform the maintenance tasks. This can often involve a formal 'permit to work' scheme to ensure that the correct sequence of safety critical tasks has been performed and all necessary precautions taken
➤ correct tools and safety equipment being available to perform the maintenance work without risks to people. For example special lighting or ventilation may be required.

13.5.2 *Inspection under PUWER*

Complex equipment and/or high-risk equipment will probably need a maintenance log and may require a more rigid inspection regime to ensure continued safe operation. This is covered by PUWER regulation 6.

PUWER requires that where safety is dependent on the installation conditions and/or the work equipment is exposed to conditions causing deterioration, which may result in a significant risk and a dangerous situation developing, the equipment is inspected by a competent person. In this case the competent person would normally be an employee, but there may be circumstances involving specialized earth moving equipment, where an outside competent person would be used.

The inspection must be done:

➤ after installation for the first time
➤ after assembly at a new site or in a new location and thereafter
➤ at suitable intervals and
➤ each time exceptional circumstances occur which could affect safety.

The inspection under PUWER will vary from a simple visual inspection to a detailed comprehensive inspection, which may include some dismantling and/or testing. The level of inspection required would normally be less rigorous and intrusive than thorough examinations under the Lifting Operations and Lifting Equipment Regulations (LOLER) for certain lifting equipment. In the case of boilers and air receivers the inspection is covered by the Pressure Systems Safety Regulations, which involves a thorough examination (see the summary in Chapter 21). The inspection under PUWER would only be needed in these cases if the examinations did not cover all the significant health and safety risks that are likely to arise.

13.5.3 *Examination of boilers and air receivers*

(a)

Figure 13.6 (a) Typical diesel powered compressor with air receiver and pneumatic chisel and air receiver;

(b)

Figure 13.6 (b) Electrically powered compressor with air receiver.

Under the Pressure Systems Safety Regulations, a wide range of pressure vessels and systems require thorough examination by a competent person to an agreed, specifically written, scheme. In construction work the most common piece of equipment is the mobile air receiver for concrete breaking and the like.

The Regulations place duties on designers and manufacturers but this section is only concerned with the duty on users to have the vessels examined. An employer who operates a steam boiler or an air receiver must ensure:

➤ that it is supplied with the correct written information and markings
➤ that the equipment is properly installed
➤ that it is used within its operating limits
➤ there is a written scheme for periodic examination of the equipment certified by a competent person (in the case of standard steam boilers and air receivers the scheme is likely to be provided by the manufacturer)
➤ that the equipment is examined in accordance with the written scheme by a competent person within the specified period
➤ that a report of the periodic examination is held on file giving the required particulars
➤ the actions required by the report are carried out
➤ any other safety critical maintenance work is carried out, whether or not covered by the report.

In these cases the competent person is usually a specialist inspector from an external inspection organization. Many of these organizations are linked to the insurance companies who cover the financial risks of the pressure vessel.

In construction work many such pieces of equipment are hired from plant hire organizations, who must supply a copy of the relevant, in-date inspection report when the equipment is supplied to site.

13.6 Operation and working environment

To operate work equipment safely it must be fitted with easily reached and operated controls; kept stable properly lit; kept clear; and provided with adequate markings and warning signs. These are covered by PUWER, which applies to all types of work equipment.

Equipment should be operated in clear unobstructed work areas, particularly if they are high-risk machines. Debris and rubbish given off from the equipment should be removed quickly.

Sometimes there will be a need for special systems to remove waste to prevent accumulations of dust and/or substances hazardous to health, for example, wood dust from wood working machinery.

13.6.1 *Controls*

Equipment should be provided with efficient means of:

➤ starting or making a significant change in operating conditions
➤ stopping in normal circumstances and
➤ emergency stopping as necessary to prevent danger.

All controls should be well positioned, clearly visible and identifiable, so that it is easy for the operator to know

Equipment controls should:

➢ be easily reached from the operating positions
➢ not permit accidental starting of equipment
➢ move in the same direction as the motion being controlled
➢ vary in mode, shape and direction of movement to prevent inadvertent operation of the wrong control
➢ incorporate adequate red stop buttons
➢ incorporate adequate red emergency stop buttons, of the mushroom headed type with lock-off
➢ have shrouded or sunk green start buttons to prevent accidental starting of the equipment
➢ be clearly marked to show what they do.

Figure 13.7 Equipment Controls – design features.

what each control does. Markings should be clearly visible and remain so under the conditions met on construction sites. See Figure 13.7 for more design information on controls.

a) Start controls

It should only be possible to start the work equipment by using the designed start control. Equipment may well have a start sequence which is electronically controlled to meet certain conditions before starting can be achieved, for example preheating a diesel engine, purge cycle for gas fed equipment. Restarting after a stoppage will require the same sequence to be performed.

Stoppage may have been deliberate or as a result of opening an interlocked guard or tripping a switch accidentally. In most cases it should not be possible to restart the equipment simply by shutting the guard or resetting the trip. Operation of the start control should be required.

Any other change to the operating conditions such as speed, pressure or temperature should only be done by using a control designed for the purpose.

b) Stop controls

The action of normal stopping controls should bring the equipment to a safe condition in a safe manner. In some cases immediate stopping may cause other risks to occur. The stop controls do not have to work instantaneously and can bring the equipment to rest in a safe sequence or at the end of an operating cycle. It is only the parts necessary for safety, i.e. accessible dangerous parts, that have to be stopped. So, for example, suitably guarded cooling fans may need to run continuously and be left on.

In some cases where there is, for example, stored energy in hydraulic systems, it may be necessary to insert physical scotches to prevent movement and/or to exhaust residual hydraulic pressure. These should be incorporated into the stopping cycle, which should be designed to dissipate or isolate all stored energy to prevent danger.

It should not be possible to reach dangerous parts of the equipment until it has come to a safe condition, e.g. stopped, cooled, electrically safe.

c) Emergency stop controls

Emergency stop must be provided where the other safeguards in place are not sufficient to prevent danger to operatives and any other persons who may be affected. Where appropriate there should be an emergency stop at each control point and at other locations around the equipment so that action can be taken quickly. Emergency stops should bring the equipment to a halt rapidly but this should be controlled where necessary so as not to create any additional hazards. Crash shutdowns of complex systems have to be carefully designed to optimize safety without causing additional risks.

Emergency stops are not a substitute for effective guarding of dangerous parts of equipment and should not be used for normal stopping of the equipment.

Emergency stop buttons should be easily identified, reached and operated. Common types are mushroom-headed buttons, bars, levers, kick-plates or pressure-sensitive cables. They are normally red and should need to be reset after use. With stop buttons this is done either by twisting or using a security key. (See Figure 13.8.)

Mobile work equipment is normally provided with effective means of stopping the engine or power source. In some cases large-scale equipment may need emergency stop controls away from the operators position.

Figure 13.8 Emergency stop buttons.

Isolation of equipment
Equipment should be provided with efficient means of isolating it from all sources of energy. The purpose is to make the equipment safe under particular conditions, for example when maintenance is to be carried out or where adverse weather conditions may make it unsafe to use. On static equipment isolation will usually be for mains

electrical energy, however in some cases there may be additional or alternative sources of energy. The isolation should cover all sources of energy such as diesel and petrol engines, LPG, steam, compressed air, hydraulics, batteries, heat. In some cases special consideration is necessary where, for example, hydraulic pumps are switched off, so as not to allow heavy pieces of equipment to fall under gravity. An example would be the loading shovel on an excavator.

Stability
Stability is important and is achieved by, for example, bolting equipment in place, using clamps, outriggers, all terrain tyres on wide axles, and only operating on level ground. Some equipment can be tied down, counterbalanced, weighted, or provided with self-levelling or cut-off devices so that it remains stable under all operating conditions. If portable equipment is weighted or counterbalanced it should be reappraised when the equipment is moved to another position. If outriggers are needed for stability in certain conditions, for example, to stabilize mobile access towers, they should be employed whenever conditions warrant the additional support. In severe weather conditions it may be necessary to stop using the equipment or reappraise the situation to ensure stability is maintained.

The quality of general, local or temporary lighting will need to be considered to ensure the safe operation of the equipment. The level of lighting and its position relative to the working area are often critical to the safe use of work equipment. Poor levels of lighting, glare and shadows can be dangerous when operating equipment. Some types of lighting, for example, sodium lights, can change the colour of equipment, which may increase the level of risk. This is particularly important if the colour coding of pipe work or cables is essential for safety. All temporary lighting should be properly installed so that it is protected against mechanical damage or water ingress. Further information is contained in Chapter 14 on electrical hazards.

Markings
Markings on equipment must be clearly visible and durable. They should follow international conventions for some hazards like radiation and lasers and, as far as possible, conform to the Health and Safety (Safety Signs and Signals) Regulations (see Chapter 21 for a summary). The contents, or the hazards of the contents, as well as controls, will need to be marked on some equipment. Warnings or warning devices are required in some cases to alert operators or people nearby to any dangers, for example, 'wear hardhats'; a flashing light on an airport vehicle; or a reversing horn on a truck.

Users' and hirers' responsibilities

13.7.1 *Users' responsibilities*

The responsibilities on users of work equipment are covered in section 7 of HSW Act and regulation 14 of the Management Regulations. Section 7 requires employees to take reasonable care for themselves and others who may be affected and to cooperate with the employer. Regulation 12 requires employees to use equipment properly in accordance with instructions and training. They must also inform employers of dangerous situations and shortcomings in protection arrangements. The self-employed who are users have similar responsibilities to employers and employees combined.

13.7.2 *Hirers' responsibilities*

Those who provide work equipment for use of an employed person include those who hire, lease or rent equipment for construction work. Hirers need to ensure that the equipment meets the standards required by the Provision and Use of Equipment Regulations and other relevant legislation. The equipment must be properly maintained and provided with any in-date Statutory Examination Reports.

13.8 ## Hand-held tools

13.8.1 *Introduction*

Work equipment includes hand tools and hand-held power tools. This section deals with hand tools. These tools need to be correct for the task, well maintained and properly used by trained people.

Five basic safety rules can help prevent hazards associated with the use of hand-held tools:

1. Keep all tools in good condition with regular maintenance.
2. Use the right tool for the job.
3. Examine each tool for damage before use and do not use damaged tools.
4. Use tools according to the manufacturers' instructions.
5. Provide and use properly the right personal protective equipment (PPE).

Employees and employers should work together to establish safe working procedures. If a hazardous situation is encountered, it should be brought immediately to the attention of the proper individual for hazard abatement.

Figure 13.9 Typical range of hand tools.

13.8.2 *Hazards of hand tools*

Hazards from the misuse or poor maintenance of hand tools include:

➤ broken handles on files/chisels/screwdrivers/hammers which can cause cut hands or hammer heads to fly off
➤ incorrect use of knives, saws and chisels with hands getting injured in the path of the cutting edge
➤ tools that slip causing stab wounds
➤ poor quality uncomfortable handles that damage hands
➤ splayed spanners that slip and damage hands or faces
➤ chipped or loose hammer heads that fly off or slip
➤ incorrectly sharpened or blunt chisels or scissors that slip and cut hands. Dull tools can cause more hazards than sharp ones. Cracked saw blades must be removed from service.
➤ flying particles that damage eyes from breaking up stone or concrete

➤ electrocution or burns by using incorrect or damaged tools for electrical work

➤ use of poorly insulated tools for hot work in catering or food industry

➤ use of pipes or similar equipment as extension handles for a spanner which is likely to slip causing hand or face injury

➤ mushroomed-headed chisels or drifts which can damage hands or cause hammers (not suitable for chisels) and mallets to slip

➤ use of spark-producing or percussion tools in flammable atmospheres

➤ painful wrists and arms (upper limb disorders) from frequent twisting when using screwdrivers

➤ when using saw blades, knives, or other tools, they should be directed away from aisle areas and away from other people working in close proximity.

13.8.3 *Hand tools safety considerations*

Hand tools should be properly controlled including those tools owned by employees. The following controls are important.

Suitability – all tools should be suitable for the purpose and location in which they are to be used. Using the correct tool for the job is the first step in safe hand tool use. Tools are designed for specific needs. That's why screwdrivers have various lengths and tip styles and pliers have different head shapes. Using any tool inappropriately is a step in the wrong direction. To avoid personal injury and tool damage, select the proper tool to do the job well and safely.

Quality professional hand tools will last many years if they are taken care of and treated with respect. Manufacturers design tools for specific applications. Use tools only for their intended purpose.

Suitability will include:

➤ specially protected and insulated tools for electricians

➤ non-sparking tools for flammable atmospheres and the use of non-percussion tools and cold cutting methods

➤ tools made of suitable quality materials which will not chip or splay in normal use

➤ the correct tools for the job, for example, using the right-sized spanner and the use of mallets not hammers on chisel heads. The wooden handles of tools must not be splintered

➤ safety knives with enclosed blades for regular cutting operations

➤ impact tools such as drift pins, wedges and cold-chisels must be kept free of mushroomed heads

➤ spanners must not be used when jaws are sprung to the point that slippage occurs.

Inspection – all tools should be maintained in a safe and proper condition. This can be achieved through:

➤ the regular inspection of hand tools

➤ discarding or prompt repair of defective tools

➤ taking time to keep tools in the proper condition and ready for use

➤ proper storage to prevent damage and corrosion

➤ locking tools away when not in use to prevent them being used by unauthorized people.

Training – all users of hand tools should be properly trained in their use. This may well have been done through apprenticeships and similar training. This will be particularly important with specialist working conditions or work involving young people.

Always wear approved eye protection when using hand tools, particularly when percussion tools are being used. However, pieces metal and wood will fly out when they are cut or planed. Other workers in the vicinity should wear eye protection, as well.

Use Well-Designed, High-Quality Tools – Finally, investing in high-quality tools makes the professional's job safer and easier:

➤ if extra leverage is needed, use high-leverage pliers, which give more cutting and gripping power than standard pliers. This helps, in particular, when making repetitive cuts or twisting numerous wire pairs.

➤ serrated jaws provide sure-gripping action when pulling or twisting wires.

➤ some side-cutting and diagonal-cutting pliers are designed for heavy-duty cutting. When cutting screws, nails and hardened wire, only use pliers that are recommended for that use.

➤ pliers with hot riveting at the joint ensure smooth movement across the full action range of the pliers, which reduces handle wobble, resulting in a positive cut. The knives align perfectly every time.

➤ induction hardening on the cutting knives adds to long life, so the pliers cut cleanly day after day.

➤ sharp cutting knives and tempered handles also contribute to cutting ease.

➤ some pliers are designed to perform special functions. For example, some high-leverage pliers have features that allow crimping connectors and pulling fish tapes.

➤ tool handles with dual-molded material allow for a softer, more comfortable grip on the outer surface and a harder, more durable grip on the inner surface and handle ends.

➤ well-designed tools often include a contoured thumb area for a firmer grip or colour-coded handles for easy tool identification.

➤ insulated tools reduce the chance of injury where the tool may make contact with an energized source.

Well-designed tools are a pleasure to use. They save time, give professional results and help to do the job more safely.

13.9 Hand-held power tools

Figure 13.10 Typical range of hand-held power tools.

13.9.1 *Introduction*

The electrical hazards of portable hand-held tools and portable appliance testing (PAT) are covered in more detail in Chapter 14. This section deals mainly with other physical hazards and safeguards relating to hand-held power tools.

The section covers, in particular, pneumatic drills/chisels, electric drills, disc-cutters, sanders, cartridge and pneumatic nail guns and chainsaws which are commonly used on construction sites and covered in the Construction Certificate syllabus.

13.9.2 *General hazards of hand-held power tools*

The general hazards involve:

➤ mechanical entanglement in rotating spindles or sanding discs
➤ waste material flying out of the cutting area
➤ coming into contact with the cutting blades or drill bits
➤ risk of hitting electrical, gas or water services when drilling into building surfaces

➤ electrocution/electric shock from poorly maintained equipment and cables or cutting the electrical cable
➤ manual handling problem with a risk of injury if the tool is heavy or very powerful
➤ hand–arm vibration especially with pneumatic drill and chainsaws, disc cutters and petrol driven units
➤ tripping hazard from trailing cables, hoses or power supplies
➤ eye hazard from flying particles
➤ injury from poorly secured or clamped work-pieces
➤ fire and explosion hazard with petrol driven tools or when used near flammable liquids, explosive dusts or gases
➤ high noise levels with pneumatic chisels, planes and saws in particular (see Chapter 17)
➤ dust and fumes given off during the use of the tools (but see Chapter 16).

13.9.3 *Typical safety controls and instructions*

Guarding

The exposed moving parts of power tools need to be safeguarded. Belts, gears, shafts, pulleys, sprockets, spindles, drums, flywheels, chains, or other reciprocating, rotating, or moving parts of equipment must be guarded.

Machine guards, as appropriate, must be provided to protect the operator and others from the following:

➤ point of operation
➤ in-running nip points
➤ rotating parts
➤ flying chips and sparks.

Safety guards must never be removed when a tool is being used. Portable circular saws must be equipped at all times with guards. An upper guard must cover the entire blade of the saw. A retractable lower guard must cover the teeth of the saw, except where it makes contact with the work material. The lower guard must automatically return to the covering position when the tool is withdrawn from the work material.

Operating controls and switches

Most hand-held power tools should be equipped with a constant-pressure switch or control that shuts off the power when pressure is release. On off switches should be easily accessible without removing hands from the equipment.

Handles should be designed to protect operators from excessive vibration and keep their hands away from danger areas. In some cases handles are also designed to activate a brake of the cutting chain or blade, for example in a chainsaw. Equipment should be designed to reduce lifting and manual handling problems with special harnesses being used as necessary, for example when using large strimmers.

Means of starting engines and holding equipment should be designed to minimize any musculoskeletal problems.

Safe operations/instructions

When using power tools, the following basic safety measures should be observed to protect against electrical shock, personal injury, ill-health and risk of fire. Also see more detailed electrical precautions in Chapter 14. Operators should read these instructions before using the equipment and ensure that they are followed:

➤ maintain a clean and tidy working area that is well lit and clear of obstructions.

➤ never expose power tools to rain. Do not use power tools in damp or wet surroundings.

➤ do not use power tools in the vicinity of combustible fluids, dusts or gases unless they are specially protected and certified for use in these areas.

➤ protect against electrical shock (if tools are electrically powered) by avoiding body contact with grounded objects such as pipes, scaffolds, metal ladders (see Chapter 14 for more electrical safety precautions).

➤ keep children away.

➤ do not let other persons handle the tool or the cable. Keep them away from the working area.

➤ store tools in a safe place when not in use where they are in a dry, locked area, which is inaccessible to children.

➤ tools should not be overloaded as they operate better and safer in the performance range for which they were intended.

➤ use the right tool. Do not use small tools or attachments for heavy work. Do not use tools for purposes and tasks for which they were not intended; for example, do not use a hand-held circular saw to cut down trees or cut-off branches.

➤ wear suitable work clothes. Do not wear loose fitting clothing or jewellery. They can get entangled in moving parts. For outdoor work, rubber gloves and non-skid footwear are recommended. Long hair should be protected with a hair net.

➤ use safety glasses.

➤ also use filtering respirator mask for work that generates dust.

➤ do not abuse the power cable.

➤ do not carry the tool by the power cable and do not use the cable to pull the plug out of the power socket. Protect the cable from heat, oil and sharp edges.

➤ secure the work piece. Use clamps or a vice to hold it. This is safer than using hands and it frees both hands for operating the tool.

➤ do not overreach the work area. Avoid abnormal body postures. Maintain a safe stance and maintain a proper balance at all times.

➤ maintain tools with care. Keep your tools clean and sharp for efficient and safe work. Follow the maintenance regulations and instructions for the changing of tools. Check the plug and cable regularly and in case of damage, have it repaired by a qualified service engineer. Also inspect extension cables regularly and replace if damaged.

➤ keep the handle dry and free of oil or grease.

➤ disconnect the power plug when not in use, before servicing and when changing the tool, that is blade, bits, cutter, sanding disc.

➤ do not forget to remove the key. Check before switching on that the key and any tools for adjustment are removed.

➤ avoid unintentional switch-on. Do not carry tools that are connected to power with your finger on the power switch. Check that the switch is turned off before connecting the power cable.

➤ when working outdoors, use only extension cables which are intended for such use and marked accordingly.

➤ stay alert, keep eyes on the work. Use common sense. Do not operate tools when there are significant distractions.

➤ check the equipment for damage. Before further use of a tool, check carefully the protection devices or lightly damaged parts for proper operation and performance of their intended functions. Check movable parts for proper function, for binding or for damaged parts. All parts must be correctly mounted and meet all conditions necessary to ensure proper operation of the equipment.

➤ damaged protection devices and parts should be repaired or replaced by a competent service centre unless otherwise stated in the operating instructions. Damaged switches must be replaced by a competent service centre. Do not use any tool which cannot be turned on and off with the switch.

➤ only use accessories and attachments that are described in the operating instructions or are provided or recommended by the tool manufacturer. The use of tools other than those described in the operating instructions or in the catalogue of recommended tool inserts or accessories can result in a risk of personal injury.

➤ use engine driven power tools in well-ventilated areas. Store petrol in a safe place in approved storage cans. Stop and let engines cool before refuelling.

13.9.4 *Specific hazards and control measures for specified hand-held power tools*

The following hand-held power tools have been put in the Construction Certificate syllabus: pneumatic drills/chisels, electric drills, disc-cutters/cut-off saws, sanders, cartridge and pneumatic nail guns and chainsaws, all of which are commonly used on construction sites. The hazards and safety control measures in addition to the general ones covered earlier are set out for each type of equipment.

a) Pneumatic drills/chisels

Figure 13.11 Pneumatic hammer/chisel.

Hazards include:

➤ entanglement, particularly of loose clothing or long hair in rotating drill bits
➤ accidental disconnection of the airline
➤ fracture of airline, particularly when internal diameter exceeds 12 mm
➤ accidental expulsion of attachments, particularly on chisels
➤ high noise levels from the machine or air compressor
➤ eye injury from flying particles and chips, particularly from chisels
➤ injury from poorly secured work-pieces
➤ electric shock from drilling into a live hidden cable
➤ hand–arm vibration hazard in hammer mode
➤ dust given off from material being worked on
➤ tripping hazard from trailing airlines
➤ upper limb disorder from powerful machines with a strong torque, particularly if they jam and kick back

➤ foot injury hazards from dropping heavy units onto unprotected feet
➤ manual handling hazards, particularly with heavy machines and intensive use
➤ hazards associated with a portable air supply compressor such as refuelling, noise, fumes, and dangerous moving parts of the engine, e.g. the fan and belts
➤ explosion risk if sparks are given off from chisels near flammable liquids and explosive dusts and gases
➤ using equipment in poor weather conditions with wet, slippery surfaces, poor visibility and cold conditions.

Specific control measures include:

➤ pneumatic tools must be checked to see that the tools are fastened securely to the air hose to prevent them from becoming disconnected. A short wire or positive locking device attaching the air hose to the tool must also be used and will serve as an added safeguard;
➤ if an air hose is more than 12.5mm (1/2 inch) in diameter, a safety excess flow valve must be installed at the source of the air supply to reduce pressure in case of hose failure;
➤ in general, the same precautions should be taken with an air hose that are recommended for electric cables, because the hose is subject to the same kind of damage or accidental striking, and because it also presents tripping hazards;
➤ when using pneumatic tools, a safety clip or retainer must be installed to prevent attachments such as chisels on a chipping hammer from being ejected during tool operation;
➤ eye protection is required, and head and face protection is recommended for employees working with pneumatic tools;
➤ screens must also be set up to protect nearby workers from being struck by flying fragments around chippers, chisels or air drills;
➤ use of heavy jackhammers (pneumatic chisels) can cause fatigue and strains. Heavy rubber grips reduce these effects by providing a secure handhold. Workers operating a jackhammer must wear safety glasses and safety shoes that protect them against injury if the jackhammer slips or falls. A face shield also should be used;
➤ noise is another hazard associated with pneumatic tools. Working with noisy tools such as jackhammers (heavy pneumatic chisels) requires proper, effective use of appropriate hearing protection;

b) Electric drills

Figure 13.12a & b   Electric drills.

Hazards are:

➤ entanglement, particularly of loose clothing or long hair in rotating drill bits
➤ high noise levels from the drill or attachment
➤ eye injury from flying particles and chips, particularly from chisels
➤ injury from poorly secured work-pieces

➤ electrocution/electric shock from poorly maintained equipment
➤ electric shock from drilling into a live hidden cable
➤ hand–arm vibration hazard in hammer mode
➤ dust given off from material being worked on
➤ tripping hazard from trailing cables
➤ upper limb disorder from powerful machines with a strong torque, particularly if they jam and kick back.
➤ foot injury hazards from dropping heavy units onto unprotected feet
➤ manual handling hazards, particularly with heavy machines and intensive use or using at awkward heights and/or reaches
➤ fire and explosion hazard when used near flammable liquids, or explosive dusts or gases
➤ using equipment in poor weather conditions with wet, slippery surfaces, poor visibility and cold conditions.

Specific control measures include:

➤ Use double-insulated tools or earthed reduced-voltage tools with a residual current device (see Chapter 14 for electrical safeguards in detail).
➤ Use a pilot hole or punch to start holes whenever possible.
➤ Select the correct drill bit for the material being drilled.
➤ Secure small pieces to be drilled to prevent spinning.
➤ Protect against damage or injury on the far side if the bit is long enough to pass through the material or there are buried services in, say, plaster walls.
➤ Take care to prevent loose sleeves, or long hair from being wound round the drill bit; for example wear short or close fitting sleeves.
➤ Wear suitable eye protection.

c) Disc-cutters/cut-off saws

Figure 13.13   Disc-cutter/cut-off saw.

Hazards include:

➤ cutting hazard from high speed rotating discs
➤ high noise levels during cutting operations
➤ overheating of discs due to lack of lubricant such as water
➤ eye injury from flying hot particles
➤ eye and puncture injuries from disintegrating, badly mounted discs
➤ injury from poorly secured work-pieces
➤ electrocution/electric shock from poorly maintained equipment if electrical
➤ electric shock from cutting a live hidden cable
➤ hand–arm vibration hazard
➤ dust given off from material being worked on
➤ tripping hazard from trailing cables
➤ upper limb disorder from powerful machines with a strong torque particularly if they jam and kick back
➤ foot injury hazards from dropping heavy units onto unprotected feet
➤ manual handling hazards, particularly with heavy machines and intensive use or using at awkward heights and/or reaches
➤ fire and explosion hazard with petrol fuelled tools
➤ fire and explosion hazard when using engine or electricity driven tools near flammable liquids, or explosive dusts or gases
➤ using equipment in poor weather conditions with wet, slippery surfaces, poor visibility and cold conditions.

Specific control measures include:

➤ the blade or disc must be fitted with a close fitting hood guard adjusted to give the maximum protection for the task.
➤ inspect wheels regularly and discard any cracked wheels immediately before using the cutter again.
➤ wse a wheel with the proper rpm (revolutions per minute) rating and cutting characteristics for the task and material.
➤ hold the tool firmly with both hands to avoid recoil caused by jamming and wedging.
➤ wear eye protection.
➤ do not apply excessive force as this will overheat the work-piece and stress the wheel or disc.
➤ if conditions are dusty suitable filtering face masks should be worn.
➤ operators should be adequately trained to fit new cutting discs.
➤ if water is used for a cooling lubricant the machine system should be used with the correct fittings to ensure that the water is correctly directed at the cutting edge of the blade.

➤ operators should wear suitable clothing avoiding loose garments, long hair and jewellery which could catch in the equipment. Protective gloves and footwear are recommended.

d) Sanders

Figure 13.14 Rotary drum floor sander.

Figure 13.15 Orbital finishing sander.

There are a large range of hand-held sanders on the market; rotating discs, random orbital, rectangular orbital, belt sanders and heavy duty floor sanders of the rotating drum type and recently the orbital floor sander. The high-speed rotating discs and drum types are the most hazardous but they all need using with care.

Hazards include:

➤ high noise levels from the sander in operation
➤ injury from poorly secured work-pieces
➤ electrocution/electric shock from poorly maintained electrical equipment
➤ potential of entanglement with rotating disc and drum sanders
➤ sanding attachments becoming loose in the chuck and flung off
➤ injury from contact with abrasive surfaces particularly with coarse abrasives and high speed rotating sanding discs and drums
➤ hand–arm vibration hazard, particularly from reciprocating equipment
➤ health hazards from extensive dust given off from material being worked on
➤ fire and health hazards from overheating of abraded surfaces, particularly if plastics are being sanded
➤ tripping hazard from trailing cables
➤ large powerful sanders suddenly gripping the surface and pulling the operator off their feet
➤ foot injury hazards from dropping heavy units onto unprotected feet
➤ manual handling hazards, particularly with heavy machines, and intensive use or using at awkward heights and/or reaches
➤ fire and explosion hazard when used near flammable liquids, or explosive dusts or gases
➤ using equipment in poor weather conditions with wet, slippery surfaces, poor visibility and cold conditions.

Figure 13.16 Disc sander.

Specific control measures include:

➤ The work pieces must be securely clamped or held in position during sanding. In some cases a jig will

be necessary. The direction of spin of disc sanders (normally anti-clockwise) is important, to ensure that a small work-piece is pushed towards a stop or fence which is normally at the left side of the work-piece, particularly when clamping is impossible;
➤ abrasive sanding belts, discs and sheets should be properly and firmly attached to the machine without any torn parts or debris underneath. The manufacturer's instructions and fixing accessories should be used to ensure correct attachment of the abrasive;
➤ old nails and fixings should be sunk below the surface or removed to prevent the sander snagging;
➤ always hold the equipment by the proper handles, and particularly on large disc and floor sanders always use both hands. Excessive pressure should not be used as the surface will be rutted and the machine may malfunction;
➤ ensure that the dust extract or is working properly and has been emptied when about 1/3 full. Some extraction systems draw dust and air through the sanding sheet, which must have correctly pre-punched holes to allow the passage of air;
➤ operators should wear suitable dust respirators, eye protection and, where necessary, hearing protection;
➤ operators should wear suitable clothing, avoiding loose garments, long hair and jewellery which could catch in the equipment. Protective gloves and footwear are recommended.

e) Cartridge and pneumatic nail guns

Figure 13.17 Cartridge powered nail gun.

These are used for driving nails either by an explosive powder charge in a cartridge, or compressed air or gas propolsion. They can be extremely dangerous if used incorrectly. Although the use of a powder cartridge does not require a firearms licence, correct training of operatives is essential.

Hazards include:

➤ flying debris or particles discharged from the work surface that the nail enters
➤ using too heavy a charge for the material resulting in the nail being shot right through the material/surface being fixed
➤ nails ricocheting if the tool is not held properly or the surface material is too hard
➤ fire and explosion hazard near flammable liquids and explosive dusts or gases
➤ using tool explosive charges in firearms or vice versa
➤ high noise levels, particularly from cartridge nail guns
➤ electric shock from driving into a live hidden cable
➤ dust given off from material being worked on and fumes from spent cartridges
➤ tripping hazard from trailing airlines
➤ upper limb disorder from powerful machines with a strong kick back
➤ foot injury hazards from dropping heavy units onto unprotected feet
➤ manual handling hazards, particularly with heavy machines and intensive use
➤ hazards associated with a portable air supply compressor such as refuelling, noise, fumes, dangerous moving parts of the engine, e.g. the fan and belts
➤ using equipment in poor weather conditions with wet, slippery surfaces, poor visibility and cold conditions.

Specific control measures for cartridge powered nail guns include:

➤ cartridge powder-actuated tools operate like a loaded gun and must be treated with extreme caution. In fact, they are so dangerous that they must be operated only by specially trained employees.
➤ when using cartridge-actuated tools, an employee must wear suitable ear, eye and face protection. The user must select a cartridge level – high or low velocity – that is appropriate for the cartridge-actuated tool and necessary to do the work without excessive force.
➤ the muzzle end of the tool must have a protective shield or guard centred perpendicular to and concentric with the barrel to confine any fragments or particles that are projected when the tool is fired. A tool

containing a high-velocity load must be designed not to fire unless it has this kind of safety device.
➤ to prevent the tool from firing accidentally, two separate motions are required for firing. The first motion is to bring the tool into the firing position, and the second motion is to pull the trigger. The tool must not be able to operate until it is pressed against the work surface with a force at least 2.2 kg greater than the total weight of the tool.
➤ if a powder-actuated tool misfires, the user must hold the tool in the operating position for at least 30 seconds before trying to fire it again. If it still will not fire, the user must hold the tool in the operating position for another 30 seconds and then carefully remove the load in accordance with the manufacturer's instructions. This procedure will make the faulty cartridge less likely to explode. The bad cartridge should then be put in water immediately after removal. If the tool develops a defect during use, it should be *tagged* and must be *taken out of service immediately* until it is properly repaired.
➤ always keep powder-actuated tools, studs and cartridges in a safe place when not in use, preferably under lock and key. Don't leave tools or accessories unattended, even for a short period of time.
➤ safety precautions that must be followed when using powder-actuated tools include the following:
  ➤ do not use a tool in an explosive or flammable atmosphere;
  ➤ inspect the tool before using it to determine that it is clean, that all moving parts operate freely, and that the barrel is free from obstructions and has the proper shield, guard, and attachments recommended by the manufacturer;
  ➤ test the tool each day before loading to see that the safety devices are working. Follow the manufacturer's test methods;
  ➤ do not load the tool unless it is to be used immediately;
  ➤ do not leave a loaded tool unattended, especially where it would be available to unauthorized persons;
  ➤ keep hands clear of the barrel end;
  ➤ never carry a loaded tool from one job to another;
  ➤ don't rest the tool against the body when loading or making adjustments;
  ➤ never point the tool at anyone;
  ➤ don't let bystanders stand too close to the operator. Clear workers from the area, particularly from the other side of any partitions being worked on. Screens must also be set up to protect nearby workers from being struck by flying fragments around nail guns, riveting guns or staplers;
  ➤ hold the tool perpendicular to the work surface.

244

➤ when using powder-actuated tools to apply fasteners, several additional procedures must be followed:

➤ make a thorough study of each job. Know the types of materials you'll be driving into, so that you can select the proper stud and cartridge;

➤ also know what is on the other side of a wall as well as what is inside it, such as electric wires and gas or water pipes;

➤ do not fire fasteners into material that would allow the fasteners to pass through to the other side;

➤ do not drive fasteners into very hard or brittle material that might chip or splatter or make the fasteners ricochet;

➤ always use an alignment guide when shooting fasteners into existing holes;

➤ when using a high-velocity tool, do not drive fasteners less than 75mm (3 inches) from an unsupported edge or corner of material such as brick or concrete;

➤ when using a high-velocity tool, do not place fasteners in steel any closer than 12.5mm (1/2 inch) from an unsupported corner edge unless a special guard, fixture, or jig is used;

➤ never point the tool at anyone;

➤ don't fire studs into cast iron, high carbon or tempered steel, armour plate, rock, glazed brick, tile, or glass.

Figure 13.18 Pneumatic powered nail gun.

Specific control measures for pneumatic powered nail guns include:

➤ pneumatic nail guns must be checked to see that the tools are fastened securely to the air hose to prevent them from becoming disconnected. A short wire or positive locking device attaching the air hose to the tool must also be used and will serve as an added safeguard.

➤ If an air hose is more than 12.5mm (1/2 inch) in diameter, a safety excess flow valve must be installed at the source of the air supply to reduce pressure in case of hose failure.

➤ In general, the same precautions should be taken with an air hose that are recommended for electric cables, because such hose is subject to the same kind of damage or accidental striking, and because it also presents tripping hazards.

➤ Pneumatic tools that shoot nails, rivets, staples, or similar fasteners and operate at pressures more than 6,890kPa (100 pounds per square inch) must be equipped with a special device to keep fasteners from being ejected, unless the muzzle is pressed against the work surface.

➤ Eye protection is required, and head and face protection is recommended for employees working with pneumatic tools.

➤ Screens must also be set up to protect nearby workers from being struck by flying fragments around nail guns, riveting guns or staplers.

➤ Compressed-air guns should never be pointed toward anyone. Workers should never 'dead-end' them against themselves or anyone else. A chip guard must be used when compressed air is used for cleaning.

➤ Noise is another hazard associated with pneumatic tools. Appropriate hearing protection should be provided.

f) Chainsaws

Hazards include:

➤ very serious cutting by contact with the high speed cutting chain

➤ kick-back due to being caught on the wood being cut or contact with the top front corner of the chain in motion, with the saw chain being kicked upwards – towards the face in particular

➤ pull-in when the chain is caught and the saw is pulled forward

➤ push-back when the chain on the top of the saw bar is suddenly pinched and the saw is driven straight back towards the operator

➤ burns from hot parts of the engine

➤ high noise levels

➤ hand–arm vibration causing white finger and other problems

➤ fire and explosion hazards from the use of highly flammable petrol as a fuel

➤ eye and face puncture wounds from ejected particles

➤ back strain while supporting the weight of the chainsaw when operating it;

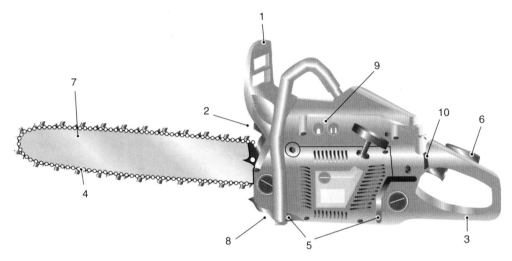

Figure 13.19 Typical chainsaw with rear handle. These have the rear handle projecting from the back of the saw. They are designed to always be gripped with both hands, with the right hand on the rear handle. It may be necessary to have a range of saws with different guide bar lengths available. As a general rule, choose a chainsaw with the shortest guide bar suitable for the work. 1. Hand guard with integral chain brake; 2. Exhaust outlet directed to the RHS away from the operator; 3. Chain breakage guard at bottom of rear handle; 4. Chain designed to have low kickback tendency; 5. Rubber anti-Vibration mountings; 6. Lockout for the throttle trigger; 7. Guide bar (should be protected when transporting chainsaw); 8. Bottom chain catcher; 9. PPE hand/eye/ear defender signs; 10. On/off switch.

➤ fire and explosion hazard when using engine or electrical driven tools near flammable liquids, explosive dusts or gases
➤ electric shock if electrically powered
➤ trailing cables on electrically powered equipment
➤ falls from height if using the chainsaw in trees and the like
➤ lone working and the risk of serious injury
➤ contact with overhead power lines if felling trees
➤ being hit by falling branches or whole trees while felling
➤ possible health hazards from cutting due to wood dust, particularly if wood has been seasoned
➤ using chainsaw in poor weather conditions with wet, slippery surfaces, poor visibility and cold conditions
➤ health hazards from engine fumes (carbon dioxide and carbon monoxide), particularly if used inside a shed or other building.

Specific control measures include:

➤ Chainsaws should only be operated by fully trained, fit and competent people. Using chainsaws in tree work requires a relevant certificate of competence or national competence award, unless the users are undergoing such training and are adequately supervised. However, in the agricultural sector, this requirement only applies to first-time users of a chainsaw.

Figure 13.20 Kevlar gloves, over trousers and over shoes providing protection against chainsaw cuts. Helmet, ear and face shields protect the head. Apprentice under instruction – first felling.

This means everyone working with chainsaws on or in trees should hold such a certificate or award **unless**:

➤ it is being done as part of agricultural operations (e.g. hedging, clearing fallen branches, pruning trees to maintain clearance for machines)

➤ the work is being done by the occupier or their employees; and

➤ they have used a chainsaw before 5 December 1998.

In any case, operators using chainsaws for any task in agriculture or any other industry must be competent under PUWER 1998.

➤ **Competence assessment**

Lantra Awards, Stoneleigh Park, Kenilworth, Warwickshire CV8 2LG Tel: 02476 419703 Fax: 02476 411655. web site: www.pantra.co.uk

NPTC, Stoneleigh Park, Kenilworth, Warwickshire CV8 2LG Tel. 024 7685 7300 web site: www.npcc.org.uk

➤ Avoid working alone with a chainsaw. Where this is not possible, establish procedures to raise the alarm if something goes wrong. These may include:

➤ regular contact with others using either a radio or telephone

➤ someone regularly visiting the worksite

➤ carrying a whistle to raise the alarm

➤ an automatic signalling device which sends a signal at a preset time unless prevented from doing so

➤ checks to ensure operators return to base or home at an agreed time.

➤ Moving engine parts should be enclosed

➤ Electrical units should be double insulated and cables fitted with residual current devices

➤ Must be fitted with a top handle and effective brake mechanism

➤ Chainsaws expose operators to high levels of noise and hand–arm vibration which can lead to hearing loss and conditions such as vibration white finger. These risks can be controlled by good management practice including:

➤ purchasing policies for low-noise/low-vibration chainsaws (e.g. with anti-vibration mounts and heated handles)

➤ providing suitable hearing protection

➤ proper maintenance schedules for chainsaws and protective equipment

➤ giving information and training to operators on the health risks associated with chainsaws and use of PPE, etc.

➤ Proper maintenance is essential for safe use and protection against ill-health from excessive noise and vibration. The saw must be maintained in its manufactured condition with all the safety devices in efficient working order and all guards in place. It should be regularly serviced by someone who is competent to do the job.

➤ Operators need to be trained in the correct chain-sharpening techniques and chain and guide bar maintenance to keep the saw in safe working condition.

➤ Make sure petrol containers are in good condition and clearly labelled, with securely fitting caps. Use containers which are specially designed for chain-saw fuelling and lubrication. Fit an auto-filler spout to the outlet of a petrol container to reduce the risk of spillage from over-filling.

➤ Do not allow operators to use discarded engine oil as a chain lubricant – it is a very poor lubricant and may cause cancer if it is in regular contact with an operator's skin.

➤ When starting the saw, operators should maintain a safe working distance from other people and ensure the saw chain is clear of obstructions.

➤ Kickback is the sudden uncontrolled upward and backward movement of the chain and guide bar towards the operator. This can happen when the saw chain at the nose of the guide bar hits an object. Kickback is responsible for a significant pro-portion of chainsaw accidents, many of which are to the face and parts of the upper body where it is difficult to provide protection. A properly maintained chain brake and use of low-kickback chains (safety chains) reduces the effect, but cannot entirely pre-vent it. Make sure operators use the saw in a way which avoids kickback by:

➤ not allowing the nose of the guide bar to acci-dentally come into contact with any obstruction, for example branches, logs, stumps

➤ not over-reaching

➤ keeping the saw below chest height

➤ keeping the thumb of the left hand around the back of the front handle

➤ using the appropriate chain speed for the mate-rial being cut.

➤ To avoid pull-in always hold the spiked bumper securely against the tree or limb

➤ To avoid push-back be alert to conditions that may cause the top of the guide bar to be pinched and do not twist the guide bar in the cut

➤ Training in good manual handling techniques and using handling aids/tools should reduce the risk of back injuries.

➤ To avoid overhead and other service hazards before felling starts on the worksite:

➤ contact the owners of any overhead power lines within a distance equal to twice the height of any tree to be felled to discuss whether the lines need to be lowered or made dead

➤ do not start work until agreement has been reached on the precautions to be taken

➤ check whether there are underground services such as power cables or gas pipes which could be damaged when the tree strikes the ground

➤ if there are roads or public rights of way within a distance equal to twice the height of the tree to be felled, ensure that road users and members of the public do not enter the danger zone. You may need to arrange warning notices, diversions or traffic control.

➤ Suitable PPE should always be worn, no matter how small the job. European standards for chainsaw PPE are published as part of EN 381 *Protective clothing for users of hand-held chainsaws*.

Protective clothing complying with this standard should provide a consistent level of resistance to chainsaw cut-through. Other clothing worn with the PPE should be close fitting and non-snagging.

NB No protective equipment can ensure 100% protection against cutting by a hand-held chainsaw.

Safety helmet – to EN 397. (Arborists working from a rope and harness may use a suitably adapted rock-climbing helmet.)

Hearing protection – to EN 352-1.

Eye protection – mesh visors to EN 1731 or safety glasses to EN 166.

Upper body protection – chainsaw jackets to BS EN 381-11.

Gloves – to EN 381-7. The use of appropriate gloves is recommended under most circumstances. The type of glove will depend on a risk assessment of the task and the machine. Consider the need for protection from cuts from the chainsaw, thorny material and cold/wet conditions.

Leg protection – to EN 381-5. (All-round protection is recommended for arborists working in trees and occasional users such as those working in agriculture.)

Chainsaw boots – to BS EN ISO 20345: 2004 and bearing a shield depicting a chainsaw to show compliance with EN 381-3. (For occasional users working in even ground where there is little risk of tripping or snagging on undergrowth or brash, protective gaiters conforming to EN 381-9 worn in combination with steel-toe-capped safety boots.)

➤ Chainsaw jackets can provide additional protection where operators are at increased risk (e.g. trainees, unavoidable use of a chainsaw above chest height). However, this needs to be weighed against increased heat stress generated by physical exertion (e.g. working from a rope and harness).

➤ If conditions are dusty suitable filtering face masks should be worn.

13.10 Mechanical machinery hazards

Most machinery has the potential to cause injury to people, and machinery accidents figure prominently in official accident statistics. These injuries may range in severity from a minor cut or bruise, through various degrees of wounding and disabling mutilation, to crushing, decapitation or other fatal injury. Nor is it solely powered machinery that is hazardous, for many manually operated machines (e.g. hand-operated guillotines and fly presses) can still cause injury if not properly safeguarded.

Machinery movement basically consists of rotary, sliding or reciprocating action, or a combination of these. These movements may cause injury by entanglement, friction or abrasion, cutting, shearing, stabbing or puncture, impact, crushing, or by drawing a person into a position where one or more of these types of injury can occur. The hazards of machinery are set out in BS EN150 12100 Part 2: 2003, which covers the classification of machinery hazards and how harm may occur. The following machinery hazards follow this standard classification.

A person may be injured at machinery as a result of:

➤ a **crushing hazard** through being trapped between a moving part of a machine and a fixed structure, such as a wall or a raised body of a dumper truck and its chassis

➤ a **shearing hazard** which traps part of the body, typically a hand or fingers, between moving and fixed parts of the machine

➤ a **cutting or severing hazard** through contact with a cutting edge, such as a band saw or rotating cutting disc

➤ **entanglement hazard** with the machinery which grips loose clothing, hair or working material, such as emery paper, around revolving exposed parts of the machinery. The smaller the diameter of the revolving part the easier it is to get a wrap or entanglement

➤ a **drawing-in or trapping hazard** such as between in-running gear wheels or rollers or between belts and pulley drives

➤ an **impact hazard** when a moving part directly strikes a person, such as with the accidental

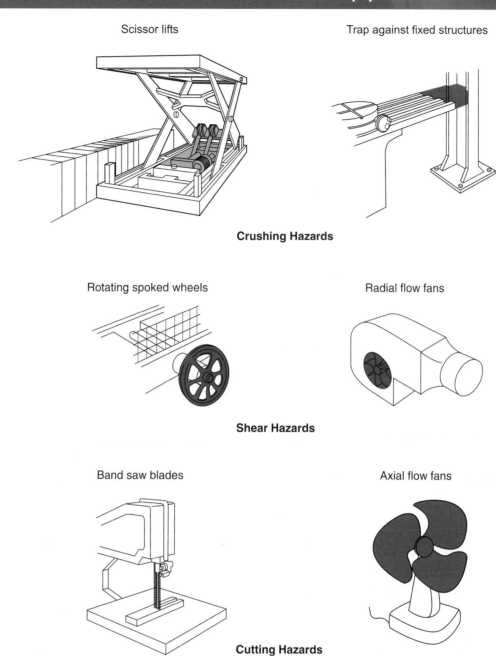

Scissor lifts

Trap against fixed structures

Crushing Hazards

Rotating spoked wheels

Radial flow fans

Shear Hazards

Band saw blades

Axial flow fans

Cutting Hazards

Figure 13.21 Range of mechanical hazards.

movement of a robot's working arm when maintenance is taking place
➤ a **stabbing or puncture hazard** through ejection of particles from a machine or a sharp operating component like a small drill
➤ contact with a **friction or abrasion hazard**, for example, on grinding wheels or sanding machines

➤ **high pressure fluid ejection hazard**, for example, from a hydraulic system leak.

In practice, injury may involve several of these at once, for example, contact, followed by entanglement of clothing, followed by trapping. Reference should also be made to PD5304: 2005, Guidance on Safe Use of Machinery from BSI.

Chains and sprockets

Rack and pinion gears

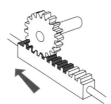

Pulley belts

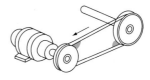

Belt conveyor

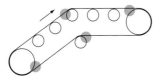

Meshing gears

Counter-rotating rolls

Drawing-in Hazards

Abrasive wheels

**Abrasion and
Ejection Hazard**

Shaft with projections

Entanglement Hazard

Figure 13.21   Continued.

13.11 Non-mechanical machinery hazards

Non-mechanical hazards include:

- access: slips, trips and falls; falling and moving objects; obstructions and projections
- lifting and handling
- electricity (including static electricity): shock, burns
- fire and explosion
- noise and vibration
- pressure and vacuum
- high/low temperature
- inhalation of dust/fumes/mist
- suffocation
- radiation: ionizing and non-ionizing
- biological: viral or bacterial
- harsh environment with wind, snow and lightning.

In many cases it will be practicable to install safeguards which protect the operator from both mechanical and non-mechanical hazards.

For example, a guard may prevent access to hot or electrically live parts as well as to moving ones. The use of guards which reduce noise levels at the same time is also common.

As a matter of policy machinery hazards should be dealt with in this integrated way instead of dealing with each hazard in isolation.

13.12 Examples of machinery hazards

The following examples are given to demonstrate a small range of machines found in the construction industry, which are included in the Construction Certificate syllabus.

13.12.1   Office – photocopier

The hazards include:

➤ contact with moving parts when clearing a jam
➤ electrical – when clearing a jam, maintaining the machine or through poorly maintained plug and wiring
➤ heat through contact with hot parts when clearing a jam
➤ health hazard from ozone or lack of ventilation in the area.

13.12.2   Office – document shredder

The hazards include:

➤ drawing in between the rotating cutters when feeding paper into the shredder
➤ contact with the rotating cutters when emptying the waste container or clearing a jam
➤ electrical through faulty plug and wiring or during maintenance
➤ noise from the cutting action of the machine
➤ possible dust from the cutting action.

13.12.3   Workshop machinery – bench top grinding machine

The hazards include:

➤ contact with the rotating wheel causing abrasion
➤ drawing in between the rotating wheel and a badly adjusted tool rest
➤ bursting of the wheel, ejecting fragments which puncture the operator
➤ electrical through faulty wiring and/or earthing or during maintenance
➤ fragments given off during the grinding process causing eye injury
➤ hot fragments given off which could cause a fire
➤ noise produced during the grinding process
➤ possible health hazard from dust/particles/fumes given off during grinding.

13.12.4   Workshop machinery – pedestal drill

The hazards include:

➤ entanglement around the rotating spindle and chuck
➤ contact with the cutting drill or work-piece
➤ being struck by the work-piece if it rotates
➤ being cut or punctured by fragments ejected from the rotating spindle and cutting device
➤ drawing in to the rotating drive belt and pulley
➤ contact or entanglement with the rotating motor
➤ electrical from faulty wiring and/or earthing or during maintenance
➤ possible health hazard from cutting fluids or dust given off during the process.

13.12.5   Workshop machinery – bench-mounted circular saw

The hazards include:

➤ contact with the cutting blade above and below the bench
➤ ejection of the workpiece or timber as it closes after passing the cutting blade
➤ drawing-in between chain and sprocket, V belt drives or powered feed rollers
➤ contact and entanglement with moving parts of the drive motor
➤ likely noise hazards from the cutting action and motor
➤ health hazards from wood dust given off during cutting
➤ electric shock from faulty wiring and/or earthing or during maintenance.

13.12.6   Workshop machinery – hand-fed power planer

The hazards include:
➤ contact with the cutting cylinder above and below the bench
➤ contact with the cutting blades under the bridge guard when hands are placed on the work-piece to push it past the planing blades
➤ ejection of the work-piece, timber or incorrectly fitted blades
➤ drawing-in between chain and sprocket, V belt drives or powered feed rollers
➤ contact and entanglement with moving parts of the drive motor
➤ likely noise hazards from the cutting action and motor
➤ health hazards from wood dust given off during cutting
➤ electric shock from faulty wiring and/or earthing or during maintenance.

Also see section 13.9.

13.12.7   Workshop machinery – spindle moulding machine

The hazards include:
➤ contact with the cutting blade above the bench
➤ contact with the cutting blades when hands are placed on the work-piece to push it past the blades and no holding jigs are used
➤ ejection of the work-piece, timber or incorrectly fitted blades
➤ drawing-in between chain and sprocket or V belt drives
➤ contact and entanglement with moving parts of the drive motor
➤ likely noise hazards from the cutting action and motor
➤ health hazards from wood dust given off during cutting
➤ electric shock from faulty wiring and/or earthing or during maintenance.

13.12.8 *Site machinery – compressor*

The hazards include:

➤ contact and entanglement with moving parts of the drive motor
➤ flying debris and blast from an exploding air receiver
➤ drawing-in between chain and sprocket drives
➤ electrical, if electrically powered
➤ burns from hot parts of petrol or diesel engine
➤ fire if highly flammable liquids used as fuel
➤ possible noise hazards from the motor and tools
➤ contamination of respirable air if used for a supplied air respirator
➤ possible health hazards from leaking oil
➤ possible hand–arm vibration causing white finger and other problems if used for road or concrete breaking
➤ eye puncture wounds from ejected chips from concrete or similar when using a pneumatic chisel.

13.12.9 *Site machinery – cement/concrete mixer*

The hazards include:

➤ contact and entanglement with moving parts of the drive motor
➤ crushing between loading hopper (if fitted) and drum
➤ drawing-in between chain and sprocket drives
➤ electrical, if electrically powered
➤ burns from hot parts of the engine
➤ fire if highly flammable liquids used as fuel
➤ possible noise hazards from the motor and dry mixing of aggregates
➤ eye injury from splashing cement slurry
➤ possible health hazard from cement dust and cement slurry while handling.

13.12.10 *Site machinery – plate compactor*

The hazards include:

➤ noise hazards from the action of the plate and a petrol driven engine
➤ vibration from the action of the plate
➤ contact with foot or lower leg
➤ flying splinters/chippings from the material being compacted
➤ manual handing injuries as they are generally handled by a single person.
➤ dust from dry material being compacted.

13.12.11 *Site machinery – ground consolidation equipment*

The hazards include:

➤ noise hazards from the action of the vibrating roller and a petrol or diesel driven engine

➤ contact with moving parts of the engine
➤ vibration from the action of the roller
➤ operative or person standing by being run over by a self-propelled roller
➤ manual handing injuries, as it is generally handled by a single person
➤ dust from dry material being consolidated
➤ possible health hazard from lubricants or fuels in the engine
➤ possible health hazards from exhaust
➤ contact with hot parts of the engine
➤ fire if highly flammable liquids are used as fuel.

13.12.12 *Site machinery – road marking equipment*

There is a variety of road and line marking equipment from the simple roller unit used in car-parks or sports fields to walk-behind line markers to lorry-mounted specialist units for large high-way projects. The hazards here relate to a lorry-mounted unit with a driver and single operative.
The hazards include:

➤ the danger of being struck by vehicles using the existing road unless it is new and not yet opened to the public
➤ injury to fellow workers, who may be hit by the vehicle while travelling or manoeuvring, or fall off the side of the vehicle
➤ contact with dangerous moving parts in the vehicle or equipment engine(s)
➤ fire hazard from burning off existing markings
➤ fire and explosion hazard if LPG is used as a fuel for burning
➤ health hazard from any toxic fumes, aerosols or dust given off from the heated products or the line marking sprayed paint
➤ health hazard from exposure to the UV from the sun
➤ hazards caused by cold or hot atmospheric conditions because of working outside
➤ potential burns from heated materials
➤ injury from damaged airlines.

13.12.13 *Site machinery – road making equipment*

There are various ranges of road making machinery including rollers, bulldozers, graders and tarmac layers. Some of the hazards are similar but those of a tarmac/asphalt laying machine will be covered here.
The hazards include:

➤ noise hazards from the action of a petrol or diesel driven engine
➤ contact with moving parts of the engine
➤ falling off the equipment when climbing on or off
➤ people standing by or working alongside being run over by a self-propelled machine

➤ being crushed between vehicle loading hopper and machine

➤ possible health hazards from lubricants or fuels in the engine

➤ possible health hazards from exhaust gases and hot tarmac/asphalt

➤ health hazards from long exposure to sunlight in the summer months

➤ contact with hot parts of engine or tarmac in hopper

➤ fire if highly flammable liquids used as fuel

➤ possible hazards of crushing or manual handling when loading a machine onto a transporter/low loader truck.

13.13 Control measures for machinery hazards

PUWER requires that access to dangerous parts of machinery should be prevented in a preferred order or hierarchy of control methods. The standard required is a 'practicable' one, so that the only acceptable reason for non-compliance is that there is no technical solution. Cost is not a factor. (See Chapter 1 for more details on standards of compliance.)

The levels of protection required are, in order of implementation:

➤ fixed enclosing guarding

➤ other guards or protection devices, such as interlocked guards and pressure-sensitive mats

➤ protection appliances, such as jigs, holders and push-sticks and

➤ the provision of information, instruction, training and supervision.

Since the mechanical hazard of machinery arises principally from someone coming into contact or entanglement with dangerous components, risk reduction is based on preventing this contact occurring.

This may be by means of:

➤ a physical barrier between the individual and the component (e.g. a fixed enclosing guard)

➤ a device which only allows access when the component is in a safe state (e.g. an interlocked guard which prevents the machine starting unless a guard is closed and acts to stop the machine if the guard is opened) or

➤ a device which detects that the individual is entering a risk area and then stops the machine (e.g. certain photoelectric guards and pressure-sensitive mats).

The best method should, ideally, be chosen by the designer as early in the life of the machine as possible.

It is often found that safeguards which are 'bolted on' instead of 'built in' are not only less effective in reducing risk, but are also more likely to inhibit the normal operation of the machine. In addition, they may in themselves create hazards and are likely to be difficult and hence expensive to maintain.

13.13.1 *Fixed guards*

Fixed guards have the advantage of being simple, always in position, difficult to remove and almost maintenance free. Their disadvantage is that they do not always properly prevent access, they are often left off by maintenance staff and they can create difficulties for the operation of the machine.

A fixed guard has no moving parts and should, by its design, prevent access to the dangerous parts of the machinery. It must be of robust construction sufficient to withstand the stresses of the process and environmental conditions. If visibility or free air flow (e.g. for cooling) is necessary, this must be allowed for in the design and construction of the guard. If the guard can be opened or removed, this must only be possible with the aid of a tool.

An alternative fixed guard is the distance fixed guard, which does not completely enclose a hazard, but which reduces access by virtue of its dimensions and its distance from the hazard. Where perimeter-fence guards are used, the guard must follow the contours of the machinery as far as possible, thus minimizing space between the guard and the machinery. With this type of guard it is important that the safety devices and operating systems prevent the machinery being operated with the guards closed and someone inside the guard, that is in the danger area. Figure 13.22 shows a range of fixed guards for some of the examples shown in Figure 13.21.

13.13.2 *Adjustable guards*

User adjusted guard
These are fixed or moveable guards, which are adjustable for a particular operation during which they remain fixed. They are particularly used with machine tools where some access to the dangerous part is required (e.g. drills, circular saws, milling machines) and where the clearance required will vary (e.g. with the size of the cutter in use on a horizontal milling machine or with the size of the timber being sawn on a circular saw bench).

Adjustable guards may be the only option with cutting tools, which are otherwise very difficult to guard, but they have the disadvantage of requiring frequent re-adjustment.

By the nature of the machines on which they are most frequently used, there will still be some access to the dangerous parts, so these machines must only be used by suitably trained operators. Jigs, push sticks and false tables must be used wherever possible to minimize

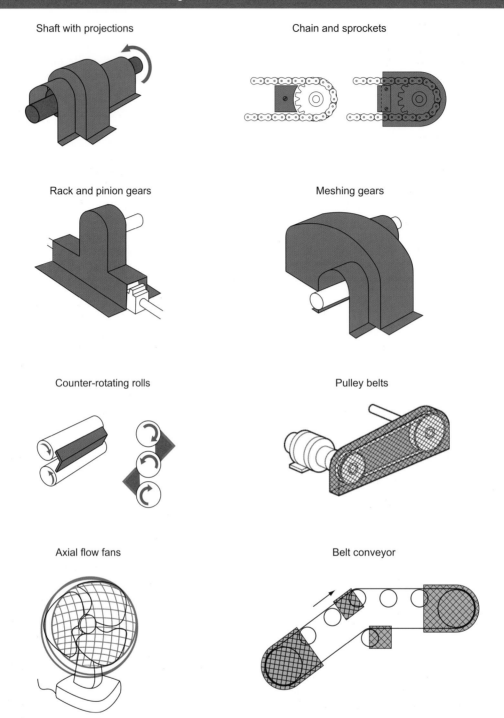

Figure 13.22   Range of fixed guards.

hazards during the feeding of the work-piece. The working area should be well lit and kept free of anything which might cause the operator to slip or trip.

Self-adjusting guard
A self-adjusting guard is one which adjusts itself to accommodate, for example, the passage of material.

A good example is the spring-loaded guard fitted to many portable circular saws (see Figure 13.24).

As with adjustable guards (see above) they only provide a partial solution in that they may well still allow access to the dangerous part of the machinery. They require careful maintenance to ensure they work to the best advantage.

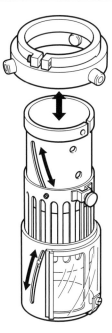

Figure 13.23 Adjustable guard for a rotating shaft, such as a pedestal drill.

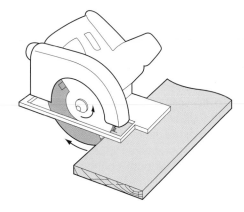

Figure 13.24 Self-adjusting guard on a wood saw.

13.13.3 *Interlocking guard*

The advantages of interlocked guards are that they allow safe access to operate and maintain the machine without dismantling the safety devices. Their disadvantage stems from the constant need to ensure that they are operating correctly and designed to fail safe. Maintenance and inspection procedures must be very strict. This is a guard which is movable (or which has a movable part), whose movement is connected with the power or control system of the machine or equipment.

An interlocking guard must be so connected to the machine controls that:

➤ until the guard is closed the interlock prevents the machinery from operating by interrupting the power medium

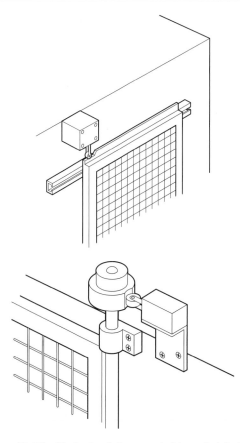

Figure 13.25 Typical sliding and hinged interlocking guards.

➤ either the guard remains locked closed until the risk of injury from the hazard has passed or opening the guard causes the hazard to be eliminated before access is possible.

A passenger lift or hoist is a good illustration of these principles: the lift will not move unless the doors are closed, and the doors remain closed and locked until the lift is stationary and in such a position that it is safe for the doors to open.

Special care is needed with systems which have stored energy. This might be the momentum of a heavy moving part, stored pressure in a hydraulic or pneumatic system, or even the simple fact of a part being able to move under gravity even though the power is disconnected. In these situations, dangerous movement may continue or be possible with the guard open, and these factors need to be considered in the overall design. Braking devices (to arrest movement when the guard is opened) or delay devices (to prevent the guard opening until the machinery is safe) may be needed. All interlocking systems must be designed to minimize the risk of failure-to-danger and should not be easy to defeat.

13.14 Other safety devices

13.14.1 *Trip devices*

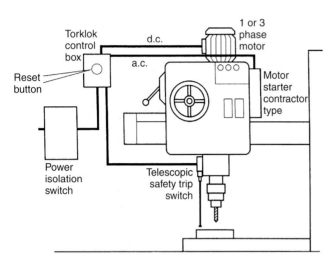

Figure 13.26 Schematic diagram of a telescopic trip device fitted to a radial drill.

A trip device does not physically keep people away but detects when a person approaches close to a danger point. It should be designed to stop the machine before injury occurs. A trip device depends on the ability of the machine to stop quickly and in some cases a brake may need to be fitted. Trip devices can be:

➤ mechanical in the form of a bar or barrier
➤ electrical in the form of a trip switch on an actuator rod, wire or other mechanism
➤ photoelectric or other type of presence-sensing device
➤ pressure-sensitive mat.

They should be designed to be self-resetting so that the machine must be restarted using the normal procedure.

13.14.2 *Two-handed control devices*

These are devices which require the operator to have both hands in a safe place (the location of the controls) before the machine can be operated. They are an option on machinery that is otherwise very difficult to guard but they have the drawback that they only protect the operator's hands; it is therefore essential that the design does not allow any other part of the operator's body to enter the danger zone during operation. More significantly, they give no protection to anyone other than the operator.

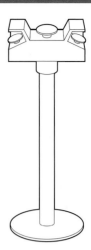

Figure 13.27 Pedestal mounted free-standing two-handed control device.

Where two-handed controls are used, the following principles must be followed:

➤ the controls should be so placed, separated and protected as to prevent spanning with one hand only, being operated with one hand and another part of the body, or being readily bridged;
➤ it should not be possible to set the dangerous parts in motion unless the controls are operated within approximately 0.5 seconds of each other. Having set the dangerous parts in motion, it should not be possible to do so again until both controls have been returned to their off position;
➤ movement of the dangerous parts should be arrested immediately or, where appropriate, arrested and reversed if one or both controls are released while there is still danger from movement of the parts;
➤ the hand controls should be situated at such a distance from the danger point that, on releasing the controls, it is not possible for the operator to reach the danger point before the motion of the dangerous parts has been arrested or, where appropriate, arrested and reversed.

13.14.3 *Hold-to-run control*

This is a control which allows movement of the machinery only as long as the control is held in a set position. The control must return automatically to the stop position when released. Where the machinery runs at crawl speed, this speed should be kept as low as practicable.

Hold-to-run controls give even less protection to the operator than two-handed controls and have the same main drawback in that they give no protection to anyone other than the operator.

However, along with limited movement devices (systems which permit only a limited amount of machine movement on each occasion that the control is operated and are often called 'inching devices'), they are extremely relevant to operations such as setting, where access may well be necessary and safeguarding by any other means is difficult to achieve.

13.15 Application of controls to the range of machines

The application of the controls or safeguards to the Construction Certificate range of machines is as follows.

Figure 13.28 Typical photocopier.

13.15.1 Office – photocopier

Application of controls:

➤ the machines are provided with an all-enclosing case which prevents access to the internal moving, hot or electrical parts;

➤ the access doors are interlocked so that the machine is automatically switched off when gaining access to clear jams or maintain the machine. It is good practice to switch off when opening the machine;

➤ internal electrics are insulated and protected to prevent contact;

➤ regular inspection and maintenance should be carried out;

➤ the machine should be on the PAT schedule;

➤ good ventilation in the machine room should be maintained.

13.15.2 Office – document shredder

Application of controls:

➤ enclosed fixed guards surround the cutters with restricted access for paper only, which prevents fingers reaching the dangerous parts;

Figure 13.29 Typical office paper shredder.

➤ interlocks are fitted to the cutter head so that the machine is switched off when the waste bin is emptied;

➤ a trip device is used to start the machine automatically when paper is fed in;

➤ machine should be on PAT schedule and regularly checked;

➤ general ventilation will cover most dust problems except for very large machines where dust extraction may be necessary;

➤ noise levels should be checked and the equipment perhaps placed on a rubber mat if standing on a hard reflective floor.

13.15.3 Workshop – bench top grinder

Application of controls:

➤ wheel should be enclosed as much as possible in a strong casing capable of containing a burst wheel;

➤ an adjustable tool rest should be adjusted as close as possible to the wheel;

➤ an adjustable screen should be fitted over the wheel to protect the eyes of the operator. Goggles should also be worn;

➤ only competent and registered people should mount an abrasive wheel;

Figure 13.30 Typical bench-mounted grinder.

➤ the maximum speed should be marked on the machine so that the abrasive wheel can be matched to the machine speed;

➤ noise levels should be checked and attenuating screens used if necessary;

➤ the machine should be on the PAT schedule and regularly checked;

➤ if necessary extract ventilation should be fitted to the wheel encasing to remove dust at source.

Outline procedure for mounting an abrasive wheel:

➤ wheel mounting should be carried out by an appropriately trained person;

➤ the wheel should be mounted only on the machine for which it was intended;

➤ before mounting, all wheels should be closely inspected to ensure that they have not been damaged in storage or transit;

➤ the speed marked on the machine should not exceed the speed marked on the wheel, blotter or identification label;

➤ the bush, if any, should not project beyond the sides of the wheel and blotters. The wheel should fit freely but not loosely on the spindle;

➤ flanges should not be smaller than their specified minimum diameter, and their bearing surfaces should be true and free from burrs;

➤ with the exception of the single flange used with threaded-hole wheels, all flanges should be properly recessed or undercut;

➤ flanges should be of equal diameter and have equal bearing surfaces (the shape of some wheels may not allow this rule to be followed, for example certain internal and cup wheels);

➤ protection flanges should have the same degree of taper as the wheel. Blotters, slightly larger than the flanges, should be used with all abrasive wheels (except: mounted wheels and points; abrasive discs; plate-mounted wheels; cylinder wheels mounted on chucks; rubber-bonded cutting-off wheels 0.5 mm or less in thickness; taper-sided wheels; wheels with threaded inserts). Wrinkles in blotters should be avoided;

➤ wheels, blotters and flanges should be free from foreign matter;

➤ clamping nuts should be tightened only sufficiently to hold the wheel firmly. When flanges are clamped by a series of screws they should be tightened uniformly in a criss-cross sequence. Screws for inserted nut mounting of discs, cylinders and cones should be long enough to engage a sufficient length of thread, but not so long that they contact the abrasive;

➤ when mounting the wheels and points, the overhang appropriate to the speed, diameter of the mandrel and size of the wheel should not be exceeded, and there should be sufficient length of mandrel in the collet or chuck;

➤ all safeguards should be reinstalled and properly adjusted including, where appropriate, the wheel enclosure, tool rests and adjustable front visor.

For more details on mounting abrasive wheels see HSG 17 *Safety in the Use of Abrasive Wheels* (3rd edn) from HSE Books.

13.15.4 *Workshop – pedestal drill*

Application of controls:

➤ motor and drive should be fitted with a fixed guard;.

➤ the spindle should be guarded by an adjustable guard, which is fixed in position during the work;

➤ a clamp should be available on the pedestal base to secure work-pieces;

➤ the machine should be on the PAT schedule and regularly checked;

➤ futting fluid, if used, should be contained and not allowed to get onto clothing or skin. A splash guard may be required but this is unlikely;

➤ goggles should be worn by the operator.

13.15.15 *Workshop machinery – bench-mounted circular saw*

Application of controls:

➤ a fixed guard should be fitted to fully enclose the blade below the bench;

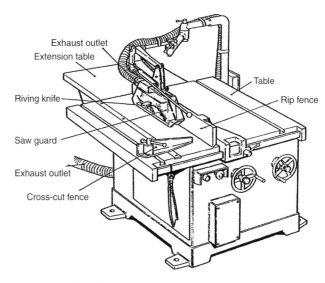

Figure 13.32 Bench-mounted circular saw.

should also be fitted. Machines should be fitted with a braking device that brings the blade to rest within 10 seconds;.

➤ blades should be removed from the bench regularly for cleaning off resins and other residues with a scraper;

➤ the diameter of the smallest saw blade that can safely be used should be marked on the saw bench;

➤ a riving knife (slightly thicker than body of blade but thinner than width of cut) should be fitted behind the blade to keep the cut timber apart and prevent ejection;

➤ there should be adequate support for the workpiece going through the saw and a push stick should be used when making any cut less than 300 mm or when feeding the last 300 mm of a longer cut;

➤ noise attenuation should be applied to the machine, for example damping, use of special quiet saw blades and if necessary fitting in an enclosure. Hearing protection may have to be used;

➤ protection against wet weather should be provided;

➤ the electrical parts should be regularly checked plus all the mechanical guards;

➤ extraction ventilation will be required for the wood dust and shavings both above and below the saw table;

➤ suitable dust masks should be worn;

➤ use vacuum for cleaning.

13.15.16 *Workshop machinery – hand-fed power planer*

Application of controls:

➤ fixed guards should be fitted to the motor and drives;

Figure 13.31 Typical pedestal drill.

➤ fixed guards should be fitted to the motor and drives;

➤ an adjustable top guard should be fitted to the blade above the bench which encloses as much of the blade as possible. An adjustable front section

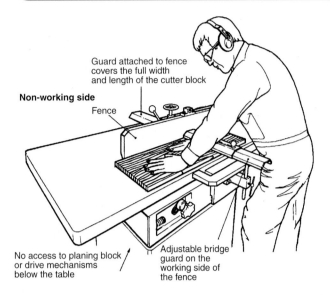

Guard attached to fence
covers the full width
and length of the cutter block

Non-working side

Fence

No access to planing block
or drive mechanisms
below the table

Adjustable bridge
guard on the
working side of
the fence

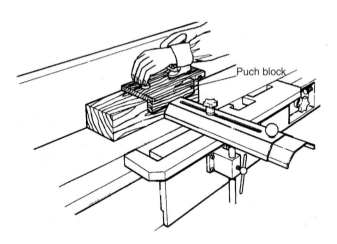

Puch block

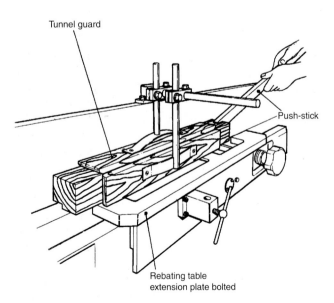

Tunnel guard

Push-stick

Rebating table
extension plate bolted

Figure 13.33 Hand-fed planing machine safeguards.

➤ an adjustable bridge guard should be fitted to the working side of the fence above the bench. It should be strong and rigid, and wide enough to cover the cutter block and, on larger tables, telescopic;

➤ a further bridge guard should be attached to the fence on the non-working side to cover the cutter block as the fence is moved across the table;

➤ machines should be fitted with a braking device to bring the cutter block to rest within 10 seconds;

➤ only cylindrical (or 'round form') cutter blocks should be used with maximum cutter projection of 1.1 mm. Keep gaps between the two sides of the table as small as possible;

➤ a push block with well-designed handles should be used when planing short pieces. In some cases a push stick will be required;

➤ work-pieces should be supported properly. Tunnel/ shaw guards should be used where appropriate;

➤ noise attenuation should be applied to the machine, for example damping, use of special quiet saw blades and if necessary fitting in an enclosure. Hearing protection may have to be used;

➤ protection against wet weather should be provided;

➤ the electrical parts should be checked regularly plus all the mechanical guards;

➤ extraction ventilation will be required for the wood dust and shavings;

➤ suitable dust masks should be worn;

➤ areas around the machine should be kept clear;

➤ use vacuum for cleaning.

13.15.7 *Workshop machinery – spindle moulding machine*

Application of controls:

➤ machines should be fitted with a braking device that brings the cutter block to rest within 10 seconds;

➤ table rings should be fitted to reduce the gap between the spindle and the table to a minimum;

➤ for most work the cutters should be effectively guarded. Where this is not possible use jigs or work holders and stops;

➤ wherever possible only feed the work-piece to the tool against the direction of spindle rotation;

➤ only tools marked 'MAN' should be used, meaning hand-free;

➤ false fences should be used for straight work;

➤ adequate work-piece support should be used;

➤ a demountable power feed should be used for straight cuts where possible. If not the work area should be enclosed with pressure pads in the form of a tunnel;

➤ jigs should be used where appropriate with, for example, stopped cuts;

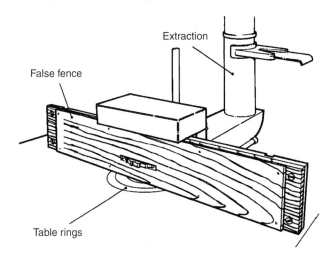

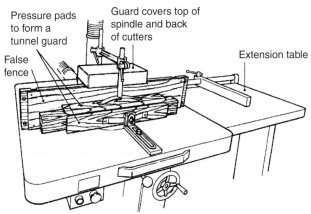

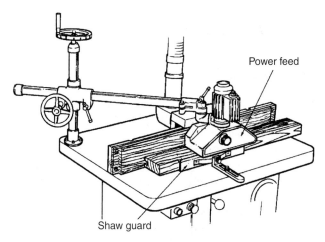

Figure 13.34 Spindle moulding machines with various forms of safeguards/controls.

➤ a ring guard or fence should be used for curved cuts;
➤ dust and chippings should be removed by efficient extraction ventilation and use vacuum for cleaning;

➤ noise attenuation should be applied to the machine, for example damping, use of special quiet saw blades and if necessary fitting in an enclosure. Hearing protection may have to be used;
➤ protection against wet weather should be provided;
➤ the electrical parts should be regularly checked plus all the mechanical guards;
➤ suitable dust masks should be worn;
➤ areas around the machine should be kept clear.

13.15.8 *Site machinery – compressor*

Portable air compressors are very common pieces of machinery on construction sites. For an example see Figure 13.6.

The application of controls:

➤ the moving, dangerous parts of the engine should be fenced locally to prevent access. This will probably be achieved through total enclosure within the engine casing for most components and use of a separate guard for the cooling fan and drive belts;
➤ for electrically driven compressors, components should be properly constructed and designed to prevent contact with dangerous live currents. RCDs should be used and, where possible, reduced voltages at 110 V. See Chapter 14 for more electrical hazards information. The equipment should be kept dry and not used in very damp or wet conditions. Extension cables should be fully unwound to avoid overheating and short circuits due to the current flow required for many units;
➤ compressor air receivers/tanks should be regularly drained of any residual water and the oil levels in the compressor checked;
➤ if using for spray painting or similar keep the compressor at least 3 m away from the work area to reduce the risk of fire and explosion;
➤ the air receiver and associated pressure pipe-work should be thoroughly examined to ensure their integrity. See Section 13.5.3;
➤ machines and tools with reduced noise and/or vibration characteristics should be used. This will involve use of acoustic enclosures around compressors and anti-vibration engine mountings. For larger units, mounting on pneumatic tyres reduces vibration considerably;
➤ all hoses should be used carefully and regularly checked. Compressors should never be moved by pulling them by a hose. They should be supplied with claw-type quick action couplings or similar to prevent accidental decoupling. Airlines should always be connected to an air tool if they are being supplied with compressed air. A mechanism to cut-off air if a hose becomes disconnected or fails should be fitted;

➤ compressors are fitted with air pressure gauges so that the pressure in the unit can be monitored and matched to the rating of a particular air tool;

➤ protective clothing including eye and ear protection should be worn by people using air tools;

➤ except for small units, moving compressors should be done by site vehicle/tractor. Only use on the road if properly equipped for road use with lights, tow bar and brakes (in most cases).

13.15.9 *Site machinery – cement mixer*

Figure 13.35 Typical cement/concrete mixer.

Application of controls:

➤ operating position for the hopper hoist should be designed so that anyone in the trapping area is visible to the operator. The use of the machine should be restricted to designated operators only. As far as possible the trapping point should be designed out. The hoist operating location should be fenced off just allowing access for barrows, etc., to the unloading area;

➤ drives and rotating parts of the engine should be enclosed;

➤ the drum gearing should be enclosed and persons kept away from the rotating drum, which is normally fairly high on large machines;

➤ no one should be allowed to stand on the machine while it is in motion;

➤ goggles should be worn to prevent cement splashes;

➤ if petrol driven, care is required with flammable liquids and refuelling;

➤ engines must only be run in the open air;

➤ electric machines should be checked regularly and be on the PAT schedule;

➤ noise levels should be checked and noise attenuation, for example, silencers and damping, fitted if necessary.

13.15.10 *Site machinery – plate compactor*

Figure 13.36 Plate compactor.

Application of controls:

➤ select reduced/anti-vibration models;

➤ have hold to run controls;

➤ always push away from body when operating and wear steel-toe-capped boots;

➤ engines must only be run in the open air;

➤ drives and rotating parts of the engine should be enclosed;

➤ if petrol driven, care is required with flammable liquids and refuelling;

➤ protection from the sun should be provided in the form of PPE for operator;

➤ noise levels should be checked and noise attenuation, for example silencers and damping, fitted if necessary;

➤ suitable eye protection should be worn;

➤ respiratory protection to be worn where dusty ground is being compacted;

➤ manual handling assessments should be carried out. Minimize handling and use two people for lifting or mechanized equipment.

13.15.11 Site machinery – ground consolidation equipment

Application of controls:

➤ select reduced/anti-vibration models;

➤ engines must only be run in the open air;

➤ drives and rotating parts of engine should be enclosed;

➤ operator seat restraints should be fitted and worn;

➤ hold to run controls should be fitted;

➤ suitable access and egress steps and hand holds are required;

➤ rollover or turnover protection (ROPS) should be fitted to protect the driver;

➤ if petrol driven, care is required with flammable liquids and refuelling;

➤ noise levels should be checked and noise attenuation, for example silencers and damping, fitted if necessary;

➤ respiratory protection to be worn where dusty ground is being consolidated, unless the cab is enclosed with air filtration/air conditioning;

➤ protection from the sun should be fitted with a canopy or cab;

➤ suitable foot protection to be worn.

Figure 13.37 Ground consolidation vibrating roller.

13.15.12 Site machinery – road marking equipment

Application of controls:

➤ protective clothing to include high visibility overalls or jackets with reflective strips for all personnel. In addition where appropriate and risk assessments require it:

　➤ goggles

　➤ helmets – when working in hazardous conditions, when workers are overhead, where there is

Figure 13.38 High-way line marking lorry mounted unit.

Figure 13.39a

Figure 13.39b Walk behind line marking equipment.

overhanging debris or where working under anything that could strike from overhead

➤ suitable respirators when working in dusty conditions or when fumes and vapours are present.

➤ roadworks may not commence until sufficient road warning signs as required by the Department for Transport Code of Practice 'Safety at street works and roadworks' (ISBN 0115523103) are in position,

263

and, where appropriate footpaths have been provided for pedestrians. All road signs must be removed at the completion of the work. Many road marking schemes will use moving signs attached to vehicles as the work proceeds. These should all be in accordance with the Code of Practice. See Section 11.4 for more details;

➤ yellow warning lights with amber lenses must be placed around hazards left overnight. Check that lamps are recharged/refilled and lit before dark;

➤ operatives of mechanically propelled road marking equipment must be over 18, hold a current driving licence for the category of vehicle, and have been trained and authorized to drive the equipment;

➤ all moving parts of the engine must be properly safeguarded. Engines are usually totally enclosed except for cooling fans, alternators and belts which need to be guarded;

➤ equipment, flexible pipes, tools, pressure vessels should all be checked for leaks or other defects before using;

➤ no air guns or paint sprays should be used on people even in jest to prevent compressed air entering the bloodstream;

➤ no smoking or naked flames should be used with flammable road marking paints;

➤ heat sources should be carefully controlled with thermostatic controls on electrical systems. LPG cylinders should be fixed in an upright position with effective on–off valves, hoses in date and in good order.

13.15.13 *Site machinery – road making laying equipment*

Application of controls:

➤ engines must only be run in the open air;
➤ drives and rotating parts of engine should be enclosed;
➤ operator seat restraints should be fitted and worn;
➤ hold to run controls should be fitted;
➤ suitable access and egress steps and hand holds are required;
➤ ROPS may be fitted to protect the driver, although the risk is very small;
➤ if petrol driven, care is required with flammable liquids and refuelling;
➤ noise levels should be checked and noise attenuation, for example silencers and damping, fitted if necessary;
➤ respiratory protection to be worn where hot asphalt fumes are a problem, unless the cab is enclosed with air filtration/air conditioning;
➤ for protection from the sun, the machine should be fitted with a canopy or cab;
➤ safe system should be devised for reversing trucks up to the hopper to ensure that no one is standing in the vicinity. Trucks to have reversing alarms. Driver should have an unrestricted view of the hopper area;
➤ suitable eye protection should be worn;
➤ suitable foot protection to be worn.

Figure 13.40 Road making laying machines.

13.16 Guard construction

The design and construction of guards must be appropriate to the risks identified and the mode of operation of the machinery in question.

The following factors should be considered:

➤ strength – guards should be adequate for the purpose, able to resist the forces and vibration involved, and able to withstand impact (where applicable)
➤ weight and size – in relation to the need to remove and replace the guard during maintenance
➤ compatibility with materials being processed and lubricants, etc.
➤ hygiene and the need to comply with food safety regulations
➤ visibility – it may be necessary to see through the guard for both operational and safety reasons
➤ noise attenuation – guards can often be used to reduce the noise levels produced by a machine. Conversely the resonance of large undamped panels may exacerbate the noise problem
➤ enabling a free flow of air – where necessary (e.g. for ventilation)
➤ avoidance of additional hazards – for example, free of sharp edges
➤ ease of maintenance and cleanliness
➤ openings – the size of openings and their distance from the dangerous parts should not allow anyone to be able to reach into a danger zone. These values can be determined by experiment or by reference to standard tables. If doing so by experiment, it is essential that the machine is first stopped and made safe (e.g. by isolation). The detailed information on openings is contained in EN 294:1992, EN 349:1993 and EN 811:1997.

13.17 Practice NEBOSH questions for Chapter 13

1. **Outline** the sources and possible effects of four non-mechanical hazards commonly encountered in a woodworking shop.

2. **Describe** the hazards associated with the use of a pedestrian-operated plate compactor being used on a highway.

3. (a) **Identify TWO** types of equipment used in the construction industry which are fitted with an abrasive wheel and describe the hazards associated with their use.
   (b) **Outline** the correct method of mounting an abrasive wheel.
   (c) **Prepare** a **10-point** checklist that can be used to monitor safe working practices when abrasive wheels are being used.

4. **Identify FOUR** mechanical hazards presented by pedestal drills and outline in each case how injury may occur.

5. (a) In relation to machine safety, **outline** the principles of:
      (i) interlocked guards
      (ii) trip devices.
   (b) Other than contact with dangerous parts, **identify FOUR** types of danger against which fixed guards on machines may provide protection.

6. **Outline EIGHT** factors that may be important in determining the maintenance requirements for an item of work equipment.

7. **Provide** sketches to show clearly the nature of the following mechanical hazards from moving parts of machinery:
   (a) entanglement
   (b) crushing
   (c) drawing-in
   (d) shear.

8. **List** the main requirements of the Provision and Use of Work Equipment Regulations 1998.

9. With reference to an accident involving an operator who comes into contact with a dangerous part of a machine, **describe**:
   (a) the possible immediate causes
   (b) the possible root (underlying) causes.

Electrical hazards and control

14

Introduction

Figure 14.1 Beware of electricity – typical sign.

Electricity is a widely used, efficient and convenient, but potentially hazardous, method of transmitting and using energy. It is in use in every factory, workshop, laboratory and office in the country. Any use of electricity has the potential to be very hazardous with possible fatal results. Legislation has been in place for many years to control and regulate the use of electrical energy and the activities associated with its use. Such legislation provides a framework for the standards required in the design, installation, maintenance and use of electrical equipment and systems, and the supervision of these activities to minimize the risk of injury. Electrical work from the largest to the smallest installation must be carried out by people known to be competent to undertake such work. New installations always require expert advice at all appropriate levels to cover design aspects of the system and its associated equipment. Electrical systems and equipment must be properly selected, installed, used and maintained.

Approximately 8% of all fatalities at work are caused by electric shock. Over the last few years, there have been between 12 and 16 employee deaths due to electricity, between 210 and 258 major accidents and about 500 over 3-day accidents each year. The majority of the fatalities occur in the agriculture, extractive and utility supply and service industries, while the majority of the major accidents happen in the manufacturing, construction and service industries.

Only voltages up to and including **mains voltage** (220/240V) will be considered in detail in this chapter and the three principal electrical hazards – electric shock, electric burns and electrical fires and explosions.

14.2
Principles of electricity and some definitions

14.2.1 *Basic principles and measurement of electricity*

In simple terms, electricity is the flow or movement of electrons through a substance which allows the transfer of electrical energy from one position to another. The substance through which the electricity flows is called a **conductor**. This flow or movement of electrons is known as the **electric current**. There are two forms of electric current – direct and alternating. **Direct current (DC)** involves the flow of electrons along a conductor from one end to the other. This type of current is mainly restricted to batteries and similar devices. **Alternating current (AC)** is produced by a rotating alternator and causes an oscillation of the electrons rather than a flow of electrons so that energy is passed from one electron to the adjacent one and so on through the length of the conductor.

It is sometimes easier to understand the basic principles of electricity by comparing its movement with that of water in a pipe flowing downhill. The flow rate of water through the pipe (measured in litres/second) is similar to the current flowing through the conductor, which is measured in amperes, normally abbreviated to **amps**. Sometimes very small currents are used and these are measured in milliamps (mA).

The higher the pressure drop is along the pipeline, the greater will be the flow rate of water and, in a similar way, the higher the electrical 'pressure difference' along the conductor, the higher the current will be. This electrical 'pressure difference' or potential difference is measured in **volts**.

The flow rate through the pipe will also vary for a fixed pressure drop as the roughness on the inside surface of the pipe varies – the rougher the surface, higher the resistance to flow becomes the and the slower the flow. Similarly, for electricity, the poorer the conductor, the higher the resistance is to electrical current and the lower the current becomes. Electrical resistance is measured in **ohms**.

The voltage (V), the current (I) and the resistance (R) are related by the following formula, known as Ohm's law:

$$V = I \times R \text{ (volts)}$$

and, electrical power (P) is given by:

$$P = V \times I \text{ (watts)}$$

These basic formulae enable simple calculations to be made so that, for example, the correct size of fuse may be ascertained for a particular piece of electrical equipment.

Conductors and insulators

Conductors are nearly always metals, copper being a particularly good conductor, and are usually in wire form but they can be gases or liquids, water being a particularly good conductor of electricity. 'Superconductors' is a term given to certain metals which have a very low resistance to electricity at low temperatures.

Very poor conductors are known as **insulators** and include materials such as rubber, timber and plastics. Insulating material is used to protect people from some of the hazards associated with electricity.

Short circuit

Electrical equipment components and an electrical power supply (normally the mains or a battery) are joined together by a conductor to form a **circuit**. If the circuit is broken in some way so that the current flows directly to earth rather than to a piece of equipment, a **short circuit** is made. Since the resistance is greatly reduced but the voltage remains the same, a rapid increase in current occurs which could cause significant problems if suitable protection were not available.

Earthing

An electricity supply company has one of its conductors solidly connected to the earth and every circuit supplied by the company must have one of its conductors connected to earth. This means that if there is a fault, such as a break in the circuit, the current, known as the earth fault current, will return directly to earth, which forms the circuit of least resistance, thus maintaining the supply circuit. This process is known as **earthing**. Other devices, such as fuses and residual current devices (RCDs), which will be described later, will also be needed within the circuit to interrupt the current flow to earth so as to protect people from electric shock and equipment from overheating. Good and effective earthing is absolutely essential and must be connected and checked by a competent person. Where a direct contact with earth is not possible, for example, in a motor car, a common voltage reference point is used, such as the vehicle chassis.

Where other potential metallic conductors exist near to electrical conductors in a building, they must be connected to the main earth terminal to ensure **equipotential bonding** of all conductors to earth. This applies to gas, water and central heating pipes and other devices such as lightning protection systems. **Supplementary bonding** is required in bathrooms and kitchens where, for example, metal sinks and other metallic equipment surfaces are present. This involves the connection of a conductor from the sink to a water supply pipe which has been earthed by equipotential bonding. There have been several fatalities due to electric shocks from 'live' service pipes or kitchen sinks.

14.2.2 *Some definitions*

Certain terms are frequently used with reference to electricity and the more common ones are defined here.

Low voltage – a voltage normally not exceeding 600 V AC between conductors and earth or 1,000 V AC between phases. Mains voltage falls into this category.

High voltage – a voltage normally exceeding 600 V AC between conductors and earth or 1,000 V AC between phases.

Mains voltage – this is the common voltage available in domestic premises and many workplaces and is normally taken from three-pin socket points. In the UK it is distributed by the national grid and is usually supplied between 220 and 240 V, AC and at 50 cycles/second.

Maintenance – a combination of any actions carried out to maintain an item of electrical equipment in, or restore it to, an acceptable and safe condition.

Testing – a measurement carried out to monitor the conditions of an item of electrical equipment without physically altering the construction of the item or the electrical system to which it is connected.

Inspection – a maintenance action involving the careful scrutiny of an item of electrical equipment, using, if necessary, all the senses to detect any failure to meet an acceptable and safe condition. An inspection does not include any dismantling of the item of equipment.

Examination – an inspection together with the possible partial dismantling of an item of electrical equipment, including measurement and non-destructive testing as required, in order to arrive at a reliable conclusion as to its condition and safety.

Isolation – involves cutting off the electrical supply from all or a discrete section of the installation by separating the installation or section from every source of electrical energy. This is the normal practice so as to ensure the safety of persons working on or in the vicinity of electrical components which are normally live and where there is a risk of direct contact with live electricity.

Competent electrical person – a person possessing sufficient electrical knowledge and experience to avoid the risks to health and safety associated with electrical equipment and electricity in general.

14.3   Electrical hazards and injuries

Electricity is a safe, clean and quiet method of transmitting energy. However, this apparently benign source of energy when accidentally brought into contact with conducting material, such as people, animals or metals, permits releases of energy which may result in serious damage or loss of life. Constant awareness is necessary to avoid and prevent danger from accidental releases of electrical energy.

The principal hazards associated with electricity are:

➤ electric shock
➤ electric burns
➤ electrical fires and explosions
➤ arcing
➤ portable electrical equipment
➤ secondary hazards.

14.3.1 *Electric shock and burns*

Electric shock is the convulsive reaction by the human body to the flow of electric current through it. This sense of shock is accompanied by pain and, in more severe cases, by burning. The shock can be produced by low voltages, high voltages or lightning. Most incidents of electric shock occur when the person becomes the route to earth for a live conductor. The effect of electric shock and the resultant severity of injury depend upon the size of the electric current passing through the body which, in turn, depends on the voltage and the electrical resistance

of the skin. If the skin is wet, a shock from mains voltage (220/240 V) could well be fatal. The effect of shock is very dependent on conditions at the time but it is always dangerous and must be avoided. Electric burns are usually more severe than those caused by heat, since they can penetrate deep into the tissues of the body.

The effect of electric current on the human body depends on its pathway through the body (e.g. hand to hand or hand to foot), the frequency of the current, the length of time of the shock and the size of the current. Current size is dependent on the duration of contact and the electrical resistance of body tissue. The electrical resistance of the body is greatest in the skin and is approximately 100,000 ohm; however, this may be reduced by a factor of 100 when the skin is wet. The body beneath the skin offers very little resistance to electricity due to its very high water content and, while the overall body resistance varies considerably between people and during the lifetime of each person, it averages at 1,000 ohms. Skin that is wounded, bruised or damaged will considerably reduce human electrical resistance and work should not be undertaken on electrical equipment if damaged skin is unprotected.

An electric current of 1 mA is detectable by touch and one of 10 mA will cause muscle contraction which may prevent the person from being able to release the conductor, and if the chest is in the current path, respiratory movement may be prevented causing asphyxia. Current passing through the chest may also cause fibrillation of the heart (vibration of the heart muscle) and disrupt the normal rhythm of the heart, though this is likely only within a particular range of currents. The shock can also cause the heart to stop completely (cardiac arrest) and this will lead to the cessation of breathing. Current passing through the respiratory centre of the brain may cause respiratory arrest that does not quickly respond to the breaking of the electrical contact. These effects on the heart and respiratory system can be caused by currents as low as 25 mA. It is not possible to be precise on the threshold current because it is dependent on the environmental conditions at the time, as well the age, sex, body weight and health of the person.

Burns of the skin occur at the point of electrical contact due to the high resistance of skin. These burns may be deep, slow to heal and often leave permanent scars. Burns may also occur inside the body along the path of the electric current causing damage to muscle tissue and blood cells. Burns associated with radiation and microwaves are dealt with in Chapter 17.

14.3.2 *Treatment of electric shock and burns*

Electric shock

There are many excellent posters available which illustrate a first-aid procedure for treating electric shock and such posters should be positioned close to electrical junction boxes or isolation switches. The recommended procedure for treating an unconscious person who has received a **low voltage** electric shock is as follows:

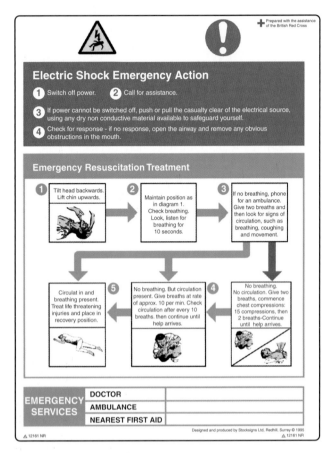

Figure 14.2 Typical electric shock poster.

1. On finding a person suffering from electric shock, raise the alarm by calling for help from colleagues (including a trained first aider).
2. Switch off the power if it is possible and/or the position of the emergency isolation switch is known.
3. Call for an ambulance.
4. If it is not possible to switch off the power, then push or pull the person away from the conductor using an object made from a good insulator, such as a wooden chair or broom. Remember to stand on dry insulating material, for example, a wooden pallet, rubber mat or wooden box. If these precautions are not taken, then the rescuer will also be electrocuted.
5. If the person is breathing, place them in the recovery position so that an open airway is maintained and the mouth can drain if necessary.
6. If the person is not breathing apply mouth-to-mouth resuscitation and, in the absence of a pulse, chest compressions. When the person is breathing normally place them in the recovery position.
7. Treat any burns by placing a sterile dressing over the burn and secure with a bandage. Any loose skin or blisters should not be touched nor any lotions or ointments applied to the burn wound.
8. If the person regains consciousness, treat for normal shock.
9. Remain with the person until they are taken to a hospital or local surgery.

It is important to note that electrocution by high voltage electricity is normally instantly fatal. On discovering a person who has been electrocuted by high voltage electricity, the police and electricity supply company should be informed. If the person remains in contact with or within 18 m of the supply, they should not be approached to within 18 m by others until the supply has been switched off and clearance has been given by the emergency services. High voltage electricity can 'arc' over distances less than 18 m, thus electrocuting the would-be rescuer.

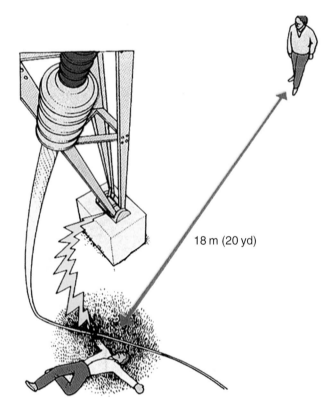

18 m (20 yd)

Figure 14.3 Keep 18 m clear of high voltage lines.

14.3.3 *Electrical fires and explosions*

Over 25% of all fires have a cause linked to a malfunction of either a piece of electrical equipment or wiring or both. Electrical fires are often caused by a lack of reasonable care in the maintenance and use of electrical installations and equipment. The electricity that provides heat and light and drives electric motors is capable of igniting insulating or other combustible material if the equipment is misused, is not adequate to carry the electrical load, or is not properly installed and maintained. The most common causes of fire in electrical installations are short circuits, overheating of cables and equipment, the ignition of flammable gases and vapours, and the ignition of combustible substances by static electrical discharges.

Short circuits happen, as mentioned earlier, if insulation becomes faulty, and an unintended flow of current between two conductors or between one conductor and earth occurs. The amount of the current depends, among other things, upon the voltage, the condition of the insulating material and the distance between the conductors. At first the current flow will be low but as the fault develops the current will increase and the area surrounding the fault will heat up. In time, if the fault persists, a total breakdown of insulation will result and excessive current will flow through the fault. If the fuse fails to operate or is in excess of the recommended fuse rating, overheating will occur and a fire will result. A fire can also be caused if combustible material is in close proximity to the heated wire or hot sparks are ejected. Short circuits are most likely to occur where electrical equipment or cables are susceptible to damage by water leaks or mechanical damage. Twisted or bent cables can also cause breakdowns in insulation materials.

Inspection covers and cable boxes are particular problem areas. Effective steps should be taken to prevent the entry of moisture as this will reduce or eliminate the risk. Covers can themselves be a problem especially in dusty areas where the dust can accumulate on flat insulating surfaces resulting in tracking between conductors at different voltages and a subsequent insulation failure. The interior of inspection panels should be kept clean and dust free by using a suitable vacuum cleaner.

Overheating of cables and equipment will occur if they become overloaded. Electrical equipment and circuits are normally rated to carry a given safe current which will keep the temperature rise of the conductors in the circuit or appliance within permissible limits and avoid the possibility of fire. These safe currents define the maximum size of the fuse (the fuse rating) required for the appliance.

A common cause of circuit overloading is the use of equipment and cables which are too small for the imposed electrical load. This is often caused by the addition of more and more equipment to the circuit thus taking it beyond its original design specification. In offices, the overuse of multisocket unfused outlet adaptors can create overload problems (sometimes

(a)

(b)

Figure 14.4 Typical (a) transformer and (b) RCD device.

known as the Christmas tree effect). The more modern multiplugs are much safer as they lead to one fused plug and cannot be easily overloaded (see Figure 14.7). Another cause of overloading is mechanical breakdown or wear of an electric motor and the driven machinery. Motors must be maintained in good condition with particular attention paid to bearing surfaces. Fuses do not always provide total protection against the overloading of motors and, in some cases, severe heating may occur without the fuses being activated.

Loose cable connections are one of the most common causes of overheating and may be readily detected (as well as overloaded cables) by a thermal imaging survey (a technique which indicates the presence of hot spots). The bunching of cables together can also cause excessive heat to be developed within the inner cable leading to a fire risk. This can happen with cable extension reels, which have only been partially unwound, used for high-energy appliances like an electric heater.

Ventilation is necessary to maintain safe temperatures in most electrical equipment and overheating is liable to occur if ventilation is in any way obstructed or reduced. All electric equipment must be kept free of any obstructions that restrict the free supply of air to the equipment and, in particular, to the ventilation apertures.

Most electrical equipment either sparks in normal operation or is liable to spark under fault conditions. Some electrical appliances, such as electric heaters, are specifically designed to produce high temperatures. These circumstances create fire and explosion hazards, which demand very careful assessment in locations where processes capable of producing flammable concentrations of gas or vapour are used, or where flammable liquids are stored.

It is likely that many fires are caused by static electrical discharges. Static electricity can, in general, be eliminated by the careful design and selection of materials used in equipment and plant, and the materials used in products being manufactured. When it is impractical to avoid the generation of static electricity, a means of control must be devised. Where flammable materials are present, especially if they are gases or dusts, then there is a great danger of fire and explosion, even if there is only a small discharge of static electricity. The control and prevention of static electricity is considered in more detail later.

The use of electrical equipment in potentially flammable atmospheres should be avoided as far as possible. However, there will be many cases where electrical equipment must be used and, in these cases, the standards for the construction of the equipment should comply with the Equipment and Protective Systems Intended for Use in Potentially Explosive Atmospheres Regulations (known as ATEX) and details on the classification or zoning of areas are published by the British Standards Institution and the Health and Safety Executive (HSE).

Before electrical equipment is installed in any location where flammable dusts, vapours or gases may be present, the area must be zoned in accordance with the Dangerous Substances and Explosive Atmospheres Regulations and records of the zoned areas must be marked on building drawings and revised when any zoned area is changed. The installation and maintenance of electrical equipment in potentially flammable atmospheres is a specialized task. It must only be undertaken by electricians or instrument mechanics who have an understanding of the techniques involved.

In the case of a fire involving electrical equipment, the first action must be the isolation of the power supply so that the circuit is no longer live. This is achieved by switching off the power supply at the mains isolation switch or at another appropriate point in the system. Where it is not possible to switch off the current, the fire must be attacked in a way which will not cause additional danger. The use of a non-conducting extinguishing medium, such as carbon dioxide or powder, is necessary. After extinguishing such a fire careful watch should be kept for renewed outbreaks until the fault has been rectified. Re-ignition is a particular problem when carbon dioxide extinguishers are used, although less equipment may be damaged than is the case when powder is used.

Finally, the chances of electrical fires occurring are considerably reduced if the original installation was undertaken by competent electricians working to recognized standards, such as the Institution of Electrical Engineers Regulations. It is also important to have a system of regular testing and inspection in place so that any remedial maintenance can take place.

14.3.4 *Electric arcing*

A person who is standing on earth too close to a high voltage conductor may suffer flash burns as a result of arc formation. Such burns may be extensive and lower the resistance of the skin so that electric shock may add to the ill effects. Electric arc faults can cause temporary blindness by burning the retina of the eye and this may lead to additional secondary hazards. The quantity of electrical energy is as important as the size of the voltage since the voltage will determine the distance over which the arc will travel. The risk of arcing can be reduced by the insulation of live conductors.

Strong electromagnetic fields induce surface charges on people. If these charges accumulate, skin sensation is affected and spark discharges to earth may cause localized pain or bruising. Whether prolonged exposure to strong fields has any other significant effects on health has not been proven. However, the action of an implanted cardiac pacemaker may be disturbed by the close proximity of its wearer to a powerful electromagnetic field. The health effects of arcing and other non-ionizing radiation are covered in Chapter 17.

14.3.5 *Static electricity*

Static electricity is produced by the build-up of electrons on weak electrical conductors or insulating materials. These materials may be gaseous, liquid or solid and may include flammable liquids, powders, plastic films and granules. The generation of static may be caused by the rapid separation of highly insulated materials by friction or by transfer from one highly charged material to another in an electric field by induction.

Discharges of static electricity may be sufficient to cause serious electric shock and are always a potential source of ignition when flammable liquid, dusts or powders are present. Flour dust in a mill, for example, has been ignited by static electricity.

Static electricity may build up on both materials and people. When a charged person approaches flammable gases or vapours and a spark ignites the substance, the resulting explosion or fire often causes serious injury. See Fig. 14.5.

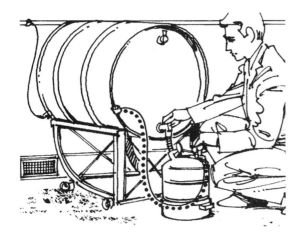

Figure 14.5 Prevention of static discharge – container connected to earthed drum.

Lightning strikes are a natural form of static electricity and result in large amounts of electrical energy being dissipated in a short time in a limited space with a varying degree of damage. The current produced in the vast majority of strikes exceeds 3,000 amps over a short period of time. Before a strike, the electrical potential between the cloud and earth might be about 100 million volts and the energy released at its peak might be about 100 million watts per metre of strike.

The need to provide lightning protection depends on a number of factors, which include:

➤ the risk of a strike occurring
➤ the number of people likely to be affected
➤ the location of structure and the nearness of other tall structures in the vicinity
➤ the type of construction, including the materials used
➤ the contents of structure or building (including any flammable substances)
➤ the value of the building and its contents.

Expert advice will be required from a specialist company in lightning protection, especially when flammable substances are involved. Lightning strikes can also cause complete destruction and/or significant disruption of electronic equipment.

Figure 14.6 Portable hand-held electric power tools.

14.3.6 *Portable electrical equipment*

Portable and transportable electrical equipment is defined by the HSE as 'not part of a fixed installation but may be connected to a fixed installation by means of a flexible cable and either a socket and plug or a spur box or similar means. It may be hand held or hand operated while connected to the supply, or is intended or likely to be moved while connected to the supply'. The auxiliary equipment, such as extension leads, plugs and sockets, used with portable tools, is also classified as portable equipment. The term 'portable' means both portable and transportable.

Almost 25% of all reportable electrical accidents involve portable electrical equipment (known as portable appliances). While most of these accidents were caused by electric shock, over 2,000 fires each year are started by faulty cables used by portable appliances, caused by a lack of effective maintenance. Portable electrical tools often present a high risk of injury, which is frequently caused by the conditions under which they are used. These conditions include the use of defective or unsuitable equipment and, indeed, the misuse of equipment. There must be a system to record the inspection, maintenance and repair of these tools.

Where plugs and sockets are used for portable tools, sufficient sockets must be provided for all the equipment and adaptors should not be used. Many accidents are caused by faulty flexible cables, extension leads, plugs and sockets, particularly when these items become damp or worn. Accidents often occur when contact is made with some part of the tool which has become live (probably at mains voltage), while the user is standing on, or in contact with, an earthed conducting surface. If the electrical supply is at more than 50 V AC, then the electric shock that a person may receive from such defective equipment is potentially lethal. In adverse environmental conditions, such as humid or damp atmospheres, even lower voltages can be dangerous. Portable electrical equipment should not be used in flammable atmospheres if it can be avoided and it must also comply with any standard relevant to the particular environment. Air-operated equipment should also be used as an alternative whenever it is practicable.

Some portable equipment requires substantial power to operate and may require voltages higher than those usually used for portable tools, so that the current is kept down to reasonable levels. In these cases, power leads with a separate earth conductor and earth screen must be used. Earth leakage relays and earth monitoring equipment must also be used, together with substantial plugs and sockets designed for this type of system.

Electrical equipment is safe when properly selected, used and maintained. It is important, however, that the environmental conditions are always carefully considered. The hazards associated with portable appliances increase with the frequency of use and the harshness of the environment (construction sites are often particularly hazardous in this respect). These factors must be considered when inspection, testing and maintenance procedures are being developed.

14.3.7 *Secondary hazards*

It is important to note that there are other hazards associated with portable electrical appliances, such as abrasion and impact, noise and vibration. Trailing leads used for portable equipment and raised socket points offer serious trip hazards and both should be used with great care near pedestrian walkways. Power drives from electric motors should always be guarded against entanglement hazards.

Secondary hazards are those additional hazards which present themselves as a result of an electrical hazard. It is very important that these hazards are considered during a risk assessment. An electric shock could lead to a fall from height if the shock occurred on a scaffold or it could lead to a collision with a vehicle if the victim collapsed on to a roadway.

Similarly, an electrical fire could lead to all the associated fire hazards outlined in Chapter 15 (e.g. suffocation, burns and structural collapse) and electrical burns can easily lead to infections.

14.4   General control measures for electrical hazards

The principal control measures for electrical hazards are contained in the statutory precautionary requirements covered by the Electricity at Work Regulations, the main provisions of which are outlined in Chapter 21. They are applicable to all electrical equipment and systems found at the workplace and impose duties on employers, employees and the self-employed.

The Regulations cover the following topics:

➤ the design, construction and maintenance of electrical systems, work activities and protective equipment
➤ the strength and capability of electrical equipment
➤ the protection of equipment against adverse and hazardous environments
➤ the insulation, protection and placing of electrical conductors
➤ the earthing of conductors and other suitable precautions
➤ the integrity of referenced conductors
➤ the suitability of joints and connections used in electrical systems
➤ means for protection from excess current
➤ means for cutting off the supply and for isolation
➤ the precautions to be taken for work on equipment made dead
➤ working on or near live conductors
➤ adequate working space, access and lighting
➤ the competence requirements for persons working on electrical equipment to prevent danger and injury.

Detailed safety standards for designers and installers of electrical systems and equipment are given a code of practice published by the Institution of Electrical Engineers, known as the IEE Regulations. While these Regulations are not legally binding, they are recognized as a code of good practice and widely used as an industry standard.

The risk of injury and damage inherent in the use of electricity can only be controlled effectively by the introduction of employee training, safe operating procedures (safe systems of work) and guidance to cover specific tasks.

Training is required at all levels of the organization ranging from simple on-the-job instruction to apprenticeship for electrical technicians and supervisory courses for experienced electrical engineers. First-aid training related to the need for cardiovascular resuscitation and treatment of electrical burns should be available to all people working on electrical equipment and their supervisors.

A **management system** should be in place to ensure that the electrical systems are installed, operated and maintained in a safe manner. All managers should be responsible for the provision of adequate resources of people, material and advice to ensure that the safety of electrical systems under their control is satisfactory and that **safe systems of work** are in place for all electrical equipment. (Chapter 6 gives more information on both safe systems of work and permits to work.)

For small factories and office or shop premises where the system voltages are normally at mains voltage, it may be necessary for an external competent person to be available to offer the necessary advice. Managers must set up a high voltage permit-to-work system for all work at and above 600 V. The system should be appropriate to the extent of the electrical system involved. Consideration should also be given to the introduction of a permit system for voltages under 600 V when appropriate and for all work on live conductors.

The additional control measures that should be taken when working with electricity or using electrical equipment are summarized by the following topics:

➤ the selection of suitable equipment
➤ the use of protective systems
➤ inspection and maintenance strategies.

These three groups of measures will be discussed in detail.

14.5 The selection and suitability of equipment

Many factors which affect the selection of suitable electrical equipment, such as flammable, explosive and damp atmospheres and adverse weather conditions, have already been considered. Other issues include high or low temperatures, dirty or corrosive processes or problems associated with vegetation or animals (e.g. tree roots touching and displacing underground power

cables, farm animals urinating near power supply lines and rats gnawing through cables). Temperature extremes will affect, for example, the lubrication of motor bearings and corrosive atmospheres can lead to the breakdown of insulating materials. The equipment selected must be suitable for the task demanded or either it will become overloaded or running costs will be too high.

The equipment should be installed to a recognized standard and capable of being isolated in the event of an emergency. It is also important that the equipment is effectively and safely earthed. Electrical supply failures may affect process plant and equipment. These are certain to happen at some time and the design of the installation should be such that a safe shutdown can be achieved in the event of a total mains failure. This may require the use of a battery backed shutdown system or emergency stand-by electric generators (assuming that this is cost effective).

Electrical equipment that is to be used on a construction site must be constructed to withstand the site conditions in terms of weather, water and dust. Site distribution units should be designed and manufactured to a suitable standard, for example BS EN 60439-4, particular to requirements for assemblies for construction sites (ACS). Units will be suitable if they are:

➤ flexible in application for repeated use on various contracts
➤ designed for ease of transport and storage
➤ robust in construction that will resist damage
➤ provided with lockable switches and means of isolation.

Socket outlets should be suitable for use on a construction site. Sockets designed for domestic usage are not suitable.

When electrical equipment is selected for use on construction sites, it must be robust and suitable for the conditions likely to be found on the site. Such conditions include damp weather and the presence of construction materials which produce dust.

Finally, it is important to stress that electrical equipment must only be used within the rating performance given by the manufacturer and any accompanying instructions from the manufacturer or supplier must be carefully followed.

14.5.1 The planning and installation of a progressively extending electrical system on site

A detailed account of this topic is given in the HSE publication *Electrical Safety on Construction Sites* (HSG141) and only a summary will be given here.

The site distribution system is the cabling system and electrical equipment installed to supply electricity to various points across the site during the construction phase. The supervision, maintenance and development of the system must be the responsibility of a qualified and competent electrician. Any extensions to the system must be agreed beforehand with the electrician.

It is a temporary system which will be removed at the end of the contract and replaced with the permanent site distribution system. Even though the system is temporary, it must be robust enough to withstand the harsh conditions on site and be adequately protected against damage and contamination. Switchgear and metering equipment should be positioned away from traffic routes in secure and sheltered accommodation. All wiring, no matter how temporary, must be installed to appropriate and recognized standards. Distribution cables must be located where they are least likely to be damaged by site activity. They should also be kept clear of walkways, ladders and other services. If they need to cross a roadway or walkway, they should be placed in ducts with a marker at each end of the duct. If the roadway is used by vehicles, the duct should be buried at least 0.5 m below the surface. A record must be kept of the location of all buried cables to avoid damage as the work progresses. Alternatively, the cables may be carried overhead using a goalpost type arrangement.

All fixed distribution cables carrying 400 or 230 V must be protected with a metal sheath and/or armour, which are continuous, effectively earthed and protected against corrosion. The HSE strongly recommend that any existing or new permanent fixed supply should not be used to supply contractors' equipment during the construction work. Any movable plant such as lifts or hoists should be supplied by armoured cable.

Finally any work on live electrical installations must be subject to a permit to work which must be signed on and off at either end of the work. The detailed requirements of permits to work are given in Chapter 6.

14.5.2 *The advantages and limitations of protective systems*

There are several different types of protective systems and techniques that may be used to protect people, plant and premises from electrical hazards, some of which, for example earthing, have already been considered earlier in this chapter. However, only the more common types of protection will be considered here.

Fuse

A fuse will provide protection against faults and continuous marginal current overloads. It is basically a thin strip of conducting wire which will melt when current in excess of the rated value passes through it, thus breaking the circuit. A fuse rated at 13 amps will melt when a current in excess of 13 amps passes through the fuse thus stopping the flow of current. A **circuit breaker** throws a switch off when excess current passes and is similar in action to a fuse. Protection against overload is provided by fuses which detect a continuous marginal excess flow of current and energy. This overcurrent protection is arranged to operate before damage occurs, either to the supply system or to the load, which may be a motor or heater. When providing protection against overload, consideration needs to be made as to whether tripping the circuit could give rise to an even more dangerous situation, such as with fire fighting equipment.

The prime objective of a fuse is to protect equipment or an installation from overheating and becoming a fire hazard. It is not an effective protection against electric shock due to the time that it takes to cut the current flow.

The examination of fuses is a vital part of an inspection programme to ensure that the correct size or rating is fitted at all times.

Insulation

Insulation is used to protect people from electric shock, and to prevent the short circuiting of live conductors and the dangers associated with fire and explosions. Insulation is achieved by covering the conductor with an insulating material. Insulation is often accompanied by the enclosure of live conductors so that they are out of reach to people. A breakdown in insulation can cause electric shock, fire, explosion or instrument damage.

Isolation

The isolation of an electrical circuit involves more than 'switching off' the current in that the circuit is made dead and cannot be accidentally re-energized. It, therefore, creates a barrier between the equipment and the electrical supply which only an authorized person should be able to remove. When it is intended to carry out work, such as mechanical maintenance or a cleaning operation on plant or machinery, isolation of electrical equipment will be required to ensure safety during the work process. Isolators should always be locked off when work is to be done on electrical equipment.

Before earthing or working on an isolated circuit, checks must be made to ensure that the circuit is dead and that the isolation switch is 'locked off' and clearly labelled.

Reduced low voltage systems

When the working conditions are relatively severe due to either wet conditions or heavy and frequent usage of equipment, reduced voltage systems should be used.

All portable tools used on construction sites, vehicle washing stations or near swimming pools, should operate on 110V or less, preferably with a centre tapped to earth at 55V. This means that while the full 110V is available to power the tool, only 55V is available should the worker suffer an electric shock. At this level of voltage, the effect of any electric shock should not be severe. For lighting, even lower voltages can be used and are even safer. Another way to reduce the voltage is to use battery (cordless) operated hand tools.

Residual current devices

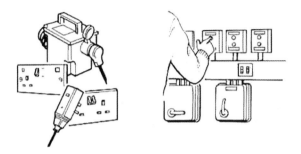

Figure 14.7 Transformer, RCD and isolators.

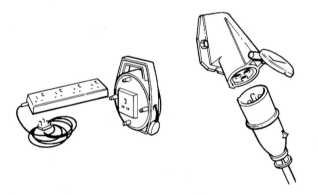

Figure 14.8 Multiplugs, extension lead and special plugs and sockets.

If electrical equipment must operate at mains voltage, the best form of protection against electric shock is the RCD. RCDs, also known as earth leakage circuit breakers, monitor and compare the current flowing in the live and neutral conductors supplying the protected equipment. Such devices are very sensitive to differences of current between the live and neutral power lines and will cut the supply to the equipment in a very short period of time when a difference of only a few milliamps occurs. It is the speed of the reaction which offers the protection against electric shock.

RCDs can be used to protect installations against fire since they will interrupt the electrical supply before

sufficient energy to start a fire has been dissipated. For protection against electric shock, the RCD must have a rated residual current of 30mA or less and an operating time of 40 milliseconds or less at a residual current of 250mA. The protected equipment must be properly protected by insulation and enclosure in addition to the RCD. The RCD will not prevent shock or limit the current resulting from an accidental contact, but it will ensure that the duration of the shock is limited to the time taken for the RCD to operate. The RCD has a test button which should be tested frequently to ensure that it is working properly. On construction sites the size and type of the RCD will depend on the particular application and should never be used in place of reduced low voltage equipment.

If mains voltage has to be used, then RCDs must also be used. In this event, they should be installed in a dustproof and weatherproof enclosure or be specifically designed to work under these conditions. They should be protected against mechanical damage and vibration and checked daily by operating the test button. They should be inspected weekly together with the equipment that being supplied with power and given an electrical test every 3 months by a competent electrician.

Double insulation

To remove the need for earthing on some portable power tools, double insulation is used. Double insulation employs two independent layers of insulation over the live conductors, each layer alone being adequate to insulate the electrical equipment safely. Since such tools are not protected by an earth, they must be inspected and maintained regularly and must be discarded if damaged.

Figure 14.9 shows the symbol which is marked on double insulated portable power tools.

Figure 14.9 Double insulation sign.

14.6 Inspection and maintenance strategies

14.6.1 *Maintenance strategies*

Regulation 4(2) of the Electricity at Work Regulations requires that 'as may be necessary to prevent danger, all systems shall be maintained so as to prevent so far as is reasonably practicable, such danger'. Regular maintenance is, therefore, required to ensure that a serious risk of injury or fire does not result from installed electrical equipment. Maintenance standards should be set as high as possible so that a more reliable and safe electrical system will result.

Inspection and maintenance periods should be determined by reference to the recommendations of the manufacturer, consideration of the operating conditions and the environment in which equipment is located. The importance of equipment within the plant, from the plant safety and operational viewpoint, will also have a bearing on inspection and maintenance periods. The mechanical safety of driven machinery is vital and the electrical maintenance and isolation of the electrically powered drives is an essential part of that safety.

The particular areas of interest for inspection and maintenance are:

➤ the cleanliness of insulator and conductor surfaces
➤ the mechanical and electrical integrity of all joints and connections
➤ the integrity of mechanical mechanisms, such as switches and relays
➤ the calibration, condition and operation of all protection equipment, such as circuit breakers, RCDs and switches.

Safe operating procedures for the isolation of plant and machinery during both electrical and mechanical maintenance, must be prepared and followed. All electrical isolators must, wherever possible, be fitted with mechanisms which can be locked in the 'open/off' position and there must be a procedure to allow fuse withdrawal wherever isolators are not fitted.

Working on live equipment with voltages in excess of 110V must not be permitted except where fault finding or testing measurements cannot be done in any other way. Reasons such as the inconvenience of halting production are not acceptable.

Part of the maintenance process should include an appropriate system of visual inspection. By concentrating on a simple, inexpensive system of looking for visible signs of damage or faults, many of the electrical risks can be controlled, although more systematic testing may be necessary at a later stage.

All fixed electrical installations should be inspected and tested periodically by a competent person, such as a member of the National Inspection Council for Electrical Installation Contracting (NICEIC).

14.6.2 *Inspection strategies*

Regular inspection of electrical equipment is an essential component of any preventative maintenance programme and, therefore, regular inspection is required under Regulation 4(2) of the Electricity at Work Regulations, quoted previously. Any strategy for the inspection of electrical equipment, particularly portable appliances, should involve the following considerations:

➤ a means of identifying the equipment to be tested
➤ the number and type of appliances to be tested
➤ the competence of those who will undertake the testing (whether in-house or bought-in)
➤ the legal requirements for portable appliance testing (PAT) and other electrical equipment testing and the guidance available
➤ organizational duties of those with responsibilities for PAT and other electrical equipment testing
➤ test equipment selection and re-calibration
➤ the development of a recording, monitoring and review system
➤ the development of any training requirements resulting from the test programme.

14.7 Portable electrical appliances testing

Portable appliances should be subject to three levels of inspection – a user check, a formal visual inspection and a combined inspection and test.

14.7.1 *User checks*

When any portable electrical hand tool, appliance, extension lead or similar item of equipment is taken into use, and at least once each week or, in the case of heavy work, before each shift, the following visual check and associated questions should be asked:

➤ Is there a recent PAT label attached to the equipment?
➤ Are any bare wires visible?
➤ Is the cable covering undamaged and free from cuts and abrasions (apart from light scuffing)?
➤ Is the cable too long or too short? (Does it present a trip hazard?)
➤ Is the plug in good condition, for example, the casing is not cracked and the pins are not bent (see Figure 14.10)?
➤ Are there no taped or other non-standard joints in the cable?
➤ Is the outer covering (sheath) of the cable gripped where it enters the plug or the equipment? (The coloured insulation of the internal wires should not be visible.)

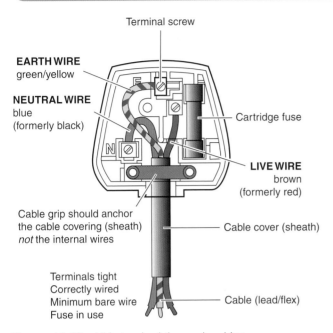

Terminal screw

EARTH WIRE
green/yellow

NEUTRAL WIRE
blue
(formerly black)

Cartridge fuse

LIVE WIRE
brown
(formerly red)

Cable grip should anchor
the cable covering (sheath)
not the internal wires

Cable cover (sheath)

Terminals tight
Correctly wired
Minimum bare wire
Fuse in use

Cable (lead/flex)

Figure 14.10 UK standard three-pin wiring.

> Is the outer case of the equipment undamaged or loose and are all screws in place?
> Are there any overheating or burn marks on the plug, cable, sockets or the equipment?
> Are the trip devices (RCDs) working effectively (by pressing the 'test' button)?

14.7.2 Formal visual inspections and tests

There should be a **formal visual inspection** routinely carried out on all portable electrical appliances. Faulty equipment should be taken out of service as soon as the damage is noticed. At this inspection the plug cover (if not moulded) should be removed to check that the correct fuse is included, but the equipment itself should not be taken apart. This work can normally be carried out by a trained person who has sufficient information and knowledge.

Some faults, such as the loss of earth continuity due to wires breaking or loosening within the equipment, the breakdown of insulation and internal contamination (e.g. dust containing metal particles may cause short circuiting if it gets inside the tool), will not be spotted by visual inspections. To identify these problems, a programme of testing and inspection will be necessary.

This formal **combined testing and inspection** should be carried out by a competent person either when there is reason to suspect the equipment may be faulty, damaged or contaminated, but this cannot be confirmed by visual inspection or after any repair, modification or similar work to the equipment, which could have affected its electrical safety. The competent person

could be a person who has been specifically trained to carry out the testing of portable appliances using a simple 'pass/fail' type of tester. When more sophisticated tests are required, a competent person with the necessary technical electrical knowledge and experience would be needed. The inspection and testing should normally include the following checks:

> that the polarity is correct
> that the correct fuses are being used
> that all cables and cores are effectively terminated
> that the equipment is suitable for its environment.

Testing need not be expensive in many low risk premises like shops and offices, if an employee is trained to perform the tests and appropriate equipment is purchased.

14.7.3 Frequency of inspection and testing

The frequency of inspection and testing should be based on a risk assessment which is related to the usage, type and operational environment of the equipment. The harsher the working environment is, the more frequent the period of inspection. Thus tools used on a construction site should be tested much more frequently than a visual display unit which is never moved from a desk. Manufacturers or suppliers may recommend a suitable testing period. Table 14.1 lists the intervals for construction and industrial equipment given by the HSE in its publication *Maintaining portable and transportable electrical equipment* (HSG107).

It is very important to stress that there is no 'correct' interval for testing – it depends on the frequency of usage, type of equipment, how and where it is used. A few years ago, a young trainee was badly scalded by a kettle of boiling water which exploded while in use. On investigation, an inspection report indicated that the kettle had been checked by a competent person and passed just a few weeks before the accident. Further investigation showed that this kettle was the only method of boiling water on the premises and was in use continuously for 24 hours each day. It was therefore unsuitable for the purpose and a plumbed-in continuous-use hot water heater would have been far more suitable.

14.7.4 Records of inspection and testing

Schedules which give details of the inspection and maintenance periods and the respective programmes, must be kept together with records of the inspection findings and the work done during maintenance. Records must include both individual items of equipment and a description of the complete system or section of the

Table 14.1 Suggested intervals for portable appliance inspection and testing

Equipment/application	Voltage	User check	Formal visual inspection	Combined inspection and test
Battery-operated power tools and torches	Less than 25V	No	No	No
25V portable hand lamps (confined or damp situations)	25V secondary winding from transformer	No	No	No
50V portable hand lamps	Secondary winding centre tapped to earth (25V)	No	No	Yearly
110V portable and hand-held tools, extension leads, site lighting, moveable wiring systems and associated switchgear	Secondary winding centre tapped to earth (55V)	Weekly	Monthly	Before first use on site and then 3 monthly
230V portable and hand-held tools, extension leads and portable floodlighting	230V mains supply through 30mA RCD	Daily/every shift	Weekly	Before first use on site and then monthly
230V equipment such as lifts, hoists and fixed floodlighting	230V supply fuses or MCBs	Weekly	Monthly	Before first use on site and then 3 monthly
RCDs fixed*		Daily/every shift	Weekly	**Before first use on site and then 3 monthly
Equipment in site offices	230V office equipment	Monthly	6 Monthly	Before first use on site and then yearly

MCB = miniature circuit breaker.
*It is recommended that portable RCDs are tested monthly.
**Note: RCDs need a different range of tests to other portable equipment, and equipment designed to carry out appropriate tests on RCDs will need to be used.

system. They should always be kept up-to-date and with an audit procedure in place to monitor the records and any required actions. The records do not have to be paper based but could be stored electronically on a computer. It is good practice to label the piece of equipment with the date of the last combined test and inspection.

The effectiveness of the equipment maintenance programme may be monitored and reviewed if a record of tests is kept. It can also be used as an inventory of portable appliances and help to regulate the use of unauthorized appliances. The record will enable any adverse trends to be monitored and to check that suitable equipment has been selected. It may also give an indication as to whether the equipment is being used correctly.

14.7.5 *Advantages and limitations of portable appliance testing*

The advantages of PAT include:

➤ an earlier recognition of potentially serious equipment faults, such as poor earthing, frayed and damaged cables and cracked plugs

➤ discovery of incorrect or inappropriate electrical supply and/or equipment
➤ discovery of incorrect fuses being used
➤ a reduction in the number of electrical accidents
➤ the misuse of portable appliances may be monitored
➤ equipment selection procedures can be checked
➤ an increased awareness of the hazards associated with electricity
➤ a more regular maintenance regime should result.

The limitations of PAT include:

➤ Some fixed equipment is tested too often leading to excessive costs.
➤ Some unauthorized portable equipment, such as personal kettles, are never tested since there is no record of them.
➤ Equipment may be misused or overused between tests due to a lack of understanding of the meaning of the test results.
➤ All faults, including trivial ones, are included on the action list so that the list becomes very long and the more significant faults are forgotten or overlooked.

➤ The level of competence of the tester can be too low.
➤ The testing equipment has not been properly calibrated and/or checked before testing takes place.

Most of the limitations may be addressed and the reduction in electrical accidents and injuries enables the advantages of PAT to greatly outweigh the limitations.

14.8 Protection against contact with live overhead power lines

At tender or negotiation stage, the existence of any overhead lines across the site must be noted and contact made with both the power supply company and the local authority. At the site assessment stage, arrangements must be made for any possible line diversions with the electricity company or to confirm with them any safe distances, clearances or other precautions that they may require to be made. If the power line cannot be diverted, the principal contractor and all sub-contractors must be informed of the presence of live overhead power lines.

Where possible all work likely to lead to contact with overhead power lines should be done in an area well clear of the line itself. It may be possible to alter the work and eliminate or reduce the risk. As a general rule vehicles, plant or equipment should be brought no closer than:

➤ 15 m of overhead lines suspended from steel towers
➤ 9 m of overhead lines supported on wooden poles.

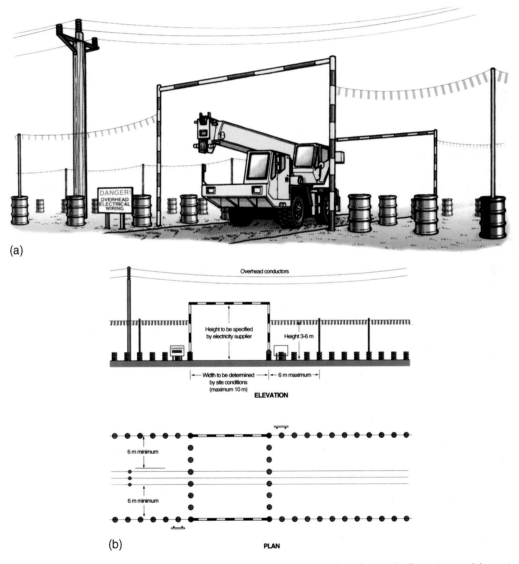

(a)

(b)   PLAN

Figure 14.11 (a) Precautions for overhead lines: 'goalpost' crossing points beneath lines to avoid contact by plant. (b) Diagram showing normal dimensions for 'goalpost' crossing points and barriers. Reproduced from HSG185 *Health and Safety in Excavations*.

Where a closer approach is likely either the lines should be made dead or barriers erected to prevent an approach. Permits to work are likely to be required if work close to the lines is necessary.

Where work is necessary directly beneath the lines, or blasting or other unusual activity has to be done adjacent to the lines, the lines may need to be made dead and a permit to work system operated. In certain situations, capacitated or induced AC voltages and even arcing can be created in fences and pipelines which run parallel to overhead lines. When there are concerns about nearby structures the advice of the safety adviser and/or the electrical supply company must be sought. Risk assessments should also be made and suitably recorded, the information being made available to all workers on the site.

On many construction sites, vehicles will need to move beneath power lines. In these cases, the roadway should be covered by goalposts covered with warning tape. Bunting should be suspended, level with the top of the goalposts and just above ground level (often using empty oil drums), between poles across the site along the length of the power line. This will ensure that vehicles can only pass under the power lines by passing through the goalposts. Suitable warning signs should be placed on either side of the roadway on each side of the power line to warn of the overhead power line.

14.9 Practice NEBOSH questions for Chapter 14

1. (a) **Outline** the dangers associated with electricity.
    (b) **Outline** the emergency action to take if a person suffers a severe electric shock.

2. (a) **Identify** the factors that may determine the severity of injury caused by contact with electricity.
    (b) **Outline** the measures that may be employed to protect against an electric shock.

3. (a) **Outline** the effects on the human body from a severe electric shock.
    (b) **Describe** how earthing can reduce the risk of receiving an electric shock.

4. (a) **Describe** the possible effects of electricity on the body.
    (b) **Outline FOUR** factors that may affect the severity of injury from contact with electricity.

5. **Outline** a range of checks that should be made to ensure electrical safety in an office environment.

6. In relation to the use of electrical cables and plugs in the workplace:
    (a) **identify FOUR** examples of faults and bad practices that could contribute to electrical accidents
    (b) **outline** the corresponding precautions that should be taken for each of the examples identified in (i).

7. **Outline** the measures to be taken to minimize the risk of shock from the use of electric hand tools.

8. **Describe** the measures to be taken to minimize the risk of electric shock from the use of portable electric tools.

9. Hand-held electric drills are commonly used on construction sites:
    (a) **Outline** the checks that should be carried out by the user of a drill to reduce the likelihood of electric shock.
    (b) Other than electricity, **outline FOUR** hazards associated with the use of such equipment.

10. **State** the items that should be included on a checklist for the routine inspection of portable electrical appliances.

11. **Outline** a regime for the inspection and testing of electrical hand tools and associated equipment used on a construction site.

12. HSG107 *Maintaining Portable and Transportable Electrical Equipment* gives guidance on the frequency of inspection for electrical hand tools and associated equipment.

    State frequencies for the inspection and testing of electrical hand tools and associated equipment used on a construction site and support your answer with appropriate examples.

13. Ground works are about to commence on a site crossed by high voltage overhead power lines. The work will involve the use of excavators and tipper lorries to remove spoil from the site.

    Outline a management strategy for the avoidance of accidents involving accidental contact with the overhead power lines.

14. **Outline** the precautions to be taken before lifting operations are carried out adjacent to high voltage overhead power lines.

Fire hazards and control

15.1 Introduction

Chapter 15 covers fire prevention on construction sites and how to ensure that people are properly protected if fire does occur. Each year UK fire and rescue services attend over 41,000 fires at work in which about 40 people are killed and over 2,500 are injured. Fire and explosions at work account for about 2% of the major injuries reported under Reporting of Injuries, Diseases and Dangerous Occurrences Regulations (RIDDOR 1995). There are over 4,000 construction fires annually and about 100 of them cause over £50,000 of damage and can result in complete dislocation of project schedules. Some, like the National Westminster Tower fire, affect large numbers of people. In this case about 600 workers were at risk when the fire broke out.

The financial costs associated with serious fires are very high including, in many cases (believed to be over 40%), the failure to start up business again. Never underestimate the potential of any fire. What may appear to be a small fire in a waste bin, if not dealt with can quickly spread through a building or structure. The Bradford City Football ground in 1985 or King's Cross Underground station in 1987 are examples of where small fires quickly became raging infernos, resulting in many deaths and serious injuries. Some construction site fires have caused many millions of pounds in losses.

Since the introduction of the Fire Services Act of 1947, the fire authorities have had the responsibility for fighting fires in all types of premises. In 1971, the Fire Precautions Act gave the fire authorities control over certain fire procedures, means of escape and basic fire protection equipment through the drawing up and issuing of Fire Certificates in certain categories of building. The Fire Certification was mainly introduced to combat a

number of serious industrial fires that had occurred, with a needless loss of life, where simple well-planned protection would have allowed people to escape unhurt.

Following a Government review in the 1990s, the Fire Precautions (Workplace) Regulations 1999 came into force in December of that year. These Regulations were made under the Health and Safety Work etc Act (HSW Act). They were amended in 1999 so as to apply to a wider range of premises including those already subject to the Fire Precautions Act 1971. In many ways these Regulations established the principles of fire risk assessment which would underpin a reformed legislative framework for fire safety. There remained a difference of opinion within the Government, as to the right home for fire legislation. The Home Office believed that process fire issues should remain with the HSW Act, while general fire safety should have a different legislative vehicle.

In 2000, the Fire Safety Advisory Board was established to reform the fire legislation to simplify, rationalize and consolidate existing legislation. It would provide for a risk-based approach to fire safety allowing more efficient, effective enforcement by the fire and rescue service and other enforcing authorities.

The Regulatory Reform (Fire Safety) Order 2005 (RRFSO), SI No 2005, 1541, was made on 7th June 2005. The Order came into force on the delayed date of 1st October 2006. A summary of the Order has been included in Chapter 21. The Fire Precautions Act 1971 is repealed and The Fire Precautions (Workplace) Regulations 1996 are revoked by the Order. Since the Construction (Design and Management) Regulations 2007 (CDM 2007) were approved the RRFSO has been amended, as the Construction (Health, Safety and Welfare) Regulations 1997 (now revoked) have been incorporated into CDM 2007.

Figure 15.1 Fire is still a major risk in many workplaces.

The RRFSO reforms the law relating to fire safety in non-domestic premises. The main emphasis of the changes is to move towards fire prevention. Fire certificates under the Fire Precautions Act 1971 are abolished by the Order and will cease to have legal status.

There is a duty to carry out fire risk assessments that should consider the safety of both employees and non-employees.

The Order imposes a number of specific duties in relation to the fire precautions to be taken. The Order provides for the enforcement of the Order, appeals, offences and connected matters. The Order also gives effect in England and Wales to parts of a number of EC Directives including the Framework, Workplace, Chemical Agents and Explosive Atmospheres Directives.

15.2 The RRFSO – requirements

15.2.1 *Outline of RRFSO*

Part 1 – The RRFSO applies to all non-domestic premises including construction sites other than those listed in Article 6. The main duty holder is the 'responsible person' in relation to the premises, defined in Article 3. The duties on the responsible person are extended to any person who has, to any extent, control of the premises to the extent of their control (Article 5).

Part 2 – imposes duties on the responsible person in relation to fire safety in premises.

Part 3 – provides for enforcement.

Part 4 – provides for offences and appeals.

Part 5 – provides for miscellaneous matters including firefighters' switches for luminous tube signs, maintenance of measures provided to ensure the safety of firefighters, civil liability for breach of statutory duty by an employer, special requirements for licensed premises and consultation by other authorities.

Schedule 1 sets out the matters to be taken into account in carrying out a risk assessment (Parts 1 and 2), the general principles to be applied in implementing fire safety measures (Part 3) and the special measures to be taken in relation to dangerous substances (Part 4).

The Order is mainly enforced by the fire and rescue authorities.

15.2.2 *Part 1 – General*

Meanings

The Order defines a responsible person who may be the person in control of the premises or the owner.

The meaning of general fire precautions is set out, and covers:

- reduction of fire risks and fire spread
- means of escape
- keeping means of escape available for use
- firefighting
- fire detection and fire warning
- action to be taken in the event of fire
- instruction and training of employees.

But it does not cover **process-related fire precautions**. These include the use of plant or machinery or the use or storage of any dangerous substances. Process-related fire precautions still come under either the Health and Safety Executive (HSE) or local authority, and are covered by the general duties imposed by the HSW Act or other specific regulations.

Duties

Duties are placed on a 'responsible person' who is:

- the employer in a workplace, to the extent they have control or
- the other person who has control of the premises or
- the owner of the premises.

The obligations of a particular responsible person relate to matters within their control. It is therefore advisable that arrangements between responsible persons, where there may be more than one, such as contracts or tenancy agreements, should clarify the division of responsibilities to avoid unnecessary breaches of the Order, where one responsible person simply assumed the other would satisfy a particular requirement.

Premises *not* covered

The RRFSO does not cover domestic premises, offshore installations, a ship (normal shipboard activities under a master), remote fields, woods or other land forming part of an agricultural, or forestry operation (it does cover the buildings), means of transport, a mine (it does cover the buildings at the surface) and a borehole site.

Other alternative provisions cover premises such as sports grounds.

15.2.3 *Part 2 – Fire safety duties*

The responsible person must take appropriate general fire precautions to protect both employees and non-employees to ensure that the premises are safe. Details are given in Chapter 21.

Risk assessment and arrangements
The responsible person must:

> make a suitable and sufficient risk assessment to identify the general fire precautions required (usually known as a fire risk Assessment)
>> if a dangerous substance is or is liable to be present, this assessment must include the special provisions of Part 1 of Schedule 1 to the Order
> review the risk assessment regularly or when there have been significant changes
> with regard to young persons, take into account Part 2 of Schedule 1
> record the significant findings where more than four people are employed or where there is a licence or Alterations Notice in place for the premises
> apply the principles of prevention, Part 3 of Schedule 1
> set out appropriate fire safety arrangements for planning, organization, control, monitoring and review
> eliminate or reduce risks from dangerous substances in accordance with Part 4 of Schedule 1.

Firefighting and fire detection
The responsible person must ensure that the premises are provided with appropriate:

> firefighting equipment
> fire detectors and alarms
> measures for firefighting which are adapted to the size and type of the undertaking
> trained and equipped competent persons to implement firefighting measures
> contacts with external emergency services, particularly as regards firefighting, rescue work, first aid and emergency medical care.

Emergency routes, exits and emergency procedures
The responsible person must ensure that routes to emergency exits and the exits themselves are kept clear and ready for use. See Chapter 21 for specific requirements for means of escape.

The responsible person must establish suitable and appropriate emergency procedures and appoint a sufficient number of competent persons to implement the procedures. This includes arrangements for undertaking fire drill and information and action regarding exposure to serious, imminent and unavoidable dangers. Where dangerous substances are used and/or stored, additional emergency measures must be set up covering the hazards. There should be appropriate visual or audible warnings and other communications systems to effect a prompt and safe exit from the endangered area.

Where necessary to protect persons' safety all premises and facilities must be properly maintained and subject to a suitable system of maintenance.

Safety assistance
The responsible person must appoint (except in the case of the self-employed or partnerships where a person has the competence themselves) one or more people to assist in undertaking the preventative and protective measures. Competent persons must be given the time and means to fulfil their responsibilities. Competent persons must be kept informed of anything relevant to their role and have access to information on any dangerous substances present on the premises.

Provision of information
The responsible person must provide:

> his employees
> the employer of any employees of an outside undertaking
> with comprehensible and relevant information on the risks, precautions taken, persons appointed for fire fighting and fire drills, and appointment of competent persons. Additional information is also required for dangerous substances used and/or stored on the premises.

Before a child is employed, information on special risks to children must be given to the child's parent or guardian.

Capabilities, training and cooperation
The responsible person must ensure that adequate training is provided when people are first employed or exposed to new or increased risks. This may occur with new equipment, changes of responsibilities, new technologies, new systems of work and new substances used. The training has to be done in working hours and repeated periodically as appropriate.

Where two or more responsible persons share duties or premises, they are required to coordinate activity in their premises and cooperate with each other including keeping each other informed of risks.

Duties of employees

These are covered in Section 21.18.17 and are similar to requirements under the HSW Act. Employees must take care of themselves and other relevant persons. They must cooperate with the employer and inform them of any situation which they would reasonably consider to present a serious and immediate danger, or a shortcoming in the protection arrangements.

15.2.4   *Part 3 – Enforcement*

Enforcing authorities

The fire and rescue authority in the area local to the premises is normally the enforcing authority. However, in stand-alone construction sites, premises which requires a Nuclear Licence, and where a ship is under repair, the HSE are the enforcing authority.

Fire Authorities officers appointed under the RRFSO have similar powers to those under the HSW Act. These include the power to enter premises, make enquiries, require information, require facilities and assistance, take samples, require an article or substance to be dismantled or subjected to a process or test and the power to issue enforcement notices.

In addition to fire safety legislation, health and safety at work legislation covers the elimination or minimization of fire risks. As well as the particular and main general duties under the HSW Act, fire risks are covered by specific rules, such as for dangerous substances and explosive atmospheres, work equipment, electricity and other hazards. Thus, environmental health officers or HSE inspectors may enforce health and safety standards for the assessment and removal or control of process-related fire risks, where it is necessary, for the protection of workers and others.

Alterations and enforcement notices

Enforcing authorities can issue (see Chapter 21 for details):

Alterations notices – where the premises constitute a serious risk to people either due to the features of the premises or hazards present in the premises. The notices require the responsible person to notify changes and may require them to record the significant findings of the risk assessment and safety arrangements; and to supply a copy of the risk assessment before making changes.

Enforcement notices – where there is a failure to comply with the requirements of the RRFS Order. The notice must specify the provisions concerned and MAY include directions on remedial action.

Prohibition notices – where the risks are so serious that the use of the premises should be prohibited or restricted. The notice MAY include directions on remedial action.

15.2.5   *Part 4 – Offences*

Cases can be tried on summary conviction in a magistrates court or on indictment in the Crown Court. The responsible person can be liable:

➤ on summary conviction to a fine not exceeding the statutory maximum
➤ on indictment to an unlimited fine and/or imprisonment for up to 2 years for failure to comply with fire safety duties where there is a risk of death or serious injury and for failure to comply with an Alterations, Enforcement or Prohibition Notice.

Any person who fails to comply with their duties under the Order as regards fire risks can be prosecuted alongside or instead of a responsible person. Fines are limited to the statutory maximum (or levels 3 or 5 on the standard scale) on summary conviction but on indictment, the fine is unlimited.

15.3   CDM 2007

Details of CDM 2007 are given in Chapter 21. Regulation 46 (see Section 21.10.11) makes the Fire and Rescue authority responsible for the non-process fire-related sections of CDM 2007 on construction sites where other people are working within or as part of the site. For example the work is being done in an occupied factory or office building.

The Regulations concerned are as, discussed, in the section below.

Regulation 39 – Emergency procedures

In so far as they concern fire emergency procedures and involve suitable procedures for evacuation, familiarization of workers and testing the arrangements.

Regulation 40 – Emergency routes and exits

In so far as they concern fire emergency routes and exits and involve the following:

➤ sufficient number of emergency routes and exits
➤ must lead directly as possible to an identified safe place

must be kept clear and free of obstruction
provided with emergency lights as necessary and
suitably signed.

Regulation 41 – Fire detection and firefighting
This involves the following:

➤ suitably located, suitable and sufficient firefighting equipment
➤ suitably located, suitable and sufficient fire detection and alarm systems
➤ these systems and equipment to be properly maintained, examined and tested
➤ these to be easily accessible unless automatically activated
➤ training for every person on the construction site in the operation of fire fighting equipment that they are likely to use
➤ where there is a particular risk of fire, people should be instructed before starting the work
➤ equipment to be suitably signed.

See Table 15.1 for the various enforcement authorities and their responsibilities on construction sites.

Basic principles of fire

15.4.1 *Fire triangle*

Fire cannot take place unless three things are present. These are shown in Figure 15.2.

The absence of any one of these elements will prevent a fire starting. Prevention depends on avoiding these

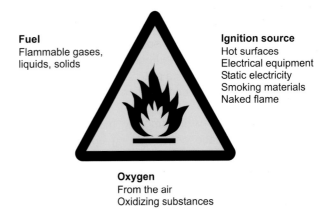

Fuel
Flammable gases, liquids, solids

Ignition source
Hot surfaces
Electrical equipment
Static electricity
Smoking materials
Naked flame

Oxygen
From the air
Oxidizing substances

Figure 15.2   Fire triangle.

Table 15.1   Enforcement in respect of fire on construction sites

Type of Premises Fire issue		Fire and Rescue Authorities Under RRFSO and CDM 2007	HSE under HSW Act, CDM 2007, and RRFSO	Local Authorities Inspectors under HSW Act, and RRFSO	Defence Fire Service	A fire inspector or person appointed by the Sec. of State
Fire emergency procedures; escape routes and exits; fire detection and fire fighting; and training issues on:	Stand-alone construction site		✓			
	Construction site where there are other activities (e.g. an occupied factory or office)	✓				
Process fire and explosion risks, where there are for example tar boilers/dangerous substances on:	Standalone sites		✓			
	Sites where there are other activities (e.g. an occupied factory or office)			✓		
Licensed Nuclear sites; and ships under construction or repair		✓				
Defence and Crown armed forces premises other than a ship					✓	✓
Sports grounds and Sports stands				✓		
Crown premises except nuclear; UK Atomic Energy premises except nuclear.						✓

three coming together. Fire extinguishing depends on removing one of the elements from an existing fire, and is particularly difficult if an oxidizing substance is present.

Once a fire starts it can spread very quickly from fuel to fuel as the heat increases.

15.4.2 *Sources of ignition*

Workplaces have numerous sources of ignition, some of which are obvious but others may be hidden inside machinery. Most of the sources may cause an accidental fire from sources inside but, in the case of arson (about 13% of industrial fires), the source of ignition may be brought from outside the workplace and will be deliberately used. The following are potential sources of ignition in the typical workplace construction site:

➤ **naked flames** – from smoking materials, cooking appliances, space heating/drying appliances, braziers and tar boilers
➤ **external sparks** – from grinding metals, welding, impact tools, electrical switch gear
➤ **internal sparking** – from electrical equipment (faulty and normal), machinery, lighting
➤ **hot surfaces** – from lighting, cooking, heating appliances, braziers, tar boilers, plant and machinery, poorly ventilated equipment, faulty and/or badly lubricated equipment, hot bearings and drive belts
➤ **static electricity** – causing significant high voltage sparks from the separation of materials such as unwinding plastic, pouring highly flammable liquids, walking across insulated floors, or removing synthetic overalls.

15.4.3 *Sources of fuel*

If it will burn it can be fuel for a fire. The things which will burn easily are the most likely to be the initial fuel, which then burns quickly and spreads the fire to other fuels. The most common things that will burn in a typical workplace are:

Solids – these include wood, paper, cardboard, wrapping materials, plastics, rubber, foam (e.g. polystyrene tiles and furniture upholstery), textiles (e.g. furnishings and clothing), wallpaper, hardboard and chipboard used as building materials, waste materials (e.g. wood shavings, dust, paper), hair.

Liquids – these include paint, varnish, thinners, adhesives, petrol, white spirit, methylated spirits, paraffin, toluene, acetone and other chemicals. Most flammable liquids give off vapours which are heavier than air so they will fall to the lowest levels. A flash flame or an explosion can occur if the vapour catches fire in the correct concentrations of vapour and air.

Gases – flammable gases include LPG (liquefied petroleum gas in [cylinders, usually butane or propane]), acetylene (used for welding) and hydrogen. An explosion can occur if the air/gas mixture is within the explosive range.

15.4.4 *Oxygen*

Oxygen is of course provided by the air all around but this can be enhanced by wind, or by natural or powered ventilation systems which will provide additional oxygen to continue burning.

Cylinders providing oxygen for medical purposes or welding can also provide an additional very rich source of oxygen.

In addition, some chemicals such as nitrates, chlorates, chromates and peroxides can release oxygen as they burn and therefore need no external source of air.

15.5 Methods of extinction

There are four main methods of extinguishing fires, which are explained as follows:

➤ **cooling** – reducing the ignition temperature by taking the heat out of the fire – using water to limit or reduce the temperature

➤ **smothering** – limiting the oxygen available by smothering, and preventing, the mixture of oxygen and flammable vapour – by the use of foam or a fire blanket

➤ **starving** – limiting the fuel supply – by removing the source of fuel by switching off electrical power, isolating the flow of flammable liquids or removing wood and textiles, etc.

➤ **chemical reaction** – by interrupting the chain of combustion and combining the hydrogen atoms with chlorine atoms in the hydrocarbon chain for example with Halon extinguishers. (Halons have been withdrawn because of their detrimental effect on the environment, as ozone depleting agents can be used on aircraft and certain fixed systems.)

15.6 Classification of fire

Fires are classified in accordance with British Standard *EN 2: 1992 Classification of Fires.* However, for all practical purposes there are *five* main classes of fire – A, B, C, D and F, plus fires involving electrical equipment. BS 7937:2000 *The Specification of Portable Fire Extinguishers for Use on Cooking Oil Fires* introduced the new class F. The categories based on fuel and the means of extinguishing are as follows:

Class A – fires which involve solid materials such as wood, paper, cardboard, textiles, furniture and plastics where there are normally glowing embers during combustion. Such fires are extinguished by cooling, which is achieved using water.

Class B – fires which involve liquids or liquefied solids such as paints, oils or fats. These can be further subdivided into:

➤ **Class B1** – fires which involve liquids that are soluble in water such as methanol. They can be extinguished by carbon dioxide, dry powder, water spray, light water and vaporizing liquids.

➤ **Class B2** – fires which involve liquids not soluble in water, such as petrol and oil. They can be extinguished by using foam, carbon dioxide, dry powder, light water and vaporizing liquid.

Class C – fires which involve gases such as natural gas or liquefied gases such as butane or propane. They can be extinguished using foam or dry powder in conjunction with water to cool any containers involved or nearby.

Class D – fires which involve metals such as aluminium or magnesium. Special dry powder extinguishers are required to extinguish these fires, which may contain powdered graphite or talc.

Class F – fires which involve high temperature cooking oils or fats in large catering establishments or restaurants.

Note:

Electrical fires – fires involving electrical equipment or circuitry do not constitute a fire class on their own, as electricity is a source of ignition that will feed a fire until switched off or isolated. But there are some pieces of equipment that can store, within capacitors, lethal voltages even when isolated. Extinguishers specifically designed for electrical use like carbon dioxide or dry powder should always be used for this type of fire hazard.

Fire extinguishers are usually designed to tackle one or more class of fire. This is discussed in Section 15.14.

15.7 Principles of heat transmission and fire spread

Fire transmits heat in several ways, which needs to be understood in order to prevent, plan escape from, and fight fires. Heat can be transmitted by convection, conduction, radiation and direct burning (Figure 15.3).

15.7.1 Convection

Hot air becomes less dense and rises drawing in cold new air to fuel the fire with more oxygen. The heat is transmitted upwards at sufficient intensity to ignite combustible materials in the path of the very hot products of combustion and flames. This is particularly important inside buildings or other structures where the shape may effectively form a chimney for the fire.

15.7.2 Conduction

This is the transmission of heat through a material with sufficient intensity to melt or destroy the material and ignite combustible materials which come into contact with or close to a hot section. Metals like copper, steel and aluminium are very effective or good conductors of heat. Other materials like concrete, brickwork and insulation materials are very ineffective or poor conductors of heat.

Poor conductors or good insulators are used in fire protection arrangements. When a poor conductor is also incombustible, it is ideal for fire protection. Care is necessary to ensure that there are no other issues such as health risks with such materials. Asbestos is a very poor

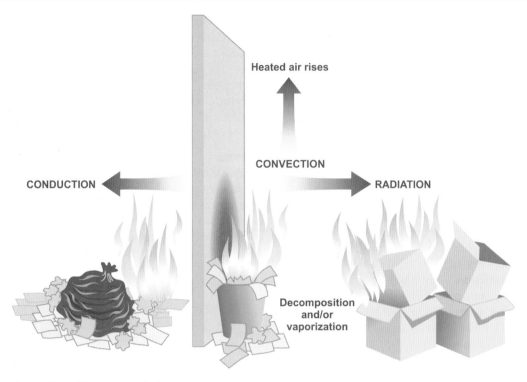

Figure 15.3 Principles of heat transmission.

conductor of heat and is incombustible. However this has, very severe health effects which now far outweigh its value as a fire protection material and it is banned in the United Kingdom, it is still found in many buildings where it was used extensively for fire protection.

15.7.3 *Radiation*

Often in a fire the direct transmission of heat through the emission of heat waves from a surface can be so intense that adjacent materials are heated sufficiently to ignite. A metal surface glowing red-hot would be typical of a severe radiation hazard in a fire.

15.7.4 *Direct burning*

This is the effect of combustible materials catching fire through direct contact with flames which causes fire to spread, in the same way that lighting an open fire, with a range of readily combustible fuels, is spread within a grate.

15.7.5 *Smoke spread in buildings*

Where fire is not contained and people can move away to a safe location there is little immediate risk to those people. However, where fire is confined inside buildings the fire behaves differently.

Figure 15.4 Fire spread in buildings.

290

The smoke rising from the fire gets trapped inside the space by the immediate ceiling then spreads horizontally across the space, deepening all the time until the entire space is filled. The smoke will also pass through any holes or gaps in the walls, ceiling or floor and get into other parts of the building. It moves rapidly up staircases or lift wells and into any areas that are left open, or rooms which have open doors connecting to the staircase corridors. The heat from the building gets trapped inside, raising the temperature very rapidly. The toxic smoke and gases are an added danger to people inside the building, who must be able to escape quickly to a safe location.

15.8 Common causes of fire and consequences

15.8.1 *Causes*

The Home Office statistics show that the causes of fires in buildings, excluding dwellings, in 2005 was as shown in Figure 15.5. The total was 35,300, that is 6% less than the previous year. This follows a 10% fall between 2003 and 2004 and follows the general trend downward since 1995.

The sources of ignition are shown in Figure 15.6. Out of the 21,200 accidental fires in 2005, it shows that cooking appliances and electrical equipment account for over 60% of the total.

The most common causes of fire on construction sites are:

➤ the changing flammability of combustible material being cut, planed, sanded, ground or filed into a finely divided form which will readily catch fire
➤ easy ignition of loose packaging and other materials left on site
➤ poor storage of highly flammable gases and other materials
➤ dismantling of tanks with flammable residues
➤ coming across buried gas or electric lines during demolition or excavation
➤ inadequate arrangements for rubbish disposal
➤ discarding of smoking materials
➤ overheating of poorly maintained plant and machinery
➤ improper use or poor maintenance of welding and cutting equipment
➤ overloaded or poorly maintained temporary electrical equipment
➤ damaged cables and improper fuses or failure of safety devices
➤ accumulation of rubbish against electrical equipment, causing overheating
➤ overheating of coiled cables used for heavy current usage, such as on heating
➤ bonfires getting out of control
➤ use of petrol or other accelerants to brighten bonfires
➤ arson through intruders or even disgruntled or bored site employees.

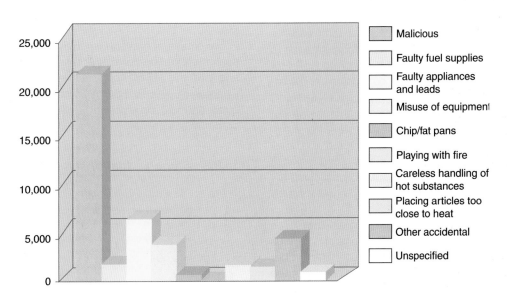

Figure 15.5 Causes of fire, 2005.

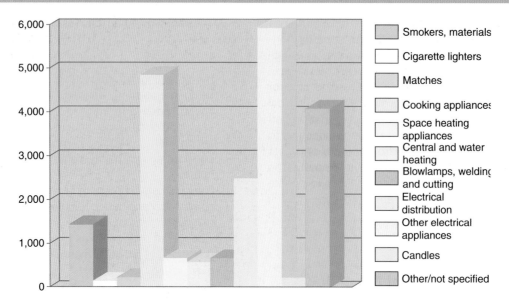

Figure 15.6 Accidental fires – sources of ignition, 2005.

15.8.2 *Consequences*

The main consequences of fire are:

➤ death – although this is a very real risk, relatively few people die in building fires that are not dwellings. In 1999, 38 (6%) people died out of a total of 663 in all fires.
 The main causes of all deaths were:
  ➤ overcome by gas or smoke 46%
  ➤ burns 27%
  ➤ burns and overcome by gas or smoke 20%
  ➤ other 7%
➤ clearly gas and smoke are the main risks
➤ personal injury – some 1900 people were injured (11% of total injuries in all fires)
➤ building damage – can be very significant, particularly if the building materials have poor resistance to fire and there is little or no built in fire protection
➤ flora and fauna damage – can be significant, particularly in a hot draught or forest fire
➤ loss of business and jobs – it is estimated that about 40% of businesses do not start up again after a significant fire. Many are under- or not insured and small companies often cannot afford the time and expense of setting up again when they probably still have the old debts to service. Many site schedules are totally disrupted by a fire
➤ transport disruption – rail routes, roads and even airports are sometimes closed because of a serious fire. The worst case was of course 11 September 2001 when airports around the world were disrupted
➤ environmental damage from the fire and/or fighting the fire – firefighting water, the products of combustion and exploding building materials, such as asbestos cement roofs, can contaminate significant areas around the fire site.

15.9 Fire risk assessment

15.9.1 *General*

What fire precautions are needed depends on the risks. There are several methods of carrying out a fire risk assessment; the one described below is based on the method contained within Fire Safety Guides published by the Department for Communities and Local Government. (See Appendix 15.1.) A systematic approach, considered in five simple stages, is generally are now the best practical method. The fire risk assessment under the RRFSO and not the HSW Act Management Regulations (see Chapter 5). The RRFSO guides were published in July 2006 and are downloadable on www.firesafetyguides.communities.gov.uk.

15.9.2 *Stage 1 – Identify fire hazards*

There are five main dangers produced by fire that should be considered when assessing the level of risk:

➤ oxygen depletion
➤ flames and heat
➤ smoke
➤ gaseous combustion products
➤ structural failure of buildings.

Of these, smoke and other gaseous combustion products are the most common cause of death in fires.

For a fire to occur it needs sources of heat and fuel. If these hazards can be kept apart, removed or reduced, then the risks to people and businesses are minimized. **Identifying fire hazards** in the workplace is the first stage as follows.

Identify any combustibles

Most worksites contain combustible materials. Usually, the presence of normal stock in trade should not cause concern, provided the materials are used safely and stored away from sources of ignition. Good standards of housekeeping are essential to minimize the risk of a fire starting or spreading quickly.

The amount of combustible materials to hand at a worksite should be kept as low as is reasonably practicable. Limit materials to half a day's supply or a single shift and return unused materials to the stores. Always choose the least flammable materials and keep site stocks down as far as possible.

Materials should not be stored in gangways, corridors or stairways or scaffolds where they may obstruct exit doors and routes. Fires often start and are assisted to spread by combustible waste or packaging. Such waste should be collected frequently and removed from the site.

Construction work such as cutting, grinding or sanding often alters the flammability of building materials. Finely devised dust or crumbs can often be ignited easily.

Highly combustible materials, such as flammable liquids, paints or plastic foams, ignite very easily and quickly produce large quantities of heat and/or dense toxic smoke. Such materials should be stored outside the buildings under construction in secure storage areas.

If combustible materials are stored inside buildings they should be kept where the means of escape in case of fire would not be affected. In some cases, stores will need to be constructed of fire-resistant materials giving at least 30 minutes' fire resistance and separated from working areas.

Identify any sources of heat

All work sites will contain heat/ignition sources. These may be heaters, boilers, engines, bonfires, smoking materials or heat from processes, or electrical circuits and equipment, whether in normal use, through misuse or accidental failure.

Where possible, sources of ignition should be removed from the site or replaced with safer forms. Where this cannot be done the ignition source should be kept well away from combustible materials or made the subject of management controls.

Particular care should be taken in areas where portable heaters are used or where smoking is permitted. Where heat is used as part of a process, it should be used carefully to reduce the chance of a fire as much as possible. Good security both inside and outside the site will help to combat the risk of arson.

Refuelling of vehicles (especially with petrol) should take place in the open air or in well-ventilated spaces away from sources of ignition.

Lights should be securely fastened to a solid back or tripods made secure so that they cannot be easily dislodged. Make sure electrical equipment cannot be inadvertently covered or that combustible materials are not close to halogen lights (in particular, as they get very hot) or heaters.

Plant should be properly maintained to avoid overheating, particularly in dusty conditions when filters become easily blocked.

Electrical systems, especially temporary ones, should be of sufficient capacity, and inspected and maintained by competent people. For more information see Chapter 14.

Bonfires should not normally be allowed on construction sites. There should be alternative arrangements for the proper disposal of rubbish and other waste materials. If under exceptional circumstances a bonfire is needed it should be separated from combustible materials by at least 10 m, confined in incinerators, never left unattended and checked for dangerous materials. Extinguishers should be available nearby. Bonfires should not be lit on windy days should be avoided (see Figure 15.7).

Smoking should be discouraged on all sites and where there are high risks it should be banned altogether. Under smoke-free legislation, smoking is not permitted in enclosed or significantly enclosed areas. Outside, designated safe areas should be provided for those who still require to smoke. The smoking rules should be rigorously enforced.

Demolition work can involve a high risk of fire and explosion. In particular:

➤ Dismantling tanks structures can cause the ignition of flammable residues. This is especially dangerous if hot methods are used to dismantle tanks before residues are thoroughly cleaned out. The work should only be done by specialists.

➤ Disruption and ignition of buried gas and electrical services is a common problem. It should always be assumed that buried services are present unless it is positively confirmed that the area is clear. A survey using buried service detection equipment must be carried out by a competent person to identify any services. The services should then be marked, and competently purged or made dead, before

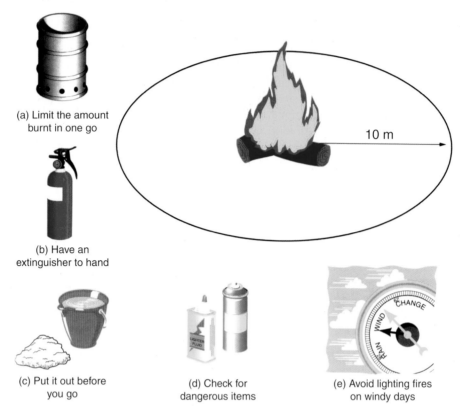

(a) Limit the amount burnt in one go

(b) Have an extinguisher to hand

(c) Put it out before you go

(d) Check for dangerous items

(e) Avoid lighting fires on windy days

Figure 15.7 Avoid lighting bonfires unless you really need to. If you do, make sure you follow points (a) to (e)

any further work is done. A permit to excavate or dig is the normal formal procedure to cover buried services.

Arson is a high risk on construction sites, particularly where trespassers can enter the site fairly easily. Measures should be in place to prevent unauthorized access, especially by children. Flammables should be securely stored or if necessary removed from the site. In some particularly vulnerable sites, security patrols, flood lighting, liaison with local police or even closed-circuit television (CCTV) may be needed.

Identify any unsafe acts
Persons undertaking unsafe acts, such as smoking next to combustible materials, etc.

Identify any unsafe conditions
These are hazards that may assist a fire to spread in the workplace, e.g. if there are large areas of hardboard or polystyrene tiles, or open stairs that can enable a fire to spread quickly, trapping people and enveloping the whole building in smoke and fire.

An ideal method of identifying and recording these hazards is by means of a simple single line plan. Checklists may also be used.

15.9.3 *Stage 2 – Identify location and persons who are at significant risk*

Consider the risk to any people who may be present. In many instances and particularly for most small sites the risk(s) identified will not be significant and specific measures for persons in this category will not be required. There will, however, be some occasions when certain people may be particularly at risk from the fire, because of their specific role, disability, or place of sleeping, or the location or general site activity. Special consideration is needed if:

➤ sleeping accommodation is provided
➤ persons are physically, visually or mentally challenged
➤ people are unable to react quickly
➤ people are isolated.

People may come into the site from outside, such as visitors, the public or sub-contractors. The assessor must decide whether the current arrangements are satisfactory or if changes are needed.

Because fire is a dynamic event, which, if unchecked, may spread throughout a building under construction, all people present will eventually be at risk if fire occurs. Where people are at risk, adequate means of escape from fire should be provided together with arrangements for detecting and giving warning of fire. Firefighting equipment suitable for the hazards throughout the site should be provided.

Some people may be at significant risk because they work in areas where fire is more likely or where rapid fire growth can be anticipated. Where possible, the hazards creating the high level of risk should be reduced. Specific steps should be taken to ensure that people affected are made aware of the danger and the action they should take to ensure their safety and the safety of others.

(a)

15.9.4 Stage 3 – Evaluate and reduce the risks

Having identified the hazards and the persons at risk, the next stage is to reduce the chance of harm to persons on the site. The principles of prevention laid down in the RRFSO should be followed at this stage (see Section 21.18). These are based on EC Directive requirements and are therefore the same as those used in the Management Regulations (see Chapter 5).

Evaluate the risks

Attempt to classify each area as 'high risk', 'normal risk' or 'low risk'. If 'high risk' it may be necessary to reconsider the principles of prevention, otherwise additional compensatory measures will be required.

Low risk – areas where there is minimal risk to people's lives, where the risk of fire occurring is low, or the potential for fire, heat and smoke spreading is negligible and people would have plenty of time to react to an alert of fire.

Normal risk – areas will account for nearly all parts of many sites; where an outbreak of fire is likely to remain confined or spread slowly, with an effective fire warning allowing persons to escape to a place of safety.

High risk – areas where the available time needed to evacuate the area is reduced by the speed of development of a fire, for example highly flammable or explosive materials stored or used (other than small quantities under controlled conditions). Also where the reaction time to the fire alarm is slower because of the type of person present or the activity on the site, for example complex scaffolding on high-rise buildings and persons sleeping on the premises.

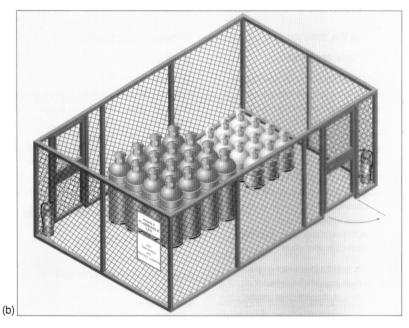

(b)

Figure 15.8 Storage arrangements: (a) for highly flammable liquids; (b) for Outside storage for LPG.

Determine if the existing arrangements are adequate, or need improvement.

Matters that will have to be considered are:

➤ **means for detecting and giving warning in case of fire** – can it be heard by all site occupants?
➤ **means of escape** – are they adequate in size, number, location, well lit, unobstructed, safe to use and so on?
➤ **signs** – for exits, fire routines and so on.
➤ **firefighting equipment** – should be wall mounted or in a cradles at fire exits; are suitable types for hazards present and sufficient in number?

15.9.5 Stage 4 – the findings (always recommended, see Stage 5 review)

The findings of the assessment and the actions (including maintenance) arising from it should be recorded. If five or more people are employed, or an Alterations Notice is required, a formal record of the significant findings and any measures proposed to deal with them must be recorded. See Appendix 15.2 and summary of the Order in Chapter 21. The record should indicate:

➤ the date the assessment was made
➤ the hazards identified
➤ any staff and other people especially at risk
➤ what action needs to be taken, and by when (action plan)
➤ the conclusions arising.

The above guidelines are to be used with caution. Each part of the worksite must be looked at and a decision made on how quickly persons would react to an alert of fire in each area. Adequate safety measures will be required if persons are identified as being at risk. Where maximum travel distances (see later) cannot be achieved, extra fire safety precautions will be needed.

Where persons are at risk or an unacceptable hazard still exists, additional fire safety precautions will be required to compensate for this, or alternatively repeat previous stages to manage risk to an acceptable level.

15.9.6 Stage 5 – Monitor and review on a regular basis

The fire risk assessment is not a one-off procedure. Construction sites are continually changing as work progresses. The site should be continually monitored to ensure that the existing fire safety arrangements and risk assessment remains realistic. The assessment should be reviewed frequently as could-significant changes affect its validity.

15.9.7 Structural features

The construction site may contain features that could promote the rapid spread of fire, heat or smoke and affect escape routes. These features may include ducts or flues, openings in floors or walls, or combustible wall or ceiling linings. Where people are put at risk from these features, appropriate steps should be taken to reduce the potential for rapid fire spread by, for example, fitting non-combustible temporary partitions and/or providing an early warning of fire so that people can leave the workplace before their escape routes become unusable.

Combustible wall or ceiling linings should not be used on escape routes and large areas should be removed wherever they are found. Other holes in fire-resisting floors, walls or ceilings should be filled in with fire-resisting material to prevent the passage of smoke, heat and flames.

During the course of construction, escape routes are likely to change and possibly become unavailable. It is important that replacement routes are provided and identified in good time.

Building designs often incorporate fire escape routes for the finished building. These should be finished as early as possible in the project. In buildings being refurbished, existing escape routes should be kept available as far as possible.

In an emergency, escape using a scaffold is difficult and should not be relied on wherever possible. Provide access from a scaffold to escape routes in the main building where possible.

15.9.8 Stairways, maintenance and refurbishment

Careful consideration should be given to means of escape from above or below ground level. Stairways and ladders must be located or protected so that any fire will not prevent people using them.

Protected stairways will be a feature in many buildings so it makes sense to install these and make them available early in the construction project, before fire risks increase when finishes are installed. Flame retardant surfaces should be used in stairways. In buildings over four storeys high, alternative protected stairways will be required.

Doors giving access to protected stairways should be 30-minute fire resistant and fitted with effective self-closing devices. They should open in the direction of escape. Revolving or sliding doors are not suitable.

If internal stairways are impracticable external temporary stairways should be provided. These can be constructed from scaffolding using wooden treads and platforms (see Figure 15.9).

Sources of heat or combustible materials may be introduced into the workplace during periods of maintenance or refurbishment. Where the work involves the

Figure 15.9 In this fire, leaking gas ignited and a man died despite the escape stairway.

introduction of heat, such as welding, this should be carefully controlled by a safe system of work, for example, a Hot Work Permit (see Chapter 6 for details). All materials brought into the workplace in connection with the work being carried out should be stored away from sources of heat and not obstruct exit routes.

15.9.9 *Fire plans*

Fire plans should be produced and attached to the fire risk assessment. A copy should be posted in the site. A single-line plan of the area or floor should be produced or an existing plan should be used which needs to show:

➤ escape routes, numbers of exits, number of stairs, fire-resisting doors, fire-resisting walls and partitions, places of safety and the like
➤ fire safety signs and notices including pictorial fire exit signs and fire action notices
➤ the location of fire warning call points and sounders or rotary gongs
➤ the location of emergency lights
➤ the location and type of firefighting equipment
➤ names of designated fire/emergency team.

15.10 Dangerous substances

15.10.1 *Introduction*

The Dangerous Substances and Explosive Atmosphere Regulations (DSEAR) apply at most workplaces where a dangerous substance is present or could be present.

The employer must:

➤ carry out a risk assessment of any work activities involving dangerous substances
➤ provide a way of eliminating or reducing risks as far as is reasonably practicable
➤ provide procedures and equipment to deal with accidents and emergencies
➤ provide training and information to employees
➤ classify places where explosive atmospheres may occur into zones and mark the zones where necessary.

The Regulations give a detailed definition of 'dangerous substance', which should be referred to for more information. They include any substance or preparation, which because of its properties or the way it is used could be harmful because of fires and explosions. The list includes petrol, LPG, paints, varnishes, solvents and some dusts. These are dusts which, when mixed with air, can cause an explosive atmosphere. Dusts from milling and sanding operations are examples of this. Most workplaces contain a certain amount of dangerous substances.

An explosive atmosphere is an accumulation of gas, mist, dust or vapour, mixed with air, which has the potential to catch fire or explode. Although an explosive atmosphere does not always result in an explosion (detonation), if it catches fire, flames can quickly travel through the workplace. In a confined space (e.g. in plant or equipment) the rapid spread of the flame front or rise in pressure can itself cause an explosion and rupture of the plant and/or building. For a detailed summary of the Regulations see Chapter 21.

15.10.2 *Risk assessment*

This is the process of identifying and carefully examining the dangerous substances present or likely to be present in the workplace, the work activities involving them and how they might fail and cause fire, explosion and similar events that could harm employees and the public. The purpose of a risk assessment is to enable the employer to decide what needs to be done to eliminate or reduce the safety risks from dangerous substances as far as is reasonably practicable. It should take account of the following:

➤ what hazardous properties the substance has
➤ the way they are used and stored
➤ the possibility of hazardous explosive atmospheres occurring
➤ any potential ignition sources.

Regardless of the quantity of dangerous substance present, the employer must carry out a risk assessment.

This will enable them to decide whether existing measures are sufficient or whether they need to make any additional controls or precautions. Non-routine activities need to be assessed as well as the normal activities within the workplace. For example, in maintenance work, there is often a higher potential for fire and explosion incidents to occur.

Unless the employer has already carried out a detailed assessment under the Management Regulations, there must be an assessment of the risks from fire, explosion and other events arising from dangerous substances, including addressing requirements specified by DSEAR.

Employers are required to ensure that the safety risks from dangerous substances are eliminated or, where this is not reasonably practicable, to take measures to control risks **and** to reduce the harmful effects of any fire, explosion or similar event, so far as is reasonably practicable (i.e. mitigation).

15.10.3 *Substitution*

Substitution is the best solution. It is much better to replace a dangerous substance with a substance or process that totally eliminates the risk. In practice this is difficult to achieve so it is more likely that the dangerous substance will be replaced with one that is less hazardous (e.g. by replacing a low-flashpoint solvent with a high-flashpoint one). Designing the process so that it is less dangerous is an alternative solution.

For example, a change could be made from a batch production to a continuous production process; or the manner or sequence in which the dangerous substance is added could be altered. However, care must be taken when carrying out these steps to make sure that no other new safety or health risks are created or increased, as this would outweigh the improvements implemented as a result of DSEAR.

The fact is that where a dangerous substance is handled or stored for use as a fuel, there is often no scope to eliminate it and very little chance to reduce the quantities handled.

Where risk cannot be entirely eliminated control and mitigation measures should be applied. This should reduce risk as follows:

15.10.4 *Control measures*

Control measures should be applied in the following order of priority:

➤ Reduce the amount of dangerous substances to a minimum.
➤ Avoid or minimize releases.
➤ Control releases at source.
➤ Prevent the formation of an explosive atmosphere.
➤ Use a method such as ventilation to collect, contain and remove any releases to a safe place.
➤ Avoid ignition sources.
➤ Avoid adverse conditions (e.g. exceeding the limits of temperature or other control settings) that could lead to danger.
➤ Keep incompatible substances apart.

15.10.5 *Mitigation measures*

Choose mitigation measures, which are consistent with the risk assessment and appropriate to the nature of the activity or operation. These can include:

➤ preventing fires and explosions from spreading to other plant and equipment or to other parts of the workplace
➤ making sure that a minimum number of employees is exposed
➤ in the case of process plant, providing plant and equipment that can safely contain or suppress an explosion, or vent it to a safe place.

In workplaces where explosive atmospheres may occur employers should ensure that:

➤ areas where hazardous explosive atmospheres may occur are classified into zones based on their likelihood and persistence
➤ areas classified into zones are protected from sources of ignition by selecting; equipment and protective systems suitably protected
➤ where necessary, areas classified into zones are marked with a specified "EX" sign at their points of entry
➤ employees working in zoned areas must be provided with appropriate clothing that does not create a risk of an electrostatic discharge igniting the explosive atmosphere
➤ before coming into operation for the first time, areas where hazardous explosive atmospheres may be present are confirmed as being safe (verified) by a person (or organization) competent in the field of explosion protection; the person carrying out the verification must be competent to consider the particular risks at the workplace and the adequacy of control and other measures put in place.

15.10.6 *Storage*

Dangerous substances should be kept in a safe place in a separate building or the open air. Only small quantities of dangerous substances should be kept in a workroom or area as follows:

➤ For flammable liquids that have a flashpoint above the maximum ambient temperature (normally taken as 32 °C), the small quantity that may be stored in a workroom or area is considered to be an amount up to **250 litres**.

➤ For extremely and highly flammable liquids and those flammable liquids with a flashpoint below the maximum ambient temperature the small quantity is considered to be up to **50 litres** and this should be held in a special metal cupboard or container.

Any larger amounts should be kept in special fire-resisting store, which should be:

➤ properly ventilated
➤ provided with spillage retaining arrangements such as sills
➤ free of sources of ignition, such as unprotected electrical equipment, sources of static electrical sparks, naked flames or smoking materials
➤ arranged so that incompatible chemicals do not become mixed together either in normal use or in a fire situation
➤ of fire-resisting construction
➤ used for empty as well as full containers – all containers must be kept closed
➤ kept clear of combustible materials such as cardboard or foam plastic packaging materials.

15.10.7 *Flammable gases*

Flammable Gas cylinders also need to be stored and used safely. The following guidance should be adopted:

➤ Both full and empty cylinders should be stored outside. They should be kept in a separate secure compound at ground level with sufficient ventilation. Open mesh is preferable.
➤ Valves should be uppermost during storage to retain them in the vapour phase of the LPG.
➤ Cylinders must be protected from mechanical damage. Unstable cylinders should be together, for example; and make sure cylinders are protected from the heat of the summer sun.
➤ The correct fittings must be used. These include hose, couplers, clamps and regulators.
➤ Gas valves must be turned off after use at the end of the shift.
➤ Take precautions against welding flame 'flash back' into the hoses or cylinders. People need training in

the proper lighting up and safe systems of work procedures; non-return valves and flame arrestors also need to be fitted.
➤ Change cylinders in a well-ventilated area remote from any sources of ignition.
➤ Test joints for gas leaks using soapy/detergent water – never use a flame.
➤ Flammable material must be removed or protected before welding or similar work.
➤ Cylinders should be positioned outside buildings with gas piped through in fixed metal piping.
➤ Make sure that both high and low ventilation is maintained where LPG appliances are being used.
➤ Flame failure devices are necessary to shut off the gas supply in the event of flame failure.

15.11 Fire detection and warning

In the event of fire it is vital that everyone in the workplace is alerted as soon as possible. The earlier the fire is discovered, the more likely it is that people will be able to escape before the fire takes hold and before it blocks escape routes or makes escape difficult.

Every site should have detection and warning arrangements. Usually the people who work there will detect the fire and in many worksites nothing further will be needed.

It is important to consider how long a fire is likely to burn before it is discovered. Fires are likely to be discovered quickly if they occur in places that are frequently visited by employees, or in occupied areas of a building. For example, employees are likely to smell burning or see smoke if a fire breaks out in an office or site hut.

The extent or sophistication of the fire warning will vary from site to site. For example:

➤ On a small open-air site, or those involving small buildings and structures, 'word of mouth' may well be adequate.
➤ On larger open-air sites, or those involving buildings and structures with a limited number of rooms, such that a shout of 'fire' might not be heard or could be misunderstood, a klaxon, whistle, gong or small self-contained proprietary fire alarm unit may well be needed.
➤ On sites for complex multi-storey buildings, it is likely that a wired-in system of call points and sounders will be required to provide an effective fire warning system that meets the relevant British Standard.

Figure 15.10 This means of escape has been blocked off and will be useless in the confusion of a fire.

The operation and effectiveness of the fire alarm should be:

➤ regularly checked and tested and
➤ periodically serviced.

Where there is concern that fire may break out in an unoccupied part of the premises, for example, in a basement, some form of automatic fire detection should be fitted. Commercially available heat or smoke detection systems can be used. In small premises a series of interlinked domestic smoke alarms that can be heard by everyone present will be sufficient. In most cases, staff can be relied upon to detect a fire.

If it is thought that there might be some delay in fire being detected, automatic fire detection should be considered, linked into an electrical fire alarm system. Where a workplace provides sleeping accommodation or where fires may develop undetected, automatic detection must be provided. If a workplace provides sleeping accommodation for fewer than six people, interlinked domestic smoke alarms (wired to the mains electricity supply) can be used, provided that they are audible throughout the workplace while people are present.

Figure 15.11 A temporary, wired-in fire alarm during major renovation of a large and complex multi-storey building.

15.12 Means of escape in case of fire

15.12.1 *General*

It is essential to ensure that everyone can escape quickly from the site if there is a fire. On open-air sites and unenclosed single-storey structures, escape routes may be obvious and plentiful. In more complex structures which have workplaces above and/or below ground, more detailed provisions will have to be made.

Proper provision is required for everyone on site wherever they are and however temporary their stay on site (e.g. workers on roofs in cellars or plant rooms near the roof).

Escape routes will change during the construction project and may become unusable. Replacement routes must be identified early on when changes occur. Building designs usually incorporate fire escape routes for the finished building and its occupants. In new buildings these should be installed as early as possible in the contract. For refurbished buildings existing routes should be maintained as far as possible.

In an emergency, escape via a scaffold can be very difficult particularly if there are large quantities of smoke or the weather is poor or it is dark. Where possible provide good access to escape routes inside the main building.

There should normally be at least two escape routes in different directions. They need to be clear, uncomplicated passageways leading to a place of safety. They must be properly maintained and kept free of obstructions.

A person confronted by a fire or its effects needs to be able to turn away from it or be able to pass the area and reach a place of safety unaided. Where this is not possible it is important to ensure that the risk of being trapped in a dead-end situation is minimized. This can be achieved by, for example, keeping combustible materials out of the area and travel distances to a minimum.

15.12.2 Doors

Some doors may need to open in the direction of travel:

➤ doors from a high-risk area, such as a paint spraying room or large kitchen
➤ doors that may be used by more than 50 persons
➤ doors at the foot of stairways where there may be a danger of people being crushed
➤ some sliding doors may be suitable for escape purposes provided that they do not put people using them at additional risk, slide easily and are marked with the direction of opening
➤ doors which only revolve and do not have hinged segments are not suitable as escape doors.

15.12.3 Escape routes

Escape routes should meet the following criteria:

➤ Where two or more escape routes are needed they should lead in different directions to places of safety.
➤ Escape routes need to be short and to lead people directly to a place of safety, such as the open air or an area of the workplace where there is no immediate danger.
➤ It should be possible for people to reach the open air without returning to the area of the fire. They should then be able to move well away from the building.
➤ Ensure that the escape routes are wide enough for the volume of people using them. A 750 mm door will allow up to 40 people to escape in 1 minute, so most doors and corridors will be wide enough. If the routes are likely to be used by people in wheelchairs, the minimum width will need to be 800 mm.

Make sure that there are no obstructions on escape routes, especially on corridors, scaffolds and stairways where people who are escaping could dislodge stored items or be caused to trip. Any fire hazards must be removed from exit routes as a fire on an exit route could have very serious consequences.

Escape routes need regular checks to make sure that they are not obstructed and that exit doors are not locked. Self-closing fire-resisting doors should be checked to ensure doors close fully, including those fitted with automatic release mechanisms.

15.12.4 Lighting

Escape routes must be well lit. If the route has only artificial lighting or if it is used during the hours of darkness, alternative sources of lighting should be considered in case the power fails during a fire. Check the routes when it is dark as, for example, there may be street lighting outside that provides sufficient illumination. In small workplaces it may be enough to provide the staff with torches that they can use if the power fails. However, it may be necessary to provide battery-operated emergency lights so that if the mains lighting fails the lights will operate automatically. Candles, matches and cigarette lighters are not adequate forms of emergency lighting.

15.12.5 Signs

Exit signs on doors or indicating exit routes should be provided where they will help people to find a safe escape route. Signs on exit routes should have directional arrows, 'up' for straight on and 'left, right or down' according to the route to be taken. Advice on the use of all signs including exit signs can be found in Chapters 6 and 21.

15.12.6 Escape times

Everyone in the building should be able to get to the nearest place of safety in between 2 and 3 minutes. This means that escape routes should be kept short. Where there is only one means of escape, or where the risk of fire is high, people should be able to reach a place of safety, or a place where there is more than one route available, in 1 minute.

The way to check this is to pace out the routes, walking slowly and noting the time. Start from where people work and walk to the nearest place of safety. Remember that the more people there are using the route, the longer the time they will take. People take longer to negotiate stairs and they are also likely to take longer if they have a disability.

Where fire drills are held, check how long it takes to evacuate each floor in the workplace. This can be used as a basis for assessment. If escape times are too long, it may be worth rearranging the workplace so that people are closer to the nearest place of safety, rather than undertake expensive alterations to provide additional escape routes.

Table 15.2 gives maximum travel distances which experience has shown are acceptable for a variety of situations in construction.

Table 15.2 Travel distances

	Fire hazard		
	Low	Medium	High
Enclosed structures:			
Alternative	60 m	45 m	25 m
Dead end	18 m	18 m	12 m
Semi-open structures (e.g. scaffolds):			
Alternative	200 m	100 m	60 m
Dead end	25 m	18 m	12 m

Reaction time needs to be considered. This is the amount of time people will need for preparation before they escape. It may involve, for example, closing down machinery, issues of security or helping visitors or members of the public out of the premises. Reaction time needs to be as short as possible to reduce risk to staff. Assessment of escape routes should include this. If reaction times are too long, additional routes may need to be provided. It is important that people know what to do in case of fire as this can speed up the time needed to evacuate the premises.

15.13 Principles of fire protection in buildings

15.13.1 *General*

The design of all new buildings and the design of extensions or modifications to existing buildings must be approved by the local planning authority.

Design data for new and modified buildings must be retained throughout the life of the structure.

Building Legislation Standards are concerned mainly with safety to life and therefore consider the early stage of fire and how it affects the means of escape, but also aim to prevent eventual spread to other buildings.

Asset protection requires extra precautions that will have effect at both early and later stages of the fire growth by controlling fire spread through and between buildings and preventing structural collapse. However,

this extra fire protection will also improve life safety, not only for those escaping at the early stages of the fire but also for fire fighters who will subsequently enter.

If a building is carefully designed and suitable materials are used to build it and maintain it, then the risk of injury or damage from fire can be substantially reduced. Three objectives must be met:

1. It must be possible for everyone to leave the building quickly and safely.
2. The building must remain standing for as long as possible.
3. The spread of fire and smoke must be reduced.

These objectives can be met through the selection of materials and design of buildings.

15.13.2 *Fire loading*

The fire load of a building is used to classify types of building use. It may be calculated simply by multiplying the weight of all combustible materials by their calorific values and dividing by the floor area under consideration. The higher the fire load the more effort needs to be made to offset this by building to higher standards of fire resistance.

15.13.3 *Surface spread of fire*

Combustible materials, when present in a building as large continuous areas, such as for lining walls and ceilings, readily ignite and contribute to spread of fire over their surfaces. This can represent a risk to life in buildings, particularly where walls of fire escape routes and stairways are lined with materials of this nature.

Materials are tested by insurance bodies and fire research establishments. The purpose of the tests is to classify materials according to the tendency for flame to spread over their surfaces. As with all standardized test methods, care must be taken when applying test results to real applications.

In the United Kingdom, a material is classified as having a surface in one of the following categories:

➤ Class 1 – Surface of very low flame spread
➤ Class 2 – Surface of low flame spread
➤ Class 3 – Surface of medium flame spread
➤ Class 4 – Surface of rapid flame spread.

The test shows how a material would behave in the initial stages of a fire.

As all materials tested are combustible, in a serious fire they would burn or be consumed. Therefore, there is an additional Class 0 of materials which are non-combustible throughout, or, under specified conditions, non-combustible on one face and combustible on the other.

Figure 15.12 Steel structures will collapse in the heat of a fire.

The spread of flame rating of the combined Class 0 product must not be worse than that for Class 1.

Internal partitions of walls and ceilings should be Class 0 materials wherever possible and must not exceed Class 1 specifications.

15.13.4 *Fire resistance of structural elements*

If elements of construction such as walls, floors, beams, columns and doors are to provide effective barriers to fire spread and to contribute to the stability of a building, they should be of a required standard of fire resistance.

In the United Kingdom, tests for fire resistance are made on elements of structure, full size if possible, or on a representative portion having minimum dimensions of 3 m long for columns and beams and 1 m^2 for walls and floors. All elements are exposed to the same standard fire provided by furnaces in which the temperature increases with time at a set rate. The conditions of exposure are appropriate to the element tested. Freestanding columns are subjected to heat all round, and walls and floors are exposed to heat on one side only. Elements of structure are graded by the length of time they continue to meet three criteria:

1. The element must not collapse.
2. The element must not develop cracks through which flames or hot gases can pass.
3. The element must have enough resistance to the passage of heat such that the temperature of the unexposed face does not rise by more than a prescribed amount.

The term 'fire resistance' has a precise meaning. It should not be applied to such properties of materials as resistance to ignition or resistance to flame propagation. For example, steel has a high resistance to ignition and

flame propagation but will distort quickly in a fire and allow the structure to collapse – it therefore has poor 'Fire Resistance'. It must be insulated to provide good fire protection. This is normally done by encasing steel frames in concrete.

In the past, asbestos has been made into a paste and plastered onto steel frames, giving excellent fire protection, but it has caused major health problems and its use in new work is banned.

Building materials with high fire resistance are, for example, brick, stone, concrete, very heavy timbers (the outside chars and insulates the inside of the timber) and some specially made composite materials used for fire doors.

15.13.5 *Insulating materials*

Building materials used for thermal or sound insulation could contribute to the spread of fire. Only approved fire-resisting materials should be used.

15.13.6 *Fire compartmentation*

A compartment is a part of a building that is separated from all other parts by walls and floors, and is designed to contain a fire for a specified time.

The principal object is to limit the effect of both direct fire damage and consequential business interruption caused by not only fire spread but also smoke and water damage on the same floor and other storeys.

Buildings are classified in purpose groups, according to their size. To control the spread of fire, any building whose size exceeds that specified for its purpose group must be divided into compartments that do not exceed the prescribed limits of volume and floor area. Otherwise, they must be provided with special fire protection. In the United Kingdom, the normal limit for the size of a compartment is 7,000 m^3. Compartments must be separated by walls and floors of sufficient fire resistance. Any openings needed in these walls or floors must be protected by fire-resisting doors to ensure proper fire-tight separation.

Ventilation and heating ducts must be fitted with fire dampers where they pass through compartment walls and floors. Firebreak walls must extend completely across a building from outside wall to outside wall. They must be stable; they must be able to stand even when the part of the building on one side or the other is destroyed.

No portion of the wall should be supported on unprotected steelwork nor should it have the ends of unprotected steel members embedded in it. The wall must extend up to the underside of a non-combustible roof surface, and sometimes above it. Any openings

303

must be protected to the required minimum grade of fire resistance. If an external wall joins a firebreak wall and has an opening near the join, the firebreak wall may need to extend beyond the external wall.

An important function of external walls is to contain a fire within a building, or to prevent fire spreading from outside.

The fire resistance of external walls should be related to the:

➤ purpose for which the building is used
➤ height, floor area and volume of the building
➤ distance of the building from relevant boundaries and other buildings
➤ extent of doors, windows and other openings in the wall.

A wall, which separates properties from each other, should have no doors or other openings in it.

15.14   Provision of firefighting equipment

15.14.1 *General*

If fire breaks out in the workplace and trained staff can safely extinguish it using suitable firefighting equipment, the risk to others will be removed. Therefore, all workplaces where people are at risk from fire should be provided with suitable firefighting equipment.

The most useful form of firefighting equipment for general fire risks is the water-type extinguisher or suitable alternative. One such extinguisher should be provided for around each 200 m² of floor space with a minimum of one per floor. If each floor has a hose reel, which is known to be in working order and of sufficient length for the floor it serves, there may be no need for water-type extinguishers to be provided.

Areas of special risks involving the use of oil, fats or electrical equipment may need carbon dioxide, dry powder or other types of extinguisher.

Fire extinguishers should be sited on exit routes, preferably near to exit doors or, where they are provided for specific risks, near to the hazards they protect.

Notices indicating the location of firefighting equipment should be displayed where the location of the equipment is not obvious or in areas of high fire risk where the notice will assist in reducing the risk to people in the workplace.

All halon fire extinguishers should have been decommissioned as from 31 December 2003.

Those carrying out hot work should have appropriate fire extinguishers with them and know how to use them.

The primary purpose of fire extinguishers is to tackle fires at a very early stage to enable people to make their escape. Putting out larger fires is the role of the fire and rescue services.

Extinguishers should conform to a recognized standard BS EN 3:7 *Portable Fire Extinguishers. Characteristics, performance requirements and test methods.*

15.14.2 *Advantages and limitations of the main extinguishing media*

Fire extinguishers are all red with 5% of the cylindrical area taken up by the colour code.

Water extinguishers (red band)
This type of extinguisher can only be used on Class A fires. They allow the user to direct water onto a fire from a considerable distance.

A 9-litre water extinguisher can be quite heavy and some water extinguishers with additives can achieve the same rating, although they are smaller and therefore considerably lighter. This type of extinguisher is not suitable for use on live electrical equipment, liquid or metal fires.

Water extinguishers with additives (red band)
This type of extinguisher is suitable for Class A fires. They can also be suitable for use on Class B fires and where appropriate, this will be indicated on the extinguisher. They are generally more efficient than conventional water extinguishers.

Foam extinguishers (cream band)
This type of extinguisher can be used on Class A or B fires and is particularly suited to extinguishing liquid fires such as petrol and diesel. They should not be used on free-flowing liquid fires unless the operator has been specially trained, as these have the potential to rapidly spread the fire to adjacent material. This type of extinguisher is not suitable for deep-fat fryers or chip pans. They should not be used on electrical or metal fires.

Powder extinguishers (blue band)
This type of extinguisher can be used on most classes of fire and achieve a good 'knock down' of the fire. They can be used on fires involving electrical equipment but will almost certainly render that equipment useless.

Because they do not cool the fire appreciably it can reignite. Powder extinguishers can create a loss of visibility and may affect people who have breathing problems, and are not generally suitable for confined spaces. They should not be used on metal fires.

	Old Colour BS 5406 / New Colour BS EN3	Class A Paper or Wood etc.	Class B Flammable Liquids	Class C Flammable Gas Fires	Class D Metal Fires	Electrical Fires
RED	WATER	✔	Do Not Use ✘			Do Not Use ✘
ROD	Fire hose reel	✔	Do Not Use ✘			Do Not Use ✘
CREAM	FOAM	Note: Multi-Purpose Foams may be used ✔	Note: Specialist Foams required for industrial slcohol ✔			Do Not Use ✘
BLACK	CO₂ GAS		Secondary ✔			Primary ✔
BLUE	POWDER	✔	Note: Specialist DP required for Solvents & Esters ✔	✔	Note: Specialist Dry Powders may be required ✔	✔
ROD	Fire blanket		Primary ✔	General Note – May be used in conjunction with other extinguishing agents or fire extinguishing techniques		
CANARY YELLOW		F — SPECIALIST HOT COOKING OIL FIRES ONLY Specifically for dealing with high temperature (360°C+) cooking oils used in large industrial size catering kitchens, restaurants and takeaway establishments with deep fat frying facilities				

Figure 15.13 Types of fire extinguishers and labels. (NB Main colour of all extinguishers is red with 5% for label.)

Carbon dioxide extinguishers (black band)

This type of extinguisher is particularly suitable for fires involving electrical equipment as they will extinguish a fire without causing any further damage (except in the case of some electronic equipment, e.g. computers). As with all fires involving electrical equipment, the power should be disconnected if possible. These extinguishers not be used on metal fires.

305

Wet Chemical-Class 'F' extinguishers (yellow band)

This type of extinguisher is particularly suitable for commercial catering establishments with deep-fat fryers. The intense heat in the fluid generated by fat fires means that when standard foam or carbon dioxide extinguishers stop discharging, reignition tends to occur.

Wet chemical extinguishers starve the fire of oxygen by sealing the burning fluid, which prevents flammable vapour reaching the atmosphere.

15.15 Maintenance and testing of firefighting equipment

It is important that equipment is fit for its purpose and is properly maintained and tested. One way in which this can be achieved is through companies that specialize in the testing and maintenance of firefighting equipment.

All equipment provided to assist escape from the premises, such as fire detection and warning systems and emergency lighting, and all equipment provided to assist with fighting fire, should be regularly checked and maintained by a suitably competent person in accordance with the manufacturer's recommendations. Table 15.3 gives guidance on the frequency of test and maintenance and provides a simple guide to good practice.

Figure 15.14 Fire extinguishers trolley on construction site.

Table 15.3 Maintenance and testing of fire equipment

Equipment	Period	Action
Fire-detection and fire-warning systems including self-contained smoke alarms and manually operated devices	Weekly	➤ Check all systems for state of repair and operation. ➤ Repair or replace defective units. ➤ Test operation of systems, self-contained alarms and manually operated devices.
	Annually	➤ Full check and test of system by competent service engineer. ➤ Clean self-contained smoke alarms and change batteries.
Emergency lighting including self-contained units and torches	Weekly	➤ Operate torches and replace batteries as required. ➤ Repair or replace any defective unit.
	Monthly	➤ Check all systems, units and torches for state of repair and apparent function.
	Annually	➤ Full check and test of systems and units by competent service engineer. ➤ Replace batteries in torches.
Fire-fighting equipment including hose-reels	Weekly	➤ Check all extinguishers including hose-reels for correct installation and apparent function.
	Annually	➤ Full check and test by competent service engineer.

15.16 Fire emergency plans

15.16.1 *Introduction*

Each workplace should have an emergency plan. The plan should include the action to be taken by staff in the event of fire, the evacuation procedure and the arrangements for calling the fire and rescue authority.

For small workplaces this could take the form of a simple fire action notice posted in positions where staff can read it and become familiar with it.

High-fire-risk or larger workplaces will need more detailed plans, which take account of the findings of the risk assessment, for example, the staff significantly at risk and their location. For large workplaces, notices giving clear and concise instructions of the routine to be followed in case of fire should be prominently displayed. The notice should include the method of raising an alarm in the case of fire and the location of an assembly point to which staff escaping from the workplace should report.

15.16.2 *Fire routines and fire notices*

Site managers must make sure that all employees and sub-contractors are familiar with the means of escape in case of fire and their use, and with the routine to be followed in the event of fire. As construction work progresses this will constantly change and will need to be regularly updated.

To achieve this, routine procedures must be set up and made known to all employees and sub-contractors, generally outlining the action to be taken in case of fire and specifically laying down the duties of certain nominated persons. Notices should be posted throughout the site.

While the need in individual sites may vary, there are a number of basic components which should be considered when designing any fire routine procedures:

➤ the action to be taken on discovering a fire
➤ the method of operating the fire alarm or alerting people to an emergency (e.g. by klaxon or even shouting on a small site)
➤ the arrangements for calling the external fire and rescue services
➤ the stopping of machinery and plant
➤ first-stage firefighting by employees
➤ evacuation of the structure or site
➤ assembly of staff, sub-contractors and visitors, and carrying out a roll-call to account for everyone on the site.

The procedures must take account of those people who may have difficulty in escaping quickly from a building or partially completed structure because of their location or a disability. Insurance companies and other responsible people may need to be consulted where special procedures are necessary to protect buildings and plant during or after people have been evacuated. The duties of various managers should include any necessary actions.

15.16.3 *Supervisory duties*

A member of the site staff should be nominated to supervise and co-ordinate all fire and emergency arrangements. This person should be in a senior position or at least have direct access to a senior manager. Senior members of the staff should be appointed as departmental fire wardens, with deputies for every occasion of absence, however brief. They should ensure the following precautions are taken, i.e. that:

➤ everyone on site can be alerted to an emergency
➤ everyone on site knows what signal will be given for an emergency and knows what to do
➤ someone who has been trained in what to do is on site and ready to co-ordinate activities
➤ emergency routes are kept clear, signed and adequately lit
➤ there are arrangements for calling the fire and rescue services and to give them special information about high hazard work, for example in tunnels or confined spaces
➤ there is adequate access to the site for the emergency services and this is always kept clear
➤ suitable arrangements for treating and recovering injured people are set up
➤ someone is posted to the site entrance to receive and direct the emergency services.

Under normal conditions fire wardens should check that good standards of housekeeping exist in their areas, that exits and escape routes are kept free from obstruction, that all fire fighting appliances are available for use and fire points are not obstructed, that smoking is rigidly controlled, and that all employees and sub-contractors under their control are familiar with the emergency procedure and know how to use the fire alarm and fire fighting equipment.

15.16.4 *Assembly and roll-call*

Assembly points should be established for use in the event of evacuation. They should be at positions, preferably undercover, which are unlikely to be affected at the time of fire. In some cases it may be necessary to make mutual arrangements with the client or occupiers of nearby premises.

In the case of small sites, a complete list of the names of all staff should be maintained so that a roll-call can be made if evacuation becomes necessary.

In those premises where the number of staff would make a single roll-call difficult, each area fire warden should maintain a list of the names of employees and sub-contractors in their area. Roll-call lists must be updated regularly.

A recent serious warehouse fire in Warwickshire, where four fire fighters died, demonstrates how important it is to ensure that the buildings have been fully evacuated and all people accounted for. The fire and rescue services will always go into a building putting themselves in grave danger, when they believe people are trapped.

15.16.5 Fire notices

Printed instructions for the action to be taken in the event of fire should be displayed throughout the site and regularly updated as work progresses. The information contained in the instructions should be stated briefly and clearly. The staff and their deputies to whom specific duties are allocated should be identified.

Instruction for the immediate calling of the fire brigade in case of fire should be displayed at the site office, welfare amenity areas and around the site as appropriate.

A typical fire notice is given in Appendix 15.3

15.16.6 Testing

The alarm system (where required) should be tested every week, while the construction project is under way. Where relevant the test should be carried out by activating a different call point each week at a fixed time.

15.16.7 Fire drills

Once a fire routine has been established, it must be tested at regular intervals in order to ensure that all employees and sub-contractors are familiar with the action to be taken in an emergency.

The most effective way of achieving this is by carrying out fire drills at prescribed intervals. A programme of fire drills should be planned to ensure that all employees and sub-contractors are covered.

15.17 Fire procedures and people with a disability

People with a disability (including members of the public) need special consideration, when planning for emergencies. People with a disability seldom work on construction sites

but where necessary their needs must be considered. Employers should:

➤ identify everyone who may need special help to get out
➤ allocate responsibility to specific staff to help people with a disability in emergency situations
➤ consider possible escape routes
➤ enable the safe use of lifts
➤ enable people with a disability to summon help in emergencies
➤ train staff to be able to help their colleagues.
➤ consider safe havens.

Advice on the needs of people with a disability, including sensory impairment, is available from the organizations which represent various groups. (Names and addresses can be found in the telephone directory.)

People with impaired vision must be encouraged to familiarize themselves with escape routes, particularly those not in regular use. A 'buddy' system would be helpful. But, to take account of absences, more than one employee working near anyone with impaired vision should be taught how to help them.

Staff with impaired hearing may not hear alarms in the same way as those with normal hearing but may still be able to recognize the sound. This may be tested during the weekly alarm audibility test. There are alternative means of signalling, such as lights or other visual signs, vibrating devices or specially selected sound signals. The Royal National Institute for Deaf People (19–23 Featherstone St, London EC1Y 8SL) can advise. Ask the fire brigade before installing alternative signals.

Wheelchair users or others with impaired mobility may need help to negotiate stairs, etc. Anyone selected to provide this help should be trained in the correct methods. Advice on the lifting and carrying of people can be obtained from the Fire Service, Ambulance Service, British Red Cross Society, St John Ambulance Brigade or certain disability organizations.

Employees with a mental handicap may also require special provision. Management should ensure that the colleagues of any employee with a mental handicap know how to reassure them and lead them to safety.

15.18 Practice NEBOSH questions for Chapter 15

1. (a) **Identify** the legal requirements and any relevant industry standards with respect to the control of fire in construction work.
   (b) **Describe** the factors to be considered in the selection and training of personnel given

special responsibilities for fire safety on a large refurbishment project in a city centre.

2. Timber sections are stored in an enclosed compound.
   (a) **Describe** the precautions that should be taken to prevent a fire occurring.
   (b) **Identify** the extinguishing media that may be used for dealing with a fire in such a situation, and **explain** how **EACH** works to extinguish the fire.

3. (a) **Explain** the dangers associated with LPG.
   (b) **Describe** the precautions needed for the storage, use and transportation of LPG in cylinders on a construction site.

4. **Outline** the requirements for the safe storage of LPG cylinders on a construction site.

5. A major hazard on a refurbishment project is fire.
   (a) **Identify THREE** activities that represent an increased fire risk in such a situation.
   (b) **Outline** the precautions that may be taken to prevent a fire from occurring.

6. **Prepare notes** for a briefing session ('toolbox talk') associated with the prevention and control of fire on a construction site.

7. (a) **Explain**, using a suitable sketch, the significance of the 'fire triangle'.
   (b) **List** the types of ignition source that may cause a fire to occur, and give an example of **EACH** type.

8. **Identify** the **FOUR** methods of heat transfer and explain how each can cause the spread of fire.

9. **List EIGHT** ways of reducing the risk of a fire starting in a workplace.

10. **Outline** the measures that should be taken to minimize the risk of fire from electrical equipment.

11. **Outline** the main requirements for a safe means of escape from a building in the event of a fire.

12. (a) With reference to the fire triangle, **outline TWO** methods of extinguishing fires.
    (b) **State** the ways in which persons could be harmed by a fire in work premises.

13. **List EIGHT** features of a safe means of escape from a building in the event of fire.

14. **Outline** reasons for undertaking regular fire drills in the workplace.

15. (a) **Outline** the main factors to be considered in the siting of fire extinguishers.
    (b) **Outline** the inspection and maintenance requirements for fire extinguishers in the workplace.

16. (a) **Identify FOUR** different types of ignition source that may cause a fire.
    (b) For **EACH** type of ignition source identified in (a), **outline** the precautions that could be taken to prevent a fire starting.

17. (a) **Explain TWO** ways in which electricity can cause a fire on a construction site.
    (b) **Outline** the various ways in which workers might be physically harmed by a fire on a construction site.

18. A major hazard on a refurbishment project is fire.
    (a) **Identify THREE** activities that represent an increased fire risk in such a situation.
    (b) **Outline** the precautions that may be taken to prevent a fire from occurring.

Appendix 15.1 – Fire risk assessment as recommended in Fire Safety Guides published by the Department for Communities and Local Government in 2006

CHECKLIST

Follow the five key steps – Fill in the checklist – Assess your fire risk and plan fire safety

1. Fire safety risk assessment

Fire starts when heat (source of ignition) comes into contact with fuel (anything that burns) and oxygen (air).
You need to keep sources of ignition and fuel apart.

How could a fire start? (ignition source)

Think about heaters, lighting, naked flames, electrical equipment, hot processes such as welding or grinding, cigarettes, matches and anything else that gets very hot or causes sparks.

What could burn? (fuel)

Packaging, rubbish and furniture could all burn just like the more obvious fuels such as petrol, paint, varnish and white spirit. Also think about wood, paper, plastic, rubber and foam. Do the walls or ceilings have hardboard, chipboard, or polystyrene? Check outside, too.

Check:

➢ Have you found anything that could start a fire?

Make a note of it.

➢ Have you found anything that could burn?

Make a note of it.

2. People at risk

People at risk

Everyone is at risk if there is a fire. Think whether the risk is greater for some because of when or where they work, such as night staff, or because they're not familiar with the premises, such as visitors or customers.
Children, the elderly or disabled people are especially vulnerable.

Check:

Have you identified:

➢ who could be at risk?
➢ who could be especially at risk?

Make a note of what you have found.

3. Evaluate and act

Evaluate

First, think about what you have found in Steps 1 and 2: what are the risks of a fire starting, ignition source and fuel coming and what are the risks to people in the building and nearby?

Remove and reduce risk

How can you avoid accidental fires? Could a source of heat or sparks fall, or be knocked or pushed into something that would on to combustible materials burn? Could that happen the other way round?

Protect

Take action to protect your premises and people from fire.

Check:

➢ Have you assessed the risks of fire in your workplace?
➢ Have you assessed the risk to staff and visitors?
➢ Have you kept any source of fuel and heat/sparks apart?
➢ If someone wanted to start a fire deliberately, is there anything around they could use?
➢ Have you removed or secured any fuel an arsonist could use?
➢ Have you protected your premises from accidental fire or arson?
➢ **How can you make sure everyone is safe in case of fire?**
   ➢ Will you know there is a fire?
   ➢ Do you have a plan to warn others?
   ➢ Who will make sure everyone gets out?
   ➢ Who will call the fire service?
   ➢ Could you put out a small fire quickly and stop it spreading?
➢ **How will everyone escape?**
   ➢ Have you planned escape routes?
   ➢ Have you made sure people will be able to safely find their way out, even at night if necessary?
   ➢ Does all your safety equipment work?
   ➢ Will people know what to do and how to use equipment?

Make a note of what you have found.

4. Record, plan and train

Record

Keep a record of any fire hazards and what you have done to reduce or remove them. If your premises are small, a record is a good idea. If you have five or more staff or have a licence then you must keep a record of what you have found and what you have done.

Plan

You must have a clear plan of how to prevent fire and how you will keep people safe in case of fire. If you share a building with others, you need to coordinate your plan with them.

Train

You need to make sure your staff know what to do in case of fire, and if necessary, are trained for their roles.

Check:

➤ Have you made a record of what you have found, and action you have taken?

➤ Have you planned what everyone will do if there is a fire?

➤ Have you discussed the plan with all staff?

➤ **Have you**

  ➤ informed and trained people (practised a fire drill and recorded how it went)?

➤ nominated staff to put in place your fire prevention measures, and trained them?

➤ made sure everyone can fulfil their role?

➤ informed temporary staff?

➤ consulted others who share a building with you, and included them in your plan?

5. Review

Keep your risk assessment under regular review. Over time, the risks may change.

If you identify significant changes in risk or make any significant changes to your plan, you must tell others who share the premises and where appropriate retrain staff.

Have you

➤ made any changes to the building inside or out?

➤ had a fire or near miss?

➤ changed work practices?

➤ begun to store chemicals or dangerous substances?

➤ significantly changed your stock, or stock levels? Have you planned your next fire drill?

The checklist above can help you with the Fire Risk Assessment but you may need additional information especially if you have large or complex premises.

Source: Department for Community and Local government.

Appendix 15.2 – Example form for recording significant findings as published in 2006 by the Department for Communities and Local Government in their Fire Safety Guides

Risk Assessment – Record of significant findings		
Risk assessment for	**Assessment undertaken by**	
Company	Date	
Address	Completed by	
	Signature	
Sheet number	**Floor/area**	**Use**
Step 1 – identify fire hazards		
Sources of ignition	**Sources of fuel**	**Sources of oxygen**
Step 2 – People at risk		
Step 3 – Evaluate, remove, reduce and protect from risk		
(3.1) Evaluate the risk of the fire occurring		
(3.2) Evaluate the risk to people from a fire starting in the premises		
(3.3) Remove and reduce the hazards that may cause a fire		
(3.4) Remove and reduce the risks to people from a fire.		
Assessment review		
Assessment review date	Completed by Signature	
Review outcome (where substantial changes have occurred a new record sheet should be used)		

Notes:
1. The risk assessment record should refer to other significant plans, records or other documents as necessary.
2. The information in this record should assist employers to develop an emergency plan; to co-ordinate measures with other 'responsible persons' in the building; and to inform and train staff and inform other relevant persons.

Appendix 15.3 – Typical fire notice

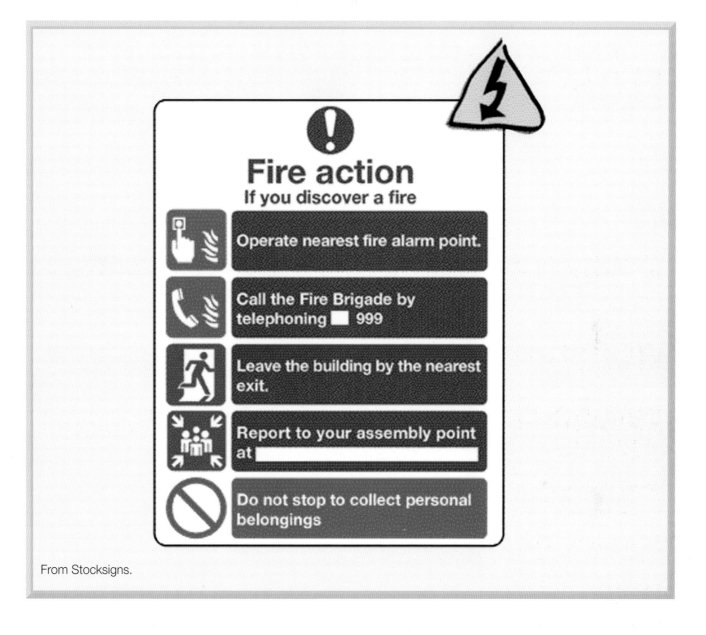

From Stocksigns.

Chemical and biological health hazards and control

16

16.1 Introduction

Occupational health is as important as occupational safety but generally receives less attention from managers. Every year twice as many people suffer ill-health caused or exacerbated by the workplace than suffer workplace injury. Although these illnesses do not usually kill people, they can lead to many years of discomfort and pain. Such illnesses include respiratory disease, hearing problems, asthmatic conditions and back pain. Furthermore, it has been estimated that 30% of all cancers probably have an occupational link – that linkage is known for certain in 8% of cancer cases.

Work in the field of occupational health has been taking place for the last four centuries and possibly longer. The main reason for the relatively low profile for occupational health over the years has been the difficulty in linking the ill-health effect with the workplace cause. Many illnesses, such as asthma or back pain, can have a workplace cause but can also have other causes. Many of the advances in occupational health have been as a result of statistical and epidemiological studies (one well-known such study linked the incidence of lung cancer to cigarette smoking). While such studies are invaluable in the assessment of health risk, there is always an element of doubt when trying to link cause and effect. The measurement of gas and dust concentrations is also subject to doubt when a correlation is made between a measured sample and the workplace environment from which it was taken. Occupational health, unlike occupational safety, is generally more concerned with probabilities than certainties.

In this chapter, chemical and biological health hazards will be considered – other forms of health hazard will be covered in Chapter 17.

The chemical and biological health hazards described in this chapter are covered by the following health and safety regulations:

➤ Control of Substances Hazardous to Health Regulations (COSHH)
➤ Control of Lead at Work Regulations
➤ Control of Asbestos Regulations.

16.2 Forms of chemical agent

Chemicals can be transported by a variety of agents and in a variety of forms. They are normally defined in the following ways.

Dusts are solid particles slightly heavier than air but often suspended in it for a period of time. The size of the particles ranges from about $0.4\,\mu m$ (fine) to $10\,\mu m$ (coarse). Dusts are created either by mechanical processes (e.g. grinding or pulverizing) or construction processes (e.g. concrete laying, demolition or sanding), or by specific tasks (e.g. furnace ash removal). The fine dust is much more hazardous because it penetrates deep into the lungs and remains there – known as **respirable dust**. In rare cases, respirable dust enters the bloodstream directly causing damage to other organs. Examples of such fine dust are cement, granulated plastic materials and silica dust produced from stone or concrete dust. Repeated exposure may lead to permanent lung disease. Any dusts which are capable of entering the nose and mouth during breathing are known as **inhalable dusts.**

Gases are any substances at a temperature above their boiling point. Steam is the gaseous form of water. Common gases include carbon monoxide, carbon dioxide, nitrogen and oxygen. Gases are absorbed into the bloodstream where they may be beneficial (oxygen) or harmful (carbon monoxide).

Vapours are substances which are at or very close to their boiling temperatures. They are gaseous in form. Many solvents, such as cleaning fluids, fall into this category. The vapours, if inhaled, enter the bloodstream and some can cause short-term effects (dizziness) and long-term effects (brain damage).

Liquids are substances which normally exist at a temperature between freezing (solid) and boiling (vapours and gases). They are sometimes referred to as fluids in health and safety regulations.

Mists are similar to vapours in that they exist at or near their boiling temperature but are closer to the liquid phase. This means that there are suspended, very small, liquid droplets present in the vapour. A mist is produced during a spraying process (such as paint spraying). Many industrially produced mists can be very damaging if inhaled producing similar effects to vapours. It is possible for some mists to enter the body through the skin or by ingestion with food.

Fume is a collection of very small metallic particles (less than 1 μm) which have condensed from the gaseous state. They are most commonly generated by the welding process. The particles tend to be within the respirable range (approximately 0.4–1.0 μm) and can lead to long-term permanent lung damage. The exact nature of any harm depends on the metals used in the welding process and the duration of the exposure.

16.3 Forms of biological agent

As with chemicals, biological hazards may be transported by any of the following forms of agent.

Fungi are very small organisms, sometimes consisting of a single cell, and were formerly regarded as plants (e.g. mushrooms and yeast). Unlike plants, they cannot produce their own food but either live on dead organic matter or on living animals or plants as parasites. Fungi reproduce by producing spores which can cause allergic reactions when inhaled. The infections produced in man by fungi may be mild, such as athlete's foot, or severe such as ringworm. Many fungal infections can be treated with antibiotics.

Moulds are a particular group of very small fungi which, under damp conditions, will grow on things such as walls, bread, cheese, leather and canvas. They can be beneficial (penicillin) or cause allergic reactions (asthma). Asthma attacks, athlete's foot and farmer's lung are all examples of fungal infections.

Blue green algae are another biological hazard, which contaminate drinking water, and provide food for *Legionella* bacteria.

Bacteria are very small single-celled organisms which are much smaller than cells within the human body. They can live outside the body and be controlled and destroyed by antibiotic drugs. There is evidence that bacteria are developing which are becoming resistant to most antibiotics. This has been caused by the widespread misuse of antibiotics. It is important to note that not all bacteria are harmful to humans. Bacteria aid the digestion of food and babies would not survive without their aid to break down the milk in their digestive systems. Legionaires', tuberculosis and tetanus are all bacterial diseases.

Viruses are minute non-cellular organisms which can only reproduce within a host cell. They are very much smaller than bacteria and cannot be controlled by antibiotics. They appear in various shapes and are continually developing new strains. They are usually only defeated by the defence and healing mechanisms of the body. Drugs can be used to relieve the symptoms of a viral attack but cannot cure it. The common cold is a viral infection as are hepatitis, AIDS (HIV) and influenza.

16.4 Classification of hazardous substances and their associated health risks

A hazardous substance is one which can cause ill-health to people at work. Such substances may include those used directly in the work processes (glues and paints), those produced by work activities (welding fumes) or those which occur naturally (dust). Hazardous substances are classified according to the severity and type of hazard which they may present to people who may come into contact with them. The contact may occur while working or transporting the substances or might occur during a fire or accidental spillage. There are several classifications but here only the five most common will be described.

Irritant is a non-corrosive substance which can cause skin (dermatitic) or lung (bronchial) inflammation after repeated contact. People who react in this way to a particular substance are **sensitized** or **allergic** to that substance. In most cases, it is likely that the concentration of the irritant may be more significant than the exposure time. Many household substances, such as wood preservatives, bleaches and glues are irritants. Many chemicals used as solvents are also irritants (white spirit, toluene and acetone). Formaldehyde and ozone are other examples of irritants.

A **corrosive** substance is one which will attack, normally by burning, living tissue. It is usually a strong acid or alkali and examples include sulphuric acid and caustic soda. Many tough cleaning substances, such as kitchen oven cleaners, are corrosives as are many dishwasher crystals.

Harmful is the most commonly used classification and describes a substance which, if it is swallowed, inhaled or penetrates the skin, **may** pose limited health risks. These risks can usually be minimized or removed by following the instruction provided with the substance (e.g. by using personal protective equipment (PPE)). Many household substances fall into this category including bitumen-based paints and paint brush restorers. Many chemical cleansers, are categorized as harmful. It is very common for substances labelled harmful also to be categorized as irritant.

Toxic substances are ones which impede or prevent the function of one or more organs within the body, such as the kidneys, liver and heart. A toxic substance is, therefore, a poisonous one. Lead, mercury, pesticides and the gas carbon monoxide are toxic substances. The effect on the health of a person exposed to a toxic substance depends on the concentration and toxicity of the substance, the frequency of the exposure and the effectiveness of the control measures in place. The state of health and age of the person and the route of entry into the body have influence on the effect of the toxic substance.

Carcinogenic substances are ones which are known or suspected of promoting abnormal development of body cells to become cancers. Asbestos, hard wood dust, creosote and some mineral oils are carcinogenic. It is very important that the health and safety rules accompanying the substance are strictly followed.

Mutagenic substances are those which damage genetic material within cells causing abnormal changes that can be passed from one generation to another.

Each of the classifications may be identified by a symbol and a symbolic letter – the most common of these are shown in Figure 16.1.

The effects on health of hazardous substances may be either acute or chronic:

Acute effects are of short duration and appear fairly rapidly, usually during or after a single or short-term exposure to a hazardous substance. Such effects may be severe and require hospital treatment but are usually reversible. Examples include asthma-type attacks, nausea and fainting.

Chronic effects develop over a period of time which may extend to many years. The word 'chronic' means 'with time' and should not be confused with 'severe' as its use in everyday speech often implies. Chronic health effects are produced from prolonged or repeated exposures to hazardous substances resulting in a gradual, latent and often irreversible illness, which may remain undiagnosed for many years. Many cancers and mental diseases fall into the chronic category. During the development stage of a chronic disease, the individual may experience no symptoms.

16.4.1 *The role of COSHH*

The Control of Substances Hazardous to Health Regulations 1988 (COSHH) were the most comprehensive and significant piece of Health and Safety legislation to be introduced since the Health and Safety at Work Act 1974. They were enlarged to cover biological agents in 1994. A detailed summary of these Regulations appears in Chapter 21. The Regulations impose duties on employers to protect employees and others who may be exposed to substances hazardous to their health and require employers to control exposure to such substances.

The COSHH Regulations offer a framework for employers to build a management system to assess health risks and to implement and monitor effective controls. Adherence to these Regulations will provide the following benefits to the employer and employee:

➤ improved productivity due to lower levels of ill-health and more effective use of materials
➤ improved employee morale
➤ lower numbers of civil court claims
➤ better understanding of health and safety legal requirements.

Organizations which ignore COSHH requirements will be liable for enforcement action, including prosecution, under the Regulations.

Harmful (Xn)   Irritant (Xi)   Corrosive (C)   Toxic (T)   Carcinogenic (T)

Figure 16.1 Classification symbols.

16.5 Routes of entry to the human body

There are three principal routes of entry of hazardous substances into the human body:

1. **Inhalation** – breathing in the substance with normal air intake. This is the main route of contaminants into the body. These contaminants may be chemical (e.g. solvents or welding fume) or biological (e.g. bacteria or fungi) and become airborne by a variety of modes, such as sweeping, spraying, grinding and bagging. They enter the lungs where they have access to the bloodstream and many other organs.
2. **Absorption through the skin** – the substance comes into contact with the skin and enters either through the pores or a wound. Tetanus can enter in this way as can toluene, benzene and various phenols.
3. **Ingestion** – through the mouth and swallowed into the stomach and the digestive system. This is not a significant route of entry to the body. The most common occurrences are due to airborne dust or poor personal hygiene (not washing hands before eating food).

Another very rare entry route is by **injection**. The abuse of compressed air lines by shooting high pressure air at the skin can lead to air bubbles entering the bloodstream. Accidents involving hypodermic syringes in a health or veterinary service setting are rare but illustrate this form of entry route.

The most effective control measures which can reduce the risk of infection from biological organisms are disinfection, proper disposal of clinical waste (including syringes), good personal hygiene and, where appropriate, PPE. Other measures include vermin control, water treatment and immunization.

There are five major functional systems within the human body – respiratory, nervous, cardiovascular (blood), urinary and integumentary (skin).

16.5.1 *The respiratory system*

This comprises the lungs and associated organs (e.g. the nose). Air is breathed in through the nose, passes through the trachea (windpipe) and the bronchi into the two lungs. Within the lungs, the air enters many smaller passageways (bronchioli) and thence to one of 300,000 terminal sacs called alveoli. The alveoli are approximately 0.1 mm across, although the entrance is much smaller. On arrival in the alveoli, there is a diffusion of oxygen into the bloodstream through blood capillaries and an effusion of carbon dioxide from the bloodstream. While soluble dust which enters the alveoli will be absorbed into the bloodstream, insoluble dust (respirable dust) will remain permanently, leading to possible chronic illness.

The whole of the bronchial system is lined with hairs, known as cilia. The cilia offer some protection against insoluble dusts. These hairs will arrest all non-respirable dust (above 5 μm) and, with the aid of mucus, pass the dust from one hair to a higher one and thus bring the dust back to the throat. (This is known as the ciliary escalator.) It has been shown that smoking damages this action. The nose will normally trap large particles (greater than 20 μm) before they enter the trachea.

Respirable dust tends to be long thin particles with sharp edges which puncture the alveoli walls. The puncture

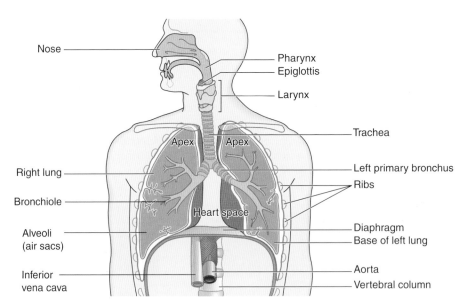

Figure 16.2 The upper and lower respiratory system.

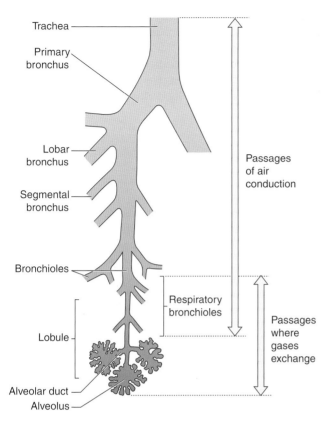

Trachea
Primary bronchus
Lobar bronchus
Segmental bronchus
Bronchioles
Lobule
Alveolar duct
Alveolus
Passages of air conduction
Respiratory bronchioles
Passages where gases exchange

Figure 16.2 Continued.

heals producing scar tissues which are less flexible than the original walls – this can lead to fibrosis. Such dusts include asbestos, coal, silica, some plastics and talc.

Acute effects on the respiratory system include bronchitis and asthma and chronic effects include fibrosis and cancer. Hardwood dust, for example, can produce asthma attacks and nasal cancer.

Finally, asphyxiation, due to a lack of oxygen, is a problem in confined spaces particularly when MIG (metal inert gas) welding is taking place.

16.5.2 *The nervous system*

The nervous system consists primarily of the brain, the spinal cord and nerves extending throughout the body. Any muscle movement or sensation (e.g. hot and cold) is controlled or sensed by the brain through small electrical impulses transmitted through the spinal cord and nervous system. The effectiveness of the nervous system can be reduced by **neurotoxins** and lead to changes in mental ability (loss of memory and anxiety), epilepsy and narcosis (dizziness and loss of consciousness). Organic solvents (trichloroethylene) and heavy metals (mercury) are well-known neurotoxins. The expression 'mad hatters' originated from the mental deterioration of top hat polishers in the 19th century who used mercury to produce a shiny finish on the top hats.

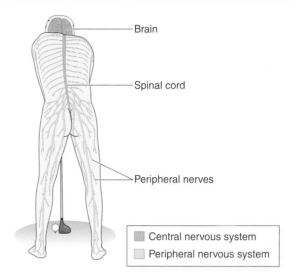

Brain
Spinal cord
Peripheral nerves

Central nervous system
Peripheral nervous system

Figure 16.3 The nervous system.

16.5.3 *The cardiovascular system*

The blood system uses the heart to pump blood around the body through arteries, veins and capillaries. Blood is produced in the bone marrow and consists of a plasma within which are red cells, white cells and platelets. The system has three basic objectives:

1. to transport oxygen to vital organs, tissues and the brain and carbon dioxide back to the lungs (red cell function)
2. to attack foreign organisms and build up a defence system (white cell function)
3. to aid the healing of damaged tissue and prevent excessive bleeding by clotting (platelets).

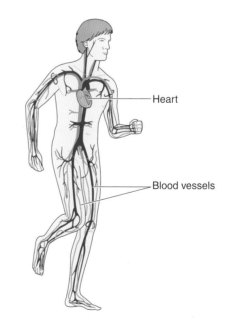

Heart
Blood vessels

Figure 16.4 The cardiovascular system.

There are several ways in which hazardous substances can interfere with the cardiovascular system. Benzene can affect the bone marrow by reducing the number of blood cells produced. Carbon monoxide prevents the red cells from absorbing sufficient oxygen and the effects depend on its concentration. Symptoms begin with headaches and end with unconsciousness and possibly death.

16.5.4 *The urinary system*

The urinary system extracts waste and other products from the blood. The two most important organs are the

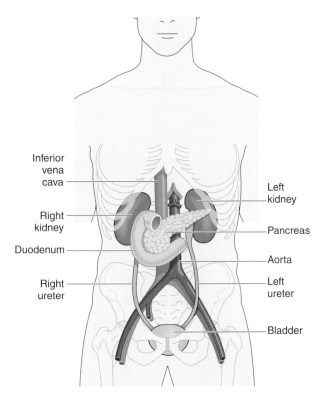

Figure 16.5 Parts of the urinary system.

liver (normally considered part of the digestive system) and the kidneys, both of which can be affected by hazardous substances within the bloodstream.

The liver removes toxins from the blood, maintains the levels of blood sugars and produces protein for the blood plasma. Hazardous substances can cause the liver to be too active or inactive (e.g. xylene), or lead to liver enlargement (e.g. cirrhosis caused by alcohol) or liver cancer (e.g. vinyl chloride).

The kidneys filter waste products from the blood as urine, regulate blood pressure and liquid volume in the body and produce hormones for making red blood cells. Heavy metals (e.g. cadmium and lead) and organic solvents (e.g. glycol ethers used in screen printing) can restrict the operation of the kidneys, possibly leading to failure.

16.5.5 *The skin*

The skin holds the body together and is the first line of defence against infection. It regulates body temperature, is a sensing mechanism, provides an emergency food store (in the form of fat) and helps to conserve water. There are two layers – an outer layer called the epidermis (0.2 mm) and an inner layer called the dermis (4 mm). The epidermis is a tough protective layer and the dermis contains the sweat glands, nerve endings and hairs.

The most common industrial disease of the skin is **dermatitis** (non-infective dermatitis). It begins with a mild irritation on the skin and develops into blisters which can peel and weep, becoming septic. It can be caused by various chemicals, mineral oils and solvents. There are two types:

1. **irritant contact dermatitis** – occurs soon after contact with the substance and the condition reverses after contact ceases (detergents and weak acids)
2. **allergic contact dermatitis** – caused by a sensitizer such as turpentine, epoxy resin, solder flux and formaldehyde.

For many years, dermatitis was seen as a 'nervous' disease which was psychological in nature. Nowadays, it is recognized as an industrial disease which can be controlled by good personal hygiene, PPE, use of barrier creams and health screening of employees. Dermatitis can appear anywhere on the body but it is normally found on the hands. Therefore, gloves should always be worn when there is a risk of dermatitis.

The risks of dermatitis occurring increases with the presence of skin cuts or abrasions, which allow chemicals to be more easily absorbed, and the type, sensitivity and existing condition of the skin.

16.6   Health hazards of specific agents

The health hazards associated with hazardous substances can vary from very mild (momentary dizziness or a skin irritation) to very serious such as a cancer.

Cancer is a serious body cell disorder in which the cells develop into tumours. There are two types of tumour – benign and malignant. Benign tumours do

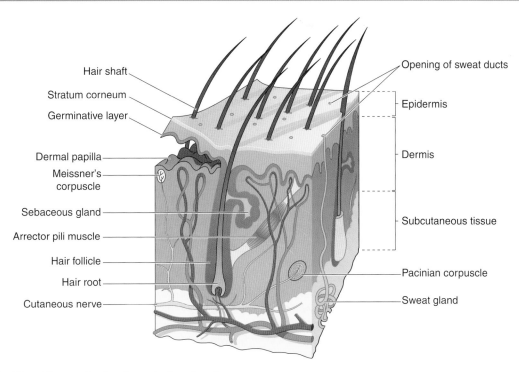

Figure 16.6 The skin – main structures in the dermis.

not spread but remain localized within the body and grow slowly. Malignant tumours are called cancers and often grow rapidly, spreading to other organs using the bloodstream and lymphatic glands. Survival rates have improved dramatically in recent years as detection methods have improved and the tumours can be found in their early stages of development. A minority of cancers are believed to be occupational in origin (20%).

Occupational asthma has approximately 4 million sufferers in the UK and it is estimated that 13 million working days are lost each year as a result of it. It is mainly caused by breathing in respiratory sensitizers, such as wood dusts, organic solvents, solder flux fumes or animal hair. The symptoms are coughing, wheezing, tightness of the chest and breathlessness due to a constriction of the airways. It can be a mild attack or a serious one that requires hospitalization. There is some evidence that stress can trigger an attack.

The following common agents of health hazard will be described together with the circumstances in which they may be found:

Ammonia is a colourless gas with a distinctive odour which, even in small concentrations, causes the eyes to smart and run and a tightening of the chest. It is a corrosive substance which can burn the skin, burn and seriously damage the eye, cause soreness and ulceration of the throat and severe bronchitis and oedema (excess of fluid) of the lungs. Good eye and respiratory protective equipment (RPE) is essential when maintaining equipment containing ammonia. Any such equipment should be tested regularly for leaks and repaired promptly if required. Ammonia is also used in the production of fertilizers and synthetic fibres. Most work on ammonia plant should require a permit-to-work procedure.

Chlorine is a greenish, toxic gas with a pungent smell which is highly irritant to the respiratory system, producing severe bronchitis and oedema of the lungs and may also cause abdominal pain, nausea and vomiting. It is used as a disinfectant for drinking water and swimming pool water and in the manufacture of chemicals.

Organic solvents are used widely in industry as cleansing and degreasing agents. There are two main groups – the hydrocarbons (includes the aromatic and aliphatic hydrocarbons) such as toluene and white spirit, and the non-hydrocarbons (such as trichloroethylene and carbon tetrachloride). All organic solvents are heavier than air and most are sensitizers and irritants. Some are narcotics whilst others can cause dermatitis and after long exposure periods liver and kidney failure. It is very important that the hazard data sheet accompanying the particular solvent is read and the recommended PPE is worn at all times. Solvents are used extensively in construction in varnishes, paints, adhesives, strippers and as thinners. They are at highest risk when used as sprays. One of the most hazardous is dichloromethane (DCM) also known as methylene chloride. It is used as a paint stripper, normally as a gel. It can produce narcotic effects and has been classified as a category 3 carcinogen in the European Community. The minimum PPE requirements are impermeable overalls, apron, footwear, long gloves and gauntlets and chemically resistant goggles or visor. Respiratory protection equipment is also required if it cannot be demonstrated that exposure is below the workplace exposure limit (WEL).

Carbon dioxide is a colourless and odourless gas which is heavier than air. It represses the respiratory system, eventually causing death by asphyxiation. At low concentrations it will cause headaches and sweating followed by a loss of consciousness. The greatest hazard occurs in confined spaces particularly where the gas is produced as a by-product.

Nitrogen is a colourless, tasteless and odourless gas but is much more benign than many other gases. It comprises 78% by volume of air – the earth's atmosphere. It is an essential constituent of all plants and animals and is used as an ingredient in the manufacture of agricultural fertilizer. It neither burns nor supports combustion – it is an inert gas. Due to these properties, it is used to pressurize systems, particularly hot water systems. In liquid form, it is used both as a refrigerant and freezing agent (see Chapter 19 (Section 19.2.1) for an example in excavation work). Since it has a melting point of −196°C, it must be used very carefully to avoid frostbite.

Carbon monoxide is a colourless, tasteless and odourless gas which makes it impossible to detect without special measuring equipment. As explained earlier, carbon monoxide enters the blood (red cells) more readily than oxygen and restricts the supply of oxygen to vital organs. At low concentrations in the blood (less than 5%), headaches and breathlessness will occur whilst at higher concentrations, unconsciousness and death will result. The most common occurrence of carbon monoxide is as an exhaust gas either from a vehicle or a heating system. In either case, it results from inefficient combustion and, possibly, poor maintenance.

Isocyanates are volatile organic compounds widely used in industry for products such as printing inks, adhesives and two-pack paints (particularly in vehicle body shops) and in the manufacture of plastics (polyurethane products). They are irritants and sensitizers. Inflammation of the nasal passages and the throat and bronchitis are typical reactions to many isocyanates. When a person becomes sensitized to an isocyanate, very small amounts of the substance often provoke a serious reaction similar to an extreme asthma attack. Isocyanates also present a health hazard to fire fighters. They are subject to a WEL, and RPE should normally be worn. Two-pack polyurethane paints and varnishes are used in construction as surface coatings.

Asbestos appears in three main forms – crocidolite (blue), amosite (brown) and chrysotile (white). The blue and brown asbestos is considered to be the most dangerous and may be found in older buildings where it was used as a heat insulator around boilers and hot water pipes and as fire protection of structure. White asbestos has been used in asbestos cement products and brake linings. It is difficult to identify an asbestos product by its colour alone – laboratory identification is usually required. Asbestos produces a fine fibrous dust of respirable dust size which can become lodged in the lungs. The fibres can be very sharp and hard causing damage to the lining of the lungs over a period of many years. This can lead to one of the following diseases:

➤ asbestosis or fibrosis (scarring) of the lungs
➤ lung cancer
➤ mesothelioma – cancer of the lining of the lung or, in rarer cases, the abdominal cavity (this is confined to blue asbestos).

If asbestos is discovered during the performance of a contract, work should cease immediately and the employer be informed. Typical sites of asbestos include ceiling tiles, asbestos cement roof and wall sheets, sprayed asbestos coatings on structural members, loft insulation and asbestos gaskets. Asbestos has its Regulations (Control of Asbestos Regulations) and a summary of these is given in Chapter 21. These cover the need for a risk assessment, a method statement covering the removal and disposal, air monitoring procedures and the control measures (including PPE and training) to be used. A summary of these controls is given later in this chapter.

The most recent asbestos Regulations have removed textured coatings (decorative products, such as Artex, Wondertex and Pebblecoat) from the asbestos licensing regime. Until 1992, these products contained white asbestos. This amendment by the Health and Safety Executive (HSE) followed research work by the Health and Safety Laboratory. The new Regulations introduce an additional training requirement for asbestos awareness training. Such training should include:

➤ the health risks caused by exposure to asbestos
➤ the materials that are likely to contain asbestos and where they are likely to be found
➤ the methods to reduce asbestos risks during work and
➤ the action to take in an emergency, such as an uncontrolled release of asbestos dust.

Asbestos is responsible for at least 4,000 deaths in the UK each year and the HSE felt that there was a need to increase awareness amongst the workforce of the risks associated with this material.

Lead is a heavy, soft and easily worked metal. It is used in many industries but is most commonly associated with plumbing and roofing work. Lead enters the body normally by inhalation but can also enter by ingestion and skin contact. The main targets for lead are the central nervous system (and the brain) and the blood (and blood production). The effects are normally chronic and develop as the quantity of lead builds up. Headaches and nausea are the early symptoms followed by anaemia, muscle weakening and (eventually) coma. Regular blood tests are a legal and sensible requirement as are good ventilation and the use of appropriate PPE. High personal hygiene standards and adequate welfare (washing) facilities are essential and must be used before smoking or food is consumed. The reduction in the use of leaded petrol was an acknowledgement of the health hazard represented by lead in the air. Lead is covered by its own set of Regulations – the Control of Lead at Work Regulations (summarized in Chapter 21). These Regulations require risk assessments to be undertaken and engineering controls to be in place. They also recognize that lead can be transferred to an unborn child through the placenta and, therefore, offers additional protection to women of reproductive capacity. Medical surveillance, in the form of a blood test, of all employees who come into contact with lead operations, is required by the Regulations. Such tests should take place at least once a year. Lead is used in construction by roofers as a roofing and guttering material and it is also used (to a lesser extent these days) by plumbers.

Silica is the main component of most rocks and is a crystalline substance made of silicon and oxygen. It occurs in quartz (found in granite), sand and flint, which are present in a wide variety of construction materials. Harm is caused by the inhalation of silica dust, which can lead to silicosis (acute and chronic), fibrosis and pneumoconiosis. Activities which can expose workers and members of the public to silica dust include:

➤ cutting building blocks and other stone masonry work
➤ cutting and/or drilling paving slabs and concrete paths
➤ demolition work
➤ sand blasting of buildings
➤ tunnelling.

In general, the use of power tools to cut or dress stone and other silica-containing materials will lead to very high exposure levels while the work is occurring. In most cases, exposure levels are in excess of WELs by factors greater than 2 and in some cases as high as 12.

The dust which causes the most harm is respirable dust which becomes trapped in the alveoli. This type of dust is sharp and very hard and, probably, causes wounding and scarring of lung tissue. The inhalation of very fine silica dust can lead to the development of silicosis. As silicosis develops, breathing becomes more and more difficult and eventually as it reaches its advanced stage, lung and heart failure occur. It has also been noted that silicosis can result in the development of tuberculosis as a further complication. Hard rock miners,

quarrymen, and stone and pottery workers are most at risk. Health surveillance is recommended for workers in these occupations at initial employment and at subsequent regular intervals. Prevention is best achieved by the use of good dust extraction systems and respiratory PPE. HSE have produced a detailed information sheet on silica – CIS No 36 (Rev 1).

Cement dust and wet cement is an important construction material and is also a hazardous substance. Contact with wet cement can cause serious burns or ulcers which will take several months to heal and may need a skin graft. Dermatitis, both irritant and allergic, can be caused by skin contact with either wet cement or cement powder. Allergic dermatitis is caused by an allergic reaction to hexavalent chromium (chromate) which is present in cement. Cement powder can also cause inflammation and irritation of the eye, irritation of the nose and throat and, possibly, chronic lung problems. Research has shown that between 5% and 10% of construction workers are probably allergic to cement. And plasterers, concreters and bricklayers or masons are particularly at risk.

Manual handling of wet cement or cement bags can lead to musculoskeletal health problems and cement bags weighing more than 25 kg should not be carried by a single worker. PPE in the form of gloves, overalls with long sleeves and full length trousers and waterproof boots must be worn on all occasions. If the atmosphere is dusty, goggles and RPE must be worn. An important factor in the possibility of dermatitis occurring is the sensitivity of the worker to the chromate in the cement and the existing condition of the skin including cuts and abrasions. Finally, adequate welfare facilities are essential so that workers can wash their hands at the end of the job and before eating, drinking or using the toilet. If cement is left on the skin for long periods without being washed off, the risk of an allergic reaction to hexavalent chromium will increase.

An amendment to the COSHH Regulations prohibits the supply of cement which has a concentration of more than two parts per million of chromium VI. This measure is designed to prevent allergic contact dermatitis when wet cement comes into contact with the skin. However, since the strong alkalinity of cement will remain, there is still the potential for skin burns.

The Approved Code of Practice (ACoP) gives useful advice on possible control measures by offering two routes for employers to comply with the amended COSHH Regulations. They can either use the generic advice given in COSHH Essentials HSG193 or design a solution themselves with the help of competent advice. In any event, the controls should be proportionate to the health risk. The Regulations stress the need to provide adequate washing, changing, eating and drinking facilities. Details of these welfare facilities are given in Chapter 10.

Wood dust can be hazardous, particularly when it is hard wood dust, which is known, in rare cases, to lead

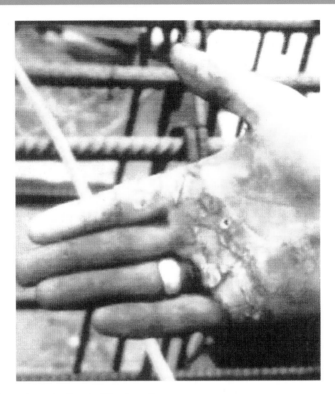

Figure 16.7 Dermatitis on the palmar surface of the hand.

to nasal cancer. Composite boards, such as medium-density fibreboard (MDF), are hazardous due to the resin bonding material used which also can be carcinogenic. There are three types of wood-based boards available: laminated board, particle board and fibreboard. The resins used to bond the fibreboard together contain formaldehyde (usually urea formaldehyde). It is generally recognized that formaldehyde is 'probably carcinogenic to humans' and is subject to a WEL. At low exposure levels, it can cause irritation to the eyes, nose and throat and can lead to dermatitis, asthma and rhinitis. The main problems are most likely to occur when the MDF is being machined and dust is produced. A suitable risk assessment should be made and gloves and appropriate masks should always be worn when machining MDF. However, it is important to stress that safer materials are available which do not contain formaldehyde and these should be considered for use in the first instance.

Wood dust is produced whenever wood materials are machined, in particular sawed, sanded, bagged as dust from dust extraction units or during cleaning operations especially if compressed air is used. The main hazards associated with all wood dusts are skin disorders, nasal problems such as rhinitis, and asthma. There is also a hazard from fire and explosion. A COSHH assessment is essential to show whether the particular wood dust is hazardous. When the wood dust is created inside a woodworking shop a well-designed extraction system is essential. PPE in the form of gloves, suitable RPE, overalls and eye protection may also be necessary as a result of the assessment. Other hazards associated with woodworking include:

➤ wood paints, varnishes, stains and preservatives
➤ stripping chemicals
➤ various solvents and
➤ adhesives.

Finally, good washing and welfare facilities are also essential.

Tetanus is a serious, sometimes fatal, disease caused by a bacterium that lives in the soil. It usually enters the human body through a wound from an infected object, such as a nail, wood splinter or thorn. On entering the wound, it produces a powerful toxin which attacks the nerves that supply muscle tissue. It is commonly known as lockjaw because after an incubation period of approximately a week, stiffness around the jaw area occurs. Later the disease spreads to other muscles including the breathing system, and this can be fatal. The disease has been well controlled with anti-tetanus immunization and it is important that all construction workers are so immunized. Booster shots should be obtained every few years. Any flesh wound should be thoroughly cleaned immediately an antiseptic cream applied and the wound covered.

Leptospirosis and Weil's disease is caused by a bacterium found in the urine of rats. In humans, the kidneys and liver are attacked causing high temperatures and headaches followed by jaundice and can, in up to 20% of cases, being fatal. It enters the body either through the skin or by ingestion. The most common source is contaminated water in a river sewer or ditch, and workers, such as canal or sewer workers, are most at risk. Leptospirosis is always a risk where rats are present, particularly if the associated environment is damp. Good impervious protective clothing, particularly wellington boots, is essential in these situations and the covering of any skin wounds. For workers who are frequently in high-risk environments (sewer workers), immunization with a vaccine may be the best protection. It is important for construction workers to be aware of this hazard when working beside or over rivers, canals or streams.

Weil's disease is, strictly, a severe form of leptospirosis. The symptoms of leptospirosis are similar to influenza but those for Weil's disease are anaemia, nose bleeds and jaundice. While the most common source of infection is from the urine of rats, it has been found in other animals, such as cattle, so farm and veterinary workers may also be at risk.

Legionella is an airborne bacterium and is found in a variety of water sources. It produces a form of pneumonia caused by the bacteria penetrating to the alveoli in the lungs. This disease is known as Legionnaires' disease, named after the first documented outbreak at a State Convention of the American Legion held at Pennsylvania in 1976. During this outbreak, 200 men were affected, of whom 29 died. That outbreak and many subsequent ones were attributed to air-conditioning systems. It is most common in those over 45 years of age and rare in the under-20s and men seem more susceptible than women. The *Legionella* bacterium cannot survive at temperatures above 60°C but grows between 20°C and 45°C, being most virulent at 37°C. It also requires food in the form of algae and other bacteria. Control of the bacteria involves the avoidance of water temperatures between 20°C and 45°C, avoidance of water stagnation and the build up of algae and sediments and the use of suitable water treatment chemicals. This work is often done by a specialist contractor.

The most common systems at risk from the bacterium are:

➤ water systems incorporating a cooling tower
➤ water systems incorporating an evaporative condenser
➤ hot and cold water systems and other plant where the water temperature may exceed 20°C.

An ACoP (*Legionnaires' disease – The control of legionella bacteria in water systems – L8*) was produced by the HSE in 2000. Where plant at risk of the development of *Legionella* exists, the following is required:

➤ a written 'suitable and sufficient' risk assessment
➤ the preparation and implementation of a written control scheme involving the treatment, cleaning and maintenance of the system

➤ appointment of a named person with responsibility for the management of the control scheme
➤ the monitoring of the system by a competent person
➤ record keeping and the review of procedures developed within the control scheme.

The code of practice also covers the design and construction of hot and cold water systems and cleaning and disinfection guidance. There have been several cases of members of the public becoming infected from a contaminated cooling tower situated on the roof of a building. It is required that all cooling towers are registered with the local authority. People are more susceptible to the disease if they are older or weakened by some other illness. It is, therefore, important that residential and nursing homes and hospitals are particularly vigilant. The most common source of isolated outbreaks of *Legionella* is showerheads, particularly when they remain unused for a period of time. Showerheads should be cleaned and descaled at least every 3 months.

Hepatitis is a disease of the liver and can cause high temperatures, nausea and jaundice. It can be caused by hazardous substances (some organic solvents) or by a virus. The virus can be transmitted from infected faeces (hepatitis A) or by infected blood (hepatitis B and C). The normal precautions include good personal hygiene particularly when handling food and in the use of blood products. Hospital workers who come into contact with blood products are at risk of hepatitis as are drug addicts who share needles. It is also important that workers at risk regularly wash their hands and wear protective disposable gloves.

16.7 Requirements of the COSHH Regulations

When hazardous substances are considered for use (or in use) at a place of work, the COSHH Regulations impose certain duties on employers and requires employees to cooperate with the employer by following any measures taken to fulfil those duties. The principal requirements are as follows:

1. Employers must undertake a suitable and sufficient assessment of the health risks created by work which is liable to expose their employees to substances hazardous to health and of the steps that need to be taken by employers to meet the requirements of these Regulations (Regulation 6).
2. Employers must prevent, or where this is not reasonably practicable, adequately control, the exposure of their employees to substances hazardous to health. Workplace Exposure Limits (WELs), which should not

be exceeded, are specified by the Health and Safety Executive (HSE) for certain substances. As far as inhalation is concerned, control should be achieved by means other than PPE. If, however, respiratory equipment, for example, is used, then the equipment must conform to HSE standards (Regulation 7).

The control of exposure can only be treated as adequate if the principles of good practice (discussed under 16.8.2 and 16.9) are applied (Schedule 2A).

3. Employers and employees must make proper use of any control measures provided (Regulation 8).
4. Employers must maintain any installed control measures on a regular basis and review systems of work and keep suitable records (Regulation 9).
5. Monitoring must be undertaken of any employee exposed to items listed in schedule 5 of the Regulations or in any other case where monitoring is required for the maintenance of adequate control or the protection of employees. Records of this monitoring must be kept for at least 5 years, or 40 years where employees can be identified (Regulation 10).
6. Health surveillance must be provided to any employees who are exposed to any substances listed in schedule 6. Records of such surveillance must be kept for at least 40 years after the last entry (Regulation 11).
7. Employees who may be exposed to substances hazardous to their health must be given information, instruction and training sufficient for them to know the health risks created by the exposure and the precautions which should be taken (Regulation 12).

16.8 Details of a COSHH assessment

Not all hazardous substances are covered by the COSHH Regulations. If there is no warning symbol on the substance container or it is a biological agent which is not directly used in the workplace (such as an influenza virus), then no COSHH assessment is required. The COSHH Regulations do not apply to those hazardous substances which are subject to their own individual regulations (asbestos, lead or radioactive substances). The COSHH Regulations do apply to the following substances:

➤ substances having WELs as listed in the HSE publication EH40 (Workplace Exposure Limits)
➤ substances or combinations of substances listed in the Chemicals (Hazard, Information and Packaging for Supply) Regulations, better known as the CHIP Regulations
➤ biological agents connected with the workplace
➤ substantial quantities of airborne dust (more than $10 \, mg/m^3$ of total inhalable dust or $4 \, mg/m^3$ of respirable dust, both 8 hour time-weighted average, when there is no indication of a lower value)

➤ any substance creating a comparable hazard but which for technical reasons may not be covered by CHIP.

16.8.1 Assessment requirements

A COSHH assessment is very similar to a risk assessment but is applied specifically to hazardous substances. There are six stages to a COSHH assessment:

1. Identify the hazardous substances present in the workplace and those who could be affected by them.
2. Gather information about the hazardous substances.
3. Evaluate the risks to health.
4. Decide on the control measures, if any, required including information, instruction and training.
5. Record the assessment.
6. Review the assessment.

It is important that the assessment is conducted by somebody who is competent to undertake it. Such competence will require some training, the extent of which will depend on the complexities of the workplace. For large organizations with many high-risk operations, a team of competent assessors will be needed.

If the assessment is simple and easily repeated, a written record is not necessary. In other cases, a concise and dated record of the assessment together with recommended control measures should be made available to all those likely to be affected by the hazardous substances. The assessment should be reviewed on a regular basis particularly when there are changes in work process or substances or when adverse ill-health is reported.

16.8.2 Workplace exposure limits

One of the main purposes of a COSHH assessment is to control adequately the exposure of employees and others to hazardous substances. This means that such substances should be reduced to levels which do not pose a health threat to those exposed to them day after day at work. Under the amendments to the COSHH Regulations 2002, the Health and Safety Commission (HSC) has assigned WELs to a large number of hazardous substances and publishes an annual update in a publication called 'Workplace Exposure Limits' EH40.

Before the introduction of WELs, there were two types of exposure limit published – the maximum exposure limit (MEL) and the occupational exposure standard (OES).

The COSHH (Amendment) Regulations replaced the OES/MEL system with a single WEL. This removed the concern of HSC that the OES was seen as a 'safe' limit rather than a 'likely safe' limit. Hence, the WEL must **not** be exceeded.

Hazardous substances which have been assigned a WEL fall into two groups:

1. Substances which are carcinogenic or mutagenic (having a risk phase R45, R46 or R49) or could cause occupational asthma (having a risk phase R42, or R42/43 or listed in section C of the HSE publication 'Asthmagen? Critical assessment for the agents implicated in occupational asthma' as updated from time to time) or are listed in Schedule 1 of the COSHH Regulations. These are substances which were assigned an MEL before 2005. The level of exposure to these substances should be reduced as far as is reasonably practicable.
2. All other hazardous substances which have been assigned a WEL. Exposure to these substances by inhalation must be controlled adequately to ensure that the WEL is not exceeded. These substances were previously assigned an OES before 2005. For these substances, employers should achieve adequate control of exposure by inhalation by applying the principles of good practice outlined in the ACoP and listed below.

The principles of good practice are to:

➤ minimize the emission, release and spread of substances hazardous to health
➤ take into account all relevant routes of exposure – inhalation, skin and ingestion
➤ control exposure by measures that are proportionate to the health risk
➤ choose the most effective and reliable control options
➤ provide personal protective equipment when exposure cannot be controlled by other means
➤ review regularly all elements of the control measures for their continuing effectiveness
➤ inform and train all employees on the hazards and risks from substances with which they work, and the use of control measures developed to minimize the risks
➤ ensure that the control measures do not increase the overall risk to health and safety.

These principles and their implications is discussed later in this chapter.

The WELs are subject to time-weighted averaging. There are two such time-weighted averages (TWA) – the long-term exposure limit (LTEL) or 8 hour reference period; and the short-term exposure limit (STEL) or 15 minute reference period. The 8 hour TWA is the maximum

exposure allowed over an 8 hour period so that if the exposure period was less than 8 hours the WEL is increased accordingly with the proviso that exposure above the LTEL value continues for no longer than 1 hour.

For example, if a person was exposed to a hazardous substance with a WEL of 100 mg/m^3 (8 hour TWA) for 4 hours, no action would be required until an exposure level of 200 mg/m^3 was reached. (Exposure at levels between 100 and 200 mg/m^3 should be restricted to 1 hour.)

If, however, the substance has an STEL of 150 mg/m^3, then action would be required when the exposure level rose above 150 mg/m^3 for more than 15 minutes.

The STEL always takes precedence over the LTEL. When a STEL is not given, it should be assumed that it is three times the LTEL value.

Table 16.1 Examples of Workplace Exposure Limits (WEL)

Group 1 WELs	LTEL (8 hour TWA)	STEL (15 minutes)
All isocyanates	0.02 mg/m^3	0.07 mg/m^3
Styrene	430 mg/m^3	1,080 mg/m^3
Group 2 WELs		
Ammonia	18 mg/m^3	25 mg/m^3
Toluene	191 mg/m^3	574 mg/m^3

The publication EH40 is a valuable document for the health and safety professional since it contains much additional advice on hazardous substances for use during the assessment of health risks, particularly where new medical information has been made public.

It is important to stress that if a WEL from Group 1 is exceeded, the process and use of the substance should cease immediately. In the longer term, the process, control and monitoring measures should be reviewed and health surveillance considered.

The over-riding requirement for any hazardous substance, which has a WEL from Group 1, is to reduce exposure as low as is reasonably practicable.

Finally, there are certain limitations on the use of the published workplace limits:

➤ They are specifically quoted for an 8 hour period (with an additional STEL for many hazardous substances). Adjustments must be made when exposure occurs over a continuous period longer than 8 hours.

➤ They can only be used for exposure in a workplace and not to evaluate or control non-occupational exposure (e.g. to evaluate exposure levels in a neighbourhood close to the workplace, such as a playground).

➤ WELs are only approved where the atmospheric pressure varies from 900 to 1,100 millibars. This could exclude their use in mining and tunnelling operations.

➤ They should also not be used when there is a rapid build-up of a hazardous substance due to a serious accident or other emergency. Emergency arrangements should cover these eventualities.

HSC has published a revised ACOP and EH40 to include these changes. More information on the latest COSHH Regulations is given in Chapter 21.

The fact that a substance has not been allocated a WEL does not mean that it is safe. The exposure to these substances should be controlled to a level to which nearly all of the working population could be exposed all the time without any adverse effects to their health. Detailed guidance and references are given for such substances in the HSE ACOP and guidance to the COSHH Regulations (L5).

16.8.3 *Sources of information*

There are other important sources of information available for a COSHH assessment in addition to the HSE Guidance Note EH40.

Product labels include details of the hazards associated with the substances contained in the product and any precautions recommended. They may also bear one or more of the CHIP hazard classification symbols.

Product safety data sheets are another very useful source of information for hazard identification and associated advice. Manufacturers of hazardous substances are obliged to supply such sheets to users giving details of the name, chemical composition and properties of the substance. Information on the nature of the health hazards and any relevant exposure standard (WEL) should also be given together with recommended exposure control measures and personal protective equipment. The sheets contain useful additional information on first aid and fire fighting measures and handling, storage, transport and disposal information. The data sheets should be stored in a readily accessible and known place for use in the event of an emergency, such as an accidental release (See Fig. 16.8).

Other sources of information include trade association publications, industrial codes of practice and specialist reference manuals.

16.8.4 *Survey techniques for health risks*

An essential part of the COSHH assessment is the measurement of the quantity of the hazardous substance in the atmosphere surrounding the workplace. This is known as air sampling. There are four common types of air sampling techniques used for the measurement of air quality:

1. **Stain tube detectors** use direct reading glass indicator tubes filled with chemical crystals which change colour when a particular hazardous substance passes through them. The method of operation is very similar to the breathalyser used by the police to check alcohol levels in motorists. The glass tube is opened at each end and fitted into a pumping device (either hand or electrically operated). A specific quantity of contaminated air, containing the hazardous substance, is drawn through the tube and the crystals in the tube change colour in the direction of the air flow. The tube is calibrated such that the extent of the colour change along the tube indicates the concentration of the hazardous substance within the air sample.

This method can only be effective if there are no leakages within the instrument and the correct volume of sampled air is used. The instrument should be held within 30 cm of the nose of the person whose atmosphere is being tested. A large range of different tubes is available. This technique of sampling is known as **grab** or **spot** sampling since it is taken at one point. (See Fig. 16.9)

The advantages of the technique are that it is quick, relatively simple to use and inexpensive. There are, however, several disadvantages:

➤ the instrument cannot be used to measure concentrations of dust or fumes

➤ the accuracy of the reading is approximately $\pm 25\%$

➤ it will yield false reading if other contaminants present react with the crystals

➤ it can only give an instantaneous reading not an average reading over the working period (TWA)

➤ the tubes are very fragile with a limited shelf life.

2. **Passive sampling** is measured over a full working period by the worker wearing a badge containing absorbent material. The material will absorb the contaminant gas and at the end of the measuring period, the sample is sent to a laboratory for analysis. The advantages of this method over the stain tube are that there is less possibility of instrument errors and it gives a time-weighted average (TWA) reading.

3. **Sampling pumps and heads** can be used to measure gases and dusts. The worker, whose breathing zone is being monitored, wears a collection head as a badge and a battery-operated pump on his back at waist level. The pump draws air continuously through a filter, fitted in the head, which will either absorb the contaminant gas or trap hazardous dust particles. After the designated testing period, it is sent to a laboratory for analysis. This system is more accurate than stain tubes and gives a TWA result but can be

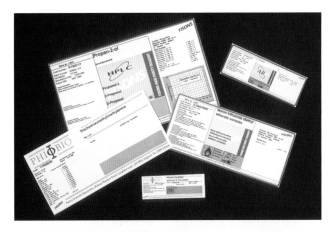

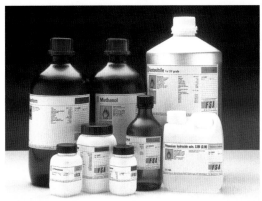

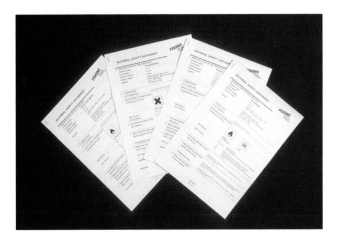

Figure 16.8 Typical product labels and data sheets.

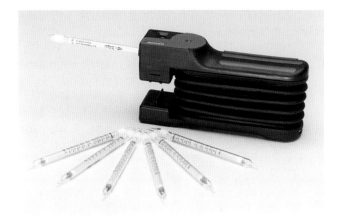

Figure 16.9 Hand pump and stain detector.

Other common monitoring instruments include **vane anemometers**, used for measuring air flow speeds and **hygrometers**, which are used for measuring air humidity.

Qualitative monitoring techniques include smoke tubes and the dust observation lamp. **Smoke tubes** generate a white smoke which may be used to indicate the direction of flow of air – this is particularly useful when the air speed is very low or when testing the effectiveness of ventilation ducting. **A dust observation lamp** enables dust particles which are normally invisible to the human eye to be observed in the light beam. This dust is usually in the respirable range and, although the lamp does not enable any measurements of the dust to be made, it will illustrate the operation of a ventilation system and the presence of such dust.

Regulation 10 requires routine sampling or monitoring of exposure where there could be serious health effects if the controls failed, WELs might be exceeded or the control measures may not be working properly. Air monitoring should also be undertaken for any hazardous substances listed in Schedule 5 of the COSHH Regulations. Records, of this monitoring should be kept for 5 years unless an employee is identifiable in the records, in which case they should be kept for 40 years.

uncomfortable to wear over long periods. Such equipment can only be used by trained personnel.

4. **Direct reading instruments** are available in the form of sophisticated analysers which can only be used by trained and experienced operatives. Infra-red gas analysers are the most common but other types of analyser are also available. They are very accurate and give continuous or TWA readings. They tend to be very expensive and are normally hired or used by specialist consultants.

16.9 The control measures required under the COSHH Regulations

16.9.1 *The principles of good practice for the control of exposure to substances hazardous to health*

The objective of the COSHH Regulations is to prevent ill-health due to exposure to hazardous substances.

Employers are expected to develop suitable and sufficient control measures by:

1. identifying hazards and potentially significant risks
2. taking action to reduce and control risks and
3. keeping control measures under regular review.

In order to assist employers with these duties, the HSE have produced the following principles of good practice:

(a) Design and operate processes and activities to minimize the emission, release and spread of substances hazardous to health.
(b) Take into account all relevant routes of exposure – inhalation, skin absorption and ingestion – when developing control measures.
(c) Control exposure by measures that are proportionate to the health risk.
(d) Choose the most effective and reliable control options which minimize the escape and spread of substances hazardous to health.
(e) When adequate control of exposure cannot be achieved by other means, provide, in combination with other control measures, suitable PPE.
(f) Check and review regularly all elements of control measures for their continuing effectiveness.
(g) Inform and train all employees on the hazards and risks from the substances with which they work and the use of control measures developed to minimize the risks.
(h) Ensure that the introduction of control measures does not increase the overall risk to health and safety.

All of these principles are embodied in the following sections on COSHH control measures.

16.9.2 *Hierarchy of control measures*

The COSHH Regulations require the prevention or adequate control of exposure by measures other than PPE, so far as is reasonably practicable, taking into account the degree of exposure and current knowledge of the health risks and associated technical remedies. The hierarchy of control measures are as follows:

➤ elimination
➤ substitution
➤ provision of engineering controls
➤ provision of supervisory (people) controls
➤ provision of PPE.

Examples where engineering controls are not reasonably practicable include emergency and maintenance work,

short-term and infrequent exposure and where such controls are not technically feasible.

Measures for preventing or controlling exposure to hazardous substances include one or a combination of the following:

➤ elimination
➤ substitution
➤ total or partial enclosure of the process
➤ local exhaust ventilation (LEV)
➤ dilute or general ventilation
➤ reduction of the number of employees exposed to a strict minimum
➤ reduced time exposure
➤ housekeeping
➤ training
➤ PPE
➤ welfare (including first aid)
➤ medical records
➤ health surveillance.

16.9.3 *Preventative control measures*

Prevention is the safest and most effective of the control measures and is achieved either by changing the process completely or by substituting for a less hazardous substance (the change from oil-based to water-based paints is an example of this). It may be possible to use a substance in a safer form, such as a brush paint rather than a spray.

The EU has introduced chemical safety regime REACH (Registration, Evaluation and Authorisation of Chemicals) Regulations which restrict the use of high-risk substances or substances of very high concern and require that safer substitute must be used. Manufacturers and importers of chemicals are responsible for understanding and managing the risks associated with their products.

16.9.4 *Engineering controls*

The simplest and most efficient engineering control is the segregation of people from the process; a chemical fume cupboard is an example of this as is the handling of toxic substances in a glove box. Modification of the process is another effective control to reduce human contact with hazardous substances.

More common methods, however, involve the use of forced ventilation – LEV and dilute ventilation.

Local exhaust ventilation
LEV removes the hazardous gas, vapour or fume at its source before it can contaminate the surrounding atmosphere and harm people working in the vicinity.

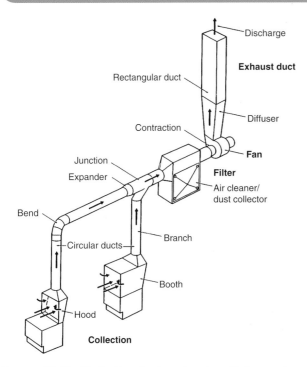

Figure 16.10 Typical local exhaust and ventilation system.

Such systems are commonly used for the extraction of welding fumes and dust from woodworking machines. All exhaust ventilation systems have the following basic components:

1. **A collection hood and intake** – sometimes this is a nozzle-shaped point which is nearest to the work piece, while at other times it is simply a hood placed over the workstation. The speed of the air entering the intake nozzle is important: if it is too low then hazardous fume may not be removed (air speeds of up to 1 m/s are normally required).

2. **Ventilation ducting** – this normally acts as a conduit for the contaminated air and transports it to a filter and settling section. It is very important that this section is inspected regularly and any dust deposits are removed. It has been known for ventilation ducting attached to a workshop ceiling to collapse under the added weight of metal dust deposits. It has also been known for them to catch fire.

3. **Filter or other air cleaning device** – normally located between the hood and the fan, the filter removes the contaminant from the air stream. The filter requires regular attention to remove contaminant and to ensure that it continues to work effectively.

4. **Fan** – this moves the air through the system. It is crucial that the correct type and size of fan is fitted to a given system and it should only be selected by a competent person. It should also be positioned so

that it can easily be maintained but does not create a noise hazard to nearby workers.

5. **Exhaust duct** – this exhausts the air to the outside of the building. It should be checked regularly to ensure that the correct volume of air is leaving the system and that there are no leakages. The exhaust duct should also be checked to ensure that there is no corrosion due to adverse weather conditions.

An Introduction to Local Exhaust Ventilation, HSG37, HSE Books, is a very useful document on ventilation systems.

The COSHH Regulations require that such ventilation systems must be inspected at least every 14 months by a competent person to ensure that they are still operating effectively.

The effectiveness of a ventilation system will be reduced by damaged ducting, blocked or defective filters and poor fan performance. More common problems include the unauthorized extension of the system, poor initial design, poor maintenance, incorrect adjustments and a lack of inspection or testing.

Routine maintenance should include repair of any damaged ducting, checking filters, examination of the fan blades to ensure that there has been no dust accumulation, tightening all drive belts and a general lubrication of moving parts.

The local exhaust ventilation system will have an effect on the outside environment in the form of noise and odour. Both these problems can be reduced by regular routine maintenance of the fan and filter. The waste material from the filter may be hazardous and require the special disposal arrangements described later in the Chapter (Section 16.15.3).

Dilution (or general) ventilation

Dilution (or general) ventilation uses either natural ventilation (doors and windows) or fan-assisted forced ventilation system to ventilate the whole working room by inducing a flow of clean air using extraction fans fitted into the walls and the roof, sometimes assisted by inlet fans. It operates by either removing the contaminant or reducing its concentration to an acceptable level. It is used either when airborne contaminants are of low toxicity, low concentration and low vapour density or contamination occurs uniformly across the workroom. Paint spraying operations often use this form of ventilation as does the glass reinforced plastics (GRP) boat-building industry – these being instances where there are not discrete points of release of the hazardous substances. It is also widely used in kitchens and bathrooms. It is not suitable for dust extraction and where it is reasonably practicable to reduce levels by other means.

There are limitations to the use of dilution ventilation. Certain areas of the work room (e.g. corners and beside cupboards) will not receive the ventilated air and a build-up of hazardous substances occurs. These areas are known as 'dead areas'. The flow patterns are also significantly affected by doors and windows being opened or the rearrangement of furniture or equipment.

16.9.5 *Supervisory or people controls*

Many of the supervisory controls required for COSHH purposes are part of a good safety culture and were discussed in detail in Chapter 4. These include items such as systems of work, arrangements and procedures, effective communications and training. Additional controls when hazardous substances are involved are as follows:

➤ Reduced time exposure – thus ensuring that workers have breaks in their exposure periods. The use of this method of control depends very much on the nature of the hazardous substance and its STEL.
➤ Reduced number of workers exposed – only persons essential to the process should be allowed in the vicinity of the hazardous substance. Walkways and other traffic routes should avoid any area where hazardous substances are in use.
➤ Eating, drinking and smoking must be prohibited in areas where hazardous substances are in use.
➤ Any special rules, such as the use of personal protective equipment, must be strictly enforced.

16.9.6 *Personal protective equipment*

Personal protective equipment is to be used as a control measure only as a last resort. It does not eliminate the hazard and will present the wearer with the maximum health risk if the equipment fails. Successful use of personal protective equipment relies on good user training, the availability of the correct equipment at all times and good supervision and enforcement.

The 'last resort' rule applies in particular to respiratory protective equipment (RPE) within the context of hazardous substances. There are some working conditions when RPE may be necessary:

➤ during maintenance operations
➤ as a result of a new assessment, perhaps following the introduction of a new substance
➤ during emergency situations, such as fire or plant breakdown
➤ where alternatives are not technically feasible.

All personal protective equipment, except respiratory protection equipment (which is covered by specific Regulations, such as COSHH, lead, etc.) is controlled by its own set of regulations – the Personal Protective Equipment at Work Regulations. A detailed summary of these Regulations is given in Chapter 21 and needs to be studied in detail with this section. The principal requirements of these regulations are as follows:

1. personal protective equipment which is suitable for the wearer and the task

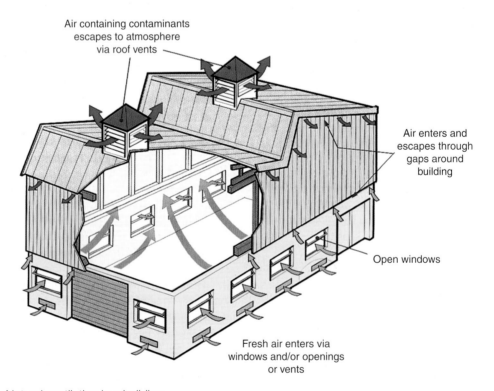

Figure 16.11 Natural ventilation in a building.

2. compatibility and effectiveness of the use of multiple personal protective equipment
3. a risk assessment to determine the need and suitability of proposed personal protective equipment
4. a suitable maintenance programme for the personal protective equipment
5. suitable accommodation for the storage of the personal protective equipment when not in use
6. information, instruction and training for the user of personal protective equipment
7. the supervision of the use of personal protective equipment by employees and a reporting system for defects.

Types of personal protective equipment

There are several types of personal protective equipment, such as footwear, hearing protectors and hard hats, which are not primarily concerned with protection from hazardous substances; those which are used for such protection, include:

➤ respiratory protection
➤ hand and skin protection
➤ eye protection
➤ protective clothing.

For all types of personal protective equipment, there are some basic standards that should be reached. The personal protective equipment should fit well, be comfortable to wear and not interfere either with other equipment being worn or present the user with additional hazards (e.g. impaired vision due to scratched eye goggles). Training in the use of particular personal protective equipment is essential, so that it is not only used correctly, but the user knows when to change an air filter or to change a type of glove. Supervision is essential with disciplinary procedures invoked for non-compliance with personal protective equipment rules.

It is also essential that everyone who enters the proscribed area, particularly senior managers, wear the specified personal protective equipment.

Respiratory protection equipment
Respiratory protection equipment can be subdivided into two categories – respirators (or face masks), which filter and clean the air; and breathing apparatus, which supplies breathable air.

Respirators should not be worn in air which is dangerous to health, including oxygen deficient atmospheres. They are available in several different forms but the common ones are:

➤ **filtering half mask**, often called disposable respirator – made of the filtering material. It covers the nose and mouth and removes respirable-size dust particles. It is normally replaced after 8–10 hours of use. It offers protection against some vapours and gases

➤ **half mask respirator** – made of rubber or plastic and covers the nose and mouth air is drawn through a replaceable filter cartridge. This apparatus can be used for vapours, gases or dusts but it is very important that the correct filter is used (a dust filter will not filter vapours)
➤ **full face mask respirator** – similar to the half mask type but covers the eyes with a visor
➤ **powered respirator** – a battery-operated fan delivers air through a filter to the face mask, hood, helmet or visor.

Breathing apparatus is used in one of the following three forms:

1. **self-contained breathing apparatus** – where air is supplied from compressed air in a cylinder and forms a completely sealed system
2. **fresh air hose apparatus** – fresh air is delivered through a hose to a sealed face mask from an uncontaminated source. The air may be delivered by the wearer, by natural breathing or mechanically by a fan
3. **compressed air line apparatus** – air is delivered through a hose from a compressed air line. This can be either continuous flow or on demand. The air must be properly filtered to remove oil, excess water and other contaminants and the air pressure must be reduced. Special compressors are normally used.

The selection of appropriate respiratory protective equipment and correct filters for particular hazardous substances is best done by a competent specialist person.

There are several important technical standards which must be considered during the selection process. Respiratory protective equipment must be either CE marked or HSE approved (HSE approval ceased on 30th June 1995 but such approved equipment may still be used). Other standards include the minimum protection required (MPR) and the assigned protection factor (APF). The CE mark does not indicate that the equipment is suitable for a particular hazard. The following information will be needed before selection can be made:

➤ details of the hazardous substance in particular whether it is a gas, vapour or dust or a combination of all three
➤ presence of a beard or other facial hair which could prevent a good leak-free fit (a simple test to see whether the fit is tight or not is to close off the air supply, breathe in and hold the breath. The respirator should collapse onto the face. It should then be possible to check to see if there is a leak)
➤ the size and shape of the face of the wearer and physical fitness
➤ compatibility with other personal protective equipment, such as ear defenders
➤ the nature of the work and agility and mobility required.

Filtering half masks to protect against particles

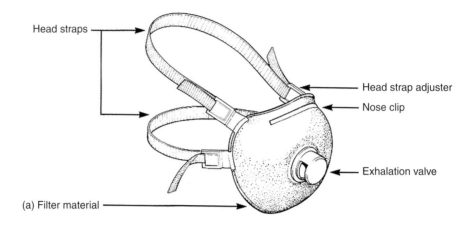

Head straps

Head strap adjuster

Nose clip

Exhalation valve

(a) Filter material

Half masks reusable with filters

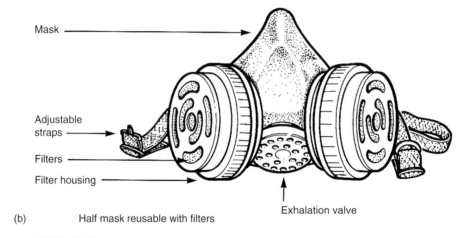

Mask

Adjustable straps

Filters

Filter housing

Exhalation valve

(b)　　Half mask reusable with filters

Figure 16.12　Types of RPE. (a) Filtering half mask; (b) half mask reusable with filters

Filters and masks should be replaced at the intervals recommended by the supplier or when taste or smell is detected by the wearer.

All respiratory protective equipment should be examined at least once a month except for disposable respirators. A record of the inspection should be kept for at least 5 years. There should be a routine cleaning system in place and proper storage arrangements.

Respiratory protective equipment – a practical guide for users, HSG53, HSE Books, contains comprehensive advice and guidance on respiratory protective equipment selection, use, storage, maintenance and training and should be consulted for more information.

Hand and skin protection
Hand and skin protection is mainly provided by gloves (arm shields are also available). A wide range of safety

gloves is available for protection from chemicals, sharp objects, rough working and temperature extremes. Many health and safety catalogues give helpful guidance for the selection of gloves. For protection from chemicals, including paints and solvents, impervious gloves are recommended. These may be made of PVC, nitrile or neoprene. For sharp objects, such as trimming knives, a Kevlar based glove is the most effective. Gloves should be regularly inspected for tears or holes since this will obviously allow skin contact to take place.

Another effective form of skin protection is the use of barrier creams and these come in two forms – pre-work and after-work. Pre-work creams are designed to provide a barrier between the hazardous substance and the skin. After-work creams are general purpose moisteners which replace the natural skin oils removed either by solvents or by washing.

Compressed-air line BA with full-face mask fitted with demand valve

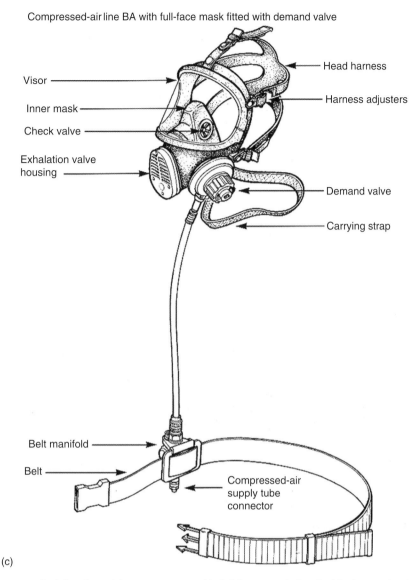

Visor

Inner mask

Check valve

Exhalation valve housing

Head harness

Harness adjusters

Demand valve

Carrying strap

Belt manifold

Belt

Compressed-air supply tube connector

(c)

Figure 16.12 (c) compressed-air line breathing apparatus with full face mask fitted with demand valve.

Eye protection

Eye protection comes in three forms – spectacles (safety glasses), goggles and face visors. Eyes may be damaged by chemical and solvent splashes or vapours, flying particles, molten metals or plastics, non-ionizing radiation (arc welding and lasers) and dust. Spectacles are suitable for low risk hazards (low speed particles such as machine swarf). Some protection against scratching of the lenses can be provided but this is the common reason for replacement. Prescription lenses are also available for people who normally wear spectacles.

Goggles are best to protect the eyes from dust or solvent vapours because they fit tightly around the eyes. Visors offer protection to the face as well as the eyes and do not steam up so readily in hot and humid

Figure 16.13 Variety of eye protection goggles.

environments. For protection against very bright lights, special light filtering lenses are used (e.g. in arc welding). Maintenance and regular cleaning are essential for the efficient operation of eye protection.

When selecting eye protection, several factors need to be considered. These include the nature of the hazard (the severity of the hazard and its associated risk will determine the quality of protection required), comfort and user acceptability, compatibility with other PPE, training and maintenance requirements and costs.

Protective clothing

Protective clothing includes aprons, boots and headgear (hard hats and bump caps). Aprons are normally made of PVC and protect against spillages but can become uncomfortable to wear in hot environments. Other lighter fabrics are available for use in these circumstances. Safety footwear protects against falling objects, collision with hard or sharp objects, hot or molten materials, slippery surfaces and chemical spills. It has metal toe-caps and comes in the form of shoes, ankle boots or knee-length boots and is made of a variety of materials dependent on the particular hazard (e.g. thermally insulated against cold environments). It must be used with care near live unprotected electricity. Specialist advice is needed for use with flammable liquids.

It is important to note that appropriate personal protective equipment should be made available to visitors and other members of the public when visiting workplaces where hazardous substances are being used. It is also important to stress that managers and supervisors must lead by example, particularly if there is a legal requirement to wear particular personal protective equipment. Refusals by employees to wear mandatory personal protective equipment must lead to some form of disciplinary action.

16.10 Health surveillance and personal hygiene

Health surveillance enables the identification of those employees most at risk from occupational ill-health. It should not be confused with health monitoring procedures such as pre-employment health checks or drugs and alcohol testing but it covers a wide range of situations from a responsible person looking for skin damage on hands to medical surveillance by a medical doctor. Health surveillance detects the start of an ill-health problem and collects data on ill-health occurrences. It also gives an indication of the effectiveness of the control procedures. Health surveillance is needed to protect workers, identify as early as possible any health changes related to exposure and warn of any lapses in control arrangements.

Simple health surveillance is normally sufficient for skin problems and takes the form of skin inspections by a 'responsible person'. A 'responsible person' is a person who has been trained by a competent medical practitioner and may well be the site supervisor.

It is required when there appears to be a reasonable chance that ill-health effects are occurring in a particular workplace as a result of reviewing sickness records or when a substance listed in schedule 6 under Regulation 11 of the COSHH Regulations is being used. Schedule 6 lists the substances and the processes in which they are used. There are a limited number of such substances. The health surveillance includes medical surveillance by an employment medical adviser or appointed doctor at intervals not exceeding 12 months. Records of the health surveillance must contain approved particulars and be kept for 40 years. The need for health surveillance is not that common and further advice on the necessary procedures is available from the Employment Medical Advisory Service.

Personal hygiene has already been covered under supervisory controls. It is very important for workers exposed to hazardous substances to wash their hands thoroughly before eating, drinking or smoking. Protection against biological hazards can be increased significantly by vaccination (e.g. tetanus). Finally, contaminated clothing and overalls need to be removed and cleaned on a regular basis.

For plumbers and roofers who use lead in their work, medical surveillance by a doctor will be required if exposure is significant and the surveillance will include blood lead tests. Similarly workers who come into contact with asbestos will require medical surveillance as described later in this chapter.

An example of a health questionnaire for persons working with respiratory sensitizers is shown in Appendix 16.1

16.11 Maintenance and emergency controls

Engineering control measures will only remain effective if there is a programme of preventative maintenance available. Indeed the COSHH Regulations require that systems of adequate control are in place and the term 'adequate control' spans normal operations, emergencies and maintenance. Maintenance will involve the cleaning, testing and, possibly, the dismantling of equipment. It could involve the changing of filters in extraction plant or entering confined spaces. It will almost certainly require hazardous substances to be handled and waste material to be safely disposed of. It may also require a permit-to-work procedure to be in place since the control equipment will be inoperative during the maintenance operations. Records of maintenance should be kept for at least 5 years.

Emergencies can range from fairly trivial spillages to major fires involving serious air pollution incidents.

The following points should be considered when emergency procedures are being developed:

➤ the possible results of a loss of control (e.g. lack of ventilation)
➤ dealing with spillages and leakages (availability of effective absorbent materials)
➤ raising the alarm for more serious emergencies
➤ evacuation procedures including the alerting of neighbours
➤ fire fighting procedures and organization
➤ availability of respiratory protective equipment
➤ information and training.

The Emergency Services should be informed of the final emergency procedures and, in the case of the Fire Service, consulted for advice during the planning of the procedures. See Chapter 6 for more details on emergency procedures.

To provide an effective occupational health programme.

Figure 16.14 Health commitment.

16.12 Control of asbestos

A recent study published by the British Medical Journal has found that there are 1,800 mesothelioma deaths each year in Britain. Since this disease can take between 15 and 60 years to develop, the peak of the epidemic has still to be reached. In the construction industry, those at risk are asbestos removal workers and those, such as electricians, plumbers and carpenters, who are involved in refurbishment, maintenance or repair of buildings.

As discussed earlier in this chapter, asbestos can occur in three forms – crocidolite (blue), amosite (brown) and chrysotile (white). It was used widely as a building material until the mid-1980s. Although much asbestos has been removed from buildings, it has been estimated that over half a million non-domestic buildings still have asbestos in them amounting to many thousands of tons.

The strategy of the HSE is to ensure that those involved in the repair, removal or disturbance of asbestos-containing materials (ACMs), such as insulation, coatings or insulation boards, are licensed, competent and working to the strict requirements of the Control of Asbestos Regulations. This requires the identification of ACMs and the planning of any subsequent work. This should prevent inadvertent exposure to asbestos and minimize the risks to those who have to work with it.

The Regulations bring together the three previous sets of Regulations covering the prohibition of asbestos, the control of asbestos at work and asbestos licensing.

The Control of Asbestos Regulations prohibit the importation, supply and use of all forms of asbestos. They continue the ban introduced for blue and brown asbestos in 1985 and for white asbestos in 1999. They also continue the ban on the second-hand use of asbestos products such as asbestos cement sheets and asbestos boards and tiles; including panels which have been covered with paint or textured plaster containing asbestos. The ban applies to new uses of asbestos. If existing ACMs are in good condition, they may be left in place and their condition monitored and managed to ensure that they are not disturbed. The Regulations only cover the safe management of asbestos in industrial and commercial premises. Domestic premises are covered by the Defective Premises Act and the Civic Government (Scotland) Act. The Regulations also include the duty to manage asbestos in non-domestic premises. Guidance on the duty to manage asbestos can be found in the HSE ACOP – 'The Management of Asbestos in Non-Domestic Premises', L127.

The Regulations require mandatory training for anyone liable to be exposed to asbestos fibres at work, including maintenance workers and others who may come into contact with or who may disturb asbestos (e.g. cable installers) as well as those involved in asbestos removal work.

When work with asbestos or which may disturb asbestos is being carried out, the Regulations require employers and the self-employed to prevent exposure to asbestos fibres. Where this is not reasonably practicable, they must make sure that exposure is kept as low as reasonably practicable by measures other than the use of respiratory protective equipment. The spread of asbestos must be prevented. The Regulations specify the work methods and controls that should be used to prevent exposure and spread.

Worker exposure must be below the airborne exposure limit (the Control Limit) of 0.1 fibres per cm^3 for all types of asbestos. The Control Limit is the maximum concentration of asbestos fibres in the air (averaged over any continuous 4-hour period) and must not be exceeded. Short-term exposures must be strictly controlled and worker exposure should not exceed 0.6 fibres per cm^3 of air averaged over any continuous

10-minute period using respiratory protective equipment if exposure cannot be reduced sufficiently using other means. Respiratory protective equipment is an important part of the control regime but it must not be the sole measure used to reduce exposure and should only be used to supplement other measures.

Most asbestos removal work must be undertaken by a licensed contractor but any decision on whether particular work needs to be licensed is based on the level of risk. Details are given in Chapter 21 of the six possible reasons for work with asbestos being exempt from licensing.

The Control of Asbestos Regulations requires those in control of premises and the duty holders to:

➤ take reasonable steps to determine the location and condition of materials likely to contain asbestos
➤ presume materials contain asbestos unless there is strong evidence that they do not
➤ make and keep an up-to-date record of the location and condition of the ACMs or presumed ACMs in the premises
➤ assess the risk of the likelihood of anyone being exposed to fibres from these materials
➤ prepare a plan setting out how the risks from the materials are to be managed
➤ take the necessary steps to put the plan into action
➤ review and monitor the plan periodically
➤ provide such information and asbestos awareness training to anyone who is liable to work on these materials or otherwise disturb them.

In addition, the regulations include the following main provisions:

1. a single tighter control limit for all types of asbestos
2. specific training requirements for those working with asbestos
3. a clear hierarchy of controls to be used to reduce exposure.

There are **three** types of survey. The HSE name these as type 1, type 2 and type 3 that can be made to identify the presence of asbestos.

Type 1 is the location and assessment survey, sometimes called the **presumptive survey**. This type of survey locates as far as is reasonably practicable, the identifies and condition of ACMs. Where a material cannot be accurately assessed as asbestos but is suspected of being an ACM, this survey presumes that it is and any required remedial action is taken.

Type 2 type of survey is a **sampling survey** which is similar to the presumptive survey but a representative number of samples are sent away for analysis. If the results of the analysis are positive, then it is assumed that material with a similar appearance to the sample material is also an ACM.

Type 3 type of survey is the most invasive one and could involve the destruction of material. It is known as a **full access and sampling survey** and is normally used prior to demolition or major refurbishment work.

The person who undertakes any of these surveys must be suitably trained and experienced in such work.

Some types of work, of an intermittent and low intensity nature, will not have to be done by a licensed contractor – artex work is an example of this. As mentioned earlier, detailed information on work which is exempt from licensing is given in Chapter 21.

There are several publications available from HSE Books which cover all these stages in some detail and the reader should refer to them for more information. Here a brief summary of the various stages will be given.

Identification of the presence of asbestos is the first action. Asbestos is commonly found as boiler and pipe lagging, insulation panels around pillars and ducting for fire protection and heat insulation, ceiling tiles and asbestos cement products, including asbestos cement sheets. Initial investigations will involve the examination of building plans, the determination of the age of the building and a thorough examination of the building. Advice is available from a number of reputable specialist consultants and details may be obtained from the Local Authority, who often offers such a service. If the specialist is in any doubt, a sample of the suspect material will be sent to a specialist laboratory for analysis. It is important for a specialist to take the sample because the operation is likely to expose loose fibres. When asbestos has been identified, a record, possibly electronic, of its location must be made so that it is available should any future maintenance be necessary.

Assessment is an evaluation as to whether the location or the condition could lead to the asbestos being

Figure 16.15 AIB wall panels badly damaged during poorly planned and uncontrolled work.

disturbed. If it is in good condition, undamaged and not likely to be disturbed, then it is usually safer to leave it in place and manage it. However, if it is in a poor condition, it may need to be repaired, sealed, enclosed or removed. If there is doubt, then specialist advice should be sought.

Removal must only be done by a licensed contractor. A detailed plan of work is essential before work begins. The plan should give details of any equipment to be used for the protection and decontamination of employees and others. This process will also require an assessment to be made to ensure that people within the building and neighbours are properly protected. At the planning stage, generally, the HSE or the local authority must be notified of the intention to remove asbestos (at least 14 days' notice) and again when the work begins. This is particularly important when the exposure to asbestos is likely to exceed the action level. The assessment should include details of the type and location of the asbestos, the number of people who could be affected, the controls to be used to prevent or control exposure, the nature of the work, the removal methods, the procedures for the provision of personal and respiratory protective equipment and details of emergency procedures. If asbestos cement sheeting is to be removed the following procedure is recommended:

➤ Where reasonably practicable, remove the asbestos sheets before any other operation, such as demolition.
➤ Avoid any breaking of the sheets.
➤ Dampen the sheets while working on them.
➤ Lower the sheets on to a clean hard surface.
➤ Remove all waste and debris from the site as soon as possible to prevent its spread around the site.
➤ Do not bulldoze broken asbestos cement or sheets into piles.
➤ Do not dry sweep asbestos cement debris.
➤ Dispose of the waste and debris safely, separate from general waste as hazard, us waste.

Control measures during the removal of asbestos include the provision of personal and RPE including overalls, good ventilation arrangements in and the segregation or sealing of the working area, suitable method statements and air monitoring procedures. A decontamination unit should also be provided. The sealed area will need to be tested for leaks. Good supervision and induction of the workforce are also essential. A high level of personal hygiene must be expected for all workers and the provision of welfare amenity arrangements, particularly washing and catering facilities and the separation of working and personal clothing. Suitable warning signs must be displayed and extra controls provided if the work is taking place at height. After the work is completed, the area must be thoroughly cleaned and a clean air certificate provided after a successful air test.

Medical surveillance in the form of a regular medical examination by an appointed doctor should be given

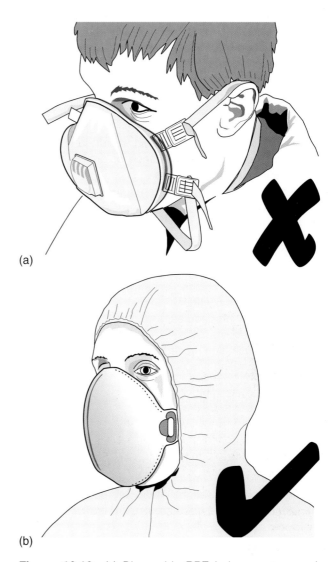

(a)

(b)

Figure 16.16 (a) Disposable RPE being worn wrongly; (b) if your disposable RPE looks like this then you are wearing it correctly.

to any employee who has been exposed to asbestos at levels above the action level. The first medical examination should take place within 2 years of exposure and further examinations at intervals of not more than 2 years. A health record of such surveillance should be kept for a period of at least 40 years after the last entry.

Awareness training is an important feature of the Control of Asbestos Regulations and they require that adequate information, instruction and training are given to those of his employees:

(a) who are or who are liable to be exposed to asbestos, or who supervise such employees
(b) who carry out work in connection with the employer's duties under these Regulations, so that they can carry out that work effectively.

Such training should cover the following topics:

(i) the properties of asbestos and its effects on health, including its interaction with smoking

(ii) the types of products or materials likely to contain asbestos

(iii) the uses and location of ACMs in buildings and plant

(iv) the operations which could result in asbestos exposure and the importance of preventive controls to minimize exposure

(v) the presence of other hazards such as working at height

(vi) the requirements of the Control of Asbestos Regulations

(vii) safe work practices, control measures, and protective equipment

(viii) the purpose, choice, limitations, proper use and maintenance of RPE

(ix) emergency procedures

(x) hygiene requirements

(xi) decontamination procedures

(xii) waste handling procedures

(xiii) medical examination requirements and

(xiv) the control limit and the need for air monitoring.

Refresher training should be given at regular intervals (at least annually) and adapted to take account of significant changes in the type of work carried out or methods of work used by the employer. It should be provided in a manner appropriate to the nature and degree of exposure identified by the risk assessment, and so that the employees are aware of:

(i) the significant findings of the risk assessment and

(ii) the results of any air monitoring carried out with an explanation of the findings.

Disposal of asbestos waste is subject to the Hazardous Waste Regulations which require it to be consigned to an authorized asbestos waste site only. The waste container must be strong enough to securely contain the waste and not become punctured; it must be easily decontaminated, kept securely on the site until required and properly labelled. The waste must only be carried by a licensed carrier.

Accidental exposure to ACMs can occur even when all reasonable precautions have been taken. If an ACM has been worked on by a worker who did not realize that it was an ACM or has accidentally damaged an ACM, then all work must stop immediately and nobody should be allowed to enter the area in question. The work supervisor must be informed and a sample of the material sent for analysis. If the result is positive, then a specialist licensed contractor should be employed unless the work

is exempt from the need for a licence. Any contaminated clothing must be removed and placed in a plastic bag. The affected worker should shower as soon as possible or wash thoroughly.

More information on the legislative requirements is given in Chapter 21.

16.13 The transport of hazardous substances by road

Although this topic is not in the NEBOSH National Construction Certificate syllabus, a brief mention of the main precautions required to safeguard the health and safety of those directly involved in the transport of hazardous substances, and general members of the public, is important.

Data sheets from the manufacturer of the hazardous substance should indicate the safest method of handling it and will give information on emergency procedures (e.g. for spillages and fire). These sheets should be available to all concerned with the transportation of the substance, in particular those responsible for loading/unloading and the driver. The hazardous substance should be loaded correctly on the vehicle in suitable containers and segregated from incompatible materials. There must be adequate emergency information provided along with the substance containers and attached to the vehicle. Drivers of the vehicles must receive special training which covers issues such as emergency procedures and route planning. There should also be emergency provisions for first aid and PPE on the vehicle.

Finally, the Transport of Dangerous Goods (Safety Adviser) Regulations require the appointment of a trained and competent safety adviser to ensure the safe loading, transportation and unloading of the hazardous substance. HSE have produced several guidance publications which offer more detailed advice on this topic.

16.14 An Illustrative example using COSHH controls

Organic solvents are widely used throughout industry and commerce in paints, inks, glues and adhesives. The COSHH hierarchy discussed in section 16.9 should be applied to minimize the health risks from the use of these solvents. The top of this hierarchy is to eliminate or substitute the use of the organic solvent by using a less volatile or water-based alternative. If this is not possible, then some form of engineering control should be applied, such as dilution or local exhaust ventilation. Alternatively, the workplace, where the solvents are being used, could be enclosed or isolated from the main work activities. Other engineering type controls could include the use of properly labelled anti-spill containers, the use of covered disposal units for any used cloths and the transfer of large quantities using a pumping/pipe arrangement rather than simply pouring the solvent.

Supervisory controls include the reduction in the length of time any employee is exposed to the solvent and the provision of good housekeeping, such as ensuring that containers are kept closed when not in use and any spills are quickly removed.

The provision of barrier creams and personal protective equipment (eye protection, gloves and aprons) and respiratory protective equipment may also be required.

Welfare issues will include first-aid provision, washing facilities and the encouragement of high levels of personal hygiene. Smoking and the consumption of food and drink should be prohibited where there might be contamination from organic solvents. All these supervisory items should be reinforced in training sessions and employees given appropriate information on the risks associated with the solvents.

Finally, some form of health surveillance will be needed so that employees who show allergies to the solvents can be treated and, possibly, assigned other duties.

16.15 Environmental considerations

Organizations must also be concerned with aspects of the environment. There will be an interaction between the health and safety policy and the environmental policy which many organizations are now developing. Many of these interactions will be concerned with good practice, the reputation of the organization within the wider community and the establishment of a good health and safety culture. The health and safety data sheet, used for a COSHH assessment, also contains information of an environmental nature covering ecological information and disposal considerations.

There are three environmental issues which place statutory duties on employers and are directly related to the health and safety function. These are:

➤ air pollution
➤ water pollution
➤ waste disposal.

The statutory duties are contained in the Environmental Protection Act 1990 (EPA) and several of its subsequent regulations. The Act is enforced by various state agencies (the Environment Agency and the Local Authorities) and these agencies have very similar powers to the HSE (e.g. enforcement and prohibition

notices and prosecution). The Act is divided into nine parts but this chapter will only be concerned with Part 1 (the control of pollution into the air, water or land) and Part 2 (waste disposal).

Pollution is a term that covers more than the effect on the environment of atmospheric emissions, effluent discharges and solid waste disposal from industrial processes. It also includes the effect of noise, vibration, heat and light on the environment. This was recognized in the 1996 EU Directive on Integrated Pollution Prevention and Control (IPCC), and The Pollution Prevention and Control Act 1999 replaced the Integrated Pollution Control Regulations made under Part 1 of the EPA by extending those powers to cover waste minimization, energy efficiency, noise and site restoration. Thus the Pollution Prevention and Control Act has now replaced Part 1 of the EPA.

The Solvents Emissions Directive (SED) has produced some tougher rules on the use of solvents in industry. Local authorities and the Environment Agency are beginning to incorporate SED provisions into Integrated Pollution Prevention and Control Permits (IPPCP). Organizations that use solvents on an industrial scale need an IPPCP to operate.

16.15.1 *Air pollution*

The most common airborne pollutants are carbon monoxide, benzene, 1,3-butadiene, sulphur dioxide, nitrogen dioxide and lead. Air pollution is monitored by Integrated Pollution Prevention and Control (IPPC). This is a system which extends the Integrated Pollution Control (IPC) system established by Part 1 of the EPA by introducing three tiers of pollution control:

➤ **Regime A1 processes**, which are certain large scale manufacturing processes with a potential to cause serious environmental damage to air, water or land. In England and Wales this is enforced by the Environment Agency. In Scotland there is a parallel system enforced by the Scottish Environmental Protection Agency.
➤ **Regime A2 processes**, which produce emissions to air, water and land with a much smaller potential to pollute than regime A1 processes. The local authority is the regulator for these processes.
➤ **Part B processes**, which may be classified as those from less polluting industries with only emissions released to air being subject to regulatory control. For such processes local authorities are the enforcing body through Environmental Health Officers. The system is known as Local Air Pollution Control (LAPC) in Northern Ireland and Scotland. In England and Wales it is known as the Local Pollution Prevention and Control LAPPC.

Figure 16.17   A clean environment.

This division has lead to some anomalies in that some Part A processes create less pollution than some Part B processes. However, the grouping of three pollution destinations under one arrangement tends to a more holistic approach.

The aim of IPPC is to control pollution of the whole environment under a single enforcement system and offers three principles to prevent and control pollution. These are:

1. The 'Best Practicable Environmental Option (BPEO)' which considers both the environmental and economic costs and benefits of the possible options available to deal with the pollution problem. BPEO is a legal requirement for Part A processes. It normally requires a technical solution.
2. The 'Best Available Techniques', similar to BATNEEC (Best Available Techniques Not Entailing Excessive Cost) introduced by the EPA to minimize the overall environmental impact of a process. Part B processes only need satisfy the BATNEEC requirement which is not restricted to pollution control technology but can include employee training and competence and building design and maintenance.
3. 'As low as reasonably practicable' applies the same test to an environmental problem as is applied to a health and safety problem. Any high or unacceptable environmental risk should be reduced to as low as is reasonably practicable.

The EPA prescribes certain listed substances from being released to air, water or land. All prescribed processes must have authorization. An operator of a prescribed process (such as a vehicle spray booth) must apply to the Environment Agency for prior authorization to operate the process. If the application is granted, the operator must monitor emissions and report them to the Environment Agency on a yearly basis. The Agency has the power to revoke the authorization, enforce the terms of the authorization or prohibit the operation of the process.

Further information on the authorization process is given in Chapter 21.

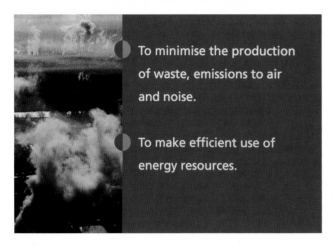

To minimise the production of waste, emissions to air and noise.

To make efficient use of energy resources.

Figure 16.18 Environmental protection commitment.

16.15.2 *Water pollution*

Pollution of rivers and other water courses can produce very serious effects on the health of plants and animals which rely on that water supply. The Environment Agency is responsible for coastal waters, inland fresh water and ground waters (known as 'controlled waters'). The EC Groundwater Directive seeks to protect groundwater from pollution since this is a source of drinking water. Such sources can become polluted by leakage from industrial soakaways. Discharges to a sewer are controlled by the Water Industry Act that defines trade effluent and those substances which are prohibited from discharge (e.g. petroleum spirit) and the Water Resources Act that covers discharge consent to controlled waters. It is an offence to pollute any controlled waters or sewage system. If hazardous substances are being used by the organization, safety data sheets give advice on the safe disposal of any residues that remain after the particular process has been completed.

The local water company has a right to sample discharges into its sewers because it is required to keep a public trade effluent register. There are two lists of prescribed substances which can only be discharged into a public sewer with the permission of the water company.

Finally, if oil is stored on site, a retaining bund wall should surround the oil store. This will not only ensure that any oil leakage is contained but will also stop the contamination of ground water by fire fighting foam in the event of a fire. This is a requirement of the Control of Pollution (Oil Storage) Regulations 2001.

16.15.3 *Waste disposal*

The statutory duty of care for the management of waste derives from Part 2 of the EPA. The principal requirements are as follows:

➤ to handle waste so as to prevent any unauthorized escape into the environment

➤ to pass waste only to an authorized person as defined by the EPA
➤ to ensure that a written description accompanies all waste. The Environmental Protection (Duty of Care) Regulations 1991 require holders or producers of waste to complete a 'Transfer Note' giving full details of the type and quantity of waste for collection and disposal. Copies of the note should be kept for at least 2 years
➤ to ensure that no person commits an offence under the Act.

The EPA is concerned with controlled waste. Controlled waste comprises household, industrial or commercial waste. All construction waste is controlled waste. It is a criminal offence to deposit controlled waste without a waste management licence and/or in a manner likely to cause environmental pollution or harm to human health.

The EPA also covers 'hazardous wastes' which can only be disposed of using special arrangements. These are sometimes substances which are life threatening (toxic, corrosive or carcinogenic) or highly flammable. Clinical waste falls within this category. A consignment note system accompanies this waste at all the stages to its final destination. Before hazardous waste is removed from the originating premises, a contract should be in place with a licensed carrier. Hazardous waste should be stored securely prior to collection to ensure that the environment is protected.

The Hazardous Waste Regulations with the exclusion of Scotland replace the Special Waste Regulations and cover many more substances, for example computer monitors, fluorescent tubes, end-of-life vehicles and television sets. Hazardous waste is waste which can cause damage to the environment or to human health. Such waste is defined in the List of Waste Regulations and producers of such waste may need to notify the Environment Agency. The Regulations seek to ensure that hazardous waste is safely managed and its movement is documented. The following points are important for construction sites:

➤ Sites that produce more than 200 kg of hazardous waste each year for removal, treatment or disposal must register with the Environment Agency.
➤ Different types of hazardous waste must not be mixed.
➤ Producers must maintain registers of their hazardous wastes.

Some form of training may be required to ensure that employees segregate hazardous and non-hazardous wastes on site and fully understand the risks and necessary safety precautions which must be taken. Personal protective equipment, including overalls, gloves and eye protection, must be provided and used. The storage site should be protected against trespassers, fire and adverse weather conditions. If flammable or combustible wastes are being stored, adequate fire protection systems must be in place. Finally, in the case of liquid wastes, any drains must be protected and bunds used to restrict spreading of the substance as a result of spills.

A hierarchy for the management of waste streams has been recommended by the Environment Agency:

1. Prevention – by changing the process so that the waste is not produced (e.g. substitution of a particular material)
2. Reduction – by improving the efficiency of the process (e.g. better machine maintenance)
3. Reuse – by recycling the waste back into the process (e.g. using reground waste plastic products as a feed for new products)
4. Recovery – by releasing energy through the combustion, recycling or composting of waste (e.g. the incineration of combustible waste to heat a building)
5. Responsible disposal – by disposal in accordance with regulatory requirements.

In 1998, land disposal accounted for approximately 58% of waste disposal, 26% was recycled and the remainder was incinerated with some of the energy recovered as heat. The Producer Responsibility Obligations (Packaging Waste) Regulations 1997 placed legal obligations on employers to reduce their packaging waste by either recycling or recovery as energy (normally as heat from an incinerator attached to a district heating system). A series of targets have been stipulated which will reduce the amount of waste progressively over the years. These Regulations are enforced by the Environment Agency, who have powers of prosecution in the event of non-compliance.

There is more information on the EPA, IPPC and waste disposal in Chapter 21.

The Environment Agency has similar powers following the introduction of the Waste Electrical and Electronic Equipment (WEEE) directive, the Restrictions of the use of certain Hazardous Substances in electrical and electronic equipment (RoHS) and the End of Life Vehicle directive (ELV). The aim of the WEEE directive is to minimize the environmental impact of electrical and electronic equipment both during their lifetimes and when they are discarded. All electrical and electronic equipment must be returned to the retailer from its end user and reused or reprocessed by the manufacturer. Manufacturers must register with the Environment Agency, who will advise on these obligations.

The Department of Farming and Rural Affairs (DEFRA) are to introduce legislation on 'Site waste management plans'. This legislation will provide a legally enforceable framework for waste management at all stages of a construction project. The plans require clients and/or principal contractors to identify:

➤ a person responsible for resource management
➤ the types of waste that are likely to be generated
➤ resource management options for these wastes
➤ the use of licensed waste management contractors and
➤ a plan for the monitoring and reporting of resource use and the amount of waste.

The minimum qualifying value for a project is £250,000 with extra requirements if the value exceeds £500,000. DEFRA proposes that local authorities and the Environment Agency will be the enforcement agencies for the plans.

This is part of the environmental permitting programme; see Chapter 21 (Section 21.5.4) for more details.

16.16 Practice NEBOSH questions for Chapter 16

1. For **EACH** of the following types of hazardous substance, give a typical example and state its primary effect on the body:
   (a) toxic
   (b) corrosive
   (c) carcinogenic
   (d) irritant.

2. (a) **Describe** the differences between acute and chronic health effects.
   (b) **Identify** the factors that could affect the level of harm experienced by an employee exposed to a toxic substance.
   (c) **Give TWO** acute and **TWO** chronic health effects on the body from exposure to lead.

3. For **EACH** of the following agents, **outline** the principal health effects **AND identify** a typical workplace situation in which a person might be exposed:
   (a) silica
   (b) asbestos
   (c) *leptospira* bacteria
   (d) lead.

4. (a) **Identify FOUR** possible ill-health effects that can be caused from working with cement.
   (b) **Outline** ways in which the ill-health effects in (a) can be prevented.

5. (a) **Identify TWO** respiratory diseases that may be caused by exposure to asbestos.
   (b) **Identify** the common sources of asbestos in buildings that should be considered when conducting an asbestos survey of work premises.

6. An employee is engaged in general cleaning activities in a large veterinary practice.
   (a) **Identify FOUR** specific types of hazard that the cleaner might face when undertaking the cleaning.
   (b) **Outline** the precautions that could be taken to minimize the risk of harm from these hazards.

7. (a) **Identify** possible routes of entry of biological organisms into the body.
   (b) **Outline** control measures that could be used to reduce the risk of infection from biological organisms.

8. (a) **List FOUR** four respiratory diseases that could be caused by exposure to dust at work.
   (b) **Identify** the possible indications of a dust problem in a workplace.

9. A general operative is about to use an epoxy resin filler with a peroxide hardener. **Outline** the health and safety measures to be considered and communicated before the operative starts work.

10. **Identify EIGHT** possible health hazards to which construction workers may be exposed **AND** in **EACH** case **give** an example of a likely source.

11. **Identify FOUR** categories of occupational health hazard and, in **EACH** case, give an example of a specific hazard prevalent within the construction industry.

12. (a) **Describe** the typical symptoms of occupational dermatitis.
    (b) **State** the factors that could affect the likelihood of dermatitis occurring in workers handling dermatitic substances.

13. **Identify** the factors that will influence the likelihood of dermatitis occurring amongst site operatives handling wet cement.

14. (a) **Explain** the meaning of the term 'workplace exposure limit' (WEL).
    (b) **Outline FOUR** actions management could take when a WEL has been exceeded.

15. Local exhaust ventilation (LEV) systems must be thoroughly examined at least every 14 months. **Outline** the routine maintenance that should be carried out between statutory examinations in order to ensure the continuing efficiency of an LEV system.

16. (a) **Explain** the meaning of the term 'dilution ventilation'.
    (b) **Outline** the circumstances in which the use of dilution ventilation may be appropriate.

17. (a) **Describe**, by means of a labelled sketch, a chemical indicator (stain detector) tube suitable for atmospheric monitoring.
    (b) **List** the main limitations of chemical indicator (stain detector) tubes.

18. The manager of a company is concerned about a substance to be introduced into one of its manufacturing processes. **Outline FOUR** sources of information that might be consulted when assessing the risk from this substance.

19. **List** the information that should be included in a manufacturer's safety data sheet for a hazardous substance used on a construction site.

20. **Identify FOUR** different types of hazard that may necessitate the use of special footwear, explaining in each case how the footwear affords protection.

21. (a) **Identify** the types of hazard against which gloves could offer protection.
    (b) **Outline** the practical limitations of using gloves as a means of protection.

22. (a) **Identify THREE** types of hazard for which personal eye protection would be required.
    (b) **Outline** the range of issues that should be addressed when training employees in the use of personal eye protection.

23. **Outline** the types of personal protective equipment required when:
    (a) cutting and shaping medium density fibreboard using an electric saw in a site workshop
    (b) cutting up very old painted metalwork using gas cutting equipment.

24. **Outline** the types of personal protective equipment required when:
    (a) chasing concrete and brick walls by mechanical means.
    (b) using solvent-based adhesives in a poorly ventilated room.

25. A furniture factory uses solvent-based adhesives in its manufacturing process.
    (a) **Identify** the possible effects on the health of employees using the adhesives.
    (b) **State FOUR** control measures to minimize such health effects.

26. (a) **Explain** the health and safety benefits of restricting smoking in the workplace.
    (b) **Outline** the ways in which an organization could effectively implement a no-smoking policy.

27. In relation to the spillage of a toxic substance from a ruptured drum stored in a warehouse:
    (a) **Identify THREE** ways in which persons working in the close vicinity of the spillage might be harmed.
    (b) **Outline** a procedure to be adopted in the event of such a spillage.

28. An employee is required to install glass-fibre insulation in a loft.
    (a) **Identify TWO** hazards connected with this activity.
    (b) **Outline** the precautions that might be taken to minimize harm to the employee carrying out this operation.

29. An essential raw material for a process is delivered in powdered form and poured by hand from bags into a mixing vessel. **Outline** the control measures that might be considered in this situation in order to reduce employee exposure to the substance.

30. On a fairly long-term contract, joiners are working with medium density fibreboard using hand-held circular saws and portable electric planers in a temporary site unit.
    Explain how you would ensure that the joiners' health is not put at risk during this activity.

31. (a) **Identify** the **THREE** main types of asbestos and where **EACH** type is likely to be found in a building during renovation work.
    (b) **Outline** the principal health effects from exposure to asbestos fibres.
    (c) **Outline** the notification requirements, and the specific issues that must be included in an assessment and plan of work, when ACM is to be removed from a building.
    (d) **Describe** the precautions necessary when stripping asbestos lagging from old pipework.

32. **Outline**, with reasons, the particular health issues that should be addressed during routine health surveillance examinations for construction workers.

33. Health surveillance is beginning to become more commonly used as a check to ensure workplace control measures are working effectively.
    Identify the particular health issues that should be addressed during routine health surveillance examinations for construction workers and for each issue identified, **give** a relevant hazard.

34. **Outline** a system for the management and disposal of waste from a construction site.

35. **Identify EIGHT** safe practices to be followed when using a skip for the collection and removal of waste from a construction site.

Appendix 16.1 – Health questionnaire for on-going surveillance of persons working with respiratory sensitizers

To be completed by the responsible person

Employee's name   Works no

The questionnaire should be completed 6 weeks, 6 months and annually after employment commences or as advised by the company occupational health adviser.

Further advice will be required from the company occupational health adviser if any Yes box is ticked.

Since starting your present job have you had any of the following symptoms either at work or at home? (Do not include isolated colds, sore throats or flu)

(a)   Recurring soreness of or watering of eyes   Yes ☐   No ☐

(b)   Recurring blocked or running nose   Yes ☐   No ☐

(c)   Bouts of coughing   Yes ☐   No ☐

(d)   Chest tightness   Yes ☐   No ☐

(e)   Wheeze   Yes ☐   No ☐

(f)   Breathlessness   Yes ☐   No ☐

(g)   Have you consulted your doctor about chest problems since   Yes ☐   No ☐
the last questionnaire?

To be completed by the responsible person

(a)   No further action required   ☐

(b)   Refer to company occupational health adviser   ☐

Signature of responsible person   Date

I confirm that the responses given by me are correct and that I have received a copy of the completed questionnaire.

Signed   Date

Physical and psychological health hazards and control

17

17.1 Introduction

Occupational health is concerned with physical and psychological hazards as well as chemical and biological hazards. The physical occupational hazards have been well known for many years and the recent emphasis has been on the development of lower risk workplace environments. Physical hazards include topics such as electricity and manual handling, which were covered in earlier chapters and noise, display screen equipment (DSE) and radiation, which are discussed in this chapter.

However, it is only really in the last 20 years that psychological hazards have been included among the occupational health hazards faced by many workers. This is now the most rapidly expanding area of occupational health and includes topics such as mental health and workplace stress, violence to staff, passive smoking, drugs and alcohol.

The physical and psychological hazards discussed in this chapter are covered by the following health and safety regulations:

➤ CDM 2007 Part 4 and Schedule 2
➤ Health and Safety (Display screen Equipment) Regulations
➤ Manual Handling Operations Regulations
➤ Control of Noise at Work Regulations
➤ Ionising Radiations Regulations
➤ Control of Vibration at Work Regulations.

17.2 Task and workstation design

17.2.1 The principles and scope of ergonomics

Ergonomics is the study of the interaction between workers and their work in the broadest sense, in that it encompasses the whole system surrounding the work process. It is, therefore, as concerned with the work organization, process and design of the workplace and work methods as it is with work equipment. The common definitions of ergonomics, the 'man–machine interface' or 'fitting the man to the machine rather than vice versa' are far too narrow. It is concerned with the physical and mental capabilities of an individual as well as their understanding of the job under consideration. Ergonomics includes the limitations of the worker in terms of skill level, perception and other personal factors in the overall design of the job and the system supporting and surrounding it. It is the study of the relationship between the worker, the machine and the environment in which it operates and attempts to optimize the whole work system, including the job, to the capabilities of the worker so that maximum output is achieved for minimum effort and discomfort by the worker. Cars, buses and lorries are all ergonomically designed so that all the important controls, such as the steering wheel, brakes, gear stick and instrument panel are easily accessed by most drivers within a wide range of sizes. Ergonomics is sometimes described as human engineering and as working practices become more and more automated, the need for good ergonomic design becomes essential.

The scope of ergonomics and an ergonomic assessment is very wide, incorporating the following areas of study:

➤ personal factors of the worker, in particular, physical, mental and intellectual abilities, body dimensions and competence in the task required
➤ the machine and associated equipment under examination
➤ the interface between the worker and the machine – controls, instrument panel or gauges and any aids including seating arrangements and hand tools

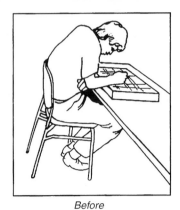

Before   *After*

Figure 17.1 Workstation egronomic design improvements.

➤ environmental issues affecting the work process such as lighting, temperature, humidity, noise and atmospheric pollutants

➤ the interaction between the worker and the task, such as the production rate, posture and system of working

➤ the task or job itself – the design of a safe system of work, checking that the job is not too strenuous or repetitive and the development of suitable training packages

➤ the organization of the work, such as shift work, breaks and supervision.

The reduction of the possibility of human error is one of the major aims of ergonomics and an ergonomic assessment. An important part of an ergonomic study is to design the workstation or equipment to fit the worker. For this to be successful, the physical measurement of the human body and an understanding of the variations in these measurements between people are essential. Such a study is known as **anthropometry**, which is defined as the scientific measurement of the human body and its movement. Since there are considerable variations in, for example, the heights of people, it is common for some part of the workstation to be variable (e.g. an adjustable seat).

17.2.2 *The ill-health effects of poor ergonomics*

Ergonomic hazards are those hazards to health resulting from poor ergonomic design. They generally fall within the physical hazard category and include the manual handling and lifting of loads, pulling and pushing loads, prolonged periods of repetitive activities and work with vibrating tools. The condition of the working environment, such as low lighting levels, can present health hazards to the eyes. It is also possible for psychological conditions such as occupational stress to result from ergonomic hazards.

The common ill-health effects of ergonomic hazards are musculoskeletal disorders (back injuries, covered in Chapter 12, and work-related upper limb disorders (WRULDs) including repetitive strain injury being the main disorders) and deteriorating eyesight.

Work-related upper limb disorders

WRULDs describe a group of illnesses which can affect the neck, shoulders, arms, elbows, wrists, hands and fingers. **Tenosynovitis** (affecting the tendons), **carpal tunnel syndrome** (affecting the tendons which pass through the carpal bones in the hand) and **frozen shoulder** are all examples of WRULDs, which differ in the manifestation and site of the illness. The term **repetitive strain injury (RSI)** is commonly used to describe WRULDs.

WRULDs are caused by repetitive movements of the fingers, hands or arms which involve pulling, pushing, reaching, twisting, lifting, squeezing or hammering. These disorders can occur to workers in offices as well as in factories or on construction sites. Typical occupational groups at risk include painters and decorators, riveters and pneumatic drill operators and desktop computer users.

During construction, structural engineers and others can suffer from RSI due to the large torques that they exert on spanners when bolt fixing and tightening nuts and screws.

The main symptoms of WRULDs are aching pain to the back, neck and shoulders, swollen joints and muscle fatigue accompanied by tingling, soft tissue swelling similar to bruising and a restriction in joint movement. The sense of touch and movement of fingers may be affected. The condition is normally a chronic one in that it gets worse with time and may lead eventually to permanent damage. The injury occurs to muscle, tendons and/or nerves. If the injury is allowed to heal before being exposed to the repetitive work again, no long-term damage should result. However, if the work is repeated

Figure 17.2 Poor workstation layout may cause WRULD.

again and again, healing cannot take place and permanent damage can result leading to a restricted blood flow to the arms, hands and fingers.

The risk factors which can lead to the onset of WRULDs are repetitive actions of lengthy duration, the application of significant force and unnatural postures possibly involving twisting and over-reaching and the use of vibrating tools. Cold working environments, work organization and worker perception of the work organization have all been shown in studies to be risk factors as is the involvement of vulnerable workers such as those with pre-existing ill-health conditions and pregnant women.

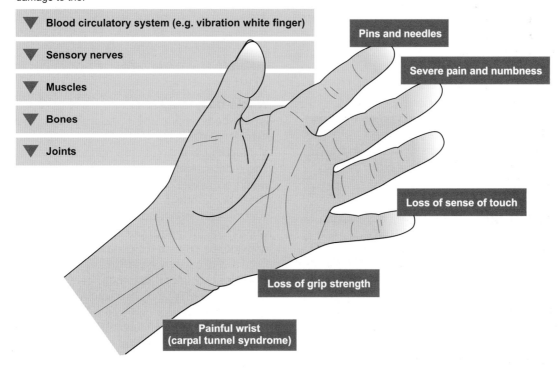

Figure 17.3 Injuries which can be caused by hand–arm vibration (also known as "white finger").

Ill-health due to vibration

Hand-held vibrating machinery (such as pneumatic drills, sanders and grinders, powered lawnmowers and strimmers and chainsaws) can produce health risks from hand–arm or whole-body vibration (WBV).

Hand–arm vibration syndrome

Hand–arm vibration syndrome (HAVS) describes a group of diseases caused by the exposure of the hand and arm to external vibration. Some of these have been described under WRULDs, such as carpal tunnel syndrome.

However, the best known disease is **vibration white finger** (VWF), in which the circulation of the blood, particularly in the hands, is adversely affected by the vibrations. The early symptoms are tingling and numbness felt in the fingers, usually some time after the end of the working shift. As exposure continues, the tips of the fingers go white and then the whole hand may become affected. This results in a loss of grip strength and manual dexterity. Attacks can be triggered by damp and/or cold conditions and, on warming, 'pins and needles' are experienced. If the condition is allowed to persist, more serious symptoms become apparent including decolorization and enlargement of the fingers. In very advanced cases, gangrene can develop leading to the amputation of the affected hand or finger. VWF was first detailed as an industrial disease in 1911.

The risk of developing HAVS depends on the frequency of vibration, the length of exposure and the tightness of the grip on the machine or tool. Some typical values of vibration measurements for equipment used in the construction industry are given in Table 17.1.

Table 17.1 Examples of vibration exposure values measured by HSE on work equipment

Equipment	Condition	Vibration reading (m/s^2)
Road breakers	Typical	12
	Modern design and trained operator	5
	Worst tools and operating conditions	20
Demolition hammers	Modern tools	8
	Typical	15
	Worst tools	25
Hammer drills	Typical	9
	Best tools and operating conditions	6
	Worst tools and operating conditions	25
Large angle grinders	Modern vibration-reduced designs	4
	Other types	8
Small angle grinders	Typical	2–6
Chainsaws	Typical	6
Brush cutter or strimmer	Typical	4
	Best	2
Sanders (orbital)	Typical	7–10

Figure 17.4 Mounted breaker reduces vibrations.

Whole-body vibration

Whole-body vibration is caused by vibration from machinery passing into the body through either the feet of standing workers or the buttocks of sitting workers. The most common ill-health effect is severe back pain which, in severe cases, may result in permanent injury. Other acute effects include reduced visual and manual control, and increased heart rate and blood pressure. Chronic or long-term effects include permanent spinal damage, damage to the central nervous system, hearing loss, and circulatory and digestive problems.

The two most common occupations which are affected by WBV are pneumatic drilling in construction and in agricultural or horticultural machinery work. There is growing concern throughout the European Community about this problem. Control measures include the proper use of the equipment including correct adjustments of air or hydraulic pressures, seating and, in the case of vehicles, correct suspension and tyre pressures, and appropriate speeds to suit the terrain. Other control measures include the selection of suitable equipment with low vibration characteristics, work rotation, good maintenance and fault reporting procedures.

Health and Safety Executive (HSE) commissioned measurements of WBV on several machines and some of the results are shown in Table 17.2.

Table 17.2 Machines which could produce significant whole-body vibrations

Machine	Activity	Vibration reading (m/s^2)
3 tonne articulated site dumper truck	Removal of soil	0.78
	Transport of soil	1.13
25 tonne articulated dumper truck	Transport of soil	0.91
Bulldozer	Dozing	1.16
4 tonne twin-drum articulated roller	Finishing tarmac	0.86
1.5 tonne 360° excavator	Excavating	0.92
80 tonne rigid dumper truck	Transport of soil	1.03
Wheeled scraper	Cut and fill	1.33

Preventative and precautionary measures
The control strategy outlined in Chapter 6 can certainly be applied to ergonomic risks. The common measures used to control ergonomic ill-health effects are to:

➤ examine results of task analysis and identification of repetitive activities
➤ eliminate vibration or hazardous tasks by performing the job in a different way
➤ ensure that the correct equipment (properly adjusted) is always used
➤ introduce job rotation so that workers have a reduced time exposure to the hazard
➤ during the design of the job ensure that poor posture is avoided
➤ undertake a risk assessment
➤ introduce a programme of health surveillance
➤ ensure that employees are given adequate information on the hazards and develop a suitable training programme

➤ ensure that a programme of preventative maintenance is introduced and include the regular inspection of items such as vibration isolation mountings
➤ keep up to date with advice from equipment manufacturers, trade associations and health and safety sources (more and more low vibration equipment is becoming available).

A very useful and extensive checklist for the identification and reduction of WRULDs is given in Appendix 2 of the HSE guide to Work-related upper limb disorders HSG60. HSE have also produced two very useful guides – (1) Hand–arm vibration, The Control of Vibration at Work Regulations 2005 Guidance on Regulations L140 and (2) Whole-body vibration, The Control of Vibration at Work Regulations 2005 Guidance on Regulations L141. These guidance documents offer detailed advice on the implementation of the regulations.

17.2.3 *The Control of Vibration at Work Regulations 2005*

The Control of Vibration at Work Regulations introduce, for both hand–arm and whole-body vibration, a daily exposure limit and action values. These values are as follows:

1. For hand–arm vibration
   (a) the daily exposure limit value normalized to an 8-hour reference period is 5 m/s^2
   (b) the daily exposure action value normalized to an 8-hour reference period is 2.5 m/s^2.
2. For whole-body vibration
   (a) the daily exposure limit value normalized to an 8-hour reference period is 1.15 m/s^2
   (b) the daily exposure action value normalized to an 8-hour reference period is 0.5 m/s^2.

An exposure limit value must not be exceeded. If an exposure action value is exceeded, then action must be taken to reduce the value. The designation A(8) is added to the exposure limit or action value to denote that it is an average value spread over an 8-hour working day. Thus the daily exposure limit value for hand–arm vibration is 5 m/s^2 A(8).

Hand–arm vibration
Many machines and processes used in construction produce hand–arm vibration. The high risk processes include:

➤ grinding, sanding and polishing wood and stone
➤ cutting stone, metal and wood
➤ riveting, caulking and hammering
➤ compacting sand, concrete and aggregate

➤ drilling and breaking rock, concrete and road surfaces

➤ surface preparation, including de-scaling and paint removal.

There are several ways to ascertain the size of the vibration generated by equipment and machines. Manufacturers must declare vibration emission values for portable hand-held and hand-guided machines and provide information on risks. Other important sources for vibration information include scientific and technical journals, trade associations and on-line databases. HSE experience has shown that the vibration level is higher in practice than that quoted by many manufacturers. The reasons for this discrepancy may be that:

➤ the equipment is not well maintained

➤ it is not suitable for the material being worked

➤ accessories are not appropriate or badly fitted

➤ the operative is not using the tool properly.

In view of these problems, it is recommended that the declared value should be doubled when comparisons are made with exposure limits.

Since the exposure limit or action value is averaged over 8 hours, it is possible to work with higher values for a reduced exposure time. Table 17.3 shows the reduction in exposure time as the size of the vibration increases.

The Guidance (L140) gives very useful advice on the measurement of vibrations, undertaking a suitable and sufficient risk assessment, control measures, health surveillance and training of employees. The following points summarize the important measures which should be taken to reduce the risks associated with hand–arm vibration:

➤ Avoid, whenever possible, the need for vibration equipment.

➤ Undertake a risk assessment which includes a soundly based estimate of the employees' exposure to vibration.

➤ Develop a good maintenance regime for tools and machinery. This may involve ensuring that tools are regularly sharpened, worn components are replaced or engines are regularly tuned and adjusted.

➤ Introduce a work pattern that reduces the time of exposure to vibration.

➤ Issue employees with gloves and warm clothing. There is a debate as to whether anti-vibration gloves are really effective but it is agreed that warm clothing helps with blood circulation, which reduces the risk of VWF. Care must be taken so that the tool does not cool the hand of the operator.

➤ Introduce a reporting system for employees to use so that concerns and any symptoms can be recorded and investigated.

Finally, information and training for operators and supervisors is required by the Regulations. A training course for the use of pneumatic drills to break up a concrete platform would include the following advice:

➤ Select the smallest correct tool capable of completing the task.

➤ Hold the drill with a light grip and keep the handles in a horizontal position.

➤ Do not press downwards on the drill but let the weight of the tool do the work.

➤ Stop the drill when lifting it to change position.

➤ Cut the concrete in small pieces so that the drill bit does not jam.

Whole-body vibration

WBV in construction arises from driving vehicles over rough terrain. It is highly unlikely that driving vehicles on smooth roads will produce WBV problems. As explained earlier, the most common health problem associated with WBV is back pain. This pain may well have been caused by other activities but WBV will aggravate it. The reasons for back pain in drivers include:

➤ poor posture while driving

➤ incorrect adjustment of the driver's seat

Table 17.3 The change in exposure times as vibration size increases

Value of vibration (m/s^2)	2.5	3.5	5	7	10	14	20
Exposure time to reach action value (h)	8	4	2	1	30 min	15 min	8 min
Exposure time to reach limit value (h)	Over 24	16	8	4	2	1	30 min

➤ difficulty in reaching all relevant controls due to poor design of the control layout
➤ frequent manual handling of loads
➤ frequent climbing up and down from a high cab.

The Regulations require that where there is a likelihood of WBV, the employer must undertake a risk assessment. The HSE Guidance document, L141, gives detailed advice to help with this risk assessment and on estimating daily exposure levels. WBV risks are low for exposures around the action value and usually only simple control measures are necessary.

The Regulations allow a transitional period for the limit value until July 2010 for vehicles supplied before July 2007.

The measurement of whole-body vibration is very difficult and can only be carried out accurately by a specialist competent person. If the risk assessment has been made and the recommended control actions are in place, there is no need to measure the exposure of employees to vibration. However, the HSE have suggested that employers can use the following checklist to estimate whether exposure to WBV is high. Check whether:

➤ there is a warning in the machine manufacturer's handbook that there is a risk of WBV
➤ the task is not suitable for the machine or vehicle being used
➤ operators or drivers are using excessive speeds or operating the machine too aggressively
➤ operators or drivers are working too many hours on machines or vehicles that are prone to WBV
➤ road surfaces are too rough and potholed
➤ drivers are being continual jolted or when going over bumps rising from their seats
➤ vehicles designed to operate on normal roads are used on rough or poorly repaired roads
➤ operators or drivers have reported back problems.

If one or more of the above applies, then exposure to WBV may be high.

The actions for controlling the risks from WBV include the following:
Ensure that:

➤ the driver's seat is correctly adjusted so that all controls can be reached easily and that the driver weight setting on their suspension seat, if available, is correctly adjusted; the seat should have a backrest with lumbar support
➤ anti-fatigue mats are used if the operator has to stand for long periods
➤ the speed of the vehicle is such that excessive jolting is avoided; speeding is one of the main causes of excessive whole-body vibration

➤ all vehicle controls and attached equipment are operated smoothly
➤ only established site roadways are used
➤ only suitable vehicles and equipment are selected to undertake the work and cope with the ground conditions
➤ the site roadway system is regularly maintained
➤ all vehicles are regularly maintained, with particular attention being paid to tyre condition and pressures, vehicles' suspension systems and the driver's seat
➤ work schedules are regularly reviewed so that long periods of exposure on a given day are avoided and drivers have regular breaks
➤ prolonged exposure to WBV is avoided for at-risk groups (older people, young people, people with a history of back problems and pregnant women)
➤ employees are aware of the health risks from WBV, the results of the risk assessment and the ill-health reporting system; they should also be trained to drive in away that reduces excessive vibration.

A simple health monitoring system should be agreed with employees or their representatives that includes a questionnaire checklist (available on the HSE website) to be completed once a year by employees at risk.

More information on the Control of Vibration at Work Regulations is given in Chapter 21.

17.2.4 *Display screen equipment DSE*

Display screen equipment, which includes visual display units, is a good example of a common work activity which relies on an understanding of ergonomics and the ill-health conditions which can be associated with poor ergonomic design.

Legislation governing DSE is covered by the Health and Safety (Display Screen Equipment) Regulations and a detailed summary of them is given in Chapter 21. The Regulations apply to a user or operator of DSE and define a user or operator. The definition is not as tight as many employers would like it to be, but usage in excess of approximately 1 hour continuously each day would define a user. The definition is important since users are entitled to free eye tests and, if required, a pair of spectacles. The basic requirements of the regulations are:

➤ a suitable and sufficient risk assessment of the workstation, including the software in use, trip and electrical hazards from trailing cables and the surrounding environment
➤ workstation compliance with the minimum specifications laid down in the schedules appended to the Regulations
➤ a plan of the work programme to ensure that there are adequate breaks in the work pattern of workers

Seating and posture for typical office tasks

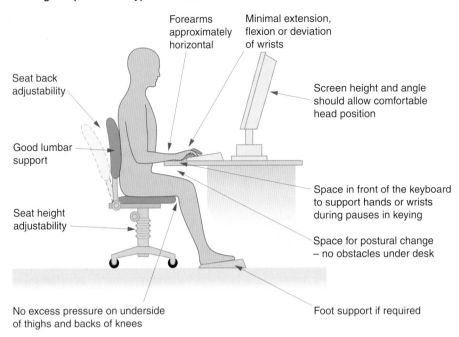

Forearms approximately horizontal

Minimal extension, flexion or deviation of wrists

Seat back adjustability

Screen height and angle should allow comfortable head position

Good lumbar support

Space in front of the keyboard to support hands or wrists during pauses in keying

Seat height adjustability

Space for postural change – no obstacles under desk

No excess pressure on underside of thighs and backs of knees

Foot support if required

Figure 17.5 Workstation design.

➤ the provision of free eyesight tests and, if required, spectacles to users of DSE

➤ a suitable programme of training and sufficient information given to all users.

There are three basic ill-health hazards associated with DSE. These are:

➤ musculoskeletal problems
➤ visual problems
➤ psychological problems.

A fourth hazard, radiation, has been shown from several studies to be very small and is now no longer normally considered in the risk assessment.

Similarly, in the past, there have been suggestions that DSE could cause epilepsy and there were concerns about adverse health effects on pregnant women and their unborn children. All these risks have been shown in various studies to be very low.

Musculoskeletal problems

Tenosynovitis is the most common and well-known problem which affects the wrist of the user. The symptoms and effects of this condition have already been covered. Suffice it to say that if the condition is ignored, then the tendon and tendon sheath around the wrist will become permanently injured. Tenosynovitis, better known as RSI, is caused by the continual use of a keyboard and can be relieved by the use of wrist supports. Other WRULDs are caused by poor posture and can produce pains in the back, shoulders, neck or arms. Less commonly,

pain may also be experienced in the thighs, calves and ankles. These problems can be mitigated by the application of ergonomic principles in the selection of working desks, chairs, footrests and document holders. It is also important to ensure that the desk is at the correct height and the computer screen is tilted at the correct angle to avoid putting too much strain on the neck. (Ideally the user should look down on the screen at a slight angle.)

The keyboard should be detachable so that it can be positioned anywhere on the desktop and a correct posture adopted while working at the keyboard. The chair should be adjustable in height, stable and have an adjustable backrest. If the knees of the user are lower than the hips when seated, then a footrest should be provided. The surface of the desk should be non-reflecting and uncluttered but ancillary equipment (e.g. telephone and printer) should be easily accessible.

Visual problems

There does not appear to be much medical evidence that DSE causes deterioration in eyesight, but many users suffer from visual fatigue which results in eye strain, sore eyes and headaches. Less common ailments are skin rashes and nausea.

The use of DSE may indicate that reading spectacles are needed and the Regulations make provision for this. It is possible that any prescribed lenses may only be suitable for DSE work since they will be designed to give optimum clarity at the normal distance at which screens are viewed (50–60 cm).

Eye strain is a particular problem for people who spend a large proportion of their working day using display screen equipment. A survey has indicated that up to 90% of display screen equipment users complain of eye fatigue. Unnecessary eye strain can be reduced by the following steps additional to those already identified in this section:

➤ Train staff in the correct use of the equipment.
➤ Ensure that a font size of at least 12 is used on the screen.
➤ Ensure that users take regular breaks away from the screen (up to 10 mins every hour).

The screen should be adjustable in tilt angle and screen brightness and contrast. Finally, the lighting around the workstation is important. It should be bright enough to allow documents to be read easily but not too bright so that either headaches are caused or there are reflective glares on the computer screen.

Psychological problems

These are generally stress-related problems. They may have environmental causes, such as noise, heat, humidity or poor lighting, but they are usually due to high speed working, lack of breaks, poor training and poor workstation design. One of the most common problems is the lack of understanding of all or some of the software packages being used.

There are several other processes and activities where ergonomic considerations are important. These include the assembly of small components (microelectronics assembly lines) and continually moving assembly lines (car assembly plants).

Many of these processes have some of the chemical hazards (such as soldering fumes) described in the previous chapter.

17.3 Work environment issues

Work environment issues are covered by the CDM 2007 Regulations together with an Approved Code of Practice.

The issues governing the workplace environment are ventilation, heating and temperature, lighting, workstations and seating.

Ventilation

Ventilation of the workplace should be effective and sufficient and free of any impurity and air inlets should be sited clear of any potential contaminant (e.g. a chimney flue). Care needs to be taken to ensure that workers are not subject to uncomfortable draughts. The ventilation plant should have an effective visual or audible warning device fitted to indicate any failure of the plant. The plant should be properly maintained and records kept. The supply of fresh air should not normally fall below 5 to 8 litres per second per occupant.

Smoking is now forbidden in enclosed or substantially enclosed areas. See Appendix 17.1

Heating and temperature

During working hours, the temperature in all workplaces inside buildings shall be reasonable (not uncomfortably high or low). 'Reasonable' is defined in the Approved Code of Practice as at least 16°C, unless much of the work involves severe physical effort, in which case the temperature should be at least 13°C. These temperatures refer to readings taken close to the workstation at working height and away from windows. The Approved Code of Practice recognizes that these minimum temperatures cannot be maintained where rooms open to the outside or where food or other products have to be kept cold. A heating or cooling method must not be used in the workplace which produces fumes, injurious or offensive to any person. Such equipment needs to be regularly maintained.

A sufficient number of thermometers should be provided and maintained to enable workers to determine

Figure 17.6 Well-lit construction site at night.

the temperature in any workplace inside a building (but need not be provided in every workroom).

Where, despite the provision of local heating or cooling, the temperatures are still unreasonable, suitable protective clothing and rest facilities should be provided.

Lighting

Every workplace shall have suitable and sufficient lighting and this shall be natural lighting so far as is reasonably practicable. Suitable and sufficient emergency lighting must also be provided and maintained in any room where workers are particularly exposed to danger in the event of a failure of artificial lighting (normally due to a power cut and/or a fire). Windows and skylights should be kept clean and free from obstruction, but the shading of windows or skylights may be necessary to prevent excessive heat or glare.

When deciding on the suitability of a lighting system, the general lighting requirements will be affected by the following factors:

➤ the availability of natural light
➤ the specific areas and processes, in particular any colour rendition aspects or concerns over stroboscopic effects (associated with fluorescent lights)
➤ the type of equipment to be used and the need for specific local lighting
➤ the lighting characteristics required (type of lighting, its colour and intensity)
➤ the location of visual display units and any problems of glare
➤ structural aspects of the workroom and the reduction of shadows
➤ the presence of atmospheric dust
➤ the heating effects of the lighting
➤ lamp and window cleaning and repair (and disposal issues)
➤ the need and required quantity of emergency lighting.

Light levels are measured in illuminance, having units of lux (lx), using a light meter. A general guide to lighting levels in different workplaces is given in Table 17.4.

Poor lighting levels will increase the risk of accidents such as slips, trips and falls. More information is available on lighting from *Lighting at work*, HSG38, HSE Books.

Workstations and seating

Workstations should be arranged so that work may be done safely and comfortably. The worker should be at a suitable height relative to the work surface and there should be no need for undue bending and stretching. Workers must not be expected to stand for long periods

Table 17.4 Typical workplace lighting levels

Workplace or type of work	Illuminance (lx)
Warehouses and stores	150
General factories or workshops	300
Offices	500
Drawing offices (detailed work)	700
Fine working (ceramics or textiles)	1,000
Very fine work (watch repairs or engraving)	1,400

of time, particularly on solid floors. A suitable seat should be provided when a substantial part of the task can or must be done sitting. The seat should, where possible, provide adequate support for the lower back and a footrest provided for any worker whose feet cannot be placed flat on the floor. It should be made of materials suitable for the environment, be stable and, possibly, have arm rests.

It is also worth noting that sitting for prolonged periods can present health risks, such as blood circulation and pressure problems, and vertebral and muscular damage.

Seating at work, HSG57, HSE Books, provides useful guidance on how to ensure that seating in the workplace is safe and suitable.

17.4 Noise

There was considerable concern for many years over the increasing cases of occupational deafness and this led to the introduction of Noise at Work Regulations in 1989 and the revised Control of Noise at Work Regulations in 2005. HSE have estimated that an additional 1.1 million workers will be covered by the revised Regulations. These Regulations, which are summarized in Chapter 21, require the employer:

➤ to assess noise levels and keep records
➤ reduce the risks from noise exposure by using engineering controls in the first instance and the provision and maintenance of hearing protection as a last resort
➤ provide employees with information and training
➤ if a manufacturer or supplier of equipment, to provide relevant noise data on that equipment (particularly if any of the three action levels is likely to be reached).

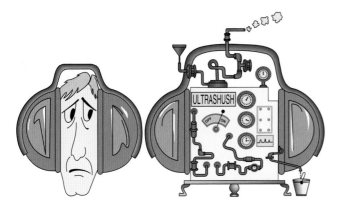

Figure 17.7 Better to control noise at source rather than wear ear protection.

The main purpose of the Noise Regulations is to control noise levels rather than measure them. This involves the better design of machines, equipment and work processes, ensuring that personal protective equipment is correctly worn and that employees are given adequate training and health surveillance.

Sound is transmitted through the air by sound waves, which are produced by vibrating objects. The vibrations cause a pressure wave which can be detected by a receiver, such as a microphone or the human ear. The ear may detect vibrations which vary from 20 to 20,000 (typically 50–16,000) cycles each second (or hertz (Hz)). Sound travels through air at a finite speed (342 m/s at 20°C and sea level). The existence of this speed is shown by the time lag between lightning and thunder during a thunderstorm. Noise normally describes loud, sudden, harsh or irritating sounds.

Noise may be transmitted directly through the air, by reflection from surrounding walls or buildings or through the structure of a floor or building. In construction, the noise and vibrations from a pneumatic drill will be transmitted from the drill itself, the ground being drilled and from the walls of surrounding buildings.

17.4.1 *Health effects of noise*

The human ear
There are three sections of the ear – the outer (or external) ear, the middle ear and the inner (or internal) ear. The sound pressure wave passes into and through the outer ear and strikes the eardrum causing it to vibrate. The eardrum is situated approximately 25 mm inside the head. The vibration of the eardrum causes the proportional movement of three interconnected small bones in the middle ear thus passing the sound to the cochlea situated in the inner ear.

Within the cochlea the sound is transmitted to a fluid causing it to vibrate. The motion of the fluid induces a membrane to vibrate which in turn causes hair cells attached to the membrane to bend. The movement of the hair cells causes a minute electrical impulse to be transmitted to the brain along the auditory nerve. Those hairs nearest to the middle ear respond to high frequency whilst those at the tip of the cochlea respond to lower frequencies.

There are about 30,000 hair cells within the ear and noise-induced hearing loss causes irreversible damage to these hair cells.

Ill-health effects of noise
Noise can lead to ear damage on a temporary (acute) or permanent (chronic) basis.

There are three principal **acute** effects:

➤ Temporary threshold shift – caused by short excessive noise exposures; this affects the cochlea by reducing the flow of nerve impulses to the brain. The result is a slight deafness which is reversible when the noise is removed.

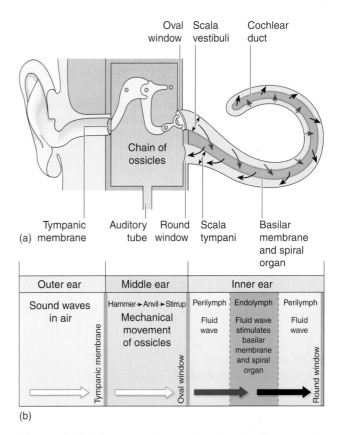

Figure 17.8 Passage of sound waves: (a) the ear with cochlea uncoiled and (b) summary of transmission.

➤ Tinnitus – is a ringing in the ears caused by an intense and sustained high noise level. It is caused by the over-stimulation of the hair cells. The ringing sensation continues for up to 24 hours after the noise has ceased.

➤ Acute acoustic trauma – caused by a very loud noise such as an explosion. It affects either the eardrum or the bones in the middle ear and is usually reversible. Severe explosive sounds can permanently damage the eardrum.

Occupational noise can also lead to one of the following three **chronic** hearing effects:

➤ Noise-induced hearing loss – results from permanent damage to the cochlea hair cells. It affects the ability to hear speech clearly but the ability to hear is not lost completely.

➤ Permanent threshold shift – this results from prolonged exposure to loud noise and is irreversible due to the permanent reduction in nerve impulses to the brain. This shift is most marked at the 4,000 Hz frequency, which can lead to difficulty in hearing certain consonants and some female voices.

➤ Tinnitus – is the same as the acute form but becomes permanent. It is a very unpleasant condition which can develop without warning.

It is important to note that, if the level of noise exposure remains unchanged, noise-induced hearing loss will lead to a permanent threshold shift affecting an increasing number of frequencies.

Presbycusis is the term used for hearing loss in older people which may have been exacerbated by occupational noise earlier in their lives.

17.4.2 Noise assessments

The Control of Noise at Work Regulations specify exposure action levels at which the hearing of employees must be protected. The conclusion as to whether any of those levels have been breached is reached after an assessment of noise levels has been made. However, before noise assessment can be discussed, noise measurement and the statutory action levels must be described.

Noise measurement

Sound intensity is measured by a unit known as a pascal (Pa or N/m^2), which is a unit of pressure similar to that used when inflating a tyre. If noise were measured in this way, a large scale of numbers would be required ranging from 1 at one end to 1 million at the other. The sound pressure level (SPL) is a more convenient scale because:

➤ it compresses the size of the scale by using a logarithmic scale to the base 10

➤ it measures the ratio of the measured pressure, p, to a reference standard pressure, p_0, which is the pressure at the threshold of hearing (2×10^{-5} Pa).

The unit is called a decibel (dB) and is defined as:

$$SPL = 20 \log_{10} (p/p_0) \, dB$$

It is important to note that since a logarithmic scale to the base 10 is used, each increase of 3 dB is a doubling in the sound intensity. Thus if a sound reading changes from 75 to 81 dB, the sound intensity or loudness has increased by four times.

Since the human ear tends to distort its sensitivity to the sound it receives by being less sensitive to lower frequencies, the scale used by sound meters is weighted so that readings mimic the ear. This scale is known as the A scale and the readings are known as dB(A). There are also three other scales known as B, C and D. Table 17.5 gives some typical dB readings for common activities and Table 17.6 gives some typical readings for various construction activities.

Table 17.5 Some typical sound pressure levels (SPL) dB(A) values

Activity or environment	SPL in dB(A)
Threshold of pain	140
Jet aircraft taking off at 25 m	140
Pneumatic drill	125
Pop group or disco	110
Heavy lorry	93
Street traffic	85
Domestic vacuum cleaner	80
Conversational speech	65
Business office	60
Living room	40
Bedroom	25
Threshold of hearing	0

Originally the A scale was used for sound pressure levels up to 55 dB, the B scale for levels between 55 and 85 dB and the C scale for values above 85 dB. However, today, the A scale is used for nearly all levels except for very high sound pressure levels, when the C scale is

Table 17.6 Some typical sound pressure levels (SPL) dB(A) values for construction processes

Activity or environment	SPL in dB(A)
Nail gun	130–140
Piling work	100+
Blasting	100+
Shovelling hardcore	94
Sandblasting	85–90
Carpenter	86–96
Excavator driver	85
Dumper truck driver	85
Asphalt paving	85
Bricklayer	81–85

used. The B scale is rarely used and the D scale is mainly used to monitor jet aircraft engine noise.

Noise is measured using a sound level meter which reads sound pressure levels in dB(A) and the **peak sound pressure** in pascals (Pa), which is the highest noise level reached by the sound. There are two basic types of sound meter – integrated and direct reading meters. Meters which integrate the reading provide an average over a particular time period, which is an essential technique when there are large variations in sound levels. This value is known as the **continuous equivalent noise level** (L_{Eq}), which is normally measured over an 8-hour period.

Direct reading devices, which tend to be much cheaper, can be used successfully when the noise levels are continuous at a near constant value.

Another important noise measurement is the **daily personal exposure level** of the worker, $L_{EP,d}$, which is measured over an 8-hour working day. Hence, if a person was exposed to 87 dB(A) over 4 hours, this would equate to an $L_{EP,d}$ of 84 dB(A) since a reduction of 3 dB(A) represents one-half of the noise dose.

The HSE Guidance on the Noise Regulations, L108, offers some very useful advice on the implementation of the Regulations and should be read by anyone who suspects that they have a noise problem at work. The guidance covers 'equipment and procedures for noise surveys' and contains a noise exposure ready reckoner which can be used to evaluate $L_{EP,d}$ when the noise occurs during several short intervals and/or at several different levels during the 8-hour period.

Noise action levels

Regulation 6 of the Control of Noise at Work Regulations places a duty on employers to reduce the risk of damage to the hearing of his employees from exposure to noise to the lowest level reasonably practicable.

The Regulations introduce exposure **action** level values and exposure **limit** values.

An exposure action value is a level of noise at which certain action must be taken.

An exposure limit value is a level of noise at the ear above which an employee must not be exposed. Therefore if the workplace noise levels are above this value, any ear protection provided to the employee must reduce the noise level to the limit value at the ear.

These exposure action and limit values are given below:

1. The lower exposure action levels are:
    (a) a daily or weekly personal noise exposure of **80 dB(A)**
    (b) a peak sound pressure of **135 dB(C)**
2. The upper exposure action levels are:
    (a) a daily or weekly personal noise exposure of **85 dB(A)**
    (b) a peak sound pressure of **137 dB(C)**
3. The exposure limit values are
    (a) a daily or weekly personal noise exposure of **87 dB(A)**
    (b) a peak sound pressure of **140 dB(C)**.

The peak exposure action and limit values are defined because high-level peak noise can lead to short- and long-term hearing loss. Explosives, guns (including nail guns), cartridge tools, hammers and stone chisels can all produce high peak sound pressures.

If the daily noise exposure exceeds the lower action level, then a noise assessment should be carried out and recorded by a competent person. There is a very simple test which can be done in any workplace to determine the need for an assessment.

Table 17.7 gives other information on simple tests to determine the need for a noise risk assessment.

If there is a marked variation in noise exposure levels during the working week, then the Regulations allow a weekly rather than daily personal exposure level, $L_{EP,w}$, to be used. It is only likely to be significantly different from the daily exposure level if exposure on 1 or 2 days in the working week is 5 dB(A) higher than on the other days or the working week has 3 or fewer days of exposure. The weekly exposure rate is not a simple arithmetic average of the daily rates. If an organization is considering the use of a weekly exposure level, then the following provisions must be made:

➤ Hearing protection must be provided if there are very high noise levels on any one day.
➤ The employees and their representatives must be consulted on whether weekly averaging is appropriate.

Table 17.7 Simple observations to determine the need for a noise risk assessment

Observation at the workplace	Likely noise level	A noise risk assessment must be made if this noise level persists for:
The noise is noticeable but does not interfere with normal conversation – equivalent to a domestic vacuum cleaner	80 dB(A)	6 hours
People have to shout to be heard if they are more than 2 m apart	85 dB(A)	2 hours
People have to shout to be heard if they are more than 1 m apart	90 dB(A)	45 minutes

➤ An explanation must be given to the employees on the purpose and possible effects of weekly averaging.

Finally, if the working day is 12 hours, then the action levels must be reduced by 3 dB(A) because the action levels assume an 8-hour working day.

The HSE Guidance document L108 gives detailed advice on noise assessments and surveys. The most important points are that the measurements should be taken at the working stations of the employees closest to the source of the noise and over as long a period as possible, particularly if there is a variation in noise levels during the working day. Other points to be included in a noise assessment are:

➤ details of the noise meter used and the date of its last calibration
➤ the number of employees using the machine, time period of usage and other work activities
➤ an indication of the condition of the machine and its maintenance schedule
➤ the work being done on the machine at the time of the assessment
➤ a schematic plan of the workplace showing the position of the machine being assessed
➤ other noise sources, such as ventilation systems, should be considered in the assessment; the control of these sources may help to reduce overall noise levels
➤ recommendations for future actions, if any.

Other actions which the employer must undertake when the lower action level is exceeded are:

➤ inform, instruct and train employees on the hearing risks

➤ supply hearing protection to those employees requesting it
➤ ensure that any equipment or arrangements provided under the Regulations are correctly used or implemented.

The additional measures which the employer must take if the upper action level is reached are:

➤ reduce and control exposure to noise by means other than hearing protection
➤ establish hearing protection zones, marked by notices, and ensure that anybody entering the zone is wearing hearing protection
➤ supply hearing protection and ensure that it is worn.

Figure 17.9 Typical ear protection zone sign.

The Control of Noise at Work Regulations place a duty on the employer to undertake health surveillance for employees whose exposure regularly exceeds the upper action level irrespective of whether ear protection was worn. The recommended health surveillance is a hearing test at induction, followed by an annual check and review of hearing levels. The checks may be extended to every 3 years if no adverse effects are found during earlier tests. If exposure continues over a long period, health surveillance of employees is recommended using a more substantial audiometric test. This will indicate whether there has been any deterioration in hearing ability. Where exposure is between the lower and upper exposure action levels or occasionally above the upper action values, health surveillance will only be required if information becomes available, perhaps from medical records, that the employee is particularly sensitive to noise-induced hearing loss.

The Regulations also place the following statutory duties on employees:

➤ For noise levels above the lower action level, they must use any control equipment (other than hearing protection), such as silencers, supplied by the employer and report any defects.

➤ For noise levels above the upper action level, they fulfil the obligations given above and wear the hearing protection provided.

➤ Take care of any equipment provided under these Regulations and report any defects.

➤ See their doctor if they feel that their hearing has become damaged.

The Management of Health and Safety Regulations prohibits the employment of anyone under the age of 18 years where there is a risk to health from noise.

17.4.3 *Noise control techniques*

In addition to **reduced time exposure** of employees to the noise source, there is a simple hierarchy of control techniques. These are to:

➤ reduction of noise at source

➤ reduction of noise levels received by the employee (known as attenuation)

➤ personal protective equipment which should only be used when the above two remedies are insufficient.

Reduction of noise at source

There are several means by which noise could be reduced at source. These are to:

➤ change the process or equipment (e.g. replace solid tyres with rubber tyres or replace diesel engines with electric motors)

➤ change the speed of the machine

➤ improve the maintenance regime by regular lubrication of bearings, tightening of belt drives.

Attenuation of noise levels

There are many methods of attenuating or reducing noise levels and these are covered in detail in the guide to the Regulations. The more common ones will be summarized here.

Orientation or re-location of the equipment – turn the noisy equipment away from the workforce or locate it away in separate and isolated areas.

Enclosure – surrounding the equipment with a good sound insulating material can reduce sound levels by up to 30 dB(A). Care will need to be taken to ensure that the machine does not become overheated.

Screens or absorption walls – can be used effectively in areas where the sound is reflected from walls. The walls of the rooms housing the noisy equipment are lined with sound-absorbent material, such as foam or mineral wool, or sound-absorbent (acoustic) screens are placed around the equipment.

Damping – the use of insulating floor mountings to remove or reduce the transmission of noise and vibrations through the structure of the building such as girders, wall panels and flooring.

Lagging – the insulation of pipes and other fluid containers to reduce sound transmission (and, incidentally, heat loss).

Silencers – normally fitted to engines which are exhausting gases to atmosphere. Silencers consist of absorbent material or baffles.

Isolation of the workers – the provision of sound-proofed workrooms or enclosures isolated away from

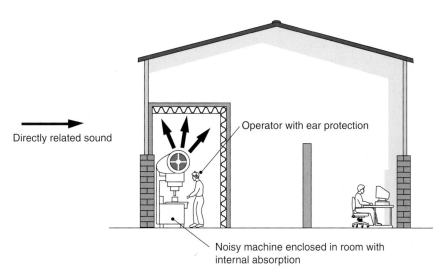

Directly related sound

Operator with ear protection

Noisy machine enclosed in room with internal absorption

Figure 17.10 Segregation of noisy operation to benefit the whole workplace.

noisy equipment (a power station control room is an example of worker isolation).

17.4.4 *Personal ear protection*

The provision of personal hearing protection should only be considered as a last resort. There is usually resistance from the workforce to use them and they are costly to maintain and replace. They interfere with communications, particularly alarm systems, and they can present hygiene problems.

The following factors should be considered when selecting personal ear protection:

➤ suitability for the range of sound spectrum of frequencies to be encountered
➤ noise reduction (attenuation) offered by the ear protection
➤ pattern of the noise exposure
➤ acceptability and comfort of the wearer, particularly if there are medical problems
➤ durability
➤ hygiene considerations
➤ compatibility with other personal protective equipment
➤ ease of communication and able to hear warning alarms
➤ maintenance and storage arrangements
➤ cost.

There are two main types of ear protection – earplugs and ear defenders (earmuffs).

Earplugs are made of sound-absorbent material and fit into the ear. They can be reusable or disposable, are able to fit most people and can easily be used with safety glasses and other personal protective equipment. Their effectiveness depends on the quality of the fit in the ear which, in turn, depends on the level of training given to the wearer. Permanent earplugs come in a range of sizes so that a good fit is obtained. The effectiveness of earplugs decreases with usage and they should be replaced at the intervals specified by the supplier. A useful simple rule to ensure that the selected earplug reduces the noise level at the ear to 87 dB(A) is to choose one with a manufacturer's rating of 83 dB(A). This should compensate for any fitting problems. The main disadvantage of earplugs is that they do not reduce the sound transmitted through the bone structure which surrounds the ear and they often work loose with time.

Ear defenders (earmuffs) offer a far better reduction of all sound frequencies. They are generally more acceptable to workers because they are more comfortable to wear and they are easy to monitor since they are clearly visible. They also reduce the sound intensity transmitted through the bone structure surrounding the ear. A communication system can be built into earmuffs. However, they may be less effective if the user has long hair or is wearing spectacles or large earrings. They may also be less effective if worn with helmets or face shields and uncomfortable in warm conditions. Maintenance is an important factor with earmuffs and should include checks for wear and tear and general cleanliness.

Selection of suitable ear protection is very important since it should not just reduce sound intensities below the statutory action levels but also reduce those intensities at particular frequencies. Normally advice should be sought from a competent supplier, who will be able to advise on ear protection to suit a given spectrum of noise using 'octave band analysis'.

Finally, it is important to stress that the use of ear protection must be well supervised to ensure that not only is it being worn correctly but that it is, in fact, being actually worn.

17.4.5 *Noise in woodworking machines*

Woodworking machines are used widely in the construction industry and the noise levels are usually in excess of the upper exposure action level. There are several Woodworking Information Sheets, published by the HSE, which give a useful indication of some practical solutions to reduce noise levels that do not simply rely on ear defenders. Noise assessments made on typical woodworking machines showed that noise levels ranged from 97 dB for beam panel saws and sanding machines to 105 dB for edge banders and multi-cutter moulding machines. The machines are often used intermittently during the day so that a worker's daily noise exposure will normally be lower than these values. An electronic spreadsheet on the HSE website (www.hse.gov.uk/noise) can be used to calculate the daily exposure level for a worker who uses any woodworking machine for any period during the day.

Noise in woodworking machines is affected by:

➤ the type, size and moisture content of the timber used
➤ the condition, speed and balance of the tooling used
➤ the machine setting and
➤ the design and effectiveness of the extraction unit used.

Effective action on any of these elements could produce a significant reduction in noise levels.

17.5 Heat and radiation hazards

17.5.1 *Extremes of temperature*

The human body is very sensitive to relatively small changes in external temperatures. Food not only provides energy and the build-up of fat reserves, but also generates heat, which needs to be dissipated to the surrounding environment. The body also receives heat from its surroundings. The body temperature is normally around 37°C and the body will attempt to maintain this temperature irrespective of the temperature of the surroundings. Therefore, if the surroundings are hot, sweating will allow heat loss to take place by evaporation caused by air movement over the skin. On the other hand, if the surroundings are cold, shivering causes internal muscular activity, which generates body heat.

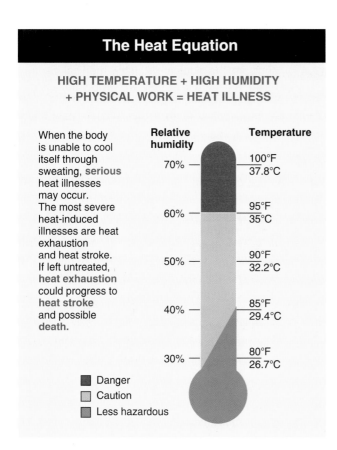

Figure 17.11 The heat equation.

At high temperatures, the body has more and more difficulty in maintaining its natural temperature unless sweating can take place and therefore water must be replaced by drinking. If the surrounding air has high humidity, evaporation of the sweat cannot take place and the body begins to overheat. This leads to heart strain and, in extreme cases, heat stroke. It follows that when working is required at high temperatures, a good supply of drinking water should be available and, further, if the humidity is high, a good supply of ventilation air is also needed. On construction sites, weather is a relevant factor and is dealt with in detail later on. Temperature and heat exhaustion are very relevant to working in a confined space, as described in Chapter 19.

At low temperatures, the body will lose heat too rapidly and the extremities of the body will become very cold leading to frostbite and possibly the loss of limbs. Under these conditions, thick warm clothing and external heating will be required.

In summary, extremes of temperature require special measures, particularly if accompanied by extremes of humidity. Frequent rest periods will be necessary to allow the body to acclimatize to the conditions.

17.5.2 *Ionizing radiation*

Ionizing radiation is emitted from radioactive materials, either in the form of directly ionizing alpha and beta particles or indirectly ionizing X- and gamma rays or neutrons. It has a high energy potential and an ability to penetrate, ionize and damage body tissue and organs.

Figure 17.12 Typical ionizing sign.

All matter consists of atoms within which is a nucleus, containing protons and neutrons, and orbiting electrons. The number of electrons within the atom defines the element – hydrogen has 1 electron and lead has 82 electrons. Some atoms are unstable and will change into atoms of another element, emitting ionizing radiation in the process. The change is called radioactive decay and the ionizing radiations most commonly emitted are alpha and beta particles and gamma rays. X-rays are produced by bombarding a metal target with electrons at very high speeds using very high voltage electrical discharge equipment. Neutrons are released by nuclear fission and are not normally found in manufacturing processes.

Alpha particles consist of two protons and two neutrons and have a positive charge. They have little power to penetrate the skin and can be stopped using

365

very flimsy material, such as paper. Their main route into the body is by ingestion.

Beta particles are high speed electrons whose power of penetration depends on their speed, but penetration is usually restricted to 2 cm of skin and tissue. They can be stopped using aluminium foil. There are normally two routes of entry into the body – inhalation and ingestion.

Gamma rays, which are similar to **X-rays**, are electromagnetic radiations and have far greater penetrating power than alpha or beta particles. They are produced from nuclear reactions and can pass through the body.

There are two principal units of radiation – the becquerel (Bq), which measures the amount of radiation in a given environment and the millisievert (mSv), which measures the ionizing radiation dose received by a person.

Ionizing radiation occurs naturally as well as from man-made processes and about 87% of all radiation exposure is from natural sources. The Ionising Radiations Regulations specify a range of dose limits, some of which are given in Table 17.8.

Table 17.8 Typical radiation dose limits

	Dose (mSv)	Area of body
Employees aged 18 years+	20	Whole body per year
Trainees 16–18 years	6	Whole body per year
Any other person	1	Whole body per year
Women employees of child-bearing age	13	The abdomen in any consecutive 3-month period
Pregnant employees	1	During the declared term of the pregnancy

Harmful effects of ionizing radiation

Ionizing radiation attacks the cells of the body by producing chemical changes in the cell DNA by ionizing it (thus producing free radicals), which leads to abnormal cell growth. The effects of these ionizing attacks depend on the following factors:

➤ the size of the dose – the higher the dose then the more serious will be the effect
➤ the area or extent of the exposure of the body – the effects may be far less severe if only a part of the body (e.g. an arm) receives the dose

➤ the duration of the exposure – a long exposure to a low dose is likely to be less harmful than a short exposure to the same quantity of radiation.

Acute exposure can cause, dependent on the size of the dose, blood cell changes, nausea and vomiting, skin burns and blistering, collapse and death. Chronic exposure can lead to anaemia, leukaemia and other forms of cancer. It is also known that ionizing radiation can have an adverse effect on the function of human reproductive organs and processes. Increases in the cases of sterility, stillbirths and malformed foetuses have also been observed.

The health effects of ionizing radiation may be summarized into two groups – **somatic effects**, which refer to cell damage in the person exposed to the radiation dose and **genetic effects**, which refer to the damage done to the children of the irradiated person.

Sources of ionizing radiation

The principal workplaces which could have ionizing radiation present are the nuclear industry, medical centres (hospitals and research centres) and educational centres. Radioactive processes are used for the treatment of cancers and radioactive isotopes are used for many different types of scientific research. X-rays are used extensively in hospitals, but they are also used in industry for non-destructive testing (e.g. crack detection in welds). Smoke detectors, used in most workplaces, also use ionizing radiations.

Ionizing radiations can also occur naturally – the best example being radon, which is a radioactive gas that occurs mainly at or near granite outcrops where there is a presence of uranium. It is particularly prevalent in Devon and Cornwall. The gas enters buildings normally from the substructure through cracks in flooring or around service inlets. The Ionising Radiations Regulations have set two action levels above which remedial action, such as sumps and extraction fans, have to be fitted to lower the radon level in the building. The first action level is 400 Bq/m^3 in workplaces and 200 Bq/m^3 in domestic properties. At levels above 1,000 Bq/m^3, remedial action should be taken within 1 year.

Site radiography equipment is sometimes used on construction sites usually as a form of non-destructive examination. Such equipment uses ionizing radiations.

Personal radiation exposure can be measured using a film badge, which is worn by the employee over a fixed time interval. The badge contains a photographic film which, after the time interval, is developed and an estimate of radiation exposure is made. A similar device, known as a radiation dose meter or detector, can be positioned on a shelf in the workplace for 3 months, so that a mean value of radiation levels may be measured.

Instantaneous radiation values can be obtained from portable hand-held instruments, known as Geiger counters, which continuously sample the air for radiation levels. Similar devices are available to measure radon levels.

17.5.3 *Non-ionizing radiation*

Non-ionizing radiation includes ultraviolet, visible light (this includes lasers which focus or concentrate visible light), infra-red and microwave radiations. Since the wavelength is relatively long, the energy present is too low to ionize atoms which make up matter. The action of non-ionizing radiation is to heat cells rather than change their chemical composition.

Other than the Health and Safety at Work Act 1974, there is no specific set of Regulations governing non-ionizing radiation. However, the Personal Protective Equipment at Work Regulations are particularly relevant since the greatest hazard is tissue burning of the skin or the eyes.

Ultraviolet radiation occurs with sunlight and with electric arc welding. In both cases, the skin and the eyes are at risk from the effect of burning. The skin will burn (as in sunburn) and repeated exposure can lead to skin cancer. Skin which is exposed to strong sunlight should be protected either by clothing or sun creams. This problem has become more common with the reduction in the ozone layer (which filters out much ultraviolet light). The eyes can be affected by a form of conjunctivitis which feels like grit in the eye, and is called by a variety of names dependent on the activity causing the problem. Arc welders call it 'arc eye' or 'welder's eye' and skiers 'snow blindness'. Cataracts caused by the action of ultraviolet radiation on the eye lens are another possible outcome of exposure. Most construction work takes place in the open air and therefore protection from strong sunlight is essential although it is seldom witnessed.

With growing concern following the rise in skin cancers, the HSE has suggested the following hierarchy of controls for outdoor working:

➤ Relocate some jobs inside a building or to a shady location.
➤ Undertake some outdoor work earlier or later in the day.
➤ Provide personal protection (hats and/or sun cream).
➤ Provide suitable education and training to outdoor workers.

Lasers use visible light and light from the invisible wavelength spectrum (infra-red and ultraviolet). As the word laser implies, they produce '**l**ight **a**mplification by **s**timulated **e**mission of **r**adiation'. This light is highly concentrated and does not diverge or weaken with distance and the output is directly related to the chemical composition of the medium used within the particular laser. The output beam may be pulsating or continuous – the choice being dependent on the task of the laser. Lasers have a large range of applications including bar code reading at a supermarket checkout, the cutting and welding of metals and accurate measurement of distances and elevations required in land and mine surveying. They are also extensively used in surgery for cataract treatment and the sealing of blood vessels.

Lasers are classified into five classes (1, 2, 3a, 3b and 4) in ascending size of power output. Classes 1 and 2 are relatively low hazard and only emit light in the visible band. Classes 3a, 3b and 4 are more hazardous and the appointment of a laser safety officer is recommended. All lasers should carry information stating their class and any precautions required during use.

The main hazards associated with lasers are eye and skin burns, toxic fumes, electricity and fire. The vast majority of accidents with lasers affect the eyes. Retinal damage is the most common and is irreversible. Cataract development and various forms of conjunctivitis can also result from laser accidents. Skin burning and reddening (erythema) are less common and are reversible.

Infra-red radiation is generated by fires and hot substances and can cause eye and skin damage similar to that produced by ultraviolet radiation. It is a particular problem to fire fighters and those who work in foundries or near furnaces. Eye and skin protection are essential.

Microwaves are used extensively in cookers and mobile telephones and there are ongoing concerns about associated health hazards (and several inquiries are currently underway). The severity of any hazard is proportional to the power of the microwaves. The principal hazard is the heating of body cells, particularly those with little or no blood supply to dissipate the heat. This means that tissues such as the eye lens are most at risk from injury. However, it must be stressed that any risks are higher for items such as cookers, than for low-powered devices, such as mobile phones.

The measurement of non-ionizing radiation normally involves the determination of the power output being received by the worker. Such surveys are best performed by specialists in the field, since the interpretation of the survey results requires considerable technical knowledge.

17.5.4 *Radiation protection strategies*

Ionizing radiation

Protection is obtained by the application of shielding, time and distance either individually or, more commonly, a mixture of all three.

Shielding is the best method because it is an 'engineered' solution. It involves the placing of a physical shield, such as a layer of lead, steel and concrete, between the worker and the radioactive source. The thicker the shield the more effective it is.

Time involves the use of the reduced time exposure principle and thus reduces the accumulated dose.

Distance works on the principle that the effect of radiation reduces as the distance between the worker and the source increases.

Other measures include the following:

➤ effective emergency arrangements
➤ training of employees
➤ the prohibition of eating, drinking and smoking adjacent to exposed areas
➤ a high standard of personal cleanliness and first-aid arrangements
➤ strict adherence to personal protective equipment arrangements, which may include full body protection and respiratory protection equipment
➤ procedures to deal with spillages and other accidents
➤ prominent signs and information regarding the radiation hazards
➤ medical surveillance of employees.

The Ionising Radiations Regulations specify a range of precautions which must be taken, including the appointment of a Radiation Protection Supervisor and a Radiation Protection Adviser.

The Radiation Protection Supervisor must be appointed by the employer to advise on the necessary measures for compliance with the Regulations and its Approved Code of Practice. The person appointed, who is normally an employee, must be competent to supervise the arrangements in place and have received relevant training.

The Radiation Protection Adviser is appointed by the employer to give advice to the Radiation Protection Supervisor and employer on any aspect of working with ionizing radiation including the appointment of the Radiation Protection Supervisor. The Radiation Protection Adviser is often an employee of a national organization with expertise in ionizing radiation.

Non-ionizing radiation

For ultraviolet and infra-red radiation, eye protection in the form of goggles or a visor is most important, particularly when undertaking arc welding or furnace work. Skin protection is also likely to be necessary for the hands, arms and neck in the form of gloves, sleeves and a collar. For construction and other outdoor workers, protection from sunlight is important, particularly for the head and nose. Sun creams should also be used.

For laser operations, engineering controls such as fixed shielding and the use of non-reflecting surfaces around the workstation are recommended. For laser in

the higher class numbers, special eye protection is recommended. A risk assessment should be undertaken before a laser is used.

Engineering controls are primarily used for protection against microwaves. Typical controls include the enclosure of the whole microwave system in a metal surround and the use of an interlocking device that will not allow the system to operate unless the door is closed.

The non-ionizing radiation hazards caused by arc welding are not the only hazards associated with welding operations. The hazards from fume inhalation, trailing cables and pipes and the manual handling of cylinders are also present. There have also been serious injuries resulting from explosions and fires during welding processes. Many of these accidents occur on farms where welding equipment is used to make on-the-spot repairs to agricultural machinery.

17.6 The causes and prevention of workplace stress

In 2001, the HSE estimated that stress in the workplace cost approximately 6.7 million days lost each year and society between £3.7 billion and £3.8 billion. This has been accompanied by an increase in civil claims resulting from stress at work.

Stress is not a disease – it is the natural reaction to excessive pressure. Stress can lead to an improved performance, but is not a good thing, since it is likely to lead to both physical and mental ill-health, such as high blood pressure, peptic ulcers, skin disorders and depression.

Most people experience stress at some time during their lives, e.g. during an illness or death of a close relative or friend. However, recovery normally occurs after the particular crisis has passed. The position is, however, often different in the workplace because the underlying causes of the stress, known as work-related stressors, are not relieved but continue to build up the stress levels until the employee can no longer cope.

The basic workplace stressors are:

➤ the job itself – boring or repetitive, unrealistic performance targets, insufficient training, job insecurity or fear of redundancy
➤ individual responsibility – ill-defined roles and too much responsibility with too little power to influence the job outputs
➤ working conditions – cramped, dirty and untidy workplace; unsafe practices; lack of privacy or security; inadequate welfare facilities; threat of violence; excessive noise, vibration or heat; poor lighting; lack of flexibility in working hours to meet domestic

requirements; and adverse weather conditions for those working outside

➤ management attitudes – poor communication, consultation or supervision, negative health and safety culture, lack of support in a crisis

➤ relationships – unhappy relationship between workers, bullying, sexual and racial harassment.

Possible solutions to all these stressors have been addressed throughout this book and involve the creation of a positive health and safety culture, effective training and consultation procedures and a set of health and safety arrangements which work on a day-to-day basis.

The HSE have produced their own generic stress audit survey tool, which is available free of charge on their website and advises the following action plan:

➤ Identify the problem.
➤ Identify the background to the problem and how it was discovered.
➤ Identify the remedial action required and give reasons for that action.
➤ Identify targets and reasonable target dates.
➤ Agree a review date with employees to check that the remedial action is working.

The following additional measures have also been found to be effective by some employers, i.e. to:

➤ Take a positive attitude to stress issues by becoming familiar with the causes and controls.
➤ Take employees' concerns seriously and develop a counselling system which will allow a frank, honest and confidential discussion of stress-related problems.
➤ Develop an effective system of communication and consultation and ensure that periods of uncertainty are kept to a minimum.
➤ Set out a simple policy on work-related stress and include stressors in risk assessments.
➤ Ensure that employees are given adequate and relevant training and realistic performance targets.
➤ Develop an effective employee appraisal system which includes mutually agreed objectives.
➤ Discourage employees from working excessive hours and/or missing break periods (this may involve a detailed job evaluation).
➤ Introduce job rotation and increase job variety.
➤ Develop clear job descriptions and ensure that the individual is matched to them.
➤ Encourage employees to improve their lifestyle (e.g. many local health authorities provide smoking cessation advice sessions).
➤ Monitor incidents of bullying, sexual and racial harassment and, where necessary, take disciplinary action.

➤ Train supervisors to recognize stress symptoms among the workforce.
➤ Avoid a blame culture over accidents and incidents of ill-health.

The individual can also take action if he feels that he is becoming over-stressed. Regular exercise, change of job, review of diet and talking to somebody, preferably a trained counsellor, are all possibilities.

Workplace stress has no specific health and safety regulations but is covered by the duties imposed by the Health and Safety at Work Act and the Management of Health and Safety at Work Regulations to:

1. ensure so far as is reasonably practicable that workplaces are safe and without risk to health
2. carry out a risk assessment relating to the risks to health
3. introduce and maintain appropriate control measures.

There have been several successful civil actions for compensatory claims resulting from the effects of workplace stress. However, the Court of Appeal in 2002 redefined the guidelines under which workplace stress compensation claims may be made. Their full guidelines should be consulted and consist of sixteen points. In summary, these guidelines are that:

➤ no occupations should be regarded as intrinsically dangerous to mental health
➤ it is reasonable for the employer to assume that the employee can withstand the normal pressures of the job, unless some particular problem or vulnerability has developed
➤ the employer is only in breach of duty if they have failed to take the steps that are reasonable in the circumstances
➤ the size and scope of the employer's operation, its resources and the demands it faces are relevant in deciding what is reasonable; including the interests of other employees and the need to treat them fairly in, for example, the redistribution of duties
➤ if a confidential counselling advice service is offered to the employee, the employer is unlikely to be found in breach of his duty
➤ if the only reasonable and effective action is to dismiss or demote the employee, the employer is not in breach of his duty by allowing a willing worker to continue in the same job
➤ the assessment of damages will take account of any pre-existing disorder or vulnerability and the chance that the claimant would have succumbed to a stress-related disorder in any event.

369

Stress usually occurs when people feel that they are losing control of a situation. In the workplace, this means that the individual no longer feels that they can cope with the demands made on them. Many such problems can be partly solved by listening to, rather than talking at, people.

The causes and prevention of workplace violence

Violence at work, particularly from dissatisfied customers, clients, claimants or patients, causes a lot of stress and in some cases injury. This is not only physical violence as people may face verbal and mental abuse, discrimination, harassment and bullying. Fortunately, physical violence is still rare, but violence of all types has risen significantly in recent years. Violence at work is known to cause pain, suffering, anxiety and stress, leading to financial costs due to absenteeism and higher insurance premiums to cover increased civil claims. It can be very costly to ignore the problem.

Figure 17.13 Security access and surveillance CCTV camera.

In 1999 the Home Office and the HSE published a comprehensive report entitled *Violence at Work: Findings from the British Crime Survey.* This was updated with a joint report *Violence at Work: New Findings of the British Crime Survey 2000,* which was published in July 2001. This report shows the extent of violence at work and how it has changed during the period 1991–1999 (Table 17.9).

Table 17.9 Trend in physical assaults and threats at work, 1991–1999 (based on working adults of working age)

Number of incidents (000s)	1991	1993	1995	1997	1999
All violence	947	1,275	1,507	1,226	1,288
Assaults	451	652	729	523	634
Threats	495	607	779	703	654
Number of victims (000s)					
All violence	472	530	570	649	604
Assaults	227	287	290	275	304
Threats	264	286	352	395	338

Source: British Crime Survey 1999.

The report defines violence at work as:

All assaults or threats which occurred while the victim was working and were perpetuated by members of the public.

Physical assaults include the offences of common assault, wounding, robbery and snatch theft. Threats include both verbal threats, made to or against the victim and non-verbal intimidation. These are mainly threats to assault the victim and, in some cases, to damage property.

Excluded from the survey are violent incidents where there was a relationship between the victim and the offender and also where the offender was a work colleague. The latter category was excluded because of the different nature of such incidents.

The British Crime Survey shows that the number of incidents and victims rose rapidly between 1991 and 1995 but then declined to 1997. Unfortunately, since 1997 the decline seems to have reversed as the number of incidents has increased by 5%.

It is interesting that almost half of the assaults and a third of the threats happened after 18:00 hours, which suggests that the risks are higher if people work at night or in the late evening. Sixteen per cent of the assaults involved offenders under the age of 16 and were mainly against teachers or other education workers.

Violence at work is defined by the HSE as:

any incident in which an employee is abused, threatened or assaulted in circumstances relating to their work.

In recognition of this, the HSE has produced a useful guide to employers which includes a four stage action plan and some advice on precautionary measures (also see 'work-related violence case studies', 2002, HSG 226, ISBN 9780 7176 2358 0 (*Violence at Work – a guide for employers*, INDG69 (rev)). The employer is just as responsible, under health and safety legislation, for protecting employees from violence as they are for any other aspects of employee safety.

The Health and Safety at Work Act 1974 puts broad general duties on employers and others to protect the health and safety of staff. In particular, section 2 of the HSW Act gives employers a duty to safeguard, so far as is reasonably practicable, the health, safety and welfare at work of their staff.

Employers also have a common law general duty of care towards their staff, which extends to the risk of violence at work. Legal precedents (see *West Bromwich Building Society v Townsend* [1983] IRLR 147 and *Charlton v Forrest Printing Ink Company Limited* [1980] IRLR 331) show that employers have a duty to take reasonable care to see that their staff are not exposed to unnecessary risks at work, including the risk of injury by criminals. In carrying out their duty to provide a safe system of work and a safe working place, employers should, therefore, have regard to, and safeguard their staff against, the risk of injury from violent criminals.

The HSE recommends the following four-point action plan:

1. Find out if there is a problem.
2. Decide on what action to take.
3. Take the appropriate action.
4. Check that the action is effective.

17.7.1 *Find out if there is a problem*

This involves a risk assessment to determine what the real hazards are. It is essential to ask people at the workplace and, in some cases, a short questionnaire may be useful. Record all incidents to get a picture of what is happening over time, making sure that all relevant detail is recorded. The records should include:

➤ a description of what happened
➤ details of who was attacked, the attacker and any witnesses
➤ the outcome, including how people were affected and how much time was lost
➤ information on the location of the event.

Due to the sensitive nature of some aggressive or violent actions, employees may need to be encouraged to report incidents and protected from future aggression.

All incidents should be classified so that an analysis of the trends can be examined.

Consider the following:

➤ fatalities
➤ major injury
➤ less severe injury or shock which requires first-aid treatment, outpatient treatment, time off work or expert counselling
➤ threat or feeling of being at risk or in a worried or distressed state.

17.7.2 *Decide on what action to take*

It is important to evaluate the risks and decide who may be harmed and how this is likely to occur. The threats may be from the public or co-workers at the workplace or may be as a result of visiting the homes of customers. Consultation with employees or other people at risk will improve their commitment to control measures and will make the precautions much more effective. The level of training and information provided, together with the general working environment and the design of the job, all have a significant influence on the level of risk.

Those people at risk could include those working in:

➤ reception or customer service points
➤ enforcement and inspection
➤ lone working situations and community based activities
➤ front line service delivery
➤ education and welfare
➤ catering and hospitality
➤ retail petrol and late night shopping operations
➤ leisure facilities, especially if alcohol is sold
➤ healthcare and voluntary roles
➤ policing and security
➤ mental health units or in contact with disturbed people
➤ cash handling or control of high value goods.

Consider the following issues:

➤ quality of service provided
➤ design of the operating environment
➤ type of equipment used
➤ designing the job.

Some violence may be deterred if measures are taken which suggest that any violence may be recorded. Many public bodies use the following measures:

➤ informing telephone callers that their calls will be recorded

➤ displaying prominent notices that violent behaviour may lead to the withdrawal of services and prosecution

➤ using closed circuit television (CCTV) or security personnel.

Quality of service provided

The type and quality of service provision has a significant effect on the likelihood of violence occurring in the workplace. Frustrated people whose expectations have not been met and who are treated in an unprofessional way may believe they have the justification to cause trouble.

Sometimes circumstances are beyond the control of the staff member and potentially violent situations need to be defused. The use of correct skills can turn a dissatisfied customer into a confirmed supporter simply by careful response to their concerns. The perceived lack of or incorrect information can cause significant frustrations.

Design of the operating environment

Personal safety and service delivery are very closely connected and have been widely researched in recent years. This has resulted in many organizations altering their facilities to reduce customer frustration and enhance sales. It is interesting that most service points experience less violence when they remove barriers or screens, but the transition needs to be carefully planned in consultation with staff and other measures adopted to reduce the risks and improve their protection.

The layout, ambience, colours, lighting, type of background music, furnishings including their comfort, information, things to do while waiting and even smell all have a major impact. Queue jumping causes a lot of anger and frustration and needs effective signs and proper queue management, which can help to reduce the potential for conflict.

Wider desks, raised floors and access for special needs, escape arrangements for staff, carefully arranged furniture and screening for staff areas can all be utilized.

Type of security equipment used

There is a large amount of equipment available and expert advice is necessary to ensure that it is suitable and sufficient for the task. Some measures that could be considered include the following:

➤ **Access control** to protect people and property. There are many variations from staffed and friendly receptions, barriers with swipe cards and simple coded security locks. The building layout and design may well partly dictate what is chosen. People inside the premises need access passes so they can be identified easily.

➤ **Closed circuit television** is one of the most effective security arrangements to deter crime and violence. Because of the high cost of the equipment, it is essential to ensure that proper independent advice is obtained on the type and the extent of the system required.

➤ **Alarms** – there are three main types:
  ➤ intruder alarms fitted in buildings to protect against unlawful entry, particularly after hours
  ➤ panic alarms used in areas such as receptions and interview rooms covertly located so that they can be operated by the staff member threatened
  ➤ personal alarms carried by an individual to attract attention and to temporarily distract the attacker.

➤ **Radios and pagers** can be a great asset to lone workers in particular, but special training is necessary as good radio discipline with a special language and codes is required.

➤ **Mobile phones** are an effective means of communicating and keeping colleagues informed of people's movements and problems such as travel delays. Key numbers should be inserted for rapid use in an emergency.

Job design

Many things can be done to improve the way in which the job is carried out to improve security and avoid violence. These include:

➤ using cashless payment methods
➤ keeping money on the premises to a minimum
➤ careful check of customer or client's credentials
➤ careful planning of meetings away from the workplace
➤ teamwork where suspected aggressors may be involved
➤ regular contact with workers away from their base; there are special services available to provide contact arrangements
➤ avoidance of lone working as far as is reasonably practicable
➤ thinking about how staff who have to work shifts or late hours will get home; safe transport and/or parking areas may be required
➤ setting up support services to help victims of violence and, if necessary, other staff who could be affected; they may need debriefing, legal assistance, time off work to recover or counselling by experts.

17.7.3 *Take the appropriate action*

The arrangements for dealing with violence should be included in the safety policy and managed like any other aspect of the health and safety procedures. Action plans should be drawn up and followed through using the consultation arrangements as appropriate. The police should also be consulted to ensure that they are happy with the plan and are prepared to play their part in providing back-up and the like.

17.7.4 *Check that the action is effective*

Ensure that the records are being maintained and any reported incidents are investigated and suitable action taken. The procedures should be regularly audited and changes made if they are not working properly.

Victims should be provided with help and assistance to overcome their distress, through debriefing, counselling, time off to recover, legal advice and support from colleagues.

17.8 The effects of alcohol and drugs

Alcohol and drug abuse damages health and causes absenteeism and reduced productivity. Both forms of abuse can lead to serious accidents in the construction industry particularly when working occurs at height. The HSE is keen to see employers address the problem and offers advice in two separate booklets.

Alcohol abuse is a considerable problem when vehicle driving is part of the job especially if driving is required on public roads. Misuse of alcohol can reduce productivity, increase absenteeism, increase accidents at work and, in some cases, endanger the public. The HSE has estimated that between 3% and 5% of all absences from work is due to alcohol and results in approximately 14 million working days lost each year. Employers need to adopt an alcohol policy following employee consultation. The following matters need to be considered:

➤ how the organization expects employees to restrict their level of drinking
➤ how problem drinking can be recognized and help offered and
➤ at what point and under what circumstances will an employee's drinking be treated as a disciplinary rather than a health problem.

Prevention of the problem is better than remedial action after a problem has occurred. On construction sites, there should be no drinking, and drinking during break periods should be discouraged. Induction training

Figure 17.14   Help needed here.

should stress this policy and managers should set guidelines. Posters can also help to communicate the message. However, it is important to recognize possible symptoms of an alcohol problem, such as lateness and absenteeism, poor work standards, impaired concentration, memory and judgement, and deteriorating relations with colleagues. It is always better to offer counselling rather than dismissal. The policy should be monitored to check on its effectiveness.

Drug abuse presents similar problems to those found with alcohol abuse – absenteeism, reduced productivity and an increase in the risk of accidents. A recent study undertaken by Cardiff University found that '...although drug use was lower among workers than the unemployed. One in four workers under the age of 30 years reported having used drugs in the previous year.... There are well documented links between drug use and impairments in cognition, perception, and motor skills, both at the acute and chronic levels. Associations may therefore exist between drug use and work performance'. The following conclusions were also drawn from the study:

➤ 13% of working respondents reported drug use in the previous year. The rate varied with age, from 3% of those over 50 to 29% of those under 30

➤ drug use is strongly linked to smoking and heavy drinking in that order

➤ there is an association between drug use and minor injuries among those who are also experiencing other minor injury risk factors

➤ the project has shown that recreational drug use may reduce performance efficiency and safety at work.

A successful drug misuse policy will benefit the organization and employees by reducing absenteeism, poor productivity and the risk of accidents. There is no simple guide to the detection of drug abuse, but the HSE has suggested the following signals:

➤ sudden mood changes
➤ unusual irritability or aggression
➤ a tendency to become confused
➤ abnormal fluctuations in concentration and energy
➤ impaired job performance
➤ poor time-keeping
➤ increased short-term sickness leave
➤ a deterioration in relationships with colleagues, customers or management
➤ dishonesty and theft (arising from the need to maintain an expensive habit).

A policy on drug abuse can be established by the following:

1. Investigation of the size of the problem.
   Examination of sicknesses, behavioural and productivity changes and accident and disciplinary records is a good starting point.

2. Planning actions.
   Develop an awareness programme for all staff and a special training programme for managers and supervisors. Employees with a drug problem should be encouraged to seek help in a confidential setting.

3. Taking action.
   Produce a written policy that includes everyone in the organization and names the person responsible for implementing the policy. It should include details of the safeguards to employees and the confidentiality given to anyone with a drug problem. It should also clearly outline the circumstances in which disciplinary action will be taken (the refusal of help, gross misconduct and possession/dealing in drugs).

4. Monitoring the policy.
   The policy can be monitored by checking for positive changes in the measures made during the initial investigation (improvements in the rates of sickness and accidents). Drug screening and testing is a sensitive issue and should only be considered with the agreement of the workforce (except in the case of pre-employment testing). Screening will only be acceptable if it is seen as part of the health policy of the organization and its purpose is to reduce risks to the misusers and others. Laboratories are accredited by the United Kingdom Accreditation Service (UKAS) and are able to undertake reliable testing.

It is important to stress that some drugs are prescribed and controlled. The side effects of these can also affect performance and pose risks to colleagues. Employers should encourage employees to inform them of any possible side effects from prescribed medication and be prepared to alter work programmes accordingly.

17.9 Practice NEBOSH questions for Chapter 17

1. (a) **Define** the term 'ergonomics'.
   (b) **List SIX** observations made during an inspection of a machine operation which may suggest that the machine has not been ergonomically designed.

2. **Outline** the factors that could contribute towards the development of WRULDs among employees working at a supermarket checkout.

3. In relation to WRULDs:
   (a) **identify** the typical symptoms that might be experienced by affected individuals
   (b) **outline** the factors that would increase the risk of developing WRULDs.

4. A computer operator has complained of neck and back pain. **Outline** the features associated with the workstation that might have contributed towards this condition.

5. (a) **Outline** the possible health risks associated with working in a seated position for prolonged periods of time.
   (b) **Outline** the features of a suitable seat for sedentary work.

6. Inadequate lighting in the workplace may affect the level of stress among employees. **Outline EIGHT** other factors associated with the physical environment that may increase stress at work.

7. (a) **Identify** the possible effects on health that may be caused by working in a hot environment such as a foundry.
   (b) **Outline** measures that may be taken to help prevent the health effects identified in (a).

8. (a) **Explain** the meaning of the term 'daily personal noise exposure' ($L_{EP,d}$).
   (b) **Outline** the actions required when employees' exposure to noise is found to be in excess of 85 dB(A) $L_{EP,d}$.

9. In relation to the Noise at Work Regulations 2005:
   (a) **state**, in dB(A), the first and second action levels
   (b) **outline** the measures that should be taken when employees are exposed to noise levels in excess of the second action level.

10. (a) **Describe** the **TWO** main types of personal hearing protection.
    (b) **Identify FOUR** reasons why personal hearing protection may fail to provide adequate protection against noise.

11. A pneumatic drill is to be used during extensive repair work to the floor of a busy warehouse.
    (a) **Identify**, by means of a labelled sketch, **THREE** possible transmission paths the noise from the drill could take.
    (b) **Outline** appropriate control measures to reduce the noise exposures of the operator and the warehouse staff.

12. **Explain** the meaning of the following terms in relation to noise control:
    (a) silencing
    (b) absorption
    (c) damping
    (d) isolation.

13. Whilst conducting a safety inspection on a civil engineering contract, you observe an employee using a jackhammer but wearing no personal protection. You further discover from an on-the-spot reading that the noise level exceeds 85 dB(A).
    Outline:
    (a) the key requirements of health and safety regulations, other than those placed on the employee, that are relevant to this situation
    (b) the actions that you would recommend to be taken.

14. **Prepare notes** for a briefing session (toolbox talk) on the prevention of ill-health effects from the use of hand-held vibrating equipment.

15. (a) **Describe** the effects that vibration may have on the body, identifying **TWO** common sources of vibration encountered in the construction industry.
    (b) In order to reduce the effects described in (a), **outline** the measures that might be taken by:
    (a) manufacturers of equipment
    (b) employers
    (c) employees.

16. (a) For each of the following types of non-ionizing radiation, **identify** a source and **state** the possible ill-health effects on exposed individuals:
    (a) infra-red radiation
    (b) ultraviolet radiation.
    (b) **Identify** the general methods for protecting people against exposure to non-ionizing radiation.

17. (a) **Identify TWO** types of non-ionizing radiation, giving an occupational source of each.
    (b) **Outline** the health effects associated with exposure to non-ionizing radiation.

18. (a) **Outline** the practical measures that should be adopted to prevent fires and explosions during welding operations on a construction site.
    (b) **Identify TWO** types of radiation associated with welding activities.

19. **Outline** the health and safety risks associated with welding operations.

20. **Outline** the precautions that should be taken on a construction site in order that welding may be carried out safely.

21. **Outline** the factors that may lead to unacceptable levels of occupational stress among employees.

22. **Outline** the health, safety and welfare issues that a company might need to consider before introducing a night shift to cope with an increased demand for its products.

23. **Outline** the actions that site management might consider in order to reduce levels of occupational stress amongst workers on site.

24. **Outline** the practical measures that might be taken to reduce the risk of violence to employees who deal with members of the public as part of their work.

25. **Outline FIVE** possible indications that an employee may be suffering from alcohol or drug abuse.

Appendix 17.1 – Smokefree workplaces

1. Introduction

Second-hand tobacco smoke is a major cause of heart disease and lung cancer amongst non-smokers who work with people who smoke. It is estimated that around 700 workers a year die as a direct result of second-hand tobacco smoke in their workplace.

Second-hand smoke is also responsible for many thousands of episodes of illness. For example, Asthma UK reports that it is the second most common asthma trigger in the workplace. Eighty-two per cent of people with asthma say that other people's smoke worsens their asthma and one in five people with asthma feel excluded from parts of their workplace where people smoke.

In Summer 2007, legal restrictions on smoking in workplaces and public places were introduced in England. Similar legislation had been implemented in Wales and Scotland already. The English legislation comes about through the Health Act 2006, which effectively bans smoking in all enclosed workplaces and public places, with some exemptions. The following information comes from the Department of Health website:

2. A quick guide to the new smokefree law is as follows:

➤ England became smokefree on Sunday, 1 July 2007. The new law was introduced to protect employees and the public from the harmful effects of second-hand smoke.

➤ From 1 July 2007 it is against the law to smoke in virtually all enclosed public places, workplaces and public and work vehicles. There will be very few exemptions from the law.

➤ Indoor smoking rooms in virtually all public places and workplaces is no longer allowed.

➤ Managers of smokefree premises and vehicles have legal responsibilities to prevent people from smoking.

➤ The new law requires no-smoking signs to be displayed in all smokefree premises and vehicles.

➤ The new law applies to anything that can be smoked. This includes cigarettes, pipes (including water pipes such as shisha and hookah pipes), cigars and herbal cigarettes.

Failure to comply with the new law is a criminal offence.

Penalties and fines for smokefree offences are set out below (there are some discounted amounts for quick payment):

➤ **Smoking in smokefree premises or work vehicles**: a fixed penalty notice of £50 imposed on the person smoking. Or a maximum fine of £200 if prosecuted and convicted by a court.

➤ **Failure to display no-smoking signs**: a fixed penalty notice of £200 imposed on whoever manages or occupies the smokefree premises or vehicle. Or a maximum fine of £1,000 if prosecuted and convicted by a court.

➤ **Failing to prevent smoking in a smokefree place**: a maximum fine of £2,500 imposed on whoever manages or controls the smokefree premises or vehicle if prosecuted and convicted by a court.

Local councils will be responsible for enforcing the new law. They will offer information and support to help businesses meet their legal obligations under the new law.

You can find out more information on the new law on the Smokefree England website at **smokefreeengland.co.uk**. You can also contact your local council for information.

Working at height – hazards and control

18

Introduction

Work at height accounts for about 50–60 deaths – more than any other workplace activity – and 4,000 injuries each year. During the first 2 weeks of June 2003, inspectors from the Health and Safety Executive (HSE) visited 1,446 construction sites and stopped all work at one-quarter of them due to concerns about the level of risks of falls from height. Another 5% of sites visited were issued with improvement notices. Other problems included a lack of or inadequate toe boards and intermediate guard rails on scaffolding and working platforms.

Attempts have been made to address these concerns by the introduction of the Work at Height Regulations which apply to all operations carried out at height not just construction work so that they are also relevant to, for example, window cleaning, tree surgery, maintenance work at height and the changing of street lamps.

The Work at Height Regulations affect approximately 3m workers where working at height is essential to their work. Amongst other measures, the height of guard rails on scaffolds is increased by 40 to 950 mm thus recognizing that the average height of workers has increased over the last 40 years. The design of construction projects can also reduce health and safety hazards and accident rates. A study of accident data on falls from height over the 5-year period from 1996 until 2001 emphasized the importance of building design in the elimination of work at height hazards. HSE recently published the following information:

- In 2001/2002, 254 painters and decorators were seriously injured due to falls from height, 89 of which were falls over 2 m.
- In 2002/2003, 223 painters and decorators were seriously injured due to falls from height, 103 of which were falls under 2 m.

- In 2003/2004, 245 painters and decorators were seriously injured due to falls from height; 6 of the falls were fatal.

To protect workers at height from serious injury, the Work at Height Regulations gives the following hierarchy of control:

1. Avoid working at height if possible.
2. Use an existing safe place of work.
3. Provide work equipment to prevent falls.
4. Mitigate distance and consequences of a fall.
5. Instruction, training and supervision.

This hierarchy is discussed in more detail later in this chapter.

The Work at Height Regulations 2004 (WAHR)

The WAHR have no minimum height requirement for work at height. They include all work activities where there is a need to control a risk of falling a distance liable to cause personal injury. This is regardless of the work equipment being used, the duration of the work at height involved or the height at which the work is performed. It includes access to and egress from a place of work, and would therefore include:

- working on a scaffold or from a mobile elevated work platform (MEWP)
- sheeting a lorry or dipping a road tanker
- working on the top of a container in docks or on a ship or storage area
- tree surgery and other forestry work at height
- using cradles or rope for access to a building or other structure like a ship under repair

Figure 18.1 Working at height – mast climbing platforms.

- climbing permanent structures like a gantry or telephone pole
- working near an excavation area or cellar opening if a person could fall into it and be injured
- painting or pasting and erecting bill posters at height
- work on staging or trestles, for example for filming or events
- using a ladder/stepladder or kick stool for shelf filling, window cleaning and the like
- using man-riding harnesses in ship repair, off-shore or steeplejack work
- working in a mine shaft or chimney
- work carried out at a private house by a person employed for the purpose, for example painter and decorator (but not if the private individual carries out work on their own home).

However it would not include:

- slips, trips and falls on the same level
- falls on permanent stairs if there is no structural or maintenance work being done

- work in the upper floor of a multi-storey building where there is no risk of falling (except separate activities like using a stepladder).

At the centre of the Regulations (regulation 6), the employer is expected to apply a three-stage hierarchy to all work which is to be carried out at height. The three steps are the avoidance of work at height, the prevention of workers from falling and the mitigation of the effect on workers of falls should they occur.

The Regulations require that:

- work is not carried out at height when it is reasonably practicable to carry the work out safely other than at height (e.g. the assembly of components should be done at ground level)
- when work is carried out at height, the employer shall take suitable and sufficient measures to prevent, so far as is reasonably practicable, any person falling a distance liable to cause injury (e.g. the use of guard rails)

Figure 18.2 Working platform, tower scaffolds and bridging units.

➤ the employer shall take suitable and sufficient measures to minimize the distance and consequences of a fall (collective measures, e.g. air bags or safety nets, must take precedence over individual measures, e.g. safety harnesses).

The risk assessment and action required to control risks, from using a kick stool to collect books from a shelf, should be simple (e.g. not overloading, not overstretching). However, the action required for a complex construction project would involve significantly greater consideration and assessment of risk. A summary of the Regulations is given in Chapter 21.

18.3 Construction hazards and controls from working at height

18.3.1 Safe place of work

Safe access to and egress from the place of work is essential when working at height. All working platforms, ladders, scaffolds, gangways and passenger and materials hoists must be safe for use and be subject to regular inspection. Working platforms should also be kept as clear and clean as possible to avoid slip and trips and the accidental loss of material to the ground below.

18.3.2 Typical work activities and injuries associated with falls from height

Falls from height can result in fractures, serious head injuries and, in some cases, death. It is, therefore, essential that special care is taken to protect workers when they are working at height.

Common construction work activities which are done at height include brick or block laying, roofing, steel work erection, rendering, cladding, high pressure water-jetting, grit-blasting, concrete repairs, painting and some demolition work. For all these activities safe systems of work are essential. General hazards, such as dust, hazardous substances, electricity, vibrations and noise can also be present due to the nature of the work being carried out at height.

Figure 18.3 Industrial roof work: use of safety nets.

One particular form of such activity which can be particularly hazardous is the use of false-work. False-work is a temporary structure used to support a non-self-supporting structure during its construction or refurbishment. An example is the wooden structure to support a brickwork arch which is being built. Only competent persons should plan, erect, load or dismantle false-work. The collapse of false-work structures is common and, when occurring at height, can cause serious injury. Most accidents caused by the collapse of false-work result from the lack of liaison between the various trades using it and its suppliers or erectors.

The cleaning of buildings and monuments involves the use of hazardous techniques such as grit-blasting or high pressure water-jets, which is often done from scaffolding or even ladders. With both these techniques, care is needed to protect the workers, the occupiers of the building and passers-by from hazards such as noise, dust, falling debris and possible flooding of walkways. The equipment must be maintained and inspected regularly and operatives provided with training, supervision and suitable personal protective equipment (goggles, ear defenders and gloves) and waterproof clothing. Other measures will be necessary, such as the boarding up of windows, which may require the provision of temporary lighting inside the building. In the case of water-jetting, the scaffold should be enclosed in tarpaulin sheets and a channel system should be used to enable water to run off.

Finally, stairwells and other holes in floors are another source of falls from height and measures must be taken to address these hazards. Such measures include handrails on stairs, guard rails to stairwells and lift shafts, and ensuring that other holes in floors are similarly protected or covered. However, adequate levels of lighting and good housekeeping are also important as is a high standard of supervision and monitoring control.

18.3.3 *Protection against falls*

Falls from a height are the most common cause of serious injury or death in the construction industry. The WAHR require that suitable and sufficient steps be taken to prevent any person falling a distance that would be liable to cause personal injury and specify that the maximum unprotected gap between the toe and guard rail of a scaffold is 470 mm. This implies the use of an intermediate guard rail or infill. They also specify requirements for personal suspension equipment and means of arresting falls (such as safety nets, Fig. 18.3).

When working at height, a hierarchy of measures should then be followed to prevent falls from occurring. These measures are:

➤ avoid working at height, if possible
➤ the provision of a properly constructed working platform, complete with toe boards and guard rails
➤ if this is not practicable or where the work is of short duration, suspension equipment should be used and only when this is impracticable
➤ collective fall arrest equipment (air bags or safety nets) may be used
➤ where this is not practicable, individual fall restrainers (safety harnesses) should be used

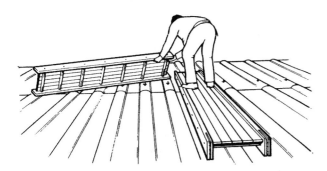

Figure 18.4 Proper precautions must always be taken when working on or near fragile roofs.

➤ only when none of the above measures are practicable, should ladders or stepladders be considered.

18.3.4 *Fragile roofs and surfaces*

Work on or near fragile surfaces is also covered by the WAHR (see summary in Chapter 21). Roof work, particularly work on pitched roofs, is hazardous and requires a specific risk assessment and method statement (see Chapter 6 for a definition) prior to the commencement of work. Particular hazards are fragile roofing materials, including those materials which deteriorate and become more brittle with age and exposure to sunlight, exposed

edges, unsafe access equipment and falls from girders, ridges or purlins. There must be suitable means of access such as scaffolding, ladders and crawling boards. Suitable barriers, guard rails or covers must be provided where people work near to fragile materials and roof lights. Suitable warning signs indicating that a roof is fragile, should be on display at ground level.

There are other hazards associated with roof work – overhead services and obstructions, the presence of asbestos or other hazardous substances, the use of equipment such as gas cylinders and bitumen boilers and manual handling hazards.

It is essential that only trained and competent persons are allowed to work on roofs and that they wear footwear having a good grip. It is good practice to ensure that a person does not work alone on a roof.

18.3.5 *Protection against falling objects*

This topic is covered in Chapter 10 because it is a general site issue. However, it is particularly relevant when working at height and is also now covered by WAHR (see summary in Chapter 21). A study by the HSE has indicated that over a 5-year period, 44 construction workers and members of the public (including children) were killed by insecure loads, unsecured equipment and pieces of plant falling on them. Over the same period 59

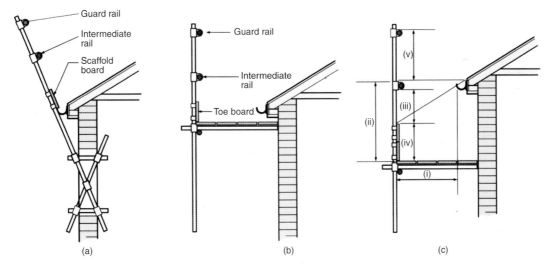

Sloping roof edge protection; typical arrangement in conventional tube and fittings
(a) Supported from window opening
(b) Working platform below the eaves
(c) Top lift of a scaffold. Dimensions should be as follows:
    (i) Working platform minimum width 600 mm
    (ii) Minimum 950 mm
    (iii) Maximum gap 470 mm
    (iv) To rise to the line of the roof slope with a minimum height of 150 mm
    (v) Gap between rails no more than 470 mm

Figure 18.5 Typical sloping roof edge protection. Barriers shown in (a) can be useful where spaces is limited, but they are not capable of sustaining loads so large as (b) and (c) which also provide a working platform.

were fatally injured by the collapse of structures or parts of structures. Whilst 37 of these deaths were caused by accidents during demolition work, the remainder occurred during routine construction work. As mentioned in Chapter 10, workers and members of the public need to be protected from the hazards associated with falling objects by the use of covered walkways or suitable netting to catch falling debris. Waste material should be brought to ground level by the use of chutes or hoists and not thrown from height. Minimal quantities of building materials should be stored on working platforms.

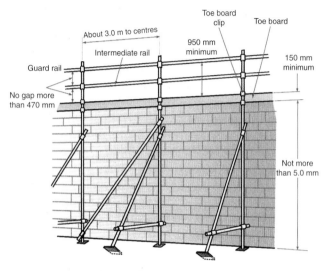

Figure 18.6 Flat roof edge protection supported at ground level. Ground-level support allows work up to the roof edge without obstruction.

Head protection is mandatory on site for workers and visitors. Several of the lives lost due to accidents from falling materials could have been saved if hard hats had been worn.

18.3.6 *The effect of deteriorating materials and weather*

As many materials age, they become brittle, weaker and less stable. This has already been mentioned as a hazard associated with fragile roofs. It is also a problem in other situations found in construction work, such as refurbishment work and the various cladding materials (including asbestos) used inside and outside buildings. The well publicized concrete 'cancer', where concrete degenerates to a powder leading to serious structural weaknesses, has occurred in several relatively young reinforced concrete structures, causing either very expensive repairs or total rebuilds. Whilst all materials will deteriorate to some extent with age, good design, maintenance, construction methods (using the correct mix ratios for mortars and concrete)

and type of cement should significantly reduce the rate of deterioration.

Weather is also a significant factor which can increase the hazards associated with working at height. High winds can affect the stability of scaffolding and ladders as has been mentioned elsewhere in this chapter. They can affect the work process itself by displacing tools and materials and cause injuries to persons working at height or those below the workings. Such winds, if strong enough, can also lift roofing tiles or sheets particularly if the roof is only partially completed. Such problems often occur outside normal working hours causing injuries to members of the public (including trespassers) and damage to neighbouring properties. It is very important to ensure that the construction work is protected after working hours especially if bad weather is forecast. Lightning is another weather hazard and it is inadvisable for work to continue on roofs or scaffolds when thunderstorms are approaching. The effects of lightning are discussed in more detail in Chapter 14.

Even apparently benign weather can present health and safety hazards. Skin cancer rates due to excessive exposure to sunlight has increased significantly over the last two decades. Even if skin cancer does not occur, serious damage to the skin will result from regular exposure to strong sun over several years. Typical damage is wrinkled skin due to the breakdown of skin elasticity and the formation of red patches of skin. There are also concerns, with the depletion of the ozone layer, that strong sunlight may cause serious deterioration in eye health, leading to possible problems such as arc eye and long-term eye damage. Precautions such as the use of protective creams, sunglasses and reduced exposure by wearing suitable clothing are all essential controls. This topic is covered in more detail in Chapter 17.

Finally, cold weather can also present hazards when working at height particularly when the cold weather is accompanied by wind, which will produce a 'chill factor' and an apparent temperature that is lower than the actual one. Guard rails will become colder and the reactions of workers much slower. Of course, cold weather will have an effect on construction processes, such as the mixing of cement and the laying of bricks. Workers should wear appropriate clothing and, if the temperature is too low, work at height should cease.

18.3.7 *Working over water*

Where construction work takes place over water, steps should be taken to prevent people falling into the water and rescue equipment should be available at all times. If the work takes place adjacent to a river or canal, special precautions are needed for persons using a towpath. The CDM Regulations cover any work where there is a risk to operatives of drowning. Working on or adjacent to water or when people have to pass near or across water on their way to work gives rise to a risk of drowning. People can also drown in other liquids such as slurries, chemicals and some free flowing foodstuffs in vats and silos.

At the planning stage of any work over or beside water, the following points need to be considered:

- the provision of suitable fencing to ensure that persons do not fall into the water, i.e. the provision of barriers or fences with toe boards
- the provision of life jackets, buoyancy aids and grab lines (fast flowing water such as rivers)
- the provision of rescue equipment, lifebelts, harnesses, safety lines and, in some cases, a rescue boat
- where transport by water is required, arrangements must be made for safe landing stages and life jackets
- when boats are used they should be of suitable construction, under the control of a competent person, not overloaded and properly maintained
- the provision of rescue teams and procedures
- specific training for boat operatives, rescue teams and supervisors
- all precautions must also protect the safety of the public, especially children
- suitable security measures will be necessary to prevent the theft of rescue equipment and boats outside working hours
- weather, tides and flooding conditions must be taken into account when planning rescue measures and using boats as transport
- all openings, drains and vats must be covered, where possible.

Site supervision is required to ensure that all barriers, fencing and rescue equipment is provided before any work commences which could place personnel at risk from drowning. The site supervisor will ensure that only authorized personnel alter barriers and operate rescue equipment and boats. The supervisor will also ensure that all rescue equipment is checked regularly and that any defective equipment is repaired or replaced immediately. He will ensure that personnel appointed as rescue teams are available during working periods and that replacement personnel are available when necessary.

There are several general hazards associated with working at height over or adjacent to water which are similar to those for any work done at height. These include falling from a height, scaffold stability, falling material, pedestrian safety and possible collision with the scaffolding by boats. Many of these issues can be addressed by the provision of a properly designed scaffold, erected by competent persons and securely tied

and provided with guard rails, toe boards and ladder access to working platforms. Other precautions include fall arrest equipment such as safety nets or harnesses, the non-overcrowding of the scaffold, the provision of rescue equipment such as lifebelts and the availability of a rescue boat under the control of a competent person, should someone fall into the water. Harnesses must be worn by those working on the scaffold in addition to the use of appropriate personal protective equipment. There should be a current register of workers and nobody should work alone on the scaffold.

For pedestrians, using perhaps a towpath, a protected access or safe diversion route must be provided. Signs, lighting and protective barriers will also be required to minimize the risk of boats colliding with the scaffold.

Finally, consideration must be given to health hazards which are associated with working near water. These include biological hazards such as leptospirosis and tetanus and general hazards to health associated with working on any construction site such as noise, vibration, musculoskeletal problems, electricity, mechanical hazards, flying particles, dust and contact with cement.

Appropriate control measures include disinfection, good standards of personal hygiene, the use of appropriate personal protective equipment, adequate washing and first-aid facilities, and the provision of information, instruction, training and supervision.

18.3.8 *Methods of avoiding working at height*

Much of the work which is done at height could often be done or partly done at ground level – thus avoiding the hazards of working at height. The partial erection of scaffolding or edge protection at ground level and the use of cranes to lift it into place at height are examples of this. The manufacture of complete window frames in a workshop and then the final installation of a frame into the building is another. By the use of suitable extension equipment high windows can be cleaned from the ground and high walls can be painted from the ground. However, in most construction work at height, the work cannot be done at ground level and suitable control measures to address the hazards of working at height will be required.

18.4 Working above ground or where there is a risk of falling

18.4.1 *Hazards and controls associated with working at height*

Work at any height involves a risk of falling. The significance of injuries resulting from falls from height, such as fatalities and other serious major injuries, have been dealt with earlier in this chapter as have the importance and legal requirements for head protection. Also covered were the many hazards involved in working at height including fragile roofs/surfaces and the deterioration of materials, unprotected edges and falling materials. Additional hazards include the weather and unstable or poorly maintained access equipment, such as ladders and various types of scaffold. Scaffolding must only be erected or modified by people who are trained and competent to do so. It should be constructed in accordance with recognized standards from components which are of adequate design and strength, and which are inspected at regular intervals. The scaffolding itself should be inspected by a competent person regularly.

The principal means of preventing falls of people or materials include the use of fencing, guard rails, toe boards, working platforms, access boards, ladder hoops, safety nets and safety harnesses. Safety harnesses arrest the fall by restricting it to a given distance due to the fixing of the harness to a point on an adjacent rigid structure. They should only be used when all other possibilities are not practical.

18.4.2 *Access equipment*

There are many different types of access equipment but only the following four categories will be considered here:

➤ ladders
➤ fixed scaffolds
➤ mobile scaffold towers
➤ mobile elevated work platforms.

Ladders

The Work at Heigh Regulations place a duty on employers to ensure that a ladder is used for work at height only if a risk assessment has demonstrated that the use of more suitable work equipment is not justified because of the low risk and:

➤ short duration of use or
➤ existing features on the site which cannot be altered.

Falls from ladders caused 13 deaths in 2005 and a third of all major injuries caused by falls at work were due to falls from ladders (1,200). Ten ladder accidents are reported to the HSE every day.

The main cause of accidents involving ladders is ladder movement whilst in use. This occurs when they have not been secured to a fixed point, particularly at the base. Other causes include over-reaching by the worker, slipping on a rung, ladder defects and, in the case of

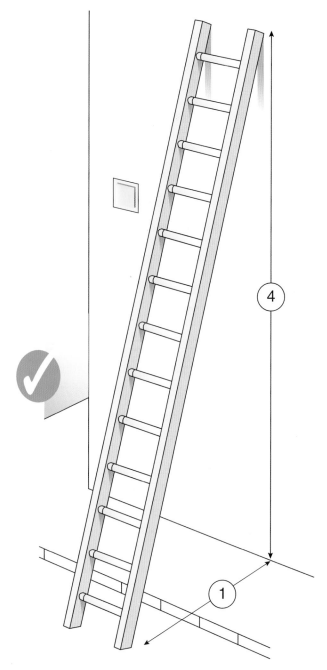

Figure 18.7 Ladder showing correct 1 in 4 angle (means of securing omitted for clarity).

metal ladders, contact with electricity. The main category of ladder accidents is falls.

There are two common materials used in the construction of ladders – aluminium and timber. Aluminium ladders have the advantage of being light but should not be used in high winds or near live electricity. Timber ladders need regular inspection for damage and should not be painted since this could hide cracks and other defects, such as knots.

The following factors should be considered when using ladders:

➤ Ensure that the use of a ladder is the safest means of access given the work to be done and the height to be climbed.

➤ The location itself needs to be checked. The supporting wall and supporting ground surface should be dry and slip free. Extra care will be needed if the area is busy with pedestrians or vehicles.

➤ The ladder needs to be stable in use. This means that the inclination should be as near the optimum as possible (1 in 4 ratio of distance from the wall to distance up the wall). The foot of the ladder should be secured to a rigid support. Ideally the ladder should be secured by making sure that the stiles are tied:
   ➤ at the top
   ➤ part way down
   ➤ near the base and
   ➤ at the base.

Footing a ladder should only be used as a last resort and should be avoided, where reasonably practicable, by the use of other access equipment. More information on the safe use of ladders and stepladders is given in INDG402 – an employers' guide. Weather conditions must be suitable (no high winds or heavy rain). The proximity of live electricity should also be checked. (This last point is important when ladders are to be carried beneath power lines).

➤ The ladder should be fitted with non-slip feet.

➤ There should be at least 1 m of ladder above the stepping off point.

➤ The work activity must be considered in some detail. Over-reaching must be eliminated and consideration given to the storage of paints or tools which are to be used from the ladder and any loads to be carried up the ladder. The ladder must be matched to work required.

➤ Workers who are to use ladders must be trained in the correct method of use and selection. Such training should include the use of both hands during climbing, clean non-slippery footwear, clean rungs and an undamaged ladder.

➤ Ladders should be inspected (particularly for damaged or missing rungs) and maintained on a regular basis and they should only be repaired by competent persons.

➤ The transportation and storage of ladders is important since much damage can occur at these times. They need to be handled carefully and stored in a dry place.

➤ When a ladder is left secured to a structure during non-working hours, a plank should be tied to the rungs to prevent unauthorized access to the structure.

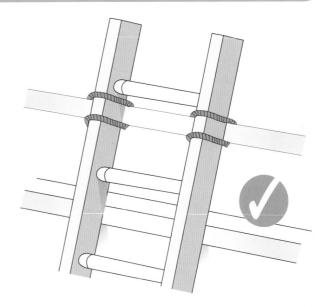

Figure 18.8a Ladder tied at top stiles (correct for working on, not for access).

Figure 18.8b Tying part way down.

There are a variety of products available to ensure that ladders are stable during use by the provision of anchors for the ladder base or fixing mechanisms to the building. However, certain work should not be attempted using ladders, particularly when the user cannot maintain a safe handhold while carrying a load. This includes work where:

➤ a secure handhold is not available
➤ the work is at an excessive height

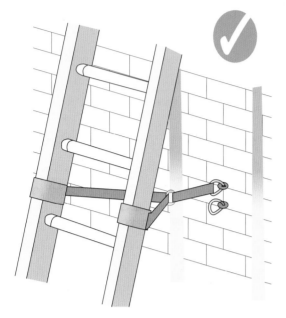

Figure 18.8c Tying at base.

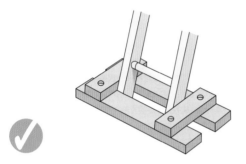

Figure 18.8d Securing at the base.

➤ where the ladder cannot be secured or made stable
➤ the work is of long duration
➤ the work area is very large
➤ the equipment or materials to be used are heavy or bulky
➤ the weather conditions are adverse
➤ there is no protection from vehicles.

A particular type of timber ladder is the wooden pole ladder, which is used by many scaffolders. Pole ladders should be suitable for the job by being of the appropriate length and having complete integrity of the steel reinforcement. Prior to use, the ladder should be inspected as outlined earlier in this section and any damage to stiles, rungs or steel reinforcement must be identified as should any unauthorized repairs.

There have been several rumours that the Work at Height Regulations have banned the use of ladders. This is **not** true. Ladders may be used for access and it is legal to work from ladders. Ladders may be used when a risk assessment shows that the risk of injury is low and the task is of short duration or there are unalterable features of the work site and that it is not reasonably practicable to use potentially safer alternative means of access. Provided that the ladder is checked before each use, the work is of short duration, only light loads are to be carried, the ladder is stable and secured and users are trained and competent, then ladders may be used. More information on ladders and their use within the requirements of the WAHR is available

Figure 18.9 Attach paint cans and the like to the ladder.

from the British Ladder Manufacturers Association. Ladders for industrial work in the UK should be marked to:

➤ Timber BS1129: Kite marked Class 1 Industrial
➤ Aluminium BS2037: 1994 Kite marked Class 1 Industrial
➤ Glass fibre BSEN131: 1993 Kite marked Industrial
➤ Step stools BS7377: 1994.

Stepladders, trestles and staging

Many of the points above for ladders apply to stepladders and trestles where stability and over-reaching are the main hazards. Neither of these should be used as a workplace above about 2 m in height unless proper edge protection is provided.

All equipment must be checked by the supervisor before use to ensure that there are no defects and must be checked at least weekly whilst in use on site. If a defect is noted, or the equipment is damaged, it must be taken out of use immediately. Any repairs must only be carried out by competent persons.

Supervisors must also check that the equipment is being used correctly and not being used where a safer method should be provided.

Where staging, such as a 'Youngmans' staging platform, is being used in roof areas, supervisors must ensure that only experienced operatives are permitted to carry out this work and that all necessary safety harnesses and anchorage points are provided and used.

The main hazards associated with stepladders, trestles and staging are:

➤ unsuitable base (uneven or loose materials)
➤ unsafe and incorrect use of equipment (e.g. the use of staging for barrow ramps)
➤ overloading
➤ use of equipment where a safer method should be provided
➤ overhang of boards or staging at supports (trap ends)
➤ use of defective equipment.

Stepladders and trestles must be:

➤ manufactured to a recognized industrial specification
➤ stored and handled with care to prevent damage and deterioration
➤ subject to a programme of regular inspection (there should be a marking, coding or tagging system to confirm that the inspection has taken place)
➤ checked by the user before use
➤ taken out of use if damaged – and destroyed or repaired

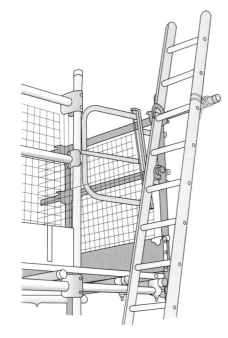

Figure 18.10 Access ladders should be tied and extend at least 1 m above the landing point to provide a secure handhold.

➤ used on a secure surface, and with due regard to ensuring stability at all times
➤ kept away from overhead cables and similar hazards.

The small platform fitted at the top of many stepladders is designed to support tools, paint pots and other working materials. It should not be used as a working place unless the stepladder has been constructed with a suitable handhold above the platform. Stepladders must not be used if they are too short for the work being undertaken, or if there is not enough space to open them out fully.

Platforms based on trestles should be fully boarded, adequately supported (at least one support for each 1.5 m of board for standard scaffold boards) and provided with edge protection when the platform is 2 m or higher.

Fixed scaffolds

It is quicker and easier to use a ladder as a means of access but it is not always the safest. Jobs such as painting, gutter repair, demolition work or window replacement should normally be done using a scaffold. Scaffolds must be capable of supporting building workers, equipment, materials, tools and any accumulated waste. A common cause of scaffold collapse is the 'borrowing' of boards and tubes from the scaffold thus weakening it. Falls from scaffolds are often caused by

✗ Wrong way

Right way ✓

✗ Stepladder too short
✗ Hazard overhead
✗ Over-reaching up and sideways
✗ No grip on ladder
✗ Sideways-on to work
✗ Foot on handrail
✗ Wearing slippers
✗ Loose tools on ladder
✗ Slippery and damaged steps
✗ Uneven soft ground
✗ Damaged stiles
✗ Non-slip rubber foot missing

Steps at right height ✓
No need to over-reach ✓
Good grip on handrail ✓
Working front-on ✓
Wearing good flat shoes ✓
Clean undamaged steps ✓
Firm level base ✓
Undamaged stiles ✓
Rubber non-slip feet all in position ✓
Meets British or European standards ✓

Figure 18.11 Working with stepladders.

badly constructed working platforms, inadequate guard rails or climbing up the outside of a scaffold. Falls also occur during the assembly or dismantling process.

There are two basic types of external scaffold:

1. independent tied – these are scaffolding structures which are independent of the building but tied to it often using a window or window recess; this is the most common form of scaffolding and
2. putlog – this form of scaffolding is usually used during the construction of a building; a putlog is a scaffold tube which spans horizontally from the scaffold into the building – the end of the tube is flattened and is usually positioned between two brick courses.

The important components of a scaffold have been defined in a guidance note issued by the HSE as follows.

Standard: an upright tube or pole used as a vertical support in a scaffold.

Ledger: a tube spanning horizontally and tying standards longitudinally.

Transom: a tube spanning across ledgers to tie a scaffold transversely. It may also support a working platform.

Bracing: tubes which span diagonally to strengthen and prevent movement of the scaffold.

Putlog: a tube which is flattened at its end and spans from a ledger to the wall of a building. A scaffold secured to a building in this way is known as a putlog scaffold.

Guard rail: a horizontal tube fitted to standards along working platforms to prevent persons from falling.

Toe boards: these are fitted at the base of working platforms to prevent persons, materials or tools falling from the scaffold.

Raker: an angled or inclined load-bearing tube used to support a cantilevered scaffold working platform.

It is fixed to the edge of the working platform at one end and the wall of the building at the other.

Base plate: a square steel plate fitted to the bottom of a standard at ground level.

Sole board: normally a timber plank positioned beneath at least two base plates to provide a more uniform distribution of the scaffold load over the ground.

Fans: specially designed platforms which prevent debris and other materials falling on people passing or working below the scaffold.

The following factors must be addressed if a scaffold is to be considered for use for construction purposes:

➤ Scaffolding must only be erected and dismantled by competent people who have attended recognized training courses. Any work carried out on the scaffold must be supervised by a competent person. For complex scaffolds, a written plan may be required. Any changes to the scaffold must be done by a competent person.

➤ In busy town centres, the erection and dismantling of the scaffold should take place during quiet times.

➤ The scaffold must be designed to carry all the required loads and use only sound fittings and materials.

➤ Adequate toe boards, guard rails and intermediate rails must be fitted to prevent people or materials from falling.

➤ The use of appropriate bay lengths and widths, bracing and tie-in arrangements together with edge and stop-end protection.

➤ The scaffold must rest on a stable surface; uprights should have base plates and timber sole boards if necessary.

➤ The scaffold must have safe access and egress.

➤ Work platforms should be fully boarded with no tipping or tripping hazards.

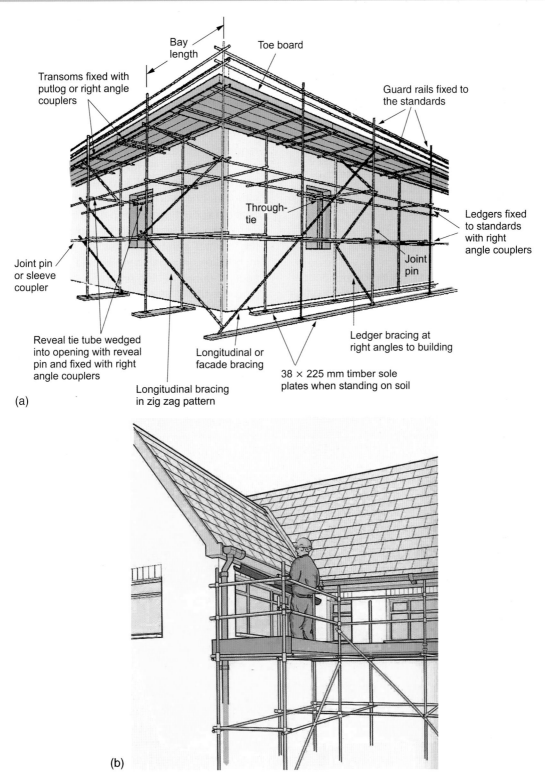

Figure 18.12 (a) Typical independent tied scaffold; (b) fixed scaffold left in place to fit the gutters.

➤ The scaffold must be maintained and cleaned, and have proper arrangements for waste disposal.
➤ The scaffold should be sited away from or protected from traffic routes so that it is not damaged by vehicles.

➤ If the scaffold must be erected near to vehicle roadways then barriers or fenders should be fitted around its base.
➤ The scaffold should be properly braced and secured to the building or structure.

389

➤ Overloading of the scaffold must be avoided.
➤ The public must be protected at all stages of the work and the appropriate highway authority contacted before a scaffold is erected beside a public thoroughfare.
➤ Regular inspections of the scaffold must be made and recorded.

There have been many examples of scaffold collapse causing serious injuries to workers and members of the public. Factors which may affect the stability of an independently tied scaffold include:

➤ poor, incompetent erection and/or a lack of regular inspections
➤ the strength of the supporting ground or foundation
➤ the proximity to the scaffold of any excavation work
➤ the effect of excessive surface water in weakening the scaffold foundation
➤ unauthorized alteration by incompetent persons (perhaps by vandals)
➤ the use of incorrect or damaged fittings during the erection or extension of the scaffold
➤ the overloading of the scaffold
➤ adverse weather conditions

➤ the sheeting in of the scaffold without the use of extra ties
➤ vehicular impact.

With all scaffolds, brick guards and debris netting should be used to prevent materials and equipment falling to the ground.

For putlog scaffolds, all the previously stated safety requirements apply and some additional ones. The putlog ends, which must be properly flattened, should be fully inserted and securely bedded into well constructed brickwork. It is recommended that a putlog scaffold should not exceed 45 m in height.

When the work is to take place in a busy area, all the normal safety features of scaffolding apply, but, in addition, special care should be taken with ladder access, debris netting and fans. Appropriate traffic signs (and possibly traffic lights), cones, barriers and lighting arrangements at night will also be necessary.

All scaffolds should have an assembly, use and dismantling plan. Before accepting a scaffold erected by a specialist scaffolding sub-contractor for use, the site supervisor must check the scaffold and obtain a handing-over certificate.

The site supervisor will ensure that all scaffolds are erected in accordance with the standards outlined above and, at the beginning of each week, he should inspect

Figure 18.13 Fan scaffold. Panels are fixed around the base of the scaffolding to prevent children from climbing.

the scaffold and ensure that any defect is rectified. A report of the inspection and action taken should be entered in the site inspection register. A similar inspection should also be carried out after high winds or other adverse weather conditions.

All materials used for scaffolding must be provided in accordance with the relevant British Standards and checked before use by a scaffolder. Such materials must be properly stored and maintained on the site.

Any scaffold that is being erected, altered or dismantled, or otherwise not suitable for use by employees, must have a warning notice erected stating that it must not be used.

All scaffolds must be checked at the end of each working day to ensure that access to the scaffold by children has been prevented.

Requirements for scaffold erectors

All scaffolds must only be erected by scaffolders who are competent and experienced to undertake the work. All the requirements, which were outlined in Chapters 3 and 10, apply equally to scaffolders. Only scaffolding firms who have a good reputation and accident record gained from earlier contracts should be employed to erect scaffolds. Scaffolders should be supervised throughout the erection of a scaffold and they should be capable of inspecting scaffolding materials and equipment before it is used. They should be equipped with personal lifelines which must be attached at all times while working at height. All scaffolders must have attended a specialist scaffolding course and be able to provide documentary evidence that they have successfully completed the course.

Design of loading platforms

A loading platform is a particular working platform on a scaffold onto which materials and equipment is loaded to further the construction project. A working platform is part of the scaffold or similar structure on which people work, walk and stand. It must be designed so that it can carry workers, their tools and working materials safely. The platform is built by laying scaffold boards across the transoms and, normally, at least three support points are required for each board. There must be a slight overhang (50–150 mm) across a support at each end of the board. Boards are normally butt jointed. Toe boards and guard rails, at least 950 mm above the working platform, must be fitted inside the standards along the edge of the working platform. Ladders used to access the loading platform must extend at least 1 m above the level of the platform.

All stacking of materials and equipment, such as bricks, must be as close to standards as possible to enable the added weight to be more evenly distributed throughout the scaffold. The boards will need some additional fixing to the structure so that they are protected from lifting during high winds.

The width of the platform will vary with the type of work being undertaken from a minimum of three boards to a maximum of seven (when trestles are to be used on the platform).

The following points are relevant to the design of all working platforms whether loading or not:

- They must be wide enough (at least 600 mm) to allow people to pass each other and for any required equipment or materials to be carried and stored.
- They must be free of openings and trip hazards.
- If members of the public are able to pass beneath, a double boarded platform with a sheet of polythene placed between the boards may be necessary to prevent debris or tools falling on them.
- The principal hazards associated with working platforms are overloading, the fall of materials, tools and equipment and slips, trips and falls of people working on the platform.
- Good housekeeping and supervision are essential on all working platforms.

Finally, if it is likely that heavy loads are to be stored on a working platform, the scaffolding firm should be informed so that special support may be designed into the scaffold structure.

Scaffold hoists

Hoists are widely used to transport materials to higher working levels. It is important that hoists are well maintained, inspected regularly and in a good working order. They should be positioned on firm and stable ground, erected following the manufacturer's instructions and properly protected at either end. The hoist should be marked with its safe working load and its control functions.

The following applies to the use of all hoists whether they are mobile or permanent. The main hazards associated with hoist operations are:

- overloading of the hoist
- overloading or incorrect use of lifting gear
- automatic safe load indicator not working
- failure of equipment due to lack of maintenance
- incorrect positioning of lifting appliance
- insecure attachment of load and falls from the platform
- unstable slinging of loads, hook not over centre of gravity
- contact with overhead electricity cables
- lack of operator training
- no banksman used when driver's view is obscured
- incorrect signals.

Only hoists which are suitable for the site and capable of lifting the required loads should be selected and all personnel working with or near lifting appliances must wear hard hats. All hoists must be secured and left in a safe condition at the end of each working period so that any child trespassers are protected.

Figure 18.14 Hoist with interlocked gates.

The following additional precautions should also normally be taken:

➤ The controls should be set so that the hoist can be operated from one position only (normally ground level) and that all the landing levels are clearly visible by the operator from the operating position thus preventing people from being struck by the platform or other moving parts.

➤ The hoistway must be enclosed at places where people might be struck; for example, working platforms or window openings.

➤ Gates must be provided at all landings and at ground level.

➤ The hoistway must be fenced wherever people could fall down it.

➤ The gates at landings are kept closed except during loading and unloading and they should be secure and not free to swing into the hoistway.

➤ The edges of the hoist platform and the landing should be as close together as possible so that there is no significant gap between for debris to fall through.

➤ Loads, such as wheelbarrows, should be securely chocked and not overfilled.

➤ Lloose loads, such as bricks, should be carried in proper containers or a hoist with an enclosed platform should be used.

➤ The platform must not be overloaded and should be clearly marked with its safe working load.

➤ The hoist must be erected by trained and experienced people who will follow the manufacturer's instructions and ensure that it is properly secured to the supporting structure.

➤ The hoist operator must be properly trained and be competent.

➤ Loads must be evenly distributed on the hoist platform.

➤ The hoist must be thoroughly examined and tested after erection, substantial alteration or repair and at 6-month intervals.

➤ Regular inspections should be carried out at least on a weekly basis and the results recorded.

Finally, no one should be allowed to ride on a goods hoist and a notice, which states this prohibition, should always be posted.

Prefabricated mobile scaffold towers

Mobile scaffold towers are frequently used throughout industry. It is essential that the workers are trained in their use since recent research has revealed that 75% of lightweight mobile tower scaffolding is erected, used, moved or dismantled in an unsafe manner.

The following points must be considered when mobile scaffold towers are to be used:

➤ Towers should only be used on level, firm and stable ground.

➤ The selection, erection, dismantling and inspection of mobile scaffold towers must be undertaken by competent and trained persons with maximum height-to-base ratios not being exceeded.

➤ No person must be permitted to erect, alter or dismantle any mobile tower scaffold unless authorized by the site supervisor.

➤ Diagonal bracing and stabilizers should always be used.

➤ Access ladders must be fitted to (not leaning on) the narrowest side of the tower or inside the tower and persons should not climb up the frame of the tower.

➤ All wheels must be locked whilst work is in progress and all persons must vacate the tower before it is moved.

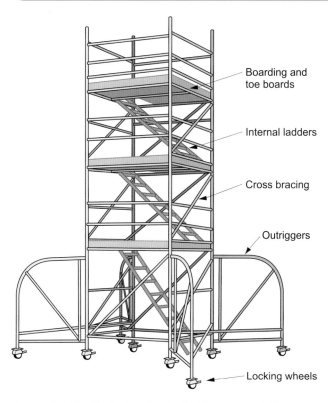

Boarding and toe boards

Internal ladders

Cross bracing

Outriggers

Locking wheels

Figure 18.15 Typical prefabricated tower scaffold.

➤ All operatives who are required to use mobile tower scaffolds will be instructed in their safe use and movement.

➤ Outriggers or stabilizers must be extended where applicable.

➤ Tower scaffolds must not be used or moved on sloping, uneven or obstructed surfaces.

➤ Tower scaffolds must only be used in the vertical position.

➤ The tower working platform must be fully boarded, fitted with guard rails and toe boards and not overloaded.

➤ Towers must be tied to a rigid structure if exposed to windy weather or to be used for work such as jet blasting.

➤ Persons working from a tower must not over-reach or use ladders from the work platform.

➤ Safe distances must be maintained between the tower and overhead power lines both during working operations and when the tower is moved.

➤ The tower should be inspected on a regular basis and a report made.

There are two approved methods for the erection of a prefabricated tower scaffold – an advance guard rail system and the 'through the trap' (3T) system. These systems have been developed by HSE and the Prefabricated Access Suppliers' and Manufacturers' Association and are described in detail on the HSE information sheet 'Tower scaffolds' CIS No. 10 (Rev 4).

The main reasons why a mobile tower scaffold may become unstable are that:

➤ the brakes are defective

➤ it was not erected by a competent person or unauthorized adjustments were made

➤ it was erected on ground that was neither firm nor level

➤ the height of the tower exceeded the recommended base to height ratio

➤ the failure to fit and/or use outriggers when they were required

➤ the safe working load was exceeded

➤ the movement of the tower either with persons and/ or materials on the working platform or under windy conditions

➤ collision with vehicles or other plant.

The recommended height to base ratio varies from 3:1 to 3.5:1, with a base not less than 1 m, dependent on whether the tower is to be used outside or inside the building. The ratio is lower for outside use due to possible wind effects. The ratios quoted should be treated as a guide and the manufacturer's instruction should always be consulted. Mobile scaffold towers should be inspected regularly as outlined later in this chapter.

Mobile elevated work platforms (MEWP)

Mobile elevated work platforms are very suitable for many different types of high-level work such as changing light bulbs in a warehouse. Examples of such platforms are scissor lifts and boom type hydraulic platforms (cherry pickers). The following factors must be considered when using a:

➤ The MEWP must only be operated by trained and competent persons and safety harnesses should be available.

➤ It should be inspected before use.

➤ The working platform should not be overloaded.

➤ It must never be moved in the elevated position

➤ It must be operated on level and stable ground with consideration being given to the stability and loading of floors.

➤ The tyres must be properly inflated and the wheels normally immobilized when the platform is in use.

➤ Outriggers, if fitted, should be fully extended and locked in position.

➤ Warning signs should be displayed and barriers erected to avoid collisions.

➤ Due care must be exercised with overhead power supplies and obstructions and adverse weather conditions.

Figure 18.16 Mobile elevating work platform – scissor lift.

> It should be maintained regularly and procedures should be in place in the event of machine failure.
> Drivers of MEWPs must be instructed in emergency procedures particularly to cover instances of power failure.

When working on a MEWP, there is a danger that the operator may become trapped against an overhead or adjacent object, preventing him from releasing the controls. There have also been accidents caused when a MEWP is reversed into areas where there is poor pedestrian segregation and the driver has limited visibility. During any manoeuvring operation, a dedicated banksman should be used.

Other techniques

Sometimes it is necessary to work from a man-riding skip. It is important that the skip is of sound construction and made from good quality material which has adequate strength. It should be at least 1 m deep and have means to prevent tipping, spinning and persons from falling out. The skip should be properly maintained and be marked with the safe working loads. The crane and lifting tackle used to lift the skip must be regularly inspected and thoroughly examined with appropriate certification being obtained. Only competent persons

should undertake this type of work and there must be proper communication with other interested parties, such as the building occupiers, the police and the local authority.

When maintenance operations take place on bridges or similar inaccessible structures, suspended access cradles or platforms may be used. The failure of access cradles has led to serious injuries and fatalities. There are many reasons for such incidents including:

> poor equipment selection procedures in that the cradle is not suitable for the purpose
> unsafe access to the cradle
> overloading of the cradle or non-uniform loading across the platform
> holes or cracks in the platform floor allowing material to fall below
> inadequate guard rails and toe boards
> the structure to which the cradle is attached is not capable of carrying the additional load
> insecure counterweights and/or braking system
> the failure of essential components of the cradle structure
> the failure of the winching and climbing devices
> poor erection, maintenance and dismantling by incompetent and untrained personnel
> inadequate emergency procedures.

The safest access to a cradle is at ground level. If access at height is unavoidable, then the cradle must be secured so that it is prevented from swinging away from the structure. It is also important that suitable anchorage points are available for safety harnesses and a secondary rope is fitted. Regular visual inspections should be made at least before each time of use and a weekly detailed inspection should be made by a competent person.

A boatswain's chair is often used for light work of a short-term nature. Whenever it is used, the worker should be protected by a harness and lanyard in case of a possible fall. Abseiling or rope access techniques are also used for inspection work. Devices such as these should only be used when a properly constructed working platform is not practicable.

Equipment, such as cradles, boatswain's chairs and rope access techniques, must be erected either by or under the supervision of a competent person and the worker must be properly trained in the use of the technique. The main rope and safety access rope should always be attached to separate anchorage points. Any tools or equipment used must be secured to the worker to prevent them falling and possibly injuring people below. Whenever possible, the area beneath the work should be fenced off.

Figure 18.17 Vehicle mounted mobile elevating work platform providing access to a large steel structure.

Fall arrest equipment
The three most common types of fall arrest equipment are safety harnesses, safety nets and air bags.

Safety harnesses should only be used alone when conventional protection, using guard rails, is no longer practicable. Such conditions occur when it is possible to fall 2 m or more from an open edge. It is important that the following points are considered when safety harnesses are to be used:

1. The length of fall only is reduced by a safety harness. The worker may still be injured due to the shock load applied to him when the fall is arrested. A free fall limit of about 2 m is maintained to reduce this shock loading. Lanyards are often fitted with shock absorbers to reduce the effect of the shock loading.
2. The worker must be attached to a secure anchorage point before they move into an unsafe position. The lanyard should always be attached above the worker, whenever possible.

3. Only specifically trained and competent workers should attach lanyards to anchorage points and work in safety harnesses. Those who wear safety harnesses must be able to undertake safety checks and adjust the harness before it is used.

Safety nets are widely used to arrest falls of people, tools and materials from height but competent installation is essential. The correct tensioning of the net is important and normally specialist companies are available to fit nets. The popularity of nets has grown since the Construction Regulations came into force in 1996 and the subsequent advocacy of their use by the HSE. Nets are used for roofing work and for some refurbishment work. Nets, however, have a limited application since they are not suitable for use in low-level construction where there is insufficient clearance below the net to allow it to deflect the required distance after impact. Nets should be positioned so that workers will not fall more than 2 m, in case they hit the ground or other obstructions.

Figure 18.18 Mast cradle.

Air bags are used when it is either not possible or practical to use safety nets. Therefore they are used extensively in domestic house building or when it is difficult to position anchorage points for safety harnesses. When air bags are used, it is important to ensure that the bags are of sufficient strength and the air pressure high enough to ensure that any falling person does not make contact with the ground. Only reputable suppliers should be employed for the provision of air bags. Air bags or bean bags are known as soft landing systems and are used to protect workers from the effects of inward falls. Other possible solutions to this problem of the inward fall are the use of internal scaffolding or lightweight 'crash deck systems'.

Air bags may be linked together to form an inflated crash deck system. Such a system is suitable when a safety net is not viable. It consists of a series of interlinked air mattresses that are positioned beneath the working area and is suitable for working at height inside a building where safety harnesses would not be practicable. The mattresses are interconnected by secure couplings and inflated in position on site using an air pump. The mattresses are made in various sizes so that any floor area configuration can be covered. A similar form of crash deck can be made from bean bags that are clipped together.

After the completion of the construction project, cable-based fall arrest systems are normally suitable for on-going building maintenance work. They may also be used for the construction of complex roof structures, such as parabolic or dome structures, where a safety net may be more than 2 m below the highest point of the structure.

Emergency procedures (including rescue)

A suitable emergency and rescue procedure needs to be in place for situations that could be expected to occur on the construction site. Such foreseeable situations could include a crane driver trapped in his cab due to a power failure or a serious ill-health problem, or the rescue of a person who has fallen into a safety net. The method of rescue may be simple and straightforward, such as putting a ladder up to a net and allowing the fallen person to descend or the use of a MEWP to lower the person to the ground.

Proprietary rescue systems are available and may well be suitable for many more complex rescues. The system selected needs to be proportionate to the risk and there should not be undue reliance on the emergency services. However, arrangements should be in

Figure 18.19 Fall arrest harness and device.

place to inform the emergency services of any serious incident. It is important that rescue teams are available at all times with a designated leader. Instruction and training are essential for the teams, preferably by simulated rescue exercises.

18.4.3 *Roof work*

Before any roof work is started, a risk assessment should be undertaken. This should include an assessment of the structural integrity of the roof and the methods to be used to repair the roof and a COSHH (Control of Substances Hazardous to Health) assessment of any hazardous substances to be used. The risk assessment should be followed by the provision of a method statement by the sub-contractor. This should be examined to ensure a safe method of working is proposed and which does not conflict with other sub-contractors. Only competent and trained workers should be employed for roof work and any equipment used should be suitable for the job.

The precautions needed when repairing the flat roof of a building include the provision of safe access to the roof (using scaffolding and/or ladders), the use of crawling boards, and the provision of safety harnesses and edge protection, using toe boards, to prevent the fall of persons or materials. Roof battens should be checked to ensure they provide safe hand and foot holds. If they do not then crawling ladders or boards will be used. Crawling boards should always be provided when work on fragile roofing materials cannot be avoided.

The perimeter of the roof requires a two-rail scaffold guard rail fitting together with a toe board where the gutter would not prevent materials slipping from the roof. Protection of the leading edge must also be provided. Where this is also used as a working platform, it is recommended that the 'Youngman' trolley system or similar fitted with guard rails is used. Suitable safety nets may also need to be provided to prevent people and materials from falling.

Access to the area immediately below the work should be restricted using suitable barriers, netting, safety signs and safety helmets. There should also be no danger to employees from fragile roof lights, voids, overhead obstructions and services.

All walkways for the passage of people and materials should be a minimum of 600 mm wide and protected with double guard rails and toe boards on the exposed edge. Where staging is used care should be taken to ensure that the ends are supported and not cantilevered over the steel work. Where a gutter is used as a walkway either along the perimeter or along a valley of the roof the need for similar protection still applies. Walkways should be kept clear of materials, tools and debris as far as possible.

Consideration should also be given to the way in which materials are transported to and from the working area, involving lifts, hoists, manual handling and/or using chutes and covered waste skips. When they are to be used on the roof, such materials should be secured in demarcated areas. Materials stacked on the roof should be securely tied down when not required, to avoid the possibility of being blown down from the roof. Where materials are stacked on purlins on the open part of the roof, a suitable form of walkway as described above should be provided around each pack and also for the working area.

Good housekeeping procedures are essential for safe roof work.

All openings left in the roof for lights, vents or similar should either be protected with guard rails or have some form of temporary covers fitted.

Where an inclined hoist is used to lift materials onto the roof this should have protective barriers around the base, an overhead mesh guard fitted to protect the operator and around the top a suitable walkway with guard

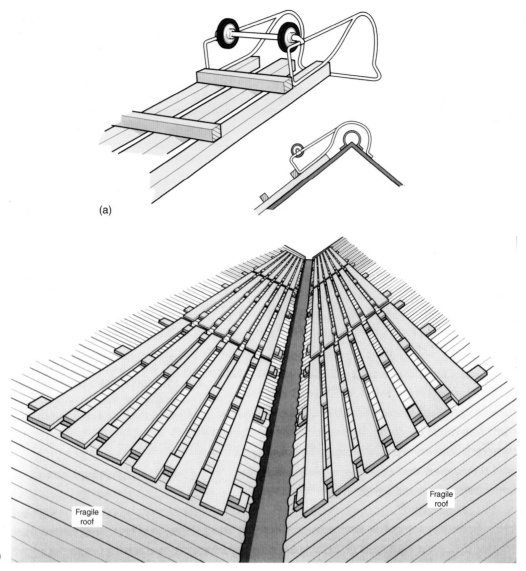

(a)

(b)

Figure 18.20 (a) Roof ladder. The ridge iron should be large enough to be clear of the ridge tile. (b) Permanent protection installed at valley gutter (the protection should be supported by at least three rafters beneath the roof sheets).

rails to provide a safe means of off-loading the materials onto the roof.

Ladders securely tied should be situated as close as possible to the working area on the roof and extend 1 m above the roof level. Where the height to the roof exceeds 9 m, it is normally a requirement that a scaffold tower with ladders be provided such that the height between intermediate platforms does not exceed 9 m.

If asbestos is present, then special precautions will be required including the provision of overalls and respiratory protection, the need for damping down and the avoidance of breaking asbestos sheets. Safe disposal of waste asbestos is particularly important.

Consideration should also be given to the wearing of a safety harness where other forms of protection

are not adequate. This applies for example when fixing the gutter or the scaffold guard rail at the start of work. Safety helmets should be worn by all persons in accordance with the Construction Head Protection Regulations whilst work is in progress overhead and/or there is a risk of head injury.

During the construction of large industrial roofs, a purlin trolley system is usually used together with a safety harness. Such systems have a double handrail on the leading edge positioned on the opposite side to the working side. As the working side is open to the roof, a safety harness is needed to protect against falls from this edge. A second line of defence has recently been designed into these trolleys by the attachment of a horizontal barrier beneath.

Figure 18.21 Soft landing system.

Other issues include the use of netting and protection against adverse weather conditions, by issuing suitable clothing (and possibly sun cream) and limiting exposure. The sensible positioning of bitumen boilers and gas cylinders will also reduce the risk of fire, and electrical risks will be reduced by the use of residual current devices or reduced low voltage for portable electrical equipment.

One final aspect of roof work which must be considered is work that takes place immediately below the roof in the attic or roof space, where access may be gained using MEWPs and 'Youngman' staging platforms. An example of this is the painting of roof girders in a warehouse. As with similar work, continual liaison with the client is essential during the planning stage and after the work begins. At the planning stage, any hazards existing in the roof space must be assessed including any lighting, heating and electrical and other services that may be present. Often other work activities may be taking place in the warehouse while the painting work is in progress. It is essential, therefore, that there is agreement with the client on the sequence of work, working periods and working areas so that there is no accidental contact between painters and high reach pallet forklift trucks. The painters will need to be briefed on the operational methods and routes used by the pallet trucks.

Similarly, the warehouse workers will require information on the materials and systems of work used by the painters. Other related aspects of the work, such as safe use of the mobile and stationary platforms, the use of harnesses and debris netting, storage arrangements for paints and solvents and the provision of respiratory protective equipment have been dealt with elsewhere.

Three examples of safe systems of roof work are given in Appendix 18.4.

18.4.4 *Protection of other persons*

Members of the public must be protected at all stages of the construction project and at all times of day and night. Scaffold should be made very visible by painting standards with red and white stripes, masking couplers, warning signs, good illumination at night and the provision of barriers and other means of safe passage. Safety precautions, such as the boarding of ladders, should be taken during periods when the site is unoccupied. Such protection should be provided to protect anyone, including trespassers, such as children.

When construction work is being undertaken on occupied premises, such as a retail store or an office block, certain important health and safety matters would need to be discussed and agreed with store or office management. These matters would include the location and isolation of the construction working areas, supervision of the work, procedures for the evacuation of the premises in the event of an emergency, the use of welfare and canteen facilities, the protection of the occupier's employees and procedures for handing back areas of the store when the work has been completed. The construction contractor must provide protected access into the building, ensuring that the area to be occupied by him is provided with adequate fences or barriers and that all waste materials are safely removed.

Other issues which need to be addressed include the selection of equipment (preferably using reduced voltage electrical equipment) and personal protective equipment and the erection of appropriate warning signs. Other precautions and controls include the provision of:

➤ safe access for workers and materials
➤ a safe working platform
➤ clean-up procedures and good housekeeping
➤ full arrest equipment
➤ protection for all premises' occupier's employees
➤ a permit system for hot work where this is necessary
➤ fire extinguishers and procedures for emergency evacuation of the building.

Finally, some or all of existing services may need to be isolated at certain times, and these times should be agreed with the occupier; adequate levels of lighting must be provided, where necessary, to carry out the work safely.

18.4.5 *Inspections and maintenance*

Inspection

Equipment for work at height needs regular inspection to ensure that it is fit for use. A marking system is probably required to show when the next inspection is due. Formal inspections should not be a substitute for any pre-use checks or routine maintenance. Inspection does not necessarily cover the checks that are made during maintenance although there may be some common features. Inspections need to be recorded but checks do not.

Scaffolds must be inspected on a regular basis by a competent person. These inspections should take place before the scaffold is used, after any alteration is made or after adverse weather conditions may have weakened it. In any event an inspection should take place every 7 days and any faults rectified.

Under the WAHR, weekly inspections are still required for scaffolding, as previously required by the Construction (Health, Safety and Welfare) Regulations where a person could fall 2 m or more. The reporting requirements for inspection are set out in schedule 7 to the Regulations as follows:

➤ the name and address of the person for whom the inspection was carried out
➤ the location of the work equipment inspected
➤ a description of the work equipment inspected
➤ the date and time of the inspection
➤ details of any matter identified that could give rise to a risk to the health or safety of any person
➤ details of any action taken as a result of any matter identified in the previous paragraph
➤ details of any further action considered necessary
➤ the name and position of the person making the report.

Whilst there is no longer a specific statutory form to be completed, a record of the inspection should be made on the recommended form for inspections shown in Appendix 18.1, with a list of typical scaffolding faults in Appendix 18.2; a possible checklist, which could be used, is given in Appendix 18.3.

Most equipment used while working at height will need regular inspection and, in some cases, a thorough examination. Such equipment includes hoists and MEWPs, grit-blasting and water-jetting equipment and electrical appliances.

Maintenance

Inspections and even thorough examinations are not substitutes for properly maintaining equipment. The information gained in the maintenance work, inspections and thorough technical examinations should inform one another. A maintenance log should be kept and be up to date. The whole maintenance system will require proper management systems. The maintenance frequency will depend on the equipment, the conditions in which it is used and the manufacturer's instructions.

18.5 Practice NEBOSH questions for Chapter 18

1. **Outline** a hierarchy of measures to be considered when a construction worker is likely to fall while working at height.

2. **Outline** the measures to be taken to prevent falls associated with stairwells and other holes in floors during the construction of a multi-storey building.

3. A gable end wall of a three-storey building adjacent to a canal needs re-pointing. An independent tied scaffold will be used for access. The scaffold will obstruct the towpath and will restrict the width of the canal.
   Outline the health and safety issues associated with this work and how they might be addressed.

4. Sheet piles are to be driven into a river bed and the water contained by the piles pumped out before construction work starts. **Identify** the health hazards that may be encountered by workers involved in this operation and **explain** how such hazards may be controlled.

5. **Describe** the specific measures that may be necessary to ensure safety when work is to be carried out from a scaffold that overhangs a fast flowing river.

6. **Outline** the precautions that might be taken in order to reduce the risk of injury when using stepladders.

7. **Explain** the issues that would need to be addressed if work is to be carried out safely from a ladder.

8. **Outline** the main points to be considered when selecting and inspecting a wooden pole ladder prior to use.

9. A wooden pole ladder is to be used as a means of access to a scaffold.
   Outline the features of the ladder that should be considered, and the particular items that should

be inspected, in order to ensure its suitability for the job.

10. **Identify** measures that should be adopted in order to protect against the dangers of people and/or materials falling from a mobile tower scaffold.

11. Grit-blasting and concrete repairs are to be carried out on the external façade of a five-storey building. Access will be by an independent tied scaffold, which will take up the entire width of the adjoining pavement.

    Outline the possible hazards associated with this work **AND** the precautions that should be taken.

12. The exterior paintwork of a row of shops in a busy high street is due to be re-painted. **Identify** the hazards associated with the work and **outline** the corresponding precautions to be taken.

13. As part of the refurbishment of a three-storey building within a busy city centre, a section of stone façade is to be cleaned using a specialist cleaning contractor. The work will be undertaken using high pressure water-jetting.

    Describe a safe system and method of work to be adopted by the contractor to ensure the safety of:
    (a) the water-jetting operatives
    (b) the workers carrying out the internal refurbishment work
    (c) the public and users of the highway.

14. Scaffolding has been erected to the outside of a block of high-rise flats in order to undertake window replacement and repairs to external cladding.

    Outline the factors that could affect the stability of the scaffold.

15. An independent scaffold tied to a 10-storey office block has collapsed into a busy street.

    Outline the factors that may have affected the stability of the scaffold.

16. **List EIGHT** items to be examined when inspecting a putlog scaffold.

17. A four-storey building on the main street of a busy town centre requires external repairs to its fabric. The scaffold to be erected for the work will take up the entire width of the pavement.
    (a) **Sketch** the front and end elevation of a suitable scaffold for the work, clearly labelling the key safety features.
    (b) **Identify**, by means of a sketch where appropriate, the measures that will be needed to ensure safe passage for pedestrians and traffic.

    (c) **Outline** the procedures to be adopted in order to ensure the safety of employees using the scaffold as a working platform.

18. A row of terraced houses is to be built alongside an existing public road that runs up a slope.
    (a) **Sketch** the front and end elevation of suitable scaffold for this work, clearly identifying the key safety features.
    (b) **Outline** the procedures and practices required to ensure the safe erection and use of the scaffold in (a).

19. A mobile tower scaffold is to be used in the re-pointing of external brickwork on the gable end wall of a building, which is 10 m high at its highest point.
    (a) **Explain** the possible dangers associated with this operation.
    (b) **Outline**, with the aid of diagrams, the precautions to be taken to minimize the risks.

20. **Identify**, by means of a labelled sketch, the main requirements for a loading bay on a scaffold in order that materials for bricklayers and roofers can be safely placed on the bay by a forklift truck.

21. **Outline** the measures required to ensure the safe operation of an inclined hoist used to raise and lower roofing materials.

22. **Outline** the precautions to be taken when carrying out repairs to the flat roof of a building.

23. The waterproof membrane on the flat roof of a two-storey, high security office block is about to be replaced. Access for personnel not engaged in the roofing work is prohibited but an access route from inside the building is available. The only external access route for materials is via a lightweight, inclined builders' hoist, which must be erected and dismantled daily.

    Describe the procedures necessary to ensure:
    (a) the safety of the occupiers of the building
    (b) the safety of the operatives whilst working on the roof
    (c) the safe erection and dismantling, and use, of the inclined hoist.

24. Minor repairs to the external fabric of a multi-storey office block are to be carried out. The chosen method of access to upper floors is by a man-riding skip suspended from the hook of a mobile crane.
    (a) **Outline** the specific requirements relating to the safe use of the skip.
    (b) **Describe** other measures that should be in place to ensure the safety of all persons affected by the work.

25. Extensive repairs are to be carried out to the flat roof of a two-storey office building, which is to remain in use during the work. A small platform hoist is to be used for lifting materials to the roof. There is no access to the roof from inside the building.

    Outline the procedures and protective measures needed to ensure:
    (a) the safety of occupiers of the building
    (b) the safety of operatives whilst work is carried out on the roof
    (c) the safe erection and use of the hoist and other access equipment.

26. A partial upgrade of an existing ventilation system is to be undertaken in a large supermarket. The work is to be carried out on a progressive overnight basis over 10 nights, when the only other activities in the store are shelf-replenishment and cleaning. The majority of the work will involve access above a suspended ceiling at a height of 5m above the ground but plant situated on the roof of the store is also to be replaced.

    Describe the main features that must be covered to ensure the health and safety of contractors and others on site during the work.

27. A painting contractor has been contracted to paint roof girders in a racked storage warehouse. Painters will gain access to the roof area from both mobile elevating work platforms and 'Youngman' staging platforms. High reach pallet trucks will continue to be used in the warehouse during the work.

Appendix 18.1 – Inspection recording form with timing and frequency chart

Timing and frequency chart (reproduced from HSG150)

Place of work or work equipment	Timing and frequency of checks, inspection and examinations equipment								
	Inspect before work at the start of every shift (see note 1)	Inspect after any event likely to have affected its strength or stability	Inspect after accidental fall of rock, or other material	Inspect after installation or assembly in any position (see notes 2 and 3)	Inspect at suitable intervals	Inspect after exceptional circumstances which are liable to jeopardise the safety of work equipment	Inspect at intervals not exceeding 7 days (see note 3)	Check on each occasion before use (REPORT NOT REQUIRED)	LOLER Thorough Examination (if work equipment subject to LOLER) (see note 4)
Excavations which are supported to prevent any person being buried or trapped by an accidental collapse or a fall or dislodgement of material	✓	✓	✓						
Cofferdams and caissons	✓	✓							
The surface and every parapet or permanent rail of every existing place of work at height								✓	
Guard rails, toe boards, barriers and similar collective means of fall protection				✓	✓	✓			
Scaffolds and other working platforms (including tower scaffolds and MEWPs) used for construction work and from which a person could fall more than 2m				✓		✓	✓		✓
All other working platforms				✓	✓	✓			✓
Collective safeguards for arresting falls (eg nets, airbags, soft landing systems)				✓	✓	✓			
Personal fall protection systems (including work positioning, rope access, work restraint and fall arrest systems)				✓	✓	✓			✓
Ladders and stepladders					✓	✓		✓	

Notes

1. Although an excavation must be inspected at the start of every shift, only one report is needed in any seven-day period. However, if something happens to affect its strength or stability, and/or an additional inspection is carried out, a report must then be completed. A record of this inspection must be made and retained for three months.
2. 'Installation' means putting into position and 'assembly' means putting together. You are not required to inspect and provide a report every time a ladder, tower scaffold or mobile elevated work platform (MEWP) is moved on site or a personal fall protection system is clipped to a new location.
3. An inspection and a report (see image titled Inspection Report) is required for a tower scaffold or MEWP (used for construction work and from which a person could fall 2 metres) after installation or assembly and every seven days thereafter, providing the equipment is being used on the same site. A record of this inspection must be made and retained for three months. If a tower scaffold is reassembled rather than simply moved, then an additonal, pre-use inspection and report is required. It is acceptable for this inspection to be carried out by the person responsible for erecting the tower scaffold, providing they are trained and competent. A visible tag system, which supplements inspection records as it is updated following each pre-use inspection, is a way of recording and keeping the results until the next inspection.
4. All work equipment subject to LOLER regulation 9, thorough examination and inspection requirements, will continue to be subject to LOLER regulation 9 requirements.

INSPECTION REPORT

1. Name and address of person for whom inspection was carried out.

2. Site address.

3. Date and time of inspection.

4. Location and description of place of work or work equipment inspected.

5. Matters which give rise to any health and safety risks.

6. Can work be carried out safely?

Y/N

7. If not, name of person informed.

8. Details of any other action taken as a result of matters identified in 5 above.

9. Details of any further action considered necessary.

10. Name and position of person making the report.

11. Date and time report handed over.

12. Name and position of person receiving report.

Appendix 18.2 – Checklist of typical scaffolding faults

Footings	Standards	Ledgers	Bracing	Putlogs and transoms	Couplings	Bridles	Ties	Boarding	Guard rails and toe boards	Ladders
Soft and uneven	Not plumb	Not level	Some missing	Wrongly spaced	Wrong fitting	Wrong spacing	Some missing	Bad boards	Wrong height	Damaged
No base plates	Jointed at same height	Joints in same bay	Loose	Loose	Loose	Wrong couplings	Loose	Trap boards	Loose	Insufficient length
No sole plates	Wrong spacing	Loose	Wrong fittings	Wrongly supported	Damaged	No check couplers	Not enough	Incomplete	Some missing	Not tied
Undermined	Damaged	Damaged			No check couplers			Insufficient supports		

Appendix 18.3 – Checklist for a safety inspection of a scaffold

Explain how the work should be planned and undertaken in order to ensure the health and safety of both painters and warehouse staff.

➤ Is the scaffold on a firm foundation?

➤ Are standards (uprights) resting on suitable base plates and (where the scaffolding is not on hard standing) timber sole boards? Bricks, blocks and other building materials should not be used as packing.

➤ The scaffold should not be undermined by excavations close to its supports.

➤ Are the uprights vertical and the horizontals horizontal?

➤ Are the uprights close enough together and is the spacing consistent? (*Note*: the permissible width of bays will vary with the operations for which the scaffolding is intended. The greater the loading, the closer the uprights will need to be.)

➤ Are load-bearing couplers used where appropriate?

➤ Are working platforms properly supported (each board resting on at least three supports – no board should overhang the last support by a distance greater than three times its thickness), fully boarded out (no gaps for people or materials to fall through) and wide enough to allow safe access and the safe movement of materials?

➤ Has the scaffold been erected by a competent person? Are inspection records in order?

➤ Are all components in good condition?

➤ Are guard rails and toe boards fitted to all working platforms to prevent people and materials falling?

➤ Is additional protection such as wire mesh brick guards used where appropriate?

➤ Is the scaffold loaded with excessive quantities of materials?

➤ Are materials safely stacked? (Loads such as piles of bricks should be in line with the standards.)

➤ Is the scaffold structure adequately stiffened by the use of diagonal bracing?

➤ Is the scaffold adequately tied to a structure, or suitably buttressed, to prevent collapse?

➤ Are incomplete sections of the scaffold marked with suitable warning notices?

➤ Where there is sheeting or some other feature which will increase the windage of the scaffold, has this been allowed for in the design?

➤ Is there safe access to all the scaffold's working platforms?

Appendix 18.4 – Examples of safe systems of work used in roof work

1. *Slating work*

1. Materials will be delivered as close as possible to the roofing operation.
2. Materials will be loaded onto the scaffold by a mechanical conveyor or ladder.
3. Felt and battens will be laid from the eaves to the ridge in accordance with the specifications.
4. All slating will progress from right to left across the roof and, on completion, the ridge tiles and fixings will be similarly loaded and laid.
5. All waste materials and debris will be cleared from the site daily.
6. The site will be left in a tidy condition on completion.

2. *Lead work on roofs*

1. The Control of Lead at Work Regulations 2002 and associated Approved Code of Practice will be followed closely during all work involving lead.
2. Appropriate manual handling procedures will be adopted.
3. Work will be confined to clean solid metallic lead and always carried out in the open air.
4. In areas where lead work is being carried out, no food or drink will be consumed and no smoking allowed.
5. Adequate washing facilities will be provided on site. Nail brushes will be provided.
6. All employees working with lead will wear overalls which must remain on site at all times (except when being laundered at a suitably equipped facility).

3. *Working on a fragile roof (with a non-asbestos roofing material)*

A risk assessment will be provided for each job.

1. The area of roof in question which is to be repaired or renewed will be assessed from the scaffold unless a specific scaffold extension or duckboards have been provided and in place on the area in question.
2. Fixing bolts will be removed one at a time either manually or with the use of 110 volt portable equipment. No smashing or breaking of sections of sheeting is permitted unless prior authorization has been obtained from the operations manager.
3. Old sheets and material are to be placed into the skip provided via the installed rubbish chute. In the event of larger sections which will not go into the chute, being removed they will be lowered by hoist or pulley wheel to ground level where they will be received by a colleague.
4. In the event of repair of a fragile roof by walking through the valley gutters the following precautions must be taken:
   (a) The valley will be cleared of debris and water, i.e. leaves and soil. Any areas where the valley is showing dips or deflection should be reported immediately, unless they were briefed on at the start of the work, for a revision of the work method.
   (b) Both elevations of the roof either side of the gutter will be provided with either a continuous 'safe site' supergrip platform to reduce the possibility of slipping and falling through fragile sheeting or a series of good condition pallets will be used to serve the same purpose.
5. To carry out leak spotting, a safety harness secured to either an internal roof purlin by removing a ridge cap or other secure fixing near the ridge, must be used.
6. To provide a pathway on fragile sheets a series of duckboards should be used with specially adapted hook assembly on inclined roof areas.
7. Under no circumstances is anyone allowed to walk on sheets, bolts or any other bare roof coverings to gain access to an area requiring attention.
8. If workers are unable to use equipment provided they must report to the operations manager and withdraw from the area until receiving further instructions.

Excavation work and confined spaces – hazards and control

19

19.1 Introduction

Excavation work is an essential part of the construction process and one of the most hazardous. Building foundations and the installation of drainage, sewage and other services require trenches to be excavated. Because it is a hazardous operation several people are killed or seriously injured each year whilst working in excavations. Many are killed or injured by excavation wall collapses and falls of materials, others by contact with buried services. Excavation work has to be properly planned, managed, supervised and carried out to prevent accidents and ill-health to workers.

A recent survey of accidents has shown that there are on average seven fatalities each year in excavation work. Over a 5-year period, the survey showed that fatal and major injuries to workers who were working in excavations were caused by the following events:

➤ struck by falling or flying object, including earth (23%)
➤ struck by a construction vehicle or plant (15%)
➤ falling into excavation or from ladders or working platforms (14%)
➤ contact with electricity (12%)
➤ trapped by a collapse of earth or materials (11%)
➤ other (25%).

A major factor in such accidents is the type of soil which is being excavated. The soil structure may range from sand to clay and rock. Sands and gravel tend to be more unstable than clays or rock and need more support during the excavation process to protect both workers and plant from the effects of collapse. The stability of soils changes rapidly with water content and many collapses in the past have occurred after storms.

There are several major accidents caused each year by contact with buried services during excavation work.

In a recent case, a man was digging a hole with a spike when he struck a high voltage electrical cable causing a massive explosion and serious injuries for him.

The HSE has produced an excellent booklet entitled *Health and Safety in Excavations* HSG185. This offers detailed advice to those involved in the planning and design of excavations.

19.2 Excavations – hazards and control

19.2.1 *Hazards*

There are about seven deaths each year due to work in excavations and these are often related to the composition and structure of the soil which forms the walls of the excavation. Many types of soil, such as clays, are self-supporting but others, such as sands and gravel, are not. The walls of excavations can collapse without any warning resulting in death or serious injury. Many such accidents occur in shallow workings. It is important to note that, although most of these accidents affect workers, members of the public can also be injured.

The specific hazards associated with excavations are as follows:

➤ collapse of the sides
➤ materials falling on workers in the excavation
➤ dangers associated with excavation machinery
➤ falls of people and/or vehicles into the excavation
➤ workers being struck by plant
➤ specialist equipment such as pneumatic drills
➤ hazardous substances particularly near the site of current or former industrial processes
➤ influx of ground or surface water and entrapment in silt or mud

Figure 19.2 Undermining of boundary wall.

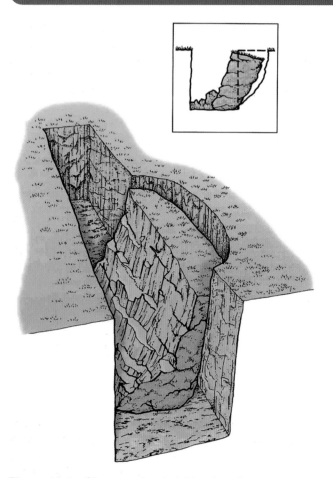

Figure 19.1 Slippage of material in a trench.

➤ proximity of stored materials, waste materials or plant
➤ proximity of adjacent buildings or structures and their stability
➤ contact with underground services
➤ contact with overhead power lines
➤ access and egress to the excavation and
➤ fumes, lack of oxygen and other health hazards (such as Weil's disease).

The build-up of fumes can be particularly hazardous since it may lead to asphyxiation, fire or explosion. When the excavation is taking place in a busy road, there will be hazards associated with passing traffic and the possibility of vehicle collision with the excavation, its workers or passing pedestrians. Clearly, alongside these specific hazards, more general hazards, such as manual handling, electricity, noise and vibrations, will also be present.

As mentioned earlier, one of the most critical factors to be considered when planning an excavation is the nature of the soil which is to be excavated. All excavations, except the most shallow, need some support using trench sheeting. When excavation work takes place in non-cohesive soil (e.g. sand, gravels and soft clays), close sheeting will be required to prevent ground movement but for more cohesive soils, such as most clays and rock, some support, such as open sheeting, will still be necessary. Since soil cohesiveness decreases as water content increases, the stability of an excavation is considerably reduced during prolonged wet weather. It is, therefore, important that a competent person should assess the likelihood of any changes in ground conditions and adjust the working method accordingly.

If the excavation work takes place below the local groundwater level or water table, then water will flow into the trench. Measures will be required to impede the inflow of water and channel the water in the trench to a sump or pond where it may be pumped to a safe discharge point. These problems will be exacerbated at times of high rainfall, such as late winter or early spring and will be even more hazardous if the soil structure is naturally weak.

Instability may also be caused by additional loading in the area adjacent to the excavation, for example by vehicles, plant or a spoil tip. This is known as **surcharging**. Possible actions to reduce this risk are given in Section 19.2.4.

If liquid nitrogen is used to freeze pipes during an excavation, additional hazards, such as frostbite and breathing difficulties due to nitrogen leakage, are presented. Permits to work should always be used for this type of work.

Finally, low levels of natural light are an additional hazard, which is particularly problematic in the winter and at the beginning and end of the working day.

19.2.2 *Planning the work*

Before starting any excavations, it is important to plan the work by following a comprehensive planning procedure (as described in HSG185) so that all significant hazards can be addressed. The principal hazards have been outlined in Section 19.2.1 and the most immediate ones for planning purposes are the following:

➤ collapse of the sides
➤ control of people and vehicles around the surface area of the excavation
➤ the proximity of adjacent structures
➤ the position of any underground services
➤ access to the excavation
➤ the protection of members of the public.

A site survey should always be undertaken during the planning process and the following points need to be considered at this stage:

➤ previous use of site
➤ location of existing buildings
➤ location of new structures
➤ results of soil investigations
➤ ground contamination
➤ level of water table and type of soil
➤ storage and disposal of excavated material
➤ amount of working space and storage required
➤ the most suitable method of temporary support of the excavation walls
➤ adequate emergency arrangements.

Much excavation work is needed to repair mains (water or gas), piped services or cables (electrical power or telecommunications) which are feeding occupied buildings. It is crucial that there is effective communication between the excavation contractors and the building occupants throughout all stages of the project. The pipe or cable must be isolated before work begins. If the work obstructs the main or emergency exits from the buildings then alternative egress arrangements must be made. All the precautions discussed in 19.2.4 will also apply.

After the site survey has been completed, meetings should be organized with the occupants of surrounding buildings and other properties so that any special arrangements can be communicated. For example, if the neighbouring property is a school, special care will be needed when the children arrive and depart from the school. Continual liaison with neighbours is an important element in the risk control strategy.

Before work begins, there must be available sufficient numbers of trained operatives and competent supervisors and suitable plant and materials for trench support. Relevant monitoring equipment and personnel trained in

its use will be required if exposure to toxic substances or lack of oxygen may occur. The location of any existing services must be completed. A method statement, which should cover external conditions, such as traffic, weather and existing structures, should also be prepared and, if necessary, Control of Substances Hazardous to Health Regulations (COSHH) assessments made.

During the final stages of planning, it will be necessary to ensure that any special personal protective equipment (PPE) or signs and equipment related to the work are ordered and available on the site. The organization of the daily inspection of the excavation before each shift and the proper administration of the inspection records will need to be made at this stage.

Arrangements will also be required for the proper induction of any sub-contractors employed on the excavation work so that they are aware of the main hazards and the site rules, including the correct use of personal protective equipment.

Finally, it is important to ensure that all the required equipment, such as trench sheets, props and walings, is available on site before the work starts.

19.2.3 *Risk assessment*

A risk assessment should be made by the contractor undertaking the work before work begins to identify those hazards that are likely to be encountered and determine the control measures required. The significant findings of an assessment should be recorded, for example in a method statement. Method statements should describe the plant and equipment and the safe methods of working required to control the risks generated. These statements should be relevant to the particular work (rather than generic) and easily understandable by those who need to use them. They should be proportionate to the level of risk and degree of complexity – if the risks and necessary precautions are straightforward, short, simple method statements are all that are needed to convey the necessary information. The documents should identify the hazards associated with the work and any factors which increase the risk of injury, such as:

➤ the nature of the ground, soil structure and groundwater regime
➤ the depth of excavation
➤ the nature of the work required to be undertaken within an excavation
➤ the location of the work (e.g. readily accessible public place, contaminated ground or heavily serviced urban area).

The method of support selected needs to take account of the work to be done and allowance made for adequate working space. It may be necessary to install support in relatively shallow excavations of less than 1.2 m depth if ground conditions are particularly poor or the nature of the work requires workers to lie or crouch in a trench. The options for the method of support may also be limited by ground conditions or the presence of services crossing the trench line. Support will also be required when plant for excavating or depositing backfill causes an extra loading on the excavation walls.

Support methods are discussed in 19.2.4, but whatever the support method selected, an essential part of the method statement is a defined safe method of work. No person should ever enter an unsupported excavation to install or remove supports. The use of proprietary support systems which can be installed safely from existing ground level may help should this situation arise. Methods of work need to take into account the following factors to avoid the risk of workers being struck by the plant itself or of the materials being placed where the visibility of the plant operator is restricted:

➤ the removal from site of surplus excavated material
➤ the stockpiling of excavated material which is to be reused
➤ the provision of storage space for materials and working space for workers and plant
➤ the type of compaction equipment to be used.

The method statement should also contain the information about the necessary emergency arrangements should an incident occur, unless they are already detailed in the health and safety plan. Details regarding any site-based personnel trained to administer first aid, and the location of the nearest hospital with accident and emergency facilities, should be included. A method for swift communication should be identified and, if there is no telephone available on site, details given on the location of a public telephone close by; otherwise a mobile telephone should be provided. Thought should be given to the actions to be taken if a person were to be injured within a trench or if any gas monitors in use identify the presence of harmful gas or oxygen deficiency. Employees and other workers should be instructed in all relevant health and safety procedures. The risk assessment also needs to categorize the level of risk. High risks include the collapse of the excavation walls, contact with existing underground or overhead services, plant and materials falling into excavations and the presence of hazardous atmospheres and contaminated soil. The possibility of persons falling into the excavations is normally considered a medium risk whereas the flooding of excavations is usually designated a low risk.

Finally, it is very important to consider the additional risks arising to members of the public, in particular to children, those with impaired mobility and the visually impaired. Work should be planned to ensure that excavations are well lit, securely fenced, and covered or

backfilled overnight and at weekends. Plant should be left in a secure compound and any engines isolated. Stored materials should be stacked carefully to prevent displacement and fenced if they could be accessed. The risk assessment should determine the precautions necessary to prevent unauthorized access to any construction site. Arrangements may be required for traffic management supervised by competent persons, involving traffic controls such as traffic lights, stop/go boards or road closures and protection for pedestrians from the work. Temporary road signs, positioned in the correct order, may be needed either side of the excavation and cones set out with lead-in and exit tapers. There should also be an indication of a safety zone between the cones and the work area. Workers who work on the carriageway must wear high visibility clothing and be trained and competent for such work under the New Road and Street Works Act. The Act also requires that any part of a street that is broken up, is open or is obstructed must be adequately guarded and lit. In addition, any traffic signs must be placed, maintained and operated as reasonably required for the guidance or direction of persons (particularly the disabled) who are using the street.

Figure 19.3 Barriers around excavation by footpath.

The New Roads and Street Works Act stipulates circumstances under which certain excavation contractors, defined as utilities and other undertakers, must give notice of their planned works. Emergency, urgent and some minor works are excluded from this duty. It also requires that contractors keep up-to-date records of the location of their apparatus and make them available, free of charge, for inspection at all reasonable hours. The exemptions applicable to the Act do not affect contractor liability under the HSW Act.

The New Roads and Street Works Act requires that the person who supervises any street works must be qualified under the Street Works (Qualifications of Supervisors and Operatives) Regulations. There must also be an operative who is qualified under the same Regulations, on site while work is in progress.

A typical risk assessment for an excavation is given in Appendix 19.2 at the end of this chapter.

19.2.4 *Precautions and controls*

Excavations must be constructed so that they are safe environments in which construction work can take place. This means that they must be fenced and suitable notices posted so that neither people nor vehicles fall into them. However, there are many other precautions and controls that must be taken and several of these should have emerged during the communication and liaison with the occupants of adjacent buildings at the planning stage of the project.

The first form of control is to decide whether the excavation is necessary. There have been many technological advances which have eliminated the need for excavation work. These trenchless technologies include micro-tunnelling, directional drilling, impact moling, auger boring, pipe relining and pipe bursting. More information on these techniques, which are not in the Construction Certificate syllabus, may be found in HSG185.

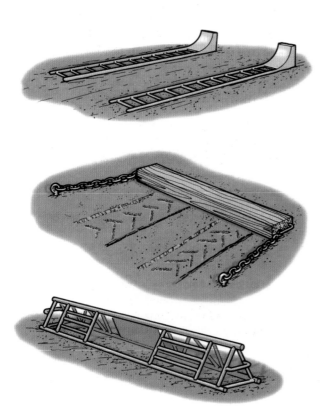

Figure 19.4 Stop blocks for dumpers.

If excavation work is the only option available, the following precautions and controls should be adopted:

➤ At all stages of the excavation, a competent person must supervise the work. Workers must be given clear instructions on working safely in the excavation.

➤ The walls of the excavation trench must be prevented from collapsing either by digging them at a safe angle (between 5° and 45° depending on soil structure and its dryness) or by shoring them up with timber, sheeting or a proprietary support system. Even shallow trenches may need support if the work involves bending or kneeling in the trench.

➤ Falls of material into the workings can also be prevented by not storing spoil material near the top of the excavation.

➤ Workers must wear hard hats and safety footwear.

➤ Where people are liable to fall into an excavation a substantial barrier, consisting of guard rails and toe boards should be provided around the surface of the workings. This is particularly important if the work is taking place in a public place where a brightly coloured barrier may be required.

➤ Workers should never enter an unsupported excavation or work ahead of supports.

➤ Vehicles should be kept away as far as possible using warning signs and barriers. Where a vehicle is tipping materials into the excavation, stop blocks should be placed behind its wheels.

➤ Whenever it is possible, workers and moving plant, such as excavators, should be kept separated.

➤ It is very important that the excavation site is well lit at night.

➤ All plant and equipment operators must be competent and non-operators should be kept away from moving plant.

➤ Personal protective equipment (gloves, overalls, boots, ear defenders and hard hats) must be worn by operators of noisy plant.

➤ Nearby structures and buildings may need to be shored up if the excavation may reduce their stability. Scaffolding could also be de-stabilized by adjacent excavation trenches. When there is doubt about this, the views of a structural engineer should be sought.

➤ The influx of water can only be controlled by the use of pumps after the water has been channelled into sumps.

➤ The presence of hazardous substances or health hazards should become apparent during the original survey work and, when possible, they should be removed or suitable control measures adopted. Any such hazards found after work has started must be reported and noted in the inspection report and remedial measures taken. Exhaust fumes can be dangerous and petrol or diesel plant should not be sited near the top of the excavation.

➤ The presence of buried services is one of the biggest hazards and the position of such services must be ascertained using all available service location drawings before work commences. Since these will probably not be accurate, service location equipment should be used by specifically trained people. Only hand tools should be used in the vicinity of underground services. Overhead services may also present risks to cranes and other tall equipment.

➤ Safe access by ladders is essential as are crossing points for pedestrians and vehicles. Whenever possible, the workings should be completely covered with steel sheeting outside working hours particularly if there is a possibility of children entering the site.

➤ Finally, care is needed during the filling-in process.

Some of these precautions will now be discussed in more detail.

A particular hazard, discussed earlier, is **surcharging**. The risk of collapse due to surcharging may be reduced by:

➤ depositing spoil and other waste material at least 1.5 m from the edge of the excavation or a distance equal to the depth of the excavation whichever is the greatest

➤ increasing the amount of temporary support to the excavation when there are buildings close by or the excavation is taking place on a slope

➤ installing suitable barriers and ensuring that all traffic routes are situated at a safe distance from the top of the excavation.

The stability of adjacent buildings and other structures will be affected by the excavation if it weakens or undermines the foundations of the structure. There have been several fatalities caused by the collapse of a retaining wall into a trench. When it is unavoidable that a trench must be dug next to a building, the building should be shored up by suitable supports. Where there is a width restriction on either side of the trench, arrangements will be needed to deposit the spoil material elsewhere at a safe distance.

Most excavations will require some form of **support**. Unsupported excavation is safe only if the sides are battered back sufficiently and the vertical depth of the trench does not exceed 1.2 m. HSG185 gives details of safe angles of slope for different soils.

There are several methods of excavation support but before they are discussed the basic components of such support and some excavation terminology must be defined.

Sheeting is the general term given to steel sheets or timber boards used to support the walls of an excavation. When vertical or horizontal sheeting is placed close together for added strength, this is known as **close sheeting**. However, **open sheeting** occurs when the sheets are only placed at intervals, normally in clay.

Trench sheets are long, narrow, rolled steel plate sections which either lap-joint or interlock at their edges with adjacent sheets. They are normally installed vertically to support the sides of the trench.

Runner is a vertical member which is driven below the floor of the excavation and supports the side of the trench, and is progressively driven as the excavation proceeds.

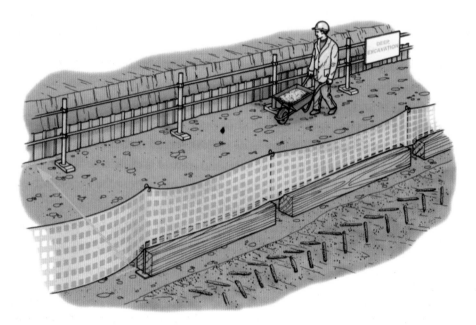

Figure 19.5 Vehicle protection at top of an excavation.

Figure 19.6 Trench sheets with timber walings, screw props, puncheons and sole plates.

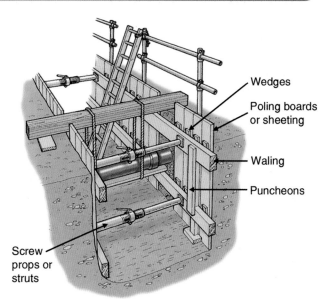

Wedges

Poling boards or sheeting

Waling

Puncheons

Screw props or struts

Figure 19.7 Timbered excavation with ladder access and supported services (guard removed on one side for clarity).

Steel sheet piles are interlocked steel sections driven into the ground to resist side pressure on the excavation trench.

Poling boards are sheeting (1 m by 5 m) which supports the sides of the trench. When poling boards are used horizontally across a gap between runners or sheeting, this is known as **cross poling**.

Walings are horizontal beams used to support sheeting or poling boards against pressure from the trench walls.

Props are sometimes called **struts** and provide extra support between pairs of sheeting, and are positioned between the walings.

Puncheons are vertical posts which support a waling or prop from below.

Well point is a long thin tube which is installed vertically into the ground and through which groundwater is pumped. A **well point system** is a system of well points situated around the excavation area and used to pump out groundwater and reduce the flooding risk in the excavation trench.

There are six basic excavation support systems in general use:

1. **Battered sides**
   This is the traditional method of trench excavation and relies on the strength of the soil to support the trench walls. There is no safe depth for this technique since all soils vary in cohesiveness and many soils consist of a mixture of sand, gravel and clay. For many soils, the angle of repose of the sides (measured from the horizontal plane) is approximately 30°. This means that such excavation covers a considerable area. It is generally recommended that this type of excavation should not exceed 1.2 m in depth.

2. **Open and close sheeting**
   These are also traditional support systems and HSG185 recommends a safe method statement for each system. Underground services can present problems with these methods of support. Such services should not themselves be used to support any part of the trench support system. It is important that sufficient struts and walings are used to give added strength to the excavation. Regular supervision is required to ensure that such support is not removed by workers to improve the size of the working space. Care is also required during the dismantling operation.

3. **Hydraulic waling frames**
   These consist of two steel beams braced apart by struts containing integral hydraulic rams. They can be used with open or close sheeting trenching and for supporting close sheeting in very deep trenches.

4. **Manhole shores**

   As the name suggests, these are supporting manhole or sump-type excavations. They are four-sided adjustable frames with integral hydraulic rams. As with waling frames, they must be secured to the trench sheeting (normally by chains).

5. **Trench boxes**

   These consist of solid side panels which are kept apart by integral hydraulic struts which may be adjusted to suit the width of the trench. The bottoms of the side panels have a tapered cutting edge. This enables them to be dug in or lowered by crane into a trench which has already been dug. Trench boxes can be fitted on top of each other when the trench is particularly deep.

6. **Drag boxes**

   These have solid sides with tapered cutting edges to their leading edges. The side sheets are braced apart by specially strengthened struts so that they may be dragged through the advancing excavation by the excavator. This technique is often used for pipe laying. The greatest problem with box systems is that they can only be used where there are no buried services or other obstructions. Their greatest advantage is that they allow trench supports to be put in place without requiring people to enter the excavation.

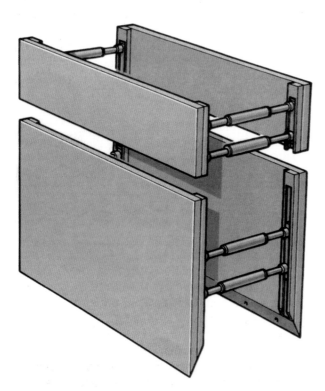

Figure 19.8 Trench box.

The presence of **underground services**, particularly electricity and gas, is a common problem. Often official knowledge of the exact location of such services is less than accurate, particularly if the underground services have been in place for many years.

The following precautions are suggested when there is uncertainty about the location of underground services in an area to be excavated:

➤ Check for any obvious signs of underground services, for example valve covers or patching of the road surface.

➤ Ensure that the excavation supervisor has the necessary service plans and is competent to use them to locate underground services.

➤ Ensure that all excavation workers are trained in safe digging practices and emergency procedures.

➤ Use locators to trace any services and mark the ground accordingly. A series of trial holes should be dug by hand to confirm the position of the pipes or cables. This is particularly important in the case of plastic pipes which cannot be detected by normal locating equipment.

➤ In areas where underground services may be present, only hand digging should be used with insulated tools. Spades and shovels should be used rather than picks and forks, which are more likely to pierce cables.

➤ Assume that all pipes or cables are 'live' unless it is known otherwise.

➤ Hand-held power tools should not be used within 0.5 m of the marked position of an electricity cable. Collars should be fitted to the tools so that initial penetration of the surface is restricted.

➤ A machine should not be used to excavate within 0.5 m of a gas pipe.

➤ Any suspected damage to services must be reported to the service providers and the health and safety enforcement authority.

➤ All exposed pipes or cables should be backfilled with fine material such as dry sand or small gravel. Backfill which is properly compacted, particularly under cast or rigid pipes, prevents settlement which could cause damage at a later date.

➤ The service plans must be updated when the new services have been laid.

Excellent guidance is available in the HSE publication *Avoiding Danger from Underground Services* HSG47.

The adoption of safe digging methods is essential if accidents are to be avoided. Several elements of such practice have already been covered and a detailed safe digging practice is given in Appendix 19.1 at the end of this chapter.

If an electricity cable is accidentally struck, all workers in the vicinity may receive an electric shock and there is also a fire risk, as there would be if a gas main were damaged. Severe flooding of the trench could result if a water main were intercepted. Any of these events could cause severe problems to the local community, particularly local hospitals. When the purpose of the excavation is to investigate or repair a cable or pipe, it must be isolated while work on it is in progress.

Figure 19.9 Using a cable locator.

An emergency plan, which includes the requirement to notify the relevant service provider, is essential to deal with damage to any underground cables or pipes. In the case of gas pipe damage, all smoking and naked flames must be banned. An evacuation procedure may be necessary which may include people in nearby properties likely to be affected by any leaks. Suitable signs should be erected to warn of the danger and the relevant health and safety enforcement authority informed.

It is important to note that **overhead power lines** can also cause serious injuries if plant, such as cranes or excavators, comes into contact with them. Although this problem is discussed in more detail in Chapter 14,

some obvious precautions will be noted here. If possible, excavation beneath power lines should be avoided. If the work is unavoidable, then highly visible barriers should be erected at least 6m away from the lines and red and white goalpost type crossing points should be positioned beneath the lines. When excavation work has to take place beneath power lines, the planning supervisor must, during the planning stage, ascertain whether the line can be diverted or made dead. The use of conductive materials, such as scaffold tubes or aluminium ladders, should be limited in these circumstances.

Access to the excavation is normally provided using strong ladders but consideration must be given to the means of rescue, perhaps using a winch or tripod, for an injured worker who is unable to use a ladder.

Fumes and exhaust gases can be hazardous, particularly if they become concentrated in an excavation. This is a particular problem for some welding gases that are heavier than air. Petrol- or diesel-engined equipment, such as generators or compressors, should not be sited in, or near the edge of, an excavation unless the exhaust gases can be ducted away or the area can be ventilated.

Where a **flooding** risk exists, cofferdams/caissons may need to be installed with pumps of suitable capacity. A cofferdam or caisson is a watertight box-like iron structure which is normally sunk in the bed of a river. It is pumped dry and filled with concrete and used in the construction of underwater foundations for structures such as bridges or quay walls. Cofferdams must be inspected by a competent person on the same day as used by a worker. A permit-to-work system may also have to be used.

Finally, if emergency routes from adjacent buildings could be obstructed by the excavation work, consideration of alternative routes will be necessary. This is a particular problem if the building is a school, surgery or hospital.

19.2.5 *Inspection requirements*

The duty to inspect and prepare a report only applies to excavations which need to be supported or battered back to prevent accidental fall of material. Only persons with a recognized and relevant competence should carry out the inspection and write the report. Inspections should take place at the following timing and frequency:

➤ after any event likely to affect the strength or stability of the excavation
➤ before work at the start of every shift
➤ after an accidental fall of any material.

Although an inspection must be made at the start of every shift, only one report is required of such inspections every 7 days. However, reports must be completed following all other inspections. Any written reports should be completed before the end of the relevant working period and a copy given to the manager responsible for the excavation within 24 hours. Reports must be kept on site until the work is completed and then retained for 3 months at an office of the organization which carried out the work. If any inspection finds that the excavation is unsafe the manager should be informed and the work stopped.

A suitable form is shown in Appendix 18.1 at the end of Chapter 18.

19.3 Confined spaces

The hazards present in confined spaces have caused the deaths of many workers and those who were trying to rescue them. On average 15 people each year are killed in confined spaces. There have been several tragedies in confined spaces due to a lack of oxygen. Several years ago in Cornwall, a young mining trainee was killed whilst attempting to rescue an unconscious colleague from a dead end underground roadway. The cause of the death was oxygen deficiency.

As with all hazardous situations, work should only take place in a confined space if it is unavoidable. Workers in confined spaces should be skilled and trained or specialist contractors.

The Confined Spaces Regulations require that:

1. no person at work shall enter a confined space for any purpose unless it is not reasonably practicable to achieve that purpose without such entry
2. other than in an emergency, no person shall enter, carry out work or leave a confined space otherwise than in accordance with a safe system of work, relevant to the specified risks.

19.3.1 *Definition of a confined space*

A confined space can be defined as any space of an enclosed nature which has limited means of access and egress, restricted natural ventilation and is not intended for continual occupancy by persons. It is, therefore, any space which, by virtue of its enclosed nature, presents a reasonably foreseeable specified risk of serious injury.

Examples of a confined space include manholes, sewers, tunnels, excavations, storage tanks, holds of ships, pits, trenches, ducts, some unventilated areas or rooms within buildings (particularly below ground level), boilers, combustion chambers in furnaces, chambers, vats, silos, pits, trenches, pipes, flues and wells.

19.3.2 *Hazards*

The principal hazards associated with a confined space are the difficult access and egress, which makes entrapment, escape and rescue more problematic; the accumulation of vapours, gases or fumes; and the lack of ventilation. These hazards may be inherent due to the nature of the confined space (e.g. the leakage of methane from the surrounding strata) or they may be introduced into the confined space due to the nature of the work (e.g. welding work or cleaning work with solvents).

Other hazards associated with confined spaces include:

➤ asphyxiation due to oxygen depletion
➤ poisoning by toxic substance or fumes
➤ explosions due to gases, vapours and dust
➤ fire due to flammable liquids, vapours and oxygen enrichment
➤ fall of materials leading to possible head injuries
➤ free flowing solid such as grain in a silo
➤ electrocution from unsuitable equipment
➤ difficulties of rescuing injured personnel
➤ drowning due to flooding
➤ fumes from plant or processes entering confined spaces
➤ excessive heat leading to heat stress
➤ claustrophobic effects due to restricted space
➤ diseases from animal wastes, infected materials or micro-organisms, e.g. fungal infections, tetanus, Weil's disease and pigeon droppings.

When oxygen levels fall to 17%, the physical and mental abilities of workers become severely reduced. As the concentration falls below 17%, unconsciousness and death follow very rapidly. In confined spaces, the act of breathing can reduce oxygen levels quickly. Oxygen depletion can be caused by the reaction of some soils to oxygen in the atmosphere, such as the action of water on chalk. Some operations, like welding and cutting, consume oxygen very rapidly.

Poisonous gases or vapours can be caused by leaching from contaminated surrounding land or by the concentration of gases in sewers or manholes. Some of the more serious hazards are introduced into the confined space. These include welding and flame cutting, painting and coating, and cleaning operations all of which can produce hazardous gases and vapours.

19.3.3 *Risk assessment*

Risk assessment is an essential requirement of the Confined Spaces Regulations and must be done so that a safe system of work may be determined. The risk assessment needs to identify the hazards present in the confined space, assess the risks and determine suitable controls to address those risks. The first part of the risk assessment will determine whether the work can only be done in the confined space.

Figure 19.10 Entering a confined space.

The risk assessment must be done by a competent person and will form the basis of a safe system of work. This will normally be formalized into a specific permit-to-work procedure which is applicable to a particular task.

The assessment will involve the examination and investigation of the following items:

➤ any previous contents in the confined space
➤ any residues that have been left in the confined space, for example sludge, rust or scale, and details of gases or vapours which may be generated if these substances are disturbed
➤ any contamination which may arise from adjacent plant, processes, services, pipes or surrounding land, soil or strata
➤ any oxygen deficiency and enrichment. There are very high risks if the oxygen content differs significantly from the normal of 20.8%
➤ the physical dimensions and layout of the space since these can affect air quality
➤ the use of cleaning chemicals and their direct effect or interaction with other substances
➤ any sources of ignition for flammable dusts, gases, vapours, plastics and the like
➤ the need to isolate the confined space from outside services or from substances held inside such as liquids, gases, steam, water, fire extinguishing media, exhaust gases, raw materials and energy sources
➤ the requirement for emergency rescue arrangements including trained people and equipment.

On completion of the risk assessment, the site manager, in conjunction with a specialist contractor if one is to be used, will draw up a method statement and a programme of work. This will detail the methods, plant and equipment to be used, general precautions and any special requirements for dealing with health hazards, and the sequencing of work. The details and procedures required for any permit-to-work systems should also be included. This method statement and programme of work will be issued to the supervisor responsible for the work on site prior to the commencement of the work. All this information should also be contained in the on-site construction phase safety plan (as required under CDM). A copy of the risk assessment and method statement should be kept on site for reference purposes. An example of a risk assessment for a confined space is given in Appendix 19.3 at the end of this chapter.

19.3.4 *Controls and precautions*

The main duty of the employer, so far as is reasonably practicable, is to avoid the need for employees or others to enter a confined space. If entry is unavoidable, he should ensure that a safe system of work is available and that adequate emergency rescue arrangements are in place including the provision of appropriate personal protective equipment such as respirators (as described in Chapter 16).

The site supervisor must ensure that all necessary equipment is available on site in accordance with the risk assessment, method statement and any other planned procedures, such as a permit to work, before any person is allowed to enter a confined space. He must also ensure that the planned procedures, including any permit-to-work systems, are adhered to strictly and that only authorized trained persons are permitted to enter the confined space. He must be notified of any changes in working methods or conditions inside the confined space which were not included in the planning procedures before any changes are implemented.

All safety equipment must be regularly checked and maintained. Records should be kept of the checks and any defects in equipment rectified immediately.

The detailed precautions required will depend on the nature of the confined space and the actual work being carried out. The main elements of a safe system of work which may form the basis of a permit to work are as follows:

➤ the type and extent of the supervision of the work
➤ the competence, training and instruction of the workforce
➤ the stipulation of a minimum gang size for large confined spaces such as reservoirs
➤ details of any information required by the workforce
➤ the methods of communication between people inside, from inside to outside and to summon help in an emergency
➤ the testing and monitoring of the atmosphere inside the confined space for hazardous gas, fumes, vapour, dust and the concentration of oxygen

Figure 19.11 Training for confined space entry.

➤ the gas-purging of toxic or flammable substances with air or inert gas such as nitrogen

➤ the provision of good ventilation, sometimes by mechanical means

➤ if appropriate, the cleaning of the confined space before the work commences

➤ the careful removal of residues using appropriate equipment which will not cause additional hazards

➤ effective protection from gases, liquids and other flowing materials by the removal of redundant piping, the blanking of pipes and the locking of valves

➤ the effective and complete isolation of electrical and mechanical equipment by the use of a lock-off and a tag system with key security. Checks will need to be made to protect the workforce against hazards such as stored energy in flywheels or the fall under gravity of heavy presses

➤ the provision of personal and respiratory protective equipment if it is not possible to enter the confined space otherwise. In this event, the supply of gas through pipes and hoses must be carefully controlled

➤ access and egress to the confined space so that there is quick and unobstructed access and escape

➤ the details of fire prevention equipment and procedures

➤ the lighting arrangements including emergency lighting

➤ the details of any behavioural rules such as the prohibition of smoking

➤ the emergency and rescue arrangements and procedures (including first aid)

➤ the maximum length of any one working period

➤ the details of any required health surveillance of the workforce

➤ monitoring and audit arrangements to ensure that the permit-to-work system is working as envisaged.

Any lighting or electrical power tools must be specially protected to the required standard in damp and flammable atmospheres.

Specific, detailed and frequent **training** is necessary for all people concerned with confined spaces, whether they are acting as rescuers, supervisors or those working inside the confined space. The training will need to cover all procedures and the use of equipment under realistic simulated conditions.

19.3.5 *Monitoring arrangements*

Before any work commences in the confined space, the site supervisor must arrange for any necessary environmental surveys and sampling needed to protect the health of the workforce. Specialists and other personnel required to test and monitor the atmosphere must be competent to take the measurements and interpret the results accurately. Continuous gas monitoring using electronic instruments should be used in preference to 'spot' detection devices which only give a measurement at a given time rather than over the complete work period. The workforce should also be given health checks periodically to ensure that they are not suffering any claustrophobic effects or problems with the wearing of breathing apparatus. An occupational health specialist will provide information on ventilation equipment, breathing apparatus, resuscitation apparatus, ropes, harnesses and monitoring equipment, and information on symptoms associated with working in a confined space.

Figure 19.12 Emergency escape breathing apparatus.

All breathing and resuscitation apparatus, rescue equipment and emergency alarms must be regularly checked and in some cases recalibrated. First aid provision must be checked regularly as must the accreditation of first aiders. The workers must also be monitored and consideration should be given to the exclusion from confined spaces of individuals who have any of the following problems:

➤ heart disorders including high blood pressure
➤ fainting, blackouts, fits or loss of balance
➤ weak eyesight or hearing
➤ asthma or chronic bronchitis
➤ poor physical size, poor mobility or back pain
➤ claustrophobia or psychological stress.

19.3.6 *Emergency arrangements*

Before people enter a confined space suitable and sufficient emergency and rescue arrangements must be in place. These arrangements must reduce the risks to rescuers so far as is reasonably practicable and include the provision and maintenance of suitable resuscitation equipment which must be designed to meet the specific risks associated with the particular confined space.

For the emergency arrangements to be suitable and sufficient, they must incorporate the following:

➤ contingency plans to deal with an emergency in the confined space. Such plans must include details of rescue teams together with individual responsibilities
➤ the contact names and telephone numbers of the local emergency services
➤ the details of the communication arrangements from inside the confined space
➤ the provision and maintenance of the rescue and resuscitation equipment
➤ the raising of the alarm, alerting the rescue team and maintaining close supervision of the workforce inside the confined space
➤ the notification of and consultation with the emergency services
➤ the protection and safeguarding of the rescue team
➤ the fire safety precautions and procedures
➤ the provision and maintenance of fire fighting equipment
➤ the control and possibly the protection of adjacent buildings, plant and equipment
➤ the arrangements for dealing with emergencies outside the normal working day
➤ the first aid arrangements including any special training required of first aiders

➤ the training of rescuers and simulations of emergencies, such as fire drills
➤ the size of access openings to permit rescue with full breathing apparatus, harnesses, fall arrest gear and lifelines, which is the normal suitable respiratory protection and rescue equipment for confined spaces.

19.4 Practice NEBOSH questions for Chapter 19

1. **Identify** the main dangers associated with excavation work on construction sites.

2. Pipe-freezing using a liquid nitrogen jacket is to be carried out in a properly supported, 2 m deep trench.
   (a) **Identify** the specific hazards associated with the pipe-freezing operation.
   (b) **Outline** the precautions that would be appropriate in such circumstances.

3. (a) **Explain** the meaning of the term 'confined space'.
   (b) **Identify FIVE** specific dangers associated with confined spaces.

4. An excavation is to be undertaken in ground which is known to contain running sand. The depth of the excavation is to be 3 m and the length 10 m to lay a 150 mm drainage pipe. There is a working width of 6 m and the time of year is February/March.
   Identify the major problems/hazards likely to be encountered and outline means of dealing with them.

5. **Sketch** a labelled diagram of a steel trench support system that clearly shows all the main components and features.

6. (a) **State** the hazards and associated risks involved with an excavation for a tank that is 2.5 m deep and is in ground previously used for landfill.
   (b) **Describe** the control measures that might be necessary to eliminate or reduce the risks identified in (a).

7. A tank that measures 4 m long, 3 m wide and 2 m deep is to be buried in a green field site as part of a surface water drainage system.
   (a) **Outline** the principal hazards that should be considered when planning the work.
   (b) **Outline** the elements of a method statement for the excavation and installation operations.

8. A sewer connection for a new housing estate is to be made in an existing 4 m deep manhole. The

manhole is 2 m from the kerbside on a 40 mph single carriageway road. The road opening notice specifies that the work must be carried out only between the hours of 9.30 am and 4.30 pm.

Describe the main features that must be covered to ensure the health and safety of the workforce and the public. Illustrate your answer with sketches of:
(a) the road layout, and
(b) the excavation.

9. The laying of a new water pipe involves excavating along and across an urban road. The width of the carriageway will be restricted to a single lane during the work and traffic flow in both directions is to be maintained.
(a) **Describe** alternative means of controlling traffic, explaining the circumstances that will determine the appropriateness of **EACH** method.
(b) **Outline** the hazards associated with such work and the precautions that will be needed to ensure the safety of workers and members of the public. Use sketches to illustrate your answer.

10. **Outline** the main precautions to be taken when carrying out excavation work.

11. The provision of main drainage to a village involves excavating to a depth of 2.5 m along the main village street.
(a) **Identify** the hazards and associated risks that may arise from this work.
(b) **Describe** the control measures that would be necessary to minimize the risks identified.

12. A leaking underground concrete reservoir has been emptied in order that it can be visually inspected prior to its subsequent repair.
(a) **Outline** the features of a safe system of work for the inspection team in order to satisfy the requirements of the Confined Spaces Regulations 1997.

(b) **Describe** the health risks faced by those repairing the reservoir when pneumatic tools are to be used to remove defective concrete and epoxy resin material used to effect internal repairs.

13. **Identify**:
(a) **FOUR** specific hazards associated with work in a confined space.
(b) **FOUR** examples of a confined space that may be encountered on a construction site.

14. It is necessary for maintenance purposes to enter a large cement silo situated on a construction site.
(a) **Identify** the likely hazards.
(b) **Outline** the precautions to be taken to reduce the risks to personnel entering the silo.

15. An underground chamber requires inspection and possible repair. A risk assessment has identified the possibility of the presence of methane within the chamber.

Describe the items of safety equipment that would be expected for such work, explaining how **EACH** item reduces the risk of harm.

16. The water main supplying a school is to be replaced. The work will be carried out in a 1.5 m deep excavation, which will be supported in order to ensure the safety of the employees working in the excavation.
(a) **Identify** when the **THREE** statutory inspections of the supported excavation must be carried out by the competent person.
(b) **State** the information that should be recorded in the excavation inspection report.
(c) Other than the provision of supports for the excavation, **outline** additional precautions to be taken during the repair work in order to reduce the risk of injury to the employees and others who may be affected by the work.

Appendix 19.1 – An example of safe digging practice

Introduction

All pipes and cables must be checked before digging begins. Service plans should be used to see whether the place intended for digging involves working near buried underground services. Proper planning and the execution of a safe system of work should enable contact with services to be avoided.

A safe system of work for digging in an excavation depends upon the use of:

➤ Cable or other service plans.
➤ Cable and service locators and.
➤ Safe digging practices.

Whenever possible, excavations should be kept well away from existing services.

Safe digging practice method statement

Before digging, ensure that:

➤ The person who is going to supervise the digging on site has service plans and is trained in the use of them.
➤ All workers involved in the digging have been trained in safe digging practice and excavation emergency procedures.
➤ All workers are properly supervised.
➤ The locator is used to trace as accurately as possible the actual line of any pipe or cable or to confirm that there are no pipes or cables in the way and the ground has been marked accordingly.
➤ There is an emergency plan to deal with damage to cables or pipes which includes the notification of the service provider in all circumstances. In the event of gas pipe damage, all smoking and naked flames must be banned. If necessary, evacuate all those at risk of injury (this may include people in nearby properties likely to be affected by leaks). Suitable signs must be erected to warn everyone of the danger.

During the excavation operation the following **safe digging practice** should be used:

➤ Check repeatedly for evidence of pipes or cables during digging using the locator. If unidentified services are found, work must cease until further checks can be made to confirm that it is safe to continue digging.
➤ Trial holes should be dug by hand to confirm the position of the pipes or cables. This is particularly important in the case of plastic pipes which cannot be detected by normal locating equipment.
➤ Always hand dig near buried pipes or cables. Use spades and shovels rather than picks and forks, which are more likely to pierce cables.
➤ Treat all pipes or cables as 'live' unless it is known otherwise. A live cable may be contained in a rusty pipe or conduit. Do not break or cut into any service until its identity is ascertained and it is known that it has been made safe.
➤ Hand-held power tools should not be used within 0.5 m of the marked position of an electricity cable. Fit check collars to the tools so that initial penetration of the surface is restricted.
➤ No machine should be used to excavate within 0.5 m of a gas pipe.
➤ Exposed services must be supported as soon as they are uncovered. This will prevent them from being damaged.
➤ Any suspected damage to services must be reported to the site supervisor immediately.
➤ Pipes or cables should be backfilled with fine material such as dry sand or small gravel. Backfill which is properly compacted, particularly under cast or rigid pipes, prevents settlement which could cause damage to them at a later date.
➤ The service plans are updated when the new services have been laid.

Appendix 19.2 – Typical excavation work risk assessment

INITIAL RISK ASSESSMENT	Work in and with excavations		
SIGNIFICANT HAZARDS	Low	Medium	High
1. Collapse of sides			✓
2. Striking existing services			✓
3. Persons falling into excavations		✓	
4. Plant and materials falling into excavations			✓
5. Flooding of excavations	✓		
6. Presence of hazardous atmospheres			✓
7. Presence of contaminated soil			✓
8.			

ACTION ALREADY TAKEN TO REDUCE THE RISKS:

Compliance with:

Provision and Use of Work Equipment Regulations (PUWER)

Control of Substances Hazardous to Health Regulations (COSHH)

Confined Spaces Regulations

CDM Regulations

HSE *Avoidance of dangers from underground services*

HSE HSG150 *Health and Safety in Construction*

Planning:

Compliance with British Standards for Earthworks. Sufficient numbers of trained operatives and competent supervision must be available before work starts. Sufficient and suitable plant and materials must be available for trench support before work starts. Suitable monitoring equipment and personnel trained in its use will be required where known exposure to toxic substances or lack of oxygen may occur. Location of existing services must be complete before work starts, also information obtained on ground conditions. A method statement should be prepared before work starts. COSHH assessments to be made as required.

Physical:

Excavations will be supported as appropriate. Where flooding risk exists, cofferdams/caissons will be installed with pumps of suitable capacity. Substantial barriers will be erected around excavations. Where poor ventilation is identified the atmosphere will be continually monitored. A permit to work will be used. Stop barriers will be used to prevent vehicle entry. Spoil and materials will be stacked at least 1.5 m from the edge of excavations. Ladders will be provided for safe access/egress. Cable location devices and local authority drawings will be used to trace buried services prior to commencement of work. Suitable signs and barriers will be provided to warn of the work.

Managerial/Supervisory:

Ensure safe system of work provided, taking account of prevailing conditions including weather, traffic and existing structures. Provide suitable PPE as required and ensure its correct use. Inspect excavations daily or before each shift. Record thorough examination. Ensure personnel selected are capable, fit and experienced unless under direct supervision. COSHH assessments are to be made of substances likely to be found or produced during the work.

Training:

Supervisors must have received training in COSHH appreciation, general site safety, and theory and practice of excavation work. Operatives must have received training in excavation support procedures and use of cable location devices. This applies to sub-contractors as well as direct employees.

Date of Assessment .... Assessment made by ....:......

Risk Re-Assessment Date .... Site Manager's Comments:

Appendix 19.3 – Typical confined spaces risk assessment

INITIAL RISK ASSESSMENT	Work in Confined Spaces		
SIGNIFICANT HAZARDS	Low	Medium	High
1. Poisoning from toxic gases			✓
2. Asphyxiation – lack of oxygen			✓
3. Explosion			✓
4. Fire			✓
5. Excessive heat		✓	
6. Drowning			✓
7.			

ACTION ALREADY TAKEN TO REDUCE THE RISKS:

Compliance with:

Confined Spaces Regulations

HSC ACoP Confined Spaces Regulations
HSE Guidance Note – Safe Work in Confined Spaces
HSE Respiratory Protective Equipment

HSE The Selection, Use and Maintenance of Respiratory
Protective Equipment

Local Authority/client safety standards, e.g. on sewer entry
CDM Regulations

Planning:

Eliminate need for entry or use of hazardous materials by selection of alternative methods of work or materials. Assessment of ventilation available and possible local exhaust ventilation requirements, potential presence of hazardous gases/atmosphere, process by-products, need for improved hygiene/welfare facility. Rescue equipment will be available.

Physical:

Documented entry permit to work system will apply. Adequate ventilation will be present or arranged. Detection equipment will be present before entry to check on levels of oxygen and presence of toxic or explosive substances. The area will be tested before entry and continually during the presence of persons in the confined space. Breathing apparatus or airlines will be provided if local ventilation is not possible. When breathing apparatus is assessed as being required, emergency BA and rescue harnesses will be provided. Rescue equipment including lifting equipment, resuscitation facilities, safety lines and harnesses will be provided. A communication system with those in confined space will be established. Air will not be sweetened with pure oxygen. Precautions for safe use of any plant or heavier-than-air gases in the confined space must be established before entry. Necessary PPE and hygiene facilities will be provided for those entering sewers. Electrical equipment will be low voltage only, suitable for the conditions in the confined space. Energy sources will be isolated to acceptable standards with, e.g., lock-off and tags on electrical equipment and valves; blanking or pipe removal is preferred.

Managerial/Supervisory:

The management role is to decide on the nature of the confined space and to put a safe system into operation, including checking the above. Flood potential and isolations must be checked. Rescuers must be in place throughout the operation in the confined spaces.

Training:

Full training required for all entering and managing confined spaces. Rescue surface party to be trained, including first aid and operation of testing equipment. All operatives must be certified as trained and supervisory staff trained to the same standard.

Date of Assessment .... Assessment made by ....
Risk Re-Assessment Date .... Site Manager's Comments:

Demolition – hazards and control

20

20.1 ## Introduction

Demolition is one of the most hazardous construction operations and is responsible for more deaths and major injuries than any other activity. Under Construction Design and Management Regulations (CDM) 2007 all demolition work requires a written plan to show how danger will be prevented. See Section 20.3 and Chapter 9 for more details. If a demolition project is well planned the risks of injury and death can be minimized. It should be emphasized that the planning and execution of a demolition project should only be done by appropriately competent persons. The work should be supervised by someone with sufficient knowledge of the particular structure being dismantled and an understanding of the demolition method statement. For complex demolition work, expert advice from structural engineers will be necessary.

All demolition work must be carried out so as to minimize, so far as is reasonably practicable, the risks to employees and others who may be affected by the work. CDM 2007 apply to all demolition work. The HSE must be notified before work begins if the construction work, including demolition, is to last for 30 days or more than 500 person days are involved.

20.2 Principal hazards of demolition work

The principal hazards associated with demolition work are:

> falls from height or on the same level
> falling debris
> premature collapse of the structure being demolished
> dust and fumes

Figure 20.1 Demolition in progress.

> the silting up of drainage systems by dust
> the problems arising from spilt fuel oils
> manual handling
> presence of asbestos and other hazardous substances
> noise and vibration from heavy plant and equipment

425

- electric shock
- fires and explosions from the use of flammable and explosive substances
- smoke from burning waste timber
- pneumatic drills and power tools
- the existence of services, such as electricity, gas and water
- collisions with heavy plant
- plant and vehicles overturning.

- the means of access to the site
- the removal of waste
- the details of any traffic or pedestrian routes through the site
- the provision of welfare facilities
- the proximity of neighbours
- the location of any public thoroughfares adjacent to the structure or building
- the name of the planning supervisor.

20.3 Pre-demolition investigation and survey

When a building is to be demolished the (non-domestic) client must provide pre-demolition information to the designer and contractor on non-notifiable projects and the CDM co-ordinator on notifiable projects. This will involve a pre-demolition investigation and survey.

Before any work is started, a full site investigation must be made by a competent person to determine the hazards and associated risks which may affect the demolition workers and members of the public who may pass close to the demolition site. The competent person is often a specialist structural engineer who will also advise on the temporary support of adjacent buildings and the correct method of dismantling or demolition.

The investigation should cover the following topics:

- the construction details of the structures or buildings to be demolished (including the materials used, fragile roofs, rot, the presence of cantilevered structures and any general weaknesses) and those of neighbouring structures or buildings
- the previous use of the premises
- the load carrying capacity of adjoining land including the presence of underground culverts
- the need for possible temporary support structures for the building being demolished and adjoining buildings
- falls of materials and people
- the location of any dangerous machinery
- the presence of asbestos, lead or other hazardous or radioactive substances and any associated health risks
- environmental issues, such as dust, water pollution and noise
- public safety including the provision of high fencing or hoardings
- manual handling issues
- the location of any underground or overhead services (water, electricity, gas and sewage)
- the location of any underground cellars, storage tanks, chimneys, balconies or bunkers particularly if flammable or explosive substances were previously stored

Figure 20.2   Asbestos removal sign.

The details of the construction structure of building to be demolished would include whether it was built of brick, pre-stressed concrete, reinforced concrete or steel. There may be certain building regulations which cover the site and the Local Authority Building Department should be contacted to ascertain whether any part of the site is affected by these regulations. It is also important to consult with legal advisers to ensure that there are no legal covenants or disputes which could affect operations on the site.

All demolition work requires those in control of the work to produce a written plan showing how danger will be prevented. This will be the responsibility of the CDM co-ordinator on notifiable projects and the contractor or designer on non-notifiable projects.

The written plan will include a risk assessment of the state and design of the structure to be demolished and the influence of that design on the demolition method proposed. This risk assessment will normally be made by the project designer who will also plan the demolition work. A further risk assessment should then be made by the contractor undertaking the demolition and a written method statement will be required before demolition takes place.

The site manager should arrange for suitable plant and equipment to be provided so that the work can be executed to the standards required by health and safety legislation, in particular the Control of Asbestos Regulations 2006. It may be necessary for the local authority and the police to be consulted about the proposed demolition so that issues of public protection, local traffic management and possible road closures can be addressed. There should be liaison with the occupiers of adjacent properties because, in some cases, they may need to be evacuated.

The provision of temporary access roads, welfare facilities, office accommodation, fuel storage and plant maintenance facilities on site will need to be considered at the planning stage.

The presence of hazardous substances and their release during the demolition process must be considered at the planning stage. Most hazardous substances create a hazard for demolition workers by being inhaled, ingested, injected or in contact with or being absorbed by the skin. Environmental monitoring may need to be carried out in certain situations. Specialist advice should be obtained from appropriately competent persons. Some of the most common hazardous substances in demolition work include:

➤ **lead** – is most dangerous when it is in the air as a fume or dust (e.g. cutting steelwork coated with lead-based paint or dismantling of tanks containing lead-based petrol)
➤ **asbestos** – where possible it should be removed before any other demolition work starts and must always be removed by a licensed contractor. Asbestos may be found in sprayed coatings, thermal and acoustic insulation materials, fire resistant walls/partitions, asbestos cement sheets or flooring materials
➤ **PCBs** – a toxic substance found in electric transformers and capacitors, refrigeration and heating equipment
➤ **silica** – occurs in stone, some bricks and concrete aggregate. Any demolition of structures constructed from these materials will give rise to dust containing silica.

Residues of hazardous substances may also create a hazard to demolition workers. Storage tanks, vessels, pipes and other confined spaces may contain flammable vapours or toxic sludges – especially those which were formerly used in industrial or chemical processes. Plans must be made to dispose of any hazardous or dangerous substance found during the demolition process in a safe way which conforms to legislative requirements.

20.4 Demolition method statement

There are two forms of demolition:

1. piecemeal – where the demolition is done using hand and mechanical tools such as pneumatic drills, cranes and demolition balls, hydraulic pusher arms or heavy duty grabs and
2. deliberate controlled collapse – where explosives are used to demolish the structure. This technique should only be used by trained, specialist competent persons. Pre-weakening of the structure, by the

removal of several load bearing elements and their replacement with temporary props, normally precedes the deliberate collapse. This is the most economic form of demolition but it is the most hazardous and everyone must be at a safe distance at the time of the collapse.

Figure 20.3(a) Long-reach hydraulic arm for piecemeal demolition.

Figure 20.3(b) Remote controlled hydraulic arm for pushing, nibbling or hammering.

Figure 20.4 Controlled collapse.

- first-aid, emergency and accident arrangements
- training and welfare arrangements
- arrangements for the separation and disposal of waste
- names of site foremen and those with responsibility for health and safety and the monitoring of the work
- the co-ordination of all work activities on the site
- the expected level of competence of site workers.

Danger
Demolition work
in progress

Figure 20.5 Sign – Danger demolition work in progress.

A risk assessment should be made by the contractor undertaking the demolition – this risk assessment will be used to draw up a method statement for inclusion in the health and safety plan. A written method statement will be required before demolition takes place. The contents of the method statement will include the following:

- details of the method of demolition to be used including the means of preventing premature collapse or the collapse of adjacent buildings and the safe removal of debris from upper levels to prevent overloading due to the debris
- details of site access and security
- details of the location of any underground or overhead services
- details of protection from falling materials arrangements
- details of equipment, including access equipment, required and any hazardous substances to be used
- arrangements for the protection of the public and the construction workforce against noise and whether hazardous substances, such as asbestos or other dust, are likely to be released
- details of the isolation methods for any services which may have been supplied to the site and any temporary services required on the site
- details of personal protective equipment, such as hard hats, which must be worn by all personnel on site

Other risk assessments, such as COSHH, personal protective equipment and manual handling, should be appended to the method statement.

The isolation of all services (gas, electricity and water) which feed the site is essential before any demolition takes place. Contact with the appropriate service provider will be necessary since sites often have a complex of feeds for many of their services. The local authority and surrounding properties also need to be informed that services are to be isolated.

Over recent years, more rules and regulations have been introduced concerning the disposal of construction waste and this topic is covered in more depth in Chapter 16. Proper arrangements must be made with a reputable waste disposal contractor for the disposal of demolition waste. If hazardous substances are included in the waste, then specialist waste contractors should be used. Only registered disposal sites should be used and records kept of each load.

The site should be made secure with relevant signs posted to warn members of the public of the dangers.

Asbestos-containing material requires special care and will only be dismantled by a licensed contractor. Indeed, if asbestos is disturbed in any way, a specialist contractor must be consulted. If possible, such work should be done before any other work is started.

It is important to develop a safe system of work for demolition which ensures that people are distanced as far as possible from the demolition area. A checklist of issues which should be incorporated into a safe system of work is given in Appendix 20.1.

Glass must be removed from all windows before demolition starts. This will avoid hazards from flying broken glass. The window spaces should be boarded up to deter trespassers. Timber can present three hazards: injuries caused by protruding nails or sharp and broken edges and pollution and fire if the timber is burnt on site.

Young persons who are undergoing training either as an apprentice or work experience trainee require close supervision on demolition sites. No young person (under 18 years old) should be permitted to operate any plant or act as banksman unless being trained under direct supervision of a competent person. Children and unauthorized persons must not be permitted to enter working areas at any time, in particular whilst plant is in use; all necessary control measures must be taken to avoid injuries to such persons on the site both during and after working hours, particularly if it is not possible to fully fence the site. The general public must also be protected by restricting access to the demolition site and by taking special precautions for walkways and other public areas.

Under CDM 2007 pre-tender information and a construction phase health and safety plan must be prepared. The have been covered extensively in Chapters 9 and 21.

For demolition work, the pre-tender plan will draw heavily on the pre-demolition survey and would take account of conditions that the client imposes, so that the work will not impinge too greatly on his business or upon the neighbourhood. It would include:

➤ the nature of the project (including the name of the client and planning supervisor)
➤ type and size of the building or structure
➤ a description of the work and its scope and timescale
➤ the previous use of the building and its proximity to other buildings
➤ any likely health risks that might arise (such as dust possibly including asbestos)
➤ the condition of the structure
➤ the position of existing services
➤ ground conditions and any possible contamination
➤ information relating to residual risks identified by the designers arising from the methods or materials of construction
➤ site rules and procedures for continuing liaison with the principal contractor.

The construction phase health and safety plan should include the following:

➤ the co-ordination of and provision of information to other contractors
➤ risk assessments and other statutory assessments such as COSHH and noise
➤ emergency procedures
➤ health and safety monitoring arrangements
➤ site rules
➤ welfare arrangements
➤ the provision of information and training to the workforce
➤ consultation arrangements with the workforce.

The following issues are specific to the demolition work and would include:

➤ the particular methods of demolition to be employed
➤ the arrangements for the control and disposal of waste
➤ the provision and use of personal protective equipment
➤ the details of permit-to-work systems
➤ the arrangements for the exclusion of unauthorized persons.

Demolition using explosives

If the demolition is to use explosives, an exclusion zone must be established at a distance from and surrounding the structure to be demolished. All persons with the possible exception of the shotfirer must be outside the exclusion zone at the time of the explosion. The HSE have produced a detailed information sheet (CIS No. 45) on the establishment of exclusion zones when using explosives in demolition.

An exclusion zone comprises four areas:

1. the plan area of the structure to be demolished
2. the designed drop area where the bulk of the structure is designed to drop
3. the predicted debris area which is beyond the design drop area and where the rest of the debris is predicted to drop
4. the buffer area between the predicted debris area and the boundary of the exclusion zone.

The size of the exclusion zone is not just related to the height of the structure – other issues must be considered such as the effect of ground vibration on buried services.

The main factors that influence the design of an exclusion zone include:

➤ the designed collapse mechanism
➤ the materials used in the structure
➤ the condition of the structure
➤ the extent of any planned pre-weakening
➤ the types of explosives to be used
➤ the topography of the site
➤ the proximity of surrounding structures
➤ any adverse effects of ground vibration or the attendant shock wave
➤ noise
➤ dust.

The competent person with prime responsibility for designing the exclusion zone is the explosives engineer. The members of the public must be kept outside the exclusion zone until the all clear is given. Any houses within the exclusion zone must be evacuated and provision for the welfare of evacuees during the evacuation period must also be made. Finally the local police should be informed of and ideally involved in the planning process.

20.5 Management of the demolition and general controls

Demolition hazard control involves implementing measures which reduce the risk at the demolition site. Where legislation requires specific methods to control the risk, these must be used.

As discussed in Chapter 6, there is a recognized hierarchy of controls in health and safety ranging from the most to the least effective. This hierarchy can be applied directly to the demolition process. The elements of the hierarchy are:

➤ elimination
➤ substitution
➤ isolation
➤ engineering controls
➤ administrative controls
➤ personal protective equipment.

Elimination is the most effective control measure. This involves removing the hazard or hazardous work practice from the site, by the following actions:

➤ disconnecting services to the demolition site
➤ ensuring there are no sparks or ignition sources, where a risk of fire or explosion exists
➤ ensuring separation between the public and demolition activities
➤ excluding unauthorized persons from the site.

Substituting or replacing a hazard or hazardous work practice with a less hazardous one could include the following:

➤ using a controlled collapse technique, in place of people working at heights
➤ using power shears in place of grinding or oxyacetylene cutting, where a risk of fire exists.

Isolating or separating the hazard or hazardous work practice from workers and others involved in the demolition area by the following:

➤ installing screens on plant to protect from dust and noise
➤ installing barriers and fences
➤ marking off hazardous areas.

If the hazard cannot be eliminated, substituted or isolated, an engineering control is the next preferred measure, by means including the following:

➤ modification of tools or equipment
➤ provision of guarding to machinery or equipment
➤ provision of shatterproof/guarded windscreens on plant
➤ provision of falling object protective structures (FOPS) and roll-over protective structures (ROPS) on vehicles and plant
➤ installation of safe working platforms on scaffolds and elevating work platforms
➤ installation of edge protection to open edges of landings, stairways and fixed platforms
➤ use of props and bracing to support loads
➤ installation of static lines and anchor points for fall-arrest systems
➤ water sprays to suppress dust
➤ use of chutes for dropping debris
➤ use of cranes for lowering loads
➤ installation of vehicle buffers where plant is exposed to an open edge
➤ use of machine mounted impact hammers, power shears, drills and saws
➤ provision of flash arresters on gauges and hoses on welding equipment.

Administrative controls, including the introduction of reduced risk work practices, are also important and include the following:

➤ limiting the amount of time a person is exposed to a particular hazard
➤ implementing and documenting safe working procedures for all hazardous tasks
➤ training and instructing all personnel

- identifying hazardous substances prior to work commencing
- implementing safe procedures for handling hazardous substances
- implementing procedures for disconnecting services to the site
- using operational observers at the site
- implementing confined space entry procedures
- ensuring that all loads to be lifted are accurately calculated
- consulting with all workers on the site.

Personal protective equipment is the least effective management control because it needs almost constant supervision and enforcement. The advice of a specialist may be required when selecting personal protective equipment. Personal protective equipment should only be considered when other control measures are not practicable or to increase the protection given to a person. Personal protective equipment includes:

- safety helmets (required by legislation)
- fall-arrest harnesses and lanyards
- safety boots
- gloves
- goggles
- respirators
- hearing protectors.

20.6 Specific issues

20.6.1 *Premature collapse*

Premature collapse of the structure is one of the main causes of serious injury resulting from demolition work. A common reason for this type of incident is the lack of effective planning before the beginning of the demolition. Such planning has already been covered earlier in this chapter – the pre-demolition survey, a detailed site risk assessment and the development of a comprehensive method statement. Premature collapse usually begins with the structural collapse of floors and is often caused by plant operating on floors which are not certified safe and/or back-propped where required. It can also be caused by poor site supervision and a lack of instruction, training and information for the workforce. This will sometimes lead to individual workers or teams being unaware of additional hazards which have been created as the demolition work progresses. A fully competent and trained workforce that has been made completely aware of the hazards associated with a particular site should lead to a successful conclusion to the demolition.

20.6.2 *Protection from falls and falling material*

A risk assessment should be undertaken in relation to falls, and fall protection should be provided for all persons exposed to the risk of a fall that could cause an injury. Having regard to the hierarchy of control, the higher order controls of elimination and isolation should be introduced where practicable, to ensure all persons work from solid construction.

Falls at demolition sites can be the result of:

- falling through fragile roofing material
- falling through penetrations or open voids (e.g. skylights and voids for air-conditioning and plumbing)
- falling from open edges (e.g. stairways, landings, fixed platforms, scaffolding and edge of roof)
- falling out of elevating work platforms (e.g. scissor lifts)
- failure of plant (e.g. elevating work platforms)
- falling whilst accessing a roof or other elevated area
- falling down lift shafts
- collapse of flooring (e.g. concrete slab and wooden decking)
- collapse of ground above cellars or pits
- tripping over debris on the ground.

The hazard of being hit, trapped or struck by an object may be present for a number of reasons, which include:

- falling debris (e.g. down service ducts and lift shafts)
- accidental or uncontrolled collapse of the structure
- use of plant and equipment (e.g. cranes lifting loads)
- failure of structural members (e.g. load bearing steelwork).

20.6.3 *Noise*

Noise can be a serious problem during the demolition process. However some of the effect might be mitigated by replacing noisy equipment with quieter machinery and ensuring that plant and equipment are regularly maintained. Noise from the processes may be difficult to control but the location of noisy plant where it would have least effect on the neighbourhood, the erection of noise barriers, the reduction of the time taken by noisy operations and the organization of the work so that noisy operations are not carried out either early in the morning or during the evening and night, would all help to reduce the risk.

The source of noise and vibration on demolition sites will normally be caused by the use of plant and equipment, falling debris and/or explosives. Compressors, pneumatic hand-held tools, front-end loaders, excavators and other equipment can create noise levels of more than the lower exposure action value of 80 dB(A) and may at times create peak noise levels of more than 135 dB(C), the lower peak sound pressure action value. A noise assessment will be required and workers will need to wear ear defenders.

20.6.4 *Dust*

During the demolition process, dust is a considerable hazard which needs to be controlled. Among the possible control measures are damping the process down with water, the sheeting of disposal lorries as they leave the site to prevent the generation of dust and the provision of filters or covers to site drainage systems to prevent the risk of them silting up. A restriction on the speed of vehicles on the site and the regular cleaning of roads will also help to reduce the dust level. Mud on the roads, which will cause dust, can be alleviated by providing 'stoned-up' vehicle routes, ensuring that vehicles pass through a wheel wash before leaving the site, and by setting up procedures for regular road cleaning and sweeping. Designating the vehicle routes to and from the site would also limit the number of affected roads.

Dust and fumes from site machinery and smoke from burning timber waste can further exacerbate the dust problem. As stressed earlier, asbestos dust is a particularly hazardous problem that can only be addressed by specialists.

20.6.5 *Siting and use of machinery*

The hazards associated with plant and equipment used on demolition sites are numerous. Only appropriately competent persons should operate plant and equipment on these sites. Examples of some of the hazards that may result from operating plant and equipment on demolition sites include:

➤ electrocution due to plant or equipment coming into contact with live electricity
➤ plant failure due to its safe working load being exceeded
➤ accidental dropping of material due to the incorrect slinging of loads
➤ plant striking or colliding with persons, particularly shared access ways
➤ excessive noise and vibration from machinery, such as front-end loaders and excavators
➤ flying particles from pneumatic tools such as impact hammers
➤ welding and cutting operations during maintenance work
➤ the structural failure of steelwork and
➤ fire and explosions.

The site supervisor must ensure that all plant and equipment when delivered to site is in good working order and fitted with all necessary safety devices, notices and guards. He must also ensure that an ongoing maintenance and defect reporting system is in place. No equipment may be used until all defects have been rectified. The site supervisor will ensure that only authorized licensed operators are permitted to operate any plant and equipment used on site.

Manual handling tasks on demolition sites should be identified, assessed and controlled using the hierarchy of control described in Chapter 12. Wherever possible, plant and other mechanical aids should be used to move debris, materials and equipment. The use of some equipment may also create a manual handling hazard. Manual handling risk assessments should be carried out as necessary. Manual handling hazards on demolition sites may result from operating plant and equipment, manual demolition, lifting materials, loading trucks and bins and tidying the site after demolition.

All wiring, except where temporary installations are required, should be disconnected before demolition work commences. Temporary electrical installations must comply with the Institution of Electrical Engineers Wiring Regulations and legislative requirements. Where there is a possibility of live wires in the structure, areas of danger should be clearly tagged and signs erected warning of the hazard. A detailed plan of these areas should be formulated.

Plant and equipment should be used with care to ensure that no part of them comes into contact with overhead or underground wires or cables. When working near any power lines or cables or other services such as gas mains, any national or local legislative requirements for such work must be followed.

20.6.6 *Environment*

Where any redundant building, structure or plant has contained flammable materials, precautions must be observed to avoid fires or explosions. Specialist advice may be required to identify any residual materials and assess whether any contamination remains. Any residue of flammable materials must be made safe.

Deliberate burning of structures must not be used as a method of demolition. The burning of any waste materials, such as timber, on demolition sites must not be undertaken without the approval of the relevant local authorities.

When welding or cutting at the demolition site, sparks will present a fire hazard. Any flammable materials should be identified and fire risks assessed before this type of work is undertaken.

The escape of explosive gases from accidental damage to pipes during the demolition process is yet another source for fire and explosions. All services, including gas, need to be identified during the pre-demolition survey.

Where services are shown on original plans for the structure and site, but cannot be found, work should cease until a further investigation determines the location of these services.

Arson, especially when the site is unattended, is a further risk of fire. Adequate site security, particularly the boarding of empty window spaces, and good housekeeping of the site, by removing and/or isolating flammable and combustible materials, will assist in reducing the risk of arson at a demolition site.

Portable fire extinguishers and access to local fire hydrants should be maintained throughout the demolition process.

Environmental airborne hazards in the form of pollution can be produced from petrol or diesel engine powered vehicles or machinery. This is a particular problem when the engines are poorly maintained or being overworked. Carbon monoxide from petrol- or diesel-driven plant is a serious health problem when such plant is operated in enclosed areas.

Finally, fuel storage tanks should be surrounded by a bund and standing vehicles and vessels should be provided with drip trays beneath them to prevent fuels from contaminating the ground and water courses.

20.6.7 Competence and training

A very important requirement for demolition work is the provision of training for all construction workers involved in the work. Specialist training courses are available for those concerned with the management of the total process. The Construction Industry Training Board (CITB) offers a training course entitled 'A Scheme for the Certification of Competence of Demolition Operatives'.

An induction training which outlines the hazards and the required control measures should be given to all workers before the start of the demolition work. This training should provide information to site workers to make them aware of the hazards that they will face, the demolition and control procedures to be adopted and the site rules. Issues such as first-aid and welfare facilities should also be covered. The raising of environmental issues at progress meetings and the inclusion of the procedures in work specifications and the health and safety plan are further actions that can be used to increase awareness. During the demolition process, communications should be continued using toolbox talks, and posters and signs around the site.

20.7 Practice NEBOSH questions for Chapter 20

1. **Outline** the main hazards that may be present during the demolition of a building.

2. A three-storey office block is to be demolished. **Outline** the likely hazards to the environment caused by the work together with the actions needed to control them.

3. **Outline** the factors to consider when undertaking a pre-demolition survey of a construction site prior to drawing up a demolition method statement.

4. **Identify** the key issues to be addressed in a pre-demolition survey of a multi-storey block of flats in a city centre.

5. A demolition contractor has been awarded a contract to demolish a large disused factory in an industrial area of a city.
   Outline the factors to consider when undertaking a pre-demolition survey of the site prior to drawing up a demolition method statement.

6. **Outline** the main areas to be addressed in a demolition method statement.

7. A contractor has been employed to demolish a disused paint factory. **Give** examples of the information that should be provided to the contractor by the owner of the premises in order to comply with the Construction (Design and Management) Regulations.

8. A large industrial site has been closed recently. The owners of the site are to have the buildings demolished before selling the site for redevelopment.
   Outline the issues the principal demolition contractor should address as part of the health and safety plan.

9. The owners of a currently disused office block are to have part of the building demolished and a large extension built. With respect to the proposed work, **outline** the areas to be addressed in the health and safety plan:
   (a) as prepared by the CDM co-ordinator
   (b) as developed by the principal contractor.

10. A large construction project in a residential area involves demolition and subsequent removal of debris and spoil.
   (a) **Outline** the potential environmental problems that might arise from this work and **explain** how the problems might be mitigated.
   (b) **Describe** the actions that could be taken by site management in order to ensure that contractors and operatives are aware of the procedures adopted.

433

Appendix 20.1 – Checklist for a safe system of work

The key to the development of an effective safe system of work for demolition is to choose a method of working with everyone as far away from the danger area as possible. After the site survey has been completed, the following issues should be addressed:

> Does the demolition method to be used, such as a long-reach machine or crane and ball, enable everyone to be distanced from the demolition?

> Will the demolition work cause nearby structures to become unstable requiring the use of temporary supports?

> Will structural elements of the building, such as intermediate floors, support the weight of demolition machinery and/or falling materials?

> Will the HSE be notified using Form 10 or its equivalent before work begins?

> Have all electric, gas, water, steam, sewer, and other service lines been shut off, capped, or otherwise controlled prior to the start of work and have temporary lines to supply power for adequate lighting during demolition work been provided?

> Have any hazards that exist from fragmentation of glass been fully controlled?

> Have all floor openings, not used as material drops, been covered?

> Have worker entrances to sites of multi-storey buildings which are to be demolished been protected through the use of walkway canopies that extend from the face of the building for a minimum of 2 m?

> Are materials to be dropped to any point lying outside the exterior walls of the structure into an area which is not effectively protected?

> Are material chutes, inclined at an angle of 45° or more to the horizontal, entirely enclosed and is there a substantial gate at the discharge end which is to be operated by a competent worker? Chute openings must be protected by a substantial guard rail approximately 1.6 m above the floor level.

> When using material chutes, is the area where the material is dropped enclosed with barricades not less than 2 m around the chute opening? Are warning signs of the hazard of falling material posted at each level?

> During mechanical demolition, is a competent person continuously inspecting the structure as work progresses to detect hazards resulting from the weakened floors and walls or from loosened materials?

Summary of the main legal requirements

21.1 Introduction

The achievement of an understanding of the basic legal requirements can be a daunting task. However, the health and safety student should not despair because at NEBOSH National Certificate in Construction level they do not need to know the full history of UK health and safety legislation, the complete range of Regulations made under the Health and Safety at Work Act 1974 or the details of the appropriate European Directives. These summaries, which are correct up to the 1st October 2007, cover those Acts and Regulations which are essential for the Certificate student and will provide the essential foundation of knowledge. They will also cover many of the requirements for similar awards at Certificate level and below. Students should check the latest edition of the NEBOSH Construction Certificate guide to ensure that the regulations summarized here are the latest for the course being undertaken. Remember that NEBOSH does not examine on legislation until 6 months after it has been published.

Many managers will find these summaries a useful quick reference to the legal requirements. However, anyone involved in achieving compliance with legal requirements should ensure that they read the Regulations themselves, any approved codes of practice (ACOP) and HSE guidance.

21.2 The legal framework

21.2.1 *General*

The Health and Safety at Work etc Act 1974 (HSW Act) is the foundation of British Health and Safety law.

It describes the general duties that employers have towards their employees and to members of the public, and also the duties that employees have to themselves and to each other.

The term '*so far as is reasonably practicable*' (SFARP) qualifies the duties in the HSW Act 1974. In other words, the degree of risk in a particular job or workplace needs to be balanced against the time, trouble, cost and physical difficulty of taking measures to avoid or reduce the risk.

The law simply expects employers to behave in a way that demonstrates good management and common sense. They are required to look at what the hazards are and take sensible measures to tackle them.

The Management of Health and Safety at Work Regulations 1999 (MHSWR, the Management Regulations) clarifies what employers are required to do to manage health and safety under the Health and Safety at Work Act. Like the Act, they apply to every work activity.

The law requires every employer to carry out *risk assessments*. If there are five or more employees in the workplace, the significant findings of the risk assessment need to be recorded.

In a place like an office, risk assessment should be straightforward; but where there are serious hazards, such as those on a construction site or when construction work is undertaken in a chemical plant or on an oil rig, it is likely to be more complicated.

The Factories Act 1961 and The Office Shops and Railway Premises Act of 1963, while still partly remaining on the Statute Book for most practical health and safety work, can be ignored. However, strangely, the notification to the authorities of premises which come within the definition of these two acts are carried out on Form 9 for factories, Form 10 for construction and OSR 1 for office shop and railway premises.

21.2.2 *The relationship between the regulator and industry*

The Health and Safety Commission (HSC) consults widely with those affected by its proposals.

The HSC/HSE works through:

➤ the HSC's Industry and Subject Advisory Committees, which have members drawn from the areas of work they cover, and focus on health and safety issues in particular industries (such as the construction industry and education or areas such as toxic substances and genetic modification)
➤ intermediaries, such as small firms' organizations
➤ Construction Industries Training Board (CITB)
➤ providing information and advice to employers and others with responsibilities under the Health and HSW Act 1974
➤ guidance to enforcers, both HSE inspectors and those of local authorities
➤ the day-to-day contact which inspectors have with people at work.

The HSC consults with small firms through Small Firms Forums. It also seeks views in detail from representatives of small firms about the impact on them of proposed legislation.

21.3 List of Acts and Regulations summarized

The following Acts and Regulations are covered in this summary:

➤ Health and Safety at Work etc Act 1974 (HSW Act)
➤ Environmental Protection Act 1990
➤ New Roads and Street Works Act 1991.

For health and safety issues the most important of these is the HSW Act 1974. Most of the relevant Regulations covering health and safety at work have been made under this Act since 1974. These are included here and are all relevant to the Construction Certificate student.

The Fire Safety Order was made under the Regulatory Reform Act 2001 and the Hazardous Waste Regulations 2005 under the Environmental Protection Act 1990.

The first list is alphabetical and some titles have been modified to allow an easier search. They are:

21.3.1 *Alphabetical list of regulations summarized*

Control of Asbestos Regulations 2006
Chemicals (Hazard Information and Packaging for Supply) Regulations 2002 plus 2005 amendment

Confined Spaces Regulations 1997
Construction (Design and Management) Regulations 2007
Construction (Head Protection) Regulations 1989
Consultation with Employees Regulations 1996
Control of Substances Hazardous to Health Regulations 2002 and 2005
Dangerous Substances and Explosive Atmospheres Regulations 2002
Display Screen Equipment Regulations 1992
Electricity at Work Regulations 1989
Employers Liability (Compulsory Insurance) Act 1969 Regulations 1998
Fire Safety Order 2005
First Aid Regulations 1981
Hazardous Waste (England and Wales) Regulations 2005
Information for Employees Regulations 1989
Ionising Radiations Regulations 1999
Lead at Work Regulations 2002
Lifting Operations and Lifting Equipment Regulations 1998
Management of Health and Safety at Work Regulations 1999
Manual Handling Operations Regulations 1992
Noise at Work Regulations 2005
Personal Protective Equipment at Work Regulations 1992
Provision and Use of Work Equipment Regulations 1998 (except Part IV – Power Presses)
Reporting of Injuries Diseases and Dangerous Occurrences Regulations 1995
Safety Representatives and Safety Committees Regulations 1997
Safety Signs and Signals Regulations 1996
Supply of Machinery (Safety) Regulations 1992
Vibration at Work Regulations 2005
Workplace (Health, Safety and Welfare) Regulations 1992
Work at Height Regulations 2005 as amended in 2007

The following Acts and Regulations have also been included even though they are not in the NEBOSH Construction Certificate syllabus. Very brief summaries only are given:

Corporate Manslaughter and Corporate Homicide Act 2007
Disability Discrimination Act 1995 and 2005
Electrical Equipment (Safety) Regulations 1994
Gas Appliances (Safety) Regulations 1992
Gas Safety (Installation and Use) Regulations 1998
Occupiers Liability Acts 1957 and 1984
Personal Protective Equipment Regulations 2002
Pesticides Regulations 1986

Pressure Systems Safety Regulations 2000
The Manufacture and Storage of Explosives Regulations 2005
Working Time Regulations 1998, 2003 and 2007

21.3.2 *Chronological list of regulations summarized*

The list below gives the correct titles (minus the year), year produced and Statutory Instrument number.

Year	SI no.	Title
1977	0500	Safety Representatives and Safety Committees Regulations
1981	0917	Health and Safety (First Aid) Regulations
1989	0635	Electricity at Work Regulations
1989	0682	Health and Safety (Information for Employees) Regulations
1989	2209	Construction (Head Protection) Regulations
1992	2792	Health and Safety (Display Screen Equipment) Regulations
1992	2793	Manual Handling Operations Regulations
1992	2966	Personal Protective Equipment at Work Regulations
1992	3004	Workplace (Health, Safety and Welfare) Regulations
1992	3073	Supply of Machinery (Safety) Regulations
1995	3163	Reporting of Injuries, Diseases and Dangerous Occurrences Regulations
1996	0341	Health and Safety (Safety Signs and Signals) Regulations
1996	1513	Health and Safety (Consultation with Employees) Regulations
1997	1713	Confined Spaces Regulations
1998	2306	Provision and Use of Work Equipment Regulations (except Part IV – Power Presses)
1998	2307	Lifting Operations and Lifting Equipment Regulations
1998	2573	Employers Liability (Compulsory Insurance) Regulations
1999	3232	Ionising Radiations Regulations
1999	3242	Management of Health and Safety at Work Regulations
2002	2677	Control of Substances Hazardous to Health Regulations
2002	3247	Chemicals (Hazard Information and Packaging for Supply) Regulations
2002	2676	Control of Lead at Work Regulations

Year	SI no.	Title
2002	2776	Dangerous Substances and Explosive Atmospheres Regulations
2005	735	Work at Height Regulations
2005	1093	Control of Vibration at Work Regulations
2005	1541	Regulatory Reform (Fire Safety) Order
2005	1643	Control of Noise at Work Regulations
2006	2739	Control of Asbestos Regulations
2007	0320	Construction (Design and Management) Regulations

Very brief summaries only for:

1957 and	Ch31 and	
1984	Ch3	Occupiers Liability Acts
1986	1510	Control of Pesticides Regulations
1992	0711	Gas Appliances (Safety) Regulations
1998	2451	Gas Safety (Installation and Use) Regulations
1994	3260	Electrical Equipment (Safety) Regulations
1998	1833	Working Time Regulations
2000	128	Pressure Systems Safety Regulations
2002	1144	Personal Protective Equipment Regulations
2005	894	The Hazardous Waste (England and Wales)
2005	895	The List of Wastes (England) Regulations
2005	Ch50	Disability Discrimination Act 1995 and 2005
2005	1082	The Manufacture and Storage of Explosives Regulations
2007	Ch19	Corporate Manslaughter and Corporate Homicide Act 2007

21.4 Health and Safety at Work Etc Act (HSW Act) 1974

The HSW Act 1974 was introduced to provide a comprehensive and integrated piece of legislation dealing with the health and safety of people at work and the protection of the public from work activities. A very small amount of pre-1974 legislation is still in place despite strenuous efforts to repeal and replace it with updated legislation under the HSW Act 1974.

The Act imposes a duty of care on everyone at work related to their roles. This includes employers, employees, owners, occupiers, designers, suppliers and manufacturers of articles and substances for use at work. It also includes self-employed people. The detailed requirements are spelt out in Regulations.

The Act basically consists of four parts:

Part 1 covers:

➤ the health and safety of people at work
➤ protection of other people affected by work activities
➤ the control of risks to health and safety from articles and substances used at work
➤ the control of some atmospheric emissions.

Part 2: sets up the Employment Medical Advisory Service
Part 3: makes amendments to the safety aspects of building regulations
Part 4: general and miscellaneous provisions.

21.4.1 *Duties of employers – section 2*

The employers' main general duties are to ensure, SFARP, the health, safety and welfare at work of all their employees, in particular:

➤ the provision of safe plant and systems of work
➤ the safe use, handling, storage and transport of articles and substances
➤ the provision of any required information, instruction, training and supervision
➤ a safe place of work including safe access and egress
➤ a safe working environment with adequate welfare facilities.

When five or more people (The Employer's health and safety Policy Statements (Exception) Regulations 1975 (SI No 1584) exempted an employer who employs less than five people) are employed the employer must:

➤ prepare a written general health and safety policy
➤ set down the organization and arrangements for putting that policy into effect
➤ revise and update the policy as necessary
➤ bring the policy and arrangements to the notice of all employees.

Employers must also:

➤ consult safety representatives appointed by recognized trade unions
➤ consult Representatives of Employee Safety (ROES) elected by employees
➤ establish a safety committee if requested to do so by recognized safety representatives.

Note that this has been enacted by the Safety Representatives and Safety Committees Regulations 1977 and enhanced by the Health and Safety (Consultation with Employees) Regulations 1996 (see later summaries).

21.4.2 *Duties of owners/occupiers – sections 3 and 4*

Every employer and self-employed person is under a duty to conduct their undertaking in such a way as to ensure, SFARP, that persons not in their employment (and themselves for self-employed) who may be affected, are not exposed to risks to their health and safety.

Those in control of non-domestic premises have a duty to ensure, SFARP, that the premises, the means of access and exit, and any plant or substances are safe and without risks to health. The common parts of residential premises are non-domestic.

Section 5 in relation to harmful emissions into the atmosphere was repealed by the Environmental Protection Act 1990.

21.4.3 *Duties of manufacturers/ suppliers – section 6*

Persons who design, manufacture, import or supply any article or substance for use at work must ensure, SFARP, that:

➤ it is safe and without risks to health when properly used (i.e. according to manufacturers' instructions)
➤ carry out such tests or examinations as are necessary for the performance of their duties
➤ provide adequate information (including revisions) to perform their duties
➤ carry out any necessary research to discover, eliminate or minimize any risks to health or safety
➤ (the installer or erector must ensure that) nothing about the way in which the article is installed or erected makes it unsafe or a risk to health.

21.4.4 *Duties of employees – section 7*

Two main duties are placed on employees:

➤ to take reasonable care for the health and safety of themselves and others who may be affected by their acts or omissions at work
➤ to co-operate with their employer and others to enable them to fulfil their legal obligations.

21.4.5 *Other duties – sections 8 and 9*

No person may misuse or interfere with anything provided in the interest of health, safety or welfare in pursuance of any of the relevant statutory provisions.

Employees cannot be charged for anything done, or provided, to comply with the relevant statutory provisions, for example personal protective equipment (PPE) required by a health and safety regulation.

21.4.6   Powers of inspectors – sections 20 and 25

Inspectors appointed under the HSW Act have the authorization to enter premises at any reasonable time (or anytime in a dangerous situation), and to:

➤ take a constable with them if necessary
➤ take with them another authorized person and necessary equipment
➤ examine and investigate
➤ require premises or anything in them to remain undisturbed for purposes of examination or investigation
➤ take measurements, photographs and recordings
➤ cause an article or substance to be dismantled or subjected to any test
➤ take possession of or retain anything for examination or legal proceedings
➤ take samples as long as a comparable sample is left behind
➤ require any person who can give information to answer questions and sign a statement. Evidence given under this Act cannot be used against that person or their spouse
➤ require information, facilities, records or assistance
➤ do anything else necessary to enable them to carry out their duties
➤ issue an **Improvement Notice(s)**, which is a notice identifying a contravention of the law and specifying a date by which the situation is remedied. There is an appeal procedure which must be triggered within 21 days. The notice is suspended pending the outcome of the appeal
➤ issue a **Prohibition Notice(s)**, which is a notice identifying and halting a situation which involves or will involve a risk of serious personal injury to which the relevant statutory provisions apply. A contravention need not have been committed. The notice can have immediate effect or be deferred, for example to allow a process to be shut down safely. Again there is provision to appeal against the notice but the order still stands until altered or rescinded by an Industrial tribunal
➤ initiate prosecutions
➤ seize, destroy or render harmless any article or substance which is a source of imminent danger.

21.4.7   Offences – section 33

Offences prosecuted under the HSW Act or Regulations attract a maximum fine of £5,000 if tried on summary conviction in a magistrates' court. Exceptionally this is a maximum of £20,000 for breaches of sections 2 and 6 (General duties) and £20,000 and/or 6 months' imprisonment for contravening a prohibition notice, an improvement notice or order of the court.

A number of cases can also be tried on indictment, in the Crown Court where the fines are unlimited and the maximum prison sentence can be up to 12 years. However, prison sentences are limited to certain cases only. Table 21.1 summarizes the position. Note that the offences relating to breaches of general health and safety requirements (shown in bold) do *not* attract a prison sentence.

If a Regulation has been contravened, failure to comply with an ACOP is admissible in evidence as failure to comply. Where an offence is committed by a corporate body with the knowledge, connivance or neglect of a responsible person, both that person and the body corporate are liable to prosecution.

In proceedings the onus of proving the limits of what is reasonably practicable rests with the accused.

21.4.8   The Health and Safety Commission/ Executive – section 10

Section 10 of the Act established these two bodies whose chief functions were to 31 March 2008:

➤ The HSC had the prime responsibility for administering the law and practice on occupational health and safety. Its nine members were appointed by the Government from industry, trade unions, local authorities and others. The HSC had wide powers to do anything (except borrow money) which is calculated to facilitate, or is conducive or incidental to, the performance of its functions.
➤ The Health and Safety Executive (HSE) is the executive and enforcement arm of the HSC. The HSC may direct its work except where it involves enforcement of the law in individual cases. Enforcement of the law in certain premises like offices and shops is the responsibility of local authority environmental health departments, whose officers have identical powers to HSE inspectors.

From 1 April 2008 the HSC/HSE have merged to form a single National Regulatory Body called The Health and Safety Executive. See press Release At 21.38.

21.4.9   References

An Introduction to Health and Safety: Guidance on a Health and Safety for Small Firms. INDG259 (rev1), reprinted 2006, HSE Books, ISBN 9780-7176-2685-7.
Health and Safety at Work etc Act 1974. Chapter 37, London, The Stationery Office.

Table 21.1   Offences and penalties under HSW Act

Offence	Summary conviction	Conviction on indictment
HSW Act s.33(1A) ➤ breach of HSW Act ss.2, 3, 4 or 6	**Max. £20,000 fine**	**Fine (unlimited)**
HSW Act s.33(2) ➤ obstructing an inspector, impersonating an inspector, contravening requirement of inspector under s.20, or similar	Max. £5,000 fine	
HSW Act s.33(2A) ➤ failure to comply with enforcement notice or court order	Max. £20,000 and/or up to 6 months' imprisonment	Fine (unlimited) and/or up to 2 years' imprisonment
HSW Act s.33(3) ➤ all other offences (except those specified in s.33(4)) ➤ breach of HSW Act ss.7, 8 or 9 ➤ breach of Regulations made under the Act ➤ making false statements or entries ➤ using, or possessing a document with the intent to deceive	**Max. £5,000 fine**	**Fine (unlimited)**
HSW Act s.33(3) – *offences specified in s.33(4)* ➤ operating without, or contravening the terms or conditions of, a licence when one is required (e.g. Under Asbestos (Licensing) Regulations 1983) ➤ acquiring or attempting to acquire, possessing or using an explosive article or substance in contravention of the relevant statutory provisions ➤ using or disclosing information in contravention of sections 27(4) and 28 (relates to information provided to HSC/HSE, etc. by notice or under relevant statutory provisions)	Max. £5,000 fine	Fine (unlimited) and/or up to 2 years' imprisonment

The Health and Safety System in Great Britain. 3rd edition, HSC, 2002 HSE Books, ISBN 9780-7176-2243-6.

What to Expect When a Health and Safety Inspector Calls: A Brief Guide for Businesses, Employees and Their Representatives. HSC 14, 1998 HSE Books.

21.5   Environmental Protection Act 1990

21.5.1   *Introduction*

The Environmental Protection Act 1990 (EPA) is still the centrepiece of current UK legislation on environmental protection. It is divided into nine parts, corresponding to the wide range of subjects dealt with by the Act.

Integrated pollution control (IPC) was a system established by Part 1 of the Act. Part 1 introduced *Part A*

Processes, which are the most potentially polluting or technologically complex processes. In England and Wales these are enforced by the Environment Agency. In Scotland there is a parallel system enforced by the Scottish Environment Protection Agency.

Less polluting industry were classified as Part B, with only emissions released to air being subject to regulatory control. For such processes local authorities are the enforcing body and the system is known as Local Air Pollution Control (LAPC).

Both IPC and LAPC have now been replaced by an Integrated Pollution Prevention and Control (IPPC) regime that implements the requirements of the EC Directive 96/61 on IPPC. This was introduced under the Pollution Prevention and Control Act of 1999 (1999 Chapter 24), which repealed Part 1 of the Environmental Protection Act.

The change is outlined in Box 21.1.

Box 21.1   Pollution Prevention and Control Regime Outline

PPC REGIME
Pollution Prevention and Control

IPC (regime A)
Integrated Pollution Control
LAPC (regime B)
Local Air Pollution Control
IPPC (regime A1 and A2)
Integrated Pollution Prevention and Control
LAPPC (regime B)
Local Air Pollution Prevention and Control

Under Environmental Protection Act 1990 – Part 1

Regime A – This is an integrated permitting regime. Emissions to the air, land and water of the potentially more polluting processes are regulated. The Environment Agency is the regulator.
    Regime B – This regime permits processes with a lesser potential for polluting emissions. Only emissions to the air are regulated. The Local Authority is the Regulator.

Under Pollution Prevention and Control Act 1999

Regime A1 – This is an integrated permitting regime. Emissions to the air, land and water of potentially more polluting processes are regulated. The Environment Agency is the Regulator.
    Regime A2 – This is an integrated permitting regime. Emissions to the air, land and water of processes with a lesser potential to pollute are regulated. The Local Authority Agency is the regulator.
    Regime B – This is the permitting of processes with a lesser potential to pollute. Only emissions to the air are regulated. The Local Authority is the Regulator.

21.5.2   *Integrated Pollution Prevention and Control*

The system of Integrated Pollution Prevention and Control (IPPC) applies an integrated environmental approach to the regulation of certain industrial activities.

This means that emissions to air, water (including discharges to sewer) and land, plus a range of other environmental effects, must be considered together. It also means that regulators must set permit conditions so as to achieve a high level of protection for the environment as a whole. These conditions are based on the use of the 'Best Available Techniques' (BAT), which balances the costs to the operator against the benefits to the environment. IPPC aims to prevent emissions and waste production and where that is not practicable, reduce them to acceptable levels. IPPC also takes the integrated approach beyond the initial task of permitting, through to the restoration of sites when industrial activities cease.

21.5.3   *Setting the legal framework*

The PPC Regulations implement the European Community (EC) Directive 96/61/EC on Integrated Pollution Prevention and Control ('the IPPC Directive'), insofar as it relates to installations in England and Wales. Separate Regulations apply the IPPC Directive in Scotland and Northern Ireland and to the offshore oil and gas industries.

➤ Prior to the PPC Regulations coming into force, many industrial sectors covered by the IPPC Directive were regulated under Part I of the EPA 1990. This introduced the systems of IPC, which controlled releases to all environmental media and LAPC, which controlled releases to air only. Other industrial sectors new to integrated permitting, such as the landfill, intensive farming and food and drink sectors were regulated, where appropriate, by separate waste management licences issued under Part II of the EPA and/or water discharge consents under the Water Resources Act 1991 or Water Industry Act 1991.

➤ The PPC Regulations create a coherent new framework to prevent and control pollution, with two parallel systems similar to the old regimes of IPC and LAPC. The first of these – the 'Part A' regime of IPPC – applies a similar integrated approach to IPC while delivering the additional requirements of the IPPC Directive. 'Part A' extends the issues that regulators must consider alongside emissions into areas such as energy use and site restoration. The main provisions of IPPC apply equally to the ex-IPC processes and the other sectors new to integrated permitting. There are also some further requirements that apply solely to waste management activities falling under IPPC.

➤ The IPPC Directive applies to those landfills receiving more than 10 tonnes per day or with a total capacity exceeding 25,000 tonnes (but excluding landfills taking only inert waste; the landfill Directive applies to all landfills. The PPC Regulations have been amended to include all landfills. For landfills the technical requirements are met through the Landfill Regulations. Department for Environment, Food and Rural Affair (DEFRA) issued separate guidance on the Landfill Regulations in 2004.

➤ The Environment Agency regulates Part A (1) installations. Part A (2) installations are regulated by the relevant local authority – usually the district, London or metropolitan borough council in England and the county or borough council in Wales. However, the local authority will always be a statutory consultee where the Environment Agency is the regulator, and vice versa. Moreover, the local authority and the Environment Agency will work together in the permitting process. Local authorities have expertise in setting standards for noise control, while the Environment Agency will ensure that permit conditions protect water adequately. Annex I describes how IPPC installations are classified into either Part A (1) or Part A (2) installations depending on what activities take place within them.

➤ The second new regime – the 'Part B' regime of Local Air Pollution Prevention and Control (LAPPC) – represents a continuation of the old LAPC regime. LAPPC is similar to IPPC from a procedural perspective, but it still focuses on controlling emissions to air only. DEFRA provides separate guidance on local authority air pollution control.

21.5.4 *Overview of the regulatory process*

The basic purpose of the IPPC regime is to introduce a more integrated approach to controlling pollution from industrial sources. It aims to achieve 'a high level of protection of the environment taken as a whole by, in particular, preventing or, where that is not practicable, reducing emissions into the air, water and land'. The main way of doing that is by determining and enforcing permit conditions based on BAT.

➤ The entire regulatory process for IPPC consists of a number of elements. These are outlined below. IPPC applies to specified 'installations', both 'existing' and 'new', requiring each 'operator' to obtain a permit from the regulator – either the Environment Agency or the local authority.

Stage 1 – Permitting

1. The procedure begins with the preparation of an application by the operator. Once the regulator receives the application, they will consult various 'statutory consultees'. Operators should be encouraged to engage the public at the earliest opportunity. Operators are required to advertise in one or more local papers and in the London Gazette details of the activity and its location together with a statement of where public representations should be made. After giving further public participation in the case of applications in respect of 'new installations', or 'substantial changes' to installations, the regulator will then determine the application, either granting a permit with conditions or refusing it.

2. The Landfill Regulations required operators of all existing landfill sites to submit a conditioning plan (CP) by 16 July 2002. If a site closed before 16 July 2002, then no plan was required. Otherwise the Environment Agency decides whether a landfill should continue operating on the basis of the CP. A PPC application building on the CP will then be required to be submitted by the date notified to the operator.

3. In making an application the operator must cover various environmental issues. These include:
   ➤ satisfactory environmental management of the installation and no significant pollution caused
   ➤ adequate compliance monitoring
   ➤ waste production is avoided and where waste is produced, it is recovered. Where that is not possible it is disposed of in a way producing the least impact on the environment, if any impact is produced at all
   ➤ assessment of polluting releases and the identification of BAT (see Box 21.2 for a definition)
   ➤ compliance with other EU directives, community and national environmental quality standards (EQSs) and domestic regulations
   ➤ energy is used efficiently
   ➤ measures are taken to avoid accidents and limit their consequences
   ➤ necessary measures are taken on the closure of an installation to avoid any pollution risk and return the site to a satisfactory condition
   ➤ for landfills, alternative requirements are specified by the Landfill Regulations.

4. The operator must also consider the condition of the site at the time of the original application. This will contribute to assessing the need for restoration when the installation closes (stage 3).

5. In determining the application, the regulator must be satisfied that the operator has addressed the above points appropriately. It is therefore the operator's responsibility to demonstrate that this is the case.

Stage 2 – Operation

1. Once the regulator has issued a permit, the operator of an IPPC installation will have to carry out monitoring

Box 21.2　Best available techniques

This term is defined as 'the most effective and advanced stage in the development of activities and their methods of operation which indicates the practicable suitability of particular techniques for providing the basis for emission limit values designed to prevent, and where that is not practicable, generally to reduce the emissions and the impact on the environment as a whole'.

This definition implies that BAT not only cover the technology used but also the way in which the installation is operated, to ensure a high level of environmental protection as a whole. BAT take into account the balance between the costs and environmental benefits (i.e. the greater the environmental damage that can be prevented, the greater the cost for the techniques).

to demonstrate compliance with the permit conditions. Regulators will also carry out their own monitoring and inspections, and have a range of enforcement powers.

2. Over time, regulators may vary permits to reflect changes in how installations are operated, or for other reasons. The regulator may vary permit conditions at either its own or the operator's instigation, with the possibility of consultation in either case. The regulator may also transfer permits from one operator to another, for example when one operator is taken over by another. More generally, regulators must review permits periodically, or whenever circumstances make a review necessary, such as when significant pollution occurs.

3. Specific conditions may apply to individual installations that the regulator considers appropriate to ensure a high level of protection to the environment as a whole. If the regulator believes that the operator is breaching the conditions of a permit, enforcement options are available where enforcement, suspension or a revocation notice can be served (the operator may appeal against this to the Secretary of State).

Stage 3 – Closure and Surrender

1. When an installation closes, an operator should apply to surrender a permit, to end regulation under IPPC and payment of the associated annual charges to the regulator. The application to surrender the permit must include a site report identifying, in particular, any changes in the condition of the site since the time at which the permit was issued. The operator is required to identify any steps that have been taken to avoid any pollution risk resulting from the operation of the installation or return it to a satisfactory state. If on closure the operator satisfies the regulator that they have removed any pollution risks and restored the site to a satisfactory state, the regulator accepts the surrender and gives the operator notice of its determination. The permit then ceases to have effect on the date specified in the notice of determination. If the regulator is not satisfied, it has to give notice of its determination stating that the application has been refused.

21.5.5　*Wider scope of IPPC*

IPPC takes a wider range of environmental impacts into account than IPC. The current system of IPC regulates emissions to land, water and air. The IPPC regime will additionally take into account waste avoidance or minimization, energy efficiency, accident avoidance and minimization of noise, heat and vibrations. These aims will achieve a higher level of protection as a whole.

IPPC applies to a wider range of industries than IPC. These industries include all installations that are currently regulated under IPC, some installations currently under LAPC, and some installations that are not currently under either regime, such as landfill sites, intensive agriculture, large pig and poultry units, and food and drink manufacturers.

Under IPPC, regulated industries are referred to as 'installations' as opposed to 'processes', which is the term used for IPC. This change in terminology enables a more integrated approach to regulation; a whole installation must be permitted rather than just individual processes within the installation.

Guidelines to establish which BAT are used published by the European Commission's IPPC Bureau. These reference notes are known as BREF notes and provide the basis for national sectoral guidance. The Environment Agency has supplementary guidance to cover many issues, some, for example energy efficiency, site remediation and noise, are new issues under IPPC. Industry sectors not previously regulated under the Environmental Protection Act 1990, such as intensive farming and food and drink installations, are also covered by guidance.

Once issued, permits for IPPC are to be reviewed periodically in addition to any updating which is made necessary by technological or other changes.

Permits have been required for all new installations and existing installations undergoing a substantial change (where there is a change in operation that

443

may have a significant negative effect on human beings or the environment) from 31 October 1999. For existing installations IPC permits will continue to be in force until IPPC permits are phased in on a sectoral basis by October 2007.

21.5.6 *Duty of care*

Waste and the duty of care

The duty of care is covered in Part II of the Environmental Protection Act 1990. The duty of care applies to anyone who produces or imports, keeps or stores, transports, treats or disposes of waste. It also applies if they act as a broker and arrange these things.

The duty holder is required to take all reasonable steps to keep the waste safe. If they give waste to someone else, the duty holder must be sure that they are authorized to take it and can transport, recycle or dispose of it safely.

The penalty for breach of this law is an unlimited fine.

Waste can be anything owned, or which a business produces, that a duty holder wants to get rid of. Controlled waste is defined in Box 21.3.

Box 21.3 Controlled Waste

Controlled waste means household, commercial or industrial waste. It includes any waste from a house, school, university, hospital, residential or nursing home, shop, office, factory or any other trade or business premises. It is controlled waste whether it is solid or liquid and even if it is not hazardous or toxic.

If the waste comes from a person's own home, the duty of care **does not** apply to them. But if the waste is not from the house they live in, for example, if it is waste from their workplace or waste from someone else's house, the duty of care **does** apply.

Animal waste collected and transported under the Animal By-Products Order 1992 is not subject to the duty of care.

Duty holders must take all reasonable steps to fulfil the duty and complete some paperwork. What is reasonable depends on what is done with the waste.

Steps to take if the duty of care applies when a duty holder has waste are as follows. They must:

➤ stop it escaping from their control and store it safely and securely. They must prevent it causing pollution or harming anyone

➤ keep it in a suitable container. Loose waste in a skip or on a lorry must be covered

➤ if the duty holder gives waste to someone else, they must check that they have authority to take it. The law says the person to whom they give the waste must be authorized to take it. Box 21.4 explains who is allowed to take waste and how the duty holder can check

➤ describe the waste in writing. The duty holder must fill in and sign a transfer note for it and keep a copy. To save on paperwork, the description of the waste can be written on the transfer note (see Box 21.5).

Box 21.4 Who has authority to take waste?

Council waste collectors

The duty holder does not have to do any checking, but if they are not a householder, there is some paperwork to complete. This is explained in Box 21.5.

Registered waste carriers

Most carriers of waste have to be registered with the Environment Agency or the Scottish Environment Protection Agency. Look at the carrier's certificate of registration or check with the Agencies.

Exempt waste carriers

The main people who are exempt are charities and voluntary organizations. Most exempt carriers need to register their exemption with the Environment Agency or the Scottish Environment Protection Agency. If someone says they are exempt, ask them why. Check with the Agencies that their exemption is registered.

Holders of waste management licences

Some licences are valid only for certain kinds of waste or certain activities. Ask to see the licence. Check that it covers the kind of waste being consigned.

Businesses exempt from waste management licences

There are exemptions from licensing for certain activities and kinds of waste. For example, the recycling of scrap metal or the dismantling of

---

scrap cars. Most exempt businesses need to register their exemption with the Environment Agency or the Scottish Environment Protection Agency. Check with the Agencies that their exemption is registered.

Authorized transport purposes

Waste can also be transferred to someone for 'Authorized transport purposes'. This means:

➤ the transfer of controlled waste between different places within the same premises
➤ the transport of controlled waste into Great Britain from outside Great Britain and
➤ the transport by air or sea of controlled waste from a place in Great Britain to a place outside Great Britain.

Registered waste brokers

Anyone who arranges the recycling or disposal of waste, on behalf of someone else, must be registered as a waste broker. Check with the Environment Agency or the Scottish Environment Protection Agency that the broker is registered.

Exempt waste brokers

Most exempt waste brokers need to register with the Environment Agency or the Scottish Environment Protection Agency. Those who are exempt are mainly charities and voluntary organizations. If someone tells you they are exempt, ask them why. You can check with the Environment Agencies that their exemption is registered.

Box 21.5   Filling in paperwork

When waste is passed from one person to another the person taking the waste must have a written description of it. A transfer note must also be filled in and signed by both persons involved in the transfer.

The duty holder can write the description of the waste on the transfer note. Who provides the transfer note is not important as long as it contains the right information. The

---

Government has published a model transfer note with the Code of Practice, which can be used if desired.

Repeated transfers of the same kind of waste between the same parties can be covered by one transfer note for up to a year. For example, weekly collections from shops.

The transfer note: The transfer note to be completed and signed by both persons involved in the transfer must include:

➤ what the waste is and how much there is
➤ what sort of containers it is in
➤ the time and date the waste was transferred
➤ where the transfer took place
➤ the names and addresses of both persons involved in the transfer
➤ whether the person transferring the waste is the importer or the producer of the waste
➤ details of which category of authorized person each one is. If the waste is passed to someone for authorized transport purposes, you must say which of those purposes applies
➤ if either or both persons is a registered waste carrier, the certificate number and the name of the Environment Agency which issued it
➤ if either or both persons has a waste management licence, the licence number and the name of the Environment Agency which issued it
➤ the reasons for any exemption from the requirement to register or have a licence
➤ where appropriate, the name and address of any broker involved in the transfer of waste.

The written description: The written description must provide as much information as someone else might need to handle the waste safely.

Keeping the papers: Both persons involved in the transfer must keep copies of the transfer note and the description of the waste for 2 years. They may have to prove in Court where the waste came from and what they did with it. A copy of the transfer note must also be made available to the Environment Agency or the Scottish Environment Protection Agency if they ask to see it.

When a person takes waste from someone else they must:

➤ be sure the law allows them to take it. Box 21.4 explains who is allowed to take waste.
➤ make sure the person giving them the waste describes it in writing. The waste receiver must fill in and sign a transfer note and keep a copy (see Box 21.5).

21.5.7 *Hazardous waste*

On 16 July 2005 the Hazardous Waste (England and Wales) Regulations 2005 and the List of Wastes (England) Regulations came into force replacing the Special Waste Regulations. The Special Waste Regulations 1996 transposed the requirements of the European Hazardous Waste Directive (91/689/EEC) which sets out requirements for the controlled management of hazardous (special) waste. The Regulations set out procedures to be followed when disposing of, carrying and receiving hazardous waste. The Special Waste Regulations 1996 were amended by the Special Waste (Amendment) Regulations 1996, the Special Waste (Amendment) Regulations 1997 and the Special Waste (Amendment) (England and Wales) Regulations 2001. These can be found on the The Office of Public Sector Information (OPSI) website at. http://www.opsi.gov.uk/legislation/.

The new regime includes a requirement for most producers of hazardous waste to notify their premises to the Environment Agency. The facility to notify premises has been available since April 2005. Preliminary guidance on notification was published by DEFRA in January 2005. Guidance on notification, including the on-line notification facility, and more general guidance on the new regime can be found on the Environment Agency's *hazardous waste* pages.

21.5.8 *Applying for a waste management licence*

Introduction
A waste management licence is a legal document issued under the Environmental Protection Act 1990. There are two types of waste management licence:

➤ a site licence – authorizing the deposit, recovery or disposal of controlled waste in or on land
➤ a mobile plant licence – authorizing the recovery or disposal of controlled waste using certain types of mobile plant.

A licence has conditions to make sure that the authorized activities do not cause pollution of the environment, harm to human health or serious detriment to local amenities.

Anyone who deposits, recovers or disposes of **controlled waste** must do so either:

➤ within the conditions of a waste management licence or
➤ within the conditions of an exemption from licensing

and must not cause pollution of the environment, harm to human health or serious detriment to local amenities. Otherwise they could be fined and sent to prison.

The conditions of the exemptions from licensing are found in the Waste Management Licensing Regulations 1994. If an activity may be exempt this should be discussed with the Environment Agency.

Before applying for a licence contact the Environment Agency for advice. If a licence is needed the Environment Agency will:

➤ meet to discuss your proposals
➤ explain the application process and provide an application package
➤ provide guidance on preparing a working plan
➤ set out in writing any additional information required specifically for the proposals
➤ help work out the application fee and
➤ say who will deal with the application.

There are financial and legal implications to holding a licence and the licensee may want to get independent advice about these.

Box 21.6 What is a working plan?

As part of the licence application a working plan will need to be prepared. This is a document describing how the licensee intends to prepare, develop, operate and restore (where relevant) the site or plant. If they write a comprehensive working plan it will help to avoid delays in processing the application. It may also mean that some of the licence conditions could be less restrictive, giving more opportunity for flexibility in the operations.

21.5.9 *Applying for a licence*

To apply for a waste management licence the licensee must send:

➤ a completed application form
➤ the fee for processing the application (this cannot be refunded)

➤ any additional information which the Agency has asked for in writing and

➤ the supporting information in the checklist on the application form to the Environment Agency at the address shown on the application form.

When all the information has been received the Environment Agency will write to confirm that the application is complete and that the 4-month period allowed in law for deciding the application has started.

The application form will be put on the public register. If the licensee feels that any information on the form should not be made public because it is commercially confidential, they can apply to have that information withheld. The application form tells them how to do this.

21.5.10   *Processing the application*

Licence applications are processed by the Environment Agency using the waste management licensing process handbook, which helps to ensure that the licensing service is efficient and consistent.

The Environment Agency will consider whether it can write licence conditions which will make sure that the proposed activities do not cause pollution of the environment, harm to human health or serious detriment to local amenities.

When processing an application the Agency will also:

➤ check that the activities to be licensed have planning permission or equivalent (where it is needed). It is the licensee's responsibility to make sure that the activities have the necessary planning permission

➤ consult with other regulatory bodies and

➤ decide whether the licensee is a fit and proper person.

Box 21.7   What is a fit and proper person?

When considering whether the licensee is a fit and proper person the Agency will look at:

➤ whether they have made adequate financial provision to cover the licence obligations. If the Agency plans to issue a licence then a draft of the licence conditions will be sent to the licensee. They should use this draft to help calculate the amount of financial provision needed

➤ whether a technically competent person will be managing the licensed activities

➤ whether the licensee, or another relevant person, has been convicted of a relevant offence.

If the Environment Agency is not satisfied on any aspect, it will write and say what it needs. If the licensee is still unable to satisfy the Agency the application will be rejected.

21.5.11   *After a licence is issued*

Once the site is licensed, the licensee must comply with the licence conditions at all times. The Environment Agency will make visits to check that the licence conditions are being met.

There will be an annual fee to cover the costs of these visits. The fees for waste management licensing are in the Waste Management Licensing (Charges) Scheme.

The licence and working plan will be put on the public register. If the licensee thinks any of the information should not be made public because it is commercially confidential, they can apply to have that information withheld.

The licensee will be responsible for the obligations arising from that licence until:

➤ the Agency accepts an application to transfer the licence to another person

for a site licence – the Agency accepts the surrender of the licence

for a mobile plant licence – the licensee surrenders the licence.

Changes to licence conditions can only be made by the Environment Agency. If the licensee wants to change their operations, this should be discussed with the Agency.

21.5.12   *Future Environmental Permitting Programme*

The Environmental Permitting Programme (EPP) is a joint initiative between the Environment Agency, the Department for Environment, Food and Rural Affairs (DEFRA), Welsh Assembly Government (WAG) and other groups to develop a new and simplified regulatory system. It starts by combining Waste Management Licensing and Pollution Prevention and Control (PPC).

The objectives of EPP can be summarized as:

➤ bringing waste and PPC regulation into one joined-up risk-based regime

➤ reducing red tape while maintaining environmental standards

➤ simplifying and clarifying supporting documents and information systems and

➤ delivering savings for both operator and regulator.

The first step will be to make a number of sets of standard rules for waste activities available from April 2008. This was the subject of consultation issued on 13 September 2007.

Standard Rules and Guidance for the introduction of the Environmental Permitting Regulations

When the new Environmental Permitting Regulations (EP Regulations) came into force in April 2008, standard permits will be available from the Environment Agency. A standard permit has one condition which refers to a fixed package of standard rules. These permits are for low- to medium-risk operations that do not require site-specific assessment of risk. The Environment Agency will be able to issue these quickly because they have no decisions to make on site-specific conditions.

For more information on the draft Standard Rules and the Environmental Permitting Regulations see the Defra website: http://www.defra.gov.uk/environment/epp/.

21.5.13 *Further information*

The law

Environmental Protection Act 1990, ISBN 9780-10-544390-5.

Hazardous Waste (England and Wales) Regulations 2005. The Stationery Office, SI 2005, No. 894.

Pollution Prevention and Control Act 1999. Stationery Office 1999, ISBN 9780-10-542499.

Waste Management Licensing Regulations 1994. SI 1994, No. 1056 (as amended), ISBN 9780-11-044056-0.

Government guidance

Integrated Pollution Prevention and Control Practical Guidance. 4th edition, DEFRA 2005.

Waste Management Paper 4. Licensing of Waste Management Facilities, ISBN 9780-11-752727-0.

DOE Circular 11/94 Environmental Protection Act 1990: Part II Waste Management Licensing. The Framework Directive on Waste (Welsh Office Circular 26/94, Scottish Office Environment Department Circular 10/94), ISBN 9780-11-752975-3.

A Guide to the Hazardous Waste Management Regulations and the List of Waste Regulations in England and Wales. Environment Agency HWR01. Environmental Permitting DEFRA website: http://www.defra.gov.uk/environment/epp/.

21.6 New Roads and Street Works Act 1991

21.6.1 *Introduction*

The prevailing legislation surrounding streetworks is primarily enshrined in The New Roads and Street Works Act 1991 (NRSWA). This is a detailed Act with 171 sections but it is Part III which applies to street works in England and Wales and Part IV which applies to road works in Scotland that most concern health and safety. A summary of the general requirements only is given.

Prior to this legislation, the Public Utilities and Street Works Act 1950 (PUSWA) gave public utilities the right to dig up roads without needing to obtain prior permission, while local authorities were generally responsible for carrying out the permanent reinstatements. As this was widely believed to be unnecessarily bureaucratic, there was a groundswell in favour of deregulation to create more favourable operating conditions for the private sector. However, the deregulation of streetworks coincided with the liberalization of the telecommunications industry, which meant that when the Act came into effect the conditions which had originally spawned it had radically altered and there was soon a plethora of companies empowered to dig up the road. In 1987 there were only two telecommunications companies licensed to dig up the road – there are now nationally over 120 and the scale of co-ordination required is daunting.

Under the terms of this Act, statutory undertakers (in general, companies or public bodies supplying gas, water, electricity and telecommunications, as well as bodies such as London Underground) have the legal right to dig up roads to either maintain or repair their existing pipes and cables or in order to install additional ones to provide service to new customers.

21.6.2 *Application of the Act to Street works/ Road works (Scotland) – Part III Section 48 and Part IV Section 107*

A street means:

➤ any highway, road, lane, footway, alley or passage,
➤ any square or court, and
➤ any land laid out as a way whether it is for the time being formed as a way or not.

Where a street passes over a bridge or through a tunnel, references in this Part to the street include that bridge or tunnel.

In section 107 the definition of 'road' is similar but here it means any way whether or not there is a public right of passage.

Street works or Road works (in Scotland) means works for any purpose (other than works for road purposes) executed in a street or road in pursuance of a statutory right or a street license (permission granted under section 109 in Scotland). They cover works of any of the following kinds:

➤ placing apparatus or
➤ inspecting, maintaining, adjusting, repairing, altering or renewing apparatus, changing the position of apparatus or removing it, or

➤ works required for or incidental to any such works (including, in particular, breaking up or opening the street, or any sewer, drain or tunnel under it, or tunnelling or boring under the street).

21.6.3 General requirements

The statutory undertakers are obliged to inform local authorities of the work but the notice required varies dramatically, depending on the nature of the work. The categories laid down in the Act are as follows in Table 21.2.

The Act requires local authorities to co-ordinate the works while obliging undertakers to co-operate with local authorities in the interests of safety, minimizing inconvenience to persons using the street and protecting the structure of the street and the integrity of the apparatus within it.

The sheer volume of streetworks is colossal – with the vast majority giving less than 7 days' notice. Clearly it would be unrealistic and unwieldy to expect local authorities to inform residents and businesses about every last piece of activity.

Table 21.2 Minimum notice periods

Categories of works	Non-traffic sensitive situations	Traffic sensitive situations
Emergency (including remedial – dangerous)	Within 2 hours of work starting	
Urgent	Within 2 hours of work starting	2 hours' notice in advance
Special cases of urgent	Within 2 hours of work starting (where immediate start is justified)	
Minor works (without excavation)	Notice not required	3 days' notice
Minor works (with excavation)	Notify by daily whereabouts	1 month advance notice and 7 days' notice of start date
Remedial works (non-dangerous)	Notify by daily whereabouts	3 days' notice
Standard works	7 days notice	1 month advance notice and 7 days' notice of start date
Major projects	1 month advance notice and 7 days' notice of start date	

There is in theory a vested interest for local authorities and statutory undertakers to notify the public in advance (through either mail-drops or signage), especially when large-scale works are carried out, yet there is no legal compulsion to do so.

Other complications are caused by the fact that the records detailing what is located underground are not always 100% accurate. Underground work can pre-date the First World War and unsurprisingly details can be patchy, or non-existent. While sophisticated equipment is available to gauge what might be lurking underground, this is not infallible either.

The introduction of the 'electronic transfer of notices' (ETON) in 1999 established a standard mechanism for undertakers to use to send their notices to highway authorities.

This promised to be far more efficient than previous paper-based systems, as well as dramatically speeding up communication between the undertakers and highway authorities.

While councils have a duty to co-ordinate the work of statutory undertakers, enforcing this is another matter. Formal co-ordination meetings are generally held on a regular basis, attended by representatives of the council, the statutory undertakers and the police. However, in central London in particular, the companies carrying out this sort of work are operating in a highly competitive environment and are therefore reluctant to disclose details of their plans to their rivals, beyond their legal obligation to give 1 month's notice of substantial works.

Although councils can try to persuade companies to work together (by laying several cables in one trench or timetabling planned work in close sequence) in order to minimize disruption, they have no authority to do anything more than encouraging them to co-ordinate their activities.

21.6.4 Code of practice under regulations 65 and 124

This Code of Practice is issued by the Secretary of State for Transport, the Scottish Executive and the National Assembly for Wales under sections 65 and 124 of the NRSWA, and by the Department for Regional Development (Northern Ireland) under article 25 of the Street Works (Northern Ireland) Order 1995. The legislation requires an undertaker, and those working on its behalf, carrying out work under the Act or the Order to do so in a safe manner as regards the signing, lighting and guarding of works.

Failure to comply with this requirement is a criminal offence. Compliance with the Code will be taken as compliance with the legal requirements to which it relates.

Highway authorities in England and Wales and roads authorities in Scotland should comply with this Code for their own works, as recommended by the respective national administrations. The Northern Ireland road authority is legally required to comply with the Code. In the application of this Code to Scotland, all references in the text to highway authorities are to be read as references to roads authorities.

Everyone on site has a personal responsibility to behave safely, to the best of their ability. Under the Health and Safety at Work etc Act 1974, employers have duties to protect their employees from dangers to their health and safety and to protect others who might be affected by the work activity (e.g. passing pedestrians and motorists). These include proper arrangements for design (including planning and risk assessment) and management (including supervision) of the works. Supervisors qualified under the NRSWA or the Order will know what to do in most situations about which they have to be consulted, and will be able to find out quickly what to do about the others. It is the employer's responsibility to ensure that these arrangements are properly carried out.

This Code applies to all highways and roads except motorways and dual carriageways with hard shoulders. More detailed advice, and advice on some situations not covered by this Code, can be found in Chapter 8 of the Traffic Signs Manual published by the Department for Transport, Local Government and the Regions in conjunction with the Scottish, Welsh and Northern Ireland administrations. This gives authoritative advice, but it does not have the status of a Code of Practice under the Act. In Northern Ireland the use of Chapter 8 is mandatory for undertakers' works on motorways or dual carriageways with hard shoulders, and elsewhere in the United Kingdom undertakers should comply with Chapter 8 when carrying out such works. On all other roads they meet their obligations under section 65 or 124 of the Act, or under article 25 of the Northern Ireland Order, if they comply with this Code, even though further relevant advice may be available in Chapter 8 and other relevant documents.

21.6.5 Changes to legislation

Prolonged occupation

As a result of the widespread perception that streetworks are often unnecessarily protracted, the government held consultations before activating section 74 of the NRSWA under The Street Works (Charges for Unreasonably Prolonged Occupation of the Highway) (England) Regulations 2001.

This provides powers for highway authorities to charge undertakers a daily fee if they fail to complete works by an agreed deadline. The Regulations define a 'prescribed' period for work (i.e. one prescribed in Regulations) and a 'reasonable' period (i.e. the period that the undertaker estimates that the work will take, if not challenged by the highway authority, or if not agreed, the period determined by arbitration). If the duration of a work exceeds both of these periods, a highway authority may levy a charge. Highway authorities have the power to waive or reduce the level of charges when they believe that circumstances warrant this.

The Government has commissioned a report from Halcrow into the effects of the scheme, which is currently awaited.

Lane rental

At the time section 74 was passed, the Government made it clear that if the legislation failed to lead to a sufficient reduction in disruption, then it would be prepared to consider making lane rental charging powers available to local authorities. The Government decided to undertake a localized test of the proposed new powers, and following the London Borough of Camden's successful application to the Secretary of State to operate a lane rental pilot scheme, this was passed under section 74A of the NRSWA as The Street Works (Charges for Occupation of the Highway) (London Borough of Camden) Order 2002. The pilot scheme was launched in March 2002 and will run until March 2004 with a similar one running in Middlesborough. Section 74 will not apply in Camden while section 74A is operational.

Under the lane rental scheme, streets in Camden are divided into 'premium routes' and 'ordinary routes' with the following charges applying per working day from the commencement of works (see Table 21.3).

The charges for premium routes and ordinary routes differ, depending on whether the jobs are works or remedial works – works in this context consist of the work originally scheduled while remedial works are streetworks which have been necessary because the local authority was dissatisfied, for instance, with the quality of reinstatement.

21.6.6 References

Crossing High-Speed Roads on Foot During Temporary Traffic Management Works. Construction Information Sheet No. 53, 2000, HSE.

New Roads and Street Works Act 1991. Chapter 22, London, The Stationery Office.

Safety at Street Works and Road Works Code of Practice. London, The Stationery Office.

The Traffic Signs Manual. Chapter 8: The traffic safety measures and signs for roadworks and temporary situations, Department for Transport.

Table 21.3 Lane rental – daily changes for workers day

Categories of work	Premium route charges		Ordinary route charges	
	Works	**Remedial works**	**Works**	**Remedial works**
Standard works	£500	£650	£100	£200
Minor works	£0	£650	£0	£200
Urgent works	£500	£650	£100	£200
Emergency works	£300	£650	£0	£200
Non-excavatory works	£0	£0	£0	£0

21.7 Control of Asbestos Regulations 2006 CAR 2006

21.7.1 *Introduction*

These Regulations implement an amendment to the Asbestos Worker Protection Directive (2003/18/EC). This Directive is designed to ensure the protection of those workers who are now considered to be most at risk from exposure to asbestos (i.e. building and maintenance workers).

Many of the requirements introduced by the amending Directive were already contained within the existing Asbestos Regulations or in the associated ACOP. However, there still remained the need to introduce a number of significant changes that include:

➤ a single lower 'Control Limit' of 0.1 fibres/cm^3 of air
➤ a new concept of sporadic and low intensity exposure to asbestos, where such work is exempt from notifying HSE and worker medical surveillance
➤ a new World Health Organisation (WHO) asbestos fibre counting method.

The Control of Asbestos Regulations 2006 came into force on 13 November 2006 (Asbestos Regulations – SI 2006/2739).

These Regulations bring together the three previous sets of Regulations covering the prohibition of asbestos, the control of asbestos at work and asbestos licensing.

The Regulations prohibit the importation, supply and use of all forms of asbestos. They continue the ban introduced for blue and brown asbestos in 1985 and for white asbestos in 1999. They also continue the ban on the second-hand use of asbestos products such as asbestos cement sheets and asbestos boards and tiles, including panels which have been covered with paint or textured plaster containing asbestos.

The ban applies to new uses of asbestos. If existing asbestos-containing materials (ACMs) are in good condition, they may be left in place, their condition monitored and managed to ensure they are not disturbed.

The Asbestos Regulations also include the 'duty to manage asbestos' in non-domestic premises. Guidance on the duty to manage asbestos can be found in the 'Approved Code of Practice. The Management of Asbestos in Non-Domestic Premises', L127, HSE Books, ISBN 9780 7176 2382 3. Domestic premises are covered by the Defective Premises Act 1972 or the Civic Government (Scotland) Act.

The Regulations require mandatory training for anyone liable to be exposed to asbestos fibres at work (see regulation 10). This includes maintenance workers and others who may come into contact with or who may disturb asbestos (e.g. cable installers) as well as those involved in asbestos removal work.

When work with asbestos or which may disturb asbestos is being carried out, the CAR 2006 require employers and the self-employed to prevent exposure to asbestos fibres. Where this is not reasonably practicable, they must make sure that exposure is kept as low as reasonably practicable by measures other than the use of respiratory protective equipment (RPE). The spread of asbestos must be prevented. The Regulations specify the work methods and controls that should be used to prevent exposure and spread.

Worker exposure must be below the airborne exposure limit (Control Limit). The CAR 2006 have a single Control Limit for all types of asbestos of 0.1 fibres/cm^3. A Control Limit is a maximum concentration of asbestos fibres in the air (averaged over any continuous 4-hour period) that must not be exceeded.

In addition, short-term exposures must be strictly controlled and worker exposure should not exceed 0.6 fibres/cm^3 of air averaged over any continuous 10-minute period using RPE if exposure cannot be reduced sufficiently using other means.

RPE is an important part of the control regime but it must not be the sole measure used to reduce exposure and should only be used to supplement other measures. Work methods that control the release of fibres such as those detailed in the *Asbestos Essentials task sheets* for non-licensed work should be used. RPE must be suitable, must fit properly and must ensure that worker exposure is reduced as low as is reasonably practicable.

Most asbestos removal work must be undertaken by a licensed contractor but any decision on whether particular work is licensable is based on the risk. Work is only exempt from licensing if:

➤ the exposure of employees to asbestos fibres is sporadic and of low intensity (but exposure cannot be considered to be sporadic and of low intensity if the concentration of asbestos in the air is liable to exceed 0.6 fibres/cm^3 measured over 10 minutes) and

➤ it is clear from the risk assessment that the exposure of any employee to asbestos will not exceed the control limit and

➤ the work involves:

➤ short, non-continuous maintenance activities. Work can only be considered as short, non-continuous maintenance activities if any one person carries out work with these materials for less than 1 hour in a 7-day period. The total time spent by all workers on the work should not exceed a total of 2 hours.

➤ removal of materials in which the asbestos fibres are firmly linked in a matrix, Such materials include: asbestos cement; textured decorative coatings and paints which contain asbestos; articles of bitumen, plastic, resin or rubber which contain asbestos where their thermal or acoustic properties are incidental to their main purpose (e.g. vinyl floor tiles, electric cables, roofing felt); and other insulation products which may be used at high temperatures but have no insulation purposes (e.g. gaskets, washers, ropes and seals).

➤ encapsulation or sealing of ACMs which are in good condition or

➤ air monitoring and control, and the collection and analysis of samples to find out if a specific material contains asbestos.

Under the CAR 2006, anyone carrying out work on asbestos insulation, asbestos coating or asbestos-insulating board (AIB) needs a licence issued by the HSE unless they meet one of the exemptions above.

Although work may not need a licence to carry out a particular job, there is still a need to comply with the rest of the requirements of the CAR 2006.

If the work is licensable there are a number of additional duties. Duty holders need to:

➤ notify the enforcing authority responsible for the site where they are working (for example HSE or the local authority)

➤ designate the work area (see Regulation 18 for details)

➤ prepare specific asbestos emergency procedures and

➤ pay for their employees to undergo medical surveillance.

The CAR 2006 require any analysis of the concentration of asbestos in the air to be measured in accordance with the 1997 WHO recommended method.

From 6 April 2007, a clearance certificate for re-occupation may only be issued by a body accredited to do so. At the moment, such accreditation can only be provided by the United Kingdom Accreditation Service (UKAS).

21.7.2 *Application regulation 3 and general note*

The Regulations and ACOP apply to all work with asbestos. They apply in particular to work on or which disturbs building materials containing asbestos, asbestos sampling and laboratory analysis with the exception of clearing asbestos-contaminated land which is not specifically covered by this ACOP. An additional ACOP entitled 'The Management of Asbestos in Non-domestic Premises' is aimed at those who have repair and maintenance responsibilities for non-domestic premises.

Most of the duties in the CAR 2006 are placed upon 'an employer', that is, the person who employs the workers who are liable to be exposed to asbestos in the course of their work. Although the Regulations always refer to an employer, regulation 3 (1) makes it clear that self-employed people have the same duties towards themselves and others as an employer has towards his or her employees and others.

There is an exemption from certain regulatory requirements for particular, specified types of work with asbestos where any worker exposure will only be sporadic and of low intensity and the exposure level is below the control limit (regulation 3 (2)). Such work will not require a licence. All other work with asbestos will require a licence (regulation 8); must be notified to the relevant enforcing authority (regulation 9); must have emergency arrangements in place (regulation 15 (1)); must have designated asbestos areas (regulation 18); and those working with the asbestos must be subject to medical surveillance and have health records (regulation 22). Some of the guidance in the ACOP is specifically aimed at this more hazardous work and, for convenience, this work has been referred to as licensable work throughout the ACOP.

If the control limit for asbestos is exceeded in the working area, this triggers particular requirements including:

(a) immediately informing employees and their representatives (regulation 11 (5) (b) (i))
(b) identification of the reasons for the control limit being exceeded and the introduction of appropriate measures to prevent it being exceeded again (regulation 11 (5) (b) (ii))
(c) stop work until adequate measures have been taken to reduce employees' exposure to below the control limit (regulation 11 (5) (b) (iii))
(d) a check of the effectiveness of the measures taken to reduce the levels of asbestos in the air by carrying out immediate air monitoring (regulation 11 (5) (b) (iv))
(e) the designation of respirator zones and
(f) the mandatory provision of RPE (regulation 11 (3)), although such equipment should always be provided if it is reasonably practicable to do so (regulation 11 (2)).

Where work with asbestos forms part of a larger project there will be a particular need to co-operate with other employers, and there may be other Regulations which must be taken into account. However, the responsibility to ensure compliance with the provisions of the Asbestos Regulations remains with the employer or self-employed person.

There are exceptions from some requirements.

Where regulation 3 (2) applies (i.e. non-licensable work):

(a) the work will not need to be notified to the relevant Enforcing Authority
(b) the work will not need to be carried out by holders of a licence to work with asbestos
(c) the workers will not need to have a current medical and a current health record
(d) the employer will not need to prepare specific asbestos emergency procedures
(e) the area around work does not need to be identified as an asbestos area.

Work with the following materials is likely only to produce sporadic and low-intensity worker exposure and can be categorized as complying with regulation 3 (2) as long as 3 (2) (b) is fulfilled, that is it is clear from the risk assessment that the control limit will not be exceeded:

(a) asbestos cement
(b) textured decorative coating which contains asbestos
(c) any article of bitumen, plastic, resin or rubber which

contains asbestos where its thermal or acoustic properties are incidental to its main purpose (e.g. vinyl floor tiles, electric cables, roofing felt) and
(d) asbestos materials such as paper linings, cardboards, felt, textiles, gaskets, washers, and rope where the products have no insulation purposes.

21.7.3 *Work with asbestos*

'Work with asbestos' includes:

(a) work which consists of the removal, repair or disturbance of asbestos
(b) work which is ancillary to such work (ancillary work) and
(c) supervising work referred to in sub-paragraphs (a) or (b) above (supervisory work).

'Ancillary work' means work associated with the main work of repair, removal or disturbance of asbestos. Work carried out in an ancillary capacity requires a licence unless the main work (i.e. the removal, repair, disturbance activity) would result in worker exposure which fulfils the conditions for regulation 3 (2) to apply.

'Supervisory work' means work involving direct supervisory control over those removing, repairing or disturbing asbestos. Work carried out in a supervisory capacity requires a licence to work with asbestos unless the work being supervised would result in worker exposure which fulfils the conditions for regulation 3 (2) to apply.

Therefore, compliance with regulations 8 (licence), 9 (notification), 15 (1) (emergency arrangements), 18 (1) (a) (designated areas) and 22 (health records and medical surveillance) are not required in such circumstance. Those other Regulations which apply to all work with asbestos must be observed.

21.7.4 *Duty to manage and identify asbestos in non-domestic premises – regulations 4 and 5*

Owners and occupiers of premises, who have maintenance and repair responsibilities for those premises, have a duty to assess them for the presence of asbestos and the condition of that asbestos. Where asbestos is present the duty holder must ensure that the risk from the asbestos is assessed, that a written plan identifying where that asbestos is located is prepared and that measures to manage the risk from the asbestos are set out in that plan and are implemented. Other parties have a legal duty to co-operate with the duty holder.

As part of the management plan required by regulation 4 of the Asbestos Regulations, occupiers or owners of premises have an obligation to inform any person liable to disturb ACMs, including maintenance workers, about the presence and condition of such materials.

If work to be carried out is part of a larger project which attracts the requirements of the Construction (Design and Management) Regulations (CDM) 1994 (Note: The 94 CDM Regulations are being revised), the health and safety plan prepared by the planning supervisor should contain information on whether the materials contain asbestos and what type they are.

The employer should not rely on the information of the other duty holders if they cannot produce reasonable evidence regarding the nature of suspect material (e.g. survey details or analytical reports).

21.7.5 *Assessment of work – regulation 6*

If work which is liable to expose employees to asbestos is unavoidable, then before starting the work, employers must make a suitable and sufficient assessment of the risk created by the likely exposure to asbestos of employees and others who may be affected by the work and identify the steps required to be taken by the Asbestos Regulations.

For non-licensable work it is not always necessary to make an assessment before each individual job. Where an employer carries out work which involves very similar jobs on a number of sites on the same type of asbestos material, for example, electrical and plumbing jobs, only one assessment for that work may be needed, although the plan of work should always be job specific.

However, for licensable work or where the degree and nature of the work varies significantly from site to site, for example in demolition or refurbishment, or where the type of asbestos material varies, a new assessment and plan of work (see regulation 7) will be necessary.

21.7.6 *Plan of work – regulation 7*

For any work involving asbestos, including maintenance work that may disturb it, the employer of the workers involved must draw up a written plan of how the work is to be carried out before work starts. Employers must make sure that their employees follow the plan of work (sometimes called a method statement) so far as it is reasonably practicable to do so. Where unacceptable risks to health and/or safety are discovered while work is in progress, for example disturbance of hidden, missed or incorrectly identified ACMs, any work affecting the asbestos should be stopped except for that necessary to render suitable control and prevent further spread. Where there is extensive damage to ACMs which causes contamination of the premises, or part of the premises, then the area should be immediately evacuated. Work should

not restart until a new plan of work is drawn up or until the existing plan is amended. Some measures, for example, should only be carried out by licensed contractors.

For licensable work in particular, the plan of work should identify procedures to adopt in emergencies and indicate clearly what remedial measures can be undertaken by staff.

21.7.7 *Licensing of work with asbestos – regulation 8*

This regulation means that an employer must not carry out work with asbestos (other than that fulfilling the conditions for regulation 3 (2) to apply), including supervisory and ancillary work and work with asbestos in their own premises with their own employees, unless the employer holds a licence issued under this regulation and complies with its terms and conditions. This includes work with asbestos insulation, asbestos coatings (excluding asbestos-containing textured decorative coatings) and AIB.

For supervisory work a licence is needed when directly supervising licensable work but not when the person concerned is:

(a) the client who has engaged a licensed contractor to do the licensable work
(b) the principal or main contractor on a construction or demolition site if the licensable work is being done by a subcontractor holding an **asbestos licence**
(c) an analyst checking that the area is clear of asbestos at the end of a job
(d) carrying out quality control work such as:
   (i) atmospheric monitoring outside enclosures while asbestos removal work is in progress or
   (ii) checking that work has been carried out to a standard which meets the terms of the contract
(e) a consultant or other preparing the method statement and
(f) a consultant or other reviewing tender submissions on behalf of the client.

For ancillary work, a licence is needed for:

(a) setting up and taking down enclosures for the asbestos work
(b) putting up and taking down scaffolding to provide access for licensable work where it is foreseeable that the scaffolding activity is likely to disturb the asbestos
(c) maintaining negative pressure units
(d) work done within an asbestos enclosure, such as sealing an electric motor in polythene and installing ducting to the motor to provide cooling air from outside the enclosure and
(e) cleaning the structure, plant and equipment inside the enclosure.

A licence holder is required to:

(a) notify the work to the appropriate enforcing authority (regulation 9)

(b) ensure medical surveillance is carried out for their employees and themselves (regulation 22)

(c) maintain health records for employees and themselves (regulation 22)

(d) prepare procedures in case of emergencies (regulation 15 (1)) and

(e) demarcate the work areas appropriately (regulation 18 (1) (a)).

All licences issued for work with asbestos are granted by HSE under the terms of this regulation. Fees are payable for issuing licences, reassessments and changes to licences. These fees are periodically updated by the Health and Safety (Fees) Regulations.

21.7.8 Notification of work with asbestos – regulation 9

If licensable work is undertaken notification has to be given to the appropriate enforcing authority with details of the proposed work. This gives the enforcing authorities the opportunity to assess your proposals for carrying out work with asbestos and to inspect the site either before or during the work.

Notification will normally be required 14 days before work begins, but the enforcing authority may allow a shorter period, for example in an emergency where there is a serious risk to the health and safety of any person. This shorter period is known as a 'waiver' or dispensation. Each individual job must normally be notified to the enforcing authority.

Form FOD ASB5 can be used for notification, available from the HSE website, local HSE offices or the Asbestos Licensing Unit.

21.7.9 Information instruction and training – regulation 10

There are three main types of information, instruction and training (simply referred to as training from now on). These are:

1. Asbestos awareness training. This is for those persons who are liable to be exposed to asbestos while carrying out their normal everyday work.

2. Training for non-licensable asbestos work. This is for those who undertake work with asbestos which is not licensable such as a roofer removing a whole asbestos cement sheet in good condition.

3. Training for licensable work with asbestos – for those working with asbestos which is licensable, such as removing asbestos lagging or insulating board.

Employers have a duty under regulation 3 (3) (a) of the Asbestos Regulations to ensure, SFARP, that adequate information, instruction and training are given to non-employees who are on the premises and could be affected by the work, as well as to their own employees.

21.7.10 Prevention or reduction of exposure to asbestos – regulation 11

Work which disturbs ACMs should only be carried out when there is no other reasonably practicable way of doing the work or the alternative method creates a more significant risk. Employers must therefore first decide whether they can prevent the exposure to asbestos SFARP, before considering how they will reduce the exposure to as low as reasonably practicable.

Where it is not reasonably practicable to prevent exposure, it must first be reduced to the lowest level reasonably practicable by means other than the use of RPE.

Airborne levels should be reduced to as low a level as reasonably practicable and exposure should be controlled so that any peak exposure is less than 0.6 fibre/cm^3 averaged over a maximum continuous period of 10 minutes by the use of appropriate RPE if exposure cannot be reduced sufficiently by other means.

Employers must ensure that the numbers of employees exposed to asbestos is kept as low as reasonably practicable. All unnecessary personnel should be excluded from the working areas if asbestos is being disturbed.

The provision of a sufficient number of suitable viewing panels in enclosures will allow managers to monitor the work of their employees without being unnecessarily exposed.

When it is not reasonably practicable to prevent exposure to asbestos the employer must choose the most effective method or combination of methods which minimizes fibre release and thereby reduces the exposure to the lowest levels reasonably practicable and document this in the written risk assessment/plan of work.

21.7.11 Use of control measures – regulation 12

Employers should have procedures in place to make sure that control measures are properly used or applied and are not made less effective by other work practices or other machinery.

These procedures should include:

(a) checks at the start of every shift and at the end of each day and

(b) prompt action when a problem is identified.

Within the general duties imposed by regulation 12 (2), employees should, in particular:

(a) use any control measures, including RPE, and protective clothing properly and keep it in the places provided

(b) follow carefully all the procedures set out in the employer's assessment and plan of work, including those for changing and decontamination, and comply with the use of control measures

(c) keep the workplace clean

(d) eat, drink and smoke only in the places provided and

(e) report any defects concerning control measures to their supervisor/manager immediately.

21.7.12 *Maintenance of control measures – regulation 13*

When working with asbestos, employers should make sure that maintenance procedures are drawn up for all control measures and for PPE. These should include the equipment used for cleaning, washing and changing facilities, and the controls to prevent the spread of contamination. The procedures should make clear:

(a) which control measures require maintenance

(b) when and how the maintenance is to be carried out and

(c) who is responsible for maintenance and for making good any defects.

21.7.13 *Provision and cleaning of protective clothing*

As part of the assessment, the employer must decide whether or not protective clothing is required for work with asbestos. The assessment should start from the assumption that protective clothing will be necessary unless exposures are extremely slight and infrequent. For work which requires a licence exposure will potentially be significant and employers will always need to provide a full set of PPE.

The protective clothing must be adequate and suitable and include footwear, whenever employees are liable to be exposed to a significant amount of asbestos debris or fibres. It should be appropriate and suitable for the job and must protect the parts of the body likely to be affected. If the assessment has concluded that a risk of contamination exists, disposable overalls (of a suitable standard fitted with a hood) and boots without laces will be required.

To be adequate and suitable and depending on the circumstances, the protective clothing must:

(a) fit the wearer

(b) be of sufficient size to avoid straining and ripping the joints

(c) be comfortable and, where appropriate, to allow for the effects of physical strain

(d) be suitable for cold environments

(e) prevent penetration by asbestos fibres

(f) be elasticated at the cuffs and ankles and on the hoods of overalls and designed to ensure a close fit at the wrists, ankles, face and neck

(g) not have pockets or other attachments which could attract and trap asbestos dust and

(h) be easily decontaminated or disposable.

Where disposable overalls are used, these should be of a suitable standard.

Non-disposable protective clothing and towels must be effectively washed after every shift. If the employer does not have the facilities and expertise for laundering asbestos-contaminated clothing, it must be sent to a specialist laundry.

Where disposable overalls are used they should be treated as asbestos waste and properly disposed of after every shift.

This may not be necessary for overalls used for occasional sampling where there is a low risk of contamination.

When working in enclosures, clothing for washing should be collected from the airlock and hygiene facility as soon as it has been discarded.

21.7.14 *Accidents incidents and emergencies – regulation 15*

Employers of people removing or repairing ACMs must have prepared procedures which can be put into effect should an accident, incident or emergency occur which could put people at risk because of the presence of asbestos unless, because of the quantity or the condition of the asbestos present at the workplace, there is only a slight risk to the health of employees.

Sufficient information should be made available to the emergency services (e.g. fire and rescue and paramedics), so that when they are attending a relevant incident they can properly protect themselves against the risks from the asbestos.

In any circumstance where there is an accidental uncontrolled release of asbestos into the workplace then measures, including emergency procedures, should be in place to limit exposure and the risks to health. Such procedures should include means to raise the alarm and procedures for evacuation, which should be tested and practised at regular intervals. The cause of the uncontrolled release should be identified, and adequate control regained as soon as possible.

Any people in the work area affected who are not wearing PPE including RPE must leave that area. Where such people have been contaminated with dust or debris then arrangements should be made to decontaminate those affected. Any clothing or PPE should be decontaminated or disposed of as contaminated waste.

21.7.15 Duty to prevent or reduce the spread of asbestos – regulation 16

Any plant or equipment which has been contaminated with asbestos should be thoroughly decontaminated before it is moved for use in other premises or for disposal. The basic decontamination procedures must be followed every time a person leaves the work area.

Asbestos materials should never be left loose or in a state where they can be trampled, tracked over by plant and machinery or otherwise spread. All asbestos-contaminated waste should be removed at regular intervals in appropriate waste containers.

For non-licensable work where a risk of significant contamination exists, the work area should be enclosed. A full enclosure will be expected where there is large-scale work, e.g. asbestos-containing textured decorative coating removal. A 'mini-enclosure' should be used where the work is minor.

It should be assumed that for most of the work which requires a licence, which is not external/remote, a full enclosure will normally be required.

21.7.16 Cleanliness of premises and plant – regulation 17

When work with asbestos comes to an end, the work area should be thoroughly cleaned before being handed over for reoccupation or for demolition. All visible traces of asbestos dust and debris should be removed and a thorough visual inspection should be carried out. Where the work is licensable then the 4-stage clearance procedure (which includes air sampling) should be carried out and a certificate of reoccupation issued. Where licensed work is performed out of doors (e.g. soffit removal), then air sampling will not be required. In this situation, the certificate of reoccupation should still be completed but without stage 3 (air monitoring). More information on clearance procedures for non-licensed work is given in Asbestos Essentials.

To aid the process of cleaning and to prevent the spread of asbestos, employers must choose work methods and equipment to prevent the build-up of asbestos waste on floors and surfaces in the working area. Wherever practicable, waste should be transferred direct into waste bags as workers remove the asbestos materials. Employers must make sure that any asbestos dust and debris is cleaned up and removed regularly to prevent it accumulating (and drying out where wet removal techniques have been used), and at least at the end of each shift.

Procedures will need to take account of the necessity for cleaning following an accidental and uncontrolled release of asbestos.

Procedures will need to be set up for cleaning:

(a) working areas including transit and waste routes
(b) plant and equipment and
(c) hygiene facilities.

Dustless methods of cleaning should be used including, wherever practicable, a type 'H' (BS 5415: 1986) vacuum cleaner with appropriate tools. Procedures for cleaning should make clear:

(a) the items and areas to be cleaned
(b) how often they need to be cleaned
(c) the cleaning methods, which should not create dust and
(d) any special precautions which need to be taken during cleaning, including the low-dust technique to be used, and the measures to be taken to reduce the spread of dust. Dry manual brushing, or sweeping or compressed air, must not be used to remove asbestos dust.

Once removal of the asbestos has been completed, the premises must be assessed to determine whether they are thoroughly clean and hence fit to be returned to the client. It is important that this includes the premises, any plant or equipment or parts of the premises where work with asbestos has taken place and the surrounding areas which may have been contaminated. The areas requiring assessment for site clearance certification for reoccupation include:

(a) the enclosed area including airlocks or the delineated work area where an enclosure has not been used
(b) the immediate surrounding area (for enclosures this will include the outside of walls and underneath polythene floors; for delineated areas this will include surfaces nearby either where asbestos may have been spread or where the pre-cleaning was not done properly)
(c) the transit route if one has been used and
(d) the waste route and area around the waste skip.

21.7.17 Designated areas – regulation 18

All areas where licensable work is being undertaken should be demarcated and identified by suitable warning notices as asbestos areas.

Any area, where an employee may be exposed to asbestos to a level which may exceed a control limit, must be designated as a respirator zone. Respirator zones, whether enclosed or not, must be demarcated and identified by suitable warning notices. Notices that RPE must be worn are also necessary.

Only employees who need to do so for their work can enter and remain in asbestos areas and respirator zones.

Only employees who are competent may enter respirator zones or supervise people working in respirator zones. To enter a respirator area, the employee must have received adequate information, instruction and training in accordance with regulation 10.

Employers should ensure the provision of suitable facilities for employees to eat and drink outside the working area and where appropriate as close as is reasonably practicable to the hygiene facilities. No one should eat, drink or smoke in the enclosure or work area, in the hygiene facilities or in any areas which have been marked as asbestos areas or respirator zones.

Employers should also ensure that toilet facilities are provided, if they are not provided elsewhere on the site.

Where hygiene facilities are not being used, personnel should wash and decontaminate themselves whenever they leave an asbestos area or respirator zone.

21.7.18 *Air monitoring – regulation 19*

Air monitoring may be required to protect the health of employees by determining or checking the concentrations of airborne asbestos to which they are exposed and to establish employee exposure records. This should be done at regular intervals for a representative range of jobs and work methods.

Air monitoring should always be done when there are any doubts about the effectiveness of the measures taken to reduce the concentration of asbestos in air (e.g. that engineering controls are working as they should to their design specification and do not need repair), and, in particular, measures taken to reduce that concentration below the control limit or below a peak level measured over 10 minutes of 0.6 fibre/cm^3. Monitoring will also be necessary to confirm that the RPE chosen will provide the appropriate degree of protection where the level of asbestos fibres in air exceeds, or is liable to exceed, the control limit or a peak level measured over 10 minutes of 0.6 fibre/cm^3.

Air monitoring will be appropriate unless:

(a) exposures are known to be low and not likely to approach the control limit or a 10 minute peak of 0.6 f/m^3

(b) the work is such that it complies with regulation 3 (2) and adequate information is available to enable the appropriate protective equipment to be provided or

(c) the protective equipment provided is of such a standard that no foreseeable measurement could indicate a need for equipment of a higher standard.

If the employer decides that monitoring is not necessary then he or she should use other sources of information about the likely concentrations of asbestos in air, for instance the guidance issued by HSE in the Licensed Contactors Guide or exposure data from previous similar work.

Monitoring of employee exposure should be by personal sampling. Static sampling can be used to check that control measures are effective. Analysis must be undertaken using the 1997 WHO recommended method.

21.7.19 *Standards for air testing, site clearance certification and analysis – regulations 20 and 21*

Those engaged to carry out air measurements and employee exposure monitoring must demonstrate that they conform with specified requirements in ISO 17025 through accreditation with a recognized accreditation body.

Employers carrying out their own air measurements or employee exposure monitoring should make sure that employees carrying out this work receive similar standards of training, supervision and quality control to those required by ISO 17025.

Those engaged to carry out site clearance certification for reoccupation must demonstrate that they conform with specified requirements in ISO 17020 and ISO 17025 through accreditation with a recognized accreditation body.

Those engaged to analyse samples of material to determine whether or not they contain asbestos must demonstrate that they conform with ISO 17025 by accreditation with a recognized accreditation body.

Employers carrying out their own analysis of samples should make sure that employees carrying out this work receive similar standards of training, supervision and quality control to those required by ISO 17025.

The UKAS is currently the sole recognized accreditation body in Great Britain.

21.7.20 *Health records and medical surveillance – regulation 22*

The employer must keep a health record for any employee who undertakes licensable work. The health record must be kept for 40 years in a safe place and should contain at least the following information:

(a) the individual's surname, forenames, sex, date of birth, permanent address, postcode and National Insurance number

(b) a record of the types of work carried out with asbestos and, where relevant, its location, with start and end dates, with the average duration of exposure in hours per week, exposure levels and details of any RPE used

(c) a record of any work with asbestos prior to this employment of which the employer has been informed and

(d) dates of the medical examinations.

Anyone who undertakes licensable work must have been medically examined within the previous 2 years. Employers will need to obtain certificates of examination for any employees who state that they have been examined under these Regulations within the previous 2 years and keep them for 4 years from the date of issue. Employers should check with the previous employer or with the examining doctor that the certificates are genuine.

Medical examinations should take place during the employee's normal working hours and be paid for by the employer. Employees should co-operate with their employer regarding attendance for medical examinations.

Where an employee is diagnosed with a condition related to exposure to asbestos then the employer must review the health of all other current employees similarly exposed, as well as reviewing his assessments and methods of work.

If the examination reveals the presence of any potentially limiting health conditions then a decision should be reached on whether a general fitness assessment is required in addition to the asbestos medical examination.

21.7.21 Washing and changing facilities – regulation 23

The type and extent of washing and changing facilities provided should be determined by the type and amount of exposure as indicated by the risk assessment.

If the work is licensable, separate facilities should be provided for the workers working with asbestos. Employers must ensure that adequate changing and showering facilities are provided so that employees can clean and decontaminate themselves completely each time they leave the work area. This includes providing shampoo, soap or gel and towels. The provision of suitable hygiene facilities (also known as a decontamination unit, DCU), should be on site and fully operational before any work (including ancillary work) commences. Maintenance records for DCUs (or copies of them) should be kept on site. The hygiene facility should not leave the site until the job is complete and the certificate of reoccupation has been issued.

The hygiene facility enables the employer to further comply with their duties to prevent the spread of asbestos and reduce the potential exposure of employees and other people to as low as reasonably practicable. The facilities will need to:

(a) have separate changing rooms for dirty, contaminated work clothing and for clean or personal clothing known as 'dirty' and 'clean' areas respectively. The showers should be located between the two changing rooms so that it is necessary to pass through them when going from one changing facility to the other. All doors between each room and those leading to the outside from the 'dirty end' should be self-closing and provide an airtight seal. The 'clean' and 'dirty' ends should be fitted with adequate seating and be of sufficient size for changing purposes

(b) be designed so that they can be cleaned easily

(c) be fitted with air extraction equipment which keeps a flow of air from the clean to the dirty areas. The extracted air should be discharged through a HEPA filter

(d) be adequately heated, lit (i.e. light switches at both the 'clean' and 'dirty' ends) and have internal vents so that air can pass through the unit

(e) be of sufficient size, including allowance for sufficient and separate storage for personal clothing and protective clothing and equipment in the 'clean' end and sufficient receptacles for contaminated clothing, towels, filters and so on in the 'dirty' end and shower area

(f) have an adequate supply of clean running hot and cold or warm water, at a suitable pressure, in the showers, and soap or gel, shampoo, nailbrushes and individual dry towels. If gas heating is provided and the heater is mounted inside the unit, it must be a room-sealed type, and not open-flued. Waste water should be filtered before being discharged to the drains. All filters should be treated as asbestos waste

(g) the shower areas should be of sufficient size to allow thorough decontamination and to have means to support the power pack of a full face respirator while it is still required to be worn (the power pack support should be out of the direct line of the shower to avoid contact with water and prevent damage to the batteries)

(h) have a wall-mounted mirror in the clean end of the unit and

(i) have the electricity supply enter it via a 30 mA residual current circuit breaker fitted at the point of entry into the unit, and the unit must be effectively earthed when in use.

21.7.22 Storage, distribution and labelling of raw asbestos and asbestos waste – regulation 24

Waste should be placed in suitable, labelled containers as it is produced. Where practicable, containers

459

should be sealed and the outside should be cleaned before removal from enclosures or the work area, and they should be taken to a suitable and clearly identified secure storage area if they are not being disposed of at once.

Any friable waste should be placed in UN-approved packaging (available in up to 2 tonnes capacity). The Licensed Contractors' Guide provides further advice.

Containers must be designed, constructed and maintained to prevent any of the contents escaping during normal handling. For most waste, double plastic sacks are suitable provided they will not split during normal use. It is important that the inner bag is not overfilled, especially when the debris is wet, and each bag should be capable of being securely tied or sealed. Air should be excluded from the bag as far as possible before sealing. Precautions will need to be taken as the exhaust air may be contaminated. Stronger packages are necessary if the waste contains sharp metal fragments or other materials liable to puncture plastic sacks.

Bags containing asbestos waste should be appropriately labelled and transported to a licensed disposal site in an enclosed vehicle, skip or freight container. The specific requirements of various Hazardous Waste Regulations in England and Wales and the Special Waste Regulations in Scotland should be adhered to, as appropriate.

Asbestos waste must be labelled:

(a) in accordance with the Carriage of Dangerous Goods and Use of Transportable Pressure Equipment Regulations 2004 where those Regulations apply

(b) where the Regulations in (a) do not apply, in accordance with schedule 2 of the Asbestos Regulations 2006.

The Licensed Contractors' Guide contains more detailed advice on waste handling.

21.7.23 *Prohibitions regulations 25, 26, 27, 28, 29*

Regulations 25 to 29 deal with the prohibitions of certain exposure to asbestos, supply and use of asbestos or any product containing asbestos. For details see the Regulations and ACOP.

21.7.24 *References*

➤ *Approved Code of Practice Work with Materials containing Asbestos.* L143, ISBN 9780 7176 6206 3.
➤ *Asbestos Essentials.* HSG 210, ISBN 9780 7176 1887 0 (Asbestos Essentials task sheets are available on the Asbestos Essentials area of the HSE website).

➤ *Asbestos Essentials Task Manual: Task Guidance Sheets for the Building Maintenance and Allied Trades.* HSG 210, HSE Books 2001, ISBN 9780 7176 1887 0.
➤ *Asbestos: The Analysts' Guide for Sampling, Analysis and Clearance Procedures.* HSG 248, ISBN 9780 7176 2875 2.
➤ *Asbestos: The Licensed Contractors Guide.* HSG 247, ISBN 9780 7176 2874 4.
➤ *The Management of Asbestos in Non-domestic Premises.* Approved Code of Practice and Guidance, L127, HSE Books 2006, ISBN 9780 7176 6209 8.

21.8 Chemicals (Hazard Information and Packaging for Supply) Regulations 2002

21.8.1 *Introduction*

CHIP is the Chemicals (Hazard Information and Packaging for Supply) Regulations 2002. The aim of CHIP is to ensure that people who are supplied with chemicals receive the information they need to protect themselves, others and the environment.

To achieve this CHIP obliges suppliers of chemicals to identify their hazards (e.g. flammability, toxicity) and to pass on this information together with advice on safe use to the people they supply the chemicals to. This is usually done by means of package labels and safety data sheets (SDSs).

CHIP applies to most chemicals but not all. The exceptions (which generally have Regulations of their own) are set out in regulation 3 of CHIP and include cosmetic products, medicinal products, foods, etc. CHIP 3, a new set of Regulations which consolidates and extends the previous ones, came into force on 24 July 2003.

CHIP is intended to protect people and the environment from the harmful effects of dangerous chemicals by making sure that users are supplied with information about the dangers. Many chemicals such as cosmetics and medicines are outside the scope of CHIP and have their own specific regulatory regimes. However, biocides and plant protection products, which have their own specific laws, have to be classified and labelled according to CHIP.

CHIP requires the supplier of a dangerous chemical to:

➤ **identify the hazards (dangers)** of the chemical (this is known as 'classification')
➤ **package the chemical safely** and

➤ **give information about the hazards** to their customers (usually by means of information on the package (e.g. a label) and, if supplied for use at work, an SDS).

These are known as supply requirements. 'Supply' is defined as making a chemical available to another person. Manufacturers, importers, distributors, wholesalers and retailers are examples of suppliers.

The Chemicals (Hazard Information and Packaging for Supply) (Amendment) Regulations 2005, entered into force on 31 October 2005. The regulations are known as CHIP 3.1. The Regulations bring into legal effect all the new entries, revisions, deletions and amendments to the classification and labelling requirements of hazardous substances set out in the 29th Adaptation to Technical Progress (29th ATP) to the Dangerous Substances Directive (European Commission Directive 2004/73/2004).

Table 21.4 outlines a description of commonly used terms. Figure 21.1 shows a summary of what needs to be done to comply.

21.8.2 *Classification – regulation 4*

The basic requirement for CHIP is for the supplier to decide whether the chemical is hazardous. CHIP with its Approved Classification and Labelling Guide (ACLG), sets out the rules for this. They tell the supplier how to:

➤ decide what kind of hazard the chemical has and
➤ explain the hazard by assigning a simple sentence that describes it (known as a 'risk phrase' or R–phrase). This is known as classification. Many commonly used substances have already been classified. They are contained in the CHIP Approved Supply List (ASL) which must be used.

21.8.3 *Information and labelling – regulations 5, 6, 8–10*

Information has to be supplied for the customers by a data sheet and a label on the package (unless the substance is provided in bulk such as a tanker or by pipeline). If the chemical is supplied for use at work an SDS must be provided. CHIP gives 16 headings for the SDS to set a standard for their quality.

Chip specifies what has to go on to the label including where it must be displayed, the size of the label, name and address of supplier, name of the substance, risk and safety phrases and indications of danger with symbols.

Table 21.4 Definitions of Commonly used terms in CHIP.

Category of danger	A description of hazard type
Classification	Precise identification of the hazard of a chemical by assigning a category of danger and a risk phrase using set criteria
Risk phrase (R-phrase)	A standard phrase which gives simple information about the hazards of a chemical in normal use
Safety phrase (S-phrase)	A standard phrase which gives advice on safety precautions which may be appropriate when using a chemical
Substance	A chemical element or one of its compounds, including any impurities
Preparation	A mixture of substances
Chemical	A generic term for substances and preparations
Tactile warning devices (TWDs)	A small raised triangle applied to a package intended to alert the blind and visually impaired to the fact that they are handling a container of a dangerous chemical
Child resistant fastenings (CRFs)	A closure which meets certain standards intended to protect young children from accessing the hazardous contents of a package
Chain of supply	The successive ownership of a chemical as it passes from manufacturer to its ultimate user

21.8.4 *Packaging of dangerous substance – regulation 7*

Packaging used for a chemical must be suitable. That means:

(a) the receptacle containing the dangerous substance or dangerous preparation is designed and constructed so that its contents cannot escape

(b) the materials constituting the packaging and fastenings are not susceptible to adverse attack by the contents or liable to form dangerous compounds with the contents

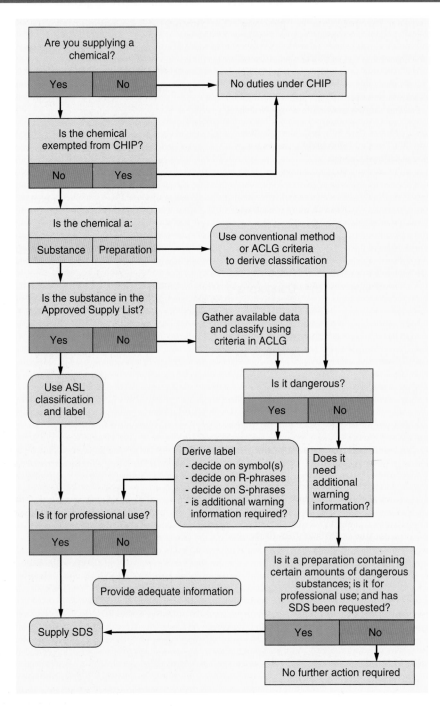

Figure 21.1 CHIP 3 summary of what needs to be done to comply.

(c) the packaging and fastenings are strong and solid throughout to ensure that they will not loosen and will meet the normal stresses and strains of handling and

(d) any replaceable fastening fitted to the receptacle containing the dangerous substance or dangerous preparation is designed so that the receptacle can be repeatedly refastened without the contents of the receptacle escaping.

21.8.5 *Child-resistant fastenings, tactile warnings and other consumer protection measures – regulation 11*

CHIP sets out special requirements for chemicals that are sold to the public.

Some have to be fitted with a child-resistant closure to a lay-down standard to prevent young children opening containers and swallowing the contents.

462

Some must have a tactile danger warning to alert the blind and partially sighted to the danger. This often a raised triangle.

21.8.6 *Retention of data – regulation 12*

Data used for classification, labelling, child-resistant fastners, and for preparing the SDS must be kept for at least 3 years after the dangerous chemical is supplied for the last time.

21.8.7 *REACH*

What is REACH?

REACH is a new EC regulation concerning the Registration, Evaluation, Authorization and restriction of CHemicals. It came into force on 1 June 2007 and replaces a patch-work of European Directives and Regulations with a single system.

DEFRA is responsible for implementing this European Regulation in the United Kingdom.

Competent Authority

Many European regulatory systems are operated at the national ('Member State') level by a Competent Authority in each Member State. In October 2006, DEFRA nominated HSE to be the **UK Competent Authority** for REACH, working closely with the Environment Agency, and other partners to manage certain key aspects of the REACH system in the UK.

The Competent Authority's responsibilities under REACH will be to:

➤ provide advice to manufacturers, importers, down-stream users and other interested parties on their respective responsibilities and obligations under REACH (Competent Authorities' help desks)
➤ conduct substance evaluation of prioritized substances and prepare draft decisions
➤ propose harmonized Classification and Labelling for CMRs and respiratory sensitizers
➤ identify substances of very high concern for authorization
➤ propose restrictions
➤ nominate candidates to membership of ECA committees on Risk Assessment and Socio-economic Analysis
➤ appoint members for the Member State Committee to resolve differences of opinion on evaluation decisions
➤ appoint a member to the Forum for Information Exchange and meet to discuss enforcement matters

➤ provide adequate scientific and technical resources to the members of the Committees that they have nominated
➤ work closely with the European Chemical Agency in Helsinki.

Aims

REACH has several aims:

➤ to provide a high level of protection of human health and the environment from the use of chemicals
➤ to allow the free movement of **substances** on the EU market
➤ to make the people who place chemicals on the market (**manufacturers** and **importers**) responsible for understanding and managing the risks associated with their use
➤ to promote the use of alternative methods for the assessment of the hazardous properties of substances
➤ to enhance innovation in and the competitiveness of the EU chemicals industry.

No data, no market

A major part of REACH is the requirement for manufacturers or importers of substances to register them with a central **European Chemicals Agency** (ECHA). A registration package will be supported by a standard set of data on that substance. The amount of data required is proportionate to the amount of substance manufactured or supplied.

If you do not register your substances, then the data on them will not be available and as a result, you will no longer be able to manufacture or supply them legally (i.e. no data, no market)!

Scope and exemptions

REACH applies to substances manufactured or imported into the EU in quantities of 1 tonne per year or more. Generally, it applies to all individual chemical substances on their own, in preparations or in articles (if the substance is intended to be released during normal and reasonably foreseeable conditions of use from an article).

Some substances are specifically excluded:

➤ radioactive substances
➤ substances under customs supervision
➤ substances being transported
➤ non-isolated intermediates
➤ waste
➤ some naturally occurring low-hazard substances.

Some substances, covered by more specific legislation, have tailored provisions, including:

➤ human and veterinary medicines
➤ food and foodstuff additives
➤ plant protection products and biocides
➤ isolated intermediates
➤ substances used for research and development.

Pre-registration

It is estimated that there are around 30,000 substances on the European Market in quantities of 1 tonne or more per year. Registering all of these at once would be a huge task for both industry and regulators. To overcome this, the registration of those substances already being manufactured or supplied is to take place in three phases. These phases are spread over 11 years. To benefit from these provisions manufacturers or suppliers should pre-register their substances between 1 June and 30 November 2008.

Once pre-registered the ECHA will identify who is intending to register the same substance and put them in contact with each other. The potential registrants can then come together and form a 'Substance Information Exchange Forum' (SIEF) where they can negotiate sharing their available data and the costs of generating any new data.

One Substance, One Registration

This is the principle that for any one substance, a single set of data is produced that is shared by all those companies that manufacture or supply that substance. The details of how this sharing is arranged (costs, etc.) and business specific (e.g. company name) and business sensitive (e.g. how it is used) information is submitted separately by each company.

Registration

Registration is a requirement on industry (manufacturers/ suppliers/importers) to collect and collate specified sets of information on the properties of those substances they manufacture or supply. This information is used to perform an assessment of the hazards and risks that a substance may pose and how those risks can be controlled. This information and its assessment is submitted to the ECHA in Helsinki.

Chemicals already existing (those on EINECS or manufactured in the EU prior to entry into force of REACH) are known as 'phase-in' substances under REACH. These will be registered in three phases according to their tonnage and/or hazardous properties.

➤ Phase 1 – substances supplied at ⩾1,000 tonnes per year; substances classified under CHIP as very toxic

to aquatic organisms supplied at ⩾100 tonnes per year; substances classified under CHIP as Category 1 or 2 carcinogens, mutagens or reproductive toxicants supplied at ⩾1 tonne per year; substances classified as very toxic to aquatic organisms must be registered in the first 3 years (by 1 December 2010).
➤ Phase 2 – substances supplied at ⩾100 tonnes per year must be registered in the first 6 years (by 1 June 2013).
➤ Phase 3 – substances supplied at ⩾1 tonne per year must be registered in the first 11 years (by 1 June 2018).

A substance can be registered at any time prior to these deadlines.

Non-phase-in substances (i.e. those not on EINECS, those which have never been manufactured previously or not pre-registered) will be subject to registration 1 June 2008. Until then, the current Notification of New Substances (NONS) Regulations continue to apply. For further details on registration please see the specific ECHA webpage or detailed guidance document.

Evaluation

Registration packages (dossiers) submitted under REACH can be evaluated for:

➤ **Compliance Check:** A quality/accuracy check of the information submitted by the industry.
➤ **Dossier Evaluation:** A check that an appropriate testing plan has been proposed for substances registered at the higher tonnage levels (⩾100 tonnes/annum)
➤ **Substance evaluation:** An evaluation of all the available data on a substance, from all registration dossiers. This is done by national Competent Authorities on substances that have been prioritized for potential regulatory action because of concerns about their properties or uses.

Substances of Very High Concern

Some substances have hazards that have serious consequences, e.g. they cause cancer, or they have other harmful properties and remain in the environment for a long time and gradually build up in animals. These are 'substances of high concern'. One of the aims of REACH is to control the use of such substances.

Authorization

Authorization is a feature of REACH that is new to the area of general chemicals management. As REACH progresses, a list of 'substances of very high concern' will be created. Substances on this list cannot be supplied or used unless an authorization has been granted. A company wishing to market or use such a substance must apply to the ECHA in Helsinki for an authorization, which may be granted or refused.

Restrictions

Any substance that poses a particular threat can be restricted. Restrictions take many forms, for example, from a total ban to not being allowed to supply it to the general public. Restrictions can be applied to any substance, including those that do not require registration. This part of REACH takes over the provisions of the Marketing and use Directive.

Classification and labelling

An important part of chemical safety is clear information about any hazardous properties a chemical has. The classification of different chemicals according to their characteristics (for example, those that are corrosive, or toxic to fish) currently follows an established system, which is reflected in REACH. Over the next few years, work will be underway to establish in the EU a classification and labelling system based on the United Nations Globally Harmonized System, or GHS. REACH has been written with GHS in mind.

Information in the supply chain

The passage of information up and down the supply chain is a key feature of REACH. Users should be able to understand what manufacturers and importers know about the dangers involved in using chemicals and how to control risks. However, in order for suppliers to be able to assess these risks they need information from the downstream users about how these chemicals are used. REACH provides a framework in which information can be passed both up and down supply chains.

REACH adopts and builds on the previous system for passing information – the SDS. This should accompany materials down through the supply chain, providing the information users need to ensure that chemicals are safely managed. In time these SDSs will include information on safe handling and use.

More detailed information about REACH can be found at the *DEFRA REACH website*, and HSE website, including impact assessment work at the national and Community levels, and consultations. A wide range of industry and commercial sources also offers commentary and advice regarding REACH.

Enforcement

Enforcing this very wide-ranging new system presents new challenges to regulators across Europe. REACH places new duties on a range of different businesses. Mostly, the new duties will be on manufacturers and importers of chemicals, but there are also requirements for downstream users of chemicals to share information with their suppliers. Although HSE will play a key role in enforcing REACH, both as the UK Competent Authority and more generally as the UK occupational health and safety regulator, enforcing REACH will fall to a number of regulatory bodies. HSE are working closely with other regulators to support DEFRA in setting up the framework for enforcing REACH in the United Kingdom.

21.8.8 *References*

Approved Classification and Labelling Guide. 5th edition, L131, HSE Books 2002, ISBN 9780 7176 2369 6.

Approved Supply List. Information Approved for the Classification and Labelling of Dangerous Substances and Preparations for Supply. 8th edition, L142, HSE Books 2005, ISBN 9780 7176 6138 5.

Chemical (Hazard Information and Packaging for Supply) Regulations 2002. SI No. 2002, 1689, The Stationery Office.

Chip for Everyone. HSG 228, HSE Books 2002, ISBN 9780 7176 2370 X.

Read the Label. INDG 352, HSE Books 2002, ISBN 9780 7176 2366 1.

The Compilation of Safety Data Sheets. Approved Code of Practice, 3rd edition, L130 HSE Books 2002, ISBN 9780 7176 2371 8.

The Idiot's Guide to CHIP. INDG 350, HSE Books 2002, ISBN 9780 7176 2333 5.

Why Do I Need a Safety Data Sheet? INDG 353, HSE Books 2002, ISBN 9780 7176 2367 X.

a. ECHA website: http://reach.jrc.it/navigator_en.htm.
b. HSE's CHIP website: http://www.hse.gov.uk/chip.
c. HSE's REACH website: http://www.hse.gov.uk/reach.

21.9 Confined Spaces Regulations 1997

21.9.1 *Introduction*

These Regulations concern any work which is carried on in a place which is substantially (but not always entirely) enclosed, where there is a reasonably foreseeable risk of serious injury from conditions and/or hazardous substances in the space or nearby. Every year about 15 people are killed and a number seriously injured across a wide range of industries ranging from simple open top pits to complex chemical plants. Rescuers without proper training and equipment, often become the victims.

21.9.2 *Definitions*

Confined space – means any place, including any chamber, tank, vat, silo, pit, trench, pipe, sewer, flue, well or similar space in which, by virtue of its enclosed nature, there arises a reasonably foreseeable specified risk.

Specified risk – means a risk to any person at work of:

- serious injury arising from a fire or explosion
- loss of consciousness arising from an increase in body temperature
- loss of consciousness or asphyxiation arising from gas, fume, vapour or the lack of oxygen
- drowning arising from an increase in the level of liquid
- asphyxiation arising from a free-flowing solid or because of entrapment by it.

21.9.3 *Employers' duties – regulation 3*

Duties are placed on employers to:

- comply regarding any work carried out by employees and
- ensure, SFARP, that other persons (e.g. use competent contractors) comply regarding work in the employer's control.

The self-employed also have duties to comply.

21.9.4 *Work in confined space – regulation 4*

1. No person at work shall enter a confined space for any purpose unless it is not reasonably practicable to achieve that purpose without such entry
2. Other than in an emergency, no person shall enter, carry out work or leave a confined space otherwise than in accordance with a safe system of work, relevant to the specified risks.

21.9.5 *Risk assessment*

Risk assessment is an essential part of complying with these Regulations and must be done (under the Management of Health and Safety at Work Regulations 1999) to determine a safe system of work. The risk assessment needs to follow a hierarchy of controls to comply. This should start with the measures, both in design and procedures, that can be adopted to enable any work to be carried out outside the confined space.

The assessment must be done by a competent person and will form the basis of a safe system of work. This will normally be formalized into a specific permit-to-work, applicable to a particular task. The assessment will involve:

(a) The general conditions to assess what may or may not be present. Consider:
- what have been the previous contents of the space
- residues that have been left in the space, for example, sludge, rust, scale and what may be given off if they are disturbed
- contamination which may arise from adjacent plant, processes, services, pipes or surrounding land, soil or strata
- oxygen deficiency and enrichment – there are very high risks if the oxygen content differs significantly from the normal level of 20.8%. If it is above this level increased flammability exists; if it is below then impaired mental ability occurs, with loss of consciousness under 16%
- that physical dimensions and layout of the space can affect air quality.

(b) Hazards arising directly from the work being undertaken. Consider:
- the use of cleaning chemicals and their direct effect or interaction with other substances
- sources of ignition for flammable dusts, gases, vapours, plastics and the like
- the need to isolate the confined space from outside services or from substances held inside such as liquids, gases, steam, water, fire extinguishing media, exhaust gases, raw materials and energy sources
- the requirement for emergency rescue arrangements including trained people and equipment.

21.9.6 *Safe system of work*

The detailed precautions required will depend on the nature of the confined space and the actual work being carried out. The main elements of a safe system of work which may form the basis of a 'permit-to-work' are:

- the type and extent of supervision
- competence and training of people doing the work
- communications between people inside, from inside to outside and to summon help
- testing and monitoring the atmosphere for hazardous gas, fume, vapour, dust, etc. and for concentration of oxygen
- gas purging of toxic or flammable substances with air or inert gas, such as nitrogen
- good ventilation, sometimes by mechanical means
- careful removal of residues using equipment which does not cause additional hazards

➤ effective isolation from gases, liquids and other flowing materials by removal of pieces of pipe, blanked off pipes, locked off valves

➤ effective isolation from electrical and mechanical equipment to ensure complete isolation with lock off and a tag system with key security. Need to secure against stored energy or gravity fall of heavy presses, etc.

➤ if it is not possible to make the confined space safe, the provision of PPE and RPE

➤ supply of gas via pipes and hoses – should be carefully controlled

➤ access and egress to give quick unobstructed and ready access and escape

➤ fire prevention

➤ lighting, including emergency lighting

➤ prohibition of smoking

➤ emergencies and rescue

➤ limiting of working periods and the suitability of individuals.

21.9.7   *Emergency arrangements – regulation 5*

Before people enter a confined space, suitable and sufficient rescue arrangements must be set up. These must:

➤ reduce the risks to rescuers SFARP

➤ include the provision and maintenance of suitable resuscitation equipment designed to meet the specified risks.

To be suitable and sufficient, arrangements will need to cover:

➤ rescue and resuscitation equipment

➤ raising the alarm, alerting rescue and watch keeping

➤ safeguarding the rescuers

➤ fire safety precautions and procedures

➤ control of adjacent plant

➤ first aid arrangements

➤ notification and consultation with emergency services

➤ training of rescuers and simulations of emergencies

➤ size of access openings to permit rescue with full breathing apparatus, harnesses, fall arrest gear and lifelines, which is the normal suitable respiratory protection and rescue equipment for confined spaces.

21.9.8   *Training*

Specific, detailed and frequent training is necessary for all people concerned with dealing with confined spaces, whether they are acting as rescuers or watchers or those carrying out the actual work inside the confined space.

The training will need to cover all procedures and the use of equipment under realistic simulated conditions.

21.9.9   *References*

Safe Work in Confined Spaces. Approved Code of Practice, Regulations and Guidance. L101, HSE Books 1997, ISBN 9780 7176 1405 0.

The Selection, Use and Maintenance of Respiratory Protective Equipment: A Practical Guide. HSG 53, HSE Books 2004, ISBN 9780 7176 2904 X.

21.10   Construction (Design and Management) (CDM) Regulations 2007

21.10.1   *Background and introduction*

The following is extracted from the new ACOP.

The revised Construction (Design and Management) Regulations 2007 (CDM 2007) came into force on 6 April 2007. They replace the Construction (design and Management) Regulations 1994 (CDM 94) and the Construction (Health, Safety and Welfare) Regulations 1996 (CHSW). The new ACOP replaces the ACOP under CDM 94.

The key aim of CDM 2007 is to integrate health and safety into the management of the project and to encourage everyone involved to work together to:

➤ improve the planning and management of projects from the very start

➤ identify risks early on so they can be eliminated or reduced at the design or planning stage and the remaining risks can be properly managed

➤ target effort where it can do the most good in terms of health and safety

➤ discourage unnecessary bureaucracy.

The Regulations are intended to focus attention on planning and management throughout construction projects, from design concepts onwards. The aim is for health and safety considerations to be treated as an essential, but normal part of a project's development – not an afterthought or bolt-on extra.

The effort devoted to planning and managing health and safety should be in proportion to the risks and complexity associated with the project. The focus should always be on action necessary to reduce and manage risks. Any paperwork produced should help with communication and risk management. Paperwork which adds little to the management of risk is a waste of effort, and can be a dangerous distraction from the real business of risk reduction and management.

Time and thought invested at the start of the project will pay dividends not only in improved health and safety, but also in:

➤ reductions in the overall cost of ownership, because the structure is designed for safe and easy maintenance and cleaning work, and because key information is available in the health and safety file
➤ reduced delays
➤ more reliable costings and completion dates
➤ improved communication and co-operation between key parties and
➤ improved quality of the finished product.

Note

Typical operating and owning cost of a building are in the ratio:

➤ 1 for construction costs
➤ 5 for maintenance and building operating costs
➤ 200 for business operating costs.

21.10.2 *Outline of Regulations, Application and Notification*

The Regulations are divided into 5 parts. Part 1 of the Regulations deals with matters of interpretation and application. The Regulations apply to all construction work in Great Britain and its territorial waters, and apply to both employers and self-employed without distinction.

Part 2 covers general management duties which apply to all construction projects, including those which are non-notifiable.

Part 3 sets out additional management duties which apply to projects above the notification threshold (projects lasting more than 3 days, or involving more than 500 person-days of construction work). These additional duties require particular appointments or particular documents which will assist with the management of health and safety from concept to completion.

Part 4 of the Regulations apply to all construction work carried out on construction sites, and covers physical safeguards which need to be provided to prevent danger. Duties to achieve these standards are held by contractors who actually carry out the work, irrespective of whether they are employers or are self-employed. Duties are also held by those who do not engage in construction work themselves, but control the way in which

the work is done. In each case, the extent of the duty is in proportion to the degree of control which the individual or organization has over the work in question.

This does not mean that everyone involved with design, planning or management of the project legally must ensure that all the specific requirements in Part 4 are complied with. They only have duties if, in practice, they exercise sufficient control over the actual working methods, safeguards and site conditions. If for example a client specifies that a particular job is done in a particular way then the client will have a duty to make sure that their instructions comply with the requirements.

Contractors must not allow work to start or continue unless the necessary safeguards are in place. Whether he is providing the safeguards or using something, for example a scaffold, supplied by someone else.

Part 5 of the Regulations covers issues of civil liability; transitional provisions which will apply during the period when the Regulations come into force; and amendments and revocations of other legislation.

The distinction in the CDM94 Regulations between their application and notification of projects was confusing. For the purposes of CDM 2007 Regulations, there should only be two types of construction projects: notifiable and non-notifiable. All of the requirements apply to notifiable projects, but the requirements relating to appointments, plans and other paperwork would not apply to non-notifiable projects.

Except where the project is for domestic clients, the HSE must be notified of projects if the construction phase is likely to involve more than (a) 30 days, or (b) 500 person-days, of construction work. An F10 (rev) may be used and is available from local HSE offices or can be completed on line. This form does not have to be used as long as all the particulars shown in schedule 1 (see Table 21.5) are provided.

21.10.3 *Definition of Construction – regulation 2*

Regulation 2 defines construction work as the following; 'construction work' means the carrying out of any building, civil engineering or engineering construction work and includes:

(a) the construction, alteration, conversion, fitting out, commissioning, renovation, repair, upkeep, redecoration or other maintenance (including cleaning which involves the use of water or an abrasive at high pressure or the use of corrosive or toxic substances), decommissioning, demolition or dismantling of a structure
(b) the preparation for an intended structure, including site clearance, exploration, investigation (but not site survey) and excavation, and the clearance or preparation of the site or structure for use or occupation at its conclusion

Table 21.5 Schedule 1 of CDM 2007 Particulars to be Notified to the HSE

SCHEDULE 1
Particulars to be notified to the executive
1. Date of forwarding.
2. Exact address of the construction site.
3. The name of the local authority where the site is located.
4. A brief description of the project and the construction work which it includes.
5. Contact details of the client (name, address, telephone number and any e-mail address).
6. Contact details of the CDM co-ordinator (name, address, telephone number and any e-mail address).
7. Contact details of the principal contractor (name, address, telephone number and any e-mail address).
8. Date planned for the start of the construction phase.
9. The time allowed by the client to the principal contractor referred to in regulation 15(b) for planning and preparation for construction work.
10. Planned duration of the construction phase.
11. Estimated maximum number of people at work on the construction site.
12. Planned number of contractors on the construction site.
13. Name and address of any contractor already appointed.
14. Name and address of any designer already engaged.
15. A declaration signed by or on behalf of the client that he is aware of his duties under these Regulations.

(c) the assembly on site of prefabricated elements to form a structure or the disassembly on site of prefabricated elements which, immediately before such disassembly, formed a structure

(d) the removal of a structure or of any product or waste resulting from demolition or dismantling of a structure or from disassembly of prefabricated elements which immediately before such disassembly formed such a structure and

(e) the installation, commissioning, maintenance, repair or removal of mechanical, electrical, gas, compressed air, hydraulic, telecommunications, computer or similar services which are normally fixed within or to a structure.

This means that there are some things which may take place on a construction site like tree planting and horticulture work, putting up marquees, positioning lightweight panels in an open plan office and surveying which are not construction work.

Domestic clients are people who have work done on their own home or the home of a family member, that does not relate to a trade or business whether for profit or not. Where an insurance company arranges construction work to be done on a domestic dwelling the insurance company becomes the client for the purposes of CDM. If however the insured arranges the work and is reimbursed by the insurance company then the insured is the client.

As with all projects, designers and contractors working for domestic clients will have to be competent and take reasonable steps to ensure that anyone they arrange for, or instruct, to manage design or construction work is also competent. They will also have to co-operate with others involved in the project to safeguard the health and safety of everyone involved. When preparing or modifying a design, designers will have to avoid risks to the health and safety of anyone constructing, maintaining, using or demolishing the structures concerned, by removing the hazards (and reducing the risks arising from any that remain). Contractors will have to plan, manage and monitor their own work; and ensure that there are suitable welfare facilities.

21.10.4 *General Duties of Clients, CDM Co-ordinators, Principal Contractors and Contractors – Part 2*

The Regulations require that appointees are competent and state that:

No person on whom these Regulations place a duty shall:

(a) appoint or engage a CDM co-ordinator, designer, principal contractor or contractor unless he has taken reasonable steps to ensure that the person to be appointed or engaged is competent

(b) accept such an appointment or engagement unless he is competent

(c) arrange for or instruct a worker to carry out or manage design or construction work unless the worker is:
  (i) competent or
  (ii) under the supervision of a competent person.

The ACOP goes into considerable detail about the assessment of organizational and individual competence. In both cases it lays down a two stage approach.

➤ Stage 1 – assessing a company's organization and arrangements for health and safety and an individual's task knowledge to assess whether they appreciate the risks

➤ Stage 2 – assessing the company's and individual's experience and track record to see if they are capable of doing the work involved.

To be competent, an organization or individual must have:

➤ sufficient knowledge of the specific tasks to be undertaken and the risks which the work will entail

➤ sufficient experience and ability to carry out their duties in relation to the project; to recognize their limitations and take appropriate action in order to prevent harm to those carrying out construction work, or those affected by the work.

The Regulations require all duty holders to co-operate and co-ordinate their work with one another to enable people to carry out their duties effectively.

Duties are placed on a number of people, which is summarized in Table 9.9 and set out as follows.

All projects require:

➤ non-domestic clients to check the competence of all their appointees; ensure that there are suitable management arrangements for the project; allow sufficient time and resources for all stages; provide pre-construction information to designers and contractors

➤ designers to eliminate hazards and reduce risks during design; and provide information about remaining risks

➤ contractors to plan, manage and monitor their own work and that of workers; check the competence of all their appointees and workers; train their own employees; provide information to their workers; comply with the requirements for health and safety on site detailed in Part 4 of the Regulations and other Regulations such as the Work at Height Regulations; and ensure that there are adequate welfare facilities for their workers

➤ everyone to assure their own competence; co-operate with others and co-ordinate work so as to ensure the health and safety of construction workers and others who may be affected by the work; report obvious risks; take account of the general principles of prevention in planning or carrying out construction work; and comply with the requirements in schedule 3, Part 4 of CDM 2007 and other Regulations for any work under their control.

To ensure the revised Regulations are proportionate to risk and the needs of small businesses, and to minimize bureaucracy, the CDM94 requirement for appointment of a Planning Supervisor and Principal Contractor and written plans for projects involving five or more workers has been withdrawn. This does not mean any lessening in the health and safety standards required by the CDM 2007 Regulations, as they have strengthened or introduced other requirements. These place the emphasis on risk management, while avoiding disproportionate bureaucracy for smaller projects.

21.10.5 *Additional duties where the project is notifiable*

As well as the above requirements under 20.10.4, a notifiable project requires:

➤ non-domestic clients to appoint a CDM co-ordinator; appoint a Principal Contractor; provide pre-construction information to CDM co-ordinator

➤ check (before construction work starts) that there is a construction phase plan and suitable welfare facilities; and retain and provide access to the health and safety file

➤ CDM co-ordinators to advise and assist clients with their duties; notify HSE; co-ordinate health and safety aspects of design work; provide pre-construction information to designers and contractors; facilitate good communications between client, designers and contractors; liaise with the Principal Contractor on ongoing design issues; prepare and update the health and safety file

➤ designers to check, before they start work, that clients are aware of their duties and a CDM co-ordinator has been appointed; check HSE has been notified; and provide any information needed for the health and safety file

➤ PCs to plan, manage and monitor the construction phase in liaison with contractors; prepare, develop and implement a written plan (the initial plan to be completed before the construction phase begins); give contractors relevant parts of the plan; make sure suitable welfare facilities are provided from the start and maintained throughout the construction phase; check the competence of all their appointees; provide site inductions; consult with the workers; liaise with the CDM co-ordinator on ongoing design issues; and secure the site

➤ contractors to confirm clients are aware of their duties and a CDM co-ordinator has been appointed; co-operate with the Principal Contractor in planning and managing work; check HSE has been notified; and provide any information needed for the health and safety file; inform Principal Contractor of problems with plan; inform Principal Contractor of any reportable accidents, diseases and dangerous occurrences.

21.10.6 *Roles of duty holders under CDM 2007*

Client's

The role of clients is one of the most difficult areas to cover in law, because of the vast range of interest and

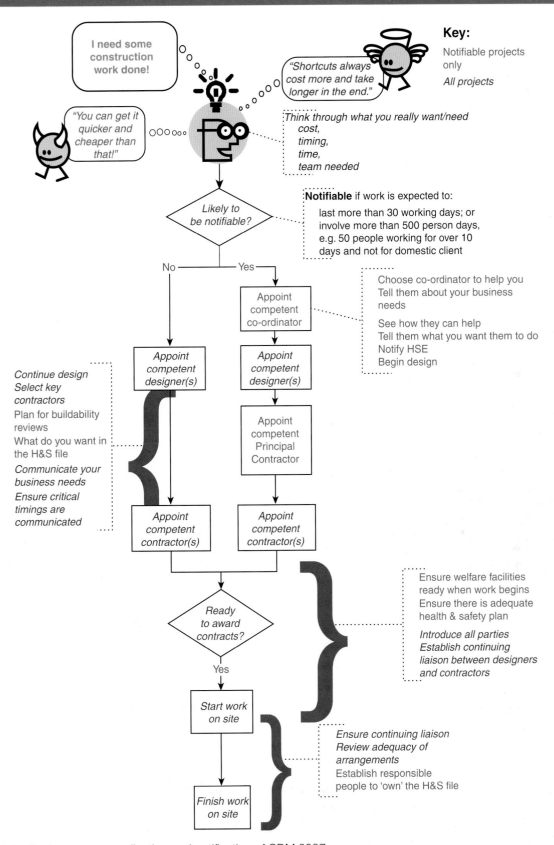

Figure 21.2 Draft summary application and notification of CDM 2007.

expertise in construction, and clients have real questions as to why they should be involved. The HSC are very conscious of the substantial influence and control that clients exert over construction projects in practice. For example they:

➤ set the tone for projects
➤ control contractual arrangements
➤ make crucial decisions (e.g. budget, time, suitability of designs); select procurement method and construction team/supply chain
➤ may have essential information about site/building.

There is a new duty on clients to ensure that suitable project management arrangements for health and safety are in place.

Clients are not expected to develop these arrangements themselves and few have the expertise to do so. They should be able to rely on the advice and support of their construction team and, in particular, the CDM co-ordinator. What the HSC expect is for them to exercise their influence and control responsibly and with due regard for those who will construct, maintain and demolish the structure.

CDM Co-ordinator

There is widespread agreement that the role of Planning Supervisor, as developed under CDM94, has not proved as effective as intended. However, views on this tend to be highly polarized. The main problems are that Planning Supervisers:

➤ are not seen as a natural part of the construction team. To be effective they need to be better integrated with the rest of the design and construction team
➤ are often appointed too late in the project so that they cannot do their job
➤ frequently have to operate at a disadvantage, due to insufficient allocation of resources by the client, in terms both of money and time
➤ have no authority to carry out their duties unless the client effectively empowers them and others co-operate and
➤ have, fairly or not, become the scapegoat for the bureaucracy linked to CDM.

To address these points CDM 2007:

➤ creates a new function – the CDM co-ordinator – to advise and assist the client
➤ places responsibility on clients to ensure that the co-ordinator's duties are carried out – only they have the information and authority to empower the CDM co-ordinator

➤ explicitly requires the CDM co-ordinator to be appointed before design work starts, with corresponding duties on designers and contractors not to begin work unless a CDM co-ordinator has been appointed and
➤ requires the client to ensure that the arrangements for managing projects include the allocation of adequate resources (including time).

Designers

The HSC fully recognize that, as well as health and safety considerations, designers need to take into account issues such as aesthetics, buildability, and cost. The challenge is to ensure that health and safety considerations are not outweighed by aesthetic and commercial priorities and, conversely, that health and safety does not inhibit aesthetics. However, it is a truth, almost universally acknowledged, that designers have considerable potential to eliminate hazards and reduce risks associated with construction work, as well as those associated with building use, maintenance, cleaning, and eventual demolition.

As part of balancing their design priorities, designers must take positive steps to use that potential and pay sufficient regard to health and safety in their designs to ensure that in the construction, use, maintenance and demolition of the resulting structures, hazards are removed **where possible** and any remaining risks reduced. Although this is already stated in the draft guidance, the HSC have explicitly acknowledged the need for such balanced decisions in CDM 2007.

The HSC also want the designers to focus on how their decisions are likely to affect those constructing, maintaining, using or demolishing the structures that they have designed and what they can do, **in the design**, to remove the hazards, for example by not specifying hazardous materials and avoiding the need for processes that create hazardous fumes, vapours, dust, noise or vibration, and reduce the resulting risks where the hazard cannot be removed. The ACOP under CDM 94 and guidance already sets out most of this and there is no plan to change these standards.

CDM 2007 is intended to require designers to eliminate hazards where they can, and then reduce those risks which remain. It does not ask designers to minimize all risks, as the HSC do not expect structures to be restricted to a height of 1 m! There are also often too many variables and no obvious *safest* design.

The duty regarding maintenance is limited in CDM94 to structural matters, but it is important that designers also consider safety during routine maintenance that is affected by their designs (e.g. how are high-level lights and ventilation systems to be maintained?).

Designers had no duty under CDM94 to ensure that their designs are safe to use. However, occupiers of workplaces have to ensure that the finished structure complies with other health and safety law, particularly the Workplace Regulations (Health, Safety and Welfare) 1992. To ensure that these issues are addressed at the design stage the CDM 2007 extends designers' duties for fixed workplaces (e.g. offices, shops, schools, hospitals and factories) to cover safe use. Com-petent designers should be doing this already – so this is likely to require minimal additional work in practice.

Principal Contractor

The role of Principal Contractor, introduced when CDM94 came into force, was built on the longstanding role of main or managing contractor and did not, there-fore, require any substantial changes in industry prac-tice. Because of this, as a role, it has worked fairly well since CDM94 came into force, and has not changed sig-nificantly in CDM 2007.

The only substantial change is to make explicit, in the Regulations, the Principal Contractor's key role in managing the construction phase, to ensure that it is carried out, SFARP, safely and without risk to health. This does not mean that the Principal Contractor has to manage the work of contractors in detail – that is the contractor's own responsibility. They do have to make sure that they themselves are competent to address the health and safety issues likely to be involved in the man-agement of the construction phase; satisfy themselves that the designers and contractors that they engage are competent and adequately resourced; and ensure that the construction phase is properly planned, managed and monitored, with adequately resourced, competent site management appropriate to the risk and activity.

The HSC did not feel that it was necessary, legally or otherwise, to specify in CDM 2007 that the Principal Contractor must be a contractor. In over 90% of projects, contractors discharge the role of Principal Contractor and those with contractors' experience and expertise are most likely to have the competence and resources to manage the work. The HSC believed that few clients have the competence or resources to manage significant construction work and do not want to encourage them to do so, although there is nothing to prevent this if they are competent – which is most likely in simple, low-risk projects.

Contractors

The only substantive change regarding contractors, is to make explicit their own duty to plan, manage and moni-tor their own work. The intention is that the management duties on PCs and contractors should complement one another, with the contractor's duty focusing on their own work and the Principal Contractor's on the co-ordination of the work of the various contractors.

21.10.7 *Health and Safety File*

Under the CDM94 a separate Health and Safety File was required for each project. The HSC believed that it would be more useful to have one file for each site, structure or, occasionally, group of structures – e.g. bridges along a road. The file can then be developed over time as informa-tion is added from different projects. The CDM co-ordinator therefore has the responsibility to prepare a suitable health and safety file or update it if one already exists.

There are also opportunities to link the health and safety file with other documents such as the Buildings Regulations Log Book. The potential practical value of the information contained in the file is also likely to increase as more clients make use of the Internet to share this information with designers and contractors. (For example, maintenance contractors could check what access equipment and parts they are likely to need to repair a fault before leaving for the site, saving them and their clients inconvenience, time and money.)

The contents of the Health and Safety File are given in the ACOP as follows:

➤ a brief description of the work carried out
➤ any residual hazards which remain and how they have been dealt with (e.g. surveys or other informa-tion concerning asbestos; contaminated land; water bearing strata; buried services)
➤ key structural principles (e.g. bracing, sources of substantial stored energy – including pre- or post-tensioned members) and safe working loads (SWL) for floors and roofs, particularly where these may preclude placing scaffolding or heavy machinery there
➤ hazardous material used (e.g. lead paint, pesticides, special coatings which should not be burnt off)
➤ information regarding the removal or dismantling of installed plant and equipment (e.g. any spe-cial arrangements for lifting, order or other special instructions for dismantling)
➤ health and safety information about equipment pro-vided for cleaning or maintaining the structure
➤ the nature, location and markings of significant services, including underground cables; gas supply equipment; firefighting services
➤ information and built drawings of the structure, its plant and equipment (e.g. the means of safe access to and from service voids, fire doors and compartmentalization).

21.10.8 *Part 4 – Health and Safety on Construction Sites*

The Constructional (Health, Safety and Welfare) 1996 have been revoked by CDM 2007. Their requirements, without the work at height provisions (regulations 6, 7 and 8), form the basis of Part 4 and schedule 2 (Welfare facilities) of the CDM 2007 Regulations. The revision is mainly intended to simplify and clarify the wording of the Regulations, without making substantive changes to what is expected in practice. One substantive change, however, has been to broaden the duty regarding explosives to cover the important issues of security and safety of storage and transport, as well as safety in use.

However, some issues already covered by the Workplace Regulations (Health,Safety and Welfare) 1992 are particularly important in construction. These include traffic management and lighting. Because these are so important CDM 2007 duplicate aspects of the Workplace Regulations (Health,Safety and Welfare) 1992 requirements, although this was not legally necessary.

The issues covered by Part 4 apply to all construction projects and are as follows.

General – regulation 25

Duties to comply with this part are placed on contractors or other person who control the way in which construction work is carried on.

Duty placed on workers to report any defect they are aware of may endanger the health and safety of themselves or another person.

Safe place of Work – regulation 26

This regulation requires SFARP:

➤ safe access and egress to places of work
➤ maintenance of access and egress
➤ that people are prevented from gaining access to unsafe access or workplaces
➤ provision of safe places of work with adequate space and suitable for workers.

Good order and site security – regulation 27

➤ requires that all parts of a construction be kept, SFARP, in good order and in a reasonable state of cleanliness
➤ requires that where necessary in the interests of health and safety the site's perimeter should be identified, signed and fenced off
➤ requires that no timber or other materials with projecting nails or similar sharp objects shall be used or allowed to remain if they could be a source of danger.

Stability of structures – regulation 28

All practical steps must be taken to ensure that:

➤ any new or existing structure which may be or become weak does not collapse accidentally

➤ temporary supporting structures are designed, installed and maintained so as to withstand all foreseeable loads
➤ a structure is not overloaded so as to be a source of danger.

Demolition or dismantling – regulation 29

➤ demolition or dismantling must be planned and carried out so as to prevent danger or reduce it to as lower a level as is reasonably possible
➤ arrangements must be recorded before the work takes place.

Explosives – regulation 30

➤ explosives, SFARP, shall be stored, transported and used safely and securely
➤ explosives may only be fired if suitable and sufficient steps have been taken to ensure no one is exposed to risk of injury.

Excavations – regulation 31

All practicable steps shall be taken to prevent danger to people, including where necessary supports or battering to:

➤ prevent excavations from collapsing accidentally
➤ prevent material from side, roof or adjacent area being dislodged or falling
➤ prevent a person from being buried or trapped by a fall of material
➤ prevent persons, work equipment or any accumulation of material falling into the excavation
➤ prevent any part of an excavation or ground adjacent being overloaded by work equipment or material.

No construction work may be carried out in an excavation where any supports or battering have been provided unless:

➤ it has been inspected by a competent person:
  ➤ at the start of the shift (only one report required every 7 days)
  ➤ after any event likely to affect the strength or stability of the excavation
  ➤ after any material accidentally falls or is dislodged
➤ the person inspecting is satisfied that it is safe.

When the person who carries out the inspection has so informed the person on whose behalf he carried out the inspection that it is not safe, work must cease until the relevant matters have been corrected.

Cofferdams and caissons – regulation 32

All cofferdams or caissons must be:

➤ suitably designed and constructed
➤ appropriately equipped so that people can gain shelter or escape if water or material enters
➤ properly maintained.

A cofferdam or caisson may be used to carry out construction only if:

➤ it and any work equipment and materials which affect its safety have been inspected by a competent person:
  ➤ at the start of the shift (only one report required every 7 days)
  ➤ after any event likely to affect its strength or stability
➤ the person inspecting is satisfied that it is safe.

When the person who carries out the inspection has so informed the person on whose behalf he carried out the inspection, work must cease until the relevant matters have been corrected.

Reports of Inspection – regulation 33

Before the end of the shift in which the report was completed the person who carries out the inspection under regulations 31 or 32 must:

➤ where he is not satisfied that construction work can be carried out safely, inform the person for whom he carried out the inspection
➤ prepare a report with the particulars set out in schedule 3 to the Regulations as shown in Table 21.6.

A copy of the report must be provided within 24 hours of the relevant inspection. If the inspector is an employee or works under someone else's control, that person must ensure that these duties are performed.

Records must be kept at the site where the inspection was carried out until the work is complete and then for 3 months. Extracts must be provided for an inspector as they require.

Energy distribution installations – regulation 34

Energy distribution installations must be suitably located, checked and clearly indicated.

Where there is a risk from electric power cables they should be:

➤ routed away from the risk areas or
➤ the power shall be cut off.

Table 21.6 Particulars for inspection report

SCHEDULE 3 Regulation 33 (1)(b)
Particulars to be included in a report of inspection
1. Name and address of the person on whose behalf the inspection was carried out.
2. Location of the place of work inspected.
3. Description of the place of work or part of that place inspected (including any work equipment and materials).
4. Date and time of the inspection.
5. Details of any matter identified that could give rise to a risk to the health or safety of any person.
6. Details of any action taken as a result of any matter identified in paragraph 5 above.
7. Details of any further action considered necessary.
8. Name and position of the person making the report.

If these safety measures are not reasonably practicable with suitable warning notices, barriers to exclude work equipment and suspended protections where vehicles have to pass underneath need to be put into place.

No construction work which is liable to create a risk to health or safety from an underground service, or from damage to or disturbance of it, shall be carried out unless suitable and sufficient steps have been taken to prevent such risk, SFARP.

Prevention of drowning – regulation 35

Where persons could fall into water or other liquid, steps must be taken to:

➤ prevent, SFARP, people from falling
➤ minimize the risk of drowning if people fall
➤ provide, use and maintain suitable rescue equipment to ensure prompt rescue.

Transport to work by water must be suitable and sufficient and any vessel used must not be overcrowded or overloaded.

Traffic routes – regulation 36

Every construction site shall be organized in such a way that, SFARP, pedestrians and vehicles can move safely.

Traffic routes shall be suitable for the persons or vehicles using them, sufficient in number, in suitable positions and of sufficient size.

Steps must be taken to ensure that:

➤ pedestrians or vehicles may use a traffic roote without causing danger to the health or safety of persons near it

➤ any door or gate for pedestrians which leads onto a traffic route is sufficiently separated from it to enable them from a place of safety to see any approaching vehicle

➤ there is sufficient separation between vehicles and pedestrians to ensure safety or, where this is not reasonably practicable:
  (i) there are provided other means for the protection of pedestrians and
  (ii) there are effective arrangements for warning any person liable to be crushed or trapped by any vehicle of its approach

➤ any loading bay has at least one exit point for the exclusive use of pedestrians and

➤ where it is unsafe for pedestrians to use a gate intended primarily for vehicles, one or more doors for pedestrians is provided in the immediate vicinity of the gate, is clearly marked and is kept free from obstruction.

Every traffic route shall be:

➤ indicated by suitable signs
➤ regularly checked and
➤ properly maintained.

No vehicle shall be driven on a traffic route unless, SFARP, that traffic route is free from obstruction and permits sufficient clearance.

Vehicles – regulation 37

The unintended movement of any vehicle must be prevented or controlled.

Suitable and sufficient steps shall be taken to ensure that, where a person may be endangered by the movement of any vehicle, the person having effective control of the vehicle shall give suitable and sufficient warning.

Any vehicle being used for the purposes of construction work shall, when being driven, operated or towed:

➤ be driven, operated or towed in a safe manner and
➤ be loaded so that it can be driven, operated or towed safely.

No person shall ride on any vehicle being used for the purposes of construction work except in a safe place provided.

No person shall remain on any vehicle during the loading or unloading of any loose material unless a safe place of work is provided and maintained.

Suitable measures must be taken to prevent any vehicle from falling into any excavation or pit, or into water, or overrunning the edge of any embankment or earthwork.

Prevention of risk from fire – regulation 38

Steps must be taken to prevent injury arising from:

➤ fires or explosions
➤ flooding or
➤ substances liable to cause asphyxiation.

Emergency procedures – regulation 39

Where necessary, emergency procedures must be prepared and, where necessary, implemented to deal with any foreseeable emergency. They must include procedures for any necessary evacuation of the site or any part thereof.

The arrangements must take account of:

(a) the type of work for which the construction site is being used
(b) the characteristics and size of the construction site and the number and location of places of work on that site
(c) the work equipment being used
(d) the number of persons likely to be present on the site at any one time and
(e) the physical and chemical properties of any substances or materials on or likely to be on the site.

Steps must be taken to make people on site familiar with the procedures and test the procedures at suitable intervals.

Emergency routes and exits – regulation 40

A sufficient number of suitable emergency routes and exits shall be provided to enable any person to reach a place of safety quickly in the event of danger.

An emergency route or exit (or as appropriate traffic route) provided must:

➤ lead as directly as possible to an identified safe area
➤ be kept clear and free from obstruction and, where necessary, provided with emergency lighting so that such emergency route or exit may be used at any time
➤ be indicated by suitable signs.

Fire detection and firefighting – regulation 41

Where necessary, duty holders must provide suitable and sufficient:

➤ firefighting equipment and
➤ fire detection and alarm systems, which shall be suitably located.

The equipment must be examined and tested at suitable intervals, properly maintained and suitably indicated by signs. If non-automatic it must be easily accessible.

Every person at work on a construction site shall, SFARP, be instructed in the correct use of any firefighting equipment which it may be necessary for him to use. Where a work activity may give rise to a particular risk of fire, a person shall not carry out such work unless he is suitably instructed.

Fresh air – regulation 42

Steps must be taken to ensure, SFARP, that every place of work or approach has sufficient fresh or purified air to ensure that the place or approach is safe and without risks to health. Any plant used for the provision of such air, must include, where necessary, an effective device to give visible or audible warning of any failure of the plant.

Temperature and weather protection – regulation 43

Steps must be taken to ensure, SFARP, that during working hours the temperature at any place of work indoors is reasonable having regard to its purpose.

Every place of work outdoors must, where necessary, SFARP and having regard to the purpose for which that place is used and any protective clothing or work equipment provided, be so arranged that it provides protection from adverse weather.

Lighting – regulation 44

Every place of work and approach thereto and every traffic route shall be provided with suitable and sufficient lighting, which shall be, SFARP, by natural light.

The colour of any artificial lighting provided shall not adversely affect or change the perception of any sign or signal provided for the purposes of health and safety.

Suitable and sufficient secondary lighting shall be provided in any place where there would be a risk to the health or safety of any person in the event of failure of primary artificial lighting.

21.10.9 *Welfare facilities – schedule 2*

Sanitary conveniences

1. Suitable and sufficient sanitary conveniences shall be provided or made available at readily accessible places. SFARP, rooms containing sanitary conveniences shall be adequately ventilated and lit.
2. SFARP, sanitary conveniences and the rooms containing them shall be kept in a clean and orderly condition.
3. Separate rooms containing sanitary conveniences shall be provided for men and women, except where and so far as each convenience is in a separate room the door of which is capable of being secured from the inside.

Washing facilities

4. Suitable and sufficient washing facilities, including showers if required by the nature of the work or for health reasons, shall SFARP be provided or made available at readily accessible places.
5. Washing facilities shall be provided:
   (a) in the immediate vicinity of every sanitary convenience, whether or not provided elsewhere and
   (b) in the vicinity of any changing rooms required by paragraph 15 whether or not provided elsewhere.
6. Washing facilities shall include:
   (a) a supply of clean hot and cold, or warm, water (which shall be running water SFARP)
   (b) soap or other suitable means of cleaning
   (c) towels or other suitable means of drying.
7. Rooms containing washing facilities shall be sufficiently ventilated and lit.
8. Washing facilities and the rooms containing them shall be kept in a clean and orderly condition.
9. Subject to paragraph 10 below, separate washing facilities shall be provided for men and women, except where and so far as they are provided in a room the door of which is capable of being secured from inside and the facilities in each such room are intended to be used by only one person at a time.
10. Paragraph 9 above shall not apply to facilities which are provided for washing hands, forearms and face only.

Drinking water

11. An adequate supply of wholesome drinking water shall be provided or made available at readily accessible and suitable places.
12. Every supply of drinking water shall be conspicuously marked by an appropriate sign where necessary for reasons of health and safety.
13. Where a supply of drinking water is provided, there shall also be provided a sufficient number of suitable cups or other drinking vessels unless the supply of drinking water is in a jet from which persons can drink easily.

Changing rooms and lockers

14. (1) Suitable and sufficient changing rooms shall be provided or made available at readily accessible places if:

   (a) a worker has to wear special clothing for the purposes of his work

   (b) a worker cannot, for reasons of health or propriety, be expected to change elsewhere

being separate rooms for, or separate use of rooms by, men and women where necessary for reasons of propriety.

(2) Changing rooms shall:

   (a) be provided with seating

   (b) include, where necessary, facilities to enable a person to dry any such special clothing and his own clothing and personal effects.

(3) Suitable and sufficient facilities shall, where necessary, be provided or made available at readily accessible places to enable persons to lock away:

   (a) any such special clothing which is not taken home

   (b) their own clothing which is not worn during working hours

   (c) their personal effects.

Facilities for rest

15. (1) Suitable and sufficient rest rooms or rest areas shall be provided or made available at readily accessible places.

(2) Rest rooms and rest areas shall:

   (a) include suitable arrangements to protect non-smokers from discomfort caused by tobacco smoke

   (b) be equipped with an adequate number of tables and adequate seating with backs for the number of persons at work likely to use them at any one time

   (c) where necessary, include suitable facilities to enable any person at work who is a pregnant woman or nursing mother to rest lying down

   (d) include suitable arrangements to ensure that meals can be prepared and eaten

   (e) include the means for boiling water

   (f) be maintained at an appropriate temperature.

21.10.10 *Civil liability – regulation 45*

The Management Regulations (Management of Health and Safety at Work Regulations 1999) were amended in 2003 to provide employees with a right of action in civil proceedings, in relation to breach of duties by their employer. To maintain consistency CDM 2007 carries forward, with three small exceptions, the rights of civil action in CDM and CHSW, and also allows employees to take action in the civil courts for injuries resulting from a failure to comply with duties under these Regulations.

21.10.11 *Enforcement*

General

The enforcement demarcation between HSE and Local Authorities (LAs) in respect of construction work was set out in the Health and Safety (Enforcing Authority) Regulations 1998, and in regulations 3 (4) and 22 of CDM94. Interpretation of these requirements was not straightforward, but the practical effect is that LAs were prevented from enforcing CDM.

The HSC wished to simplify this by omitting regulations 3 (4) and 22 from CDM 2007.

The effect of this is that, under CDM 2007, HSE are the enforcing authority for:

➤ all notifiable construction work, except that undertaken by people in LA-enforced premises who normally work on the premises

➤ work done to the exterior of the premises and

➤ work done in segregated areas.

Enforcement in respect of fire – regulation 46

This Regulation of CDM 2007 is given in full:

46. (1) Subject to paragraphs (2) and (3):

   (a) in England and Wales the enforcing authority within the meaning of article 25 of the Regulatory Reform (Fire Safety) Order 2005 or

   (b) in Scotland the enforcing authority within the meaning of section 61 of the Fire (Scotland) Act 2005

shall be the enforcing authority in respect of a construction site which is contained within, or forms part of, premises which are occupied by persons other than those carrying out the construction work or any activity arising from such work as regards regulations 39 and 40, in so far as those regulations relate to fire, and regulation 41.

(2) In England and Wales paragraph (1) only applies in respect of premises to which the Regulatory Reform (Fire Safety) Order 2005 applies.

(3) In Scotland paragraph (1) only applies in respect of premises to which Part 3 of the Fire (Scotland) Act 2005 applies.

21.10.12 *References*

Backs for the Future, HSG 149, HSE Books, ISBN 9780-7176-1122-1.

Construction (Design and Management) Regulations 2007. Approved Code of Practice, L144, 2007 HSE Books, ISBN 9780 7176 62234.

Controlling Exposure to Stonemasonry Dust: Guidance for Employers. 2001 HSE Books, ISBN 9780-7176-1760-2.

Electrical Safety on Construction Sites. HSG 141, 1995 HSE Books, ISBN 9780-7176-1000-4.

Fire Safety in Construction Work, HSG 168, 1997 HSE Books, ISBN 9780-7176-1332-1.

Health and Safety in Construction. HSG 150 (rev3), 2006 HSE Books, ISBN 9780-7176-6182-2.

Health and Safety in Excavations. HSG 185, 1999 HSE Books, ISBN 9780-7176-1563-4.

Health and Safety in Roof Work, Revised. HSG 33, 1998 HSE Books, ISBN 9780-7176-1425-5.

Managing Health and safety in Construction, Guidance. HSG 224, 2007 HSE Books, ISBN 9780-7176-21391.

Safe Use of Vehicles on Construction Sites. HSG 144, 1998 HSE Books, ISBN 9780-7176-1610-X.

Want Construction Work Done Safely: A Guide for Clients on the CDM 2007 Regulations. INDG 411, ISBN 9780-7176-6246-3.

21.11 Construction (Head Protection) Regulations 1989

21.11.1 *Application*

These Regulations apply to 'building operations' and 'works of engineering construction' as defined originally by the Factories Act 1961.

21.11.2 *Provision and maintenance*

Every employer must provide suitable head protection for each employee and shall maintain it and replace it whenever necessary. A similar duty is placed on self-employed people. The employer must make an assessment and review it as necessary, to determine whether the head protection is suitable. Accommodation must be provided for head protection when it is not in use.

21.11.3 *Ensuring that head protection is worn – regulation 4*

Every employer shall ensure, SFARP, that each employee (and any other person over whom they have control) at work wears suitable head protection, unless there is no foreseeable risk of injury to their head other than through falling.

21.11.4 *Rules and directions – regulation 5*

The person in control of a site **may** make rules regulating the wearing of suitable head protection. These must be in writing. An employer may give directions requiring their employees to wear head protection.

21.11.5 *Wearing of suitable head protection, reporting loss – regulations 6–9*

Every employee who has been provided with suitable head protection shall wear it when required to do so by rules made or directions given under regulation 5 of these Regulations. They must make full and proper use of it and return it to the accommodation.

Employees must take reasonable care of head protection, and report any loss or obvious defect.

Exemption certificates, issued by the HSE, are allowed.

21.11.6 *References*

A guide to the Construction (Head Protection) Regulations L 102 (revised). HSE Books 1998, ISBN 9780-7176-1478-6.

Head Protection for Sikhs Wearing Turbans, INDG262.

21.12 Health and Safety (Consultation with Employees) Regulations 1996

21.12.1 *Application*

These Regulations apply to all employers and employees in Great Britain except:

➤ where employees are covered by safety representatives appointed by recognized trade unions under the Safety Representatives and Safety Committees Regulations 1977
➤ domestic staff employed in private households
➤ crew of a ship under the direction of the master.

21.12.2 *Employer's duty to consult – regulation 3*

The employer must consult relevant employees in good time with regard to:

➤ the introduction of any measure which may substantially affect their health and safety
➤ the employer's arrangements for appointing or nominating competent persons under the Management of Health and Safety at Work Regulations 1999

➤ any information required to be provided by legislation

➤ the planning and organization of any health and safety training required by legislation

➤ the health and safety consequences to employees of the introduction of new technologies into the workplace.

The guidance emphasizes the difference between informing and consulting. Consultation involves listening to employees' views and taking account of what they say before any decision is taken.

21.12.3 Persons to be consulted – regulation 4

Employers must consult with either:

➤ the employees directly or

➤ one or more persons from a group of employees, who were elected by employees in that group to represent them under these regulations. They are known as 'Representatives of Employee Safety' (ROES).

Where ROES are consulted, all employees represented must be informed of:

➤ the names of ROES

➤ the group of employees represented.

Employers shall not consult a ROES if:

➤ the ROES does not wish to represent the group

➤ the ROES has ceased to be employed in that group

➤ the election period has expired

➤ the ROES has become incapacitated from carrying out their functions.

If an employer decides to consult directly with employees he must inform them and ROES of that fact. Where no ROES are elected employers will have to consult directly.

21.12.4 Duty to provide information – regulation 5

Employers must provide enough information to enable ROES to participate fully and carry out their functions. This will include:

➤ what the likely risks and hazards arising from their work may be

➤ reported accidents and diseases, etc. under Reporting of Injuries Diseases and Dangerous Occurrences Regulation (RIDDOR) 1995

➤ the measures in place, or which will be introduced, to eliminate or reduce the risks

➤ what employees ought to do when encountering risks and hazards.

An employer need not disclose information which:

➤ could endanger national security

➤ violates a legal prohibition

➤ relates specifically to an individual without their consent

➤ could substantially hurt the employers undertaking or infringe commercial security

➤ was obtained in connection with legal proceedings.

21.12.5 Functions of ROES – regulation 6

ROES have the following functions (but no legal duties):

➤ to make representations to the employer on potential hazards and dangerous occurrences related to the group of employees represented

➤ to make representations to the employer on matters affecting the general health and safety of relevant employees

➤ to represent the group of employees for which they are the ROES in consultation with inspectors.

21.12.6 Training, time off and facilities – regulation 7

Where an employer consults ROES, he must:

➤ ensure that each ROES receives reasonable training at the employer's expense

➤ allow time off with pay during the ROES working hours to perform the duties of a ROES and while a candidate for election

➤ provide such other facilities and assistance that ROES may reasonably require.

21.12.7 Civil liability and complaints – regulation 9 and schedule 2

Breach of these Regulations does not confer any right of action in civil proceedings subject to regulation 7 (3) and schedule 2 relating to complaints to employment tribunals.

A ROES can complain to an industrial tribunal that:

➤ their employer has failed to permit time off for training or to be a candidate for election

➤ their employer has failed to pay them as set out in schedule 1 to the Regulations.

21.16.8 *Elections*

The guidance lays down some ideas for the elections, although there are no strict rules.

ROES do not need to be confined to consultation related to these Regulations. Some employers have ROES sitting on safety committees and taking part in accident investigations similarly to union Safety Representatives.

21.12.9 *References*

A guide to the Health and Safety (Consultation with Employees) Regulations 1996. L95, HSE Books, ISBN 9780-7176-1234-1.

21.13 Control of Substances Hazardous to Health Regulations (COSHH) 2002 as amendmented

21.13.1 *Introduction*

The 2002 COSHH Regulations: have updated the 1999 Regulations with a few changes which include additional definitions like 'inhalable dust' and 'health surveillance'; clarify and extend the steps required under risk assessment; introduce a duty to deal with accidents and emergencies.

The 2004 Amendment Regulations replace regulation 7 (7) and (8) by substituting new requirements to observe principles of good practice for the control of exposure to substances hazardous to health introduced by schedule 2A, to ensure that workplace exposure limits are not exceeded, and to ensure in respect of carcinogens and asthmagens that exposure is reduced to as low a level as is reasonably practicable. They also introduce a single new Workplace Exposure Limit for substances hazardous to health which replaces occupational exposure standards and maximum exposure limits. The Amendment Regulations introduce a duty to review control measures other than the provision of plant and equipment, including systems of work and supervision, at suitable intervals.

COSHH covers most substances hazardous to health found in workplaces of all types. The substances covered by COSHH include:

➤ substances used directly in work activities (e.g. solvents, paints, adhesives, cleaners)
➤ substances generated during processes or work activities (e.g. dust from sanding, fumes from welding)
➤ naturally occurring substances (e.g. grain dust).

But COSHH does **not** include:

➤ asbestos and lead, which have specific regulations
➤ substances which are hazardous only because they are:
  ➤ radioactive
  ➤ simple asphyxiants
  ➤ at high pressure
  ➤ at extreme temperatures
  ➤ have explosive or flammable properties (separate Dangerous Substances Regulations cover these)
  ➤ biological agents if they are not directly connected with work and are not in the employer's control, such as catching flu from a workmate.

21.13.2 *Definition of substance hazardous to health – regulation 2*

The range of substances regarded as hazardous under COSHH are:

➤ substances or mixtures of substances classified as dangerous to health under the Chemicals (Hazard, Information and Packaging for Supply) Regulations 2002 (CHIP). These have COSHH warning labels and manufacturers must supply data sheets. They cover substances that are: very toxic, toxic, harmful, corrosive or irritant under CHIP
➤ substances with workplace exposure limits as listed in EH40 published by the HSE
➤ biological agents (bacteria and other microorganisms) if they are directly connected with the work
➤ any kind of dust in a concentration specified in the regulations, that is:
  ➤ 10 mg/m^3, as a time-weighted average over an 8-hour period, of total inhalable dust
  ➤ 4 mg/m^3, as a time-weighted average over an 8-hour period, of respirable dust
➤ any other substance which has comparable hazards to people's health but which, for technical reasons, is not covered by CHIP.

21.13.3 *Duties under COSHH – regulation 3*

The duties placed on employers under these Regulations are extended to (except for health surveillance, monitoring and information and training) persons who may be on the premises but not employed whether they are at work or not, for example, visitors and contractors.

21.13.4 *General requirements*

There are eight basic steps to comply with COSHH; they are to:

1. assess the risks to health
2. decide what precautions are needed

481

3. prevent or adequately control exposure
4. ensure that control measures are used and maintained
5. monitor the exposure of employees to hazardous substances
6. carry out appropriate health surveillance where necessary
7. prepare plans and procedures to deal with accidents, incidents and emergencies
8. ensure employees are properly informed, trained and supervised.

21.13.5 *Steps 1 and 2 – assessment of health risk – regulation 6*

No work may be carried out where employees are liable to be exposed to substances hazardous to health unless:

➤ a suitable and sufficient risk assessment, including the steps need to meet COSHH, has been made.

The assessment must be reviewed and changes made regularly and immediately if:

➤ if it is suspected that it is no longer valid
➤ there has been a significant change in the work to which it relates
➤ the results of any monitoring carried out in accordance with regulation 10 show it to be necessary.

The assessment involves identifying the substance present in the workplace; assessing the risk it presents in the way it is used; and deciding what precautions (and health surveillance) are needed. Where more than four employees are employed the significant findings must be recorded and the steps taken to comply with regulation 7.

21.13.6 *Step 3 – prevention or control of exposure – regulation 7*

Every employer must ensure that exposure to substances hazardous to health is either:

➤ prevented or, where this is not reasonably practicable
➤ adequately controlled.

Preference must be given to substituting with a safer substance. Where it is not reasonably practicable to prevent exposure, protection measures must be adopted in the following order of priority:

➤ the design and use of appropriate work processes, systems and engineering controls and the provision and use of suitable work equipment

➤ the control of exposure at source, including adequate ventilation systems and appropriate organizational measures and
➤ where adequate control of exposure cannot be achieved by other means, the additional provision of suitable PPE.

The measures must include:

➤ safe handling, storage and transport (plus waste)
➤ reducing to a minimum required for the work the number of employees exposed, the level and duration of exposure, and the quantity of substance present at the workplace
➤ the control of the working environment including general ventilation
➤ appropriate hygiene measures.

Control of exposure shall only be treated as adequate if:

➤ the principles of good practice as set out in schedule 2A are applied
➤ any workplace exposure limit approved for that substance is not exceeded and
➤ for a carcinogen (in schedule 1 or with risk phrase R45, R46 or R49) or sensitizer (Risk phrases R42 or R42/43) or asthmagen (section C of HSE publication on Asthmagens (ISBN 9780-7176-1465-4)) exposure is reduced to as low a level as is reasonably practicable.

Box 21.8

Principles of Good Practice for the Control of Exposure to Substances Hazardous to Health (schedule 2A).

➤ Design and operate processes and activities to minimize emission, release and spread of substances hazardous to health.
➤ Take into account all relevant routes of exposure – inhalation, skin absorption and ingestion – when developing control measures.
➤ Control exposure by measures that are proportionate to the health risk.
➤ Choose the most effective and reliable control options which minimize the escape and spread of substances hazardous to health.

➤ Where adequate control of exposure cannot be achieved by other means, provide, in combination with other control measures, suitable PPE.
➤ Check and review regularly all elements of control measures for their continuing effectiveness.
➤ Inform and train all employees on the hazards and risks from the substances with which they work and the use of control measures developed to minimize the risks.
➤ Ensure that the introduction of control measures does not increase the overall risk to health and safety.

PPE must conform to the Personal Protective Equipment (EC Directive) Regulations 2002.

Additionally RPE must be:

➤ suitable for the purpose
➤ comply with the EC Directive Regulations or
➤ if there is no requirement imposed by these Regulations, be of a type or shall conform to a standard approved by the HSE.

Where exposure to a Carcinogen or Biological Agent is involved, additional precautions are laid down in regulation 7.

21.13.7 *Step 4 – use, maintenance, examination and test of control measures – regulations 8 and 9*

➤ Every employer must take all reasonable steps to ensure that control measures, PPE or anything else provided under COSHH, are properly used or applied.
➤ Employees must make full and proper use of control measures, PPE or anything else provided; employees must as far as is reasonable return items to accommodation and report defects immediately.

Employers shall also:

➤ properly maintain controls and, in the case of PPE, keep them clean
➤ carry out thorough examination and tests on engineering controls:

➤ in the case of local exhaust ventilation (LEV) at least once every 14 months (except those in schedule 4, which covers blasting of castings – monthly; dry grinding, polishing or abrading of metal for more than 12 hours per week – 6 months; and jute manufacture – 6 months)
  ➤ in any other case, at suitable intervals
➤ carry out thorough examination and tests where appropriate (except on disposable items) on RPE at suitable intervals
➤ keep a record of examination and tests for at least 5 years.

21.13.8 *Step 5 – monitoring exposure – regulation 10*

Employers must monitor exposure:

➤ where this is necessary to ensure maintenance of control measures or the protection of health
➤ specifically for vinyl chloride monomer or chromium plating as required by schedule 5 and
➤ keep a record of identifiable personal exposures for 40 years and any other exposures for 5 years.

21.13.9 *Step 6 – health surveillance – regulation 11*

Employers must ensure employees, where appropriate, who are exposed or liable to be exposed are under health surveillance. It is considered appropriate when:

➤ an employee is exposed (if significant) to substances or processes in schedule 6
➤ an identifiable disease or adverse health effect may be related to the exposure, and there is a reasonable likelihood that disease may occur and there are valid disease indication or effect detection methods.

Records of health surveillance containing approved particulars must be kept for 40 years or offered to HSE if trading ceases. If a person is exposed to a substance and/or process in schedule 6, the health surveillance shall include medical surveillance by an employment medical adviser or appointed doctor at intervals not exceeding 12 months.

If a medical adviser certifies that a person should not be engaged in particular work they must not be permitted to carry out that work except under specified conditions. Health records must be available for the individual employee to see after reasonable notice.

Employees must present themselves, during working hours at the employer's expense, for appropriate health surveillance.

An employment medical adviser or appointed doctor has the power to inspect the workplace or look at records for the purpose of carrying out functions under COSHH.

21.13.10 Steps 7 and 8 – information, instruction and training and emergencies – regulations 12 and 13

Where employees are likely to be exposed to substances hazardous to health employers must provide:

➤ information, instruction and training on the risks to health and the precautions which should be taken (this duty is extended to anyone who may be affected)

➤ information on any monitoring of exposure (particularly if there is a maximum exposure limit where the employee or their representative must be informed immediately)

➤ information on collective results of health surveillance (designed so that individuals cannot be identified)

➤ procedures for accidents and emergencies including the provision of appropriate first aid facilities and relevant safety drills. Information on emergency arrangements and suitable warning and other communications must all be established to ensure suitable response and rescue operations.

21.13.11 Defence – regulation 16

It is a defence under these Regulations for a person to show that they have taken all reasonable precautions and exercised all due diligence to avoid the commission of an offence.

21.13.12 References

CHIP for Everyone. HSG 228, 2002, HSE Books, ISBN 9780-7176-2370-X.
Control of Substances Hazardous to Health Regulations 2002, SI 2002, 2677, 2002 Stationery Office, ISBN 9780-11-042919-2.
Control of Substances Hazardous to Health Approved Code of Practice and Guidance. L5, 5th edition, 2005 HSE Books, ISBN 9780-7176-2981-3.
COSHH: A Brief Guide to the Regulations. INDG136 rev3, 2005 HSE Books, ISBN 9780-7176-2982-1.
COSHH Essentials: Easy Steps to Control Chemicals. HSG 193, 2nd edition, 2003 HSE Books, ISBN 9780-7176-2737-3. COPSHH Essentials website: http://www.coshh-essentials.org.uk/.

COSHH HSE Micro website: http://www.hse.gov.uk/coshh/index.htm.
Seven Steps to Successful Substitution of Hazardous Substances. HSG 110, 1994 HSE Books, ISBN 9780-7176-0695-3.
The Idiot's Guide to CHIP3. Chemicals (Hazard Information and Packaging for Supply) Regulations, INDG350, 2003 HSE Books, ISBN 9780-7176-2333-5.
Why Do I Need a Safety Data Sheet? INDG353, 2002, HSE Books, ISBN 9780-7176-2367-X.
Workplace Exposure Limits EH40. 2005 HSE Books, ISBN 9780-7176-2083-2 (revised annually).

21.14 Dangerous Substances and Explosive Atmospheres Regulations (DSEAR) 2002

21.14.1 Introduction

These Regulations are designed to implement the safety requirements of the EU Chemical Agents and Explosive Atmospheres Directives. DSEAR deals with the prevention of fires, explosions and similar energy releasing events arising from dangerous substances. Following the introduction of DSEAR the opportunity will also be taken to modernize all the existing laws on the use and storage of petrol.

In Summary the DSEAR Regulations require employers and the self-employed to:

➤ carry out an assessment of the fire and explosion risks of any work activities involving dangerous substances

➤ provide measures to eliminate, or reduce SFARP, the identified fire and explosion risks

➤ apply measures, SFARP, to control risks and to mitigate the detrimental effects of a fire or explosion

➤ provide equipment and procedures to deal with accidents and emergencies and

➤ provide employees with information and precautionary training.

Where explosive atmospheres may occur:

➤ the workplaces should be classified into hazardous and non-hazardous places; and any hazardous places classified into zones on the basis of the frequency and duration of an explosive atmosphere, and where necessary marked with a sign

➤ equipment in classified zones should be safe so as to satisfy the requirements of the Equipment and

Protective Systems Intended for Use in Potentially Explosive Atmospheres Regulations 1996 and

➤ the workplaces should be verified by a competent person, as meeting the requirements of DSEAR.

21.14.2 Scope of regulations

Dangerous Substance

The regulations give a detailed interpretation of *dangerous substance*, which should be consulted. In summary it means that:

➤ a substance or preparation that because of its chemical and sometimes physical properties and the way it is present and/or used at work, creates a fire or explosion risk to people; for example substances like petrol, liquefied petroleum gas (LPG), paints, cleaners, solvents and flammable gases

➤ any dusts which could form an explosive mixture in air (not included in a substance or preparation); for example many dusts from grinding, milling or sanding.

Explosive Atmosphere

The Regulations give the definition as:

a mixture under atmospheric conditions, of air and one or more dangerous substances in the form of gases, vapours, mists or dusts, in which, after ignition has occurred, combustion spreads to the entire unburned mixture.

21.14.3 Application

The Regulations do not in general apply to:

➤ ships under the control of a master
➤ areas used for medical treatment of patients
➤ many gas appliances used for cooking, heating, hot water, refrigeration, etc. (except an appliance specifically designed for an industrial process), gas fittings
➤ manufacture, use, transport of explosives or chemically unstable substances
➤ mine, quarry or borehole activities
➤ activity at an offshore installation
➤ the use of means of transport (but the Regulations do cover means of transport intended for use in a potentially explosive atmosphere).

21.14.4 Risk assessment – regulation 5

The risk assessment required by regulation 5 must include:

➤ the hazardous properties of the dangerous substance
➤ supplier's information and SDS

➤ the circumstances of the work including:
  ➤ work processes and substances used and their possible interactions
  ➤ the amount of substance involved
  ➤ risks of substances in combination
  ➤ arrangements for safe handling, storage and transport and any waste which might contain dangerous substances
➤ high risk maintenance activities
➤ likelihood of an explosive atmosphere
➤ likelihood of ignition sources, including electrostatic discharges being present
➤ scale of any possible fire or explosion
➤ any places connected by openings to areas where there could be an explosive atmosphere
➤ any additional information which may be needed.

21.14.5 Elimination or reduction of risks – regulation 6

Regulation 6 concerns the reduction of risks and tracks the normal hierarchy as follows:

➤ substitution of a dangerous substance by a substance or process which eliminates or reduces the risk, for example the use of water-based paints, or using a totally enclosed continuous process
➤ reducing the quantity of dangerous substance to a minimum, for example only a half-day supply in the workroom
➤ avoiding releasing of dangerous substance or minimizing releases, for example keeping them in special closed containers
➤ controlling releases at source
➤ preventing the formation of explosive atmospheres, including the provision of sufficient ventilation
➤ ensuring that any releases are suitably collected, contained and removed; suitable LEV in a paint spray booth is an example
➤ avoiding ignition sources and adverse conditions, for example keeping electrical equipment outside the area
➤ segregating incompatible dangerous substances, for example oxidizing substances and other flammable substances.

Steps must also be taken to mitigate the detrimental effects of a fire or explosion by:

➤ keeping the number of people exposed to a minimum
➤ avoidance of fire and explosion propagation
➤ provision of explosion relief systems
➤ provision of explosion suppression equipment

➤ provision of very strong plant which can withstand an explosion

➤ the provision of suitable PPE.

21.14.6 *Classification of workplaces – schedule 2*

Where an explosive atmosphere may occur workplaces must be classified into hazardous and non-hazardous places. The following zones are specified by schedule 2 of the Regulations (Table 21.7).

Table 21.7 Classification Zones

Zone 0	A place in which an explosive atmosphere consisting of a mixture with air of dangerous substances in the form of gas, vapour or mist is present continuously or for long periods.
Zone 1	A place in which an explosive atmosphere consisting of a mixture with air of dangerous substances in the form of gas, vapour or mist is likely to occur in normal operation occasionally.
Zone 2	A place in which an explosive atmosphere consisting of a mixture with air of dangerous substances in the form of gas, vapour or mist is not likely to occur in normal operation but, if it does, will persist for a short period only.
Zone 20	A place in which an explosive atmosphere in the form of a cloud of combustible dust in air is present continuously, or for long periods.
Zone 21	A place in which an explosive atmosphere in the form of a cloud of combustible dust in air is likely to occur in normal operation occasionally.
Zone 22	A place in which an explosive atmosphere in the form of a cloud of combustible dust in air is not likely to occur in normal operation but, if it does occur, will persist for a short period only.

21.14.7 *Accidents, incidents and emergencies – regulation 8*

In addition to any normal fire prevention requirements employers must under these Regulations:

➤ ensure that procedures and first aid are in place with tested relevant safety drills

➤ provide information on emergency arrangements including work hazards and those that are likely to arise at the time of an accident

➤ provide suitable warning and other communications systems to enable an appropriate response, remedial actions and rescue operations to be made

➤ where necessary, before any explosion condition is reached provide visual or audible warnings and withdraw employees

➤ provide escape facilities where the risk assessment indicates it is necessary.

In the event of an accident immediate steps must be taken to:

➤ mitigate the effects of the event

➤ restore the situation to normal and

➤ inform employees who may be affected.

Only essential persons may be permitted in the affected area. They must be provided with PPE, protective clothing and any necessary specialized safety equipment and plant.

21.14.8 *Information instruction and training – regulation 9*

Under regulation 9, where a dangerous substance is present an employer must provide:

➤ suitable and sufficient information, instruction and training on the appropriate precautions and actions

➤ plus details of the substances any relevant data sheets and legal provisions

➤ plus the significant findings of the risk assessment.

21.14.9 *Contents of containers and pipes – regulation 10*

Regulation 10 requires that for containers and pipes (except where they are marked under legislation contained in schedule 5 to the Regulations) the content and the nature of those contents and any associated hazards be clearly identified.

21.14.10 *References*

Control and Mitigation Measures. L136, Dangerous Substances and Explosive Atmospheres Regulations 2002, Approved Code of Practice and guidance, 2003, HSE Books,ISBN 0-7176-2201-0.

Dangerous Substances and Explosive Atmosphere Regulations 2002. SI 2002 No. 2776, ISBN 9780-11-042957-5.

Dangerous Substances and Explosive Atmospheres. L138, Dangerous Substances and Explosive Atmospheres Regulations 2002, Approved Code of Practice, 2003, HSE Books, ISBN 9780-7176-2203-7.

Fire and Explosions How Safe is Your Workshop. A Short Guide to DSEAR, HSE INDG370, 2002 HSE Books, ISBN 9780-7176-2589-3.

Safe Handling of Combustible Dusts; Precautions against Explosion. HSG 103, HSE Books 2003, ISBN 9780-7176-2726-8.

Safe Maintenance, Repair and Cleaning Procedures. L137, Dangerous Substances and Explosive Atmospheres Regulations 2002, Approved Code of Practice and guidance, 2003, HSE Books, ISBN 9780-7176-2202-9.

Safe Use and Handling of Flammable Liquids. HSG 140, HSE Books, ISBN 9780-7176-0967-7.

Storage of Dangerous Substances. L135, Dangerous Substances and Explosive Atmospheres Regulations 2002, Approved Code of Practice and guidance, 2003 HSE Books, ISBN 9789780-7176-2200-2.

21.15 Health and Safety (Display Screen Equipment) Regulations 1992

21.15.1 *General*

These Regulations cover the minimum health and safety requirements for the use of display screen equipment (DSE) and are accompanied by a guidance note. They typically apply to computer equipment with either a cathode ray tube or liquid crystal monitors. But any type of display is covered with some exceptions, for example, on board a means of transport, or where the main purpose is for screening a film or use for a television. Multimedia equipment would generally be covered.

Equipment not specifically covered by these Regulations or where it is not being used by a defined 'user' is, nevertheless, covered by other requirements under the Management of Health and Safety at Work Regulations and the Provision and Use of Work Equipment Regulations (PUWER).

21.15.2 *Definitions – regulation 1*

(a) *Display screen equipment* (DSE) refers to any alphanumeric or graphic display screen, regardless of the display process involved

(b) A *user* is an employee and an *operator* is a self-employed person, both of whom habitually use DSE as a significant part of their normal work. Both would be someone to whom most or all of the following apply, i.e. a person:
➤ who depends on the DSE to do their job
➤ who has no discretion as to use or non-use

➤ who needs particular training and/or skills in the use of DSE to do their job
➤ who uses DSE for continuous spells of an hour or more at a time
➤ who does so on a more or less daily basis
➤ for whom fast transfer of data is important for the job
➤ of whom a high level of attention and concentration is required, in particular to prevent critical errors.

(c) A *workstation* is an assembly comprising:
➤ DSE with or without a keyboard, software or input device
➤ optional accessories
➤ disk drive, telephone, modem, printer, document holder, chair, desk, work surface, etc.
➤ the immediate working environment.

21.15.3 *Exemptions – regulation 1 (4)*

Exemptions include DSE used in connection with:

➤ drivers' cabs or control cabs for vehicles or machinery
➤ being on board a means of transport
➤ being mainly intended for public operation
➤ portable systems not in prolonged use
➤ calculators, cash registers and small displays related to the direct use of this type of equipment
➤ window typewriters.

21.15.4 *Assessment of risk – regulation 2*

Possible hazards associated with DSE are physical (musculoskeletal) problems, visual fatigue and mental stress. They are not unique to display screen work for an inevitable consequence of it, and indeed research shows that the risk to the individual from typical display screen work is low. However, as in other types of work, ill-health can result from poor work organization, working environment, job design and posture, and from inappropriate working methods.

Employers must carry out a suitable and sufficient analysis of users' and operators' workstations in order to assess the risks to health and safety. The guidance gives detailed information on workstation minimum standards and possible effects on health.

The assessment should be reviewed when major changes are made to software, hardware, furniture, environment or work requirements.

21.15.5 *Workstations – regulation 3*

All workstations must meet the requirements laid down in the schedule to the Regulations. This schedule lays down minimum requirements for display screen worksta-tions, covering the equipment, the working environment, and the interface between computer and user/operator.

21.15.6 *Daily work routine of users – regulation 4*

The activities of users should be organized so that their daily work on DSE is periodically interrupted by breaks or changes of activity that reduce their workload at the equipment.

In most tasks, natural breaks or pauses occur from time to time during the day. If such breaks do not occur, deliberate breaks or pauses must be introduced. The guidance requires that breaks should be taken before the onset of fatigue and must be included in the working time. Short breaks are better than occasional long ones; for example, a 5–10 minute break after 50–60 minutes con-tinuous screen and/or keyboard work is likely to be better than a 15-minute break every 2 hours. If possible, breaks should be taken **away** from the screen. Informal breaks, with time on other tasks, appear to be more effective in relieving visual fatigue than do formal rest breaks.

21.15.7 *Eyes and eyesight – regulation 5*

Initially on request, employees have the right to a free eye and eyesight test conducted by a competent person where they:

➤ are already users or
➤ before becoming a user.

The employer must provide a further eye and eyesight test at regular intervals thereafter or when a user is experiencing visual difficulties which could be caused by working with DSE.

There is no reliable evidence that work with DSE causes any permanent damage to eyes or eyesight, but it may make users with pre-existing vision defects more aware of them.

An *eye and eyesight test* means a *sight test* as defined in the Opticians Act 1989. This should be car-ried out by a registered ophthalmic optician or medi-cal practitioner (normally only those with an ophthalmic qualification do so).

Employers shall provide special corrective appli-ances to users where:

➤ normal corrective appliances cannot be used
➤ the result of the eye and eyesight test shows that such provision is necessary.

The guidance indicates that the liability of the employer extends only to the provision of corrective appliances,

which are of a style and quality adequate for their func-tion. If an employee chooses a more expensive design or multi-function correction appliance, the employer needs to pay only a proportion of the cost.

Employers are free to specify that users' eye and eyesight tests and correction appliances are provided by a nominated company or optician.

The confidential clinical information from the tests can only be supplied to the employer with the employ-ee's consent.

Vision screen tests can be used to identify people with defects but they are not a substitute for the full eye-sight test and employees have the right to opt for the full test from the outset.

21.15.8 *Training – regulation 6*

Employers shall ensure that adequate health and safety training is provided to users and potential users in the use of any workstation and refresh the training following any reorganization. The guidance suggests a range of topics to be covered in the training. In summary these involve:

➤ the recognition of hazards and risks, including the absence of desirable features and the presence of undesirable ones
➤ causes of risk and how harm may occur
➤ what the user can do to correct them
➤ how problems can be communicated to management
➤ information on the regulations
➤ the user's contribution to assessments.

21.15.9 *Information – regulation 7*

Operators and users shall be provided with adequate information on all aspects of health and safety relating to their workstation and what steps the employer has taken to comply with the Regulations (insofar as the action taken relates to that operator or user and their work). Under regulation 7 specific information should be provided as outlined in Table 21.8.

21.15.10 *References*

The Law on VDUs: An Easy Guide to the Regulations. HS (G) 90, 2003 HSE Books, ISBN 9780-7176-2602-4.
VDU Workstation Checklist, 2003 HSE Books, ISBN 9780-7176-2617-2.
Work with Display Screen Equipment, Guidance on Regulations, Health and Safety (Display Screen Equipment) Regulations 1992 as amended in 2002. L26, HSE Books, ISBN 9780-7176-2582-6.
Working with VDU's INDG36 (rev3). 2006 HSE Books, ISBN 9780-7176-6222-5.

Table 21.8   Provision of information under regulation 7

Does employer have to provide information to display screen workers who are:		Information on:					
		Risks from DSE and workstation	Risk assessment and measures to reduce the risks (Regulations 2 and 3)	Breaks and activity changes (Regulation 4)	Eye and eyesight tests (Regulation 5)	Initial training (Regulation 6(1))	Training when workstation modified (Regulation 6(2))
	users employed by the undertaking	Yes	Yes	Yes	Yes	Yes	Yes
	users employed by other employer	Yes	Yes	Yes	No	No	Yes
	operators in the undertaking	Yes	Yes	No	No	No	No

Source: HSE.

21.16   Electricity at Work Regulations 1989

The purpose of these Regulations is to require precautions to be taken against the risk of death or personal injury from electricity in work activities. The Regulations impose duties on persons ('duty holders') in respect of systems, electrical equipment and conductors and in respect of work activities on or near electrical equipment. They apply to almost all places of work and electrical systems at all voltages.

Guidance on the Regulations is contained in the *Memorandum of guidance on the Electricity at Work Regulations 1989*, published by the HSE.

21.16.1   *Definitions*

(a) **Electrical equipment** – includes anything used to generate, provide, transmit, rectify, convert, conduct, distribute, control, store, measure or use electrical energy.

(b) **Conductor** – means a conductor of electrical energy. It means any material (solid, liquid or gas), capable of conducting electricity.

(c) **System** – means an electrical system in which all the electrical equipment is, or may be, electrically connected to a common source of electrical energy, and includes the source and equipment. It also includes portable generators and systems on vehicles.

(d) **Circuit conductor** – term used in regulations 8 and 9 only, meaning a conductor in a system which is intended to carry electric current in normal conditions. It would include a combined neutral and earth conductor, but does not include a conductor provided solely to perform a protective connection to earth or other reference point and energized only during abnormal conditions.

(e) **Danger** – in the context of these Regulations, means a risk of injury from any electrical hazard.

Every year about 30 people die from electric shock or electric burns at work. Each year several hundred serious burns are caused by arcing when the heat generated can be very intense. In addition, intense ultraviolet radiation from an electric arc can cause damage to the eyes – known as arc-eye. These hazards are all included in the definition of **Danger**.

21.16.2 *Duties – regulation 3*

Duties are imposed on employers, self-employed and employees. The particular duties on employees are intended to emphasize the level of responsibility which many employees in the electrical trades and professions are expected to take on as part of their job. They are:

➤ to co-operate with their employer so far as is necessary to enable any duty placed on the employer to be complied with (this reiterates section 7(b) of the HSW Act)

➤ to comply with the provisions of these Regulations in so far as they relate to matters which are within their control. (This is equivalent to duties placed on employers and self-employed where these matters are within their control.)

21.16.3 *Systems, work activities and protective equipment – regulation*

Systems must, at all times, be of such construction as to prevent danger. Construction covers the physical condition, arrangement of components and design of the system and equipment.

All systems must be maintained so as to prevent danger.

Every work activity, including operation, use and maintenance or work near a system shall be carried out in a way which prevents danger.

Protective equipment shall be suitable, suitably maintained and used properly.

21.16.4 *Strength and capability of equipment – regulation 5*

No electrical equipment may be put into use where its strength and capability may be exceeded in such a way as may give rise to danger, in normal transient or fault conditions.

21.16.5 *Adverse or hazardous environments – regulation 6*

Electrical equipment which may be exposed to:

➤ mechanical damage
➤ the effects of weather, natural hazards, temperature or pressure
➤ the effects of wet, dirty, dusty or corrosive conditions
➤ any flammable or explosive substances including dusts, vapours or gases

shall be so constructed and protected that it prevents danger.

21.16.6 *Insulation, protection and placing of conductors – regulation 7*

All conductors in a system which may give rise to danger shall either be suitably covered with insulating material and further protected as necessary, for example, against mechanical damage, using trunking or sheathing; or have precautions taken that will prevent danger, for example, being suitably placed like overhead electric power cables, or by having strictly controlled working practices.

21.16.7 *Earthing, integrity and other suitable precautions – regulation 9*

Precautions shall be taken, either by earthing or by other suitable means, for example, double insulation, use of safe voltages and earth-free non-conducting environments, where a conductor, other than a circuit conductor, could become charged as a result of either the use of, or a fault in, a system.

An earth conductor shall be of sufficient strength and capability to discharge electrical energy to earth. The conductive part of equipment, which is not normally live but energized in a fault condition, could become a conductor.

If a circuit conductor is connected to earth or to any other reference point nothing which could break electrical continuity or introduce high impedance, for example, fuse, thyristor or transistor, is allowed in the conductor unless suitable precautions are taken. Permitted devices would include a joint or bolted link, but not a removable link or manually operated knife switch without bonding of all exposed metal work and multiple earthing.

21.16.8 *Connections – regulation 10*

Every joint and connection in a system shall be mechanically and electrically suitable for its use. This includes terminals, plugs and sockets.

21.16.9 *Excess current protection – regulation 11*

Every part of a system shall be protected from excess current, for example, short circuit or overload, by a suitably located efficient means such as a fuse or circuit breaker.

21.16.10 *Cutting off supply and isolation – regulation 12*

There should be suitably located and identified means of cutting off (switch) the supply of electricity to any electrical equipment and also isolating any electrical equipment. Although these are separate requirements, they could be effected by a single means. The isolator should be capable of being locked off to allow maintenance to be done safely.

Sources of electrical energy (accumulators, capacitors and generators) are exempt from this requirement, but precautions must be taken to prevent danger.

21.16.11 *Work on equipment made dead – regulation 13*

Adequate precautions shall be taken to prevent electrical equipment that has been made dead from becoming live while work is carried out on or near the equipment. This will include means of locking off isolators, tagging equipment, permits to work and removing fuses.

21.16.12 *Work on or near live conductors – regulation 14*

No person shall work near a live conductor, except if it is insulated, unless:

➤ it is unreasonable in all the circumstances for it to be dead
➤ it is reasonable in all the circumstances for them to be at work on or near it while it is live
➤ suitable precautions (including where necessary the provision of suitable protective equipment) are taken to prevent injury.

21.16.13 *Working space access and lighting – regulation 15*

Adequate working space, means of access and lighting shall be provided for all electrical equipment at which or near which work is being done in circumstances which may give rise to danger. This covers work of any kind. However, when the work is on live conductors the access space must be sufficient for a person to fall back out of danger and, if needed, for persons to pass one another with ease and without hazard.

21.16.14 *Competence – regulation 16*

Where technical knowledge or experience is necessary to prevent danger, all persons must possess such knowledge or experience or be under appropriate supervision.

21.16.15 *References*

Avoidance of Danger from Overhead Electric Lines. GS6, 1997 HSE Books, ISBN 9780-7176-1348-8.

Electrical Safety and You. INDG231, 1996 HSE Books. ISBN 9780-7176-1207-4.
Electrical Safety in Construction: A summary Guide for the Construction Industry on Safe Isolation Practices. HSE and Electrical safety Council, http://www.electricalsafetycouncil.org.uk.
Electrical Safety on Construction Sites. HSG 141, 1995, ISBN 9780-7176-1000-4.
Electrical Switchgear and Safety. A Concise Guide for Users. INDG372, HSE Books 2003, ISBN 9780-7176-2187-1.
Electricity at Work: Safe Working Practices. HSG 85 2003, HSE Books, ISBN 9780-7176-2164-2.
Guidance on Safe Isolation Procedures. SELECT, http://www.select.org.uk, ISBN 9780-7176-1272-4.
Maintaining Portable and Transportable Electrical Equipment. HSG 107 2004, HSE Books, ISBN 9780-7176-2805-1.
Maintaining Portable Electrical Equipment in Offices and Other Low-risk Environments. INDG236, 1996 HSE Books.
Memorandum of Guidance on the Electricity at Work Regulations, Guidance on Regulations, HSR25, 2nd edition, 2007 HSE Books, ISBN 9780-7176-6228-9.
New Wiring Colours. 2006, http://www.iee.org/cablecolours.
Safety in Electrical Testing at Work, General Guidance. INDG354, 2003 HSE Books, ISBN 9780-7176-2296-7.

21.17 Employers' Liability (Compulsory Insurance) Act 1969 and Regulations 1998 amended in 2004

21.17.1 *Introduction*

Employers are responsible for the health and safety of employees while they are at work. Employees may be injured at work, or they or former employees may become ill as a result of their work while employed. They may try to claim compensation from the employer if they believe them to be responsible. The Employers' Liability Compulsory Insurance Act 1969 ensures that an employer has at least a minimum level of insurance cover against any such claims.

Employers' liability insurance will enable employers to meet the cost of compensation for employees' injuries or illnesses whether they are caused on or off site. However, any injuries or illnesses relating to motor accidents that occur while employees are working for them, may be covered separately by motor insurance.

Public liability insurance is different. It covers for claims made against a person/company by members of the public or other businesses, but not for claims made by employees. While public liability insurance is generally voluntary, employers' liability insurance is compulsory. Employers can be fined if they do not hold a current employers' liability insurance policy which complies with the law.

21.17.2 *Application*

An employer needs employers' liability insurance unless they are exempt from the Employers' Liability Compulsory Insurance Act. The following employers are exempt:

➤ most public organizations including government departments and agencies, local authorities, police authorities and nationalized industries
➤ health service bodies, including National Health Service trusts, health authorities, Family Health Services Authorities and Scottish Health Boards and State Hospital Management Committees
➤ some other organizations which are financed through public funds, such as passenger transport executives and magistrates' courts committees
➤ family businesses, that is if employees are closely related to the employer (as husband, wife, father, mother, grandfather, grandmother, stepfather, stepmother, son, daughter, grandson, granddaughter, stepson, stepdaughter, brother, sister, half-brother or half-sister). However, this exemption does not apply to family businesses that are incorporated as limited companies except any employer which is a company that has only one employee who owns 50% or more of the share capital.

A full list of employers who are exempt from the need to have employers' liability insurance is shown at schedule 2 of the Employers' Liability (Compulsory Insurance) Regulations 1998.

21.17.3 *Coverage*

Employers are only required by law to have employers' liability insurance for people whom they employ. However, people who are normally thought of as self-employed may be considered to be employees for the purposes of employers' liability insurance.

Whether or not an employer needs employers' liability insurance for someone who works for them depends on the terms of the contract. This contract can be spoken, written or implied. It does not matter whether someone is usually called an employee or self-employed or what their tax status is. Whether the contract is called a contract of employment or a contract for services is largely irrelevant. What matters is the real nature of the employer/employee relationship and the degree of control the employer has over the work they do.

There are no hard and fast rules about who counts as employee for the purposes of Employers' liability insurance. The following paragraphs may help to give some indication.

In general, employers' liability insurance may be needed for a worker if:

➤ national insurance and income tax is deducted from the money paid to them
➤ the employer has the right to control where and when they work and how they do it
➤ most materials and equipment are supplied by the employer
➤ the employer has a right to any profit workers make even though the employer may choose to share this with them through commission, performance pay or shares in the company. Similarly, the employer will be responsible for any losses
➤ that person is required to deliver the service personally and they cannot employ a substitute if they are unable to do the work
➤ they are treated in the same way as other employees, for example, if they do the same work under the same conditions as some other employee.

In most cases employers' liability insurance is needed for volunteers. Although in general, the law may not require an employer to have insurance for:

➤ students who work unpaid
➤ people who are not employed but are taking part in youth or adult training programmes
➤ schoolchildren on work experience programmes.

However, in certain cases, these groups might be classed as employees. In practice, many insurance companies will provide cover for people in these situations.

One difficult area is domestic help. In general, an employer will probably not need employers' liability insurance for people such as cleaners or gardeners if they work for more than one person. However, if they only work for one employer, that employer may be required to take out insurance to protect them.

21.17.4 *Display of certificate*

Under the Regulations, employers must display in a suitable convenient location, a current copy of the certificate of insurance at each place of business where they employ relevant people.

21.17.5 *Retention of certificates*

An employer must retain for at least 40 years copies of certificates of insurance which have expired. This is because claims for diseases can be made many years after the disease is caused. Copies can be kept electronically if this is more convenient than paper. An employer must make these available to health and safety inspectors on request.

These requirements do not apply to policies which expired before 1 January 1999. However, it is still very important to keep full records of previous insurance policies for the employer's own protection.

21.17.6 *Penalties*

The HSE enforces the law on employers' liability insurance and HSE inspectors will check that employers have employers' liability insurance with an approved insurer for at least £5 million. They will ask to see the certificate of insurance and other insurance details.

Employers can be fined up to £2,500 for any day they are without suitable insurance. If they do not display the certificate of insurance or refuse to make it available to HSE inspectors when they ask, employers can be fined up to £1,000.

21.18 Fire Precautions Regulatory Reform (Fire Safety) Order 2005

21.18.1 *Introduction*

This order, made under the Regulatory Reform Act 2001, reforms the law relating to fire safety in non-domestic premises. It replaces fire certification under the Fire Precautions Act 1971 (which it repeals) with a general duty to ensure, SFARP, the safety of employees, a general duty, in relation to non-employees to take such fire precautions as may reasonably be required in the circumstances to ensure that premises are safe and a duty to carry out a risk assessment.

The Fire Certificate (Special premises) Regulations 1976, The Fire Precautions (Workplace) Regulations 1997 and amendment Regulations 1999 are all revoked.

The Fire Safety Order is the responsibility of the Department for Communities and Local Government. It is enforced by the fire and rescue authorities with some exceptions, which include:

➤ the HSE for:
  ➤ nuclear Installations
  ➤ ships in construction or repair and
  ➤ construction sites other than a construction site which is contained within, or forms part of, premises which are occupied by persons other than those carrying out the construction work or any activity arising from such work

➤ in Crown occupied and Crown owned buildings enforcement is carried out by the Fire Services Inspectorates appointed under the Fire Services Act 1947
➤ Local Authorities for Certificated Sports Grounds.

The order came into force on 1 October 2006 with a number of guides covering a wide range of premises. They are available on http://www.communities.gov.uk/fire/firesafety/firesafetylaw/.

Scotland
In Scotland the Fire Safety legislation is enacted through the Fire (Scotland) Act 2005, which in brief, covers the following.

The Fire (Scotland) Act 2005 received Royal Assent on 1 April 2005. Parts 1, 2, 4 and 5 of the Act commenced in August 2005. Part 3 introduces a new fire safety regime for non-domestic premises and came into force on 1 October 2006 and will replace the Fire Precautions Act 1971 and the Fire Precautions (Workplace) Regulations 1997, as amended. Fire certificates will no longer be required after 1 October 2006 and the new fire safety regime will be based on the principle of risk assessment (similar to the Fire Precautions (Workplace) Regulations).

Fire safety Regulations in Scotland have been prepared which contain detailed provisions in respect of fire safety risk assessments and fire safety measures. These Regulations came into force on 1 October 2006.

Ten guidance documents to complement the new legislation and help those with fire safety responsibilities to understand their duties is available under the Scottish law. A guidance booklet which is applicable to all premises, regardless of the size and activities undertaken on the premises, is available, free of charge, from Blackwell's Bookshop, 53 South Bridge, Edinburgh, EHI IYS. Local fire and rescue authorities also have a supply of the booklets and may be able to assist with any requests for additional copies (check the phonebook for contact details).

The guidance booklet is available for downloading from a new website dedicated to the new fire safety regime, which can be accessed at http://www.infoscotland.com/firelaw. The website includes a facility for signing up to receive email updates on progress with the new legislation and the availability of guidance documents.

If there are any queries about the new fire safety legislation, publicity campaign or guidance documents, contact the Justice Department Fire Branch on 0131 244 5338 or 0131 244 2784. Alternatively send an email to FireScotlandAct@scotland.gsi.gov.uk.

21.18.2 *Process fire risks*

HSE is mainly concerned with process fire precautions. These are the special fire precautions requireda in any workplace in connection with the work process that is being carried out there (including the storage of articles, substances and materials relating to that work process). They are to prevent or reduce the likelihood of a fire breaking out and if a fire does occur, to reduce its spread and intensity. Some examples of process fire precautions are:

➤ storage of flammable liquids in process areas, workrooms, laboratories and similar working areas
➤ ventilation systems to dilute or remove flammable gas or vapour
➤ selecting equipment that will not be a source of ignition
➤ extraction systems to remove combustible materials such as wood dust.

Process fire precautions are enforced by HSE or the local authority, under the Health and Safety at Work etc Act 1974 (HSW Act); the Management of Health and Safety at Work Regulations 1999 (MHSWR); and more specific health and safety legislation such as the DSEAR.

Part 1 General

21.18.3 *Interpretation – articles 1–7*

Here are a few of the more important definitions from the articles. For a full list consult the Order directly.

(a) 'Premises' includes any place and, in particular, includes:
➤ any workplace
➤ any vehicle, vessel, aircraft or hovercraft
➤ any installation on land (including the foreshore and other land intermittently covered by water), and any other installation (whether floating, or resting on the seabed or the subsoil thereof, or resting on other land covered with water or the subsoil thereof) and
➤ any tent or movable structure
(b) 'risk' means the risks to the safety of persons from fire
(c) 'safety' means the safety of persons in the event of fire
(d) 'workplace' means any premises or parts of premises, not being domestic premises, used for the purposes of an employer's undertaking and which are made available to an employee of the employer as a place of work and includes:
➤ any place within the premises to which such employee has access while at work and

➤ any room, lobby, corridor, staircase, road, or other place:
➤ used as a means of access to or egress from that place of work or
➤ where facilities are provided for use in connection with that place of work, other than a public road.
(e) 'responsible person' means
➤ in relation to a workplace, the employer, if the workplace is to any extent under their control
➤ in relation to any premise not falling within paragraph (a):
➤ the person who has control of the premises (as occupier or otherwise) in connection with the carrying on by them of a trade, business or other undertaking (for profit or not) or
➤ the owner, where the person in control of the premises does not have control in connection with the carrying on by that person of a trade, business or other undertaking.
(f) 'general fire precautions' in relation to premises means
➤ measures to reduce the risk of fire and the risk of the spread of fire
➤ the means of escape from the premises
➤ measures for securing that, at all material times, the means of escape can be safely and effectively used
➤ measures in relation to fighting fires
➤ means for detecting fires and giving warning in case of fire
➤ action to be taken in the event of fire, including:
➤ instruction and training of employees
➤ measures to mitigate the effects of the fire.

These issues do not cover process-related fire risks including:

➤ the use of plant or machinery or
➤ the use or storage of any dangerous substance.

Duties are placed on responsible persons in a workplace and in premises which are not workplaces to the extent that they have control over the premises.
The order does not apply (article 8) to:

➤ domestic premises
➤ an offshore installation
➤ a ship, in respect of the normal ship-board activities of a ship's crew which are carried out solely by the crew under the direction of the master
➤ fields, woods or other land forming part of an agricultural or forestry undertaking but which is not inside a building and is situated away from the undertaking's main buildings
➤ an aircraft, locomotive or rolling stock, trailer or semi-trailer used as a means of transport or a vehicle

➤ a mine other than any building on the surface at a mine

➤ a borehole site.

In addition certain provisions of the Order do not apply to groups of workers such as:

➤ occasional work which is not harmful to young people in a family undertaking
➤ armed forces
➤ members of police forces
➤ emergency services.

Part 2 Fire safety duties

21.18.4 *Duty to take general fire precautions – article 8*

The responsible person must:

(a) take such general fire precautions as will ensure, SFARP, the safety of any of his employees and
(b) in relation to relevant persons who are not his employees, take such general fire precautions as may reasonably be required in the circumstances of the case to ensure that the premises are safe.

21.18.5 *Risk assessment and fire safety arrangements – articles 9 and 11*

The responsible person must make a suitable and sufficient assessment of the risks to identify the general fire precautions he needs to take.

Where a dangerous substance is or is liable to be present in or on the premises, the risk assessment must include consideration of the matters set out in Part 1 of schedule 1 reproduced in Table 21.9.

Risk assessments must be reviewed by the responsible person regularly and if they are no longer valid or there has been significant changes. The responsible person must not employ a young person unless risks to young persons have been considered in an assessment covering the following, which is Part 2 of schedule 1.

Matters to be taken into particular account in risk assessment in respect of young persons
The matters are:

(a) the inexperience, lack of awareness of risks and immaturity of young persons
(b) the fitting-out and layout of the premises
(c) the nature, degree and duration of exposure to physical and chemical agents
(d) the form, range, and use of work equipment and the way in which it is handled
(e) the organization of processes and activities

Table 21.9 Matters to be considered in risk assessment in respect of dangerous substances

The matters are:
(a) the hazardous properties of the substance
(b) information on safety provided by the including supplier, information contained in any relevant safety data sheet
(c) the circumstances of the work including
(i) the special, technical and organizational measures and the substances used and their possible interactions
(ii) the amount of the substance involved
(iii) where the work will involve more than one dangerous substance, the risk presented by such substances in combination and
(iv) the arrangements for the safe handling, storage and transport of dangerous substances and of waste containing dangerous substances
(d) activities, such as maintenance, where there is the potential for a high level of risk
(e) the effect of measures which have been or will be taken pursuant to this Order
(f) the likelihood that an explosive atmosphere will occur and its persistence
(g) the likelihood that ignition sources, including electrostatic discharges, will be present and become active and effective
(h) the scale of the anticipated effects
(i) any places which are, or can be connected via openings to, places in which explosive atmospheres may occur and
(j) such additional safety information as the responsible person may need in order to complete the assessment.

(f) the extent of the safety training provided or to be provided to young persons and
(g) risks from agents, processes and work listed in the Annex to Council Directive 94/33/EC (**a**) on the protection of young people at work.

As soon as practicable after the assessment is made or reviewed, the responsible person must record the information prescribed where:

(a) he employs five or more employees
(b) a licence under an enactment is in force in relation to the premises or
(c) an **Alterations Notice** requiring this is in force in relation to the premises.

The prescribed information is:

(a) the significant findings of the assessment, including the measures which have been or will be taken by the responsible person and

(b) any group of persons identified by the assessment as being especially at risk.

The responsible person must make and record (as per risk assessments) arrangements as are appropriate, for the effective planning, organization, control, monitoring and review of the preventive and protective measures.

21.18.6 *Principles of prevention to be applied and fire safety arrangements – article 10*

Preventive and protective measures must be implemented on the basis of the principles specified in Part 3 of schedule 1 as follows, which are broadly the same as those in the Management Regulations:

Principles of prevention
The principles are:

(a) avoiding risks
(b) evaluating the risks which cannot be avoided
(c) combating the risks at source
(d) adapting to technical progress
(e) replacing the dangerous by the non-dangerous or less dangerous
(f) developing a coherent overall prevention policy which covers technology, organization of work and the influence of factors relating to the working environment
(g) giving collective protective measures priority over individual protective measures and
(h) giving appropriate instructions to employees.

Fire Safety arrangements must be made and put into effect by the responsible person where more than four are employed, there is a licence for the premises or an Alterations Notice requires them to be recorded. The arrangements must cover effective planning, organization, control, monitoring and review of the preventative and protective measures.

21.18.7 *Elimination or reduction of risks from dangerous substances – article 12*

Where a dangerous substance is present in or on the premises, the responsible person must ensure that risk of the substance is either eliminated or reduced SFARP, by firstly replacing the dangerous substance with a safer alternative.

Where it is not reasonably practicable to eliminate risk the responsible person must, SFARP, apply measures

including the measures specified in Part 4 of schedule 1 to the Order (see Table 21.10).

The responsible person must:

(a) arrange for the safe handling, storage and transport of dangerous substances and waste containing dangerous substances and

(b) ensure that any conditions necessary for ensuring the elimination or reduction of risk are maintained.

21.18.8 *Firefighting and fire detection – article 13*

The premises must be equipped with appropriate firefighting equipment and with fire detectors and alarms; any non-automatic firefighting equipment provided must be easily accessible, simple to use and indicated by signs.

Appropriate measures must be taken for firefighting in the premises; nominate and train competent persons to implement the measures; arrange any necessary contacts with external emergency services.

A person is to be regarded as competent where they have sufficient training and experience or knowledge and other qualities to enable them to properly implement the measures referred to.

21.18.9 *Emergency routes and exits – article 14*

Where necessary, routes to emergency exits from premises and the exits themselves must be kept clear at all times.s

The following requirements must be complied with:

(a) Emergency routes and exits must lead as directly as possible to a place of safety.

(b) In the event of danger, it must be possible for persons to evacuate the premises as quickly and as safely as possible.

(c) The number, distribution and dimensions of emergency routes and exits must be adequate having regard to the use, equipment and dimensions of the premises and the maximum number of persons who may be present there at any one time.

(d) Emergency doors must open in the direction of escape.

(e) Sliding or revolving doors must not be used for exits specifically intended as emergency exits.

(f) Emergency doors must not be so locked or fastened that they cannot be easily and immediately opened by any person who may require to use them in an emergency.

(g) Emergency routes and exits must be indicated by signs and.

(h) Emergency routes and exits requiring illumination must be provided with emergency lighting of adequate intensity in the case of failure of their normal lighting.

Table 21.10   Measures to be taken in respect of dangerous substances

1. In applying measures to control risks the responsible person must, in order of priority:
(a) reduce the quantity of dangerous substances to a minimum
(b) avoid or minimize the release of a dangerous substance
(c) control the release of a dangerous substance at source
(d) prevent the formation of an explosive atmosphere, including the application of appropriate ventilation
(e) ensure that any release of a dangerous substance which may give rise to risk is suitably collected, safely contained, removed to a safe place, or otherwise rendered safe, as appropriate
(f) avoid:
(i) ignition sources including electrostatic discharges and
(ii) such other adverse conditions as could result in harmful physical effects from a dangerous substance.
(g) segregate incompatible dangerous substances.
2. The responsible person must ensure that mitigation measures applied in accordance with article 12 (3) (b) include:
(a) reducing to a minimum the number of persons exposed
(b) measures to avoid the propagation of fires or explosions
(c) providing explosion pressure relief arrangements
(d) providing explosion suppression equipment
(e) providing plant which is constructed so as to withstand the pressure likely to be produced by an explosion and
(f) providing suitable personal protective equipment.
3. The responsible person must:
(a) ensure that the premises are designed, constructed and maintained so as to reduce risk
(b) ensure that suitable special, technical and organizational measures are designed, constructed, assembled, installed, provided and used so as to reduce risk
(c) ensure that special, technical and organizational measures are maintained in an efficient state, in efficient working order and in good repair
(d) ensure that equipment and protective systems meet the following requirements:
(i) Where power failure can give rise to the spread of additional risk, equipment and protective systems must be able to be maintained in a safe state of operation independently of the rest of the plant in the event of power failure.
(ii) Means for manual override must be possible, operated by employees competent to do so, for shutting down equipment and protective systems incorporated within automatic processes which deviate from the intended operating conditions, provided that the provision or use of such means does not compromise safety.
(iii) On operation of emergency shutdown, accumulated energy must be dissipated as quickly and as safely as possible or isolated so that it no longer constitutes a hazard.
(iv) Necessary measures must be taken to prevent confusion between connecting devices.
(e) where the work is carried out in hazardous places or involves hazardous activities, ensure that appropriate systems of work are applied including:
(i) the issuing of written instructions for the carrying out of work and
(ii) a system of permits to work, such permits being issued by a person with responsibility for this function prior to the commencement of the work concerned.

21.18.10 *Procedures for serious and imminent danger and for danger areas – article 15*

The responsible person must:

(a) establish appropriate procedures, including safety drills
(b) nominate a sufficient number of competent persons to implement evacuation procedures
(c) provide adequate safety instruction for restricted areas.

Persons who are exposed to serious and imminent danger must be informed of the nature of the hazard and of the steps taken or to be taken to protect them from it; they must be able to stop work and immediately proceed to a place of safety in the event of their being exposed to serious, imminent and unavoidable danger; and procedures must require the persons concerned to be prevented from resuming work in any situation where there is still a serious and imminent danger.

21.18.11 *Additional emergency measures in respect of dangerous substances – article 16*

In order to safeguard persons from an accident, incident or emergency related to the presence of a dangerous substance the responsible person must (unless the risk assessment shows it is unnecessary) ensure that:

(a) information on emergency arrangements is available, including:
  (i) details of relevant work hazards and hazard identification arrangements and
  (ii) specific hazards likely to arise at the time of an accident, incident or emergency.
(b) suitable warning and other communication systems are established to enable an appropriate response, including remedial actions and rescue operations, to be made immediately when such an event occurs
(c) where necessary, before any explosion conditions are reached, visual or audible warnings are given and relevant persons withdrawn and
(d) where the risk assessment indicates it is necessary, escape facilities are provided and maintained to ensure that, in the event of danger, persons can leave endangered places promptly and safely.

The information required must be:

(a) made available to accident and emergency services and
(b) displayed at the premises, unless the results of the risk assessment make this unnecessary.

In the event of a fire arising from an accident, incident or emergency related to the presence of a dangerous substance in or on the premises, the responsible person must ensure that:

(a) immediate steps are taken to:
  (i) mitigate the effects of the fire
  (ii) restore the situation to normal and
  (iii) inform those persons who may be affected.
(b) only those persons who are essential for the carrying out of repairs and other necessary work are permitted in the affected area and they are provided with:
  (i) appropriate PPE and protective clothing and
  (ii) any necessary specialized safety equipment and plant,

which must be used until the situation is restored to normal.

21.18.12 *Maintenance – article 17*

Any facilities, equipment and devices provided must be subject to a suitable system of maintenance and maintained in an efficient state, in efficient working order and in good repair.

21.18.13 *Safety assistance – article 18*

The responsible person must (except a competent self-employed person) appoint one or more competent persons to assist them in undertaking the preventive and protective measures. If more than one person is appointed, the responsible person and the appointed competent persons must make arrangements for ensuring adequate co-operation between them.

The number of persons appointed, the time available for them to fulfil their functions and the means at their disposal must be adequate having regard to the size of the premises, the risks to which relevant persons are exposed and the distribution of those risks throughout the premises.

21.18.14 *Provision of information to employees and others – articles 19 and 20*

The responsible person must provide the employees with comprehensible and relevant information on:

➤ the risks to them identified by the risk assessment
➤ the preventive and protective measures
➤ the procedures for fire drills
➤ the identities of persons nominated for firefighting or fire drills
➤ the risks notified to him regarding shared premises.

Before employing a child, a parent (or guardian) of the child must be provided with comprehensible and relevant information on the risks to that child as identified by the risk assessment; the preventive and protective measures; and the risks notified to the parent (or guardian) regarding shared premises.

Where a dangerous substance is present in or on the premises, additional information must be provided for employees as follows:

(a) the details of any such substance including:
 (i) the name of the substance and the risk which it presents
 (ii) access to any relevant SDS and
 (iii) legislative provisions which apply to the substance
(b) the significant findings of the risk assessment.

The responsible person must ensure that the employer of any employees from an outside undertaking who are working in or on the premises is provided with comprehensible and relevant information on:

(a) the risks to those employees and
(b) the preventive and protective measures taken by the responsible person.

The responsible person must ensure that any person working in his undertaking who is not his employee is provided with appropriate instructions and comprehensible and relevant information regarding any risks to that person.

21.18.15 Capabilities and training – article 21

The responsible person must ensure that his employees are provided with adequate safety training:

(a) at the time when they are first employed and
(b) on their being exposed to new or increased risks because of:
 (i) their being transferred or given a change of responsibilities
 (ii) the introduction of new work equipment into, or a change respecting work equipment already in use
 (iii) the introduction of new technology or
 (iv) the introduction of a new system of work, or a change respecting a system of work already in use.

The training must:

(a) include suitable and sufficient instruction and training on the appropriate precautions and actions to be taken by the employee
(b) be repeated periodically where appropriate
(c) be adapted to take account of any new or changed risks to the safety of the employees concerned

(d) be provided in a manner appropriate to the risk identified by the risk assessment and
(e) take place during working hours.

21.18.16 Co-operation and co-ordination – article 22

➤ Where two or more responsible persons share, or have duties in respect of, premises (whether on a temporary or a permanent basis), each such person must co-operate with, co-ordinate safety measures and inform the other responsible person concerned so far as is necessary to enable them to comply with the requirements and prohibitions imposed on them.
➤ Where two or more responsible persons share premises (whether on a temporary or a permanent basis) where an explosive atmosphere may occur, the responsible person who has overall responsibility for the premises must co-ordinate the implementation of all the measures required.

21.18.17 General duties of employees at work – article 23

Every employee must, while at work:

(a) take reasonable care for the safety of himself and of other relevant persons who may be affected by his acts or omissions at work
(b) as regards any duty or requirement imposed on his employer by or under any provision of this Order, co-operate with him so far as is necessary to enable that duty or requirement to be performed or complied with and
(c) inform his employer or any other employee with specific responsibility for the safety of his fellow employees:
 (i) of any work situation which a person with the first-mentioned employee's training and instruction would reasonably consider represented a serious and immediate danger to safety and
 (ii) of any matter which a person with the first-mentioned employee's training and instruction would reasonably consider represented a short-coming in the employer's protection arrangements for safety.

Part 3

21.18.18 Enforcement – articles 25–31

Part 3 of the order is about enforcement and penalties. An enforcing authority may be the fire and rescue authority of the area (most cases), the HSE (nuclear sites, ships and construction sites without other operations) and Fire Inspectors maintained by the Secretary of State (armed forces, UK Atomic Energy Authority and Crown premises).

Fire Inspectors or an officer of the fire brigade maintained by the fire and rescue authority, have similar powers (articles 27 and 28) to inspectors under the HSW Act.

There are differences between the HSW Act and the Fire order in the notices which can be issued. Appeals can be made against these notices within 21 days and the notices are suspended until the appeal is heard. However, a Prohibition notice stands until confirmed or altered by the court. Fire Inspectors can issue the following notices.

Alterations Notices – Article 29

These may be issued where the premises concerned constitute a serious risk to relevant people or may constitute a serious risk if any change is made to the premises or their use.

Where an Alterations Notice has been served the responsible person must notify the enforcing authority before making any specified changes, which are as follows:

➤ a change to the premises
➤ a change to the services, fittings or equipment in or on the premises
➤ an increase in the quantities of dangerous substances which are present in or on the premises
➤ a change in the use of the premises.

In addition, the Alterations Notice may also include the requirement for the responsible person to:

➤ record the significant findings of the risk assessment as per articles 9 (7) and 9 (6)
➤ record the fire safety arrangements as per articles 11 (1) and 11 (2)
➤ before making the above changes to send to the enforcing authority a copy of the risk assessment and a summary of the proposed changes to the general fire precautions.

Enforcement Notices – Article 30

Where the enforcing authority is of the opinion that the responsible person has failed to comply with the requirements of this Order or any regulations made under it they can issue an Enforcement Notice which must:

➤ state that the enforcing authority is of this opinion
➤ specify the provisions which have not been complied with; require that person to take steps to remedy the failure within a period (not less than 28 days) from the date of service of the notice.

An Enforcement Notice may include directions on the measures needed to remedy the failures. Choices of remedial action must be left open.

Before issuing an Enforcement Notice the enforcing authority must consult the relevant enforcing authorities including those under HSW Act and the Building Regulations.

Prohibition Notices – Article 31

If the enforcing authority is of the opinion that the risks, relating to escape from the premises, are so serious that the use of the premises ought to be prohibited or restricted, they may issue a Prohibition Notice. The notice must:

➤ state that the enforcing authority is of this opinion
➤ specify the provisions which give or may give rise to that risk and
➤ direct that the use to which the notice relates is prohibited or restricted as may be specified until the specified matters have been remedied.

A Prohibition Notice may include directions on the measures needed to remedy the failures. Choices of remedial action must be left open.

A Prohibition Notice takes effect immediately it is served.

Part 4

21.18.19 *Offences and appeals – articles 32–36*

Cases can be tried in a magistrates' court or on indictment in the Crown Court.

The responsible person can be liable for conviction or indictment to a fine (not limited), or to imprisonment for a term not exceeding 2 years, or both.

Any person can be liable to:

➤ on conviction or indictment to a fine (not limited) for an offence where that failure places one or more relevant people at risk of death or serious injury in case of fire.
➤ on summary conviction to a fine at standard level 3 or 5 depending on the particular offence.

In general where an offence committed by a body corporate is proved to have been committed with the consent or connivance of any director, manager, secretary or other similar officer of the body corporate they as well as the body corporate are guilty of that offence (article 32 (9)).

21.18.20 *References*

Construction Site Fire Prevention Checklist. 2007, Fire Protection Association, www.thefpa.co.uk.

Fire Precautions on Construction Sites. 6th edition, Fire Protection Association, http://www.thefpa.co.uk.

Fire Safety, An Employers Guide. HMSO, 1999 HSE Books, ISBN 9780-11-341229-0.

Fire Safety in Construction Work. HSG 168, 1997, 9780-7176-1332-1.

Regulatory Reform, England and Wales. *The Regulatory Reform (Fire Safety) Order 2005*, SI 2005 No. 1541.

There are a Total of Eleven Official Guides Produced by *The Department for Communities and Local Government*, see their website: http://www.communities.gov.uk/fire/firesafety/firesafetylaw/aboutguides/. The guides covering the following subjects are:

- Educational Premises ISBN 978-1-85112-819-8
- Factories and Warehouses ISBN 978-1-85112-816-7
- Healthcare Premises ISBN 978-1-85112-824-2
- Large Places of Assembly ISBN 978-1-85112-821-1
- Offices and Shops ISBN 978-1-85112-8150
- Open Air Events and Venues ISBN 978-1-85112-823-5
- Residential Care Premises ISBN 978-1-85112-818-1
- Sleeping Accommodation ISBN 978-1-85112-817-4
- Small and Medium Places of Assembly ISBN 13: 978-1-85112-820-4
- Theatres, Cinemas and Similar Places ISBN 978-1-85112-822-8
- Transport Premises and Facilities ISBN 978-1-85112-825-9.

21.19 Health and Safety (First Aid) Regulations 1981 as amended in 2002

21.19.1 *Introduction*

These Regulations set out employers' duties to provide adequate first-aid facilities. They define first aid as:

- treatment for the purposes of preserving life and minimizing the consequences of injury and illness until medical help is obtained
- treatment of minor injuries which would otherwise receive no treatment or which do not need treatment by a medical practitioner or nurse.

21.19.2 *Duty of the employer – regulation 3*

An employer shall provide or ensure that they are provided with:

- adequate and appropriate facilities and equipment
- qualified first aiders to render first aid

- an appointed person, being someone to take charge of situations as well as first-aid equipment and facilities, where medical aid needs to be summoned. An appointed person will suffice where:
  - the nature of the work is such that there are no specific serious hazards (offices, libraries, etc.), the workforce is small, the location makes further provision unnecessary
  - there is temporary (not planned holidays) or exceptional absence of the first aider.

There must always be at least an appointed person in every workplace during working hours.

Employers must make an assessment of the first-aid requirements that are appropriate for each workplace.

Any first-aid room provided under this regulation must be easily accessible to stretchers and to any other equipment needed to convey patients to and from the room. They must be signposted according the safety signs and signals Regulations.

21.19.3 *Employees information – regulation 4*

Employees must be informed of the arrangements for first aid, including the location of facilities, equipment and people.

21.19.4 *Self-employed – regulation 5*

The self-employed shall provide such first-aid equipment as is appropriate to render first aid to themselves.

21.19.5 *References*

Basic Advice on First Aid at Work. INDG347 (revised), 2006, ISBN 9780-7176-6193-8.

First Aid at Work. INDG214, HSE Books 1997, ISBN 9780-7176-1074-8.

First Aid at Work. Health and Safety (First Aid) Regulations 1981. Approved code of practice and guidance, L74, HSE Books 1997, ISBN 9780-7176-1050-0.

HSE First Aid Micro Site for General Information, http://www.hse.gov.uk/firstaid/index.htm.

The Training of First Aid at Work. HSG 212, HSE Books 2000, ISBN 9780-7176-1896-X.

21.20 Health and Safety (Information for Employees) Regulations 1989

21.20.1 *General requirements*

These Regulations require that the Approved Poster entitled *Health and safety – what you should know*, is displayed or the Approved Leaflet is distributed.

This information tells employees in general terms about the requirements of health and safety law.

Employers must also inform employees of the local address of the enforcing authority (either the HSE or the local authority) and the Employment Medical Advisory Service. This should be marked on the poster or supplied with the leaflet.

From 1 July 2000 the latest version of the poster must be displayed or distributed.

References to obsolete legal requirements are removed and the revised text focuses on the modern framework of general duties, supplemented by the basics of health and safety management and risk assessment. It includes two additional boxes: one for details of trade union or other safety representatives and one for competent persons appointed to assist with health and safety and their responsibilities.

21.20.2 *References*

Health and Safety (Information for Employees) Regulations 1989. SI No. 682.
Health and Safety Information for Employees (Modifications and Repeals) Regulations 1995. Poster: ISBN 9780-7176-2493-5, Leaflet: ISBN 9780-7176-1702-5. The leaflet is available in a number of languages.

21.21 Hazardous Waste (England and Wales) Regulations 2005

21.21.1 *Introduction*

The Hazardous Waste (England and Wales) Regulations 2005 were implemented on 16 July 2005. These are made under the Environmental Protection Act 1990. There were many changes to the previous Special Waste Regulations, but the two key ones are that Hazardous waste producers are now required to pre-register before any Hazardous waste can be collected from their premises and the Regulations apply the European Waste Catalogue codes of Hazardous wastes that will affect a much wider range of producers.

21.21.2 *Summary*

The list below is not exhaustive, but summarizes the main requirements. There is also a range of web links that will help to obtain more detail.

➤ From 16 July 2005, it is an offence for hazardous waste to be collected from a site that has not been registered or is exempt.

➤ All non-exempt sites that produce hazardous waste must be registered even if they are unlikely to have that waste collected for some time. Recent EA Guidance has clarified that it is an offence to produce hazardous waste on site and not be registered.

➤ The Regulations implement, through the List of Wastes (England) Regulations 2005, the European Waste Catalogue list of Hazardous wastes for the purposes of collection. This can be found at http://www.hmso.gov.uk and will mean that things like Principal Contractor monitors, Principal Contractor base units, fridges, TVs, oily rags and separately collected fluorescent tubes require collection under the new hazardous waste notification and documentation procedures.

➤ The EA will accept postal registrations, and registrations can be made on line through the website http://www.environment-agency.gov.uk.

➤ Each site producing hazardous waste has to have a separate registration although multiple sites can be registered on the same notification. Therefore, a head office could register all its sites centrally, but each site would have a separate unique registration number and require a separate fee.

➤ Some sites are exempt if they expect to produce less that 200 kg of hazardous waste a year – although they would then have to register if, part way through the year, they went over the threshold. These include agricultural premises, office premises, shops, premises where waste electrical and electronic equipment (WEEE) is collected, dental, veterinary and medical practices and ships. The EA indicate that 200 kg is approximately 10 small TVs, 500 fluorescent tubes or 5 small domestic fridges.

➤ Domestic waste is excluded from the Hazardous Waste Regulations on collection from the domestic property, but is then subject to the Regulations if it is separately collected or if it consists of asbestos. This includes Prescription Only Medicines (other than cytotoxic and cytostatic medicines) which will be hazardous waste.

➤ The Regulations require a new consignment note to be used from 16 July 2005 in place of the former section 62. Each consignment will require a fee to be paid to the EA by the consignee with their quarterly returns to the Agency. Clearly, this will be charged back to the collector, but a consignment might well attract more than one consignment fee if, for instance, it goes through a transfer station and the collector would have to ensure that this was considered in the price.

➤ Collection rounds will be possible but again, each site where the waste is collected would have to be left a copy of the consignment note, so the process of tracking and the paper trail will get quite complex especially as each collection will count as a consignment from the fee point of view.

➤ The Regulations ban the mixing of Hazardous waste and state that it must be stored separately on site. However, clarification on the interpretation of this is still awaited from the EA as it would, for instance, preclude the collection of computer systems that included a base unit and screen.

➤ Registration as a Hazardous Waste producer places a statutory duty on the EA to inspect the site where the hazardous waste arises.

21.21.3 References

A Guide to the Hazardous Waste Regulations and the List of Waste Regulations in England and Wales, Environment Agency. HWR01, see: http://www.environment-agency.gov.uk.

21.22 Ionising Radiation Regulations 1999

21.22.1 Introduction

The Ionising Radiations Regulations 1999 (IRR99) implement the majority of the Basic Safety Standards Directive 96129/Euratom (BSS Directive). From 1 January 2000, they replaced the Ionising Radiations Regulations 1985 (IRR85) (except for regulation 26 (special hazard assessments)).

The main aim of the Regulations and the supporting ACOP is to establish a framework for ensuring that exposure to ionizing radiation arising from work activities, whether from man-made or natural radiation and from external radiation (e.g. X-ray set) or internal radiation (e.g. inhalation of a radioactive substance), is kept as low as reasonably practicable and does not exceed dose limits specified for individuals. IRR99 also:

(a) replaces the Ionising Radiations (Outside Workers) Regulations 1993 (OWR93), which were made to implement the Outside Workers Directive 90/641/Euratom and

(b) implements a part of the Medical Exposures Directive 97143/Euratom in relation to equipment used in connection with medical exposures.

The guidance which accompanies the Regulations and ACOP gives detailed advice about the scope and duties of the requirements imposed by IRR99. It is aimed at employers with duties under the Regulations but should also be useful to others, such as radiation protection advisers, health and safety officers, radiation protection supervisors and safety representatives.

21.22.2 Essentially, work with ionizing radiation means

(a) a practice, which involves the production, processing, handling, use, holding, storage, transport or disposal of artificial radioactive substances and some naturally occurring sources, or the use of electrical equipment emitting ionizing radiation at more than 5 kV (see definition of practice in regulation 2 (1))

(b) work in places where the radon gas concentration exceeds the values in regulation 3 (1) (b) or

(c) work with radioactive substances containing naturally occurring radionuclides not covered by the definition of a practice.

21.22.3 Radiation employers

Radiation employers are essentially those employers who work with ionizing radiation, i.e. they carry out:

(a) a practice (see definition in regulation 2 (1)) or

(b) work in places where the radon gas concentration exceeds the values in regulation 3 (1) (b); or

(c) work with radioactive substances containing naturally occurring radionuclides not covered by the definition of a practice.

21.22.4 Duties of self-employed people

A self-employed person who works with ionizing radiation will simultaneously have certain duties under these Regulations, both as an employer and as an employee.

For example, self-employed persons may need to take such steps as:

➤ carrying out assessments under regulation 7

➤ providing control measures under regulation 8 to restrict exposure

➤ designating themselves as classified persons under regulation 20

➤ making suitable arrangements under regulation 21 with one or more approved dosimetry services (ADS) for assessment and recording of doses they receive

➤ obtaining a radiation passbook and keeping it up to date in accordance with regulation 21

➤ if they carry out services as an outside worker, making arrangements for their own training as required by regulation 14

➤ ensuring they use properly any dose meters provided by an ADS as required by regulation.

21.22.5 General requirements

Some of the major considerations are:

➤ notification to HSE of specific work unless specified in schedule 1 to the Regulations (regulation 6)

➤ carrying out prior risk assessment by a radiation employer before commencing a new activity (regulation 7)
➤ use of PPE (regulation 9)
➤ maintenance and examination of engineering controls and PPE (regulation 10)
➤ dose limitations (regulation 11)
➤ contingency plans for emergencies (regulation 12)
➤ radiation protection adviser appointment (regulation 13)
➤ information instruction and training (regulation 14).

21.22.6 Prior risk assessment

Where a radiation employer is required to undertake a prior risk assessment, the following matters need to be considered, where they are relevant:

➤ the nature of the sources of ionizing radiation to be used, or likely to be present, including accumulation of radon in the working environment
➤ estimated radiation dose rates to which anyone can be exposed
➤ the likelihood of contamination arising and being spread
➤ the results of any previous personal dosimetry or area monitoring relevant to the proposed work
➤ advice from the manufacturer or supplier of equipment about its safe use and maintenance
➤ engineering control measures and design features already in place or planned
➤ any planned systems of work; estimated levels of airborne and surface contamination likely to be encountered
➤ the effectiveness and the suitability of PPE to be provided
➤ the extent of unrestricted access to working areas where dose rates or contamination levels are likely to be significant; possible accident situations, their likelihood and potential severity
➤ the consequences of possible failures of control measures – such as electrical interlocks, ventilation systems and warning devices – or systems of work; steps to prevent identified accident situations, or limit their consequences.

This prior risk assessment should enable the employer to determine:

➤ what action is needed to ensure that the radiation exposure of all persons is kept as low as reasonably practicable (regulation 8 (1))
➤ what steps are necessary to achieve this control of exposure by the use of engineering controls, design features, safety devices and warning devices (regulation 8 (2) (a)) and, in addition, by the development of systems of work (regulation 8 (2) (b))

➤ whether it is appropriate to provide PPE and if so what type would be adequate and suitable (regulation 8 (2) (c))
➤ whether it is appropriate to establish any dose constraints for planning or design purposes, and if so what values should be used (regulation 8 (3))
➤ the need to alter the working conditions of any female employee who declares she is pregnant or is breastfeeding (regulation 8 (5))
➤ an appropriate investigation level to check that exposures are being restricted as far as reasonably practicable (regulation 8 (7))
➤ what maintenance and testing schedules are required for the control measures selected (regulation 10)
➤ what contingency plans are necessary to address reasonably foreseeable accidents (regulation 12)
➤ the training needs of classified and non-classified employees (regulation 14)
➤ the need to designate specific areas as controlled or supervised areas and to specify local rules (regulations 16 and 17)
➤ the actions needed to ensure restriction of access and other specific measures in controlled or supervised areas (regulation 18)
➤ the need to designate certain employees as classified persons (regulation 20)
➤ the content of a suitable programme of dose assessment for employees designated as classified persons and for others who enter controlled areas (regulations 18 and 21)
➤ the responsibilities of managers for ensuring compliance with these Regulations
➤ an appropriate programme of monitoring or auditing of arrangements to check that the requirements of these Regulations are being met.

21.22.7 References

Work with Ionising Radiation, Ionising Radiations Regulations, Approved Code of Practice and Guidance. HSC, L121, 2000 HSE Books, ISBN 9780-7176-1746-7.
This is a large document and need only be studied by those with a specific need to control IR. The HSE have also produced a number of information sheets which are available free on the Internet at their website.
Protection of Outsider Workers against Radiation. IRIS4, 2000 HSE Books, website: http://www.hsebooks.co.uk.
Industrial Radiography Managing radiation risk. IRIS1, 2000 HSE Books, website: htt[://www.hsebooks.co.uk.
Control of Radioactive Substances. IRIS8, HSE Books 2001, HSE Books website: www.hsebooks.co.uk.
HSE Radiation Micro Site: http://www.hse.gov.uk/radiation/index.htm.

21.23 Control of Lead at Work Regulations 2002

21.23.1 Introduction

These Regulations came into force in November 2002 and impose requirements for the protection of employees who might be exposed to lead at work and others who might be affected by the work. The Regulations:

➤ require occupational exposure levels for lead and lead alkyls
➤ require blood-lead action and suspension levels for women of reproductive capacity and others
➤ re-impose a prohibition for women of reproductive capacity and young persons in specified activities
➤ require an employer to carry out a risk assessment
➤ require employers to restrict areas where exposures are likely to be significant if there is a failure of control measures
➤ impose requirements for the examination and testing of engineering controls and RPE and the keeping of PPE
➤ impose new sampling procedures for air monitoring
➤ impose requirements in relation to medical surveillance
➤ require information to be given to employees
➤ require the keeping of records and identification of containers and pipes.

21.23.2 Interpretations – regulation 2

In these Regulations 'lead' means lead (including lead alkyls, lead alloys, any compounds of lead and lead as a constituent of any substance or material) which is liable to be inhaled, ingested or otherwise absorbed by persons except where it is given off from the exhaust of a vehicle on a road within the meaning of section 192 of the Road Traffic Act 1988.

'Occupational (workplace) exposure limit for lead' means in relation to:

➤ lead other than lead alkyls, a concentration of lead in the atmosphere to which any employee is exposed of 0.15 mg/m^3 (8-hour time-weighed average, TWA) and
➤ lead alkyls, a concentration of lead contained in lead alkyls in the atmosphere to which any employee is exposed of 0.10 mg/m^3 (8-hour TWA).

Suspension levels for both blood-lead concentrations and urinary-lead concentrations are defined in the Regulations.

21.23.3 Prohibitions – regulation 4

There are prohibitions on the use of glazes containing lead in the manufacture of pottery, and employing a young person or a women of reproductive capacity in any activity contained in schedule 1 in lead smelting and refining or in many processes in lead acid battery manufacture.

21.23.4 Assessment of risk to health – regulation 5

A suitable and sufficient risk assessment must be made and the findings implemented prior to any work which may expose people to lead. The risk assessments must include:

➤ hazardous properties of lead
➤ information on health effects from the supplier
➤ level, type and duration of exposure
➤ the circumstances of the work
➤ activities such as maintenance, where there is the potential for high exposures
➤ relevant occupational exposure levels
➤ the effect of preventative and control measures
➤ results of relevant medical surveillance
➤ results of monitoring exposure under regulation 9
➤ exposure where there are combined risks from lead and other substances
➤ whether exposure to lead is likely to be significant
➤ any other information needed to complete the risk assessment.

The risk assessment must be fully reviewed if exposure information shows it to be necessary, there are significant changes to the work, or medical surveillance shows blood-lead concentrations are exceeded.

Where there are more than four employees the employer must record the significant findings and the steps taken to meet the requirements of regulation 6.

21.23.5 Prevention or control of exposure to lead – regulation 6

Exposure must be prevented or where this is not reasonably practicable adequately controlled.

A hierarchy of controls must be considered starting with substitution with a safer alternative. Other protection measures in order of priority are:

➤ the design and use of appropriate work processes, systems and engineering controls and the provision and use of suitable work equipment and materials
➤ the control of exposure at source, including adequate ventilation systems and appropriate organizational measures and

505

➤ where adequate control of exposure cannot be achieved by other means, the provision of suitable PPE in addition to the measures required by (a) and (b).

The measures shall include:

➤ arrangements for the safe handling, storage and transport of lead and waste containing lead
➤ the adoption of suitable maintenance procedures
➤ reducing to the minimum required for the work concerned:
  ➤ the numbers of employees subject to exposure
  ➤ the level and duration of exposure
  ➤ the quantity of lead present at the workplace
➤ the control of the working environment, including appropriate general ventilation and
➤ appropriate hygiene measures including washing facilities.

Control will only be considered adequate if:

➤ the workplace exposure limit is not exceeded or
➤ if exceeded, the employer identifies the reasons and takes immediate steps to remedy the situation.

PPE must be provided in compliance with the PPE Regulations 2002 and with respiratory equipment where it is not covered with a type approved by the HSE.

Employers must take reasonable steps to make sure that control measures and facilities provided is properly used or applied.

Employees shall make full and proper use of any control measures or facility provided and report any defects discovered.

21.23.6 *Eating, drinking and smoking – regulation 7*

Eating, drinking and smoking in an area liable to be contaminated by lead is not allowed. Drinking facilities may be provided, if required for employees' welfare, and used as long as they are not liable to be contaminated by lead.

21.23.7 *Maintenance, examination and testing of control measures – regulation 8*

All control measures must be maintained by the employer in an efficient state, in efficient working order, in good repair and in a clean condition.

Engineering controls must have a thorough examination and testing at suitable intervals (at least every 14 months for exhaust ventilation plant).

Respiratory equipment must have a thorough examination and where appropriate testing at suitable intervals.

Suitable records must be kept, for at least 5 years, of the examinations and tests and of any repairs carried out.

PPE must be properly stored, checked at suitable intervals and when defective repaired or replaced. Contaminated PPE must be separated and decontaminated or destroyed.

21.23.8 *Air monitoring – regulation 9*

Where significant exposure is liable to happen the concentration of lead in air must be measured in accordance with a suitable procedure.

The monitoring must be repeated every 3 months but on complying with certain conditions can be extended to 12 months (except lead alkyls). The conditions are:

➤ no material changes have taken place in the work or conditions of exposure
➤ the lead in air concentrations for each group of employees or work area has not exceeded $0.10 \, mg/m^3$ on the two previous occasions on which monitoring was carried out.

Records must be kept for minimum of 5 years unless medical surveillance is being carried out (then 40 years).

Employees have a right to see personal monitoring record on giving reasonable notice to the employer.

21.23.9 *Medical surveillance – regulation 10*

Every employee who is liable to be exposed to lead must be under suitable medical surveillance where exposure is liable to be significant; specified blood-lead concentrations or urinary-lead concentrations are equalled or exceeded; and a relevant doctor certifies that the employee should be under such medical surveillance:

➤ The technique of investigation should be of low risk to the employees.
➤ Medical surveillance must commence before or at least within 14 days of an employee being exposed and thereafter at 12 months or as the relevant doctor prescribes.
➤ Biological monitoring must be carried out at least every 6 months except for young persons and women of reproductive capacity, when it must be done every 3 months.
➤ Records must be made and retained for at least 40 years.
➤ Access to personal health records is permitted on reasonable notice from the employee concerned.

The employer must take steps to determine the reason for blood levels having been exceeded and take steps to reduce the levels. When the blood-lead levels or urinary-lead levels reach the appropriate suspension levels a relevant doctor must certify whether the person should be suspended from this type of work; or if no suspension is certified reasons must be given.

Employers must provide suitable facilities when examinations are carried out on the premises and employees must make themselves available for examination.

21.23.10 *Information, instruction and training – regulation 11*

Employees must be provided with suitable information, instruction and training to include:

➤ details of the form of lead, risks to health, exposure limits, action level and suspension level
➤ access to relevant SDS
➤ the significant findings of the risk assessment
➤ appropriate precautions and actions to be taken
➤ results of any monitoring of exposure
➤ collective results of any medical surveillance in a form calculated to prevent individuals being identified.

These duties extend to any person whether or not they are employees who carry out work in connection with the employer's duties under these Regulations.

Where containers and pipes are not marked by any relevant legislation (given in schedule 2), the employer shall make sure they are marked with the nature of the contents and any associated hazards.

21.23.11 *References*

Control of Lead at Work Regulations 2002. SI No. 2676, Office of Public Sector Information website: http://www.opsi.gov.uk/si/si2002/uksi_20022676_en.pdf.
The Control of Lead at Work Approved Code of Practice. L132, 2002 HSE Books, ISBN 9780-7176-2565-6.
Lead and You. INDG305, 1998 HSE Books, ISBN 9780-7176-1523-3.

21.24 Lifting Operations and Lifting Equipment Regulations 1998

21.24.1 *Introduction*

This summary gives information about the Lifting Operations and Lifting Equipment Regulations 1998 (LOLER) which came into force on 5 December 1998.

In the main, LOLER replaced existing legal requirements relating to the use of lifting equipment, for example

the Construction (Lifting Operations) Regulations 1961, the Docks Regulations 1988 and the Lifting Plant and Equipment (Records of Test and Examination, etc.) Regulations 1992.

The Regulations aim to reduce risks to people's health and safety from lifting equipment provided for use at work. In addition to the requirements of LOLER, lifting equipment is also subject to the requirements of the PUWER 98.

Generally, the Regulations require that lifting equipment provided for use at work is:

➤ strong and stable enough for the particular use and marked to indicate SWL
➤ positioned and installed to minimize any risks
➤ used safely, that is the work is planned, organized and performed by competent people
➤ subject to ongoing thorough examination and, where appropriate, inspection by competent people.

21.24.2 *Definition*

Lifting equipment includes **any equipment used at work for lifting or lowering loads**, including attachments used for anchoring, fixing or supporting it. The Regulations cover a wide range of equipment including cranes, forklift trucks, lifts, hoists, mobile elevating work platforms, and vehicle inspection platform hoists. The definition also includes lifting accessories such as chains, slings and eyebolts. LOLER **does not** apply to escalators; these are covered by more specific legislation (i.e. the Workplace (Health, Safety and Welfare) Regulations 1992).

If employees are allowed to provide their own lifting equipment, then this too is covered by the Regulations.

21.24.3 *Application*

The Regulations apply to an employer or self-employed person providing lifting equipment for use at work, or who has control of the use of lifting equipment. They do not apply to equipment to be used primarily by members of the public, for example, lifts in a shopping centre. However, such circumstances are covered by the HSW Act 1974.

LOLER applies to the way lifting equipment is used in industry and commerce. LOLER applies only to work activities, for example:

➤ a crane on hire to a construction site
➤ a contract lift
➤ a passenger lift provided for use of workers in an office block
➤ refuse collecting vehicles lifting on a public road
➤ patient hoist
➤ forklift truck.

These Regulations add to the requirements of PUWER 98 and should be interpreted with them. For example, when selecting lifting equipment, PUWER regulation 4, regarding suitability, should be considered in connection with:

➤ ergonomics
➤ the conditions in which the equipment is to be used
➤ safe access and egress
➤ preventing slips, trips and falls
➤ protecting the operator.

While employees do not have duties under LOLER, they do have general duties under the HSW Act and the Management of Health and Safety at Work Regulations 1992 (MHSWR), for example, to take reasonable care of themselves and others who may be affected by their actions and to co-operate with others.

The Regulations cover places where the HSW Act applies – these include factories, offshore installations, agricultural premises, offices, shops, hospitals, hotels, places of entertainment, etc.

21.24.4 *Strength and stability – regulation 4*

Lifting equipment shall be of adequate strength and stability for each load, having regard in particular to the stress induced at its mounting or fixing point.

Every part of a load and anything attached to it and used in lifting it shall be of adequate strength.

Account must be taken of the combination of forces to which the lifting equipment will be subjected, as well as the weight of any lifting accessories. The equipment should include an appropriate factor of safety against failure.

Stability needs to take into account the nature, load-bearing strength, stability, adjacent excavations and slope of the surface. For mobile equipment, keeping rails free of obstruction and tyres correctly inflated must be considered.

21.24.5 *Lifting equipment for lifting persons – regulation 5*

To ensure safety of people being lifted, there are additional requirements for such equipment. The use of equipment not specifically designed for raising and lowering people should only be used in exceptional circumstances.

The regulation applies to all lifting equipment used for raising and lowering people and requires that lifting equipment for lifting persons shall:

➤ prevent a person using it from being crushed, trapped or struck, or falling from the carrier

➤ prevent, SFARP, persons using it while carrying out work from the carrier being crushed, trapped or struck or falling from the carrier
➤ have suitable devices to prevent the risk of the carrier falling. If a device cannot be fitted, the carrier must have:
  ➤ an enhanced safety coefficient suspension rope or chain
  ➤ the rope or chain inspected every working day by a competent person
  ➤ be such that a person trapped in any carrier is not thereby exposed to any danger and can be freed.

21.24.6 *Positioning and installation – regulation 6*

Lifting equipment must be positioned and installed so as to reduce the risks, SFARP, from:

➤ equipment or a load striking another person
➤ a load drifting, falling freely or being released unintentionally

and it is otherwise safe.

Lifting equipment should be positioned and installed to minimize the need to lift loads over people and to prevent crushing in extreme positions. It should be designed to stop safely in the event of a power failure and not release its load. Lifting equipment, which follows a fixed path, should be enclosed with suitable and substantial interlocked gates and any necessary protection in the event of power failure.

21.24.7 *Marking of lifting equipment – regulation 7*

Machinery and accessories for lifting loads shall be clearly marked to indicate their SWL, and:

➤ where the SWL depends on the configuration of the lifting equipment:
  ➤ the machinery should be clearly marked to indicate its SWL for each configuration
  ➤ information which clearly indicates its SWL for each configuration should be kept with the machinery
➤ accessories for lifting (e.g. hooks, slings) are also marked in such a way that it is possible to identify the characteristics necessary for their safe use (e.g. if they are part of an assembly)
➤ lifting equipment which is designed for lifting people is appropriately and clearly marked
➤ lifting equipment not designed for lifting people, but which might be used in error, should be clearly marked to show it is not for lifting people.

21.24.8 *Organization of lifting operations – regulation 8*

Every lifting operation, that is lifting or lowering of a load, shall be:

➤ properly planned by a competent person
➤ appropriately supervised
➤ carried out in a safe manner.

The person planning the operation should have adequate practical and theoretical knowledge and experience of planning lifting operations. The plan will need to address the risks identified by the risk assessment and identify the resources, the procedures and the responsibilities required so that any lifting operation is carried out safely. For routine simple lifts a plan will normally be left to the people using the lifting equipment. For complex lifting operations, for example where two cranes are used to lift one load, a written plan may need to be produced each time.

The planning should take account of avoiding suspending loads over occupied areas, visibility, attaching/detaching and securing loads, the environment, location, overturning, proximity to other objects, lifting of people and pre-use checks of the equipment.

21.24.9 *Thorough examination and inspection – regulation 9*

Before using lifting equipment for the first time by an employer, it must be thoroughly examined for any defect unless:

➤ the lifting equipment has not been used before
➤ an EC declaration of conformity (where one should have been drawn up) has been received or made not more than 12 months before the lifting equipment is put into service
➤ if it is obtained from another undertaking, it is accompanied by physical evidence of an examination.

A copy of this thorough examination report shall be kept for as long as the lifting equipment is used (or, for a lifting accessory, 2 years after the report is made) (regulation 11).

Where safety depends on the installation conditions, it shall be thoroughly examined:

➤ after installation and before being put into service
➤ after assembly and before being put into service at a new site or in new location

to ensure that it has been installed correctly and is safe to operate.

A copy of the thorough examination report shall be kept for as long as the lifting equipment is used at the place in which it was installed or assembled (regulation 11).

Lifting equipment which is exposed to conditions causing deterioration that which may result in dangerous situations, shall be:

➤ thoroughly examined at least every 6 months (for lifting equipment for lifting persons, or a lifting accessory); at least every 12 months (for other lifting equipment); or in accordance with an examination scheme; and each time that exceptional circumstances, liable to jeopardize the safety of the lifting equipment, have occurred, a copy of the report is kept until the next report is made, or for 2 years (whichever is longer)
➤ inspected, if appropriate, by a competent person at suitable intervals between 'thorough examinations' (and a copy of the record kept until the next record is made).

All lifting equipment shall be accompanied by physical evidence that the last 'thorough examination' has been carried out before it leaves an employer's undertaking (or before it is used after leaving another undertaking).

The user, owner, manufacturer or some other independent party may draw up examination schemes provided they have the necessary competence. Schemes should specify the intervals at which lifting equipment should be thoroughly examined and, where appropriate, those parts that need to be tested. The scheme should take account, for example, of its condition, the environment in which it is used, the number of lifting operations and the loads lifted.

The 'competent person' carrying out a thorough examination should have appropriate practical and theoretical knowledge and experience of the lifting equipment to be examined to enable them to detect defects or weaknesses and to assess their importance in relation to the safety and continued use of the lifting equipment. They should also determine whether a test is necessary and the most appropriate method for carrying it out.

21.24.10 *Reports and defects – regulation 10*

The person making a 'thorough examination' of lifting equipment shall:

➤ notify the employer forthwith of any defect which, in their opinion, is or could become, dangerous
➤ as soon as is practicable (within 28 days) write an authenticated report to:
  ➤ the employer
  ➤ any person who hired or leased the lifting equipment, containing the information specified in schedule 1

➤ send a copy (as soon as is practicable) to the relevant enforcing authority when there is, in their opinion, a defect with an existing or imminent risk of serious personal injury (this will always be HSE if the lifting equipment has been hired or leased).

Every employer notified of a defect following a 'thorough examination' of lifting equipment should ensure that it is not used:

➤ before the defect is rectified
➤ after a time specified in the schedule accompanying the report.

The person making an 'inspection' shall also notify the employer when, in his opinion, a defect is, or could become, dangerous and, as soon as is practicable, make a record of the inspection in writing.

21.24.11   *Reports – schedule 1*

Schedule 1 lists the information to be contained in a report of a thorough examination, for example name and address; identity of equipment; date of last through examination; SWL; appropriate interval; any dangerous or potentially dangerous defects; repairs required; date of next examination and test; details of the competent person.

21.24.12   *References*

Lifting Equipment and Lifting Operations Regulations 1998. SI No. OPSI website: http://www.opsi.gov.uk/si/si1998/19982307.htm.

LOLER 1998 (Lifting Operations and Lifting Equipment Regulations 1998). Open learning guidance, 1999 HSE Books, ISBN 9780-7176-2464-1.

LOLER: How the Regulations apply to agriculture. AIS28, HSE Books 1998.

LOLER: How the Regulations apply to forestry. AIS29, HSE Books 1998.

LOLER: How the Regulations apply to Arboriculture. AIS30, HSE Books 1998.

Safe Use of Lifting Equipment, Lifting Operations and Lifting Equipment Regulations 1998 Approved Code of Practice and Guidance. L113, 1998 HSE Books, ISBN 9780-7176-1628-2.

Simple Guide to the LOLER, 1999 HSE, INDG290, HSE Books, ISBN 9780-7176-2430-7.

Thorough Examination and Testing of Lifts. INDG339, HSE Books 2001, ISBN 9780-7176-2030-1.

21.25   Management of Health and Safety at work Regulations 1999 as amended in 2003 and 2006

21.25.1   *General*

These Regulations give effect to the European Framework Directive on health and safety. They supplement the requirements of the Health and Safety at Work Act etc 1974 and specify a range of management issues, most of which must be carried out in all workplaces. The aim is to map out the organization of precautionary measures in a systematic way, and to make sure that all staff are familiar with the measures and their own responsibilities.

21.25.2   *Risk assessment – regulation 3*

Every employer is required to make a 'suitable and sufficient' assessment of risks to employees, and risks to other people who might be affected by the organization, such as visiting contractors and members of the public. A systematic investigation of risks involved in all areas and operations is required, together with identification of the persons affected, a description of the controls in place and any further action required to reduce risks.

The risk assessments must take into account risks to new and expectant mothers and young people.

Significant findings from the assessments must be written down (or recorded by other means, such as on a computer) when there are five or more employees. The assessments need to be reviewed regularly and, if necessary, when there have been significant changes, they should be modified.

21.25.3   *Principles of prevention – regulation 4*

The following principles must be adopted when implementing any preventative and protective measures:

➤ avoiding risks
➤ evaluating the risks which cannot be avoided
➤ combating the risks at source
➤ adapting the work to the individual, especially as regards the design of workplaces, the choice of work equipment and the choice of working and production methods, with a view, in particular, to alleviating monotonous work and work at a predetermined work rate and to reducing their effect on health
➤ adapting to technical progress
➤ replacing the dangerous by the non-dangerous or the less dangerous

➤ developing a coherent overall prevention policy which covers technology, organization of work, working conditions, social relationships and the influence of factors relating to the working environment

➤ giving collective protective measures priority over individual protective measures

➤ giving appropriate instruction to employees.

21.25.4 *Effective arrangements for health and safety – regulation 5*

Formal arrangements must be devised (and recorded) for effective planning, organization, control, monitoring and review of safety measures. This will involve an effective health and safety management system to implement the policy. Where there are five or more employees the arrangements should be recorded.

Planning involves a systematic approach to risk assessment, the selection of appropriate risk controls and establishing priorities with performance standards.

Organization involves consultation and communication with employees; employee involvement in risk assessment; the provision of information; and securing competence with suitable instruction and training. Control involves clarifying responsibilities and making sure people adequately fulfil their responsibilities. It involves adequate and appropriate supervision.

Monitoring should include the measurement of how well the policy is being implemented and whether hazards are being controlled properly. It covers inspections of the workplace and management systems and the investigation of incidents and accidents to ascertain the underlying causes and effect a remedy.

Review is essential to look at the whole of the health and safety management system to ensure that it is effective and achieving the correct standard of risk control.

21.25.5 *Health surveillance – regulation 6*

In appropriate circumstances health surveillance of staff may be required – the ACOP describes more fully when this duty will arise. Health surveillance is considered relevant when there is an identifiable disease or poor health condition; there are techniques to detect the disease; there is a reasonable likelihood that the disease will occur; and surveillance is likely to enhance the protection of the workers concerned. A competent person, who will range from a manager, in some cases, to a fully qualified occupational medical practitioner in others, should assess the extent of the surveillance.

21.25.6 *Competent assistance – regulation 7*

Every employer is obliged to appoint one or more 'competent person(s)' to advise and assist in undertaking the necessary measures to comply with the relevant statutory requirements. They may be employees or outside consultants. The purpose is to make sure that all employers have access to health and safety expertise. Preference should be given to an in-house appointee, who may be backed up by external expertise.

The competence of the person(s) appointed is to be judged in terms of their training, knowledge and experience of the work involved; it is not necessarily dependent upon particular qualifications. In simple situations, it may involve knowledge of relevant best practice, knowing one's limitations and taking external advice when necessary. In more complex situations or risks, fully qualified and appropriately experienced practitioners will be required.

Appointed competent persons must be provided with adequate information, time and resources to do their job.

21.25.7 *Procedures for serious and imminent danger and contact with external services – regulations 8 and 9*

Procedures must be established for dealing with serious and imminent dangers, including fire evacuation plans and arrangements for other emergencies. A sufficient number of competent persons must be appointed to evacuate the premises in the event of an emergency. The procedures should allow for persons at risk to be informed of the hazards and how and when to evacuate to avoid danger. In shared workplaces employers must co-operate. Access to dangerous areas should be restricted to authorized and properly trained staff. Any necessary contact arrangements with external services for first aid, emergency medical care and rescue work must be set up.

21.25.8 *Information for employees – regulation 10*

Information must be provided to staff on the risk assessment, risk controls, emergency procedures, the identity of the people appointed to assist on health and safety matters and risks notified by others.

The information provided must take into account the level of training, knowledge and experience of the employees. It must take account of language difficulties and be provided in a form that can be understood by everyone. The use of translations, symbols and diagrams should be considered. Where children under school leaving age are at work, information on the risk assessments and control measures must be provided to the child's parent or guardians of children at work before the child starts work. It can be provided verbally or directly to the parent, guardians or school.

21.25.9 Co-operation and co-ordination – regulations 11, 12 and 15

Where two or more employers share a workplace, the following applies:

➤ Each must co-operate with other employers in health and safety matters.
➤ Each must take reasonable steps to co-ordinate their safety precautions.
➤ Each must inform the other employers of the risks to their employees (i.e. risks to neighbours' employees).
➤ Where people from outside organizations are present to do work they, and their employers, have to be provided with appropriate information on risks, health and the necessary precautions to be taken.
➤ Temporary staff and staff with fixed-term contracts as well as permanent employees must be supplied with health and safety information before starting work (regulations 12 and 15).
➤ Regulation 11 does not apply to multi-occupied premises or sites where each unit, under the control of an individual tenant employer or self-employed person, is regarded as a separate workplace. In other cases, common areas may be shared workplaces, such as a reception area or canteen or they may be under the control of a person to whom section 4 of HSW Act applies. Suitable arrangements may need to be put in place for these areas.

21.25.10 Capabilities and training – regulation 13

When giving tasks to employees, their capabilities with regard to health and safety must be taken into account.

Employees must be provided with adequate health and safety training:

➤ on recruitment
➤ on being exposed to new or increased risks
➤ on the introduction of new procedures, systems or technology; training must be repeated periodically and take place in working hours (or while being paid).

21.25.11 Duties on employees – regulation 14

Equipment and materials must be used properly in accordance with instructions and training. Obligations on employees are extended to include certain requirements to report serious and immediate dangers and any shortcomings in the employer's protection arrangements.

21.25.12 New or expectant mothers – regulations 16–18

Where work is of a kind that could present a risk to new or expectant mothers working there or their babies, the risk assessments must include an assessment of such risks. When the risks cannot be avoided the employer must alter a woman's working conditions or hours to avoid the risks; offer suitable alternative work; or suspend from work on full pay. The woman must notify the employer in writing of her pregnancy that she has given birth within the last 6 months or she is breastfeeding.

21.25.13 Young persons – regulation 19

Employers must protect young persons at work from risks to their health and safety which are the result of lack of experience, or absence of awareness of existing or potential risks or because they have not yet fully matured. Young persons may not be employed in a variety of situations enumerated in the Regulations that pose a significant risk to their health and safety. The exception to this is young persons over school leaving age:

➤ where the work is necessary for their training
➤ where they will be supervised by a competent person
➤ where the risk will be reduced to the lowest level that is reasonably practicable.

21.25.14 Provisions as to liability – regulation 21

A new provision has been added to prevent a defence for an employer by reason of any act or default by an employee or a competent person appointed under regulation 7.

21.25.15 Restriction of civil liability for breach of statutory duty – regulation 22

The Management of Health and Safety at Work Regulations 1999 were amended in 2003, by the 2003 Amendment Regulations (SI 2003 No. 2457), to enable employees to claim damages from their employer in a civil action where they suffered injury or illness as a result of the employer being in breach of those Regulations. They were also intended to enable civil claims to be brought against employees for a breach of their duties under those Regulations that resulted in injury or illness. Employees have duties under those Regulations to use any equipment, dangerous substance, etc. in accordance with any training and instruction provided by the employer. Employees are also required to alert their employer of serious and imminent danger in the workplace or any shortcomings in the health and safety arrangements.

In April 2006 the Regulations were further amended by the Management of Health and Safety at Work (Amendment) Regulations 2006.

The amendment changes the civil liability provisions in the Regulations so as to exclude the right of third parties to take legal action against employees for contraventions of their duties under these Regulations. This extends to employees the same protection against third party action as that provided for employers.

The amendment neither creates any new duties nor removes any. The practical effect will be to reduce the likelihood of claims against employees by third parties. Therefore, it is expected that there will be no additional burdens on businesses.

The wording of the 2003 amendment produced the unintended consequence of allowing claims to be brought against employees by third parties who were affected by their work activity, e.g. members of the public. This had not been the intention. One concerned group raised this unintended consequence of the 2003 amendment. HSE sought independent advice, which also concluded that there was potential for third parties to make claims against employees.

21.25.16 *References*

A Guide to Risk Assessment Requirements. INDG218, HSE Books 1996, HSE Books website only.

An Introduction to Health and Safety: Health and Safety in Small Businesses. INDG259 (rev1), 2003 HSE Books.

Directors' Responsibilities for Health and Safety at Work. INDG343, 2002, ISBN 9780-7176-2080-8.

Five Steps to Risk Assessment. HSE INDG163 (rev2), 2006 HSE Books, ISBN 9780-7176-6189-X.

Management of Health and Safety at Work: Approved Code of Practice and Guidance. HSC L21, 2nd edition, 2000 HSE Books, ISBN 9780-7176-2488-9.

The Management of Health and Safety at Work Regulations 1999. ISBN 9780-11-025051-6, OPSI website: http://www.opsi.gov.uk/si/si1999/uksi_19993242_en.pdf.

21.26 Manual Handling Operations (MHO) Regulations 1992 as amended in 2002

21.26.1 *General*

The Regulations apply to the manual handling (any transporting or supporting) of loads, i.e. by human effort, as opposed to mechanical handling by forklift truck, crane, etc. Manual handling includes lifting, putting down, pushing, pulling, carrying or moving. The human effort may be applied directly to the load, or indirectly by pulling on a rope, chain or lever. Introducing mechanical assistance, like a hoist or sack truck, may reduce but not eliminate manual handling, since human effort is still required to move, steady or position the load.

The application of human effort for purposes other than transporting or supporting a load, for example pulling on a rope to lash down a load or moving a machine control, is not a manual handling operation. A load is a discrete movable object, but it does not include an implement, tool or machine while being used.

Injury in the context of these Regulations means to any part of the body. It should take account of the physical features of the load which might affect grip or cause direct injury, for example slipperiness, sharp edges, and extremes of temperature. It does not include injury caused by any toxic or corrosive substance which has leaked from a load, is on its surface or is part of the load.

21.26.2 *Duties of employers – avoidance of manual handling – regulation 4 (1) (a)*

Employers should take steps to avoid the need for employees to carry out MHO which involves a risk of their being injured.

The guidance suggests that a preliminary assessment should be carried out when making a general risk assessment under the Management of Health and Safety at Work Regulations 1999. Employers should consider whether the operation can be eliminated, automated or mechanized.

21.26.3 *Duties of employers – assessment of risk – regulation 4 (1) (b) (i)*

Where it is not reasonably practicable to avoid MHO, employers must make a suitable and sufficient risk assessment of all such MHO in accordance with the requirements of schedule 1 to the Regulations (shown later). This duty to assess the risk takes into account the **task**, the **load**, the **working environment** and **individual capability**.

21.26.4 *Duties of employers – reducing the risk of injury – regulation 4 (1) (b) (ii)*

Where it is not reasonably practicable to avoid the MHO at which there is a risk of injury, employers must take steps to reduce the risk of injury to the lowest level reasonably practicable.

The structured approach (considering the task, the load, the working environment and the individual capability) is recommended in the guidance. The steps taken will involve ergonomics, changing the load, mechanical assistance, task layout, work routines, PPE, team working and training.

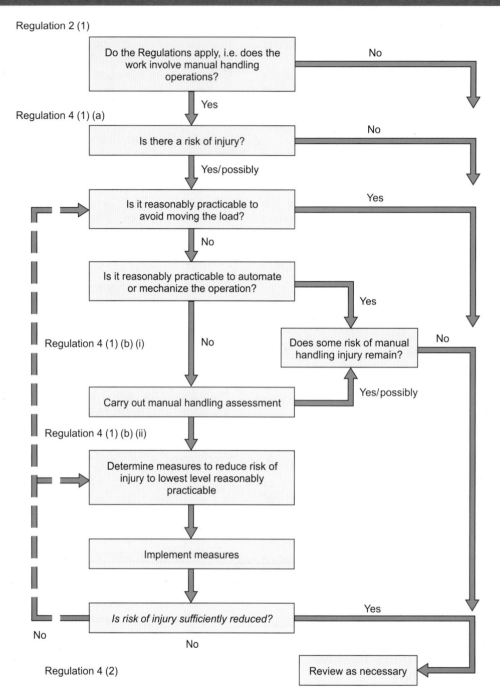

Regulation 2 (1)

Regulation 4 (1) (a)

Regulation 4 (1) (b) (i)

Regulation 4 (1) (b) (ii)

Regulation 4 (2)

Figure 21.3 Manual Handling Operations Regulations – flow chart.

21.26.5 *Duties of employers – additional information on the load – regulation 4 (1) (b) (iii)*

Employers must take appropriate steps where manual handling cannot be avoided to provide general indications and, where practicable precise information on:

➤ the weight of each load
➤ the heaviest side of any load which does not have a central centre of gravity.

The information is probably best marked on the load. Sections 3 and 6 of the HSW Act may place duties on originators of loads, like manufacturers or packers.

21.26.6 *Duties of employers – reviewing assessment – regulation 4 (2)*

The assessment should be reviewed if there is reason to suspect that it is no longer valid or there have been significant changes in the particular MHO.

Table 21.11   Schedule 1 to the Manual Handling Operations Regulations

Factors to which the employer must have regard and questions they must consider when making an assessment of manual handling operations	
Factors	**Questions**
1. The tasks	Do they involve: ➤ holding or manipulating loads at distance from trunk? Is there unsatisfactory bodily movement or posture, especially: ➤ twisting the trunk? ➤ stooping? ➤ reaching upwards? Is there excessive movement of loads, especially: ➤ excessive lifting or lowering distances? ➤ excessive carrying distances? ➤ excessive pulling or pushing of loads? ➤ risk of sudden movement of loads? ➤ frequent or prolonged physical effort? ➤ insufficient rest or recovery periods? ➤ a rate of work imposed by a process?
2. The loads	Are they: ➤ heavy? ➤ bulky or unwieldy? ➤ difficult to grasp? ➤ unstable, or with contents likely to shift? ➤ sharp, hot or otherwise potentially damaging?
3. The working environment	Are there: ➤ space constraints preventing good posture? ➤ uneven, slippery or unstable floors? ➤ variations in level of floors or work surfaces? ➤ extremes of temperature or humidity? ➤ conditions causing ventilation problems or gusts of wind? ➤ poor lighting conditions?
4. Individual capability	Does the job: ➤ require unusual strength, height, etc.? ➤ vreate a hazard to those who might reasonably be considered to be pregnant or have a health problem? ➤ require special information or training for its safe performance?
5. Other factors	Is movement or posture hindered by personal protective equipment or by clothing?

21.26.7 *Individual capability – regulation 4 (3) (a)–(f)*

A new requirement was added in 2002 by Amendment Regulations concerning the individual capabilities of people undertaking manual handling. Regard must be had in particular to:

(a) the physical suitability of the employee to carry out the operation
(b) the clothing, footwear or other personal effects they are wearing
(c) their knowledge and training
(d) results of any relevant risk assessments carried out under regulation 3 of the Management Regulations; 1999

(e) whether the employee is within a group of employees identified by that assessment as being especially at risk and
(f) the results of any health surveillance provided pursuant to regulation 6 of the Management Regulations 1999.

21.26.8 *Duty of employees – regulation 5*

Each employee, while at work, has to make proper use of any system of work provided for their use. This is in addition to other responsibilities under the HSW Act and the Management of Health and Safety at Work Regulations.

The provisions do not include well intentioned improvization in an emergency, for example, rescuing a casualty or fighting a fire (see Figure 21.3 and Table 21.11).

21.26.9 *References*

A Pain in Your Workplace: Ergonomic Problems and Solutions. HSG 121, 1994 HSE Books, ISBN 9780-7176-06668-6.

Are You Making the Best Use of Lifting and Handling Aids? INDG398, 2004 HSE Books, ISBN 9780-7176-2900-7.

Backs for the Future: Safe Manual Handling in Construction. HSG 149, 2000 HSE Books, ISBN 9780-7176-1122-1.

Getting to Grips with Manual Handling. A Short Guide for Employers Revised. INDG143, 2004 HSE Books, ISBN 9780-7176-2828-0.

Manual Handling, Manual Handling Operations Regulations 1992, Guidance on Regulations Revised. L23, 2004 HSE Books, ISBN 9780-7176-2823-X.

Manual Handling, Solutions You can Handle. HSG 115, 1994 HSE Books, ISBN 9780-7176-0693-7.

Manual Handling Assessment Chart. INDG383, 2003 HSE Books, ISBN 9780-7176-2741-1.

21.27 Control of Noise at Work Regulations 2005

21.27.1 *Introduction*

The Control of Noise at Work Regulations 2005 require employers to prevent or reduce risks to health and safety from exposure to noise at work. Employees also have duties under the Regulations.

The Regulations require employers to:

➤ assess the risks to their employees from noise at work
➤ take action to reduce the noise exposure that produces those risks
➤ provide their employees with hearing protection if they cannot reduce the noise exposure enough by using other methods
➤ make sure the legal limits on noise exposure are not exceeded
➤ provide their employees with information, instruction and training
➤ carry out health surveillance where there is a risk to health.

The Regulations do not apply to:

➤ members of the public exposed to noise from their non-work activities, or making an informed choice to go to noisy places

➤ low-level noise which is a nuisance but causes no risk of hearing damage.

Employers in the music and entertainment sectors have until 6 April 2008 to comply with the Noise Regulations 2005. Meanwhile they must continue to comply with the Noise at Work Regulations 1989, which the 2005 Regulations replace for all other workplaces.

21.27.2 *Exposure limit values and action levels – regulation 4*

The Regulations require employers to take specific action at certain action values. These relate to:

➤ the levels of exposure to noise of their employees averaged over a working day or week; and
➤ the maximum noise (peak sound pressure) to which employees are exposed in a working day.

The values are:

➤ lower exposure action values:
   ➤ daily or weekly exposure of 80 dB (A-weighted)
   ➤ peak sound pressure of 135 dB (C-weighted)
➤ upper exposure action values:
   ➤ daily or weekly exposure of 85 dB (A-weighted)
   ➤ peak sound pressure of 137 dB (C-weighted).

Figure 21.4 helps to decide what needs to be done.

There are also levels of noise exposure which must not be exceeded:

➤ exposure limit values:
   ➤ daily or weekly exposure of 87 dB (A-weighted)
   ➤ peak sound pressure of 140 dB (C-weighted).

These exposure limit values take account of any reduction in exposure provided by hearing protection.

21.27.3 *Risk assessment – regulation 5*

If employees are likely to be exposed to noise at or above the lower exposure value a suitable and sufficient assessment of the risks must be made. Employers need to decide whether any further action is needed, and plan how to do it.

The risk assessment should:

➤ identify where there may be a risk from noise and who is likely to be affected
➤ contain a reliable estimate of employees' exposures, and compare the exposure reading with the exposure action values and limit values

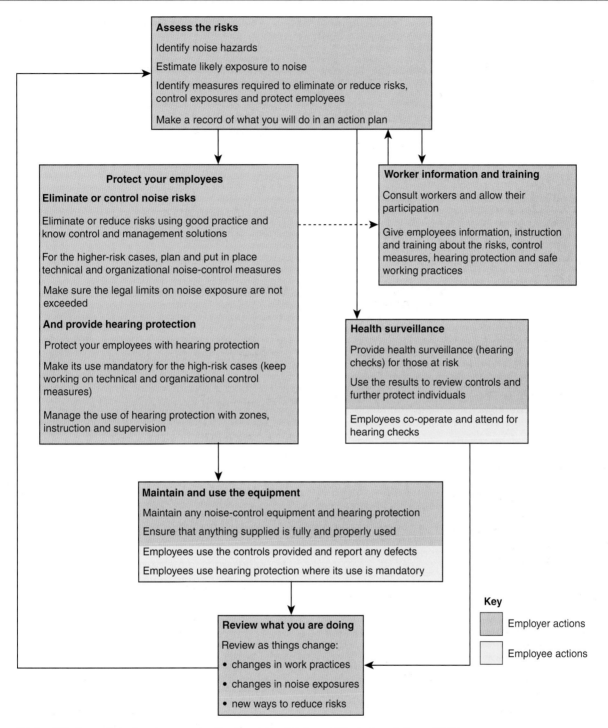

Figure 21.4 What needs to be done under the Control of Noise at Work Regulations 2005.

➤ identify what needs to be done to comply with the law, e.g. whether noise-control measures or hearing protection are needed, and, if so, where and what type and

➤ identify any employees who need to be provided with health surveillance and whether any are at particular risk.

The risk assessment should include consideration of:

➤ the level, type and duration of exposure, including any exposure to peak sound pressure

➤ the effects of exposure to noise on employees or groups of employees whose health is at particular risk

➤ SFARP any effects on the health and safety of employees resulting from the interaction, for example, between noise and vibration
➤ any indirect effects from the interaction between noise and audible warnings
➤ manufacturer's information
➤ availability of alternative equipment designed to reduce noise emissions
➤ any extension to noise exposure due to extended hours or in supervised rest facilities
➤ information following health surveillance
➤ availability of personal hearing protectors with adequate attenuation characteristics.

It is essential that employers can show that their estimate of employees' exposure is representative of the work that they do. It needs to take account of:

➤ the work they do or are likely to do
➤ the ways in which they do the work and
➤ how it might vary from one day to the next.

The estimate must be based on reliable information, e.g. measurements in their own workplace, information from other similar workplaces, or data from suppliers of machinery.

Employers must record the significant findings of their risk assessment. They need to record in an action plan anything identified as being necessary to comply with the law, setting out what they have done and what they are going to do, with a timetable and saying who will be responsible for the work.

The risk assessment should be reviewed if circumstances in the workplace change and affect noise exposures. Also, it should be reviewed regularly to make sure that the employer continues to do all that is reasonably practicable to control the noise risks. Even if it appears that nothing has changed, employers should not leave it for more than about 2 years without checking whether a review is needed.

21.27.4   *Elimination or control of exposure – regulation 6*

The purpose of the Control of Noise at Work Regulations is to make sure that people do not suffer damage to their hearing – so controlling noise risks and noise exposure should be where efforts are concentrated.

Wherever there is noise at work employers should be looking for alternative processes, equipment and/or working methods which would make the work quieter or mean people are exposed for shorter times. They should also be keeping up with what is good practice or the standard for noise control within their industry.

Where there are things that can be done to reduce risks from noise that are reasonably practicable, they should be done. However, where noise exposures are below the lower exposure action values, risks are low and so employers would only be expected to take actions which are relatively inexpensive and simple to carry out.

Where the assessment shows that employees are likely to be exposed at or above the upper exposure action values, employers must put in place a planned programme of noise control.

The risk assessment will have produced information on the risks and an action plan for controlling noise. Employers should use this information to:

➤ tackle the immediate risk (e.g. by providing hearing protection)
➤ identify what is possible to control noise, how much reduction could be achieved and what is reasonably practicable
➤ establish priorities for action and a timetable (e.g. consider where there could be immediate benefits, what changes may need to be phased in over a longer period of time and the number of people exposed to the noise in each case)
➤ assign responsibilities to people to deliver the various parts of the plan
➤ ensure the work on noise control is carried out
➤ check that what has been done has worked.

Actions taken should be based on the general principles set out in the Management Regulations and should include consideration of:

➤ other working methods
➤ choice of appropriate work equipment emitting the least possible noise
➤ the design and layout of workplaces, work stations and rest facilities
➤ suitable and sufficient information and training
➤ reduction of noise by technical means
➤ appropriate maintenance programmes
➤ limitation of the duration and intensity of exposure and
➤ appropriate work schedules with adequate rest periods.

21.27.5   *Hearing protection – regulation 7*

Hearing protection should be issued to employees:

➤ where extra protection is needed above what can been achieved using noise control
➤ as a short-term measure while other methods of controlling noise are being developed.

Hearing protection should not be used as an alternative to controlling noise by technical and organizational means. Employers should consult with their employees or their representatives on the type of hearing protection to be used.

Employers are required to:

➤ provide employees with hearing protectors if they ask for it and their noise exposure is between the lower and upper exposure action values

➤ provide employees with hearing protectors and make sure they use them properly when their noise exposure exceeds the upper exposure action values

➤ identify hearing protection zones, that is areas where the use of hearing protection is compulsory, and mark them with signs if possible; restrict access to hearing protection zones where this is practicable and the noise exposures justifies it

➤ provide employees with training and information on how to use and care for the hearing protectors

➤ ensure that the hearing protectors are properly used and maintained.

21.27.6 *Maintenance and use of equipment – regulation 8*

Employers need to make sure that hearing protection works effectively and to:

➤ check that all equipment provided in compliance with the regulations remains in good, clean condition and that there are no unofficial modifications

➤ check that hearing protection is fully and properly used (except where it is provided for employees who are exposed at or above the lower exposure but below the upper exposure level), which is likely to mean that an employer needs to:
  ➤ put someone in authority in overall charge of issuing the protection required and making sure replacements are readily available
  ➤ carry out spot checks to see that the rules are being followed and that hearing protection is being used properly
  ➤ ensure all managers and supervisors set a good example and wear hearing protection at all times when in hearing protection zones
  ➤ ensure only people who need to be there enter hearing protection zones and do not stay longer than they need to.

21.27.7 *Health surveillance – regulation 9*

Employers must provide health surveillance (hearing checks) for all their employees who are likely to be regularly exposed above the upper exposure action values, or are at risk for any reason, e.g. they already suffer from hearing loss or are particularly sensitive to damage.

The purpose of health surveillance is to:

➤ warn when employees might be suffering from early signs of hearing damage

➤ give employers an opportunity to do something to prevent the damage getting worse

➤ check that control measures are working.

Trade union safety representatives, or employee representatives and the employees concerned should be consulted before introducing health surveillance. It is important that employees understand that the aim of health surveillance is to protect their hearing. Employers will need their understanding and co-operation if health surveillance is to be effective.

Health surveillance for hearing damage usually means:

➤ regular hearing checks in controlled conditions

➤ telling employees about the results of their hearing checks (required after reasonable notice by employee)

➤ keeping health records

➤ ensuring employees are examined by a doctor where hearing damage is identified.

If the doctor considers the hearing damage is likely to be the result of exposure, the employer must:

➤ ensure that the employee is informed by a suitably qualified person

➤ review the risk assessment and control measures

➤ consider assigning employee to alternative work

➤ ensure continued surveillance.

21.27.8 *Information, instruction and training – regulation 10*

Where employees are exposed to noise which is likely to be at or above the lower exposure level, suitable and sufficient information, instruction and training must be provided and kept up to date. This includes the nature of the risks, compliance action taken, significant findings of the risk assessment, hearing protection, how to detect and report signs of hearing damage, entitlement to health surveillance, safe working practices and collective results of health surveillance.

21.27.9 *References*

Control of Noise at Work Regulations 2005 (SI No. 1643) http://www.opsi.gov.uk/si/si2005/uksi_20051643_en.pdf

Controlling Noise at Work. The Control of Noise at Work Regulations 2005, L108 (2nd edition), 2005 HSE Books, ISBN 9780-7176-6164-4.

Noise at Work. Guidance for Employers on the Control of Noise at Work Regulations 2005. INDG362 (rev1), 2005 HSE Books, ISBN 9780-7176-6165-2.

Protect Your Hearing or Lose It! INDG363 (rev1), 2005 HSE Books, ISBN 9780-7176-6166-0.

Sound Solutions Techniques to Reduce Noise at Work. HSG 138, 1995 HSE Books, ISBN 9780-7176-0791-7.

21.28 Personal Protective Equipment at Work Regulations 1992 as amended in 2002

21.28.1 *Introduction*

The effect of the Personal Protective Equipment (PPE) at Work Regulations is to ensure that certain basic duties governing the provision and use of PPE apply to all situations where PPE is required. The Regulations follow sound principles for the effective and economical use of PPE, which all employers should follow.

PPE, as defined, includes all equipment (including clothing affording protection against the weather) which is intended to be worn or held by a person at work and which protects them against one or more risks to their health and safety. Waterproof, weatherproof or insulated clothing is covered only if its use is necessary to protect against adverse climatic conditions.

Ordinary working clothes and uniforms, which do not specifically protect against risks to health and safety, and protective equipment worn in sports competitions, are not covered.

Where there is overlap in the duties in these Regulations and those covering lead, ionizing radiations, asbestos, hazardous substances (COSHH), noise, and construction head protection, then the specific legislative requirements should prevail.

21.28.2 *Provision of PPE – regulation 4*

Every employer shall ensure that suitable PPE is provided to their employees who may be exposed to risks to their health and safety except where it has been adequately or more effectively controlled by other means. (Management Regulations require PPE to be the last choice in the principles of protection.)

PPE shall not be suitable unless:

➤ it is appropriate for the risks and the conditions of use

➤ it takes account of ergonomic requirements, the state of health of the wearer and the characteristics of each workstation

➤ it is capable of fitting the wearer correctly, by adjustments if necessary

➤ it is, so far as is practicable, able to combat the risks without increasing overall risks

➤ it complies with UK legislation on design or manufacture, i.e. it has a CE marking.

Where it is necessary that PPE is hygienic and otherwise free of risk to health, PPE must be provided to a person solely for their individual use.

21.28.3 *Compatibility – regulation 5*

Where more than one health and safety risk necessitates the wearing of multiple items of PPE simultaneously then they shall be compatible and remain effective.

21.28.4 *Assessment – regulation 6*

Before choosing any PPE, employers must ensure that an assessment is made to determine whether the PPE is suitable.

The assessment shall include:

➤ assessing risks which have not been avoided by other means

➤ a definition of the characteristics that PPE must have to be effective, taking into account any risks created by the PPE itself

➤ a comparison of available PPE with the required characteristics

➤ information on whether the PPE is compatible with any other PPE which is in use, and which an employee should be required to wear simultaneously.

The assessment should be reviewed if it is no longer valid or there have been significant changes. In simple cases it will not be necessary to record the assessment but, in more complex cases, written records should be made and kept readily available for future reference.

21.28.5 *Maintenance – regulation 7*

Every employer (and self-employed person) shall ensure that any PPE provided is maintained, including replaced and cleaned, in an efficient state, in efficient working order and in good repair.

The guide emphasizes the need to set up an effective system of maintenance for PPE. This should be proportionate to the risks and appropriate to the particular PPE. It could include, where appropriate, cleaning, disinfection, examination, replacement, repair and testing.

For example, mechanical fall arrestor equipment or sub-aqua breathing apparatus will require planned preventative maintenance with examination, testing and overhaul. Records should be kept of the maintenance work. Gloves may only require periodic inspection by the user as necessary, depending on their use.

Spare parts must be compatible and be the proper part suitably CE marked where applicable. Manufacturers' maintenance schedules and instructions should be followed unless alternative schemes are agreed with the manufacturer or agent.

In some cases these requirements can be fulfilled by using disposable PPE which can be discarded after use or when their life has expired. Users should know when to discard and replace disposable PPE.

21.28.6 Accommodation – regulation 8

When an employer or self-employed person has to provide PPE they must ensure that appropriate accommodation is provided to store it when not in use.

The type of accommodation will vary and may simply be suitable hooks for special clothing and small portable cases for goggles. It should be separate from normal outer clothing storage arrangements and protect the PPE from contamination or deterioration.

21.28.7 Information, instruction and training – regulation 9

Employers shall provide employees with adequate and appropriate information, instruction and training on:

➤ the risks which the PPE will avoid or limit
➤ the purpose for which and the manner in which PPE should be used
➤ any action required of the employee to maintain the PPE.

employers are required to provide demonstrations of PPE where appropriate.

The guidance suggests the training should include:

➤ an explanation of the risks and why PPE is needed
➤ the operation, performance and limitations of the equipment
➤ instructions on the selection, use and storage of PPE
➤ problems that can affect PPE relating to other equipment, working conditions, defective equipment, hygiene factors and poor fit
➤ the recognition of defects and how to report problems with PPE
➤ practice in putting on, wearing and removing PPE
➤ practice in user cleaning and maintenance
➤ how to store safely.

21.28.8 Use and reporting of loss or defects – regulations 10 and 11

Every employer shall take all reasonable steps to ensure that PPE is properly used.

Every employee shall:

➤ use PPE provided in accordance with training and instructions
➤ return it to the accommodation provided after use
➤ report any loss or obvious defect.

21.28.9 References

A Short Guide to the Personal Protective Equipment at Work Regulations 1992. INDG174 (rev), 2005 HSE Books, ISBN 9780-7176-6139-3.

Keep Your Top on Health Risks from Working in the Sun. INDG147, 1998 HSE Books, ISBN 9780-7176-1578-2.

Personal Protective Equipment (EC Directive) Regulations 1992. SI 1992 No. 3139, http://www.opsi.gov.uk/si/si1992/Uksi_19922966_en_6.htm.

Personal Protective Equipment at Work (2nd edition): *Personal Protective Equipment at Work Regulations 1992 (as Amended). Guidance on Regulations*. L25, 2005 HSE Books, ISBN 9780-7176-6139-3.

Selecting Protective Gloves for Work with Chemicals. Guidance for Employers and Health and Safety Specialists. INDG330, HSE Books 2000, web only: http://www.hse.gov.uk/pubns/ppeindex.htm.

Sun Protection Advice for Employers of Outdoor Workers. INDG337, 2001 HSE Books, ISBN 9780-7176-1982-6.

The Selection, Use and Maintenance of Respiratory Protective Equipment: A Practical Guide. HSG53, 2005 HSE Books, ISBN 9780-7176-2904-X.

21.29 Provision and Use of Work Equipment Regulations 1998 (except Part IV) as amended in 2002

21.29.1 Introduction

The Provision and Use of Work Equipment Regulations 1998 (PUWER) are made under the HSW Act and their primary aim is to ensure that work equipment is used without risks to health and safety, regardless of its age, condition or origin. The requirements of PUWER that are relevant to woodworking machinery are set out in the *Safe use of woodworking machinery Approved Code of Practice*. PUWER has specific requirements for risk assessment which are covered under the Health and Safety Management Regulations 1999.

Part IV of PUWER is concerned with power presses and is not part of the Certificate syllabus, and is therefore not covered in this summary.

21.29.2 *Definitions*

Work equipment means any machinery, appliance, apparatus, tool or installation for use at work.

Use in relation to work equipment means any activity involving work equipment and includes starting, stopping, programming, setting, transporting, repairing, modifying, maintaining, servicing and cleaning.

21.29.3 *Duty holders – regulation 3*

Under PUWER the following groups of people have duties placed on them:

➤ employers
➤ the self-employed
➤ people who have control of work equipment, for example plant hire companies.

In addition to all places of work the Regulations apply to common parts of shared buildings, industrial estates and business parks; to temporary works sites including construction; to home working (but not to domestic work in a private household); to hotels, hostels and sheltered accommodation.

21.29.4 *Suitability of work equipment – regulation 4*

Work equipment:

➤ has to be constructed or adapted so that it is suitable for its purpose
➤ has to be selected with the conditions of use and the users' health and safety in mind
➤ may only be used for operations for which, and under conditions for which, it is suitable.

This covers all types of use and conditions and must be considered for each particular use or condition. For example: scissors may be safer than knives with unprotected blades and should therefore be used for cutting operations where practicable; risks imposed by wet, hot or cold conditions must be considered.

21.29.5 *Maintenance – regulation 5*

The regulation sets out the general requirement to keep work equipment maintained in:

➤ an efficient state
➤ efficient working order
➤ good repair.

Compliance involves all three criteria. In addition, where there are maintenance logs for machinery, they must be kept up to date.

In many cases this will require routine and planned preventive maintenance of work equipment. When checks are made priority must be given to:

➤ safety
➤ operating efficiency and performance
➤ the equipment's general condition.

21.29.6 *Inspection – regulation 6*

Where the safety of work equipment depends on the installation conditions, it must be inspected:

➤ after installation and before being put into service for the first time
➤ after assembly at a new site or in a new location to ensure that it has been installed correctly and is safe to operate.

Where work equipment is exposed to conditions causing deterioration which is liable to result in dangerous situations it must be inspected:

➤ at suitable intervals
➤ when exceptional circumstances occur.

Inspections must be determined and carried out by competent persons. An inspection will vary from a simple visual external inspection to a detailed comprehensive inspection which may include some dismantling and/or testing. However, the level of inspection would normally be less than that required for a thorough examination under, for example, LOLER for certain items of lifting equipment.

Records of inspections must be kept with sufficient information to properly identify the equipment, its normal location, dates, faults found, action taken, to whom faults were reported, who carried out the inspection, when repairs were made, date of the next inspection.

When equipment leaves an employer's undertaking it must be accompanied by physical evidence that the last inspection has been carried out.

21.29.7 *Specific risks – regulation 7*

Where the use of work equipment involves specific hazards, its use must be restricted to those persons given the specific task of using it and repairs, etc. must be restricted to designated persons.

Designated persons must be properly trained to fulfil their designated task.

Hazards must be controlled using a hierarchy of control measures, starting with elimination where this is possible, then considering hardware measures such as physical barriers and, lastly, software measures such as a safe system of work.

21.29.8 Information, instruction and training – regulations 8 and 9

Persons who use work equipment must have adequate:

➤ health and safety information
➤ where appropriate, written instructions about the use of the equipment
➤ training for health and safety in methods which should be adopted when using the equipment, and any hazards and precautions which should be taken to reduce risks.

Any persons who supervise the use of work equipment should also receive information, instruction and training. The training of young persons is especially important with the need for special risk assessments under the Management Regulations.

Health and safety training should take place within working hours.

21.29.9 Conformity with community requirements – regulation 10

The intention of this regulation is to require that employers ensure that equipment, provided for use after 31 December 1992, complies with the relevant essential requirements in various European Directives made under article 95 of the EC Treaty (formerly 100A of the Treaty of Rome). The requirements of PUWER 98 regulations 11 to 19 and 22 to 29 only apply if the essential requirements do not apply to a particular piece of equipment.

However, PUWER regulations 11–19 and 22–29 will apply if:

➤ they include requirements which were not included in the relevant product legislation
➤ the relevant product legislation has not been complied with (e.g. the guards fitted on a machine when supplied were not adequate).

Employers using work equipment need to check that any new equipment has been made to the requirements of the relevant Directive, and has a CE marking, suitable instructions and a Certificate of Conformity.

The Machinery Directive was brought into UK law by the Supply of Machinery (Safety) Regulations 1992 as amended, which duplicate PUWER regulations 11–19 and 22–29.

The employer still retains the duty to ensure that the equipment is safe to use.

21.29.10 Dangerous parts of machinery – regulation 11

Measures have to be taken which:

➤ prevent access to any dangerous part of machinery or to any rotating stock-bar
➤ stop the movement of any dangerous part of machinery or rotating stock-bar before any part of a person enters a danger zone.

The measures required follow the normal hierarchy and consist of:

➤ the provision of fixed guards enclosing every dangerous part of machinery or rotating stock-bar where and to the extent it is practicable, but where it is not
➤ the provision of other guards or protection devices, where and to the extent it is practicable, but where it is not
➤ the provision of jigs, holders, push-sticks or similar protection appliances used in conjunction with the machinery where and to the extent it is practicable
➤ provision of information, instruction, training and supervision as is necessary.

All guards and protection devices shall:

➤ be suitable for its purpose
➤ be of good construction, sound material and adequate strength
➤ be maintained in an efficient state, in efficient working order and in good repair
➤ not give rise to increased risks to health and safety
➤ not be easily bypassed or disabled
➤ be situated at sufficient distance from the danger zone
➤ not unduly restrict the view of the operating cycle of the machine where this is relevant
➤ be so constructed or adapted that they allow operations necessary to fit or replace parts and for maintenance work, if possible without having to dismantle the guard or protection device.

21.29.11 Protection against specified hazards – regulation 12

Exposure to health and safety risks from the following hazards must be prevented or adequately controlled:

➤ any article falling or being ejected from work equipment

➤ rupture or disintegration of work equipment
➤ work equipment catching fire or overheating
➤ the unintended or premature discharge of any article, or of any gas, dust, liquid, vapour or other substance which is produced, used or stored in the work equipment
➤ the unintended or premature explosion of the work equipment or any article or substance produced, used or stored in it.

21.29.12 *High or very low temperature – regulation 13*

Work equipment and any article or substance produced, used or stored in work equipment which is at a high or very low temperature must have protection to prevent injury by burn, scald or sear.

This does not cover risks such as from radiant heat or glare.

Engineering methods of control such as insulation, doors, temperature control and guards, should be used where practicable but, there are some cases, for example cooker hot plates, where this is not possible.

21.29.13 *Controls – regulations 14 to 18*

Where work equipment is provided with (regulation 14):

➤ starting controls (including restarting after a stoppage)
➤ controls which change speed, pressure or other operating condition which would affect health and safety

it should not be possible to perform any operation except by a deliberate action on the control. This does not apply to the normal operating cycle of an automatic device.

Where appropriate, one or more readily accessible **stop controls** shall be provided to bring the work equipment to a safe condition in a safe manner (regulation 15). They must:

➤ bring the work equipment to a complete stop where necessary
➤ if necessary switch off all sources of energy after stopping the functioning of the equipment
➤ operate in priority to starting or operating controls.

Where appropriate one or more readily accessible **emergency stop controls** (regulation 16) must be provided unless it is not necessary:

➤ by the nature of the hazard
➤ by reason of the time taken for the stop controls to bring the equipment to a complete stop.

Emergency stop controls must have priority over stop controls. They should be provided where other safeguards are not adequate to prevent risk when something irregular happens. They should not be used as a substitute for safeguards or the normal method of stopping the equipment.

All **controls** for work equipment shall (regulation 17):

➤ be clearly visible and identifiable including appropriate marking where necessary
➤ not expose any person to a risk to their health and safety except where necessary.

Where appropriate, employers shall ensure that:

➤ controls are located in a safe place
➤ systems of work are effective in preventing any person being in a danger zone when equipment is started
➤ an audible, visible or other suitable warning is given whenever work equipment is about to start.

Persons in a danger zone as a result of starting or stopping equipment must have sufficient time and suitable means to avoid any risks.

Control systems (regulation 18) must be safe and chosen so as to allow for failures, faults and constraints. They must:

➤ not create any increased risk to health and safety
➤ not result in additional or increased risks when failure occurs
➤ not impede the operation of any stop or emergency stop controls.

21.29.14 *Isolation from sources of energy – regulation 19*

Work equipment must be provided with readily accessible and clearly identified means to isolate it from all sources of energy.

Re-connection must not expose any person using the equipment to any risks.

The main purpose is to allow equipment to be made safe under particular circumstances, such as maintenance, when unsafe conditions occur, or when adverse conditions such as electrical equipment in a flammable atmosphere or wet conditions, occur.

If isolation may cause a risk in itself, special precautions must be taken, for example a support for a hydraulic press tool which could fall under gravity if the system is isolated.

21.29.15 *Stability – regulation 20*

Work equipment must be stabilized by clamping or otherwise as necessary to ensure health and safety.

Most machines used in a fixed position should be bolted or fastened so that they do not move or rock in use.

21.29.16 Lighting – regulation 21

Suitable and sufficient lighting, taking account of the operations being carried out, must be provided where people use work equipment.

This will involve general lighting and in many cases local lighting, such as on a sewing machine. If access for maintenance is required regularly, permanent lighting should be provided.

21.29.17 Maintenance operations – regulation 22

SFARP work equipment should be constructed or adapted to allow maintenance operations to be:

➤ conducted while they are shut down
➤ undertaken without exposing people to risk
➤ carried out after appropriate protection measures have been taken.

21.29.18 Markings and warnings – regulations 23 and 24

Work equipment should have all appropriate **markings** for reasons of health and safety made in a clearly visible manner. For example, the maximum SWL, stop and start controls, or the maximum rotation speed of an abrasive wheel.

Work equipment must incorporate **warnings or warning devices** as appropriate, which are unambiguous, easily perceived and easily understood.

They may be incorporated in systems of work, a notice, a flashing light or an audible warning. They are an active instruction or warning to take specific precautions or actions when a hazard exists.

21.29.19 Mobile work equipment – regulations 25–30

The main purpose of this section is to require additional precautions relating to work equipment while it is travelling from one location to another or where it does work while moving. If the equipment is designed primarily for travel on public roads the Road Vehicles (Construction and Use) Regulations 1986 will normally be sufficient to comply with PUWER 98.

Mobile equipment would normally move on wheels, tracks, rollers, skids, etc. Mobile equipment may be self-propelled, towed or remote controlled and may incorporate attachments. Pedestrian controlled work equipment such as lawn mowers is not covered by Part III.

21.29.20 Employees carried on mobile work equipment – regulation 25

No employee may be carried on mobile work equipment unless:

➤ it is suitable for carrying persons
➤ it incorporates features to reduce risks as low as is reasonably practicable, including risks from wheels and tracks.

21.29.21 Rolling over of mobile work equipment and forklift trucks – regulations 26 and 27

Where there is a risk of overturning it must be minimized by:

➤ stabilizing the equipment
➤ fitting a structure so that it only falls on its side (ROPS, roll over protection structure)
➤ fitting a structure which gives sufficient clearance for anyone being carried if it turns over further (ROPS)
➤ a device giving comparable protection
➤ fitting a suitable restraining system for people if there is a risk of being crushed by rolling over.

This regulation does not apply:

➤ to a fork truck fitted with ROPS
➤ where it would increase the overall risks
➤ where it would not be reasonably practicable to operate equipment
➤ to any equipment provided for use before 5 December 1998 where it would not be reasonably practicable.

21.29.22 Overturning of forklift trucks – regulation 27

Forklift trucks, which carry an employee, must be adapted or equipped to reduce the risk to safety from overturning to as low as is reasonably practicable.

21.29.23 Self-propelled work equipment – regulation 28

Where self-propelled work equipment may involve risks while in motion it shall have:

➤ facilities to prevent unauthorized starting
➤ (with multiple rail-mounted equipment) facilities to minimize the consequences of collision
➤ a device for braking and stopping
➤ (where safety constraints so require) emergency facilities for braking and stopping, in the event of

failure of the main facility, which have readily accessible or automatic controls

➤ (where the driver's vision is inadequate) devices fitted to improve vision

➤ (if used at night or in dark places) appropriate lighting fitted or otherwise it shall be made sufficiently safe for its use

➤ (if there is anything carried or towed that constitutes a fire hazard liable to endanger employees (particularly if escape is difficult such as from a tower crane)) appropriate firefighting equipment carried, unless it is sufficiently close by.

21.29.24 *Remote-controlled self-propelled work equipment – regulation 29*

Where remote-controlled self-propelled work equipment involves a risk while in motion it shall:

➤ stop automatically once it leaves its control range
➤ have features or devices to guard against the risk of crushing or impact.

21.29.25 *Drive shafts – regulation 30*

Where seizure of the drive shaft between mobile work equipment and its accessories or anything towed is likely to involve a risk to safety:

➤ the equipment must have means to prevent a seizure
➤ where it cannot be avoided, every possible measures should be taken to avoid risks
➤ the shaft should be safeguarded from contacting the ground and becoming soiled or damaged.

21.29.26 *Part IV – power presses*

Regulations 31–35 relate to power presses and are not included here as they are excluded from the NEBOSH Certificate syllabus. Details can be found in the Power Press ACOP.

21.29.27 *References*

Buying New Machinery. A Short Guide to the Law. INDG271, 1998 HSE Books, ISBN 9780-7176-1559-6.
Circular Saw Benches. WIS16, 1999 HSE Books.
Hiring and Leasing Out of Plant: Application of PUWER 98, Regulations 26 and 27. HSE MISC156, 1998 HSE Books.

Provision and Use of Work Equipment regulations 1992. SI 1992 No. 2932. http://www.opsi.gov.uk/si/si1992/Uksi_19922932_en_5.htm.
PUWER 1998 (Provision and Use of Work Equipment Regulations 1998): Open learning guidance. 1999 HSE Books, ISBN 9780-7176-2459-1.
Retrofitting of Roll-Over Protective Structures, Restraining Systems and Their Attachment Points to Mobile Work Equipment. MISC175, 1999 HSE Books.
Safe Use of Hand-Fed Planning Machines. WIS17, 2000 HSE Books.
Safe Use of Manual Powered Cross-Cut Saws. WIS36, 1998 HSE Books.
Safe Use of Power-Operated Cross-Cut Saws. WIS35, 1998 HSE Books.
Safe Use of Spindle Moulding Machines. WIS18, 2001 HSE Books.
Safe Use of Woodworking Machinery, Approved Code of practice. L114, 1998 HSE Books, ISBN 9780-7176-1630-4.
Safe Use of Work Equipment, Provision and Use of Work Equipment Regulations 1998. Approved Code of Practice and Guidance, HSC, L22, 1998 HSE Books, ISBN 9780-7176-1626-6.
Supplying New Machinery. A Short Guide to the Law. INDG270, 1998 HSE Books, ISBN 9780-7176-1560-X.
Using Work Equipment Safely. INDG229, 2002 HSE Books, ISBN 9780-7176-2389-0.

21.30 The Reporting of Injuries, Diseases and Dangerous Occurrences Regulations 1995

21.30.1 *Introduction*

These Regulations require the reporting of specified accidents, ill-health and dangerous occurrences to the enforcing authorities. The events all arise out of or in connection with work activities covered by the HSW Act. They include death, major injury and more than 3-day lost-time accidents. Schedules to the Regulations specify the details of cases of ill-health and dangerous occurrences.

For most businesses reportable events will be quite rare and so there is little for them to do under these Regulations apart from keeping the guidance and forms available and being aware of the general requirements.

21.30.2 *Definitions – regulation 2*

Accident includes an act of non-consensual physical violence done to a person at work.

This means that injuries through physical violence to people not at work are not reportable. Neither is any injury which occurs between workers over a personal matter or carried out by a visiting relative of a person at work to that person. However, if a member of the public caused injury to a person at work through physical violence, that is reportable.

Incidents involving acts of violence may well be reportable to the police, which is outside the requirements of these Regulations.

21.30.3 *Notification and reporting of major injuries and dangerous occurrences – regulation 3 (1), 4 and schedules 1 and 2*

The HSE or Local Authority shall be notified immediately by the quickest practicable means and sent a report form F2508 (or other approved means) within 10 days:

(a) following **death** of a person as a result of an accident arising out of or in connection with work. Also if death occurs within 1 year of an accident, the Authorities must be informed
(b) following **major Injury** to a person as a result of an accident arising out of or in connection with work
(c) where a person not at work suffers an injury as a result of or in connection with work, and that person is taken from the site to hospital for treatment
(d) where a person not at work suffers a major injury as a result of an accident arising out of or in connection with work at a hospital
(e) where there is a dangerous occurrence.

21.30.4 *Reporting of 3-day plus accidents – regulation 3 (2)*

Where a person at work is incapacitated for work of a kind which they might reasonably be expected to do, either under their own contract of employment or in the normal course of employment, for more than three consecutive days (excluding the day of the accident, but including any days which would not have been working days) because of an injury resulting from an accident at work, the responsible person shall within 10 days send a report on form F2508 or other approved form, unless it has been reported under regulation 3 (1) as a Major Injury, etc.

21.30.5 *Reporting of cases of disease – regulation 5*

Where a medical practitioner notifies the employer's responsible person that an employee suffers from a reportable work-related disease, a completed disease report form (F2508A) should be sent to the enforcing authority. The full list is contained in schedule 3 to the Regulations, which is summarized in this guide.

21.30.6 *Which enforcing authority?*

Local Authorities are responsible for retailing, some warehouses, most offices, hotels and catering, sports, leisure, consumer services, and places of worship.

The HSE are responsible for all other places of work.

21.30.7 *Records*

A record of each incident reported must be kept at the place of business for at least 3 years.

21.30.8 *Major injuries – schedule 1*

1. any fracture, other than to the fingers, thumbs or toes
2. any amputation
3. dislocation of the shoulder, hip, knee or spine
4. loss of sight (whether temporary or permanent)
5. a chemical or hot metal burn to the eye or any penetrating injury to the eye
6. any injury resulting from an electric shock or electrical burn (including any electrical burn caused by arcing or arcing products) leading to unconsciousness or requiring resuscitation or admittance to hospital for more than 24 hours
7. any other injury
   (a) leading to hypothermia, heat-induced illness or to unconsciousness
   (b) requiring resuscitation
   (c) requiring admittance to hospital for more than 24 hours
8. loss of consciousness caused by asphyxia or by exposure to a harmful substance or biological agent
9. either of the following conditions which result from the absorption of any substance by inhalation, ingestion or through the skin
   (a) acute illness requiring medical treatment
   (b) loss of consciousness
10. acute illness which requires medical treatment where there is reason to believe that this resulted from exposure to a biological agent or its toxins or infected material.

21.30.9 *Dangerous occurrences – schedule 2 summary*

Part I – General
1. **Lifting machinery, etc.**
   The collapse, overturning, or the failure of any load-bearing part of lifts and lifting equipment.

2. **Pressure systems**

The failure of any closed vessel or of any associated pipe work, in which the internal pressure was above or below atmospheric pressure.

3. **Freight containers**

The failure of any freight container in any of its load-bearing parts.

4. **Overhead electric lines**

Any unintentional incident in which plant or equipment comes into contact with overhead power lines.

5. **Electrical short circuit**

Electrical short circuit or overload attended by fire or explosion.

6. **Explosives**

The unintentional explosion or ignition of explosives, misfire, the failure of the shots in any demolition operation to cause the intended extent of collapse, the projection of material beyond the boundary of the site, any injury to a person resulting from the explosion or discharge of any explosives or detonator.

7. **Biological agents**

Any accident or incident which resulted or could have resulted in the release or escape of a biological agent likely to cause severe human infection or illness.

8. **Malfunction of radiation generators, etc.**

Any incident in which the malfunction of a radiation generator or its ancillary equipment used in fixed or mobile industrial radiography, the irradiation of food or the processing of products by irradiation, causes it to fail to de-energize at the end of the intended exposure period; or to fail to return to its safe position at the end of the intended exposure period.

9. **Breathing apparatus**

Any incident in which breathing apparatus malfunctions while in use, or during testing immediately prior to use.

10. **Diving operations**

In relation to a diving project the failure or the endangering of diving equipment, the trapping of a diver, any explosion in the vicinity of a diver, any uncontrolled ascent or any omitted decompression.

11. **Collapse of scaffolding**

The complete or partial collapse of any scaffold which is more than 5 m in height, erected over or adjacent to water, in circumstances such that there would be a risk of drowning to a person falling from the scaffold into the water; or the suspension arrangements of any slung or suspended scaffold which causes a working platform or cradle to fall.

12. **Train collisions**

Any unintended collision of a train with any other train or vehicle.

13. **Wells**

Any dangerous occurrence at a well other than a well sunk for the purpose of the abstraction of water.

14. **Pipelines or pipeline works**

A dangerous occurrence in respect of a pipeline or pipeline works.

15. **Fairground equipment**

The failure of any load-bearing part or the derailment or the unintended collision of cars or trains.

16. **Carriage of dangerous substances by road**

Any incident involving a road tanker or tank container used for the carriage of dangerous goods in which the tanker overturns, is seriously damaged, there is an uncontrolled release or there is a fire.

Dangerous occurrences which are reportable except in relation to offshore workplaces

17. **Collapse of building or structure**

Any unintended collapse or partial collapse of: any building or structure under construction, reconstruction, alteration or demolition which involves a fall of more than 5 tonnes of material; any floor or wall of any building; or any false-work.

18. **Explosion or fire**

An explosion or fire occurring in any plant or premises which results in the suspension of normal work for more than 24 hours.

19. **Escape of flammable substances**

The sudden, uncontrolled release inside a building:
  (i) of 100 kg or more of a flammable liquid
  (ii) of 10 kg or more of a flammable liquid above its boiling point
  (iii) of 10 kg or more of a flammable gas.

500 kg or more of any of the substances, if released in the open air.

20. **Escape of substances**

The accidental release or escape of any substance in a quantity sufficient to cause the death, major injury or any other damage to the health of any person.

21.30.10 *Reportable diseases – schedule 3 brief summary*

These include:

➤ certain poisonings
➤ some skin diseases such as occupational dermatitis, skin cancer, chrome ulcer, oil folliculitis/acne
➤ lung diseases, such as occupational asthma, farmer's lung, pneumoconiosis, asbestosis, mesothelioma
➤ infections such as leptospirosis, heptatitis, tuberculosis, anthrax, legionellosis and tetanus
➤ other conditions, such as occupational cancer, certain musculoskeletal disorders, decompression illness and hand–arm vibration (HAV) syndrome.

The details can be checked by consulting the guide, looking at the pad of report forms, checking the HSE's website, or ringing the HSE's Infoline or the Incident Control Centre (ICC).

21.30.11 References

Accident Book. BI510, 2003 HSE Books, ISBN 9780-7176-2603-2.

A Guide to the Reporting of Injuries, Diseases and Dangerous Occurrences Regulations 1995. L73, 1999 HSE Books, ISBN 9780-7176-2431-5.

RIDDOR Explained. HSE31, 1999 HSE Books, ISBN 9780-7176-2441-2.

RIDDOR Information for Doctors. HSE32, 1996 HSE Books.

The Incident Contact Centre, MISC310, HSE.

21.31 Safety Representatives and Safety Committees Regulations 1977

These Regulations, made under the HSW ACT section 2(4), prescribe the cases in which recognized trade unions may appoint safety representatives, specify the functions of such representatives, and set out the obligations of employers towards them.

Employers' obligations to consult non-union employees are contained in the Health and Safety (Consultation with Employees) Regulations 1996.

21.31.1 Appointment – regulation 3

A recognized trade union may appoint safety representatives from among employees in all cases where one or more employees are employed. When the employer is notified in writing the safety representatives have the functions set out in regulation 4.

A person ceases to be a safety representative when:

➤ the appointment is terminated by the trade union
➤ they resign
➤ employment ceases.

A safety representative should have been with the employer for 2 years or have worked in similar employment for at least 2 years.

21.31.2 Functions – regulation 4

These are functions and not duties. They include:

➤ representing employees in consultation with the employer
➤ investigating potential hazards and dangerous occurrences

➤ investigating the causes of accidents
➤ investigating employee complaints relating to health, safety and welfare
➤ making representations to the employer on health, safety and welfare matters
➤ carrying out inspections
➤ representing employees at the workplace in consultation with enforcing inspectors
➤ receiving information
➤ attending safety committee meetings.

Safety representatives must be afforded time off with pay to fulfil these functions and to undergo training.

21.31.3 Employers' duties – regulation 4a

Every employer shall consult safety representatives in good time regarding:

➤ the introduction of any measure which may affect health and safety
➤ the arrangements for appointing or nominating competent person(s) under the Management Regulations
➤ any health and safety information required for employees
➤ the planning and organizing of any health and safety training for employees
➤ the health and safety consequences of introducing new technology.

Employers must provide such facilities and assistance as safety representatives may reasonably require to carry out their functions.

21.31.4 Inspections – regulations 5 and 6

Following reasonable notice in writing, safety representatives may inspect the workplace every quarter (or more frequently by agreement with the employer). They may inspect the workplace at any time, after consultation, when there have been substantial changes in the conditions of work or there is new information on workplace hazards published by the HSC or HSE.

Following an injury, disease or dangerous occurrence subject to RIDDOR, and after notifying the employer, where it is reasonably practicable to do so, safety representatives may inspect the workplace if it is safe.

Employers must provide reasonable assistance and facilities, including provision for independent investigation and private discussion with employees. The employer may be present in the workplace during inspections.

529

21.31.5 *Information – regulation 7*

Having given reasonable notice to the employer, safety representatives are entitled to inspect and take copies of any relevant statutory documents (except any health record of an identified person).

An exempt document is one:

➤ which could endanger national security
➤ which could cause substantial commercial injury to the employer
➤ which contravenes a prohibition
➤ that relates to an individual without their consent
➤ which has been obtained specifically for legal proceedings.

21.31.6 *Safety committees – regulation 9*

When at least two safety representatives have requested in writing that a safety committee is set up, the employer has 3 months to comply. The employer must consult with the safety representatives and post a notice stating its composition and the workplaces covered by it, in a place where it can be easily read by employees. The guidance gives details on the composition and running of safety committees.

21.31.7 *Complaints – regulation 11*

If the employer fails to permit safety representatives time off or fails to pay for time off, complaints can be made to an industrial tribunal within 3 months of the incident.

21.31.8 *References*

HSC, Safety Representatives and Safety Committees. (The Brown Book) Approved Code of Practice and Guidance on the Regulations, 3rd edition, L87, 1996 HSE Books, ISBN 9780-7176-1220-1.

21.32 Health and Safety (Safety Signs and Signals) Regulations 1996

21.32.1 *Introduction*

These Regulations came into force in April 1996, but if existing signs comply with BS 5378 and fire safety signs BS 5499 no changes are required. The results of the risk assessment made under the Management of Health and Safety At Work Regulations will have identified situations where there may be a residual risk when warnings

or further information are necessary. If there is no significant risk there is no need to provide a sign.

21.32.2 *Definitions – regulation 2*

➤ 'Safety sign' means a sign referring to a specific object, activity or situation and providing information or instruction about health and safety at work by means of a signboard, a safety colour, an illuminated sign, an acoustic sign, a verbal communication or a hand signal.
➤ 'Signboard' means a sign which provides information or instructions by a combination of geometric shape, colour and a symbol or pictogram and which is rendered visible by lighting of sufficient intensity.
➤ 'Hand signal' means a movement or position of the arms or hands or a combination, in coded form, for guiding persons who are carrying out manoeuvres which create a risk to the health and safety of people at work.
➤ 'Acoustic signal' means a coded sound signal which is released and transmitted by a device designed for that purpose, without the use of a human or artificial voice.
➤ 'Verbal communication' means a predetermined spoken message communicated by a human or artificial voice.

21.32.3 *Provision and maintenance of safety signs – regulation 4*

The Regulations require employers to use and maintain a safety sign where there is a significant risk to health and safety that has not been avoided or controlled by other means, like engineering controls or safe systems of work, and where the use of a sign can help reduce the risk.

They apply to all workplaces and to all activities where people are employed, but exclude signs used in connection with transport or the supply and marketing of dangerous substances, products and equipment.

The Regulations require, where necessary, the use of road traffic signs in workplaces to regulate road traffic.

21.32.4 *Information, instruction and training – regulation 5*

Every employer shall ensure that:

➤ comprehensible and relevant information on the measures to be taken in connection with safety signs is provided to each employee
➤ each employee receives suitable and sufficient instruction and training in the meaning of safety signs.

21.32.5 *Functions of colours, shapes and symbols in safety signs*

Safety colours
Red

Red is a safety colour and must be used for any:

➤ prohibition sign concerning dangerous behaviour (e.g. the safety colour on a 'No Smoking' sign). Prohibition signs must be round, with a black pictogram on a white background with red edging and a red diagonal line (top left, bottom right). The red part must take up at least 35% of the area of the sign.

➤ danger alarm concerning stop, shutdown, emergency cut out devices, evacuate (e.g. the safety colour of an emergency stop button on equipment)

➤ firefighting equipment.

No fork-lift trucks   No smoking

Red and white alternating stripes may be used for marking surface areas to show obstacles or dangerous locations.

Yellow

Yellow (or amber) is a safety colour and must be used for any warning sign concerning the need to be careful, take precautions, examine or the like (e.g. the safety colour on hazard signs, such as for flammable material, electrical danger). Warning signs must be triangular, with a black pictogram on a yellow (or amber) background with black edging. The yellow (or amber) part must take up at least 50% of the area of the sign.

General danger   Explosive

Yellow and black alternating stripes may be used for marking surface areas to show obstacles or dangerous locations.

Yellow may be used in continuous lines showing traffic routes.

Blue

Blue is a safety colour and must be used for any mandatory sign requiring specific behaviour or action (e.g. the safety colour on a 'Safety Helmet Must Be Worn' sign or a 'Pedestrians Must Use This Route' sign). Mandatory signs must be round, with a white pictogram on a blue background. The blue part must take up at least 50% of the area of the sign.

Ear protection   Eye protection
must be worn   must be worn

Green

Green is a safety colour and must be used for: emergency escape signs (e.g. showing emergency doors, exits and routes) and first aid signs (e.g. showing location of first-aid equipment and facilities). Escape and first-aid signs must be rectangular or square, with a white pictogram on a green background. The green part must take up at least 50% of the area of the sign. So long as the green takes up at least 50% of the area, it is sometimes permitted to use a green pictogram on a white background, for example where there is a green wall and the reversal provides a more effective sign than one with a green background and white border; no danger.

Means of Escape   First Aid

Other colours
White

White is *not* a safety colour but is used: for pictograms or other symbols on blue and green signs; in alternating red and white stripes to show obstacles or dangerous locations; in continuous lines showing traffic routes.

Black

Black is *not* a safety colour but is used: for pictograms or other symbols on yellow (or amber) signs and, except for fire signs, red signs; in alternating yellow and black stripes to show obstacles or dangerous locations.

Shapes

Round signs must be used for any prohibition (red) sign; mandatory (blue) sign.

Triangular signs must be used for any warning (yellow or amber) sign.

Square or rectangular signs must be used for any emergency escape sign and any first-aid sign.

Pictograms and other symbols

The meaning of a sign (other than verbal communication) must not rely on words. However, a sign may be supplemented with words to reinforce the message provided the words do not in fact distract from the message or create a danger.

A sign (other than verbal communication, acoustic signals or hand signals) should use a simple pictogram and/or other symbol (such as directional arrows, exclamation mark) to effectively communicate its message and so overcome language barriers.

Pictograms and symbols are included in the Regulations. Employee training is needed to understand the meaning of these since many are not inherently clear, some are meaningless to anyone who has not had their meaning explained and some can even be interpreted with their opposite meaning.

Pictograms and symbols included in the Regulations do not cover all situations for which graphic representation of a hazard or other detail may be needed. Any sign used for a situation not covered in the Regulations, should include:

➤ the international symbol for general danger (exclamation mark!) if the sign is a warning sign and tests show that the sign is effective
➤ in any other case a pictogram or symbol which has been tested and shown to be effective.

The text of any words used to supplement a sign must convey the same meaning. For example, a round blue sign with a pictogram showing the white outline of a face with a solid white helmet on the head means 'Safety Helmet Must Be Worn' and so any text used must maintain the obligatory nature of the message.

21.32.6 *References*

Health and Safety (Safety Signs and Signals) Regulations 1996. SI 1996/341, http://www.opsi.gov.uk/si/si1996/Uksi_19960341_en_2.htm#end.
Safety Signs and Signals Guidance on the Regulations. L64, 1996 HSE Books, ISBN 9780-7176-0870-0.
Signpost to the Health and Safety (Safety Signs and Signals) Regulations Guidance on the Regulations 1996. INDG 184, HSE Books, ISBN 9780-7176-1139-6.

21.33 Supply of Machinery (Safety) Regulations 1992

21.33.1 *Introduction*

The Supply of Machinery (Safety) Regulations 1992 entered into force on 1 January 1993, although there was a transitional period to 31 December 1994 during which the manufacturer or importer into the EC was able to choose between **either** the Community regime described below or complying with the health and safety legislation in force on 31 December 1992.

The Supply of Machinery (Safety) (Amendment) Regulations 1994 made a number of changes to the 1992 Regulations, in particular to widen the scope to include machinery for lifting persons and safety components for machinery. The main provisions of the amending Regulations entered into force on 1 January 1995.

Therefore from **1 January 1995**:

➤ most machinery supplied in the UK, including imports, must:
  ➤ satisfy wide-ranging health and safety requirements, for example, on construction, moving parts and stability
  ➤ in some cases, have been subjected to type examination by an approved body
  ➤ carry CE marking and other information
➤ the manufacturer or the importer will generally have to be able to assemble a file containing technical information relating to the machine.

21.33.2 *Other relevant legislation to the supply of machinery*

The two sets of Regulations that will often apply are the:

➤ Electrical Equipment (Safety) Regulations 1994, which apply to most electrically powered machinery used in workplaces
➤ Electromagnetic Compatibility Regulations 1992, which cover equipment likely to cause electromagnetic disturbance, or whose performance is likely to be affected by electromagnetic disturbance.

In some cases, other laws may apply such as the Simple Pressure Vessels (Safety) Regulations 1991 or the Gas Appliances (Safety) Regulations 1995. All these Regulations implement European Directives and contain various requirements. The existence of CE marking on machinery should indicate that the manufacturer has met **all** of the requirements that are relevant.

Special transitional arrangements remain for products covered by existing Directives on roll-over and falling-object protective structures and industrial trucks, and for safety components and machinery for lifting persons.

21.33.3 *Failure to comply with these requirements*

➤ will mean that the machinery cannot legally be supplied in the United Kingdom
➤ could result in prosecution and penalties, on conviction, of a fine up to £5,000 or, in some cases, of imprisonment for up to 3 months, or both.

The same rules apply everywhere in the European Economic Area (EEA), so machinery complying with the Community regime may be supplied in any EEA State.

21.33.4 *Coverage*

Machinery, described as:

➤ an assembly of linked parts or components, at least one of which moves, with the appropriate actuators, control and power circuits, joined together for a specific application, in particular for the processing, treatment, moving or packaging of a material
➤ an assembly of machines which, in order to achieve the same end, are arranged and controlled so that they function as an integral whole
➤ interchangeable equipment modifying the function of a machine which is supplied for the purpose of being assembled with a machine (or a series of different machines or with a tractor) by the operator himself in so far as this equipment is not a spare part or a tool.

Safety components for machinery, described as:

➤ components which are supplied separately to fulfil a safety function when in use and the failure or malfunctioning of which endangers the safety or health of exposed persons.

21.33.5 *Exceptions*

The Regulations do not apply to machinery or safety components:

➤ listed in Schedule 5 (see section 21.33.6)
➤ previously used in the EC or, since 1 January 1994 the EEA (e.g. second hand)
➤ for use outside the EEA which does not carry CE marking

➤ where the hazards are mainly of electrical origin (such machinery is covered by the Electrical Equipment (Safety) Regulations 1994)
➤ to the extent that the hazards are wholly or partly covered by other Directives, from the date those other Directives are implemented into UK law
➤ relating to machinery first supplied in the EC before 1 January 1993.
➤ to safety components or machinery for lifting persons first supplied in the EEA before 1 January 1995.

Such products first supplied on or after 1 January 1995 must comply **either** with the Supply of Machinery (Safety) Regulations or the UK health and safety legislation in force relating to these items on 14 June 1993. All such products first supplied after 1 January 1997 must comply with the Supply of Machinery (Safety) Regulations.

21.33.6 *Machinery excluded – Schedule 5 to the Regulations*

➤ machinery whose only power source is directly applied manual effort, unless it is a machine used for lifting or lowering loads
➤ machinery for medical use used in direct contact with patients
➤ special equipment for use in fairgrounds and/or amusement parks
➤ steam boilers, tanks and pressure vessels
➤ machinery specially designed or put into service for nuclear purposes, which, in the event of failure, may result in an emission of radioactivity
➤ radioactive sources forming part of a machine
➤ firearms
➤ storage tanks and pipelines for petrol, diesel fuel, highly flammable liquids and dangerous substances
➤ means of transport, that is vehicles and their trailers intended solely for transporting passengers by air or on road, rail or water networks. Also, transport, which is designed for transporting goods by air, on public road or rail networks or on water. Vehicles used in the mineral extraction industry are not excluded
➤ sea-going vessels and mobile offshore units together with equipment on board, such as vessels or units
➤ cableways, including funicular railways, for the public or private transportation of people
➤ agriculture and forestry tractors, as defined by certain European Directives
➤ machines specially designed and constructed for military or police purposes
➤ certain goods and passenger lifts

➤ means of transport of people using rack and pinion rail-mounted vehicles

➤ mine winding gear

➤ theatre elevators

➤ construction site hoists intended for lifting individuals or people and goods.

21.33.7 *General requirements*

Subject to the exceptions and transitional arrangements described above, the Regulations make it an offence for a 'responsible person' to supply machinery or a safety component unless:

➤ it satisfies the essential health and safety requirements

➤ the appropriate conformity assessment procedure has been carried out

➤ an EC declaration of conformity or declaration of incorporation has been issued

➤ CE marking has been properly affixed (unless a declaration of incorporation has been issued)

➤ it is, in fact, safe.

The Regulations also make it an offence for any supplier to supply machinery or a safety component **unless it is safe**.

21.33.8 *Essential health and safety requirements*

To comply with the Regulations, machinery and safety components must satisfy the essential health and safety requirements (set out in Schedule 3 to the Regulations), which apply to it. The requirements are wide ranging, and take into account potential dangers to operators and other exposed persons within a 'danger zone'. Aspects covered in Part 1 include: the materials used in the construction of the machinery; lighting; controls; stability; fire; noise; vibration; radiation; emission of dust, gases, etc.; and maintenance. Part 2 has additional requirements for agri-foodstuffs machinery, portable hand-held machinery, and machinery for working wood and analogous materials. Part 3 deals with particular hazards associated with mobility, Part 4 with those associated with lifting, Part 5 with those associated with underground working and Part 6 with those associated with the lifting or moving of persons. The requirements also comment on instructions (including translation requirements) and marking.

When applying the essential health and safety requirements, technical and economic limitations at the time of construction may be taken into account.

21.33.9 *Standards*

Machinery and safety components manufactured in conformity with specified, published European standards which have also been published as identically worded national standards (transposed harmonized standards), will be presumed to comply with the essential health and safety requirements covered by those standards.

The European Committee for Standardization (CEN) is working to produce a complex of European standards at three levels in support of the Machinery Directive. The first (A) level comprises general principles for the design of machinery. The second (B) level covers specific safety devices and ergonomic aspects. The third (C) level deals with specific classes of machinery by calling up the appropriate standards from the first two levels and addressing requirements specific to the class.

21.33.10 *Step 1 – conformity assessment (attestation)*

The Annexes refer to the Product Standards Machinery Guide URN 95/650. (See Reference 21. 33. 65).

The responsibility for demonstrating that the machinery or safety component satisfies the essential health and safety requirements rests on the 'responsible person'.

For most machinery or safety components (other than those listed in Annex D): the 'responsible person', must be able to assemble the technical file described in the Regulations.

For machinery or safety components listed in Annex D: the 'responsible person' must follow the special procedures described below.

21.33.11 *Step 2 – declaration procedure*

Declaration of conformity: the 'responsible person' must then draw up an EC declaration of conformity, described in Annex G, for each machine or safety component supplied. This declaration is intended to be issued with the machine or safety component and declares that it complies with the relevant essential health and safety requirements or with the example that underwent type-examination.

Declaration of incorporation: alternatively, where the machinery is intended for incorporation into other machinery or assembly with other machinery to constitute machinery covered by the Regulations, the 'responsible person' may draw up a declaration of incorporation, described in Annex H, for each machine.

21.33.12 *Step 3 – marking*

Once a declaration of conformity has been issued, the 'responsible person' must affix the CE marking to the machinery.

CE markings must be affixed in a distinct, visible, legible and indelible manner.

The CE marking should not be affixed to safety components or for machinery for which a declaration of incorporation has been issued.

The Regulations make it an offence to affix a mark to machinery that may be confused with CE marking.

21.33.13 Enforcement

In Great Britain, the HSE is responsible for enforcing the Regulations in relation to machinery and safety components for use at work. Local authority Trading Standards Officers are responsible for enforcement in relation to machinery and safety components for private use.

In Northern Ireland, the Department of Economic Development and the Department of Agriculture are responsible for enforcing the Regulations in relation to machinery and safety components for use at work. District councils are responsible for enforcement in relation to machinery and safety components for private use.

The enforcement authorities have available to them various powers under the Health and Safety at Work etc Act 1974, the Health and Safety at Work (Northern Ireland) Order 1978 and the Consumer Protection Act 1987, for example, relating to suspension, prohibition and prosecution.

Where machinery bearing the CE marking is safe, but there are breaches of other obligations, the 'responsible person' will be given the opportunity to correct the breach before further enforcement action is taken.

The Machinery Directive, as amended, requires Member States to inform the European Commission of any specific enforcement action taken. The Commission will consider whether the action is justified and advise the parties concerned accordingly.

21.33.14 Penalties

The maximum penalty for contravening the prohibition on supply of non-compliant machinery and safety components is imprisonment for up to 3 months or a fine of up to £5,000 or both. The penalty for other contraventions of the Regulations is a fine of up to the same amounts. It is for the courts to decide the penalty in any given case, taking into account the severity of the offence.

The Regulations provide a defence of due diligence. They also provide for proceedings to be taken against a person other than the principal offender, if it is the other person's fault, and against officers of a company or other body corporate.

21.33.15 References

Buying New Machinery: A Short Guide to the Law, INDG 271, 1998 HSE Books, ISBN 9780-7176-1559-6.
Supplying New Machinery Advice to Suppliers, INDG270, 1998 HSE Books, ISBN 9780-7176-1560-X.
Product Standard-Machinergy Guideline Noes on UK Regulation 1995 URN 95/650 www.Berr.gov.uk/files/file1127.pdf

21.34 Control of Vibration at Work Regulations 2005

21.34.1 Introduction

These Regulations implement European Directive Vibration Directive 2002/44/EC. They came into force in July 2005 with some transitional arrangements for the exposure limits until 2010 (2014 for WBV (whole-body vibration) exposure limit value for agriculture and forestry sectors).

The Regulations impose duties on employers to protect employees who may be exposed to risk from vibration at work, and other persons who may be affected by the work.

21.34.2 Interpretation – regulation 2

'Daily exposure' means the quantity of mechanical vibration to which a worker is exposed during a working day, normalized to an 8 hour reference period, which takes account of the magnitude and duration of the vibration.

'Hand–arm vibration' (HAV) means mechanical vibration which is transmitted into the hands and arms during a work activity.

'Whole-body vibration' (WBV) means mechanical vibration which is transmitted into the body when seated or standing through the supporting surfaces, during a work activity or as described in regulation 5 (3) (f).

21.34.3 Application – regulation 3

For work equipment first provided to employees for use prior to 6th July 2007 and where compliances with the exposure limit values is not possible, employers have until 2010 to comply and in the case of agriculture and forestry, 2014 (for WBV).

However, action must be taken to use the latest technical advances and the organizational measures in accordance with regulation 6 (2)

21.34.4 Exposure limit values and action values – regulation 4

HAV: (a) 8 hour daily exposure limit value is 5 m/s^2 A (8).
(b) 8 hour daily exposure action value is 2.5 m/s^2 A (8).
(c) Daily exposure is ascertained as set out in schedule 2 Part I.

WBV: (a) 8 hour daily exposure limit value is 1.15 m/s^2 A (8).
(b) 8 hour daily exposure action value is 0.5 m/s^2 A (8).
(c) Daily exposure is ascertained as set out in schedule 2 Part I.

21.34.5 *Assessment of risk to health created by vibration at the workplace – regulation 5*

A suitable and sufficient risk assessment must be made where work liable to expose employees is carried on. The risk assessment must identify the measures which need to be taken to comply with these Regulations.

Assessment of daily exposure should be by means of:

➤ observation of specific working practices
➤ reference to relevant work equipment vibration data
➤ if necessary, measurement of the magnitude of vibration to which employees are exposed and
➤ likelihood of exposure at or above an exposure action value or above an exposure limit value.

The risk assessment shall include consideration of:

➤ the magnitude, type and duration of exposure, including any exposure to intermittent vibration or repeated shocks
➤ the effects of exposure to vibration on employees whose health is at particular risk from such exposure
➤ any effects of vibration on the workplace and work equipment, including the proper handling of controls, the reading of indicators, the stability of structures and the security of joints
➤ any information provided by the manufacturers of work equipment
➤ the availability of replacement equipment designed to reduce exposure to vibration
➤ any extension of exposure at the workplace to whole-body vibration beyond normal working hours, including exposure in rest facilities supervised by the employer
➤ specific working conditions such as low temperatures and
➤ appropriate information obtained from health surveillance including, where possible, published information.

The risk assessment shall be reviewed regularly and forthwith if:

➤ there is reason to suspect that the risk assessment is no longer valid or
➤ there has been a significant change in the work to which the assessment relates,

and where, as a result of the review, changes to the risk assessment are required, those changes shall be made.

The employer shall record:

➤ the significant findings of the risk assessment as soon as is practicable after the risk assessment is made or changed and
➤ the measures which he has taken and which he intends to take to meet the requirements of regulation 6.

21.34.6 *Elimination or control of exposure to vibration at the workplace – regulation 6*

Following the general principles of prevention in the Management Regulations, the employer shall ensure that risk from the exposure of his employees to vibration is either eliminated at source or reduced to as low as reasonably practicable.

Where this is not reasonably practicable and the risk assessment indicates that an exposure action value is likely to be reached or exceeded, the employer shall reduce exposure to as low as reasonably practicable by establishing and implementing a programme of organizational and technical measures which is appropriate.

Consideration must be given to:

➤ other working methods which eliminate or reduce exposure to vibration
➤ a choice of work equipment of appropriate ergonomic design which, taking account of the work to be done, produces the least possible vibration
➤ the provision of auxiliary equipment which reduces the risk of injuries caused by vibration
➤ appropriate maintenance programmes for work equipment, the workplace and workplace systems
➤ the design and layout of workplaces, work stations, and rest facilities
➤ suitable and sufficient information and training for employees, such that work equipment may be used correctly and safely, in order to minimize their exposure to vibration
➤ limitation of the duration and intensity of exposure to vibration
➤ appropriate work schedules with adequate rest periods and
➤ the provision of clothing to protect employees from cold and damp.

Subject to implementation dates, the employer shall:

➤ ensure that his employees are not exposed to vibration above an exposure limit value or
➤ if an exposure limit value is exceeded, he shall forthwith:
  ➤ take action to reduce exposure to vibration to below the limit value
  ➤ identify the reason for that limit being exceeded and
  ➤ modify the organizational and technical measures taken in accordance with paragraph (2) to prevent it being exceeded again.

This shall not apply where the exposure of an employee to vibration is usually below the exposure action value but varies markedly from time to time and may occasionally exceed the exposure limit value, provided that:

➤ any exposure to vibration averaged over 1 week is less than the exposure limit value

➤ there is evidence to show that the risk from the actual pattern of exposure is less than the corresponding risk from constant exposure at the exposure limit value

➤ risk is reduced to as low as reasonably practicable, taking into account the special circumstances and

➤ the employees concerned are subject to increased health surveillance, where such surveillance is appropriate.

Account must be taken of any employee whose health is likely to be particularly at risk from vibration.

21.34.7 *Health surveillance – regulation 7*

If:

➤ the risk assessment indicates that there is a risk to the health of his employees who are, or are liable to be, exposed to vibration or

➤ employees are exposed to vibration at or above an exposure action value

the employer shall ensure that such employees are under suitable health surveillance. Records must be kept and employees given access to their own records and the enforcing authorities provided with copies, as may be required. A range of specified action is required if problems are found with the health surveillance results.

21.34.8 *Information, instruction and training – regulation 8*

If:

➤ the risk assessment indicates that there is a risk to the health of employees who are, or who are liable to be, exposed to vibration or

➤ employees are exposed to vibration at or above an exposure action value

the employer shall provide those employees and their representatives with suitable and sufficient information, instruction and training.

The information, instruction and training provided shall include:

➤ the organizational and technical measures taken in order to comply with the requirements of regulation 6

➤ the exposure limit values and action values

➤ the significant findings of the risk assessment, including any measurements taken

➤ why and how to detect and report signs of injury

➤ entitlement to appropriate health surveillance

➤ safe working practices to minimize exposure to vibration; and

➤ the collective results of any health surveillance undertaken in accordance with regulation 7 in a form calculated to prevent those results from being identified as relating to a particular person.

The information, instruction and training required must be adapted to take account of significant changes in the type of work carried out or methods of work used by the employer, and cover all persons who carry out work in connection with the employer's duties under these Regulations.

21.34.9 *References*

Control of Vibration at Work Regulations 2005. SI 2005 No. 1093.

Control Back-Pain Risks from Whole-Body Vibration Advice for Employers on the Control of Vibration at Work Regulations 2005. INDG242 (rev1), 2005 HSE Books, ISBN 9780-7176-6119-9.

Drive away Bad Backs Advice for Mobile Machine Operators and Drivers. INDG404, 2005 HSE Books, ISBN 9780-7176-6120-2.

Hand Arm Vibration: Guidance on the Control of Vibration at Work Regulations 2005, L140, 2005 HSE Books, ISBN 9780-7176-6125-3.

Hand–Arm Vibration Advice for Employees. INDG296 (rev1), 2005 HSE Books, ISBN 9780-7176-6118-0.

Health risks from hand–arm vibration, Advice for employers on the Control of Vibration at Work Regulations 2005. INDG 175 (rev2), 2005 HSE Books, ISBN 9780-7176-6117-2.

OPSI: http://www.opsi.gov.uk/si/si2005/uksi_20051093_en.pdf.

Power Tools, How to Reduce Vibration Health Risks. INDG338, 2001 HSE Books, ISBN 9780-7176-2008-5.

Vibration Solutions Practical Way to Reduce the Risk of Hand–Arm Vibration. HSG 170, 1997 HSE Books, ISBN 9780-7176-0954-5.

Whole-Body Vibration: The Control of Vibration at Work Regulations 2005, Guidance on the Regulations. L141, 2005 HSE Books, ISBN 9780-7176-6126-1.

21.35 Workplace (Health, Safety and Welfare) Regulations 1992 as amended in 2002

21.35.1 *General*

These Regulations were made to implement the European Directive on the minimum safety and health requirements for the workplace. A workplace for these purposes is defined very widely to include any part of non-domestic premises to which people have access while at work and any room, lobby, corridor, staircase or other means of access to or exit therefrom. The main exceptions to these rules are construction sites, means of transport, mines and quarries and other mineral extraction sites.

The main requirements are summarized below. They are expressed in very general terms, and in each case it will be necessary to turn to the ACOP associated with these Regulations for clarification of what is necessary to meet the objectives set. The requirements of people with a disability must now be taken into account following amendments in 2002.

21.35.2 *Health – the working environment – regulations 6–10*

Ventilation
Ventilation must be effective in enclosed areas, and any plant used for this purpose must incorporate warning devices to signal breakdowns which might endanger health or safety.

A reasonable temperature
A reasonable temperature must be maintained during working hours and sufficient thermometers must be provided to enable people at work to determine the temperature in any workroom. The temperature should be comfortable without the need for special clothing. Special guidance is available for areas like food processing where it could be very hot or very cold.

Temperature should be at least 16°C or, where strenuous effort is involved, 13°C. Workrooms should be adequately thermally insulated where necessary. The excessive effects of sunlight on temperature must be avoided.

Lighting
Lighting must be suitable and sufficient. This should be natural lighting SFARP.

Emergency lighting
Emergency lighting shall be provided where persons are particularly exposed to danger if artificial light fails. Lights should avoid glare and dazzle and should not themselves cause a hazard. They should not be obscured and should be properly maintained.

Workplaces, furniture and fittings
Workplaces, furniture and fittings should be kept sufficiently clean. Surfaces inside buildings shall be capable of being kept sufficiently clean.

Floors
Floors should not be slippery and wall surfaces should not increase fire risks.

Waste
Waste should not be allowed to accumulate, except in suitable receptacles and these receptacles should be kept free from offensive waste products and discharges which should be dealt with in a specialist way.

Room dimensions
Room dimensions have to allow adequate unoccupied space to work in and to move freely. 11 m³ minimum per person is required, excluding anything over 3 m high and furniture etc.

Workstations
Workstations shall be suitable for any persons in the workplace who are likely to work at those workstations and for any work that is likely to be done there.

Outside workstations
Outside workstations shall provide, SFARP, protection from adverse weather; and adequate means of escape in emergencies; and ensure that no person is likely to slip or fall.

Seating
Seating shall be provided where work can or must be done sitting and shall be suitable for the person as well as the work. A footrest shall be provided where necessary.

21.35.3 *Safety – accident prevention – regulations 5, 12–19*

Maintenance
The workplace and equipment, devices and systems shall be maintained (including being cleaned as appropriate) in an efficient state, efficient working order and in good repair, and where appropriate subjected to a system of maintenance. This generally means planned rather than breakdown maintenance. Systems including ventilation, emergency lighting, safety fences, window cleaning devices and moving walkways are all given as examples.

Floors and traffic routes
Floors and the surface of traffic routes shall be suitably constructed for their intended purpose and free of slope and holes (unless fenced). They should not be uneven or slippery. The traffic routes should be of adequate width and height to allow people and vehicles to circulate

safely with ease and they should be kept free of obstructions.

Additional precautions are necessary where pedestrians have to cross or share vehicle routes. Open sides of staircases should be fenced with an upper rail 900 mm or higher and a lower rail. Loading bays should have exits or refuges to avoid people getting crushed by vehicles.

Falls and falling objects
Now covered by the Work at Height Regulations 2005.

Tanks and pits
Where there is a risk of falling into a tank, pit or structure containing a dangerous substance that is likely to:

➤ scald or burn
➤ be poisonous or corrosive
➤ have an asphyxiating gas, fume or vapour or
➤ have any granular or free-flowing substance likely to cause harm

measures must be taken to securely fence or cover the tank, pit or structure.

Ladders and roofs
Now covered by the Work at Height Regulations 2005.

Glazing
Windows and transparent doors and partitions must be appropriately marked and protected against breakage.

Windows
Windows and skylights must open and close safely, and be arranged so that people may not fall out of them. They must be capable of being cleaned safely.

Traffic routes
Pedestrians and vehicles must be able to circulate safely. Separation should be provided between vehicle and people at doors, gateways and common routes. Workplaces should have protection from vehicles.

Doors and gates
Doors, gates and moving walkways have to be of sound construction and fitted with appropriate safety devices.

21.35.4 *Welfare – provision of facilities – regulations 20–25*

Sanitary conveniences and washing facilities
Suitable and sufficient sanitary conveniences and washing facilities should be provided at readily accessible places. The facilities must be kept clean, adequately ventilated and lit. Washing facilities should have running hot and cold or warm water, soap and clean towels or other method of cleaning or drying. If necessary, showers should be provided. Men and women should have separate facilities unless each facility is in a separate room with a lockable door and is for use by only one person at a time.

Drinking water
An adequate supply of wholesome drinking water, with an upward drinking jet or suitable cups, should be provided. Water should only be supplied in refillable enclosed containers where it cannot be obtained directly from a mains supply.

Accommodation for clothing and facilities for changing
Adequate, suitable and secure space should be provided to store workers' own clothing and special clothing. The facilities should allow for drying clothing. Changing facilities should also be provided for workers who change into special work clothing. They must be easily accessible, of sufficient capacity and provided with seating.

Facilities to rest and to eat meals
Suitable and sufficient, readily accessible, rest facilities should be provided. Arrangements should include suitable facilities to eat meals; adequate number of seats with backrests and tables to accommodate the number of persons likely to use them at any one time; means of heating food (unless hot food is available nearby) and making hot drinks; adequate and suitable seating for people employed with a disability.

Canteens or restaurants
Canteens and restaurants may be used as rest facilities provided there is no obligation to purchase food.

Suitable rest facilities
Suitable rest facilities should be provided for pregnant women and nursing mothers. They should be near sanitary facilities and where necessary, include the facility to lie down.

Non-smokers
This is now covered by the smoke-free legislation which bans smoking in largely enclosed or enclosed work and public premises.

21.35.5   *Accommodation for people with a disability*

In 2002 a new regulation 25a was added stating that where necessary, those parts of the workplace (including in particular doors, passageways, stairs, showers, wash-basins, lavatories and workstations) used or occupied directly by disabled persons at work shall be organized to take account of such persons.

21.35.6   *References*

General Ventilation in the Workplace: Guidance for employers. HSG 202, 2000 HSE Books, ISBN 9780-7176-1793-9.

How to Deal with Sick Building Syndrome: Guidance for Employers, Building Owners and Building Managers. HSG 132, 1995 HSE Books, ISBN 9780-7176-0861-1.

Lighting at Work. HSG 38, 1996 HSE Books, ISBN 9780-7176-1232-5.

Seating at Work. HSG 57, 1997 HSE Books, ISBN 9780-7176-1231-7.

Workplace (Health, Safety and Welfare) Regulations 1992. ISBN 9780-11-025804-5 OPSI: http://www.opsi.gov.uk/si/si1992/Uksi_19923004_en_1.htm.

Workplace Health, Safety and Welfare. Workplace (Health, Safety and Welfare) Regulations 1992 Approved Code of Practice and Guidance. L24, 1992 HSE Books, ISBN 9780-7176-0413-6.

Workplace Health, Safety and Welfare. A Short Guide, INDG244, 1997 HSE Books, ISBN 9780-7176-1328-3.

21.36   Work at Height Regulations 2005 as amended in 2007

21.36.1   *Introduction*

These Regulations bring together all current requirements on work at height into one goal-based set of Regulations. They implement the requirements of the 2nd Amending Directive (2001/45/EC) to the Use of Work Equipment Directive (89/955/EEC) which sets out requirements for work at height. The 2nd Amending Directive has become known as the Temporary Work at Height Directive.

The Work at Height (Amendment) Regulations 2007 which came into force on 6 April 2007 apply to those who work at height providing instruction or leadership to one or more people engaged in caving or climbing by way of sport, recreation, team building or similar activities in Great Britain.

The WAH Regulations require a risk assessment for all work conducted at height and arrangements to be put in place for:

➤ eliminating or minimizing risks from working at height
➤ safe systems of work for organizing and performing work at height
➤ safe systems for selecting suitable work equipment to perform work at height
➤ safe systems for protecting people from the consequences of work at height.

21.36.2   *Definitions – regulation 2*

'Work at height' means:

➤ work in any place, including a place at or below ground level
➤ obtaining access to or egress from such place while at work except by a staircase in a permanent workplace.

where, if measures required by these Regulations were not taken, a person could fall a distance liable to cause personal injury.

'Working platform':

➤ means any platform used as a place of work or as a means of access to or egress from a place of work
➤ includes any scaffold, suspended scaffold, cradle, mobile platform, trestle, gangway, gantry and stairway which is so used.

21.36.3   *Organization, planning and competence – regulations 4 and 5*

Work at height must be properly planned, appropriately supervised and carried out in a manner which is, SFARP, safe. The selection of appropriate work equipment is included in the planning. Work must not be carried out if the weather conditions would jeopardize safety or health (this does not apply where members of the police, fire, ambulance or other emergency services are acting in an emergency).

All people involved in work at height activity including planning, organizing and supervising must be competent for such work, or if being trained, under competent supervision.

21.36.4   *Avoidance of risk – regulation 6*

A risk assessment carried out under the Management Regulations must be taken into account when identifying the measures required by these Regulations. Work at height should be avoided if there are reasonably practicable alternatives.

Where work at height is carried out employers must take suitable and sufficient measures to prevent persons falling a distance liable to cause personal injury. The measures include:

(a) ensuring work is carried out
   ➤ from an existing workplace
   ➤ using existing means of access and egress that comply with schedule 1 of the Regulations (assuming it is safe and ergonomic to do so)
(b) where this is not reasonably practicable, providing work equipment (SFARP) to prevent a fall occurring.

Employers must take steps to minimize the distance and the consequences of a fall, if it is not prevented.

Where the distance cannot be minimized (SFARP) the consequence of a fall must be minimized and additional training, instruction and other additional suitable and sufficient measures must be adopted to prevent a person falling a distance liable to cause personal injury.

21.36.5 *General principles for selection of work equipment – regulation 7*

Collective protection measures must have priority over personal protection measures.

Employers must take account of:

➤ working conditions and the risks to the safety of persons at the place where the work equipment is to be used
➤ the distance to be negotiated for access and egress
➤ distance and consequences of a potential fall

Work at height flow chart

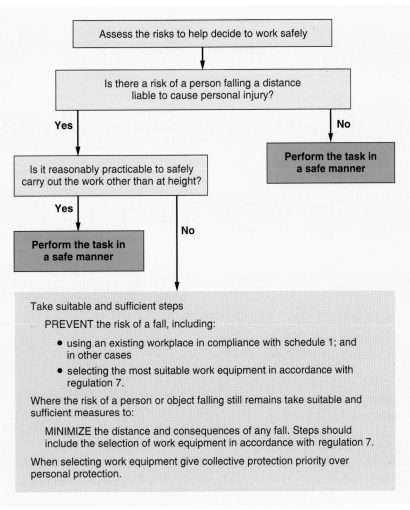

Figure 21.5 Work at height flow chart.

- duration and frequency of use
- need for evacuation and rescue in an emergency
- additional risks of using, installing and removing the work equipment used or evacuation and rescue from it
- other provisions of the Regulations.

Work equipment must:

- be appropriate for the nature of the work and the foreseeable loadings
- allow passage without risk
- be the most suitable work equipment having regard in particular to regulation 6.

21.36.6 *Requirements for particular work equipment – regulation 8*

Regulation 8 requires that various pieces of work equipment comply with the schedules to the Regulations as follows:

(a) A guard rail, toe-board, barrier or similar means of protection, must comply with schedule 2:

Schedule 2 in summary: the equipment

- must be suitable, of sufficient dimensions, of sufficient strength and rigidity
- must be so placed, secured and used to prevent accidental displacement
- must prevent fall of persons and materials
- must have supporting structure of sufficient strength and suitable for the purpose
- must have no lateral opening save at point of access to a ladder or stairway where an opening is necessary; and:
  - must be in place except for a time to gain access perform a particular task and then replaced
  - must have compensatory safety measures provided if protection removed temporarily.

(b) A working platform must comply with schedule 3 Part 1 and, in addition, where scaffolding is provided, schedule 3 Part 2.

Schedule 3 Part 1 in summary
Supporting structures must:

- be suitable, of sufficient strength and rigidity
- if wheeled, prevented, by appropriate device, from moving during work
- not slip and must have secure attachment
- be stable while being erected, used when modified, or dismantled.

Working platform to:

- be suitable and strong enough
- have no accidental displacement of components
- remain stable during dismantling
- be dismantled so as to prevent accidental displacement
- be of sufficient dimensions for safe passage and use
- have a suitable surface and be constructed to prevent people falling through
- have a suitable surface and be constructed to prevent material or objects falling through unless measures have been taken to protect other persons from falling objects
- be erected, used and maintained so that risk of slipping and tripping is prevented and no person can be caught between working platform and adjacent structure
- not be loaded so as to give risk of collapse or deformation.

Schedule 3 Part 2 additional for scaffolds:

- strength and stability calculations required unless they are available already or scaffold is assembled in conformity with a generally recognized standard configuration
- an assembly, use and dismantling plan to be drawn up; could be standard plan with supplements
- copy of the plan and instructions to be available for persons doing the work
- dimensions, form and layout to be suitable
- when not available for use to be marked with warning signs and physical barrier preventing access
- must be assembled, dismantled or significantly altered under supervision of competent person.

(c) A net, airbag or other collective safeguard for arresting falls which is not part of a personal fall protection system, must comply with schedule 4.
(d) Any personal fall protection system to comply with Part 1 of schedule 5 and
  - in the case of a work positioning system, comply with Part 2 of schedule 5
  - in the case of rope access and positioning techniques, comply with Part 3 of schedule 5
  - in the case of a fall arrest system, comply with Part 4 of schedule 5
  - in the case of a work restraint system, comply with Part 5 of schedule 5.
(e) A ladder must comply with schedule 6.

Schedule 6 in summary:

➤ only to be used for work at height if the risk assessment demonstrates that the use of more suitable equipment is not justified because of the low risk and the short duration of use or existing features on site that cannot be altered

➤ surface on which ladder rests must be stable, firm, and of sufficient strength and suitable composition such that its rungs remain horizontal

➤ must be positioned to ensure stability

➤ suspended ladder attached firmly without swing (except flexible ladder)

➤ portable ladder to be prevented from slipping by securing the stiles near their upper and lower ends and using an anti-slip device or by any other equivalent measures

➤ when used for access must protrude above place of landing unless other firm handholds provided

➤ interlocking or extension pieces must not move, relative to each other, during use

➤ mobile ladder to be prevented from moving before being stepped on

➤ sufficient safe landings provided for vertical distances of 9 m or more above base

➤ must be used so that a secure foothold and handhold always available

➤ must be used so that user can maintain a secure handhold while carrying a load (exceptions for using stepladders for low risk, short duration work).

21.36.7 *Fragile surfaces – regulation 9*

Employers must take steps to prevent people falling through any fragile surface. Steps to be taken include:

➤ avoiding, SFARP, passing across or near, or working on or near, a fragile surface

➤ where this is not reasonably practicable:
  ➤ providing suitable and sufficient platforms, covering, guard rails or other similar means of support or protection and using them
  ➤ providing suitable and sufficient guard rails to prevent persons falling through fragile surfaces
  ➤ where a risk of falling remains taking suitable and sufficient measures to minimize the distances and consequences of a fall.

➤ providing prominent warning signs at approach to the fragile surface, or if not reasonably practicable making people aware of the fragile surface

➤ where the risk of falling remains despite the precautions providing a suitable and sufficient fall arrest system.

21.36.8 *Falling objects and danger areas – regulations 10 and 11*

Suitable and sufficient steps must be taken to:

➤ prevent the fall of any material or object

➤ if not reasonably practicable, prevent the fall of any material or object, prevent persons being struck by falling objects or material (if liable to cause personal injury) by for example providing covered walkways or fan scaffolds

➤ prevent materials or objects being thrown or tipped from height where they are likely to cause injury

➤ store materials and objects in such a way as to prevent risk to any person arising from the collapse, overturning or unintended movement

➤ where danger of being struck exists, keeping unauthorized persons out of the area by suitable devices, the area being clearly indicated and

➤ store materials and objects so as to prevent them collapsing, overturning or moving in a way that could be a risk to people.

21.36.9 *Inspection of work equipment – regulation 12*

This regulation applies only to work equipment to which regulation 8 and schedules 2–6 apply. The requirements include the following:

➤ Where work equipment used for work at height depends for safety on how it is installed or assembled, the employer must ensure that it is not used after installation or assembly in any position unless it has been inspected in that place.

➤ Where work equipment is exposed to conditions causing deterioration which is liable to result in dangerous situations, it must also be inspected at suitable intervals and each time that exceptional circumstances which are liable to jeopardize the safety of the work equipment have occurred, for example a severe storm.

➤ In addition a working platform used for construction and from which a person could fall more than 2 m has to be inspected in that position (mobile equipment on the site) before use and within the previous 7 days; the particulars for this inspection are set out in schedule 7.

➤ No work equipment (lifting equipment is covered under LOLER) may either leave the undertaking or be obtained from another undertaking without evidence of in-date inspection.

➤ Results of inspections must be recorded and retained until the next due inspection is recorded; reports must be provided within 24 hours and kept at the site until construction work is complete and there-after at the office for 3 months.

21.36.10 *Inspection of places of work at height – regulation 13*

The surface and every parapet, permanent rail or other such fall protection measure of every place of work at height must be checked on each occasion before the place is used.

21.36.11 *Duties of persons at work – regulation 12*

Every person must, where working under the control of another person, report to that person any activity or defect relating to work at height which they know is likely to endanger themselves or others.

Work equipment must be used in accordance with training and instructions.

21.36.12 *References*

A Head for Heights Guidance for Working at Height in Construction. Video, 2003 HSE Books, ISBN 9780-7176-2217-7.

A Tool Box Talk on Leaning Ladders and Stepladders Safety. INDG403, 2005 HSE Books, ISBN 9780-7176-6106-7.

General Access Scaffolds and Ladders. CIS49, 1997 HSE Books.

Health and Safety in Roof Work. HSG 33, 1998 HSE Books.

Height Safe Essential Health and Safety Information for People who Work at Height. HSE's website; http://www.hse.gov.uk, ISBN 9780-7176-1425-5.

Preventing Falls from Ladders and Stepladders: An Employer's Guide, INDG402 2005 HSE Books, ISBN 9780-7176-6105-9.

Safe Use of Ladders and Stepladders. An Employers' Guide. INDG402, 2005 HSE Books, ISBN 9780-7176-6105-9.

Top Tips for Ladders and Stepladder Safety. INDG405, 2006 HSE Books, ISBN 9780-7176-6127-X.

Tower Scaffolds. CIS10, 1997 HSE Books.

Work at Height (Amendment Regulations 2007). SI 2007 No. 114, OPSI: http://www.opsi.gov.uk/si/si2007/20070114.htm.

Work at Height Regulations 2005. SI 2005/735, OPSI: http://www.opsi.gov.uk/si/si2005/20050735.

Work at Height Regulations 2005 (as amended). A Brief Guide. INDG401 (rev1), 2007 HSE Books, ISBN 9780-7176-6231-9.

Working on Roofs. INDG284, 1999 HSE Books.
Useful website, HSE's Falls from Height: http://www.hse.gov.uk/falls.

21.37 Other relevant Regulations in brief

There are a number of other Regulations which do not form part of the NEBOSH General Certificate syllabus. Nevertheless, they are important to a wider understanding of health and safety legislation. Very brief summaries are covered here.

21.37.1 *Corporate Manslaughter and Corporate Homicide Act 2007*

The Act creates a new statutory offence of corporate manslaughter which will replace the common law offence of manslaughter by gross negligence where corporations and similar entities are concerned. In Scotland the new offence will be called 'corporate homicide'.

An organization will have committed the new offence if it:

➤ owes a duty of care to another person in defined circumstances and
➤ there is a management failure by its senior managers and
➤ it amounts to a gross breach of that duty resulting in a person's death.

On conviction the offence will be punishable by an unlimited fine and the courts will be able to make remedial orders requiring organizations to take steps to remedy the management failure concerned.

It is important to note that the Act does not create a new individual liability. Individuals may still be charged with the existing offence of manslaughter by gross negligence.

Crown immunity will not apply to the offence, although a number of public bodies and functions will be exempt from it (in defined circumstances).

The Act comes into force on 6 April 2008. Available at: http://www.opsi.gov.uk/acts/acts2007/pdf/ukpga_20070019_en.pdf.

21.37.2 *Disability Discrimination Act 1995 and 2005*

The Disability Discrimination Act 1995
The Disability Discrimination Act (DDA) 1995 aims to end the discrimination that many disabled people face. This Act gives disabled people rights in the areas of:

➤ employment
➤ education

➤ access to goods, facilities and services

➤ buying or renting land or property.

The Act also allows the Government to set minimum standards so that disabled people can use public transport easily.

From 1 October 2004, Part 3 of the DDA 1995 has required businesses and other organizations to take reasonable steps to tackle physical features that act as a barrier to disabled people who want to access their services.

This may mean to remove, alter or provide a reasonable means of avoiding physical features of a building which make access impossible or unreasonably difficult for disabled people. Examples include:

➤ putting in a ramp to replace steps

➤ providing larger, well defined signs for people with a visual impairment

➤ improving access to toilet or washing facilities.

Businesses and organizations are called 'service providers' and include shops, restaurants, leisure centres and places of worship.

Available at: http://www.opsi.gov.uk/acts/acts1995/1995050.htm.

The Disability Discrimination Act 2005

In April 2005 a new DDA was passed, which amends or extends existing provisions in the DDA 1995, including:

➤ making it unlawful for operators of transport vehicles to discriminate against disabled people

➤ making it easier for disabled people to rent property and for tenants to make disability-related adaptations

➤ making sure that private clubs with 25 or more members cannot keep disabled people out, just because they have a disability

➤ extending protection to cover people who have HIV, cancer and multiple sclerosis from the moment they are diagnosed

➤ ensuring that discrimination law covers all the activities of the public sector

➤ requiring public bodies to promote equality of opportunity for disabled people

Other changes will come into force in December 2006 – the Disability Rights Commission (DRC) website has more details on these.

Available at: http://www.opsi.gov.uk/acts/acts2005/ukpga_20050013_en_1.

21.37.3 *Electrical (safety) Regulations 1994*

These Regulations came into force in January 1995 and relate to the supply of electrical equipment with a working voltage between 50 and 1000 volts and are made under the Consumer Protection Act 1987. They apply to suppliers, which include both landlords and letting agents.

The Regulations apply to all mains voltage household electrical goods and require them to be safe so that there is no risk of injury or death to humans or pets, or risk of damage to property. They do not apply to fixed electrical wiring and built-in appliances like central heating systems. The Regulations also require that instructions be provided where safety depends on the user being aware of certain issues, and equipment should be labelled with the CE marking.

There are other electrical consumer Regulations, such as The Plugs and Sockets Regulations 1994 and the Low Voltage Electrical Equipment Regulations 1989.

Available at: http://www.opsi.gov.uk/si/si1994/Uksi_19943260_en_1.htm#tcon.

21.37.4 *Gas Appliances (Safety) Regulations 1992*

The Gas Appliance Regulations cover the safety standards on new gas appliances which have to:

➤ satisfy safety and efficiency standards

➤ undergo type examination and supervision during manufacture

➤ carry the CE mark and specified information

➤ be accompanied by instructions and warnings in the language of destination.

Available at: http://www.opsi.gov.uk/si/si1995/Uksi_19951629_en_1.htm.

21.37.5 *Gas Safety (Installation and Use) Regulations 1998*

The Installation and Use Regulations place duties on gas consumers, installers, suppliers and landlords to ensure that:

➤ only competent people work on gas installations (registered with CORGI, the Council for Registered Gas Installers)

➤ no one is permitted to use suspect gas appliances

➤ landlords are responsible, in certain cases, to make sure that fittings and flues are maintained

➤ with the exception of room-sealed appliances there are restrictions on gas appliances in sleeping accommodation

➤ instantaneous gas water heaters must be room-sealed or fitted with appropriate safety devices.

Available at: http://www.opsi.gov.uk/si/si1994/Uksi_19941886_en_1.htm.

21.37.6 *Occupiers Liability Acts 1957 and 1984 – Civil Law*

The 1957 Act concerns the duty that the occupier of premises has towards visitors in relation to the condition of the premises and to things which have or have not been done to them. The Act imposes:

➤ a duty to take reasonable care to see that a visitor is reasonably safe in using the premises for the purpose for which they were invited or permitted by the occupier to be there
➤ that the common duty of care will differ depending on the visitor, so a greater duty is owed to children
➤ that an occupier can expect that a person in the exercise of his trade or profession will appreciate and guard against normal risks, for example, a window cleaner
➤ that no duty of care is owed to someone exercising a public right of way.

The 1984 Act extends the duty of care to persons other than visitors (i.e. trespassers). The occupier has to take reasonable care in all the circumstances to see that non-visitors do not get hurt on the premises because of its condition or the things done or not done to it. The occupier must cover perceived dangers and must have reasonable grounds to know that the trespassers may be in the vicinity.

Available at: http://www.opsi.gov.uk/acts/acts1984/PDF/ukpga_19840003_en.pdf.

21.37.7 *Control of Pesticides Regulations 1986 and Amendments*

These Regulations made under the Food and Environment Protection Act 1985 all came into force by 1 January 1988. The Regulations apply to any pesticide or any substance used generally for plant control and protection against pests of all types, including antifouling paint used on boats. They do not apply to substances covered by other Acts like the Medicines Acts 1968 and the Food Safety Act 1990, those Acts used in laboratories and a number of other specific applications.

No person may advertise, sell, supply, store, or use a pesticide unless it has received ministerial approval and the conditions of the approval have been complied with. The approval may be experimental, provisional or full, and the Minister has powers to impose conditions.

The Regulations also cover the need for users to be competent and to have received adequate information and training. Certificates of competence (or working under the supervision of a person with a certificate) are required where pesticides approved for agricultural use are used commercially.

Update on future consolidation

DEFRA issued a consultation document in August 2007 which covers the update of these Regulations as follows:

➤ At present there are four Regulations (plus their amendments) to control and monitor marketing and use of pesticides in England and Wales.
  ➤ The Control of Pesticides Regulations 1986 and its amendment in 1997
  ➤ The Plant Protection Products Regulations 2005 and its amendment in 2006
  ➤ The Plant Protection Products (Basic Conditions) Regulations 1997 and
  ➤ The Plant Protection Products (Fees) Regulations 2007.
➤ DEFRA's proposal is to consolidate these four Regulations (plus amendments) into one Regulation by April 2008.

Available at: http://www.pesticides.gov.uk/approvals. asp?id=2182.

21.37.8 *The Manufacture and Storage of Explosives Regulations 2005*

The Regulations (SI Number 2005/1082) came into force on 26 April 2005.

These Regulations cover the manufacture, storage and handling of all explosives, including blasting explosives, propellants, detonators and detonating cord, fireworks and other pyrotechnic articles, ammunition, and other explosive articles such as airbags and seat belt pretensioners. The Regulations cover the manufacture of explosives and intermediate products for on-site mixing and storage, and also handling operations that are not in themselves considered to 'manufacture'. These include fusing fireworks, assembling fireworks displays from components, and filling shotgun cartridges and other cartridges for small arms. The main requirements of the Regulations are as follows:

➤ Anyone manufacturing or storing explosives must take appropriate measures to prevent fire or explosion; to limit the extent of any fire or explosion should one occur; and to protect persons in the event of a fire or explosion. These are the key requirements of the Regulations and are backed up by extensive guidance in the ACOP.
➤ In most cases a separation distance must be maintained between the explosives building and neighbouring inhabited buildings. This is intended to ensure that risks to those living or working in the area are kept to an acceptable level. If there is

development in this separation zone then the quantity that may be kept must be reduced.

➤ with certain exceptions a licence is required for the manufacture or storage of explosives. HSE licenses manufacturing activities because of the greater risks involved. HSE also licenses larger explosives storage facilities. In most cases, stores holding less than 2 tonnes of explosives are either licensed or registered by the local authority or the police.

➤ the HSE may not grant a licence for a manufacturing facility or, in most cases, store until the local authority has given its assent (normally following a public hearing). This is an important safeguard in the present system that is to be retained.

Available at: http://www.opsi.gov.uk/si/si2005/uksi_20051082_en.pdf.

And the amendment Regulations are available at: http://www.opsi.gov.uk/si/si2007/20072598.htm.

Also available: *The Approved Code of Practice and guidance to the Regulations* L139 2005 HSE Books, ISBN 9780-7176-2816-7.

21.37.9 *Pressure Systems Safety Regulations 2000 (PSSR)*

These Regulations came into effect in February 2000 and replace the Pressure Systems and Transportable Gas Containers Regulations 1989. Transportable gas containers are covered by the Use of Transportable Pressure Receptacles Regulations 1996 (SI 1996 No. 2092).

The aim of PSSR is to prevent serious injury from the hazard associated with stored energy as a result of a pressure system or one of its parts failing. The Regulations cover:

➤ steam at any pressure
➤ gases which exert a pressure in excess of 0.5 bar above atmospheric pressure
➤ fluids which may be mixtures of liquids, gases and vapours where the gas or vapour phase may exert a pressure in excess of 0.5 bar above atmospheric pressure.

With the exception of scalding from steam, the Regulations do not consider the effects of the hazardous contents being released following failure. The stored contents are of concern where they can accelerate wear and cause more rapid deterioration and an increased risk of failure.

Available at: http://www.opsi.gov.uk/si/si2000/uksi_20000128_en.pdf.

21.37.10 *Personal Protective Equipment Regulations 2002*

These Regulations relate to approximation of the laws of EU Member States. They place duties on persons who place PPE on the market to comply with certain standards. These requirements are that: PPE must satisfy the basic health and safety requirements which are applicable to that class or type of PPE; the appropriate conformity assessment procedures must be carried out; CE marking must be correctly affixed; and the PPE must not compromise the safety of individuals, domestic animals or property when properly maintained and used.

Available at: http://www.opsi.gov.uk/si/si2002/uksi_20021144_en.pdf

21.37.11 *Working Time Regulations 1998 as amended by 2003 and 2007 Regulations*

These Regulations came into force on October 1998 and, for specified workers, restrict the working week to 48 hours per 7-day period. Individuals can voluntarily agree to disapply the weekly working hours limit. Employers must keep a copy of all such individual agreements. Workers whose working time is not measured or predetermined, or who can themselves determine the duration of their working day are excepted from weekly working time, night work, rest periods and breaks.

The Regulations were amended, with effect from 1 August 2003, to extend working time measures in full to all non-mobile workers in road, sea, inland waterways and lake transport, to all workers in the railway and offshore sectors, and to all workers in aviation who are not covered by the sectoral Aviation Directive. The Regulations applied to junior doctors from 1 August 2004.

Mobile workers in road transport have more limited protections. Those subject to European Drivers' hours rules 3820/85 are entitled to 4 weeks' paid annual leave and health assessments if a night worker from 1 August 2003. Mobile workers not covered by European Drivers' hours rules will be entitled to an average 48 hours per week, 4 weeks' paid holiday, health assessments if a night worker and adequate rest.

The Regulations were previously amended, with effect from 6 April 2003, to provide enhanced rights for adolescent workers.

The basic rights and protections that the Regulations provide are:

➤ a limit of an average of 48 hours a week which a worker can be required to work (though workers can choose to work more if they want to)
➤ a limit of an average of 8 hours' work in 24 which nightworkers can be required to work

➤ a right for night workers to receive free health assessments

➤ a right to 11 hours' rest a day

➤ a right to a day off each week

➤ a right to an in-work rest break if the working day is longer than 6 hours

➤ a right to 4 weeks' paid leave per year; this is extended to 5.6 weeks with the 2007 amendment from 1 October 2007; the extended leave time is phased in over about 2 years.

1998 Regulations available at: http://www.opsi.gov.uk/si/si1998/19981833.htm.

2003 amendment available at: http://www.opsi.gov.uk/si/si2003/20031684.htm.

2007 amendment available at: http://www.opsi.gov.uk/si/si2007/20072079.htm.

21.38 Health and Safety Executive from 1st April 2008

News release from the Department for Work and Pensions - 01 April 2008 – Health and Safety Commission and Health and Safety Executive merge to form a single regulatory body: The Health and Safety Executive.

The Health and Safety Commission (HSC) and the Health and Safety Executive (HSE) today merged to form a single national regulatory body responsible for promoting the cause of better health and safety at work, announced Health and Safety Minister Lord McKenzie. The merged body will be called the Health and Safety Executive and will provide greater clarity and transparency whilst maintaining its public accountability.

The decision to merge the HSC and the HSE was reached after extensive consultation with stakeholders, and through the process determined by the Legislative and Regulaory Reform Act 2006.

Welcoming the merger, Health and Safety Minister Lord McKenzie said:

"The Health and Safety Commission and the Health and Safety Executive have done an excellent job over the last 30 years in bringing about significant improvements to health and safety at work. However, to face the challenges and demands of the changing world of work, now is the right time to merge the organizations into one which can provide a platform for further improvements to health and safely at work across Great Britain."

Judith Hackitt, Chair of the new Health and Safety Executive added:

"The new Health and Safety Executive will strengthen the importance of workplace health and safety in Great Britain. With a single regulatory body we will be able to strengthen the links between strategy and delivery in order to provide the accountability expected of a public body in today's workplace climate. The merger will not fundamentally change the day-to-day operations but will set the tone for closer working throughout the organization.

The HSE will build on the independence, good relationship with stakeholders and in particular our relationship with local authorities to develop a revised strategy for health and safety in Greath Britain."

The HSE will retain its independence, reflecting the interests of employers, employees and local authorities and is committed to maintaining its service delivery. The Board of the new Executive will assume responsibility for running all aspects of the orgnaization, including setting the overall strategic direction financal and performance management and prioritization of resources. The HSE will be revisting its strategy to develop a long term view for the next five years, which will be published towards the end of 2008.

The merger will mean:

➤ There will be a single national regulatory body responsible for promoting the cause of better health and safety at work.

➤ The current Chair of the Commission becomes Chair of the Board of the new Executive.

➤ Existing Commissioners are appointed as non-executive Directors of the new Executive for the remainder of their of office with the relevant responsibilities of the new roles secretary.

➤ The potential size of the Board of the new Executive will be no more than eleven members plus the Chair and members will continue to be appointed by the Secretary of State.

➤ All the fundamental contents of the Health & Safety at Work Act remain.

➤ None of the statuatory functions of the previous Commission and Executive will be removed.

➤ There is no change in health and safety requirements, how they are enforced or how stakeholders relate to the health and safety regulator – no health and safety protections will be removed.

International aspects of health and safety in construction

22

22.1 Introduction

In 2004, an International General Certificate was introduced by NEBOSH following requests from countries around the world and international companies with cross-border operations. The main difference between the National General Certificate and the International Certificate syllabuses is the exclusion of all UK legal issues and requirements and the inclusion of three occupational health and safety management systems which are used around the world. It is likely that an International Construction Certificate to complement the International General Certificate will be developed in the future.

NEBOSH allows where appropriate that local occupational health and safety legal requirements and enforcement procedures will be covered during any International Certificate course. In some parts of the world these requirements and procedures may well be based on UK law. There has also been a significant change in reference material in that International Labour Organization (ILO) Conventions, Recommendations and Codes of Practice have been included. This change is particularly significant for Chapter 13 Work equipment hazards and control.

There have been several other minor differences in the International General Certificate compared to the UK National General Certificate, the majority of which affect Chapter 1 although there are some small changes for other chapters. Minor differences are covered in Section 22.8 of this chapter. Most of these differences will be reflected in an International Construction Certificate because the course will be unitized and have an identical management unit – Unit IGC1 Management of international health and safety. The second unit will concern the control of international construction site hazards and be based on the ILO Code of Practice – Safety and Health in Construction.

Figure 22.1 Good Scaffold – Leaning Tower.

The largest change concerns the comparison of the three major occupational health and safety management systems and this comparison covers most of this chapter.

The ILO estimates that two million women and men die each year as a result of occupational accidents and work-related diseases. Table 22.1 shows the global numbers in more detail.

Table 22.1 Number of global work-related adverse events

Event	Average daily	Annually
Work-related death	5,000	2,000,000
Work-related deaths to children	60	22,000
Work-related accidents	740,000	270,000,000
Work-related disease	438,000	160,000,000
Hazardous substance deaths	1,205	440,000
Asbestos related deaths	274	100,000

Source: ILO.

In the USA in 2002, approximately two million workers were victims of workplace violence. In the UK 1.7% of working adults (357,000 workers) were the victim of one or more incidents of workplace violence.

Figure 22.2 Local authority building site, Turkey.

Ten per cent of all skin cancers are estimated to be attributable to workplace exposure to hazardous substances. Thirty-seven per cent of miners in Latin America have silicosis, rising to fifty per cent among miners over fifty. In India 54.6% of slate pencil workers and 36.2% of stone cutters have silicosis.

In the course of the 20th century, industrialized countries saw a clear decrease in serious injuries, not least because of real advances in making the workplace healthier and safer. The challenge is to extend the benefits of this experience to the whole working world.

Figure 22.3 Poor standard of scaffold not to EU Standards in Poland.

However, 1984 saw the worst chemical disaster ever when 2,500 people were killed and over 200,000 injured in the space of a few hours at Bhopal. Not only were the workers affected but their families, their neighbours and whole communities. Twenty years later many people are still affected by the disaster and are dying as a result. The rusting remains of a once-magnificent plant remain as a reminder of the disaster.

Experience has shown that a strong safety culture is beneficial for workers, employers and governments alike. Various prevention techniques have proven themselves effective, both in avoiding workplace accidents and illnesses and in improving business performance. Today's high standards in some countries are a direct result of long-term policies encouraging tripartite social dialogue, collective bargaining between trade unions and employers, and effective health and safety legislation backed by potent labour inspection. The ILO believe that Safety Management Systems like ILO-OSH 2001 provide a powerful tool for developing a sustainable safety and health culture at the enterprise level and mechanisms for the continual improvement of the working environment.

The size of the health and safety 'problem' in terms of numbers of work-related fatalities and injuries and incidence of ill-health will vary from country to country. However, these figures should be available from the statistics branch of the national regulator as they are available in the UK from the annual report on health and safety statistics from the Health and Safety Executive.

22.2 The role and function of the ILO

The ILO is a specialized agency of the United Nations that seeks to promote social justice through establishing and safeguarding internationally recognized human

Figure 22.4   Mobile elevating work platform, France.

and labour rights. It was founded in 1919 by the Treaty of Versailles at the end of the First World War.

The motivation behind the creation of such an organization was primarily humanitarian. Working conditions at the time were becoming unacceptable to a civilized society. Long hours, unsafe, unhygienic and dangerous conditions were common in low paid manufacturing careers. Indeed, in the wake of the Russian Revolution, there was concern that such working conditions could lead to social unrest and even other revolutions. The ILO was created as a tripartite organization with governments, employers and workers represented on its governing body.

The ILO formulates international labour standards and attempts to establish minimum rights including freedom of association, the right to organize, collective bargaining, abolition of forced labour, equality of opportunity and treatment and other standards that regulate conditions across all work-related activities.

Representatives of all ILO Member States meet annually in Geneva for the International Labour Conference, which acts as a forum where social and labour questions of importance to the entire world are discussed. At this conference, labour standards are adopted and decisions made on policy and future programmes of work.

The ILO has 178 Member States but if a country is not a member, the ILO still has influence as a source of guidance when social problems occur.

The main principles on which the ILO is based are:

1. Labour is not a commodity.
2. Freedom of expression and of association are essential to sustained progress.
3. Poverty anywhere constitutes a danger to prosperity everywhere.
4. The war against want requires to be carried on with unrelenting vigour within each nation, and by continuous and concerted international effort in which

the representatives of workers and employers, enjoying equal status with those of governments, join with them in free discussion and democratic decision with a view to the promotion of the common welfare.

The most recent campaign launched by the ILO has been to seek to eliminate child labour throughout the world. In particular, the ILO is concerned about children who work in hazardous working conditions, bonded child labourers and extremely young working children. It is trying to create a worldwide movement to combat the problem by:

➤ implementing measures which will prevent child labour
➤ withdrawing children from dangerous working conditions
➤ providing alternatives and
➤ improving working conditions as a transitional measure towards the elimination of child labour.

22.2.1   *ILO Conventions and Recommendations*

The international labour standards were developed for four reasons.

The main motivation was to improve working conditions with respect to health and safety and career advancement. The second motivation was to reduce the potential for social unrest as industrialization progressed. Thirdly, the Member States want common standards so that no single country has a competitive advantage over another due to poor working conditions. Finally, the union of these countries creates the possibility of a lasting peace based on social justice.

International labour standards are adopted by the International Labour Conference. They take the form of Conventions and Recommendations. At the present time, there are 187 Conventions and 198 Recommendations, some of which date back to 1919.

International labour standards contain flexibility measures to take into account the different conditions and levels of development among Member States. However, a government that ratifies a Convention must comply with all its articles. Standards reflect the different cultural and historical backgrounds of the Member States as well as their diverse legal systems and levels of economic development.

ILO occupational safety and health standards can be divided into four groups and an example is given in each case:

1. Guiding policies for action
   The Occupational Safety and Health Convention, 1985 (No. 155) and its accompanying Recommendation (No. 164) emphasize need for preventative measures and a coherent national policy on occupational safety

and health. They also stress employers' responsibilities and the rights and duties of workers.

2. Protection in given branches of economic activity
    The Safety and Health in Construction Convention, 1988 (No. 167) and its accompanying Recommendation (No. 175) stipulate the basic principles and measures to promote safety and health of workers in construction.
3. Protection against specific risks
    The Asbestos Convention, 1986 (No. 162) and its accompanying Recommendation (No. 172) gives managerial, technical and medical measures to protect workers against asbestos dust.
4. Measures of protection
    Migrant Workers (Supplementary Provisions) Convention, 1975 (No. 143) aims to protect the safety and health of migrant workers.

ILO **Conventions** are international treaties signed by ILO Member States and each country has an obligation to comply with the standards that the Convention establishes.

In contrast, ILO **Recommendations** are non-binding instruments that often deal with the same topics as Conventions. Recommendations are adopted when the subject, or an aspect of it, is not considered suitable or appropriate at that time for a Convention. Recommendations guide the national policy of Member States so that a common international practice may develop and be followed by the adoption of a Convention.

ILO standards are the same for every Member State and the ILO has consistently opposed the concept of different standards for different regions of the world or groups of countries.

The standards are modified and modernized as needed. The Governing Body of the ILO periodically reviews individual standards to ensure their continuing relevance.

Supervision of international labour standards is conducted by requiring the countries that have ratified Conventions to present periodically a report with details of the measures that they have taken, in law and practice, to apply each ratified Convention. In parallel, employers' and workers' organizations can initiate contentious proceedings against a Member State for its alleged non-compliance with a Convention it has ratified. In addition, any member country can lodge a complaint against another Member State which, in its opinion, has not ensured in a satisfactory manner the implementation of a Convention which both of them have ratified. Moreover, a special procedure exists in the field of freedom of association to deal with complaints submitted by governments or by employers' or workers' organizations against a Member State whether or not the country concerned has ratified the relevant Conventions. Finally, the ILO has systems in place to examine the enforcement of international labour standards in specific situations.

The ILO also publishes Codes of Practice, guidance and manuals on health and safety matters. These are often used as reference material by either those responsible for drafting detailed Regulations or those who have responsibility for health and safety within an organization. They are more detailed than either Conventions or Recommendations and suggest practical solutions for the application of ILO standards. Codes of Practice indicate 'what should be done'. They are developed by tripartite meetings of experts and the final publication is approved by the ILO Governing Body.

For construction, the Safety and Health in Construction Convention, 1988 (No. 167) obliges signatory ILO member States to comply with the construction standards laid out in the Convention – the Convention is a relatively brief statement of those standards. The accompanying Recommendation (No. 175) gives additional information on the Convention statements. The Code of Practice gives more detailed information than the Recommendation. This can best be illustrated by contrasting the coverage of scaffolds and ladders by the three documents referred to in Appendix 22.1.

The ILO Codes of Practice and guidelines on health and safety matters that are relevant to the International Construction Certificate are:

➤ Safety and Health in Construction (ILO Code of Practice)
➤ Ambient factors in the Workplace (ILO Code of Practice)
➤ Safety in the Use of Chemicals at Work (ILO Code of Practice)
➤ Recording and Notification of Occupational Accidents and Diseases
➤ Ergonomic Checkpoints
➤ Work Organisation and Ergonomics
➤ Occupational safety and health management systems (ILO Guidelines).

Important ILO Conventions (C) and Recommendations (R) in the field of occupational health and safety include:

➤ C 115 Radiation Protection and (R 114), 1960
➤ C 120 Hygiene (Commerce and Offices) and (R 120), 1964
➤ C 139 Occupational Cancer and (R 147), 1974
➤ C 148 Working Environment (Air, Pollution, Noise and Vibration) and (R 156), 1977
➤ C 155 Occupational Safety and Health and (R164), 1981
➤ C 161 Occupational Health Services and (R 171), 1985
➤ C 162 Asbestos and (R 172), 1986
➤ C 167 Safety and Health in Construction and (R 175), 1988

➤ C 170 Chemicals and (R 177), 1990
➤ C 174 Prevention of Major Industrial Accidents and (R 181), 1993
➤ C 176 Safety and Health in Mines and (R 176), 1995
➤ C 184 Safety and Health in Agriculture and (R 192), 2001
➤ C 187 Promotional Framework for Occupational Safety and Health and (R 197), 2006
➤ R 97 Protection of Workers' Health Recommendation, 1953
➤ R 102 Welfare Facilities Recommendation, 1956
➤ R 31 List of Occupational Diseases Recommendation, 2002.

Copies of Conventions, Recommendations and Codes of Practice can be obtained from the ILO website www.ilo.org/safework.

22.3 Major occupational health and safety management systems

There are three major occupational health and safety management systems that are in use globally:

1. HSG 65, which has been developed by the UK Health and Safety Executive (HSE), is described in Section 1.16. The syllabuses for both Certificate courses and the management chapter headings in this text book have followed the elements of HSG 65.
2. OHSAS 18001 has been developed by the British Standards Institute from its recommended occupational health and safety management system BS 8800. It has been developed in conjunction with the ISO 9000 series for quality management and the ISO 14000 series for environmental management.
3. ILO-OSH 2001 was developed by the ILO after an extensive study of many occupational health and safety management systems used across the world. It was established as an international system following the publication of 'Guidelines on occupational safety and health management systems' in 2001. It is very similar to OHSAS 18001.

22.3.1 *Basic elements of all health and safety management systems*

All recognized occupational health and safety management systems have some basic and common elements. These are:

➤ a planning phase
➤ a performance phase
➤ a performance assessment phase
➤ a performance improvement phase.

The planning phase
The planning phase always includes a **policy statement** which outlines the health and safety aims, objectives and commitment of the **organization** and lines of responsibility. **Hazard identification and risk assessment** takes place during this phase and the significant hazards may well be included in the policy statement. It is important to note that in some reference texts, in particular those in languages other than English, the whole process of hazard identification, risk determination and the selection of risk reduction or control measures is termed 'risk assessment'. However, all three occupational health and safety management systems described in this chapter refer to the individual elements of the process separately and use the term 'risk assessment' for the determination of risk only.

At the **planning stage**, emergency procedures should be developed and relevant health and safety legal requirements and other standards identified together with appropriate benchmarks from similar industries. An **organizational** structure must be defined so that health and safety responsibilities are allocated at all levels of the organization and issues such as competent persons and health and safety training are addressed. Realistic targets should be agreed within the organization and be published as part of the policy.

The performance phase
The performance phase will only be successful if there is good communication at and between all levels of the organization. This implies employee participation as both worker representatives and on safety committees. Effective communication with the workforce, for example with clear safe systems of work and other health and safety procedures, will not only aid the implementation and operation of the plan but also produce continual improvement of performance – a key requirement of all occupational health and safety, quality and environmental management systems. There should also be effective communication with other stakeholders, such as regulators, contractors, customers and trade unions. The performance phase must be monitored on a regular basis since this will indicate whether there is an effective occupational health and safety management system and a good health and safety culture within the organization.

The performance assessment phase
The performance assessment phase may be either **active or reactive** or, ideally, a mixture of both. Active assessment includes work-based **inspections and audits**, regular health and safety committee meetings, feedback from training sessions and a constant review of risk assessments. Reactive assessment relies on records of accident, work-related injuries and ill-health

as well as near miss and any enforcement notices. Any recommended remedial or preventative actions, following an investigation, must be implemented immediately and monitored regularly.

The performance improvement phase

The performance improvement phase involves a **review** of the effectiveness of the health and safety management system and the identification of any weaknesses. The review, which should be undertaken by the management of the organization, will assess whether targets have been met and the reasons for any underperformance. Issues such as the level of resources made available, the vigilance of supervisors and the level of co-operation of the workforce should be considered at the review stage. When recommendations are made, the review process must define a time scale by which any improvements are implemented and this part of the process must also be monitored. **Continual improvement** implies a commitment to improve performance on a proactive continuing basis without waiting for a formal review to take place. Most management systems include an **audit** requirement, which may be either internal or external. The audit process examines the effectiveness of the whole management process and may be act as a control on the review process. Many inquiry reports into health and safety management issues have asserted that health and safety performance should be subject to audit in the same way that financial performance must be audited.

In the publication 'Successful health and safety management – HSG 65', the HSE recommend a similar four step approach to occupational health and safety management known as:

1. **PLAN** – establish standards for health and safety management based on risk assessment and legal requirements.
2. **DO** – implement plans to achieve objectives and standards.
3. **CHECK** – measure progress with plans and compliance with standards.
4. **ACT** – review against objectives and standards and take appropriate action.

The **Plan–Do–Check–Act** for occupational health and safety management forms the basis of the three occupational health and safety management systems HSG 65, OHSAS 18001 and ILO-OSH 2001.

22.3.2 HSG 65

In the UK HSE first published 'Successful health and safety management – HSG 65' in 1991 and has amended it since then so that it is now more concerned with continual improvement and less with the attainment of minimum health and safety standards. Unlike the other two systems under consideration, HSG 65 was developed by a regulator (HSE) and, therefore, is much more widely recognized in the UK. Although HSE inspectors may use the framework of HSG 65 when auditing the health and

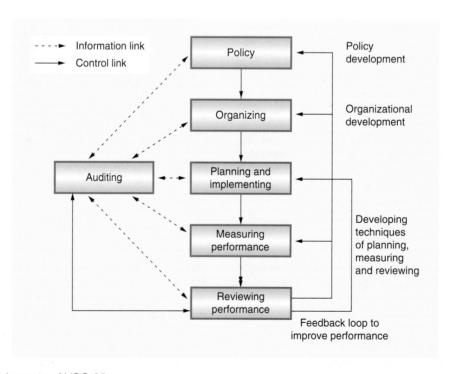

Figure 22.5 Key elements of HSG 65.

safety management arrangements of an organization, although it would be wrong, however, to assume that it is the only system recognized by the HSE.

HSG 65 is shown in Figure 22.5. An outline of the five key elements of HSG 65 was given in Chapter 1 and is given in more detail here.

1. **A clear health and safety policy** – Evidence shows that a well-considered policy contributes to business efficiency and continual improvement throughout the organization. It helps to minimize financial losses arising from avoidable accidents and demonstrates to the workforce that accidents are not necessarily the fault of any individual member of the workforce. Such a management attitude could lead to an increase in workforce co-operation, job satisfaction and productivity. This demonstration of senior management involvement offers evidence to all stakeholders that responsibilities to people and the environment are taken seriously by the organization. A good health and safety policy helps to ensure that there is a systematic approach to risk assessment and sufficient resources, in terms of people and money, have been allocated to protect the health and safety and welfare of the workforce. It can also support quality improvement programmes which are aimed at continual improvement.

2. **A well-defined health and safety organization** – The shared understanding of the organization's values and beliefs at all levels of the organization, is an essential component of a positive health and safety culture. For a positive health and safety culture to be achieved, an organization must have clearly defined health and safety responsibilities so that there is always management control of health and safety throughout the organization. The formal organizational structure should be such that the promotion of health and safety becomes a collaborative activity between the workforce, safety representatives and the managers. An effective organization will be noted for good staff involvement and participation; high quality communications; the promotion of competency; and the empowerment of all employees to make informed contributions to the work of the organization.

3. **A clear health and safety plan** – This involves the setting and implementation of performance standards and procedures using an effective occupational health and safety management system. The plan is based on risk assessment methods to decide on priorities and set objectives for controlling or eliminating hazards and reducing risks. Measuring success requires the establishing of performance standards against which achievements can be identified.

4. **The measurement of health and safety performance** – This includes both active and reactive monitoring to see how effectively the occupational health and safety management system is working. Active monitoring involves looking at the premises, plant and substances plus the people, procedures and systems. Reactive monitoring discovers through investigation of accidents and incidents why controls have failed. It is also important to measure the organization against its own long-term goals and objectives.

5. **The audit and review of health and safety performance** – The results of monitoring and independent audits should be systematically reviewed to see if the management system is achieving the right results. This is not only required by the HSW Act but is part of any company's commitment to continual improvement. Comparisons should be made with internal performance indicators and the performance of external organizations with exemplary practices and high standards.

22.3.3 *OHSAS 18001*

Figure 22.6 Sugar factory, Poland moving to 18001 accreditation.

OHSAS 18001 and ILO-OSH 2001 have many similarities in the details of the two systems. One of the biggest differences is that OHSAS 18001 is certifiable whereas ILO-OSH 2001 is not. OHSAS 18001, which is shown in Figure 22.7, was developed from the British Standard 8800 and was designed to be integrated with the two British Standards for quality and environmental management. Both occupational health and safety management systems have been shaped by internationally agreed occupational health and safety principles defined in international labour standards and the European Union

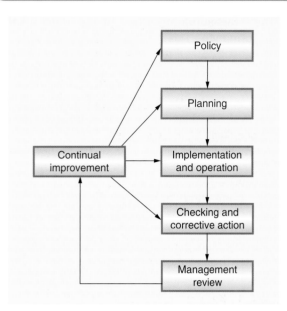

Figure 22.7 Key elements of OHSAS 18001.

Framework Directive 89/391/EEC. OHSAS 18001: 2007 requires an external audit as part of the certification process. OHSAS 18002: 2000 has also been produced by BSI as a guideline for implementation of OHSAS 18001. 18002 will be withdrawn when its contents are published as a British Standard.

There are five elements in the OHSAS 18001 occupational health and safety management system:

1. **Policy** – The general requirements are similar to for HSG 65 with the additional inclusions of a minimum compliance with national occupational safety and health laws and regulations and a willingness to consult fully with the workforce and their representatives. The policy statement should also refer to the need for continual improvement.

2. **Planning and organizing** – The organization must establish effective arrangements for the identification, elimination or control of work-related hazards and risks and to promote health, safety and welfare at work. It should also include in its plan a procedure for the maintenance of a record of all relevant and current legal and other health and safety standards. Finally, the plan must address the health and safety objectives identified in the policy statement.

3. **Implementing and operation** – As with HSG 65, there is a stress on the responsibility of senior management for the effective implementation and operation of the plan. This requires effective communication with and supervision of the workforce and the full participation of the workforce and their representatives. All procedures must be well documented and accessible by the workforce. Emergency procedures

(including prevention) must be included within the system together with arrangements for the procurement of goods and services and the selection and supervision of contractors – see Chapter 3 for more information on this latter point.

4. **Checking and action for improvement** – This is similar to the performance measurement phase of HSG 65. However, if possible, at the same time that any necessary corrective measures are noted, they should be introduced into the workplace processes and/or procedures. This is a good example of continual improvement.

5. **Management review** – There must be an initial review after the introduction of the management system into the workplace. This initial review needs to identify all relevant national laws and regulations and any significant hazards which might affect the workforce. It should check that adequate controls are in place and analyse any employee health surveillance records. It must be properly documented since it will form the base line from which continual improvement will be measured. At each subsequent management review, checks must be made on all work-related injuries, ill-health, diseases and incidents to ensure that they have been properly logged and investigated by a competent person. The conclusions of such investigations should be monitored to ensure that any recommended remedial actions have been completed. The following actions should also be included during the review:
   - an evaluation of the effectiveness of the management system in achieving the results required by the senior management of the organization
   - the identification of any changes required in the procedures or arrangements
   - an evaluation of the progress made with the policy objectives
   - a review of progress made with recommendations from earlier reviews
   - a report to the health and safety committee for communication to the workforce.

The OHSAS 18001 management system recognizes that the system is bound to require modification and change as the activities of the organization change and national regulatory requirements change.

22.3.4 *ILO-OSH 2001*

The ILO is a United Nations agency based in Geneva and has a considerable influence on the development of employment law in many countries across the world. As companies become more international in terms of both its markets and production bases, this influence of the ILO has increased and its working standards have become accepted in many parts of the world. Many of the ILO standards cover health and safety issues and are used in the International General Certificate course. As mentioned earlier, the ILO-OSH 2001 guidelines offer a recommended occupational health and safety management system based on an ILO survey of several contemporary schemes including HSG 65 and OHSAS 18001. There are, therefore, many common elements between the three schemes. The guidelines are not legally binding and are not intended to replace national laws, regulations or accepted standards.

At the national level, the guidelines should be used to establish a national framework for occupational health and safety management systems, preferably supported by national laws and regulations. They should also provide guidance for the development of voluntary arrangements to strengthen compliance with regulations and standards leading to continual improvement in occupational health and safety performance.

The ILO recognizes that a management system can usually only be successful in a country if there is some form of national policy on health and safety and occupational health and safety management systems. The ILO recommends, therefore, that the following general principles and procedures be established:

1. the implementation and integration of occupational health and safety management systems as part of the overall management of an organization
2. the introduction and improvement of voluntary arrangements for the systematic identification, planning, implementation and improvement of occupational health and safety activities at both national and organizational levels
3. the promotion of worker participation in occupational health and safety management at organizational level
4. the implementation of continual improvement without unnecessary bureaucracy and cost
5. the encouragement of national labour inspectors to support the arrangements at organizational level for a health and safety culture within the framework of an occupational health and safety management system
6. the evaluation of the effectiveness of the national policies at regular intervals
7. the evaluation of the effectiveness of occupational health and safety management systems within those organizations which are operating within the country
8. the inclusion of all those affected by the organization, such as contractors, members of the public and temporary workers, at the same level of health and safety provision as for employees.

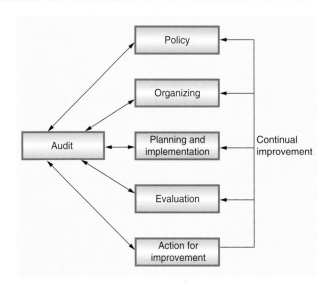

Figure 22.8 Key elements of ILO-OSH 2001.

Figure 22.8 shows the key elements of the ILO-OSH 2001 occupational health and safety management system. Each element will be discussed in turn but since it has been developed from HSG 65 and OHSAS 18001 amongst other systems, there are many similarities. The elements are:

1. **Policy** – There is a more specific emphasis on worker participation, which the ILO sees as an essential element of the management system and should be referenced in the policy statement. It is expected that workers and their health and safety representatives should have sufficient time and resources allocated to them so that they can participate actively in each element of the management system. The formation of a health and safety committee is also a recommended part of the system. The occupational health and safety management system should be compatible with or integrated with other management systems operating in the organization.
2. **Organizing** – There is much in common with both HSG 65 and OHSAS 18001 with responsibility, accountability, competence, training and communication being key parts of this element. There is a specific responsibility to provide effective supervision to ensure the protection of the health and safety of workers and to establish prevention and health promotion programmes. Workers should have access to any records, such as accident and monitoring records, which are relevant to their occupational health and safety while respecting the need for confidentiality, health and safety. Training must be available to all members of the organization and be provided during normal working hours at no cost to employees.
3. **Planning and implementation** – Unlike other systems, the ILO system combines planning and

implementation. Following an initial review of any existing health and safety management system, a plan should be developed to remedy any deficiencies found. The plan should support compliance with national laws and regulations and include continual improvement of health and safety performance. It should contain measurable objectives which are realistic and achievable and, as with the other occupational health and safety management systems, hazard identification and risk assessment. There should also be an adequate provision of resources and technical support. The plan must be capable of accommodating the impact on health and safety of any internal changes in the organization, such as new processes, new technologies and amalgamations with other organizations, or external changes due, for example, to changes in national laws or regulations. As with OHSAS 18001, emergency and procurement arrangements and detailed arrangements for the selection and supervision of contractors must be included in the health and safety plan.

4. **Evaluation** – This is very similar to the performance measurement phase of HSG 65 with a greater emphasis on the health and welfare of the worker. The recommendations concerning the investigation of work-related injuries, ill-health, diseases and incidents are identical to those for the management review element of OHSAS 18001.

5. **Action for improvement** – Arrangements should be introduced and maintained for any preventative and corrective action to be undertaken identified by performance monitoring, audits and management reviews of the health and safety management system. Arrangements should also be in place for the continual improvement of the management system. More details on the factors to be considered for continual improvement are given later in this chapter.

6. **Audit** – The ILO recommend that an audit should be performed by competent and trained personnel at agreed and regular intervals. It should cover all elements of the management system including worker participation, communication, procurement, contracting and continual improvement. The audit conclusions must state whether the health and safety management system is effective in meeting the organizational health and safety policy and objectives and promotes full worker participation. The audit should also check that there is compliance with national laws and regulations and that there has been a satisfactory response to earlier audit findings.

It is clear that there is much that is common between the three management systems discussed and any differences are differences in emphasis. HSG 65 was the first of the three systems and the emphasis was on legal

compliance. OHSAS 18001 introduced the concept of continual improvement, integration with other management systems and certification. ILO-OSH 2001 combines the features of HSG 65 and OHSAS 18001 and stresses the importance of worker participation for the system to be effective.

22.4 Other key characteristics of a health and safety management system

The four basic elements common to all occupational health and safety management systems, as described earlier in this chapter, contain the different activities of the system together with the detailed arrangements and activities required to deliver those activities. However, there are four key characteristics of a **successful** occupational health and safety management system:

1. a positive health and safety culture
2. the involvement of all stakeholders
3. an effective audit
4. continual improvement.

22.4.1 *A positive health and safety culture*

In Chapter 4, the essential elements for a successful health and safety culture were detailed and discussed. In summary, they were:

➤ leadership and commitment to health and safety throughout the organization
➤ an acceptance that high standards of health and safety are achievable
➤ the identification of all significant hazards facing the workforce and others
➤ a detailed assessment of health and safety risks in the organization and the development of appropriate control and monitoring systems
➤ a health and safety policy statement outlining short- and long-term health and safety objectives; such a policy should also include national codes of practice and health and safety standards.
➤ relevant communication and consultation procedures and training programmes for employees at all levels of the organization
➤ systems for monitoring equipment, processes and procedures and the prompt rectification of any defects found
➤ the prompt investigation of all incidents and accidents and reports made detailing any necessary remedial actions.

Some of these essential elements form part of the health and safety management system but unless all are present within the organization, it is unlikely that occupational health and safety will be managed successfully no matter which system is introduced. The chosen management system must be effective in reducing risks in the workplace else it will be nothing more than a paper exercise.

22.4.2 *The involvement of stakeholders*

There are a number of internal and external stakeholders of the organization who will have an interest and influence on the introduction and development of the occupational health and safety management system.

Internal stakeholders include:

➤ **Directors and trustees of the organization** – Following several reports on corporate governance in recent years culminating in the Combined Code of Corporate Governance 2003, the measurement of occupational health and safety performance and the attainment of health and safety targets has been recognized as being as important as other measures of business performance and targets. A report on health and safety performance should be presented at each board meeting and be a periodic agenda item for sub-committees of the board such as audit and risk management.

➤ **The workforce** – Without the full co-operation of the workforce, including contractors and temporary employees, the management of health and safety will not be successful. The workforce is best qualified to ensure and provide evidence that health and safety procedures and arrangements are actually being implemented at the workplace. So often this is not the case even though the occupational health and safety managements system is well designed and documented. Worker representatives can provide useful evidence on the effectiveness of the management system at shop floor level and they can also provide a useful channel of communication between the senior management and the workforce. One useful measure of the health and safety culture is the enthusiasm with which workers volunteer to become and continue to be representatives. For a representative to be really effective, some training is essential. Organizations which have a recognized trade union structure will have fewer problems with finding and training worker representatives. More information on the duties of worker representatives is given in Chapter 3.

➤ **Health and safety professionals** – Such professionals will often be appointed by the organization to manage the occupational health and safety management system and monitor its implementation. They will, therefore, have a particular interest in the development of the system, the design of objectives and the definition of targets or goals. Unless they have a direct line management responsibility, which is not very common, they can only act as advisers but still have a positive influence on the health and safety culture of the organisation. The appointed health and safety professional also liaises with other associated competent persons, such as for electrical appliance and local exhaust ventilation testing.

External stakeholders include:

➤ **Insurance companies** – As compensation claims increase, insurance companies are requiring more and more evidence that health and safety is being effectively managed. There is increasing evidence that insurance companies are becoming less prepared to offer cover to organizations with a poor health and safety record and/or occupational health and safety management system.

➤ **Investors** – Health and safety risks will, with other risks, have an effect on investment decisions. Increasingly, investment organizations require evidence that these risks are being addressed before investment decisions are made.

➤ **Regulators** – In many parts of the world, particularly in the Far East, national regulators and legislation require certification to a recognized international occupational health and safety management system standard. In many countries, regulators use such a standard to measure the awareness of a particular organization to health and safety issues.

➤ **Customers** – Customers and others within the supply chain are increasingly insisting on some form of formal occupational health and safety management system to exist within the organization. The construction industry is a good example of this trend. Much of this demand is linked to the need for corporate social responsibility and its associated guidelines on global best practice.

➤ **Neighbours** – The extent of the interest of neighbours will depend on the nature of the activities of the organization and the effect that these activities have on them. The control of noise, dust and other atmospheric contaminants are examples of common problem areas which can only be addressed on a continuing basis using a health and safety and environmental management system.

➤ **International organizations** – The United Nations, the ILO, the International Monetary Fund, the World Bank and the World Health Organization are all examples of international bodies which have shown a direct or indirect interest in the management of occupational health and safety. In particular, the ILO is keen to see minimum standards of health and safety established around the world. The ILO works to ensure for everyone the right to work in freedom, dignity and security – which includes the right to a safe and healthy working environment. More than 70 ILO Conventions and Recommendations relate to questions of safety and health. In addition the ILO has issued more than 30 Codes of Practice on Occupational Health and safety. For more information see their website www.ilo.org/safework.

There is a concern of many these international organizations that as production costs are reduced by relocating operations from one country to another, there is also a lowering in occupational health and safety standards. The introduction of internationally recognized occupational health and safety management systems will help to alleviate such fears.

Figure 22.9 Construction site, Sienna.

22.4.3 *An effective audit*

An effective audit is the final step in the occupational health and safety management system control cycle. The 'feedback loop' produced by audit enables the reduction of risk levels and the effectiveness of the occupational health and safety management system to be improved. Audit is a business discipline which is frequently used in finance, environmental matters and quality and can equally well be applied to health and safety. It will check on the implementation of occupational health and safety management systems and the adequacy and effectiveness of the management arrangements and risk control systems. Audit is critical to a health and safety management system but is not a substitute for the essential day-to-day management of health and safety.

The audit aims to establish that the three major components of any occupational health and safety

management system are in place and operating effectively. It should show that:

➤ appropriate management arrangements are in place
➤ adequate risk control systems exist, are implemented and consistent with the hazard profile of the organization
➤ appropriate workplace precautions are in place.

Where the organization is spread over a number of separate sites, the management arrangements linking the centre with these sites should be examined by the audit.

Some elements of the occupational health and safety management system do not need to be audited as often as others. For example, an audit to verify the implementation of critical risk control systems should be made more frequently than an audit of the management arrangements for health and safety of the whole organization. Where there are complex workplace precautions in place, such as in the chemical industry, it may also be necessary to undertake technical audits.

The audit programme should produce a comprehensive picture of the effectiveness of the health and safety management system in controlling risks. The programme must indicate when and how each component part will be audited. The audit team should include managers, safety representatives and workers. Such inclusiveness will help during the implementation of any audit recommendations.

More detailed information on the auditing process is given in Chapter 7.

When planning an audit, a decision has to be made as to whether to use internal or external auditors. When formal certification is required by either the organization or the client, competent external auditors are essential. However, it is likely that, in such cases, internal auditors will also be used.

The advantage of using internal auditors is that they know the critical areas to monitor and will help to spread good practices around the organization. The disadvantages are that clients may question their independence and they may well be unaware of external benchmarks, unless they are audited against the standard. These two disadvantages are the advantages of using external auditors. External auditors have the disadvantages that they do not know the organization, require much more documentation and can offer bland reports. It is also usually more expensive to use external rather than internal auditors. Auditor competence is an important issue and it is important that any auditors used, whether internal or external, are properly trained.

22.4.4 *Continual improvement*

Continual improvement has been mentioned several times in this chapter and is recognized as a vital element of all occupational health and safety management systems if they are to remain effective and efficient as internal and external changes affect the organization. Internal changes may be caused by business reorganization, such as a merger, new branches and changes in products, or by new technologies, employees, suppliers or contractors. External changes could include new or revised legislation, guidance or industrial standards, new information regarding hazards or campaigns by regulators.

Continual improvement need not necessarily be done at high cost or add to the complexity of the management system. Its benefits include:

➤ a decrease in the rate of injuries, ill-health and damage
➤ a possible reduction in the resources required to manage the system
➤ an acceptability of higher standards and an improved health and safety culture
➤ overall improvements in the management system itself.

The simplest way to achieve continual improvement is to implement the recommendations of audits and management reviews, and use benchmarks from similar organizations and any revised national or industrial guidelines. The workforce, managers and supervisors often have very good suggestions on ways to improve processes and procedures. Such suggestions will lead to improvements in the management system. Finally, the health and safety committee can be a very effective vehicle for continual improvement, particularly if working parties are used to investigate specific issues within an agreed timeframe.

22.5 The role of the regulatory authorities

22.5.1 *The legal framework*

The framework for regulating health and safety will vary across the world; for example, European countries use the EU framework, the Pacific Rim countries tend to use a US framework whereas the Caribbean countries follow the UK framework. The course provider should be able to describe the legal and regulatory framework appertaining to any particular country.

Most legislation is driven by a framework of Acts, Regulations and support material including Codes of Practice and standards, as illustrated in Figure 22.11. Within Europe there is another layer of legislation known as Directives, above the Member States' own legislation. These are legally binding on each state; for more information

Figure 22.10 Different solutions have to be considered in different countries – delivering furniture in Certaldo, Italy.

Figure 22.11 Typical health and safety legal framework.

see Chapter 1. The US system of federal and individual state legislation is very similar.

22.5.2 *Regulatory authorities and safety management systems*

The role of the national regulatory authority is crucial to the successful implementation of an occupational health

561

and safety management system. In many parts of the world, such as South East Asia, formal adoption of a recognized management system is required with third party auditing by government approved auditors. In the USA, organizations with approved management systems may be exempted from normal inspections by the Occupational Safety and Health Administration.

In the UK, there has been a movement from prescriptive legislation to risk assessment by the employer and this movement is now occurring in many parts of the world including the EU. A management system is an essential tool to achieve this movement and such a system is implied in the UK Management of Health and Safety at Work Regulations. Countries such as Canada, Australia, New Zealand and Norway have developed occupational health and safety management systems as an encouragement for such self-regulation. The ILO-OSH 2001 system has been adopted by Germany, Sweden, Japan, Finland, Korea, China, Mexico, Costa Rica, Brazil, Indonesia, Vietnam, Malaysia, India, Thailand, the Czech Republic, Poland and Russia.

It is likely that more countries and multi-national companies will expect occupational health and safety management systems to be adopted either as a legal duty or an implied duty following rulings from local courts of law.

22.6 The benefits and problems associated with occupational health and safety management systems

The use of occupational health and safety management systems has many benefits of which the principal ones are:

1. Legal compliance is much easier to be achieved and demonstrated; enforcement authorities have more confidence in organizations which have a health and safety management system in place.
2. It ensures that health and safety is given the same emphasis as other business objectives, such as quality and finance; it will also aid integration, where appropriate, with other management systems.
3. It enables significant health and safety risks to be addressed in a systematic manner.
4. It can be used to show legal compliance with terms such as 'practicable' and 'so far as is reasonably practicable'.
5. It indicates that the organization is prepared for an emergency.
6. It illustrates that there is a genuine commitment to health and safety throughout the organization.

There are, however, several problems associated with occupational health and safety management systems although most of them are soluble because they are caused by poor implementation of the system. The main problems are that:

1. the arrangements and procedures are not apparent at the workplace level and the audit process is only concerned with a desk top review of procedures
2. the documentation is excessive and not totally related to the organization due to the use of generic procedures
3. other business objectives, such as production targets, lead to ad hoc changes in procedures
4. integration, which should really be a benefit, can lead to a reduction in the resources and effort applied to health and safety
5. a lack of understanding by supervisors and the workforce lead to poor system implementation
6. the performance review is not implemented seriously thus causing cynicism throughout the organization.

22.7 Conclusions on the three health and safety management systems

Three occupational health and safety management systems have been described in some detail in this chapter and it has been shown that the similarities between them outweigh their differences. Any occupational health and safety management system will fail unless there is a positive health and safety culture within the organization and the active involvement of internal and external stakeholders.

A structured and well-organized occupational health and safety management system is essential for the maintenance of high health and safety standards within all organizations and countries. Some systems, such as OHSAS 18001, offer the opportunity for integration with quality and environmental management systems. This enables a sharing of resources although it is important that technical activities, such as health and safety risk assessment, are only undertaken by persons trained and competent in that area.

For an occupational health and safety management system to be successful, it must address workplace risks and be 'owned' by the workforce. It is, therefore, essential that the audit process examines shop floor health and safety behaviour to check that it mirrors that required by the health and safety management system.

Finally, whichever system is adopted, there must be continual improvement in health and safety performance if the application of the occupational health and safety management system is to succeed in the long term.

Figure 22.12 Mobile tower crane – Cinque Terre, Italy.

22.8 Other minor additions to the management unit – Unit IGC1 Management of international health and safety

The sources of health and safety information have been extended to include information provided by the websites or publicity offices of national or international agencies (e.g. International Labour Organization (ILO),

Occupational Safety and Health Administration (USA), European Agency for Safety and Health at Work (EU) and Worksafe (Western Australia)). The sources of information covered in Chapters 1 and 9 are also included.

The hazards of working in unfamiliar countries and/ or climates are important health and safety issues – snake bites, diseases (such as malaria and yellow fever) and sunstroke. The course provider will relate these and any other particular hazards to the country for which the course is being provided.

Many countries have either fault or no-fault compensation schemes for workers involved in accidents. Knowledge of these schemes is important for those who operate in several different countries.

22.8.1 *Fault and no-fault compensation*

In the UK, compensation for an injury following an accident is achieved by means of a successful legal action in a civil court (as discussed in Chapter 1). In such cases, injured employees sue their employer for negligence and the employer is found liable or at fault. This approach to compensation is adversarial, costly and can deter injured individuals of limited means from pursuing their claim. In a recent medical negligence claim in Ireland, costs were awarded against a couple who were acting on behalf of their disabled son, and they were faced with a bill for £3 million.

The spiralling cost of insurance premiums to cover the increasing level and number of compensation awards, despite the Woolfe reforms (see Chapter 8), has led to another debate on the introduction of a no-fault compensation system. It has been estimated that in medical negligence cases, it takes on average 6 years to settle a claim and only 10% of claimants ever see any compensation.

No-fault compensation systems are available in many parts of the world, in particular New Zealand and several states in the USA. In these systems, amounts of compensation are agreed centrally at a national or state level according to the type and severity of the injury. The compensation is often in the form of a structured continuous award rather than a lump sum and may be awarded in the form of a service, such as nursing care, rather than cash.

The no-fault concept was first examined in the 1930s in the USA to achieve the award of compensation quickly in motor car accident claims without the need for litigation. It was introduced first in the State of Massachusetts in 1971 and is now mandatory in nine other states although several other states have either repealed or modified no-fault schemes.

In 1978, the Pearson Commission in the UK rejected a no-fault system for dealing with clinical negligence even though it acknowledged that the existing tort system was costly, cumbersome and prone to delay. Its principal reasons for rejection were given as the difficulties in reviewing the existing tort liability system and in determining the causes of injuries.

In New Zealand, there was a general dissatisfaction with the workers, compensation scheme, which was similar to the adversarial fault-based system used in Australia and the UK. In 1974, a no-fault accident compensation system was introduced and administered by the Accident Compensation Corporation (ACC). This followed the publication of the Woodhouse Report in 1966, which advocated 24 hour accident cover for everybody in New Zealand.

The Woodhouse Report suggested the following five principles for any national compensation system:

1.   community responsibility
2.   comprehensive entitlement irrespective of income or job status
3.   compete rehabilitation for the injured party
4.   real compensation for the injured party
5.   administrative efficiency of the compensation scheme.

The proposed scheme was to be financed by channelling all accident insurance premiums to one national organization (the ACC).

The advantages of a no-fault compensation scheme include:

1.   Accident claims are settled much quicker than in fault schemes.
2.   Accident reporting rates will improve.
3.   Accidents become much easier to investigate because blame is no longer an issue.
4.   Normal disciplinary procedures within an organization or through a professional body are unaffected and can be used more quickly if the accident resulted from negligence on the part of an individual.
5.   More funds are available from insurance premiums for the injured party and less used in the judicial and administrative process.

The possible disadvantages of a no-fault compensation system are that:

1.   there is often an increase in the number of claims, some of which may not be justified
2.   there is a lack of direct accountability of managers and employers for accidents
3.   mental injury and trauma are often excluded from no-fault schemes because of the difficulty in measuring these conditions
4.   there is more difficulty in defining the causes of many injuries and industrial diseases than in a faculty system

5.   the monetary value of compensation awards tend to be considerably lower than those in fault schemes (although this can be seen as an advantage).

No-fault compensation schemes exist in many countries, including Canada and the Scandinavian countries. However, a recent attempt to introduce such a scheme in New South Wales, Australia was defeated in the legislative assembly.

22.9   A brief review of the issues facing the international construction industry

The vast majority of hazards that face the global construction industry are similar to those found in the UK. The contents and recommendations found in ILO guidance reflects this similarity and repeats most of the control measures discussed earlier in this book. The additional hazards centre on problems associated with climate (extremes of both hot and cold), site location (in densely populated areas and very remote ones) and local legal frameworks, customs and practice.

22.9.1   *ILO Code of Practice – Safety and Health in Construction*

The ILO Code of Practice – Safety and Health in Construction – contains general and specific duties. It is very comprehensive and essential reading for those involved in international construction contracts. The general duties are ascribed to the following groups:

➤   competent authorities
➤   employers
➤   self-employed persons
➤   co-operation and co-ordination
➤   workers, including their rights
➤   designers, engineers and architects
➤   project clients.

The specific duties cover all the topics described earlier in this book (Chapters 3–20) and the following additional ones:

➤   tunnelling
➤   underground pipelines
➤   cofferdams and caissons
➤   work with compressed air
➤   pile driving
➤   use of explosives
➤   work on tall chimneys, including demolition
➤   structural steel work.

Stress is placed on the need for good welfare facilities on the site, in particular a potable water supply, adequate sleeping accommodation for the workforce, good sanitary and washing facilities and suitable provision either for cooking food or obtaining cooked food.

The supply of potable water is very important and where a mains supply is not available, arrangements will be needed to ensure that a clean supply of drinking water is transported to the site on a regular basis. Any non-drinking water used on the site, must be clearly marked as unsuitable for drinking.

In many climates, vermin and mosquitoes are a serious problem, particularly at night, and effective control arrangements will need to be in place (e.g. disposal systems for waste food and mosquito nets). The provision of suitable and relevant first aid and medical assistance is also a priority. When working is to take place in high temperatures, there should be work breaks around noon and several drink breaks during the rest of the day.

22.10 Practice NEBOSH questions for the International Construction Certificate

The International General Certificate is a relatively new award and all examination papers to date (April 2008)

have used many questions taken from the question bank for the National General Certificate other than those questions directly related to UK law. Therefore, all practice questions given at the end of each chapter in this book can be used by students on the International General Certificate except those at the end of Chapter 1 which relate to UK law. The questions at the end of Chapter 1 which may be used on the International General Certificate are given below.

1. (a) **Draw** a flowchart to **identify** the main components of the health and safety management system described in the HSE publication 'Successful Health and Safety Management' (HSG 65).
   (b) **Outline any TWO** components of the health and safety management system identified in (a).

2. **Outline** the economic benefits that an organization may obtain by implementing a successful health and safety management system.

3. (a) **Draw** a flowchart to show the relationships between the **SIX** elements of the health and safety management model in HSE's 'Successful Health and Safety Management' (HSG 65).
   (b) **Outline** the part that **EACH** element of the HSG 65 model plays within the health and safety management system.

Appendix 22.1 – Scaffolds and ladders

1.1 Convention (Safety and Health in Construction) (167)

Article 14

1. Where work cannot safely be done on or from the ground or from part of a building or other permanent structure, a safe and suitable scaffold shall be provided and maintained, or other equally safe and suitable provision shall be made.
2. In the absence of alternative safe means of access to elevated working places, suitable and sound ladders shall be provided. They shall be properly secured against inadvertent movement.

3. All scaffolds and ladders shall be constructed and used in accordance with national laws and regulations.
4. Scaffolds shall be inspected by a competent person in such cases and at such times as shall be prescribed by national laws or regulations.

1.2 Recommendation (Safety and Health in Construction) (175)

Scaffolds

16. Every scaffold and part thereof should be of suitable and sound material and of adequate size and strength for the purpose for which it is used and be maintained in a proper condition.

(Continued)

Appendix 22.1 – *Continued*

17. Every scaffold should be properly designed, erected and maintained so as to prevent collapse or accidental displacement when properly used.
18. The working platforms, gangways and stairways of scaffolds should be of such dimensions and so constructed and guarded as to protect persons against falling or being endangered by falling objects.
19. No scaffold should be overloaded or otherwise misused.
20. A scaffold should not be erected, substantially altered or dismantled except by or under the supervision of a competent person.
21. Scaffolds as prescribed by national laws or regulations should be inspected, and the results recorded, by a competent person:
    (a) before being taken into use
    (b) at periodic intervals thereafter
    (c) after any alteration, interruption in use, exposure to weather or seismic conditions or any other occurrence likely to have affected their strength or stability.

1.3 Code of Practice – Safety and Health in Construction

The Code of Practice covers scaffolds and ladders under the following topics over five pages:

1. general provisions
2. materials
3. design and construction
4. inspection and maintenance
5. lifting appliances on scaffolds
6. prefabricated scaffolds
7. use of scaffolds
8. suspended scaffolds.

Study skills

23

23.1 Introduction

The NEBOSH Construction Certificate is like any other examination-based award. To be successful there are several things a student needs to do. The best results can be achieved by planning in the following way:

➤ having clear and realistic goals, both in the short and the long term
➤ having knowledge of some techniques for studying and for passing exams
➤ having a well-organized approach
➤ being motivated to learn.

It is often said that genius is 99% application and only 1% inspiration. Good organization and exam technique are secondary to academic ability when it comes to passing exams. Studying is an activity in its own right and there are a number of ways in which students can make their study much more effective to give themselves the greatest possible chance of success: choosing a suitable place for study; sensible planning of study time; adapting the type of study to suit the individual; adopting appropriate styles of note-taking; improving memory and concentration skills; choosing appropriate revision techniques; being clear about the contents of the syllabus and about what is expected by the examiner. Attention to all these details will maximize a student's available time and mental ability.

23.2 Finding a place to study

Studying does not always come naturally to people and it often has to be learned, so it is worth paying attention to the basics of study. It is easier to acquire the habit of study if the student works in a place that has been designated for the purpose. Choose somewhere reasonably quiet and free from distractions, with good ventilation (essential for alertness) and a comfortable temperature. Make sure there is enough light – a reading lamp helps to prevent eye strain and tiredness.

Choose an upright chair rather than an armchair and make sure that the workspace is large enough. The size recommended by Warwick University, for example, is a minimum of 60 cm × 1.5 m. If possible, find a place where study materials do not have to be cleared away as it is very easy to put off starting work if they are not instantly accessible.

23.3 Planning for study

Start by making a study timetable. Planned study periods need to be established so that they do not get squeezed out of the busy working week. Depending on the type of syllabus being followed, the timetable will need to include time for carrying out assignments set by tutors, time for going over lecture notes and materials, any further reading rated as 'essential' and, if possible, reading rated as 'desirable'. Time should also be allocated for revision as the course progresses because the more regularly the information is revised, the more firmly it will become fixed in 'long-term memory'. This makes it very much easier to recall information for an exam.

After every hour of study, take a short break and, if possible, have a change of atmosphere. Physical exercise is very helpful in increasing the ability to concentrate and even a brisk walk round the block can do wonders for the attention span.

Variation is another way of holding the attention, so it helps to alter the type of activity. For example, time could be spent gathering information from books, the Internet and so on, followed by drawing a diagram or a graph, writing, reading and so on.

23.4 Blocked thinking

Sometimes a student becomes 'blocked' in an area of study. If this happens, try leaving it alone for a few days and tackle a different area of the subject, since concentrating too hard on something that is too difficult is likely to produce loss of confidence and can be very demotivating. It usually happens that, having had a break, the difficulty vanishes and the problem clears. This is either because the solution has emerged in the mind from another perspective, or because the student has learnt something new that has supplied the answer. Psychologists consider that it is possible that some problems can be solved during sleep when the mind has a chance to wander and apply lateral thinking.

23.5 Taking notes

The most efficient way to store notes is in a loose-leaf folder, because items can so easily be added. Write only on one side of the sheet and use margined paper to make it easier to add information. The facing pages can be used to make summaries or extra points. If notes are clearly written and well- spaced out, they are much more straightforward to work from and more attractive to return to later. Use colour, highlighting and underlining too; notes that look good are less daunting when the time comes to revise.

During revision it will be important to be able to identify subjects quickly, so use headings, numbering, lettering, bullet points, indention and so on. It is worth spending a bit of extra time if it will make revision easier.

Key words and phrases are better than continuous prose when note-taking. While writing down information it is easy to miss essential points that are being made by the tutor. The notes should be read through within 36 hours to make sure that they are completely clear when the time comes to revise. At this point it is convenient to add in anything that has been missed, while it is fresh in the mind. Reading through notes in this way also helps to fix the information in the memory and the level of recall will be further improved by reading them again a week later. Although it seems time consuming, this technique will save a lot of time in the long term.

Many people find that they can understand and remember more effectively by making 'mind maps' (also known as 'pattern notes'). These will be discussed in more detail in the revision section.

23.6 Reading for study

It is very rarely necessary to read a whole book for study purposes. Use the index and be selective. Since most students will be working while they are studying, they will need techniques to help them make the most effective use of their time.

Skim reading is a great time-saver. First of all, flip through the book to see what it contains, looking at the contents, summaries, introduction, tables, diagrams and so on. It should be possible to identify areas that are necessary to the syllabus. These can be marked with removable adhesive strips or 'Post-it' notes.

Secondly, quickly read through the identified parts of the book. Two ways of doing this are to:

(a) run a finger slowly down the centre of a page, watching it as it moves. The eye will pick up relevant words and phrases as the finger moves down
(b) read in phrases, rather than word by word, which will increase the reading speed.

Both these methods need practice but they save a lot of time and effort, and it is better to read through a piece two or three times, quickly, than to read it once slowly. However, if there is a really difficult piece of information that is absolutely essential to understand, reading it aloud, slowly, will help.

➤ Finally, at the end of each section or chapter, make a few brief notes giving the essential points.

It is well worth learning to speed read if a course involves a large amount of reading. Tony Buzan's series *The Mind Set* explains the technique in simple terms (Buzan, 2000).

23.7 Revision

As the exam approaches, all the information from the course needs to be organized in preparation for revision. This should, ideally, have been an ongoing process throughout the course, but if not there will be extra work to be undertaken at this point. Information from lectures, reading, practical experience and assignments needs to be organized into a form that makes it readily accessible to the memory. In addition there will be the student's existing knowledge that is relevant to the syllabus. It is worth spending some time considering how this store of information relates to and can be used to expand and back up the new information from the course. The 2004 Examiners' Reports refer specifically to the advantages to be gained from this.

23.8 Organizing information

There are a number of ways to organize information for an exam:

➤ Reading back through the notes on a regular basis while the course is still in progress will help to fix the information in long-term memory.

➤ Make revision cards: condensing the information down so that it fits on a set of post cards makes it possible to carry a lot of information around in a pocket. The activity of condensing the information is revision in itself. The revision material is then available for spare moments in the day, such as queuing in the canteen, sitting in a traffic jam on the motorway, waiting for the computer to function and so on.

For example, the examiner asks the student to 'Outline the factors that may increase risks to pregnant employees'. On a card (Figure 23.1), the information could be condensed into:

```
              Pregnant Employees – Risks
Exposure to      ⎧ pesticides
chemicals        ⎨ lead
                 ⎪ mutagens (inter-cellular
                 ⎪            change)
                 ⎩ teratogens (affect embryo)

biological         (e.g. hepatitis)
Physical           (ionizing radiations
                    extremes of temperature)
manual handling
ergonomic          (prolonged standing
                    awkward body movements)
Stress
issues relating to  use/wearing of
                    protective equipment
```

Figure 23.1 Revision notes.

➤ Make mind maps (also known as pattern notes). Like revision cards, the act of making them is revision in itself. Because they are based on visual images they are easier for many people to remember and they can be made as complex or as simple as the individual requires. Using colour and imagery makes it easier to follow and to recall the information in a mind map (Buzan and Buzan, 1993). Figure 23.2 shows a mind map based on the report writing section of this book.

➤ Use list words: for example, students were asked to list eight ways of reducing the risk of a fire starting in the workplace. The examiners' report suggested the following:

Candidates could have chosen from a list including the control of smoking and smoking materials, good housekeeping to prevent the accumulation of waste paper and other combustible materials, regular lubrication of machinery, frequent inspection of electrical equipment for damage, ensuring ventilation outlets on equipment are not obstructed, controlling hot work, the provision of proper storage facilities for flammable liquids and the segregation of incompatible chemicals.

This list could be reduced to an eight-letter nonsense word, to jog the memory, as follows:

FLICSHEV:

Flammable liquids – provide proper storage facilities
Lubrication – regular lubrication of machinery
Incompatible chemicals need segregation
Control of hot work
Smoking – of cigarettes and materials, to be controlled
Housekeeping (good) – prevent accumulation of waste
Electrical equipment – needs frequent inspection for damage
Ventilation – outlets should not be obstructed

The effort involved in constructing this list and inventing the word helps to fix the information in memory.

There are other methods to aid memory but these four are probably the most practical aids to exam technique.

23.9 How does memory work?

Understanding how memory works will help the student to use it more effectively. As in any scientific subject, there is still a lot to be discovered about the workings of the mind, and a considerable amount of disagreement about what has been discovered so far; but psychologists generally seem to agree about the way that memory works.

The process of remembering is divided, roughly, into two sections – short- and long-term memory. Items

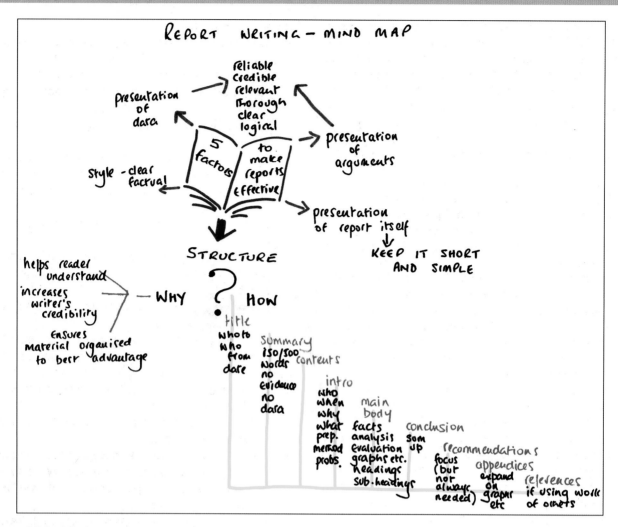

Figure 23.2 Mind map report writing.

that go into short-term memory, if no further attention is paid to them, will fade away and be forgotten. If they are rehearsed they will stay in short-term memory for a while, as, for example, when trying to remember a phone number to make a call. To put something into long-term memory demands a greater 'depth of processing', that is to say, more mental activity so:

(a) the information stays in the memory (storage) and
(b) the information can be found when it is needed (retrieval).

The techniques described in the section 23.7 entitled 'Revision' give some ways in which storage and retrieval systems in the brain can be made to work effectively.

Contrary to popular opinion, mature students have many assets when it comes to learning. Because their life and work experience is almost always more comprehensive and of a higher complexity than that of younger students, they normally possess more 'schemas' (areas of knowledge) to which new information

can be attached. In addition to this there will be more experiences already stored in the brain that can provide an explanation for the new pieces of knowledge that are being acquired. Add to this a very high level of motivation, a stronger level of incentive towards success and more determined application, and it becomes clear that the mature student has many learning advantages.

It cannot be denied, however, that mature students tend to be very nervous about exams and often need a good deal of reassurance and support to enable them to realize their full potential.

23.10 How to deal with exams

There are three stages to taking an exam:

1. planning and revision
2. the exam room
3. after the exam.

For people with **special needs**, there are provisions within NEBOSH to allow extra time or the use of special equipment. Students who think they may be eligible for help should apply to NEBOSH several weeks before the date of the exam (see sections 3.9 and 3.10 of the *Guide to the NEBOSH National Certificate in Construction Safety and Health*).

23.10.1 *Planning and revision*

1. It is absolutely essential for students to know what they are going to be examined on and what form the exam will take. Students should read through the syllabus and if they are concerned about any area of it, this should be raised with the course tutor well before the date of the exam.

2. Read the examiners' reports for the subject. After the exams every year, the examiners highlight the most common mistakes made by students (see Section 23.11). They also provide useful information about, for example, pass rates, levels attained (distinction, credit, pass, fail) time management and other hints on exam technique.

3. It is very useful to work through some recent past papers, against the clock, to get used to the 'feel' of the exam. If possible, tutors should set up mock exams and make sure that the papers are marked. Some of the shorter courses will not be able to provide this service, and, if this is the case, students should try to do at least one or two questions in their own time. The examiners' reports will give an indication of what could have gone into the answers.

Past papers, examiners' reports and syllabuses are available from NEBOSH. The website is at www.nebosh. org.uk.

4. It is vital to know where the exam is to be held and the date and time. If possible, visit the building beforehand to help build confidence about the location, availability of parking and so on.

5. Make a chart of the time leading up to the exam. Include all activities, work, leisure, social, as well as the time to be used for revision, so that the schedule is realistic.

6. Try to eat and sleep well and take some exercise.

7. Revision techniques are covered in the sections on revision and memory earlier in this chapter.

SET REALISTIC TARGETS, THEN ACHIEVE THEM. MAKE PLANS AND STICK TO THEM.

23.10.2 *In the exam room*

➤ Read through the exam paper very carefully.
➤ Check the instructions – how many questions have to be answered? From which section?
➤ Make a time plan.
➤ Underline key words, e.g. 'define', 'describe', 'explain', 'identify' and so on. NEBOSH provides definitions of the meanings of these in the examiners' reports (see 'Terminology' section later). Using some words in the question when writing the answers may help to keep the answer on track.
➤ Stick to the instructions given in the question. If the question says 'list', then list; if it says 'describe', then describe. If it says 'briefly' make sure the answer is brief.
➤ Write clearly. Illegible answers don't get marked.
➤ Look at the number of marks allocated to a question to pick up clues as to how much time should be spent on it.
➤ Mark questions which look possible and identify any that look impossible.
➤ It is rarely necessary to answer exam questions in a particular order. Start with the question that you feel most comfortable with since it will help to boost confidence. Make sure it is clearly identified by number for the examiner.
➤ Answer the question that is set, not the one you wish was on the paper.
➤ If ideas for other answers spring to mind while writing, jot down a reminder on a separate piece of paper. It is easy to forget that bit of information in the heat of the moment.
➤ Plan the use of time and plan the answers. Include some time to check over each answer.
➤ Stick to the time plan; stick to the point; make points quickly and clearly.

Early marks in an exam question are easier to pick up than the last one or two, so make sure that all the questions are attempted within the time plan. No marks are given for correct information that is not relevant to the question. Examiners are only human and usually have to work under pressure; it is quite possible that some vital and correct point may be missed if there is a mass of irrelevant information from which the point has to be extracted.

Avoid being distracted by the behaviour of other students. Someone who is requesting more paper has not necessarily written a better answer; they may simply have larger handwriting. People who start to scribble madly as soon as they turn over the question sheet are not in possession of some extra ability – they simply have not planned their exam paper properly.

Keep calm, plan carefully, don't panic.

Terminology used in NEBOSH exams

(Taken from the *Guide to the NEBOSH National Certificate in Construction Safety and Health* Appendix 3)

Action verb	Meaning
define	provide a generally recognized or accepted definition
describe	give a word picture
explain	give a clear account of, or reasons for
give	provide without explanation (used normally with the instruction to 'give an example (or examples) of' …)
identify	select and name
list	provide list without explanation
outline	give the most important features of (less depth than either 'explain' or 'describe' but more depth than 'list')
sketch	provide a simple line drawing using labels to identify specific features
state	a less demanding form of 'define', or where there is no generally recognized definition

23.10.3   *After the exam*

If there are several exams to be taken, it is important to keep confidence levels high. It is not a good idea to get into a discussion about other people's experiences of the exam. After one exam, focus on the next. If something went wrong during the exam (e.g. illness or severe family problems) the tutor and the examining board should be alerted immediately.

23.11   The examiners' reports

23.11.1   *A few points from the examiners' reports*

The latest reports available at the time of going to press are for December 2006 and June 2007. While acknowledging the achievements and hard work of candidates,

the examiners point out that, as before, avoidable mistakes are being made, and marks are being lost simply through lack of exam technique.

The examiners advise that while acquisition of knowledge and understanding across the syllabus are a vital ingredient for success, exam technique is an essential. They define this as 'the skill of reading a question, identifying the breadth of issues relevant to that question and putting them down on paper in a logical and coherent way and to the depth required'. They give some very useful information about time planning on the final page and a brief section on exam technique.

Here are a few examples from the latest reports:

Paper 1 Q3 (December 2006): The question asked candidates to 'outline the principal duties of a client under the Construction (Design and Management) Regulations 1994'. A number of candidates outlined the duties of the principal contractor or the planning supervisor.

Paper 2 Q5 (December 2006): Candidates were asked to give the precautions that should be taken on a construction site in order that welding may be carried out safely. The answers varied in quality and the examiners make the point that when making reference to the need to provide or wear personal protective equipment, details or examples of the equipment should be given.

Paper 1 Q5 (June 2007): This was about what should be taken into account when employing sub-contractors, some of whom may not have English as their first language. The examiners make the point that there was little imagination used in answering the question and they give a comprehensive account of what could have been included in the answers to gain the available 8 marks.

Paper 1 Q9 (June 2007): The question was about site safety committees and asked the candidate to outline the topics that would be included on a typical agenda. Unfortunately quite a lot of answers got no marks because instead they went into detail on standard committee procedure.

Paper 2 Q2 (June 2007): In some cases this was an example of not reading the question, which was about moving concrete blocks by hand; so discussion of the use of mechanical means to do the task was irrelevant.

Paper 2 Q3 (June 2007): This asked for a description of the features of a traffic management system but some answers were about safety devices fitted to vehicles and provision of high visibility clothing, which was not the focus of the question.

23.11.2   *Marks for NEBOSH questions*

We have been asked about the allocation of marks to each question shown throughout the book. However, since the marks awarded to each part of a question can vary, only general guidance can be given.

All one- or two-part questions are 8 mark questions with a minimum of 2 marks awarded to any part.

Questions with three or more parts are 20 mark questions with a minimum of 4 marks for each part.

23.12 Conclusion

Passing health and safety exams and assessments has a lot in common with any other subject being examined. If candidates do not apply themselves to study effectively, read the questions carefully and plan their answers, success will be very limited. As once said by an old carpenter:

'Measure twice, think twice, and cut once.'

Applied to exams this can be changed to:

'Read twice, think twice and write once.'

23.13 References

Buzan, T. and Buzan, B. (1993): *The Mind Map Book*. BBC, Revised 2003.

Buzan, T. (2000): *The Mind Set*. BBC Worldwide (includes *Use Your Head, Master Your Memory, The Speed Reading Book, and Use Your Memory*).

Elson, D. and Turnbull, G.: *How to Do Better in Exams*, AEB 1989.

Guide to the NEBOSH National Certificate in Construction Safety and Health, January 2005.

Leicester, Coventry and Nottingham Universities study skills booklets.

'NEBOSH Examiners' Report's December 2006 and June 2007, June and December 2004.

Specimen answers to NEBOSH examinations

24

24.1 Introduction

The NEBOSH National Certificate in Construction Health and Safety is assessed by two written examination papers (NGC1 – The management of health and safety; and NCC1 – Managing and controlling hazards in construction activities) and a practical assessment (NCC2 – Construction health and safety practical application). Both examination papers and the practical application must be passed so that the NEBOSH National Certificate in Construction Health and Safety may be awarded. Since the unitization of the course, students who have the recently passed NEBOSH National General Certificate will not need to re-take NGC1.

In this chapter, specimen answers are given to one long and two short questions for each of the two written papers. A specimen practical assessment and management report is also given for the practical application. It is important to stress that there are no unique answers to these questions but the answers provided should provide a useful guide to the depth and breadth expected by the NEBOSH examiners. Candidates are strongly advised to read past examiners reports which are published by NEBOSH. These are very useful documents because they not only give an indication of the expected answers to each question of a particular paper but they also indicate some common errors made by candidates.

24.2 The written examinations

The previous chapter on study skills gives some useful advice on tackling examinations and should be read in conjunction with this chapter. NEBOSH has a commendably thorough system for Questions Paper preparation to ensure that no candidates are disadvantaged by question ambiguity. Candidates should pay particular attention to the meaning of Action Verbs. Such as 'OUTLINE' used in the questions.

It is difficult to give a definitive guide on the exact length of answer to the examination questions because some expected answers will be longer than others. As a general guide, for the long answer question on the examination paper, it should take about 25 minutes to write about one and a half pages (550–620 words). Each of the ten short answer questions requires about half a page of writing (170–210 words).

While candidates may not be guaranteed 100% by following the guidance in this chapter, they should obtain a comfortable pass.

24.2.1 NGC1 – the management of health and safety

Question 1

1. (a) **Outline** the factors that should be considered when selecting individuals to assist in carrying out risk assessments in the workplace. **(5)**
   (b) **Describe** the key stages of a general risk assessment. **(5)**
   (c) **Outline** a hierarchy of measures for controlling exposures to hazardous substances. **(10)**

Answer:

(a) The most important factor is the competence and experience of the individuals in hazard identification and risk assessment. Some training in these areas should offer evidence of the required competence. They should be experienced in the process or activity under assessment and have technical knowledge of any plant or equipment used. They should have knowledge of any relevant standards, Health and Safety Executive (HSE) guidance and Regulations relating to the activity or process.

They must be keen and committed but also aware of their own limitations. They need good communication skills and the ability to write interesting and accurate reports based on evidence and the detail found in health and safety standards, codes of practice, Regulations and guidance. Some IT skills would also be advantageous. Finally, the views of their immediate supervisor should be sought before they are selected as team members.

(b) There are five key stages to a risk assessment suggested by the HSE as follows:

Stage 1 is hazard identification, which involves looking at significant hazards which could result in serious harm to people. Trivial hazards should be ignored. This will involve touring the workplace concerned looking for the hazards in consultation with workers themselves and also reviewing any accidents, ill health or incidents that have occurred.

Stage 2 is to identify the person who could be harmed-they may be employees, visitors, contractors, neighbours or even the general public. Special groups at risk like young persons, nursing or expectant mothers and people with disability should also be identified.

Stage 3 is the evaluation of the risks and deciding if existing precautions or control measures are adequate. The purpose is to reduce all residual risks after controls have been put in to as low as is reasonably practicable. It is usual to have a qualitative approach and rank risks as high, medium or low after looking at the severity of likely harm and the likelihood of it happening. A simple risk matrix can be used to get a level of risk.

The team should then consider whether the existing controls are adequate and meet any guidance or legal standards using the hierarchy of controls and the General principles of prevention set out in the Management Regulations.

Stage 4 of the risk assessment, is to record the significant findings which must be done if there are five or more people employed. The findings should include any action that is necessary to reduce risks and improve existing controls – preferably set against a time scale. The information contained in the risk assessment must be disseminated to employees and discussed at the next health and safety committee meeting.

Stage 5 is a time scale set to review and possibly revise the assessment. This must also be done if there are significant changes in the workplace or the equipment and materials being used.

(c) The various stages of the usual hierarchy of risk controls are underlined in this answer.

Elimination or substitution is the best and most effective way of avoiding a severe hazard and its associated risks. Elimination occurs when a process or activity is totally abandoned because the associated risk is too high. Substitution describes the use of a less hazardous form of the substance. There are many examples of substitution, such as the use of water-based rather than oil-based paints and the use of asbestos substitutes.

In some cases it is possible to *change the method of working* so that exposures are reduced, such as the use of rods to clear drains instead of strong chemicals. It may be possible to use a substance in a safer form, for example as liquid or pellets to prevent dust and powders. Sometimes the pattern of work can be changed so that people can do things in a more natural way, for example by encouraging people in offices to take breaks from computer screens by getting up to photocopy or fetch documents. *Reduced* or *limited time exposure* involves reducing the time during the working day during which the employee is exposed to a hazardous substance by giving the employee either other work or rest periods.

If the above measures cannot be applied, then the next stage in the hierarchy is the introduction of *engineering controls*, such as isolation (using an enclosure, a barrier or guard), insulation (used on any electrical or temperature hazard) or ventilation (exhausting any hazardous fumes or gases either naturally or by the use of extractor fans and hoods). If ventilation is to be used, it must reduce the exposure level for employees to below the workplace exposure limit.

Housekeeping is a very cheap and effective means of controlling risks. It involves keeping the workplace clean and tidy at all times and maintaining good storage systems for hazardous substances.

A *safe system of work* is a requirement of the HSW Act and describes the safe method of performing the job.

Training and information are important but should not be used in isolation. Information includes such items as signs, posters, systems of work and general health and safety arrangements.

Personal protective equipment (PPE) should only be used as a last resort. There are many reasons for this. It relies on people wearing the equipment at all times and must be used properly. *Welfare facilities*, which include general workplace ventilation, lighting and heating and the provision of drinking water, sanitation and washing facilities, are the next stage in the hierarchy.

All risk control measures, including training, and *supervision*, must be *monitored* by competent people to check on their continuing effectiveness. Periodically the risk control measures should be *reviewed*. Monitoring and other reports are crucial for the review to be useful. Reviews often take place at safety committee and/or at management meetings. A serious accident or incident should lead to an immediate review of the risk control measures in place.

Finally, special control requirements are needed for carcinogens.

Question 2

2. **Outline** ways in which employers may motivate their employees to comply with health and safety procedures. **(8)**

Answer:

Motivation is the driving force behind the way a person acts or the way in which people are stimulated to act. The best way to motivate employees to comply with health and safety procedures is to improve their understanding of the consequences of not working safely and their knowledge of good safety practices, and to promote of their ownership of health and safety. This can be done by effective training (induction, refresher and continuous) and the provision of information showing the commitment of the organization to safety, and by the encouragement of a positive health and safety culture

with good communications systems. Managers should set a good example by encouraging safe behaviour and obeying all the health and safety rules themselves even when there is a difficult conflict between production schedules and health and safety standards. A good working environment and welfare facilities will also encourage motivation. Involvement in the decision making process in a meaningful way, such as regular team briefings, the development of risk assessments and safe systems of work, health and safety meetings and effective joint consultation arrangements, will also improve motivation as will the use of incentive schemes. However, there are other important influences on motivation such as recognition and promotion opportunities, job security and job satisfaction. Self-interest, in all its forms, is a significant motivator.

Although somewhat negative, sometimes it is necessary to resort to disciplinary procedures to get people to behave in a safe way. This is rather like speed cameras on roads with the potential for fines and points on your licence.

Question 3

3. (a) **Explain** why young persons may be at a greater risk from accidents at work. **(4)**

   (b) **Outline** the measures that could be taken to minimize the risks to young employees. **(4)**

Figure 24.1 Young persons are more vulnerable.

Answer:

(a) Young workers have a lack of experience, knowledge and awareness of risks in the workplace. They tend to be subject to peer pressure and behave in a boisterous manner. They are often willing to work hard and want to please their supervisor, and can become over-enthusiastic. This can lead to the taking of risks without the realization of the consequences. Some younger workers have underdeveloped communication skills and a limited attention span. Their physical strength and capabilities may not be fully developed and so they may be more vulnerable to injury when manually handling equipment and materials. They are also more susceptible to physical agents, biological and chemical agents such as temperature extremes, noise, vibration, radiation and hazardous substances.

(b) The Management of Health and Safety at Work Regulations require that a special risk assessment must be made before a young person is employed. This should help to identify the measures which should be taken to minimize the risks to young people.

Measures should include:

➤ additional supervision to ensure that they are closely looked after, particularly in the early stages of their employment

➤ induction and other training to help them understand the hazards and risk at their workplace

➤ not allowing them to be exposed to extremes of temperature, noise or vibration

➤ not allowing them to be exposed to radiation, or to use compressed air or do diving work

➤ carefully controlling levels of exposure to hazardous materials so that exposure to carcinogens is as near zero as possible and other exposure is below the WEL limits which are set for adults

➤ not allowing them to use: highly dangerous machinery like power presses and circular saws; explosives; mechanical lifting equipment such as forklift trucks

➤ restricting the weight that young persons lift manually to well below any weights permitted for adults.

There should be clear lines of communication and regular appraisals. A health surveillance programme should also be in place.

24.2.2 NCC1 – managing and controlling hazards in construction activities

Question 1

1. It is necessary for maintenance purposes to enter a large cement silo situated on a construction site (Figure 24.2).

(i) **Identify** the likely hazards. **(5)**

(ii) **Outline** the precautions to be taken to reduce the risks to personnel entering the silo. **(15)**

Figure 24.2 Cement silo.

Answer:

(i) There are several hazards in entering and working in this silo.

The main hazards include difficult access and egress, which makes entrapment, escape and rescue more problematic; being engulfed in a free flowing solid material if the cement has blocked and bridged; the accumulation of vapours, gases or fumes; and the lack of ventilation. Some of these hazards may be introduced into the silo by the nature of the maintenance work (e.g. welding work or cleaning work with solvents) or by incoming services.

Other possible hazards are:

➤ working at height
➤ asphyxiation due to oxygen depletion
➤ poor lighting levels
➤ claustrophobic effects due to restricted space
➤ fall of materials leading to possible head injuries
➤ electrocution from unsuitable equipment
➤ fire due to flammable liquids and vapours.

Finally, the presence of cement residues is also a hazard. Contact with wet cement can cause serious burns or ulcers which will take several months to heal and may need a skin graft. Dermatitis, both irritant and allergic, can be caused by skin contact with either wet cement or cement powder.

(ii) The silo is a confined space and the health and safety requirements are covered by the Confined Spaces Regulations. The detailed precautions required will depend on the nature of the silo and the actual maintenance work being carried out.

First a risk assessment must be done so that relevant precautions may be taken and a safe system of work determined. The risk assessment needs to identify the hazards present in the silo, estimate the risks and determine suitable controls to address those risks. It must be done by a competent person and will form the basis of a safe system of work. This will normally be formalized into a specific permit-to-work procedure which will be applicable to the maintenance work. The main elements of the 'permit-to-work' are likely to be as follows:

➤ the competence, training and instruction of the workforce and the designated rescue team
➤ the effective and complete isolation of electrical and mechanical equipment and other services by the use of a lock off and a tag system with key security
➤ the methods of communication between people inside the silo, from inside to outside and to summon help in an emergency
➤ the testing and monitoring of the atmosphere inside the silo for hazardous gas, fume, vapour, dust and the concentration of oxygen
➤ the provision of good ventilation, perhaps by mechanical means and escape breathing apparatus for all persons inside the silo
➤ harness and attached lanyard for rescue purposes
➤ means of effecting rescue such as a tripod manual winch lifting unit in place during the work
➤ the careful removal of cement residues using appropriate equipment which will not cause additional hazards
➤ the provision of an outside watcher to raise the alarm and summon emergency assistance
➤ the access and egress to the silo so that there is quick and unobstructed access and escape
➤ the details of fire prevention equipment and procedures, such as method of raising the alarm
➤ the lighting arrangements
➤ the maximum length of any one working period.

It is also important that there are monitoring and audit arrangements to ensure that the permit-to-work system is working as envisaged.

The site supervisor must ensure that all necessary equipment is available on site in accordance with the risk assessment, method statement and any other planned procedures, including the permit-to-work, before any person is allowed to enter the silo. He must also ensure that the planned procedures, particularly the permit-to-work, are adhered to strictly and that only authorized trained persons are permitted to enter the silo. He must be notified of any changes in working methods or conditions inside the silo which were not included in the planning procedures before those changes are implemented.

All safety equipment must be regularly checked and maintained. Records should be kept of the checks and any defects in equipment rectified immediately. Appropriate PPE must be provided for the workforce in the silo (helmets, gloves, boots and hearing protection). Any lighting or electrical power tools must be specially protected to the required standard for damp and/or flammable atmospheres.

Specific training is necessary for all people concerned with silo maintenance, whether they are acting as rescuers, supervisors or those working inside it. The training will need to cover all procedures and the use of equipment under realistic simulated conditions.

The workers must also be monitored and consideration should be given to the exclusion from the silo of individuals who suffer from any medical condition which may be exacerbated by working in it (e.g. claustrophobia).

Before people enter the silo, suitable and sufficient emergency, rescue and first-aid arrangements must be in place. These arrangements should reduce the risks to rescuers and include the provision and maintenance of suitable resuscitation equipment.

Question 2

2. **Outline** the precautions that may be needed when carrying out repairs to the flat roof of a building. **(8)**

Answer:

Roof work is hazardous and requires a specific risk assessment and method statement prior to the commencement of work so that the required precautions may be identified. The particular hazards are fragile roofing materials, including those materials which deteriorate and become more brittle with age and exposure to sunlight, exposed edges, unsafe access equipment and falls.

There must be suitable means of access such as scaffolding, ladders and crawling boards. Suitable edge protection will be needed in the form of guard rails to prevent the fall of people or materials and access must be restricted to the area below the work using visible barriers. Warning signs indicating that a roof is fragile, should be displayed at ground level. Protection should be provided in the form of covers where people work

Figure 24.3 Flat roof.

near to fragile materials and roof lights. The means of transporting materials to and from the roof may require netting under the roof and even weather protection.

Precautions will be required for other hazards associated with roof work, such as overhead services and obstructions, the use of equipment such as gas cylinders and bitumen boilers, and manual handling.

Finally, only trained and competent persons must be allowed to work on roofs and they must wear footwear having a good grip. It is good practice to ensure that a person does not work alone on a roof.

Question 3

3. An independent scaffold tied to a 10-storey office block has collapsed into a busy street.

   Outline the factors that may have affected the stability of the scaffold. **(8)**

Answer:

The stability of the scaffold will be undermined if its original erection did not follow the intended design or the design was inadequate. Other factors may include one or more of the following:

➤ poor, incompetent erection and/or a lack of regular inspections such that the standards were bent or not plumb
➤ poor tying into the building or insecure parts of the building

➤ lack of strength of the supporting ground or foundation
➤ lack of effective sole plates as a means of spreading the load at the base of the scaffold
➤ lack of longitudinal bracing (along the length of the scaffold) or cross bracing (at right angles to the building)
➤ the proximity to the scaffold of any excavation work
➤ the effect of excessive surface water in weakening the scaffold foundation
➤ unauthorized alteration by incompetent persons (perhaps by vandals)
➤ the use of incorrect and/or damaged or corroded fittings during the erection or extension of the scaffold
➤ the overloading of the scaffold either with materials or because waste chutes became blocked
➤ adverse weather conditions including high winds
➤ the sheeting-in of the scaffold without the use of extra ties
➤ vehicular impact (either a road vehicle or a load suspended from a crane)
➤ the 'borrowing' of boards and tubes from the scaffold, thus weakening it–a common cause of scaffold collapse.

24.3 NCC2 – The practical application

The guide to the NEBOSH National Certificate in Construction Health and Safety is essential reading for all candidates before they attempt the practical application. The guide contains details on the aims of such application, the rules to be followed, the breadth and depth of observations and management report expected, such and the marking scheme.

In this section, the NEBOSH guide is summarized under 24.3.1 and a specimen practical application, with associated observations and management report, is given under 24.3.2.

24.3.1 *Requirements of the practical application*

The practical application aims to test the application of basic health and safety knowledge to a construction workplace inspection. The test assesses the understanding of key issues and the ability to communicate the findings of the inspection in an effective way to a member of the senior management team. The formal aim of the assessment is to prove a candidate's ability to complete successfully two activities:

➤ to carry out unaided a safety inspection of a construction workplace, identifying the more common hazards, deciding whether they are adequately controlled and, where necessary, suggesting appropriate and cost-effective remedial action; and

➤ to prepare a report that persuasively urges management to take appropriate action, explaining why such action is needed (including reference to possible breaches of legislation) and identifying, with due consideration of reasonable practicability, the remedial measures that should be implemented.

The inspection should take between 30 and 45 minutes and cover as wide a range of hazards as possible. Observations and comments should be noted on the observation sheets provided by NEBOSH (see Appendix 24.1).

Candidates are expected to recognize actual and potential hazards and good – as well as bad – work practices and welfare provisions. While only brief details of each hazard are required, it is important that the assessor can subsequently identify:

➤ where the hazard was located
➤ the nature of the hazard
➤ the degree of risk associated with the hazard
➤ the remedial actions, where appropriate, with relevant prioritization.

A selection from the following topics is likely to be observed at most construction sites:

1. General site issues including:
   ➤ site security
   ➤ stores
   ➤ housekeeping.

2. Management issues including:
   ➤ health and safety policy
   ➤ insurance, training
   ➤ health and safety plan
   ➤ method statements
   ➤ risk assessments
   ➤ competent workers.

3. Documentation including:
   ➤ test certificates
   ➤ inspection registers
   ➤ accident book.

4. Welfare
5. PPE
6. Scaffolds
7. Ladders
8. Roof work
9. Excavations
10. Plant and equipment
11. Power and hand tools
12. Traffic and vehicles
13. Manual Handling
14. Hazardous substances
15. Noise and vibration
16. Fire and emergency procedures.

On completion of the inspection, candidates are allowed about one hour to produce the report, which must normally be in their own handwriting but be legible to the examiner. Candidates are given a summary of the headings under which marks are allocated both for the report and observation sheets.

The whole assessment must be carried out under as near examination conditions as possible and should not normally take more than 2 hours. Candidates are not allowed to use previously prepared or organizational checklists or any other reference material.

For the inspection and **practical assessment** of the workplace, candidates are expected to do more than simply identify physical hazards such as 'work at height carried out unsafely' although this is important. In most workplaces they should find examples of chemical, fire, ergonomic, electrical, manual handling, slip, trip and fall (people and objects) hazards. Consideration should also be given to health and physical hazards (e.g. dermatitis, noise and vibration), access, emergency arrangements and welfare and environmental problems. They need to comment when there is adequate control of hazards and where safe working practices are being observed, as well as when the opposite is the case.

Medium and long-term actions must be noted as well as immediate actions to control any dangerous hazard. The distinction between immediate and longer-term action must be clear. The removal of a hazard is an example of immediate action whereas the provision of information, training or additional workers is longer term. The proposed remedial actions must not only remove or control the hazard but be realistic and cost-effective and the associated time scales realistic and appropriate.

Many candidates confuse comments with actions. The examiner looks for clear action proposals to remedy any observed weakness. The hazard recorded must be specific so that *examples* of poor housekeeping are recorded rather than a single statement of 'poor housekeeping'. For some assessments, issues of employee awareness, supervision, maintenance, inspection and testing procedure may need to be considered. There is always a requirement for a monitoring system even if no action is required for a particular hazard at the time of the inspection. For most practical assessments, the examiner will be looking for between 20 and 30 observations.

For the **management report**, candidates are marked against the following five criteria:

1. *Selection of topics for urgent management action (0 to 10 marks)*
    This requires candidates to emphasize those items on their observation sheets that they consider require urgent attention by management and to present them, together with suggested remedial actions (both short and longer term), in a logical and coherent manner. Those issues that are high risk and high priority, such as electrical safety, head protection and poor work practices, must be emphasized and highlighted. The report must differentiate between urgent and routine matters with simple cross-referencing to the observation sheets but it must not give undue weight to trivial matters.

2. *Consideration of cost implications (0 to 5 marks)*
    Candidates must demonstrate that they are aware of cost implications although they will not be judged on the value of the actual costs of a given measure.

    If training is recommended as a solution to a problem, candidates should indicate if this is likely to require a few hours of work-based instruction or several days of more costly off-the-job training. It is the assessment of the magnitude of the cost that is important, rather than precise figures. Words, such as 'cheap' and 'expensive', are better than giving no estimate. There should be an emphasis on the costs of taking no action (e.g. lower productivity, more compensation claims and possible enforcement action).

3. *Identification of breaches of legislation (0 to 5 marks)*
    Since reference material is not allowed during the application, candidates are only expected to indicate knowledge of the relevant legislation rather than the exact title and date. For example: Health and Safety at Work Act, CDM, First-Aid Regulations, PUWER

and the Workplace Regulations are acceptable abbreviations on the observation sheets and in the management report.

The following list of Acts and Regulations would be relevant to all inspections:

➤ the Health and Safety at Work etc Act
➤ the Management of Health and Safety at Work Regulations
➤ the Regulatory Reform (Fire Safety) Order and
➤ the Construction (Design and Management) Regulations.

Most inspections would also include some of the following regulations:

➤ the Construction (Head Protection) Regulations
➤ the Work at Height Regulations
➤ the Lifting and Lifting Equipment Regulations
➤ the Manual Handling Operations Regulations
➤ the Personal Protective Equipment at Work Regulations
➤ the Provision and Use of Work Equipment Regulations
➤ the Electricity at Work Regulations
➤ the Health and Safety First Aid Regulations
➤ the Control of Noise at Work Regulations
➤ the Control of Vibration at Work Regulations
➤ the Control of Asbestos Regulations
➤ the Control of Substances Hazardous to Health Regulations
➤ the Control of Lead Regulations
➤ the Health and Safety (Safety Signs and Signals) Regulations
➤ the Health and Safety Consultation with Employees Regulations
➤ the Workplace (Health, Safety and Welfare) Regulations

4. *Presentation of information (0 to 10 marks)*
    A good report should normally comprise about three sides of handwritten A4 (i.e. about 500–750 words). It should cover the following points in a logical sequence:

➤ where and when the inspection took place
➤ a brief summary of what was found
➤ a short list of issues requiring urgent action by management with convincing arguments why such action is needed and calling attention to possible breaches of health and safety legislation
➤ reference to the list of observations and recommended actions (which should be attached to the report), calling particular attention to any

recommendations which could have a high cost in terms of finance, inconvenience or time.

5. *Effectiveness in convincing management to take action (0 to 15 marks)*
Managers are unlikely to have time to plough through lengthy reports. The report should be written in such terms that a manager would be able to take reasonable action based on facts relatively quickly. Therefore, it should be concise, readable and highly selective in terms of the action required by management. It should contain balanced arguments on why action is needed and explain the effect it would have on the standards of health and safety at the workplace.

24.3.2 *Specimen practical application*

The specimen practical application contains more observations than would normally be expected from students but it does attempt to cover most of the major hazards encountered on a construction site.

The specimen practical assessment took place on a construction site, called the Beacon site, where 30 domestic houses were being built – 12 erected and being fitted out, 12 half erected and 6 with just the foundations in place. There were 5 employees, 1 trainee and several sub-contractors at the site on the day of the assessment. The 5 employees were a supervisor, 2 bricklayers, 1 carpenter and 1 building operative. In addition to the 12 houses on the site, there were a site office, welfare facilities, a stores compound and a vehicle compound. The site was surrounded by a security fence.

The following management report was submitted to the Chief Executive of A B Construction Ltd.

The management report

1. Introduction

This report follows a health and safety inspection at the Beacon construction site.

The inspection took place on 7 June 2007. The report's aim is to highlight good practice and areas of concern requiring urgent action. The observations are listed in Appendix 24.1 attached to this report.

2. Summary

There have been no serious accidents at this site since construction work began six months ago. This good record appears to have led to a relaxation in the monitoring of standards. Several inexpensive improvements in the health and safety arrangements are recommended.

There are some serious concerns involving fire precautions, the excavation work, missing toe boards on one scaffold, and plant maintenance and control of vehicles on the site, leading to possible breaches of the Health and Safety at Work Act (HSWA), the Management of Health and Safety at Work Regulations (MHSWR), the Construction (Design and Management) (CDM) Regulations, the Work at Height (WAH) Regulations, the Regulatory Reform (Fire Safety) Order (Fire Regs), the Provision and Use of Work Equipment Regulations (PUWER), the Personal Protective Equipment at Work Regulations (PPE Regs) and the Electricity at Work Regulations. Any of these breaches could attract enforcement action from the HSE.

The provision of a site induction course for all employee and sub-contractors is a further urgent requirement.

A more detailed description of the topics requiring immediate attention is given below with reference to the observation sheets given in brackets.

3. Issues requiring attention

(a) Health and safety management issues
Although the strengths outnumbered the weaknesses, many of the observed weaknesses resulted from a lack of health and safety awareness and poor supervision. It is essential that site safety induction training is introduced as a matter of urgency (7) for all workers at the site. Such a course would ensure that everyone is aware of the health and safety plan (6), risk assessments (8), the company health and safety policy (13), the need to wear hard hats (20) and the fire and emergency arrangements (39–42). The course need not be long (about half an hour) and a simple cheap booklet covering the main points would help to ensure an effective induction for all the site workers.

Improved site supervision and some delegated health and safety responsibilities would significantly reduce risks on the site. This would ensure that the stores were locked at night (4), and the site kept tidy (5) and protected from trespassers (3). Particular attention should be paid to the toilets (15) and the kitchen area (16).

Under the First Aid Regulations, the first aider needs to attend a short refresher course at a recognized training centre. This will cost about £200.

Finally, it was noted that toe boards were missing from one of the scaffolds (21) even though the scaffold had passed an inspection earlier in the week when, presumably, the boards were present. This is a serious health and safety issue and an urgent investigation should be made to discover the reason for the board removal.

The total cost of these recommendations will be small (about £300) because most of them involve a reorganization of work loads.

(b) Documentation

Most of the documentation was in good order and highly commendable (10, 11, 14).

Risk assessments have been completed except one to cover the activities of the trainee. The Management Regs require a specific risk assessment to be made for persons under the age of 18 years (8). If risk assessments are to be effective, they must be reviewed periodically so that they are relevant to the particular site being worked at the time and that no new risks have arisen. This review could take place at a short site meeting.

There were two further concerns – the absence of any COSHH assessments for the hazardous substances used on the site (37) and the lack of any excavation inspection regime (12) as required by CDM. These are both legal requirements.

Some of the workers' records were incomplete in that courses attended by individuals had not been recorded (9), so their current level of competence was not recorded.

Finally, both the company health and safety policy and the Employers' Liability Insurance certificate should be prominently displayed at the site (13).

The cost of these recommendations will be not be large and could be done by existing staff. The COSHH assessment will be the largest job but the number of hazardous substances on site is limited.

(c) Site issues

Several aspects of the site were very commendable (1, 2, 24, and 26). The main concerns were: the excavations (the lack of barriers (29) and the positioning of the spoil material (30)); the lack of exclusion zone below roof work (27); and the storage of roofing tiles on the scaffold working platform (28). Also, several ladders were not adequately secured (25) or protected from use by trespassers (3). All of these concerns require urgent attention.

The control of traffic on the site also needs urgent attention and a one-way system with pedestrian walkways (34, 35) is recommended.

The total cost of all these recommendations should not exceed £500 for various barriers which can be re-used on future sites.

(d) Other issues

The Fire Regs require that fire risk assessments, fire fighting arrangements, fire information and basic fire training be provided at the site. A simple fire risk assessment should be made as soon as possible – it is a legal requirement.

Other fire precautions that should be in place include the storing of highly flammable substances (39) in a separate store (approximate cost £300), the designation of a fire assembly point and a regular fire drill (every three months) (41). Details of emergency procedures and the location of the assembly point should be given on a posted emergency procedure notice.

No maintenance records were available for the cement mixer (31) although assurances were given that all plant is regularly serviced. A maintenance schedule for this and any other plant should be recorded and monitored.

There is a risk of back injury due to the manual handling of kerb stones. It is recommended that a mechanical lifting device should be purchased or hired for this work (£500 to £1,500).

Finally, new legislation on Whole Body Vibration (WBV) requires that an assessment must be made for drivers who may experience excessive WBV. The dumper truck driver may well require such an assessment (36). The Approved Code of Practice from the HSE offers advice on this.

The total cost of all these improvements should be between £800 and £1,800.

4. **Conclusions**

The site was found to be reasonably well run with many health and safety issues being addressed. There were, however, a number of issues which need to be addressed urgently to avoid possible legal action or compensation claims following injuries. All these problems should be rectifiable within a total cost range of £1,300–£2,000.

Signed...A Smith...................

Date...16 October 2007...

Enclosures: **Appendix 24.1   The practical assessment**

Appendix 24.1 – The practical assessment

NATIONAL CERTIFICATE IN CONSTRUCTION HEALTH AND SAFETY	Candidate's observation sheet

Unit NGC2 – Health and safety practical application

nebosh

Sheet number <u>1</u> of <u>4</u>

Candidate name   A Smith

Candidate number **C** 1416

Place inspected   A B Construction Ltd.

Date of inspection 16 / 10 / 2007

Observations List hazards, unsafe practices and good practices	Priority/risk (H, M, L)	Actions to be taken (if any) List all immediate *and* longer-term actions required	Time scale (immediate, 1 week, etc)
General site issues			
1. Site enclosed by a strong security fence.	L	Commendable – no immediate actions. Check fence every week.	Every week
2. A comprehensive set of health and safety signs on the fence.	L	Ongoing monitoring.	Every week
3. All ladders must be fitted with scaffold boards overnight to protect child trespassers.	H	Delegate responsibility for this to an employee.	Immediate
4. Stores are not secured at night.	H	Delegate responsibility for this to an employee.	Immediate
5. Site was generally untidy leading to trip and fire hazards.	H	Delegate responsibility for this to an employee.	Immediate
Management issues			
6. A health and safety plan and a complete set of method statements were available. (CDM)	L	Ensure that workers are aware of method statements. Some contractors had not seen them.	Ongoing
7. No site induction training was given to workers.	H	A simple induction course outlining the company health and safety policy and site hazards should be devised as soon as possible.	1 week
8. All risk assessments completed except for the trainee (Management Regs).	H	The young worker risk assessment should be completed as soon as possible with his/her supervisor. All other risk assessments should be reviewed and discussed at site meetings.	1 week 3 months
9. All employees and contractors were competent but some records were incomplete.	L	Records for employees and contractors should be kept up to date.	1 month
Documentation			
10. A range of test certificates for plant and equipment were available (e.g. dumper trucks, compressors, cranes, lift trucks).	–	Good practice.	
11. Register of regular scaffold inspections available.	–	Good practice.	

(Continued)

585

Appendix 24.1 – *Continued*

NATIONAL CERTIFICATE IN CONSTRUCTION HEALTH AND SAFETY		nebosh	**Candidate's observation sheet**
Unit NGC2 – Health and safety practical application			Sheet number 2 of 4

Candidate name   A Smith   Candidate number **C** 1416

Place inspected   A B Construction Ltd.   Date of inspection 16 / 10 / 2007

Observations List hazards, unsafe practices and good practices	Priority/risk (H, M, L)	Actions to be taken (if any) List all immediate **and** longer-term actions required	Time scale (immediate, 1 week, etc)
12. No register of excavation inspection kept. (CDM)	H	Inspection of excavations is required if it needs to be supported.	Immediate
13. Although a health and safety poster was displayed, no health and safety policy statement	M	The company health and safety policy is contained in the safety plan but a copy should also be on public display.	1 month
or current employers' liability insurance certificate was posted. (HSW)	H	Employers' liability insurance is a legal requirement and the certificate should be placed on a notice board. The trainee should also be covered by insurance.	Immediate
14. Accident book was available and up to date.	–	Good practice.	
Welfare			
15. Toilet and washing facilities were adequate but the paper towel bins need regular emptying and the facilities more cleaning. (CDM)	M	These facilities should be cleaned at least twice a day.	1 week
16. The rest facilities used by workers for meal breaks need attention, particularly the kitchen equipment. (CDM)	H	Regular cleaning at least twice a day especially the kitchen sink and utensils.	1 week
17. First-aid box was fully and properly stocked and appropriate notices displayed.	–	Commendable.	
18. There is a first aider on site but he must attend a refresher course. (First Aid Regs)	M	Arrange for attendance at a refresher course.	3 months
Personal protective equipment			
19. All relevant personal protective equipment was available and in good condition. (PPE Regs)	–	Commendable.	
20. Some contractors were not wearing safety helmets. (Head Protection Regs)	H	No one should be allowed on site without hard hats on. This topic should be included in the site induction training.	Immediate

NATIONAL CERTIFICATE IN CONSTRUCTION HEALTH AND SAFETY

Unit NGC2 – Health and safety practical application

nebosh

Candidate name   A Smith   Candidate number **C** 1416

Place inspected   A B Construction Ltd.   Date of inspection 16 / 10 / 2007

Observations List hazards, unsafe practices and good practices	Priority/risk (H, M, L)	Actions to be taken (if any) List all immediate **and** longer-term actions required	Time scale (immediate, 1 week, etc)
Scaffolds			
21. All guard rails were adequate but some toe boards were missing. (WAH Regs)	H	The toe boards must be replaced immediately and were present during the weekly inspection. An investigation is recommended.	Immediate
22. The scaffolding is inspected regularly and records are kept.	–	Good practice.	
23. No warning signs were fitted to an incomplete scaffold. (WAH Regs)	H	Fit signs immediately and monitor in future.	Immediate. Monitor daily when scaffold is being erected
Ladders			
24. All ladders rise to a sufficient height above landing place.	–	Good practice.	
25. Ladders are not secured to prevent slipping. (WAH Regs)	H	Secure all ladders at top and bottom.	Immediate
Roof work			
26. Suitable edge protection is provided.	–	Commendable.	
27. People are not excluded from the area below the roof work.	H	Cordon off area below roof work and post suitable signs.	Immediate
28. Too much building material (tiles) stored on the roof.	H	Reduce the number of tiles on the roof to 1 hour's supply.	Immediate
Excavations			
29. No barriers around the drainage excavations. (CDM)	H	Erect barriers and suitable warning signs.	Immediate
30. Spoil material stored too close to the excavation.	H	Store the material so that there is no danger of it re-entering the trench.	Immediate
Plant and equipment			
31. No maintenance records were available for the cement mixer. (PUWER)	M	Either obtain copies or have the mixer serviced.	1 month

(Continued)

Appendix 24.1 – *Continued*

<table>
<tr><td colspan="2">

NATIONAL CERTIFICATE IN CONSTRUCTION HEALTH AND SAFETY

Unit NGC2 – Health and safety practical application
</td><td>

nebosh
</td><td>

Candidate's observation sheet

Sheet number 4 of 4
</td></tr>
</table>

Candidate name   A Smith   Candidate number **C** 1416

Place inspected   A B Construction Ltd.   Date of inspection 16 / 10 / 2007

Observations List hazards, unsafe practices and good practices	Priority/risk (H, M, L)	Actions to be taken (if any) List all immediate **and** longer-term actions required	Time scale (immediate, 1 week, etc)
Power and hand tools			
32. Low voltage equipment used throughout site.	–	Commendable.	
33. No records of PAT testing of electrical tools. (Electricity Regs)	M	Arrange for all electrical equipment to be tested on a 3 monthly basis.	1 month
Traffic and vehicles			
34. There have been several near collisions between vehicles on the site. (CDM)	H	Urgent consideration should be given to the introduction of a one-way system on the site.	1 month
35. No pedestrian only walkways have been designated on the site. (CDM)	H	Where possible, pedestrians should be separated from vehicles. (Incorporate into one-way system?)	1 months
36. Whole body vibration of dumper truck drivers seemed excessive. (Vibration Regs)	M	Read ACOP on Whole body vibration to develop a suitable action plan. Include in induction training.	3 months
Hazardous substances			
37. No COSHH assessments have been done. (COSHH Regs)	M	COSHH assessments should be made as soon as possible with particular attention paid to cement.	3 months
Manual Handling			
38. Risk of back injury due to manual lifting of kerb stones. (Manual Handling Regs)	H	Use a mechanical lifting device.	Immediate
Fire and emergency procedures			
39. Highly flammable substances not stored separately. (Fire Regs)	H	Restrict to a maximum of 50 litres on site and store in a special metal container.	1 week
40. Fire extinguishers are adequate and checked regularly.	–	Good practice.	
41. Fire drills do not take place and emergency procedures are not posted.	M	Identify a suitable assembly point and mark it. Arrange a fire drill very 3 months and record assembly time. Include details in induction training and in emergency procedures.	1 week 3 months
42. No fire risk assessments done. (Fire Regs)	H	Complete fire risk assessment as soon as possible.	Immediate

Index

Index